Methods in Enzymology

Volume 148

PLANT CELL MEMBRANES

METHODS IN ENZYMOLOGY

EDITORS-IN-CHIEF

Sidney P. Colowick   Nathan O. Kaplan

Methods in Enzymology

Volume 148

Plant Cell Membranes

EDITED BY

Lester Packer

DEPARTMENT OF PHYSIOLOGY AND ANATOMY
UNIVERSITY OF CALIFORNIA
BERKELEY, CALIFORNIA

Roland Douce

DÉPARTEMENT DE RECHERCHE FONDAMENTALE
CENTRE D'ETUDES NUCLÉAIRES ET UNIVERSITÉ
DE GRENOBLE
GRENOBLE, FRANCE

ACADEMIC PRESS, INC.
Harcourt Brace Jovanovich, Publishers
San Diego New York Berkeley Boston
London Sydney Tokyo Toronto

Academic Press, Inc.
San Diego, California 92101

United Kingdom Edition published by
ACADEMIC PRESS INC. (LONDON) LTD.
24-28 Oval Road, London NW1 7DX

Library of Congress Catalog Card Number: 54-9110

ISBN 0-12-182048-3 (alk. paper)

PRINTED IN THE UNITED STATES OF AMERICA
87 88 89 90 9 8 7 6 5 4 3 2 1

Table of Contents

Section I. Cells, Protoplasts, and Liposomes

Section II. Vacuoles and Tonoplasts

Section III. Plastids

Section IV. Mitochondria

Section V. Peroxisomes and Glyoxysomes

Section VI. Nuclei, Endoplasmic Reticulum, and Plasma Membrane

Section VII. General Physical and Biochemical Methods

Contributors to Volume 148

Article numbers are in parentheses following the names of contributors.
Affiliations listed are current.

TAKASHI AKAZAWA (3), *Research Institute for Biochemical Regulation, School of Agriculture, Nagoya University, Chikusa, Nagoya 464, Japan*

HANS-ERIK ÅKERLUND (26), *Department of Biochemistry, University of Lund, S-221 00 Lund, Sweden*

CLAUDE ALBAN (20), *Laboratoire de Physiologie Cellulaire Végétale, Unité Associée au Centre National de la Recherche Scientifique, Département de Recherche Fondamentale, Centre d'Etudes Nucléaires et Université Scientifique, Technologique et Médicale de Grenoble, F-38041 Grenoble, France*

GILBERT ALIBERT (8), *Ecole Nationale Supérieure d'Agronomie, 31076 Toulouse, France*

F. AMBARD-BRETTEVILLE (57), *Génétique Moléculaire des Plantes, Unité Associée 115, Université Paris-Sud, Orsay 91405, France*

BERTIL ANDERSSON (26), *Department of Biochemistry, Arrhenius Laboratory, University of Stockholm, S-106 91 Stockholm, Sweden*

C. ANDING (61), *Centre de Recherche de la Dargoire, Rhône-Poulenc Agrochimie, 69263 Lyon, France*

JAEN E. ANDREWS (32), *Institut für Allgemein Botanik, Universität Hamburg, D-2000 Hamburg 52, Federal Republic of Germany*

TOM AP REES (21), *Botany School, University of Cambridge, Cambridge CB2 3EA, England*

B. ARRIO (14), *Bioénergétique Membranaire, Unité Associée 1128 au Centre National de la Recherche Scientifique, 91405 Orsay, France*

TADASHI ASAHI (46), *Laboratory of Biochemistry, School of Agriculture, Nagoya University, Chikusa, Nagoya 464, Japan*

JAMES BARBER (29), *Department of Pure and Applied Biology, Imperial College of Science and Technology, London SW7 2BB, England*

HARRY BEEVERS (49), *Department of Biology, University of California, Santa Cruz, California 95064*

DEREK S. BENDALL (27), *Department of Biochemistry, University of Cambridge, Cambridge CB2 1QW, England*

PIERRE BENVENISTE (58), *Laboratoire de Biochimie Végétale, Unité Associée au Centre National de la Recherche Scientifique, Institut de Botanique, 67083 Strasbourg, France*

P. BEYER (36), *Institut für Biologie II, Zellbiologie, D-7800 Freiburg, Federal Republic of Germany*

RICHARD BLIGNY (1), *Laboratoire de Physiologie Cellulaire Végétale, Centre d'Etudes Nucléaires de Grenoble, 38041 Grenoble, France*

MARYSE A. BLOCK (20), *Laboratoire de Physiologie Cellulaire Végétale, Unité Associée au Centre National de la Recherche Scientifique, Département de Recherche Fondamentale, Centre d'Etudes Nucléaires et Université Scientifique, Technologique et Médicale de Grenoble, F-38041 Grenoble, France*

EDUARDO BLUMWALD (12), *Department of Botany, University of Toronto, Toronto, Ontario, Canada M5S 1A1*

ALAIN M. BOUDET (8), *Centre de Physiologie Végétale, Unité Associée au Centre National de la Recherche Scientifique, Université Paul Sabatier, 31062 Toulouse, France*

JACQUES BOURGUIGNON (37), *Laboratoire de Physiologie Cellulaire Végétale, Unité Associée au Centre National de la Recherche Scientifique, Département de Recherche Fondamentale, Centre d'Etudes Nucléaires et Université Scientifique, Technologique et Médicale de Grenoble, F-38041 Grenoble, France*

CHARLES BOYER (22), *Department of Horticulture, Pennsylvania State University, University Park, Pennsylvania 16802*

DONALD P. BRISKIN (51), *Department of Agronomy, University of Illinois, Urbana, Illinois 61801*

RENAUD BROUQUISSE (37), *Laboratoire de Physiologie Cellulaire Végétale, Unité Associée au Centre National de la Recherche Scientifique, Département de Recherche Fondamentale, Centre d'Etudes Nucléaires et Université Scientifique, Technologique et Médicale de Grenoble, F-38041 Grenoble, France*

MICHEL CABOCHE (5), *Laboratoire de Biologie Cellulaire, Institut National de Recherche Agronomique, 78000 Versailles, France*

JEAN-PIERRE CARDE (56), *Laboratoire de Physiologie Cellulaire Végétale, Unité Associée au Centre National de la Recherche Scientifique, Université de Bordeaux I, F-33405 Talence, France*

ZORAN G. CEROVIC (15), *Botanical Institute and Garden, Faculty of Science, University of Belgrade, Takovska 43, 11000 Beograd, Yugoslavia*

DAVID J. CHAPMAN (29), *Department of Pure and Applied Biology, Imperial College of Science and Technology, London SW7 2BB, England*

H. CHRESTIN (10), *Biotechnology and Plant Physiology Laboratory, Orstom, 01 BPV-5 Abidjan 01, Ivory Coast*

MARCO COLOMBINI (43), *Laboratories of Cell Biology, Department of Zoology, University of Maryland, College Park, Maryland 20742*

JACQUES COVÈS (20), *Laboratoire de Physiologie Cellulaire Végétale, Unité Associée au Centre National de la Recherche Scientifique, Département de Recherche Fondamentale, Centre d'Etudes Nucléaires et Université Scientifique, Technologique et Médicale de Grenoble, F-38041 Grenoble, France*

JOHN C. CUSHMAN (16), *Waksman Institute of Microbiology, Rutgers University, Piscataway, New Jersey 08855*

COLIN DALTON (2), *Biotechnology Group, PA Technology, Melbourn, Royston, Herts SG8 6DP, England*

J. D'AUZAC (10), *Laboratoire de Physiologie Végétale Appliquée, Université des Sciences et Techniques du Languedoc, F-34060 Montpellier, France*

ROBERT P. DONALDSON (47), *Department of Biological Sciences, George Washington University, Washington, D.C. 20052*

ALBERT-JEAN DORNE (19), *Laboratoire de Physiologie Cellulaire Végétale, Unité Associée au Centre National de la Recherche Scientifique, Département de Recherche Fondamentale, Centre d'Etudes Nucléaires et Université Scientifique, Technologique et Médicale de Grenoble, F-38041 Grenoble, France*

ROLAND DOUCE (19, 20, 37), *Laboratoire de Physiologie Cellulaire Végétale, Unité Associée au Centre National de la Recherche Scientifique, Département de Recherche Fondamentale, Centre d'Etudes Nucléaires et Université Scientifique, Technologique et Médicale de Grenoble, F-38041 Grenoble, France*

A. DUPAIX (14), *Bioénergétique Membranaire, Unité Associée 1128 du Centre National de la Recherche Scientifique, 91405 Orsay, France*

YVES DUPONT (62), *Laboratoire de Biologie Moleculaire et Cellulaire, Département de Recherches Fondamentales, Centre d'Etude Nucléaires de Grenoble, F-38041 Grenoble, France*

EDGARDO ECHEVERRIA (22), *Citrus Research and Education Center, University of Florida, Lake Alfred, Florida 33850*

GERALD E. EDWARDS (39), *Department of*

Botany, Washington State University, Pullman, Washington 99164

INGEMAR ERICSON (40, 41), *Department of Biochemistry, University of Umeå, S-901 87 Umeå, Sweden*

TUNG K. FANG (47), *Department of Biochemistry, University of Missouri, Columbia, Missouri 65211*

PER GARDESTRÖM (39, 40, 41), *Department of Plant Physiology, University of Umeå, S-901 87 Umeå, Sweden*

BERNT GERHARDT (48), *Botanisches Institut, Universität Münster, D-4400 Münster, Federal Republic of Germany*

ELMA GONZÁLEZ (49), *Department of Biology, University of California, Los Angeles, California 90024*

KAREN L. GREER (28), *Department of Botany, University of Georgia, Athens, Georgia 30602*

J. L. HALL (53), *Department of Biology, The University of Southampton, Southampton S09 5NH, England*

IKUKO HARA-NISHIMURA (3), *Department of Biology, Faculty of Science, Kobe University, Rokkoudai 1-1, Nada, Kobe 657, Japan*

MARIE-ANDRÉE HARTMANN (58), *Laboratoire de Biochimie Végétale, Unité Associée au Centre National de la Recherche Scientifique, Institut de Botanique, 67083 Strasbourg, France*

JOHN L. HARWOOD (44), *Department of Biochemistry, University College, Cardiff CF1 1XL, Wales*

TSUKAHO HATTORI (46), *Laboratory of Biochemistry, School of Agriculture, Nagoya University, Chikusa, Nagoya 464, Japan*

E. HEINZ (59), *Institut für Allgemeine Botanik, Universität Hamburg, D-2000 Hamburg 52, Federal Republic of Germany*

M. HILL (14), *Bioénergétique Membranaire, Unité Associée 1128 au Centre National de la Recherche Scientifique, 91405 Orsay, France*

THOMAS K. HODGES (51), *Department of Botany and Plant Pathology, Purdue University, West Lafayette, Indiana 47907*

BENJAMIN JACOBY (11), *Department of Agricultural Botany, The Hebrew University of Jerusalem, Rehovot 76100, Israel*

ETIENNE-PASCAL JOURNET (23), *Laboratoire de Physiologie Cellulaire Végétale, Unité Associée au Centre National de la Recherche Scientifique, Département de Recherche Fondamentale, Centre d'Etudes Nucléaires et Université Scientifique, Technologique et Médicale de Grenoble, F-38041 Grenoble, France*

K. W. JOY (17), *Department of Biology, Carleton University, Ottawa, Ontario, Canada K1S 5B6*

JACQUES JOYARD (19, 20), *Laboratoire de Physiologie Cellulaire Végétale, Unité Associée au Centre National de la Recherche Scientifique, Département de Recherche Fondamentale, Centre d'Etudes Nucléaires et Université Scientifique, Technologique et Médicale de Grenoble, F-38041 Grenoble, France*

JEAN-CLAUDE KADER (60), *Laboratoire de Physiologie Cellulaire, Unité Associée au Centre National de la Recherche Scientifique, Université Pierre et Marie Curie, 75005 Paris, France*

PAUL KEIM (50), *Department of Biology, Iowa State University, Ames, Iowa 50011*

J. KESSELMEIER (59), *Botanisches Institut, Universität zu Köln, D-5000 Köln 41, Federal Republic of Germany*

PER KJELLBOM (52), *Plant Biology Laboratory, The Salk Institute, San Diego, California 92138*

HANS KLEINIG (45), *Institut für Biologie II, Zellbiologie, Universität Freiburg, D-7800 Freiburg, Federal Republic of Germany*

A. KURKDJIAN (14), *Laboratoire de Physiologie Cellulaire Végétale, Centre National de la Recherche Scientifique, 91190 Gif-sur-Yvette, France*

CHRISTER LARSSON (52), *Department of Plant Physiology, University of Lund, S-220 07 Lund, Sweden*

JEAN-JACQUES LEGUAY (1), *Laboratoire de Physiologie Cellulaire Végétale, Centre National de la Recherche Scientifique–Institut National de la Recherche Scientifique, 91190 Gif-sur-Yvette, France*

NORA W. LEM (33), *Clinical Biochemistry, Faculty of Medicine, University of Toronto, Toronto, Ontario, Canada M5S 1A1*

ROBERT T. LEONARD (9, 51), *Department of Botany and Plant Sciences, University of California, Riverside, California 92521*

HARTMUT K. LICHTENTHALER (34), *Botanisches Institut (Plant Physiology), University of Karlsruhe, D-7500 Karlsruhe, Federal Republic of Germany*

ANNIKA C. LIDÉN (41), *Department of Plant Physiology, University of Lund, S-220 07 Lund, Sweden*

BODO LIEDVOGEL (24), *Institut für Biologie II, Zellbiologie, Universität Freiburg, D-7800 Freiburg, Federal Republic of Germany*

J. MICHAEL LORD (54), *Department of Biological Sciences, University of Warwick, Warwick, Coventry CV4 7AL, England*

PAUL F. LURQUIN (5), *Program in Genetics and Cell Biology, Washington State University, Pullman, Washington 99164*

FRIEDHELM LÜTKE-BRINKHAUS (45), *Department of Biological Chemistry, School of Medicine, University of Utah, Salt Lake City, Utah 84108*

FRASER D. MACDONALD (21), *Division of Molecular Plant Biology, University of California, Berkeley, California 94720*

IAN MACKENZIE (2), *Biotechnology Group, PA Technology, Melbourn, Royston, Herts SG8 6DP, England*

MASAYOSHI MAESHIMA (46), *Laboratory of Biochemistry, School of Agriculture, Nagoya University, Chikusa, Nagoya 464, Japan*

CARMEN A. MANNELLA (42), *Wadsworth Center for Laboratories and Research, New York State Department of Health, Albany, New York 12201*

B. MARIN (10), *Orstom, F-75480 Paris, France*

PAUL MAZLIAK (60), *Laboratoire de Physiologie Cellulaire, Unité Associée au Centre National de la Recherche Scientifique, Université Pierre et Marie Curie, 75005 Paris, France*

LETICIA R. MENDIOLA-MORGENTHALER (16), *Institut für Biochemie, Universität Bern, CH-3012 Bern, Switzerland*

W. R. MILLS (17), *Department of Biology, University of Houston at Clear Lake, Houston, Texas 77058*

MICHAEL L. MISHKIND (28), *Department of Biochemistry and Microbiology, Cook College, Rutgers University, New Brunswick, New Jersey 08903*

IAN M. MØLLER (41), *Department of Plant Physiology, University of Lund, S-220 07 Lund, Sweden*

ANTHONY L. MOORE (38), *Department of Biochemistry, University of Sussex, Falmer, Brighton BN1 9QG, England*

THOMAS S. MOORE, JR. (55), *Department of Botany, Louisiana State University, Baton Rouge, Louisiana 70803*

MARIE-JOSÉ MOUTIN (62), *Laboratoire de Biologie Moleculaire et Cellulaire, Département de Recherche Fondamentale, Centre d'Etudes Nucléaires de Grenoble, F-38041 Grenoble, France*

J. BRIAN MUDD (32), *ARCO Plant Cell Research Institute, Dublin, California 94568*

SATORU MURAKAMI (25), *Department of Biology, University of Tokyo, Komaba, Meguro-ku, Tokyo 153, Japan*

TOSHIYUKI NAGATA (4), *Department of Cell Biology, National Institute for Basic Biology, 38 Nishigonaka, Myodaijicho, Okazaki 444, Japan*

MICHEL NEUBURGER (37), *Laboratoire de Physiologie Cellulaire Végétale, Unité Associée au Centre National de la Recherche Scientifique, Département de Recherche Fondamentale, Centre d'Etudes Nucléaires et Université Scientifique, Technologique et Médicale de Grenoble, F-38041 Grenoble, France*

MIKIO NISHIMURA (3), *Department of Biology, Faculty of Science, Kobe University, Rokkoudai 1-1, Nada, Kobe 657, Japan*

RONALD J. POOLE (12), *Department of Biology, McGill University, Montreal, Quebec, Canada H3A 1B1*

CARL A. PRICE (16), *Waksman Institute of Microbiology, Rutgers University, Piscataway, New Jersey 08855*

MICHAEL O. PROUDLOVE (38), *Department of Biochemistry, University of Sussex, Falmer, Brighton BN1 9QG, England*

STEPHEN K. RANDALL (13), *Department of Biochemistry, McGill University, Montreal, Quebec, Canada H3G 1Y6*

R. G. RATCLIFFE (63), *Department of Plant Sciences, University of Oxford, Oxford 0X1 3PF, England*

PHILIP A. REA (12), *Department of Biology, University of York, York Y01 5DD, England*

ELLEN M. REARDON (16), *Department of Biochemistry, University of Arizona, Tucson, Arizona 85721*

R. REMY (57), *Génétique Moléculaire des Plantes, Unité Associée 115, Université Paris-Sud, Orsay 91405, France*

SIMON P. ROBINSON (15, 18), *Division of Horticultural Research, Commonwealth Scientific and Industrial Research Organization, Adelaide, South Australia 5001, Australia*

STEPHEN A. ROLFE (27), *Department of Biochemistry, University of Cambridge, Cambridge CB2 1QW, England*

GRATTAN ROUGHAN (31), *Division of Horticulture and Processing, Department of Scientific and Industrial Research, Mt. Albert Research Centre, Auckland, New Zealand*

OTTO SCHIEDER (6), *Freie Universität Berlin, Institut für Angewandte Genetik, 1000 Berlin 33, Federal Republic of Germany*

GREGORY W. SCHMIDT (28), *Department of Botany, University of Georgia, Athens, Georgia 30602*

JACK C. SHANNON (22), *Department of Horticulture, Pennsylvania State University, University Park, Pennsylvania 16802*

JÜRGEN SOLL (35), *Botanisches Institut, Universität München, 8000 Munich 19, Federal Republic of Germany*

S. A. SPARACE (32), *Department of Plant Science, Macdonald College, Ste. Anne de Bellvue, Quebec, Canada H9X 1CO*

SALLIE G. SPRAGUE (30), *Department of Biochemistry, West Virginia University, School of Medicine, Morgantown, West Virginia 26506*

L. ANDREW STAEHELIN (30), *Department of Molecular, Cellular and Developmental Biology, University of Colorado, Boulder, Colorado 80309*

HEVEN SZE (13), *Department of Botany, University of Maryland, College Park, Maryland 20742*

CHANTAL VERGNOLLE (60), *Laboratoire de Physiologie Cellulaire, Unité Associée au Centre National de la Recherche Scientifique, Université Pierre et Marie Curie, 75005 Paris, France*

EUGENE L. VIGIL (47), *Department of Horticulture, University of Maryland, College Park, Maryland 20742*

P. VOLFIN (14), *Bioénergétique Membranaire, Unité Associée 1128 du Centre National de la Recherche Scientifique, 91405 Orsay, France*

GEORGE J. WAGNER (7), *Plant Physiology/Biochemistry/Molecular Biology Program, Department of Agronomy, University of Kentucky, Lexington, Kentucky 40546*

DAVID A. WALKER (15), *Research Institute for Photosynthesis, University of Sheffield, Sheffield S10 2TN, England*

SUSANNE WIDELL (52), *Department of Plant Physiology, University of Lund, S-220 07 Lund, Sweden*

JOHN P. WILLIAMS (33), *Department of Botany, University of Toronto, Toronto, Ontario, Canada M5S 1A1*

Preface

With the exception of the chloroplast thylakoid membrane, other plant cell membranes have been largely neglected in the *Methods in Enzymology* series. However, during the past decade, spectacular advances have been made in the isolation and characterization of almost all of the other membrane systems of the cell, plastids, mitochondria, tonoplasts, nuclei, and other intracellular organelles. The availability of these new methods and the techniques involved are extremely important for advancing our understanding of plant cell physiology and biochemistry and for applying techniques of genetic engineering and molecular biology for the solution of basic and applied research problems. The diversity of approaches employed by investigators interested in plant cell membranes indicated that this was an appropriate time to compile a methodological reference work for this area of plant biochemistry. Related methods and techniques which have been included in other volumes of this series have not been repeated unless recent developments represented significant improvements in the methods described previously.

We would like to express our deepest gratitude to all the authors for their fine contributions, advice, and suggestions which made this volume possible. We also wish to thank Françoise Bucharles for her excellent secretarial assistance. Finally, we wish to acknowledge the patience and valuable cooperation of the staff of Academic Press.

LESTER PACKER
ROLAND DOUCE

METHODS IN ENZYMOLOGY

EDITED BY

Sidney P. Colowick and Nathan O. Kaplan

VANDERBILT UNIVERSITY
SCHOOL OF MEDICINE
NASHVILLE, TENNESSEE

DEPARTMENT OF CHEMISTRY
UNIVERSITY OF CALIFORNIA
AT SAN DIEGO
LA JOLLA, CALIFORNIA

I. Preparation and Assay of Enzymes
II. Preparation and Assay of Enzymes
III. Preparation and Assay of Substrates
IV. Special Techniques for the Enzymologist
V. Preparation and Assay of Enzymes
VI. Preparation and Assay of Enzymes (*Continued*)
Preparation and Assay of Substrates
Special Techniques
VII. Cumulative Subject Index

METHODS IN ENZYMOLOGY

EDITORS-IN-CHIEF

Sidney P. Colowick and Nathan O. Kaplan

Volume VIII. Complex Carbohydrates
Edited by Elizabeth F. Neufeld and Victor Ginsburg

Volume IX. Carbohydrate Metabolism
Edited by Willis A. Wood

Volume X. Oxidation and Phosphorylation
Edited by Ronald W. Estabrook and Maynard E. Pullman

Volume XI. Enzyme Structure
Edited by C. H. W. Hirs

Volume XII. Nucleic Acids (Parts A and B)
Edited by Lawrence Grossman and Kivie Moldave

Volume XIII. Citric Acid Cycle
Edited by J. M. Lowenstein

Volume XIV. Lipids
Edited by J. M. Lowenstein

Volume XV. Steroids and Terpenoids
Edited by Raymond B. Clayton

Volume XVI. Fast Reactions
Edited by Kenneth Kustin

Volume XVII. Metabolism of Amino Acids and Amines (Parts A and B)
Edited by Herbert Tabor and Celia White Tabor

VOLUME XVIII. Vitamins and Coenzymes (Parts A, B, and C)
Edited by DONALD B. MCCORMICK AND LEMUEL D. WRIGHT

VOLUME XIX. Proteolytic Enzymes
Edited by GERTRUDE E. PERLMANN AND LASZLO LORAND

VOLUME XX. Nucleic Acids and Protein Synthesis (Part C)
Edited by KIVIE MOLDAVE AND LAWRENCE GROSSMAN

VOLUME XXI. Nucleic Acids (Part D)
Edited by LAWRENCE GROSSMAN AND KIVIE MOLDAVE

VOLUME XXII. Enzyme Purification and Related Techniques
Edited by WILLIAM B. JAKOBY

VOLUME XXIII. Photosynthesis (Part A)
Edited by ANTHONY SAN PIETRO

VOLUME XXIV. Photosynthesis and Nitrogen Fixation (Part B)
Edited by ANTHONY SAN PIETRO

VOLUME XXV. Enzyme Structure (Part B)
Edited by C. H. W. HIRS AND SERGE N. TIMASHEFF

VOLUME XXVI. Enzyme Structure (Part C)
Edited by C. H. W. HIRS AND SERGE N. TIMASHEFF

VOLUME XXVII. Enzyme Structure (Part D)
Edited by C. H. W. HIRS AND SERGE N. TIMASHEFF

VOLUME XXVIII. Complex Carbohydrates (Part B)
Edited by VICTOR GINSBURG

VOLUME XXIX. Nucleic Acids and Protein Synthesis (Part E)
Edited by LAWRENCE GROSSMAN AND KIVIE MOLDAVE

VOLUME XXX. Nucleic Acids and Protein Synthesis (Part F)
Edited by KIVIE MOLDAVE AND LAWRENCE GROSSMAN

VOLUME XXXI. Biomembranes (Part A)
Edited by SIDNEY FLEISCHER AND LESTER PACKER

VOLUME XXXII. Biomembranes (Part B)
Edited by SIDNEY FLEISCHER AND LESTER PACKER

VOLUME XXXIII. Cumulative Subject Index Volumes I–XXX
Edited by MARTHA G. DENNIS AND EDWARD A. DENNIS

VOLUME XXXIV. Affinity Techniques (Enzyme Purification: Part B)
Edited by WILLIAM B. JAKOBY AND MEIR WILCHEK

VOLUME XXXV. Lipids (Part B)
Edited by JOHN M. LOWENSTEIN

VOLUME XXXVI. Hormone Action (Part A: Steroid Hormones)
Edited by BERT W. O'MALLEY AND JOEL G. HARDMAN

VOLUME XXXVII. Hormone Action (Part B: Peptide Hormones)
Edited by BERT W. O'MALLEY AND JOEL G. HARDMAN

VOLUME XXXVIII. Hormone Action (Part C: Cyclic Nucleotides)
Edited by JOEL G. HARDMAN AND BERT W. O'MALLEY

VOLUME XXXIX. Hormone Action (Part D: Isolated Cells, Tissues, and Organ Systems)
Edited by JOEL G. HARDMAN AND BERT W. O'MALLEY

VOLUME XL. Hormone Action (Part E: Nuclear Structure and Function)
Edited by BERT W. O'MALLEY AND JOEL G. HARDMAN

VOLUME XLI. Carbohydrate Metabolism (Part B)
Edited by W. A. WOOD

VOLUME XLII. Carbohydrate Metabolism (Part C)
Edited by W. A. WOOD

VOLUME XLIII. Antibiotics
Edited by JOHN H. HASH

VOLUME XLIV. Immobilized Enzymes
Edited by KLAUS MOSBACH

VOLUME XLV. Proteolytic Enzymes (Part B)
Edited by LASZLO LORAND

VOLUME 85. Structural and Contractile Proteins (Part B: The Contractile Apparatus and the Cytoskeleton)
Edited by DIXIE W. FREDERIKSEN AND LEON W. CUNNINGHAM

VOLUME 86. Prostaglandins and Arachidonate Metabolites
Edited by WILLIAM E. M. LANDS AND WILLIAM L. SMITH

VOLUME 87. Enzyme Kinetics and Mechanism (Part C: Intermediates, Stereochemistry, and Rate Studies)
Edited by DANIEL L. PURICH

VOLUME 88. Biomembranes (Part I: Visual Pigments and Purple Membranes, II)
Edited by LESTER PACKER

VOLUME 89. Carbohydrate Metabolism (Part D)
Edited by WILLIS A. WOOD

VOLUME 90. Carbohydrate Metabolism (Part E)
Edited by WILLIS A. WOOD

VOLUME 91. Enzyme Structure (Part I)
Edited by C. H. W. HIRS AND SERGE N. TIMASHEFF

VOLUME 92. Immunochemical Techniques (Part E: Monoclonal Antibodies and General Immunoassay Methods)
Edited by JOHN J. LANGONE AND HELEN VAN VUNAKIS

VOLUME 93. Immunochemical Techniques (Part F: Conventional Antibodies, Fc Receptors, and Cytotoxicity)
Edited by JOHN J. LANGONE AND HELEN VAN VUNAKIS

VOLUME 94. Polyamines
Edited by HERBERT TABOR AND CELIA WHITE TABOR

VOLUME 95. Cumulative Subject Index Volumes 61–74, 76–80
Edited by EDWARD A. DENNIS AND MARTHA G. DENNIS

VOLUME 96. Biomembranes [Part J: Membrane Biogenesis: Assembly and Targeting (General Methods; Eukaryotes)]
Edited by SIDNEY FLEISCHER AND BECCA FLEISCHER

VOLUME 97. Biomembranes [Part K: Membrane Biogenesis: Assembly and Targeting (Prokaryotes, Mitochondria, and Chloroplasts)]
Edited by SIDNEY FLEISCHER AND BECCA FLEISCHER

VOLUME 98. Biomembranes (Part L: Membrane Biogenesis: Processing and Recycling)
Edited by SIDNEY FLEISCHER AND BECCA FLEISCHER

VOLUME 99. Hormone Action (Part F: Protein Kinases)
Edited by JACKIE D. CORBIN AND JOEL G. HARDMAN

VOLUME 100. Recombinant DNA (Part B)
Edited by RAY WU, LAWRENCE GROSSMAN, AND KIVIE MOLDAVE

VOLUME 101. Recombinant DNA (Part C)
Edited by RAY WU, LAWRENCE GROSSMAN, AND KIVIE MOLDAVE

VOLUME 102. Hormone Action (Part G: Calmodulin and Calcium-Binding Proteins)
Edited by ANTHONY R. MEANS AND BERT W. O'MALLEY

VOLUME 103. Hormone Action (Part H: Neuroendocrine Peptides)
Edited by P. MICHAEL CONN

VOLUME 104. Enzyme Purification and Related Techniques (Part C)
Edited by WILLIAM B. JAKOBY

VOLUME 105. Oxygen Radicals in Biological Systems
Edited by LESTER PACKER

VOLUME 106. Posttranslational Modifications (Part A)
Edited by FINN WOLD AND KIVIE MOLDAVE

VOLUME 107. Posttranslational Modifications (Part B)
Edited by FINN WOLD AND KIVIE MOLDAVE

VOLUME 108. Immunochemical Techniques (Part G: Separation and Characterization of Lymphoid Cells)
Edited by GIOVANNI DI SABATO, JOHN J. LANGONE, AND HELEN VAN VUNAKIS

VOLUME 109. Hormone Action (Part I: Peptide Hormones)
Edited by LUTZ BIRNBAUMER AND BERT W. O'MALLEY

VOLUME 110. Steroids and Isoprenoids (Part A)
Edited by JOHN H. LAW AND HANS C. RILLING

VOLUME 111. Steroids and Isoprenoids (Part B)
Edited by JOHN H. LAW AND HANS C. RILLING

VOLUME 112. Drug and Enzyme Targeting (Part A)
Edited by KENNETH J. WIDDER AND RALPH GREEN

VOLUME 113. Glutamate, Glutamine, Glutathione, and Related Compounds
Edited by ALTON MEISTER

VOLUME 114. Diffraction Methods for Biological Macromolecules (Part A)
Edited by HAROLD W. WYCKOFF, C. H. W. HIRS, AND SERGE N. TIMASHEFF

VOLUME 115. Diffraction Methods for Biological Macromolecules (Part B)
Edited by HAROLD W. WYCKOFF, C. H. W. HIRS, AND SERGE N. TIMASHEFF

VOLUME 116. Immunochemical Techniques (Part H: Effectors and Mediators of Lymphoid Cell Functions)
Edited by GIOVANNI DI SABATO, JOHN J. LANGONE, AND HELEN VAN VUNAKIS

VOLUME 117. Enzyme Structure (Part J)
Edited by C. H. W. HIRS AND SERGE N. TIMASHEFF

VOLUME 118. Plant Molecular Biology
Edited by ARTHUR WEISSBACH AND HERBERT WEISSBACH

VOLUME 119. Interferons (Part C)
Edited by SIDNEY PESTKA

VOLUME 120. Cumulative Subject Index Volumes 81–94, 96–101

VOLUME 121. Immunochemical Techniques (Part I: Hybridoma Technology and Monoclonal Antibodies)
Edited by JOHN J. LANGONE AND HELEN VAN VUNAKIS

VOLUME 122. Vitamins and Coenzymes (Part G)
Edited by FRANK CHYTIL AND DONALD B. MCCORMICK

VOLUME 123. Vitamins and Coenzymes (Part H)
Edited by FRANK CHYTIL AND DONALD B. MCCORMICK

VOLUME 124. Hormone Action (Part J: Neuroendocrine Peptides)
Edited by P. MICHAEL CONN

VOLUME 125. Biomembranes (Part M: Transport in Bacteria, Mitochondria, and Chloroplasts: General Approaches and Transport Systems)
Edited by SIDNEY FLEISCHER AND BECCA FLEISCHER

VOLUME 126. Biomembranes (Part N: Transport in Bacteria, Mitochondria, and Chloroplasts: Protonmotive Force)
Edited by SIDNEY FLEISCHER AND BECCA FLEISCHER

VOLUME 127. Biomembranes (Part O: Protons and Water: Structure and Translocation)
Edited by LESTER PACKER

VOLUME 128. Plasma Lipoproteins (Part A: Preparation, Structure, and Molecular Biology)
Edited by JERE P. SEGREST AND JOHN J. ALBERS

VOLUME 129. Plasma Lipoproteins (Part B: Characterization, Cell Biology, and Metabolism)
Edited by JOHN J. ALBERS AND JERE P. SEGREST

VOLUME 130. Enzyme Structure (Part K)
Edited by C. H. W. HIRS AND SERGE N. TIMASHEFF

VOLUME 131. Enzyme Structure (Part L)
Edited by C. H. W. HIRS AND SERGE N. TIMASHEFF

VOLUME 132. Immunochemical Techniques (Part J: Phagocytosis and Cell-Mediated Cytotoxicity)
Edited by GIOVANNI DI SABATO AND JOHANNES EVERSE

VOLUME 133. Bioluminescence and Chemiluminescence (Part B)
Edited by MARLENE DELUCA AND WILLIAM D. MCELROY

VOLUME 134. Structural and Contractile Proteins (Part C: The Contractile Apparatus and the Cytoskeleton)
Edited by RICHARD B. VALLEE

VOLUME 135. Immobilized Enzymes and Cells (Part B)
Edited by KLAUS MOSBACH

VOLUME 136. Immobilized Enzymes and Cells (Part C)
Edited by KLAUS MOSBACH

VOLUME 137. Immobilized Enzymes and Cells (Part D) (in preparation)
Edited by KLAUS MOSBACH

VOLUME 138. Complex Carbohydrates (Part E)
Edited by VICTOR GINSBURG

VOLUME 139. Cellular Regulators (Part A: Calcium- and Calmodulin-Binding Proteins)
Edited by ANTHONY R. MEANS AND P. MICHAEL CONN

VOLUME 140. Cumulative Subject Index Volumes 102–119, 121–134 (in preparation)

VOLUME 141. Cellular Regulators (Part B: Calcium and Lipids)
Edited by P. MICHAEL CONN AND ANTHONY R. MEANS

VOLUME 142. Metabolism of Aromatic Amino Acids and Amines
Edited by SEYMOUR KAUFMAN

VOLUME 143. Sulfur and Sulfur Amino Acids
Edited by WILLIAM B. JAKOBY AND OWEN GRIFFITH

VOLUME 144. Structural and Contractile Proteins (Part D: Extracellular Matrix)
Edited by LEON W. CUNNINGHAM

VOLUME 145. Structural and Contractile Proteins (Part E: Extracellular Matrix)
Edited by LEON W. CUNNINGHAM

VOLUME 146. Peptide Growth Factors (Part A)
Edited by DAVID BARNES AND DAVID A. SIRBASKU

VOLUME 147. Peptide Growth Factors (Part B)
Edited by DAVID BARNES AND DAVID A. SIRBASKU

VOLUME 148. Plant Cell Membranes
Edited by LESTER PACKER AND ROLAND DOUCE

VOLUME 149. Drug and Enzyme Targeting (Part B)
Edited by RALPH GREEN AND KENNETH J. WIDDER

VOLUME 150. Immunochemical Techniques (Part K: *In Vitro* Models of B and T Cell Functions and Lymphoid Cell Receptors)
Edited by GIOVANNI DI SABATO

VOLUME 151. Molecular Genetics of Mammalian Cells
Edited by MICHAEL M. GOTTESMAN

VOLUME 152. Guide to Molecular Cloning Techniques
Edited by SHELBY L. BERGER AND ALAN R. KIMMEL

VOLUME 153. Recombinant DNA (Part D)
Edited by RAY WU AND LAWRENCE GROSSMAN

VOLUME 154. Recombinant DNA (Part E)
Edited by RAY WU AND LAWRENCE GROSSMAN

VOLUME 155. Recombinant DNA (Part F)
Edited by RAY WU

VOLUME 156. Biomembranes (Part P: ATP-Driven Pumps and Related Transport: The Na,K-Pump) (in preparation)
Edited by SIDNEY FLEISCHER AND BECCA FLEISCHER

VOLUME 157. Biomembranes (Part Q: ATP-Driven Pumps and Related Transport: Calcium, Proton, and Potassium Pumps) (in preparation)
Edited by SIDNEY FLEISCHER AND BECCA FLEISCHER

VOLUME 158. Metalloproteins (Part A) (in preparation)
Edited by JAMES F. RIORDAN AND BERT L. VALLEE

Section I

Cells, Protoplasts, and Liposomes

[1] Techniques of Cell Suspension Culture

By RICHARD BLIGNY and JEAN-JACQUES LEGUAY

Plant cell suspension cultures rapidly generate large amounts of cell material that offer a remarkable tool to study plant cell physiology. Thus, it is simple and practical with cell suspensions to measure growth parameters and physiological and biochemical properties of plant cells such as respiration, energy metabolism, or synthesis of specific compounds. Cell culture in liquid medium avoids some disadvantages inherent in the culture of tissues on agar. Movement of cells in the liquid growth medium facilitates gaseous exchange, suppresses any polarity of the tissue due to gravity, and eliminates the gradients of nutrients within the medium and from cell to cell.

Cell lines can also be selected to produce a desired natural product at a high yield. Suitable fermentation technologies are now available for growing cells on a large scale. However, there is no general single method applicable to the culture of different strains under well-standardized culture conditions. Each new plant cell culture requires a systematic study of nutrient medium and growth parameters such as pH, temperature, or oxygenation. This chapter relates to culture techniques adopted in the case of sycamore cell (*Acer pseudoplatanus* L.) suspensions, a species frequently used for physiological research on plants.

Growth Pattern in Cell Cultures

A suspension culture of higher plant cells consists of cells and cell clumps dispersed and growing in agitated liquid medium. After inoculation, the cell number increases for a limited time and reaches a value of maximum density of plant material. When the culture at any stage of this growth cycle is diluted by subculture to the same initial cellular density, a similar pattern of growth is obtained. Thus, the culture can be continuously subcultured by successive transfers of inoculum at regular intervals.

The incubation of friable tissue (callus) in liquid nutrient medium is usually used for starting cell suspension cultures. The cultures can also be started from sterile seedling or plant tissues by breaking these up in a hand homogenizer and then transferring the material to liquid medium.

Initiation of Cell Suspensions

Remove calluses from the culture tubes and place four to five calluses in a 250-ml conical flask containing 50 ml of nutrient medium (the composition of different nutrient media is indicated in Table I). Place flasks on a rotary shaker at constant temperature (25–30°). When calluses are dissociated (it usually takes 7–14 days) and give rise to small clumps in suspension, filter each culture through a 500-μm sieve fitted on a sterile 100-ml measuring cylinder. Allow the contents to settle for 15 min and discard the supernatant. Adjust the volume of the cylinder to 50 ml with fresh medium and pour into a 250-ml conical flask. Replace flasks on the shaker. At the next subculture (after 7–14 days) repeat the procedure with a measuring cylinder, but take only a fraction (one-fourth to one-fifth) of the residual cells as the inoculum.

After four to five transfers at intervals of 7 days under these conditions, the cell suspensions generally contain only small cell aggregates or single cells. These cultures are now suitable to determine the growth rates.

Nutritional Aspects of Cell Culture

Nutritive Requirements. The composition of classical nutrient media is given in Table I. Media MS (Murashige–Skoog) and B_5 (Gamborg), respectively, established for tobacco[1] cells and soybean[2] cells, are used for a large number of different species.

Macro- and micronutrients. Macronutrients are supplied at millimolar concentrations (see Table I). Nitrogen is usually supplied as NO_3^- or NH_4^+ (or both) and varies, according to the different media, from 25 to 60 m*M*. Plant cells can grow in a medium containing exclusively nitrate as nitrogen source. NH_4^+ in excess may be toxic. Usually, cells cannot grow with NH_4^+ as sole source of nitrogen, unless the medium is buffered in order to counteract the alkalization of cytosol due to massive NH_4^+ absorption.[3-5] The enrichment of nutrient medium in organic acids (citrate, malate, α-ketoglutarate) allows cells to grow with ammonium as the sole nitrogen source.[6]

[1] T. Murashige and F. Skoog, *Physiol. Plant.* **15,** 473 (1962).
[2] O. L. Gamborg, R. A. Miller, and K. Ojima, *Exp. Cell Res.* **50,** 151 (1968).
[3] S. M. Martin, D. Rose, and V. Hui, *Can. J. Bot.* **55,** 2828 (1977).
[4] C. V. Givan and H. A. Collin, *J. Exp. Bot.* **18,** 321 (1967).
[5] A. Kurkdjian, J.-J. Leguay, and J. Guern, *Plant Physiol.* **64,** 1053 (1979).
[6] O. L. Gamborg and J. P. Shyluk, *Plant Physiol.* **45,** 598 (1970).

TABLE I
COMPOSITION OF MURASHIGE–SKOOG (MS), GAMBORG (B_5), AND LAMPORT'S MEDIA FOR PLANT TISSUE AND CELL CULTURE

	MS		B_5		Lamport modified by Lescure[a]	
Composition	(mg/liter)	(mM)	(mg/liter)	(mM)	(mg/liter)	(mM)
Macronutrients						
NH_4NO_3	1,650	20.6	—	—	—	—
KNO_3	1,900	18.8	2,500	25	1,960	19
$CaCl_2 \cdot 2H_2O$	440	3.0	150	1.0	360	1.4
$MgSO_4 \cdot 7H_2O$	370	1.5	250	1.0	—	—
KH_2PO_4	170	1.2	—	—	560	3.6
$(NH_4)_2SO_4$	—	—	134	1.0	—	—
$NaH_2PO_4 \cdot H_2O$	—	—	150	1.1	—	—
$Na_2HPO_4 \cdot 12H_2O$	—	—	—	—	97	0.27
KCl	—	—	—	—	65	0.087
$Ca(NO_3)_2 \cdot 4H_2O$	—	—	—	—	290	1.2
Micronutrients		(μM)		(μM)		(μM)
KI	0.83	5	0.75	4.5	0.75	4.5
H_3BO_3	6.2	100	3.0	50	1.50	24
$MnSO_4 \cdot 4H_2O$	22.3	100	—	—	—	—
$MnSO_4 \cdot H_2O$	—	—	10	60	3.38	19
$ZnSO_4 \cdot 7H_2O$	8.6	30	2.0	7.0	1.50	5.2
$Na_2MoO_4 \cdot 2H_2O$	0.25	1.0	0.25	1.0	—	—
$CuSO_4 \cdot 5H_2O$	0.025	0.1	0.025	0.1	—	—
$CoCl_2 \cdot 6H_2O$	0.025	0.1	0.025	0.1	—	—
Fe–versenate (EDTA)	43.0	100	43.0	100	2.785	1
Organic compounds						
Inositol	100		100		—	
Nicotinic acid	0.5		1.0		—	
Pyridoxine · HCl	0.5		1.0		—	
Thiamin · HCl	0.1		10.0		1.0	
IAA	1.30		—		—	
Kinetin	0.04–10		0.1		—	
2,4-D	—		0.1–2.0		1.0	
Sucrose	30,000		20,000		20,000	
pH	5.7		5.5		5.5	

[a] D. T. A. Lamport, *Exp. Cell Res.* **33,** 195 (1964); A. M. Lescure, *Physiol. Veg.* **4,** 365 (1966).

Micronutrients are supplied at micromolar concentrations (see Table I). Iron is furnished as a chelated compound, usually with ethylenediaminetetraacetic acid (EDTA).

Carbon source. A large number of carbon sources have now been tested for their ability to support the growth of plant cell cultures.[7] The most readily assimilated carbon sources are sucrose (10–20 g/liter) and glucose (10–20 g/liter) but it is possible to select cell lines adapted to galactose,[8] glycerol,[9] and lactose.[10]

Organic nitrogen. Though most suspension culture can grow on media containing strictly mineral nitrogen, casein hydrolysate (1–2 g/liter) or a mixture of amino acids enhances cell growth.[11] Addition of amino acids reactivates mitosis in isolated leaf cells.[12,13]

Cofactors and phytohormones. B-group vitamins and frequently *myo*-inositol (in tobacco cell cultures for instance) are commonly required. The most common vitamins are B_1 (thiamin), B_2 (riboflavin), and B_6 (pyridoxine). Thiamin as thiamin pyrophosphate functions as a coenzyme in several enzymatic reactions in which aldehyde groups are transferred from a donor to an acceptor molecule (for example in the reactions catalyzed by pyruvate decarboxylase). Riboflavin is a component of the flavin nucleotides which function as prosthetic groups for a class of dehydrogenases (for example in the reaction catalyzed by succinate dehydrogenase). Pyridoxal phosphate is the prosthetic group of enzymes catalyzing reactions of transamination.

Auxins and cytokinins are two phytohormones controlling cell division in plants. They are usually used at concentrations ranging from 10^{-7} to 10^{-5} *M*. The auxins commonly used are 2,4-dichlorophenoxyacetic acid (2,4-D) and 1-naphthaleneacetic acid (1-NAA). Common cytokinins derived from adenine include 6-benzyladenine (6-BAP) and isopentenyladenine (IPA). Other phytohormones such as abscissic acid or gibberellic acid are not essential to grow plant cells.

Physical Factors. pH. Initially, the pH of freshly prepared medium is acid (around 5.5). It must be noted that there is a drop of 0.5 pH units after autoclaving. During the culture cycle, pH values vary from 5.5 to 7 because culture media are poorly buffered. These variations are attributed

[7] M. W. Fowler and G. Stepman-Sarkissian, *in* "Primary and Secondary Metabolism of Plant Cell Cultures" (K. H. Neumann, W. Bartz, and E. Reinhard, eds.), p. 66. Springer-Verlag, Berlin and New York, 1985.

[8] A. Maretzki and M. Thom, *Plant Physiol.* **61,** 544 (1978).

[9] A. Jones and I. A. Veliky, *Can. J. Bot.* **59,** 2095 (1980).

[10] N. Chaubet, V. Petiard, and A. Pareilleux, *Plant Sci. Lett.* **22,** 369 (1981).

[11] O. L. Gamborg, R. A. Miller, and R. Ojima, *Exp. Cell Res.* **50,** 148 (1968).

[12] M. Jullien and J. Guern, *Physiol. Veg.* **17,** 445 (1979).

[13] L. Crépy, M. C. Chupeau, and Y. Chupeau, *Z. Pflanzenphysiol.* **107,** 123 (1982).

to the assimilation of NH_4^+ and NO_3^-. Addition of organic buffers like 5–10 mM MES [2(*N*-morpholino)ethane sulfonic acid] lessen these variations.

Temperature. Cultures are commonly maintained at a constant temperature between 20° and 30°C. The cell doubling time is directly dependent on temperatures up to 30°C. Temperatures over 35°C are lethal for a number of strains.

Oxygenation. A cell suspension less than 1 liter is conveniently oxygenated by using platform shakers with orbital movement (amplitude, 8–10 cm in diameter; 60–120 rpm) or by using simple rotators to spin culture vessels about their own longitudinal axes on an inclination.

In heterotrophic conditions, cultures grow equally well in the dark and in the light.

Minimum Effective Cell Density. The smallest inoculum from which a new suspension culture can be reproducibly grown is a function of the cell strain, of the growth phase, and of the composition of the culture medium. Below a threshold cell density (~10^3 cells/ml for sycamore cells according to Street[14]), cell growth cannot start. However, the use of culture media in which cells have already grown ("conditioned" medium according to Benbadis[15]) permits cell division at very low cell densities for an unknown reason. This could be due to unknown compounds excreted by the cells[16] or to components present at toxic concentrations in fresh culture media. For example, 1 mg auxin/liter [indole-3-acetic acid (IAA), NAA or 2,4-D] can be toxic when added to cultures containing cells at very low densities (10^3 cell/ml, or less).[17] In this case, lowering the auxin supply allows cells to divide.

Some Growth Parameters

Cell Number. Culture aliquots are macerated in 8–10% chromic acid solutions at 20°, overnight. Macerated cell suspensions are then submitted to sonication (20 sec; 20 kHz, 60 W; sonimass 259 T, Ultrason-Annemasse, France, or a similar sonicator) to improve cell dispersion. Cell numbering is done under the microscope using a Nageotte-type slide (Prieux, Joinville-le-Pont, France).

Cell Fresh Weight. Culture aliquots are collected on a glass fiber filter (15 sec; pressure of suction, 0.8 bar) and washed twice with distilled water.

[14] H. E. Street, *in* "Plant Tissue and Cell Culture" (H. E. Street, ed.), p. 191. Blackwell, Oxford, 1973.

[15] A. Benbadis, *C.R. Hebd. Seances Acad. Sci.* **261,** 4829 (1965).

[16] R. Stuart and H. E. Street, *J. Exp. Bot.* **20,** 556 (1969).

[17] M. Caboche, *Planta* **149,** 7 (1980).

TABLE II
EVOLUTION OF SOME CELL PARAMETERS IN SYCAMORE CELL CULTURES AS A FUNCTION OF CULTURE CONDITIONS[a]

Parameter	Subculture		
	Each 2 days	Each 7 days	7 days after growth plateau
Cell doubling time	40 hr	55 hr	—
Cell fresh weight (mg)	25	33	35–60
Cell dry weight (mg)	3.1	3.5	4.2
Protein (mg)	0.56	0.55	0.45
Total lipids (μg)	175	170	150
Phospholipids (μg)	48	45	35
Respiration (nmol O_2/min)	15–18	12–15	8–10

[a] Cell parameters are expressed per 10^6 cells; cell fresh weight was measured after straining on a glass fiber filter (15 sec; pressure of suction, 0.8 bar); cell dry weight was measured after lyophilization. Protein, lipids, and cell respiration were measured as described by Journet *et al.*[25]

Cell Dry Weight. This is measured after desiccation (110°, 1 hr) or preferably after lyophilization of the preceding samples.

Protein Content. Total cell protein can be measured on filtered cell aliquots, after elimination of the soluble N_2 by precipitation and washing of the proteins with 10% trichloroacetic acid, according to the method of Lance.[18] It is also possible to measure directly the protein content of cell (or sonicated cell) aliquots, using the method of Lowry *et al.*[19] with bovine serum albumin as standard. In this case a reduction coefficient (0.8 for sycamore cells) must be applied to the measurements.

Total Lipid. Lipids are extracted from filtered cell aliquots according to the method of Folch *et al.*[20] and directly weighed.

When expressed relative to cell number, the growth parameters may vary as a function of the physiological state of the cells. For example, one can observe that the most homogeneous and vigorous sycamore cell cultures are obtained after very frequent subcultures (each 2 days). Under these conditions most of the cells exhibit a roughly round shape and a rather small size (20- to 25-μm diameter) in relation with a relatively small vacuolar system. In this case, the cell doubling time is very low (40–50 hr at 25°), as indicated in Table II. In contrast, old cultures (at the growth plateau since 7 days or more) are a heterogeneous mixture of small active

[18] C. Lance, *Ann. Sci. Nat., Bot. Biol. Veg.* [12] **4,** 1 (1963).

[19] O. H. Lowry, N. J. Rosebrough, A. L. Farr, and R. Randall, *J. Biol. Chem.* **193,** 265 (1951).

[20] J. Folch, M. Lees, and G. H. S. Standley, *J. Biol. Chem.* **226,** 497 (1957).

cells and cells characterized by a lengthening and a larger volume (40- to 50-μm diameter) in relation with a more developed vacuolar system. The cell heterogeneity in routine cultures transferred every 7 days is moderate. Expressed per 10^6 cells, the fresh weight of sycamore cells vary from 25 mg (cells subcultured each 2 days) to 60 mg (cells at the growth plateau). By comparison, the cell dry weight and the protein and total lipid contents are more reliable parameters (Table II).

The physiological state of the cells can be appraised through respiration measurements. This is done polarographically using a Clark-type O_2 electrode (Hansatech, Ltd., King's Lynn, Norfolk, United Kingdom). A value of 240 μmol of O_2/ml is used for the solubility of O_2 at 25° in typical media.

Finally it is important to make frequent cell mortality measurements. The following technique by Hugh[21] is very rapid and handy. Dead cells are stained red by erythrosine B, while living cells remain colorless. The stain recipe is as follows:

Erythrosine B (GURR)	0.40 g
NaCl	0.81 g
KH_2PO_4	0.06 g
Methyl *p*-hydroxybenzoate	0.05 g
H_2O	92 ml

The mixture is boiled till dissolution of the dye. After cooling, the pH is adjusted to 7 with 1 *N* NaOH and the volume is adjusted to 100 ml.

For light microscope observation, 1 vol of erythrosine B solution is added to 9 vol of cell suspension. During the exponential growth phase, in routine cultures, rates of 2–5% dead cells are frequently observed.

Deficient Nutrient Medium

It is commonly thought that a culture medium is deficient in at least one element when the cell division stops (growth plateau). For example, the nitrogen source is the growth-limiting factor of sycamore cell cultures in Lamport's medium.[22] Similarly, a carbohydrate supply of 6–7 g/liter limits the maximum growth of tobacco cells in the modified Murashige and Skoog medium.[23,24] However, the same amount of carbohydrate is not limiting in the case of sycamore cell cultures. One reason for this is that the cell division time for tobacco (about 150 hr) is three to four times

[21] J. P. Hugh, *in* "Tissue Culture: Methods and Application" (P. F. Kruse, Jr., and M. K. Patterson, Jr., eds.), p. 406. Academic Press, New York, 1973.

[22] J. J. Leguay and J. Guern, *Plant Physiol.* **56,** 356 (1977).

[23] J. P. Jouanneau and C. Péaud-Lenoël, *Physiol. Plant.* **20,** 384 (1967).

[24] G. Gregorini and M. Laloue, *Plant Physiol.* **65,** 363 (1980).

longer than that for sycamore cells (40–45 hr). Consequently, in tobacco cell culture, more carbohydrate is consumed for respiration (10^6 cells consume about 0.4 mg sucrose/day) relative to biosynthesis of cell components. Clearly, the optimum carbohydrate supply varies with the cell strain. On the other hand, some deficiencies of microelements or hormones may have qualitative effects on the cell physiology before any quantitative effect on the cell growth.

Carbohydrate-Deficient Culture Medium

The most frequently added carbon sources (sucrose or glucose, 10–20 g/liter) are sufficient for the optimum growth of higher plant cell cultures having a cell doubling time of 2–4 days. However, during the plateau, a culture rapidly becomes sucrose (or glucose)-deficient. When all the sucrose added to the culture medium is consumed, cells hydrolyze their endogenous stock of carbohydrates. Vacuolar sucrose is first consumed. After 10 hr of sucrose starvation, the cytoplasmic phosphorylated compounds decrease and starch breakdown progressively replaces sucrose hydrolysis.[25] When 60–70% of starch is hydrolyzed, the cell respiration rates decline concomitantly with autolysis of portions of cytoplasm, accompanied by a decline in the number of mitochondria per cell and by the accumulation of phosphorylcholine in the cytoplasm.[25a] The effects of a sucrose starvation during 2–4 days are reversed within a few hours by the addition of sucrose.

Auxin Starvation

In sycamore cell cultures, the total growth is proportional to the initial 2,4-D concentration for values ranging from 0.05 to 1 μM (Fig. 1). This property was used to show that 2,4-D itself could be considered as the active form of auxin and that a threshold 2,4-D concentration was needed for cell division.[26,27] When 2,4-D is not limiting (2,4-D concentration higher than 1 μM), the total cell growth is not modified but the size of cell clumps diminishes as 2,4-D concentration increases.

Copper-Deficient Culture Medium

Copper atoms are present at the catalytic center of some plant cell enzymes like cytochrome oxidase (EC 1.9.3.1), cytosolic superoxide dis-

[25] E.-P. Journet, R. Bligny, and R. Douce, *J. Biol. Chem.* **261,** 3193 (1986).
[25a] C. Roby, J.-B. Martin, R. Bligny, and R. Douce, *J. Biol. Chem.* **262,** 5000 (1987).
[26] J.-J. Leguay and J. Guern, *Plant Physiol.* **60,** 265 (1977).
[27] A. Kurkdjian, J.-J. Leguay, and J. Guern, *Plant Physiol.* **64,** 1053 (1979).

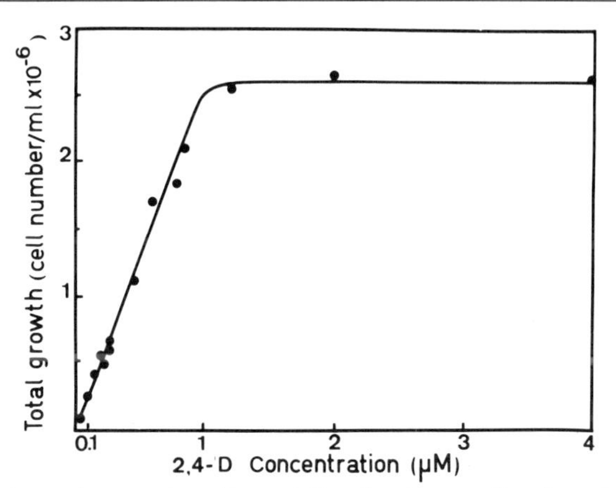

FIG. 1. Maximum increase in cell population density as a function of the initial 2,4-D concentration. The maximum increase in cell population density ($N_{max} - N_0$) is calculated from N_{max}, number of cells/ml when the population is at the stationary phase of growth, and N_0, number of cells/ml at time zero after inoculating the medium.

mutase (EC 1.15.1.1), and polyphenol oxidase [catechol oxidase (EC 1.10.3.1) and laccase (EC 1.10.3.2)]. When the culture medium contains no copper (see Lamport, Table I) there is no cell growth.[28] Below 1–2 μg Cu^{2+}/liter, the initial copper concentration limits the cell number at the growth plateau. Between 1 and 20 μg Cu^{2+}, the initial copper concentration does not limit the cell growth but the cytochrome oxidase content of the cells remains below the standard value and is directly proportional to the added copper (Table III). Similarly, the excretion of active laccase by sycamore cells is proportional to the added copper up to 100 μg Cu^{2+}/liter. Laccase is a blue copper–protein (polyphenol oxidase) continuously excreted by the cells into the culture medium.[29] Consequently, it is clear that the addition of micronutrients like copper must be considered as a function of the specific problem under investigation.

Poorly Oxygenated Culture Medium

When sycamore cells are cultivated in the presence of 20 μM O_2 or less (poor bubbling or shaking) they preferentially synthesize monounsaturated fatty acids (50–55% oleic acid) in contrast with control cells (in presence of 250 μM O_2) which synthesize 2% oleic acid and 75% polyunsaturated fatty acids (linoleic and linolenic acids). In this case, the fatty acid composition of all membrane systems is strongly modified even

[28] R. Bligny and R. Douce, *Plant Physiol.* **60,** 675 (1977).

[29] R. Bligny and R. Douce, *Biochem. J.* **209,** 489 (1983).

TABLE III
EFFECT OF THE INITIAL COPPER CONCENTRATION OF THE CULTURE MEDIUM ON CYTOCHROME OXIDASE CONTENT OF SYCAMORE CELLS AND TOTAL AMOUNT OF ACTIVE LACCASE EXCRETED BY THE CELLS[a]

	Initial copper content of the culture medium (μg/liter)						
Enzyme content	1	5	10	20	50	100	200
Cyt aa_3 per 10^6 cells (pmol)	0.6	2.8	6.8	12	13	13	13
Laccase per milliliter of sycamore cell suspensions (nmol)	0.01	0.03	0.06	0.13	0.3	0.4	0.45

[a] Cytochrome oxidase was measured spectrophotometrically by difference spectra in liquid N_2 (77 K) between dithionite-reduced and oxidized suspensions of sonicated sycamore cells, using the 598-nm absorption peak. Laccase was measured in the growth medium,[29] at the growth plateau, by O_2 consumption in the presence of 20 mM 4-methylcatechol at pH 6.7.

though the growth rate of the cell cultures and the physiological properties of the cells remain unchanged.[30]

Mass Culture

Turbidostat and Chemostat

In order to maintain nonlimiting conditions during a period of time and to establish constant culture conditions it is necessary to grow cells in a turbidostat or chemostat. In a turbidostat, the medium input system is active when the population density exceeds a preselected value and the dilution of the culture reduces the cell density to just below this value. An output system keeps the volume constant. Under such conditions the culture has a constant population density growing at a constant rate.

In a chemostat, the steady state of growth in a fixed volume of culture is obtained by continuous dilution with fresh medium at a *chosen* rate. Any chosen rate (within certain limits) gives a characteristic cell density. The initial descriptions of chemostatic conditions for bacterial cells were described independently by Monod[31] and by Novick and Szilard.[32] Monod

[30] R. Bligny, F. Rébeillé, and R. Douce, *J. Biol. Chem.* **260,** 9166 (1985).
[31] J. Monod, *Ann. Inst. Pasteur, Paris* **79,** 390 (1950).
[32] A. Novick and L. Szilard, *Science* **112,** 715 (1950).

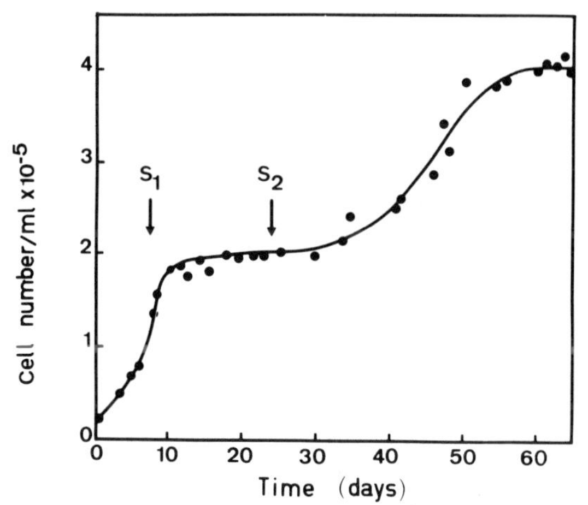

FIG. 2. Change in the biomass of a 2-liter chemostat culture of *Acer pseudoplatanus* cells following a change in the concentration of the limiting nutrient (2,4-D) in the input medium. After 8 days of growth the value of 2,4-D concentration in the inflow medium was 0.1 μM (S_1). At day 25, this value was changed to 0.2 μM (S_2). In both cases the culture medium was diluted at a constant rate of 0.069 day^{-1} (T_d = 10 days).

showed that under certain conditions the specific growth rate (μ) was dependent on the concentration of the growth-limiting nutrient (S) and follows the equation $\mu = \mu_{max} S/(K + S)$ in which (K) is a constant and (μ_{max}) is linked to the doubling time (T_d) of the cell population by the following relationship: $\mu_{max} = \log 2/T_d$.

Sycamore cells follow the equation of Monod when nitrate,[33] phosphate[34] or 2,4-D[35] are limiting factors. One example of steady state growth of sycamore cells in a chemostat with 2,4-D as the limiting nutrient is shown on Fig. 2. In these cultures, the doubling time of the cell population can be controlled by the dilution rate of the growth-limiting factor, but this does not necessarily mean that the duration of the cell cycle (the time separating two successive mitosis) is also controlled by the dilution rate. For example, Leguay[34] has shown that under chemostatic conditions with 2,4-D as limiting factor it is possible to cultivate cells with an apparent doubling time of 10 days having a cell cycle (T_c) of 3 days. According to Mendelsohn,[36] this could be explained if, at each generation, only 60% of the newly divided cells enter cell division.

[33] S. B. Wilson, P. J. King, and H. E. Street, *J. Exp. Bot.* **21,** 177 (1971).
[34] G. Wilson, Ph.D. Thesis, University of Birmingham, England (1971).
[35] J.-J. Leguay, Thesis, University of Paris VI, Paris, France (1979).
[36] M. L. Mendelsohn, *J. Natl. Cancer Inst.* (*U.S.*) **28,** 1015 (1962).

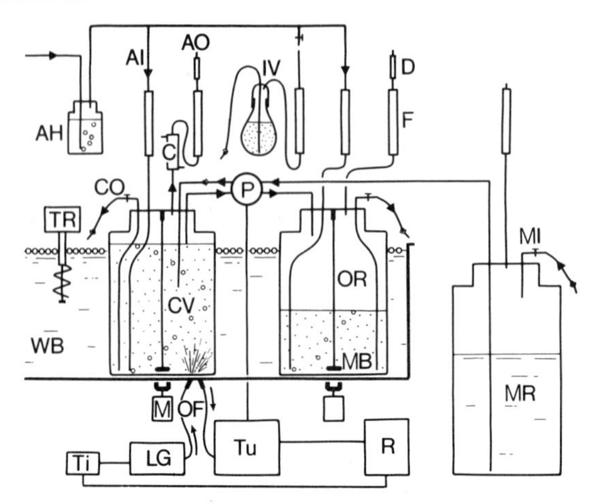

FIG. 3. Phytostat for the mass culture of higher plant cells with automatic and continuous control of culture turbidity. AH, Air humidifier; AI, air inlet; AO, air outlet; C, humidity condenser; CO, culture output; CV, culture vessel; D, flowmeter; F, air filter; IV, inoculation vessel; LG, light generator; M, magnetic stirrer; MB, suspended magnetic bar; MI, medium input; MR, medium reservoir; OF, optical fiber; OR, outflow reservoir; P, pump; R, recorder; Ti, timer; TR, thermoregulator; Tu, turbidimeter; WB, thermoregulated water bath. [Reproduced from R. Bligny, *Plant Physiol.* **59,** 502 (1977) with permission.]

Culture Apparatus

A functional diagram of a 2- to 50-liter phytostat for the culture of higher plant cell suspensions[37] is given in Fig. 3. The culture medium is sterilized by ultrafiltration (Millipore, 0.22 μm; diameter 142 mm) and introduced into the culture vessel (CV) either directly or from a reservoir of sterilized medium (MR) through a double-peristaltic pump (P). Then, an aliquot of cell suspension (0.5–1 liter) is poured into an intermediate inoculation vessel (IV) and from there into the fermentor. Culture vessels are immersed in a thermoregulated water bath (WB) allowing accurate thermoregulation between 0 and 50°. Cultures are stirred continuously by a Teflon-coated magnetic bar (MB) at 60 rpm. The magnetic bar is suspended on a flexible Teflon stalk 1 cm above the bottom of the culture vessel in order to avoid cell abrasion which occurs when the magnetic bar is lying on the bottom of the culture vessel. O_2 is supplied by bubbling air at a flow rate of 60 liter/hr. The air is sterilized by passage through a sintered nickel filter (F) having a pore size of 0.10 μm. Culture aliquots are harvested by clamping the air outlet (AO) and opening the culture output (CO).

[37] R. Bligny, *Plant Physiol.* **59,** 502 (1977).

Measurement and Control of Cell Density

Cell suspensions illuminated through their culture vessel at a given point reflect an amount of light proportional to the cell concentration. Thus the following optical measuring system is particularly convenient: a 150 W white-light generator (LG) (générateur EP 150, Fort, Paris); an optical fiber (OF) for focusing the light on a fixed point under the bottom of the culture vessel; a second optical fiber for transmitting the reflected light to a photoresistor; a measuring and monitoring apparatus, the turbidimeter (TU), for amplifying and analyzing the optical signal received by the photoresistor, and for operating automatically the dilution of the culture; a timer (TI) to modulate the frequency and the duration of action of the turbidimeter; a double-peristaltic pump (P), to transfer simultaneously new medium from the medium reservoir (MR) to the culture vessel, and the excess of culture from the fermentor to the outflow reservoir (OR).

Under these culture conditions the cell number (Fig. $4A_1$) and the culture turbidity (Fig. 4A) increase exponentially without a lag phase. Throughout the exponential phase of growth, the cell number and culture turbidity are closely correlated, indicating that the measurement of suspension turbidity is a reliable indicator of growth in different types of culture: batch or semicontinuous cultures (Fig. 4A and B), turbidostat,

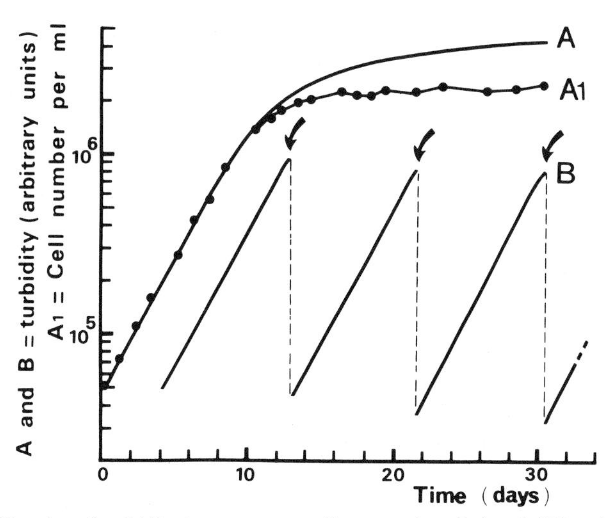

FIG. 4. Kinetics of turbidity in sycamore cell suspensions in batch (A) and semicontinuous (B) cultures of 20 liters. A_1: Kinetics of cell number. Arrows indicate successive culture dilutions. Temperature of culture, 25°C. Cell counting was by microscope with a Nageotte slide (see text). [Reproduced from R. Bligny, *Plant Physiol.* **59,** 502 (1977) with permission.]

and chemostat. In a turbidostat, the culture dilution (operated by the pump P) is controlled by the turbidimeter (TU) to maintain the cell concentration at the chosen value; in a chemostat a constant dilution rate is assigned by the timer.

[2] Culture and Characteristics of Green Plant Cells

By COLIN DALTON and IAN MACKENZIE

Introduction

Green plant cells in culture can provide a consistent supply of chlorophyllous and photosynthetic cells and the basis of experimental systems which are independent of the vagaries of climate and the influences of the whole plant. In ways to be described, cultured green plant cells have provided systems which are claimed to be superior for certain types of experiment than those achieved with the whole plant or its isolated parts. For instance, our understanding of the effect of chilling on *de novo* synthesis of membrane lipids, the regulation of chloroplast differentiation by external factors, and the excretion of photosynthetic products has been advanced by the use of such systems. Of particular relevance is the dramatic synthesis of lipids during transition from heterotrophic to photoautotrophic growth—a transition which can now be controlled in culture and triggered by different chosen events.

This short description of the methods involved and the characteristics of the green cells so produced will hopefully enable readers to culture green plant cells for their own experiments. For more general reviews readers are referred to two recent articles.[1,2]

Initiation and Culture of Green Plant Cells

There are now 25 species which have been reported as green and photosynthetic cultures (Table I). The quickest and easiest way of obtaining a green cell culture is to request an axenic sample from the author of a

[1] M. E. Horn and J. M. Widholm, *in* "Application of Genetic Engineering to Crop Improvement" (G. B. Collins and J. F. Petolino, eds.), p. 113. Martinus Nijhoff/Dr. W. Junk Publishers, Boston, Massachusetts, 1984.

[2] W. Husemann, *in* "Cell Culture and Somatic Cell Genetics of Plants" (I. K. Vasil, ed.), Vol. 2, pp. 182–191. Academic Press, New York, 1985.

TABLE I
SPECIES REPORTED AS GREEN AND PHOTOSYNTHETIC CULTURES

Genes and species	Culture type	Comments
Amaranthus retroflexus[a]	Batch flask (suspension)	
Arachis hypogaea[b]	Batch fermenter	Peanut
Asparagus officinalis[c]	Continuous culture	Monocot
Brassica napus[d]	Batch flask (suspension)	Rape
Chamaecereus sylvestrii[e]	Batch flask (suspension)	Cactus[f]
Chenopodium rubrum[b]	Continuous fermenter	
Cytisus scoparius[g]	Batch flask (callus)	Scotch broom
Datura stramonium[b]	Batch flask (callus)	
Daucus carota[b]	Batch fermenter	Carrot
Digitalis purpurea[h]	Batch flask (suspension)	
Froelichia gracilis[i]	Batch flask (callus)	C_4
Glycine max[b]	Batch flask (suspension)	Soybean
Gossypium spp.[j]	Batch flask (suspension)	Cotton
Hyoscyamus niger[b]	Batch flask (callus)	
Kalanchoe crenata[a]	Batch flask (callus)	CAM
Marchantia polymorpha[b]	Batch flask (suspension)	Liverwort
Morinda lucida[b]	Batch flask (suspension)	
Nicotiana tabacum[b]	Batch fermenter	Tobacco
Ocimum basilicum[e]	Continuous fermenter	Sweet basil
Peganum harmala[b]	Batch flask (suspension)	
Petroselinum crispum[k]	Batch flask (suspension)	Parsley
Psoralea bituminosa[c]	Batch flask (callus)	Leguminosae
Ruta graveolens[b]	Batch flask (callus)	Rue
Solanum tuberosum[l]	?	Potato
Spinacia oleracea[b]	Continuous fermenter	Spinach

[a] I. McLaren and D. R. Thomas, *New Phytol.* **66,** 683 (1967).
[b] Cited in Horn and Widholm.[1]
[c] C. C. Dalton and E. Peel, *in* "Progress in Industrial Microbiology" (M. E. Bushell, ed.), p. 109. Am. Elsevier, New York, 1983.
[d] N. Weber and H. K. Mangold, *Planta* **158,** 111 (1983).
[e] C. C. Dalton, *J. Exp. Bot.* **167,** 1008 (1985).
[f] A plant which exhibits crassulacean acid metabolism.
[g] S. Seeni and A. Gnanam, *Physiol. Plant.* **49,** 465 (1980).
[h] M. Hagimori, T. Matsumoto, and Y. Mikiami, *Plant Cell Physiol.* **25**(6), 1099 (1984).
[i] W. M. Laetsch and H. P. Kortschak, *Plant Physiol.* **49,** 1021 (1972).
[j] Horn and Widholm,[1] p. 135.
[k] F. Ellenbracht, W. Barz, and H. K. Mangold, *Planta* **150,** 114 (1980).
[l] P. C. La Rosa, P. M. Hasegawa, and R. A. Bressan, cited in Chaumont and Gudin.[14]

recent publication. If this is not possible then the next easiest approach is to request an axenic sample of a white cell culture and to use this to obtain a green culture according to Method I (Selection at Subculture) outlined below.

Alternatively a new culture can be initiated, but it generally takes 6 to 12 months from the start of initiation before a quantity of callus sufficient and suitable for experimentation is obtained. Essentially two methods for initiation of green cultures from nonaxenic whole plant material have been developed: *Method 1:* indirect, via nonphotosynthetic cells and *Method 2:* direct, from photosynthetic cells. For both methods aseptic technique should be used.[3] [Readers not experienced in the art are best advised to obtain a few hours of instruction.]

Method 1 (Indirect)

Procurement. Obtain authentic samples of seeds of the species of interest.

Surface Sterilization. Clean and surface sterilize the seeds (one method is to gently remove the seed coat with sand paper, sterilize the "naked" seed with 20% domestic bleach for 5 min, then to wash the seeds several times in sterile distilled water).

Germination. Germinate the surface-sterilized seeds on damp sterile filter paper in sterile Petri dishes.

Callus Initiation. Cut the axenic seedling into pieces (explants) and place these on to the surface of a nutrient medium solidified with 0.8% agar.

If a published protocol (for the species of interest) is available then this should be followed precisely. Where this is not the case Murashige and Skoog[4] medium (MS) should be tried with different auxins [e.g., 1-naphthaleneacetic acid (NAA) or 2,4-dichlorophenoxyacetic acid (2,4-D): 1–10 mg/liter]. Other possible trials could include the addition of organic factors (e.g., coconut water, casamino acids, or vitamins). Although it is possible to make up this medium from pure chemicals it is more convenient to obtain a prepared pack of basic medium from a supplier (e.g., Gibco or Flow Laboratories). There are numerous different media to be found in the literature but most species will produce callus on MS. In our experience more benefit can be gained by varying the physiological state of the tissue and limiting the number of different media involved.

Explant cultures are incubated at constant temperature (usually 25 ± 2°). It is advisable to incubate some cultures in the dark and some in continuous low light provided by warm white fluorescent tubes (80 ± 40 $\mu mol/m^2 \cdot sec$).

[3] J. Reinert and M. M. Yeoman, "Plant Cell and Tissue Culture: A Laboratory Manual." Springer-Verlag, Berlin and New York, 1982.

[4] T. Murashige and F. Skoog, *Physiol. Plant.* **15,** 473 (1962).

Some typical problems encountered during callus initiation with possible solutions are given in Table II.

A variation of this method is to use explants from nonaxenic whole plants which have been grown in the field or greenhouse. However, this variation invariably results in higher losses because of explants being contaminated with microorganisms, being killed by overexposure to the surface sterilant, or not producing callus due to the lack of juvenility in the source tissue.

Callus Culture. Examine the explants at weekly intervals and discard any contaminated by microorganisms. After about 4 weeks the first sign of callus growth should be apparent as a swelling of the explant. Eventu-

TABLE II
PROBLEMS ENCOUNTERED ON INITIATING CALLUS CULTURES AND SOME SOLUTIONS FOUND

Typical problems	Solutions
Contamination by microorganisms	Increase duration or concentration of the sterilant
	Change sterilant (e.g., $HgCl_2$, 0.1%)
	Change sample tissue
	Modify procedure (e.g., use several vacuum cycles to remove bubbles)
	Use antibiotics[a] or fungicides[b]
Explant dies or goes brown	Reduce concentration of surface sterilant or duration of sterilization
	Change sterilant
	Change source tissue (e.g., use stems not roots; use fresh tissue)
	Modify procedure (e.g., wax cut ends of explant to prevent uptake of sterilant)
	Incubate in the dark
Callus does not form or organized tissue forms	Increase the auxin activity (e.g., change to 2,4-D)
	Add organic factors (e.g., amino acids, vitamins, coconut water)
	Change the incubation conditions (e.g., temperature, dark/light)
	Use a different explant (e.g., meristematic tissue: buds, embryos, stem tips, or root tips)

[a] K. Pollock, D. G. Barfield, and R. Shields, *Plant Cell Rep.* **2,** 36 (1983).
[b] R. Shields, S. J. Robinson, and P. A. Anslow, *Plant Cell Rep.* **3,** 33 (1984).

ally this growth manifests itself as the spongy mass of cells (called callus) which usually appears at the cut ends of the explant. When about 5 mm in diameter this primary callus is cut from the explant and then placed on fresh medium and incubated as before.

Callus Bulking Up. The growth of callus on the explant tends to be slow but growth in the subsequent two or three subculture periods can increase. At each subculture the fastest growing and most friable calluses should be selected but at this stage it is not important to have or select for green calluses. During this period callus is bulked up until sufficient has been obtained to provide a supply of material for experimental trials and selection of green cultures.

Selection at Subculture. The conditions chosen or required for the initiation of a fast-growing, friable, nonorganized callus culture may not be optimal for greening. Experimental trials now have to be undertaken to initiate greening, then green parts of the callus are selected at subculture.

If a white culture has been procured from another worker this is the stage at which it can be introduced into the program.

Greening in culture requires white or blue light which can be provided as continuous light (see above, Callus Initiation). The choice of medium for the induction of greening of callus cultures will probably be different to that chosen for callus initiation. Trials should be undertaken to determine an appropriate choice; for guidance the changes noted in Table III have been found to induce greening. One medium based on MS which we have found to be suitable for several dicotyledonous species is described in Table IV. The only monocotyledonous species which we know has been

TABLE III
VARIABLES WHICH HAVE BEEN FOUND TO INCREASE GREENING OF CALLUS CULTURES

Variables	Action
Light	Put culture in white fluorescent light (at about 100 $\mu mol/m^2 \cdot sec$)
Phytohormones	Reduce auxin activity (i.e., change from 2,4-D to NAA and/or reduce the concentration of auxin); increase cytokinin activity
Carbohydrate	Reduce the concentration of readily utilized sugars like glucose (to about 5 g/liter) or replace them with poorly utilized carbohydrates like starch, inulin, raffinose, or lactose

TABLE IV
MODIFICATION OF MS MEDIUM FOUND TO STIMULATE GREENING IN A WIDE RANGE OF SPECIES

Stock	Amount of stock for 1 liter of medium	Ingredients of stock
Salts[a]	100 ml	KH_2PO_4 (8.5 g), $CaCl_2 \cdot 2H_2O$ (22 g), $MgSO_4 \cdot 7H_2O$ (18.5 g), NH_4NO_3 (82.45 g), KNO_3 (95 g), microelement stock (50 ml),[b] distilled water to 5 liters
Organics[a]	5 ml	Mesoinositol (20 g), glutamine (40 g), D-biotin (10 mg), thiamin (20 mg), nicotinic acid (10 mg), pyridoxine (10 mg), glycine (40 mg), distilled water to 1 liter
Iron[a]	5 ml	$Na_2EDTA \cdot 2H_2O$ (7.45 g), $FeSO_4 \cdot 7H_2O$ (5.57 g), distilled water to 1 liter (dissolve $Na_2EDTA \cdot 2H_2O$ and heat to 90° before mixing in the $FeSO_4 \cdot 7H_2O$)
Kinetin[a]	0.5 to 5 ml	20 mg/100 ml
NAA[a]	1 to 10 ml	10 mg/100 ml
Sugar	5 to 15 g	One of the following: sucrose, glucose, starch, lactose, etc.
pH	6.2	
Agar[c]	8 g	

[a] The stock solution is filter sterilized and kept at room temperature.
[b] Microelement stock containing $CuSO_4 \cdot 5H_2O$ (0.025 g/liter), $ZnSO_4 \cdot 7H_2O$ (8.6 g/liter), $MnSO_4 \cdot 4H_2O$ (22.3 g/liter), $CoCl_2 \cdot 6H_2O$ (0.025 g/liter), $Na_2Mo \cdot O_4 \cdot 2H_2O$ (0.25 g/liter), KI (0.83 g/liter), and H_3BO_4 (6.2 g/liter). Stored frozen at −15° in 50-ml aliquots.
[c] Solid medium only.

maintained as a green and photosynthetic culture is *Asparagus officinalis*. In this case the medium of Table IV was modified by omitting biotin and sugar while including 1 mg/liter NAA and 0.5 mg/liter kinetin.

Selection by Plating. A white suspension culture may have been procured. Alternatively a suspension culture can be initiated from a friable callus culture by placing pieces of callus into 150 ml of liquid medium in a 250-ml conical flask and incubating on an orbital shaker at 100 rpm. After the culture has grown, a 10- to 20-ml aliquot of cells is subcultured into 90 ml of fresh medium every 14 days.

"Plating" permits individual cells or very small clumps of cells to be cultured at a density low enough to prevent confluent growth. In this way individual green minicalluses can be selected and sublines obtained.

In practice a low density of about 500 cells/ml is required but growth of minicalluses from a culture at this density requires a "conditioned medium" which can be provided by a nurse layer of cells plated at higher density. This procedure is as follows:

1. Vacuum filter the parent cell suspension through 0.6-mm nylon bolting cloth and use the filtrate to establish a high-density (20,000 cells/ml) layer of cells in solid medium (10 ml in a 90-mm Petri dish). Seal the Petri dish with wax tape and incubate in the dark for 170 hr to establish an actively growing nurse layer.
2. Repeat step (1) but place a sterile filter paper on top of the nurse layer before establishing the low-density (500 cells/ml) layer of cells. Reseal the Petri dish and incubate as before in the dark for 500 hr then transfer to continuous white fluorescent light (50 $\mu mol/m^2 \cdot sec$) for 500 hr.
3. Pick the green minicalluses and place on 5 ml and subsequently 10 ml of solid agar in test tubes. These sublines can then be used to initiate new suspension cultures or they can be maintained as callus cultures.

Initiation and Culture of a Green Suspension Culture. Callus is a convenient way of maintaining a bank of material but it does not provide the basis of a good system for experimentation. This is because of variation within and between different calluses and because of sampling difficulties. A suspension culture is preferred but it is often found that when a suspension is initiated from a green callus culture, the suspension is less green or white when compared to the parent callus.

Even if a green suspension can be initiated without difficulty the culture of the same cells in an aerated fermenter has always been reported to result in degreening.

It is now known that the greening of plant cells, when cultured in the presence of sugar (photoheterotrophic conditions), is influenced by the gas phase. The principle effect is associated with oxygen partial pressure. When the partial pressure of oxygen is above 5% (air is 21%) degreening can occur. In addition there can also be a degreening effect of ethylene (which is often produced by plant cells) above 1 vpm. A photoheterotrophic system consumes oxygen while producing carbon dioxide and ethylene. The rates of consumption or production and the limiting resistance to mass transfer are the principal influences on the partial pressure of these gases at any time during the batch growth cycle.

To obtain greening in suspension culture and in particular a fermenter culture, it is necessary to reduce the partial pressure of oxygen below that found in air. This can be done in several ways: by increasing the resistance offered to mass transfer by the flask neck closure (e.g., using polythene film stuck to the flask neck) or inoculating at high density (e.g.,

50%, v/v). However, the ideal way is to flush the gas phase of a 100-ml mixed suspension culture with 5–10% O_2 balance N_2 at about 25 ml/flask/minute. In this way the operator determines the gas phase; it is not determined by the cells or uncontrolled cultural factors.

In practice gas flushing with 5–10% O_2 is only used for fermenter cultures (preferably in conjunction with a dissolved oxygen probe) or for special experiments in shake flasks. For routine maintenance of suspension cultures in shake flasks it is usually possible to develop a more convenient means of roughly controlling the gas phase.

For each species the exact system has to be determined by trials involving variations in the flask closure, culture volume to flask volume ratio, and use of gas absorbents such as mercuric perchlorate (which reduces the partial pressure of ethylene).

To initiate fermenter cultures special attention must be paid to the method of mixing. Plant cells do not grow when mixed by high shear agitators; slowly rotating large paddles (marine-type impellors) or air-lift fermenters have been found most suitable. It is our experience that bubble mixing with an appropriately placed baffle or paddle at 60 rpm is adequate for photoheterotrophic cell suspensions.[5]

In fermenter cultures dissolved oxygen (dO_2) should be measured and is best controlled by varying the proportion of air in an air–nitrogen mixture (this can be achieved by regular manual adjustments). The best system for experimentation is regarded by some authors to be the continuous culture. However, experiments invariably take 3–9 months and for this reason we prefer to use a fed batch fermenter in which all the sugar is supplied as a glucose syrup in a continuously pumped stream.[6] In this way the two key factors (dO_2 and sugar supply rate) which influence greening in the fermenter can be controlled by the operator.

Development and Maintenance of a Photosynthetic Culture. Even the greenest cultures of plant cells may show little if any photosynthesis. Development of a photosynthetic culture from a green culture can be achieved by gradual or stepwise removal of the supplied sugar while maintaining a low partial pressure of oxygen (5–10% or a dO_2 of 65–125 nmol O_2/ml). As sugar is removed a supply of carbon dioxide at 2–5% is started (mixotrophic conditions). Carbon dioxide is best supplied by gas flushing[1] but the more convenient system is to use a shake flask which has been modified to take a sample of bicarbonate (1.3 *M*)–carbonate (0.7 *M*) buffer.[7]

[5] C. Dalton, *in* "Heliosynthèse et aquaculture" (CNRS, ed.), Semin Martigues (France), 1978, p. 119. AFEDES, Paris, 1980.

[6] C. C. Dalton, *Plant Sci. Lett.* **32,** 263 (1983).

[7] M. Hagimori, T. Matsomoto, and Y. Mikami, *Plant Cell Physiol.* **25,** 1099 (1984).

The growth rate of photosynthetic cultures in the absence of supplied sugar is often limited by the photon flux density. To increase the growth rate the photon flux density can be increased (from 100 to 300 μmol/$m^2 \cdot sec$) or the surface area to volume ratio of the culture system can be increased (e.g., by using a tubular reactor).

Truly photoautotrophic cultures can be obtained by the removal from the medium of all organic factors (vitamins, amino acids, hormones, etc.). This has been achieved for a few species (i.e., *Spinacia oleracea, Chenopodium rubrum, Ruta graveolens*). However, for many species truly autotrophic growth has not been achieved and for some the complete absence of organic factors has been found to be suboptimal for growth.

Method 2 (Direct)

The following method, originally developed by Rossini,[8] has been used successfully to obtain suspensions of mesophyll cells with many species, including *Calystegia sepium* and *Asparagus officinalis*, and has been described in detail by Jullien and Rossini.[9]

Preparation of Material. Approximately 2 g of fresh leaf material is taken from the species of choice and surface sterilized. The main veins are removed, and the remaining tissue cut into small fragments of approximately 25 mm^2.

Cell Separation. The fragments are placed in one or more 25-ml Potter–Elvehjem homogenizers with approximately 10 ml of a liquid medium. The composition of the liquid medium may be varied but typically contains mineral salts, vitamins, and a sugar. For example we have used a modification of the medium of Jullien and Rossini[9] which contained the following components in mg/liter: KNO_3 (950), NH_4NO_3 (725), $MgSO_4 \cdot 7H_2O$ (190), $CaCl_2 \cdot 2H_2O$ (170), KH_2PO_4 (70), Na_2EDTA (37), $FeSO_4 \cdot 7H_2O$ (25), $MnSO_4 \cdot 4H_2O$ (20), H_3BO_3 (10), $ZnSO_4 \cdot 7H_2O$ (0.025), $CuSO_4 \cdot 5H_2O$ (0.025), glycine (2), *myo*-inositol (100), nicotinic acid (5), pyridoxine-HCl (0.5), thiamin-HCl (0.5), folic acid (0.5), biotin (0.05), and sucrose (10,000). The medium is adjusted to a pH of 5.5.

The leaf pieces are gently ground in the medium in the homogenizer until a green suspension is obtained. The resulting mixture will contain pieces of vascular tissue, cell clumps, single cells, and quantities of various cellular debris.

Cell Purification. The mixture is then passed successively through two metal filters with mesh sizes of 0.061- and 0.038-mm diameter

[8] L. Rossini, *C.R. Hebd. Seances Acad. Sci., Ser. D* **268,** 683 (1969).

[9] M. Jullien and L. Rossini, *Ann. Amelior. Plant.* **27,** 87 (1977).

and the filtrate is collected. This removes the larger undissociated leaf pieces.

The filtrate, which contains the isolated cells plus various small particles, is then centrifuged (400 *g*) for 5 min, the supernatant discarded, and the pellet resuspended in a similar volume of medium. This operation is repeated a second time after which a clean preparation of isolated cells is obtained.

In the case of species such as *Asparagus officinalis,* between 55 and 65% of the isolated cells obtained by this method are considered to be intact and in good condition. However, the amount of isolated cells obtained and their viability vary markedly with different species and conditions. The above method has been evaluated[9] on some 200 species of which 27 from 20 different genera gave good results; it can readily be carried out under sterile conditions to yield an axenic suspension of cells.[8]

Applicability of the Method. Not all species will yield intact mesophyll cells following mechanical dissociation of leaf tissue. Even in those which do give high yields of isolated cells, the viability of the cells can be very variable depending upon both the physiological state of the tissue used and the exact method employed. Workers wishing to use these methods will need to optimize and standardize all aspects of the technique for their own material before consistent results can be guaranteed. For certain species mechanical methods have to be combined with enzymatic treatment.

Other Direct Methods. Gnanam and Kulandaivelu[10] used mild grinding in a smooth-surfaced porcelain pestle and mortar, and a different medium to obtain mechanically isolated leaf cell suspensions from leaves of a variety of species. They reported 24 species including both monocots and dicots which gave mesophyll cell suspensions using this method. Rajendrudu *et al.*[11] also used a pestle and mortar and a medium containing both salts and 0.7 *M* mannitol to screen 146 species of angiosperms from 35 families for the ability of their leaf tissue to yield living mesophyll cells on mild maceration. They found that 73 species from 22 families yielded intact mesophyll cells by this method. Miksch and Beiderbeck[12] tested 96 dicotyledonous plants for their yield of single cells after grinding in a Potter–Elvehjem homogenizer. They reported that about 30% of species tested yielded more than 2×10^6 cells/g fresh weight of leaves used.

Axenic Culture. Although isolated leaf cells have been used to initi-

[10] A. Gnanam and G. Kulandaivelu, *Plant Physiol.* **44,** 1451 (1969).

[11] G. Rajendrudu, I. Madhusudana Rao, A. S. Raghavendra, and V. S. Rama Das, *Proc.—Indian Acad. Sci., Sect. B* **88B,** Part II (2), 143 (1979).

[12] U. Miksch and R. Beiderbeck, *Biochem. Physiol. Pflanz.* **169,** 191 (1976).

TABLE V

GENERALIZED COMPARISON BETWEEN GREEN CULTURED CELLS AND LEAF CELLS (C_3 PLANTS)

Characteristic	Green cultured cells	Leaf cells
Doubling time (*d*)[a]	1 to 7	0
Chlorophyll content (mg/g dry biomass)[a]	1 to 3	3 to 8
Photosynthetic rate (μmol/mg Chl.h)[b]	60 to 500	100 to 300
Quantum flux density which saturates photosynthesis[b]	Low	High
CO_2 requirement (%)[b]	2 to 5	0.03 to 0.1
K_m HCO_3^- (m*M*)[b]	0.3 to 1.2	0.2 to 1.0
Photorespiration[b]	Suppressed	Active
PEP-carboxylase activity[b]	High	Low
Membrane input resistance (ohms)[c]	270	28
Membrane potential (mV)[c]	−180	−180

[a] Husemann[2] and Chaumont and Gudin.[14]

[b] Horn and Widholm[1] and Husemann.[2]

[c] Cited in Ref. 1 as T. Ohkawa, K. Kohler, and F.-W. Bentrup, *Planta* **151,** 88 (1981).

TABLE VI

LIPID CLASSES OF GREEN CULTURED CELLS AND LEAF CELLS FOR TWO SPECIES

	Glycine max[a]		*Petroselinum crispum*[b]	
Lipid class	Green culture	Leaf	Green culture	Leaf
Phosphatidylcholines	21.4	17.9	39.2	15.5
Phosphatidylethanolamines	2.3	9.6	12.8	5.9
Phosphatidylglycerols	7.1	7.4	0	5.4
Monogalactosyldiacylglycerols	27.8	29.6	11.4	16.8
Digalactosyldiacylglycerols	28.1	30	6.8	29.9
Sulfoquinovosyldiacylglycerols	—	—	2.4	3.4
Acylsterylglycosides	—	—	8.3	18.1
Steryl esters	—	—	2.1	1.4
Triacylglycerols	5.4	0.3	10.2	3.6
Free fatty acids	1.4	0.6	—	—
Diacylglycerol	4.4	0.2	—	—
Phosphatidic acid	0.4	1.4	—	—
Total lipids (percentage fresh biomass)	—	—	0.51	0.67
Total lipids (μmol/g fresh biomass)	14.7	26.4	—	—

[a] Expressed as mol% total lipids after B. A. Martin, M. E. Horn, J. M. Widholm, and R. W. Rinne, *Biochim. Biophys. Acta* **796,** 146 (1984).

[b] Expressed as wt% total lipids after F. Ellenbracht, W. Barz, and H. K. Mangold, *Planta* **150,** 114 (1980).

ate several white cultures,[9,13] only a few recent reports[14,15] describe the use of this method for the selection of green and photosynthetic cell cultures.

Using a variation of this direct method, Yasuda *et al.*[16] placed axenic leaf segments directly onto sugar-free medium in the presence of 1% CO_2. Photoautotrophic cells were selected at high efficiency using this method.

Characteristics of Green Plant Cells

A general comparison is made between green cultured cells and leaf cells (Table V). The lipid composition of cultured cells has been reviewed by Radwan and Mangold[17]; a comparison of lipid classes for green cultured cells and leaf cells is presented in Table VI. In general the similarity between green cultured cells and leaf cells has been found to increase concomitant with the photosynthetic development of the cultured cells. The notable differences are that photosynthetic cultured cells require a higher partial pressure of CO_2 for maximum photoautotrophic growth, they have a high phosphoenolpyruvate (PEP) carboxylase activity, and are actively dividing.

[13] H. Lang and H. W. Kohlenbach, *in* "Production of Natural Compounds by Cell Culture Methods" (A. W. Alfermann and E. Reinhard, eds.), p. 274. Gesellschaft fur Strahlen-und Umweltforschung mbH, Munich, 1978.

[14] D. Chaumont and C. Gudin, *Biomass* **8,** 41 (1985).

[15] S. Seeni, M. Krishnan, and A. Gnanam, *Plant Cell Physiol.* **24,** 823 (1983).

[16] T. Yasuda, T. Hashimoto, F. Sato, and Y. Yamada, *Plant Cell Physiol.* **21,** 929 (1980).

[17] S. S. Radwan and H. K. Mangold, *Adv. Biochem. Eng.* **16,** 109 (1980).

[3] Preparation of Protoplasts from Plant Tissues for Organelle Isolation

By MIKIO NISHIMURA, IKUKO HARA-NISHIMURA, and TAKASHI AKAZAWA

Isolated protoplasts offer some advantages as an experimental system for studies of plant cell biology,[1] one of which is their use as a starting material for the preparation of constituent organelles. When intact tissues are used, even with favorable material and propitious disintegration pro-

[1] E. Galun, *Annu. Rev. Plant Physiol.* **32,** 237 (1981).

cedures, the forces necessary to bring about rupture of the cell wall severely damage the organelles. The use of protoplasts removes this impediment and cell breakage can then be achieved by procedures that minimize this damage. Since cell wall materials are enzymatically degraded, the protoplasts are surrounded only by the plasma membrane, which is easily ruptured.

With the availability of potentially powerful microbial hydrolases, the preparation of protoplasts from various types of plant cells has been reported.[1,2] Commercially available hydrolases are listed in Table I. The most popular is cellulase Onozuka, which is derived from the fungus *Trichoderma viride,* which attacks the mushroom *Lentinus edodes*. Cellulase Onozuka RS is derived from a mutant strain of the fungus that has a stronger cellulase activity and shortens the time required for preparation of protoplasts. Macerozyme and pectolyase are frequently used pectinases (polygalacturonase). The different combinations of cellulases and pectinases, as well as the osmolarity of the enzyme solution, are important factors for the preparation of protoplasts from plant tissues. As these hydrolases are not purified and are contaminated by many different enzymes (e.g., proteases) and even toxic substances, we recommend that investigators purchase these enzymes in a small lot, and check protoplast formation and intactness of the prepared protoplasts as described in Nishimura *et al.*[3,14] before purchasing a large amount. Most of these hydrolases are products of Japanese companies, but the same enzymes are available from Calbiochem and Sigma (Table I). Here we describe the methods for preparing protoplasts from three different plant tissues.

Preparation of Protoplasts

Protoplasts from Grapevine (Vitas vinifera) Leaves[3]

Reagents

Medium A: 1% (w/v) cellulase Onozuka RS and 1% (w/v) Driselase dissolved in 0.7 *M* mannitol containing 5 m*M* dithiothreitol, 0.4% (w/v) polyethylene glycol (6000), 5 m*M* KCl, 2 m*M* $CaCl_2$, and 25 m*M* 2-[*N*-morpholino]ethanesulfonic acid (MES)–KOH (pH 5.8). Before the start of digestion, medium A is centrifuged at 500 *g* for 10 min in order to remove the insoluble materials present in Driselase and then the pH is adjusted to 5.8

[2] Y. P. S. Bajaj, *in* "Plant Cell, Tissue, and Organ Culture" (J. Reinert and Y. P. S. Bajaj, eds.), p. 467. Springer-Verlag, Berlin and New York, 1977.

[3] M. Nishimura, I. Hara-Nishimura, and S. P. Robinson, *Plant Sci. Lett.* **37,** 171 (1984).

TABLE I
HYDROLASES FOR ISOLATION OF PROTOPLASTS

Name	Major components	Origin	Source[a]
Cellulase			
Cellulase Onozuka R-10	Cellulase	*Trichoderma viride*	Yakult
Cellulysin	—	—	Calbiochem
Cellulase	—	—	Sigma
Cellulase Onozuka RS	Cellulase	*Trichoderma viride* (mutant)	Yakult
Cellulase	—	—	Sigma
Meicelase	Cellulase	*Trichoderma koninge*	Meiji Seika
Cellulase	Cellulase	*Aspergillus niger*	Sigma
Cellulase	Cellulase	*Penicillium funiculosum*	Sigma
Driselase	Cellulase Hemicellulase Polygalacturonase	*Basidiomycetes*	Kyowa Hakko
Driselase	—	—	Sigma
Pectinase			
Macerozyme R-10	Polygalacturonase	*Rhizopus* sp.	Yakult
Macerase	—	—	Calbiochem
Pectinase	—	—	Sigma
Pectolyase Y-23	Endopolygalacturonase Endopectinylase	*Aspergillus japonicus*	Seishin
Pectolyase	—	—	Sigma
Pectinase	Polygalacturonase	*Aspergillus niger*	Sigma
Other			
Hemicellulase	Hemicellulase	*Aspergillus niger*	Sigma

[a] Yakult Honsha Co., Ltd., Higashishinbashi, 1-1-19, Minato, Tokyo 105, Japan; Calbiochem-Behring, P.O. Box 12087, San Diego, CA 92112, USA; Sigma, P.O. Box 14508, St. Louis, MO 63178, USA; Meiji Seika, Ltd., Kyobashi-2, Chuo, Tokyo 104, Japan; Kyowa Hakko Kogyo Co., Ltd., Otemachi 1-6-1, Chiyoda, Tokyo 100, Japan; Seishin Pharm. Co., Ltd., Kaomi 4-13, Nihonbashi, Chuo, Tokyo 103, Japan.

Medium B: 0.7 *M* mannitol containing 10 m*M* HEPES–KOH (pH 7.5)
70% Percoll (Pharmacia), containing medium B (pH should be adjusted to 7.5). Solutions of 50% and 25% Percoll in medium B are prepared by dilution of the 70% Percoll stock solution with medium B

Young leaves (leaf length <7 cm) were harvested from the grapevines and left in the dark for 1–2 hr in order to decrease the starch content in the leaves. The leaves (0.4 g) were cut into 1-mm slices with a razor blade and placed in a Petri dish containing 10 ml of medium A. After vacuum infiltration (100 mm Hg) for 1 min, enzymatic digestion was carried out on a reciprocal shaker (40 strokes/min) at room temperature. After 3–4 hr, many protoplasts were released from the leaf segments.

The medium A containing protoplasts was collected by filtration through nylon bolting cloth (35 mesh) and was centrifuged at 100 *g* for 3 min. The supernatant fluid was discarded and the pellet was resuspended in 1 ml of 70% Percoll containing medium B. One ml of 50, 25, and 0% Percoll (each in medium B) were layered on top of the suspension. This Percoll step gradient was centrifuged at 200 *g* for 15 min. After centrifugation, the protoplasts were floated at the interface between the 50 and 25% Percoll layers. The protoplast fraction was carefully collected with a Pasteur pipet and washed with 3 ml of medium B. The purified protoplasts were suspended in 1 ml of medium B.

The isolated protoplasts showed photosynthetic activities of 20–30 μmol O_2 evolved/mg chlorophyll/hr and respiratory activities of 7–8 μmol O_2 uptake/mg chlorophyll/hr. The protoplasts kept at 4° in darkness did not lose the original photosynthetic activities during at least 2 hr of storage.

Protoplasts from Castor Bean (Ricinus communis) Endosperm[4]

Reagents

Medium A: 0.5% (w/v) Macerozyme R-10 and 2% (w/v) cellulase Onozuka R-10 dissolved in 0.7 *M* mannitol (pH adjusted to 5.5)
Medium B: 0.7 *M* mannitol (pH adjusted to 5.5)

Protoplasts can be readily prepared from 3- to 7-day-old germinating castor bean endosperm (grown at 25°) by this method. However, using this procedure, protoplasts cannot be prepared from dry seed or 1- to 2-day-old endosperm. Sixteen endosperm halves (about 7 g) were cut into 1-mm slices with a razor blade and placed in a 50-ml Erlenmeyer flask containing 8 ml of medium A. After vacuum infiltration (70 mm Hg) for 30 sec, enzymatic digestion was carried out on a reciprocal shaker (72 strokes/min) at room temperature. As no protoplasts are released during the first 1-hr treatment, the medium A was removed and 7 ml of fresh medium A was substituted. After a further 2-hr treatment, the medium A containing protoplasts was collected by filtration through nylon bolting cloth (35 mesh). Seven milliliters of fresh medium A was added to the slices and the whole mixture was incubated for an additional 2-hr period.

[4] M. Nishimura and H. Beevers, *Plant Physiol.* **62,** 40 (1978).

The protoplast suspensions were combined and collected by centrifugation at about 200 *g* for 3 min. The pellet was resuspended and washed with 3 ml of medium B. The isolated protoplast showed a stable O_2 uptake (18 μmol $O_2/10^7$ protoplasts · hr) and did not lose the activity during at least 6 hr when the protoplasts were kept at 4°.

Protoplasts from Suspension-Cultured Cells of Sycamore (Acer pseudoplatanus)[5]

Reagents

Cell culture medium: see Bligny and Leguay[6] in this volume

Medium A: 1% (w/v) cellulase Onozuka RS and 0.1% (w/v) Pectolyase dissolved in the culture medium containing 0.5 *M* mannitol (pH adjusted to 5.8)

Medium B: Culture medium containing 0.5 *M* mannitol

Medium C: 0.5 *M* mannitol containing 2% sucrose

Sycamore cells were grown in a 500-ml Erlenmeyer flask containing 200 ml of culture medium. Before the enzymatic digestion, the cells were suspended twice for 25 min in medium B for plasmolysis, and then collected using Miracloth. Usually 40–50 g of cells (fresh weight) was suspended in 150 ml of medium A and distributed into three 100-ml Erlenmeyer flasks. The enzymatic digestion was carried out using a reciprocal shaker (72 strokes/min) at room temperature (25°) for 1.5 to 2 hr. The formation of protoplasts was time-sequentially monitored by light microscope. After the appearance of a large number of protoplasts, the whole preparation was filtered through Miracloth to remove cells; the protoplasts were collected by centrifugation (100 *g*, 3 min), and finally washed twice with medium C.

Purification of the protoplasts was achieved following the methods described above for *Vitas vinefera* except the concentration of mannitol was 0.5 *M* instead of 0.7 *M*.

Intactness of Prepared Protoplasts

It is crucial that the protoplasts isolated are intact and the constituent organelles are not damaged. An additional requirement is the stability of protoplasts, since it takes a certain length of time to finally isolate the

[5] M. S. Ali, M. Nishimura, T. Mitsui, T. Akazawa, and K. Kojima, *Plant Cell Physiol.* **26,** 1119 (1985).

[6] R. Bligny and J.-J. Leguay, this volume [1].

organelles after rupturing protoplasts. Several dye staining techniques[7,8] and some cellular activities such as respiration[4,9] and photosynthesis[10,11] have been employed as markers of potential activities of isolated protoplasts. However, these methods not only do not provide a quantitative measure of intactness, but also do not secure the intrinsic potential activities of the original cell. Protoplasts have been isolated from oat and barley aleurone layers,[9,12] but only the oat protoplasts were shown to produce α-amylase in response to gibberellin A_3 (GA_3). The preparation of intact protoplasts from mature barley aleurone layers, capable of producing GA_3-responsive α-amylase, has been reported.[13]

A simple and rapid method was developed for the quantitative analysis of intactness of isolated protoplasts using glycolate oxidase activity as an index (Fig. 1).[14] This method is generally applicable to green leaf tissues and, in fact, has been applied to grapevine protoplasts (see above).

Reagents

Isolated protoplasts suspended in 10 m*M* potassium phosphate (pH 7.2) and suitable osmoticum
KCN, 0.1 *M*
Sodium glycolate, 0.3 *M*
Triton X-100, 10% (w/v)

Throughout the analysis, experiments are performed in the dark to prevent photosynthesis with chloroplast-containing protoplasts. One milliliter of the protoplast suspension is placed in an oxygen electrode chamber (Hansatech O_2-Electrode, King's Lynn, Norfolk, UK) and the rate of respiration is measured (V_1). To suppress protoplast respiration and inhibit catalase activity, KCN (2 μl) is added to the preparation. Consumption of O_2 may not be completely inhibited by 200 μM KCN (V_2), because some plant mitochondria possess cyanide-insensitive respiration. The addition of glycolate (5 μl) causes an increase of O_2 consumption ($V_3 - V_2$) ascribable to glycolate oxidase either released from protoplasts or to damaged ("leaky") protoplasts, as glycolate is impermeable to intact protoplasts under these conditions. The total activity of glycolate oxidase in protoplasts can be measured after addition of 5 μl of 10% Triton X-100

[7] K. Glimelius, A. Wallin, and T. Eriksson, *Physiol. Plant.* **31,** 225 (1974).
[8] P. J. Larkin, *Planta* **128,** 213 (1976).
[9] L. Taiz and R. L. Jones, *Planta* **101,** 95 (1971).
[10] G. E. Edwards and S. C. Huber, *Encycl. Plant Physiol., New Ser.* **6,** 102 (1978).
[11] M. Nishimura and T. Akazawa, *Plant Physiol.* **55,** 712 (1975).
[12] R. Hooley, *Planta* **154,** 29 (1982).
[13] J. V. Jacobsen, J. A. Zwar, and P. M. Chandler, *Planta* **163,** 430 (1985).
[14] M. Nishimura, R. Douce, and T. Akazawa, *Plant Physiol.* **78,** 343 (1985).

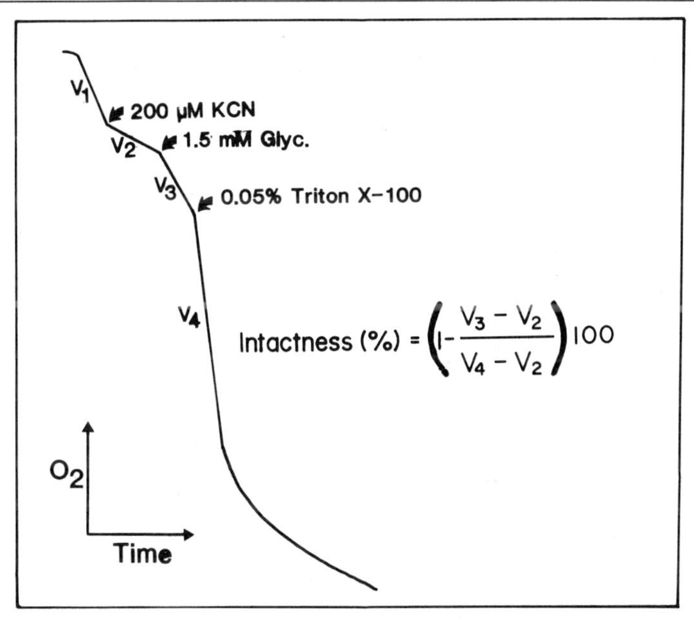

FIG. 1. Principle of calculating intactness of protoplasts using glycolate oxidase activities.

$(V_4 - V_2)$. From the ratio of glycolate oxidase activity before and after the addition of Triton X-100, percentage of intactness of protoplasts can be calculated as:

$$\text{Intactness (\%)} = \{1 - [(V_3 - V_2)/(V_4 - V_2)]\}100$$

From our experiences, intactness of 80–90% can readily be sustained in preparations of protoplasts isolated from leaves of grapevine[3] and spinach.[13]

Disruption of Protoplasts

Osmotic shock is generally considered to be the most reliable method for gently disrupting protoplasts. However, it is frequently encountered that a simple osmotic shock is not sufficient to completely release cellular organelles. A gentle mechanical device such as passage through nylon mesh or Miracloth[4,15] or Teflon homogenizer[16] has been successfully used

[15] M. Nishimura, D. Graham, and T. Akazawa, *Plant Physiol.* **58,** 309 (1976).

[16] M. Nishimura, R. Douce, and T. Akazawa, *Plant Physiol.* **69,** 916 (1982).

to isolate chloroplasts,[15] mitochondria,[16] microbodies,[4,15] vacuoles,[17–19] Golgi apparatus,[5] plastids,[4] and amyloplasts[20] from plant protoplasts. The details of these organelle preparations should be addressed in their original papers.

Generally speaking, refined separation techniques are required in this step and each individual investigator needs to develop and/or improve the method most suitable and satisfactory for a given research purpose. Most plant organelles are structurally labile compared to the protoplasts. One of the critical factors is the change of osmolarity during organelle preparation. For example, it is difficult for the high osmolarity of sucrose encountered to isolate "photosynthetically competent" chloroplasts using sucrose gradient centrifugation, even though the individual organelles, e.g., chloroplasts, mitochondria, and peroxisomes, are well separated by this technique.[12] To overcome this problem, Percoll and Ficoll are better gradient solvents, and allow the isolation of functionally active "intact" mitochondria[16] and chloroplasts.[21]

[17] M. Nishimura and H. Beevers, *Plant Physiol.* **62,** 44 (1978).
[18] S. Asami, I. Hara-Nishimura, M. Nishimura, and T. Akazawa, *Plant Physiol.* **77,** 963 (1985).
[19] I. Hara-Nishimura, M. Hayashi, M. Nishimura, and T. Akazawa, *Protoplasma* **136,** 49 (1987).
[20] D. Macherel, H. Kobayashi, T. Akazawa, S. Kawano, and T. Kuroiwa, *Biochem. Biophys. Res. Commun.* **133,** 140 (1985).
[21] T. Takabe, M. Nishimura, and T. Akazawa, *Agric. Biol. Chem.* **43,** 2137 (1979).

[4] Interaction of Plant Protoplast and Liposome

By TOSHIYUKI NAGATA

The study of the interaction between plant protoplasts and liposomes relies on the elucidation of the biological characteristics of plant membranes. The reasons for using this combination of protoplast and liposome are that (1) in the plant system the protoplast is the sole material whose cell membrane is exposed to the environment and can be directly manipulated *in situ* from the outside and (2) the liposome retains characteristics of model membranes and its mutual interactions have been well established.[1,2] This approach therefore could resolve the dynamic aspects of

[1] D. Papahadjopoulos, ed., *Ann. N.Y. Acad. Sci.* **308,** 1 (1978).
[2] F. Szoka, Jr., and D. Papahadjopoulos, *Annu. Rev. Biophys. Bioeng.* **9,** 467 (1980).

the cell membranes of plant cells. When both materials are mixed in an appropriate solvent, an electrostatic interaction between both surfaces is observed.[3] By controlling environmental conditions, molecular interactions between them are expected. As described below, the endocytotic uptake of liposomes by protoplasts has been observed.

In this chapter we will discuss the interaction of plant protoplasts with liposomes and describe the preparation of protoplasts and liposomes based on our laboratory studies. As there are several comprehensive reviews on these subjects,[4,5] only minimal information directly related to the present subject is described.

Preparation of Protoplasts. In our study of the interaction between protoplasts and liposomes, protoplasts from the suspension culture cell line of tobacco BY-2 (*Nicotiana tabacum* L. cv. BY-2) were used most frequently. This cell line is maintained by a weekly interval subculture at 50 times dilution with a fresh medium consisting of modified Linsmaier–Skoog's medium[6] (KH_2PO_4 and thiamine–HCl are increased to 370 and 1 mg/liter, respectively), supplemented with 3% sucrose, 0.2 mg/liter, 2,4-dichlorophenoxyacetic acid (2,4-D), pH 5.8, on a gyratory shaker at 120 rpm at 26° in the dark. Protoplasts were prepared from the cells cultured for 3–4 days after subculture by treating with an enzyme mixture of 0.1% pectolyase Y23 and 1% cellulase YC dissolved in 0.4 *M* D-mannitol, pH adjusted to 5.5. Both enzymes were purchased from Seishin Pharmaceutical Company, Ltd. (Nihonbashi-Koamicho, Chuoku, Tokyo). Incubation for 1 hr at 30° with gentle occasional swirlings was sufficient to convert all of cells into protoplasts.[7] The enzymatic removal of cell walls was monitored by observation under a fluorescence microscope after staining with Calcofluor White ST (American Cyanamid Company, Wayne, NJ),[8] which has specific affinity to cellulose. The naked protoplasts do not show any fluorescence under ultraviolet light. After enzyme treatment protoplasts were collected by centrifugation at 150 *g* for 2 min.

Preparation of Liposomes. There are several types of liposomes dependent on the preparation procedures—multilamellar vesicles (MLV), small unilamellar vesicles (SUV), and large unilamellar vesicles (LUV)[1]—but LUV is the most suitable for studying the interaction with cells. LUV

[3] T. Nagata, *Encycl. Plant Physiol., New Ser.* **17,** 491 (1984).

[4] G. Gregoriadis, ed., "Liposome Technology," Vols. 1–3. CRC Press, Boca Raton, Florida, 1984.

[5] I. K. Vasil, ed., "Cell Culture and Somatic Cell Genetics of Plants," Vol. 1. Academic Press, Orlando, Florida, 1984.

[6] E. M. Linsmaier and F. Skoog, *Physiol. Plant.* **18,** 100 (1965).

[7] T. Nagata, K. Okada, I. Takebe, and C. Matsui, *Mol. Gen. Genet.* **184,** 161 (1981).

[8] T. Nagata and I. Takebe, *Planta* **92,** 12 (1970).

can encapsulate appropriate marker materials, whereby interaction with cells can be explicitly monitored. Although LUV liposomes can be prepared according to several methods, the reversed-phase evaporation vesicles (REV)[9] method is described here because of its versatility in further applications.

In our study tobacco mosaic virus (TMV) RNA was encapsulated in liposomes. The introduction of TMV RNA into cells can be easily assessed by staining with a fluorochrome-labeled antibody to TMV, if this RNA retains its biological function.

REV liposomes are prepared according to the procedure described in Nagata *et al.*[7] L-α-Phosphatidyl-L-serine (PS) (Pharmacia P-L Biochem., Inc.) (5 μmol) and cholesterol (Chol) (5 μmol) dissolved in chloroform are mixed in a 50-ml round-bottom centrifuge tube and the solvent is evaporated by a stream of nitrogen gas. The phospholipids are redissolved in 1.0 ml of diethyl ether, from which peroxide was removed by repeated distillation. The resolved phospholipids are transferred to a test tube with a screw cap (13 mm × 105 mm). TMV RNA (100 μg) dissolved in 0.33 ml of TES buffer containing 2 m*M* *N*-[tris(hydroxymethyl)methyl-2-amino]-ethanesulfonic acid, 2 m*M* L-histidine, 150 m*M* NaCl, and 0.1 m*M* EDTA, pH 7.4, is added to the phospholipid solution. The resulting two-phase solution is shaken vigorously by hand for 1 min and then sonicated three times for 5 sec at 3-sec intervals by a low-power sonicator (28 kHz and 20 W, UR-20 P of Tomy Seiko Company, Ltd., Tokyo) to make the water phase opalescent. The centrifuge tube containing this mixture is placed in a round-glass joint to a rotary evaporator (#5030-11 of Shibata Chem. Apparatus Mfg. Co., Ltd., Tokyo). Rotary evaporation is continued at 45 rpm at 400 mmHg until the suspension becomes a gel and is followed by brief mixing with a Vortex mixer (Vortex-Genie, Sci. Ind., Inc., Bohemia, NY). Evaporation is then carried out at 730 mmHg until a homogeneous suspension is obtained. The prepared liposomes are washed once with TES buffer by centrifugation at 48,000 *g* for 20 min and are finally suspended in 0.5 ml of 0.4 *M* D-mannitol. According to the same procedure other liposomes of phosphatidylcholine (PC) and of PC and stearylamine (SA) (9 : 1), which have neutral and positive charges, respectively, are prepared. Liposomes of PS and cholesterol give negative charges.

Treatment of Protoplasts with Liposomes. Protoplasts (3×10^6) are pelleted in a 10-ml, round-bottom, screw-capped centrifuge tube and the supernatant is discarded. To this pellet is added 0.1 ml suspension of liposomes (2 μmol of lipids) encapsulating TMV RNA and the mixture is allowed to stand for 10 min. Then an equal volume of either 25% (w/w)

[9] F. Szoka, Jr., and D. Papahadjopoulos, *Proc. Natl. Acad. Sci. U.S.A.* **75,** 4194 (1978).

polyethylene glycol (PEG) 1540 (Koch-Light Lab., Ltd., Colnbrook, Buchs, England or Nippon Oil & Fat Co., Ltd., Yurakucho, Tokyo) or 20% (w/w) poly(vinyl alcohol) (PVA) #203 (degree of polymerization 300, 88% hydrolyzed, Kuraray Co., Kurashiki, Okayama, Japan) dissolved in the protoplast culture medium shown below is added dropwise. The mixture is incubated at room temperature for 10 min in the case of PEG and for 20 min in the case of PVA. Thereafter, PEG or PVA is diluted with 10 vol of the high pH–high Ca medium and the protoplasts are collected by centrifugation at 50 *g* for 5 min. The high pH–high Ca medium is prepared just before use by mixing 0.1 *M* glycine–NaOH buffer, pH 10.5, with an equal volume of 0.1 *M* $CaCl_2$ containing 0.8 *M* D-mannitol. The protoplasts are washed once and finally suspended in 10 ml of the protoplast culture medium. The composition of this medium is the modified Linsmaier–Skoog medium[6] supplemented with 1% sucrose, 0.2 mg/liter 2,4-D, and 0.4 *M* D-mannitol, pH 5.8. The protoplasts are cultured at a concentration of 3×10^5/ml at 28° in diffuse light in 10-cm plastic Petri dishes (Falcon #1001) sealed with Parafilm M.

Assessment of the Association of Protoplasts with Liposomes. The association of protoplasts with liposomes encapsulating TMV RNA can be assessed with a fluorescent antibody technique. If TMV RNA is introduced into protoplasts mediated by liposomes, viral coat proteins are produced to form virus particles, which can be detected by an FITC-labeled antibody to TMV. After 24 hr in the culture medium aliquots of the treated protoplasts are smeared on glass slides previously coated with Meyer's albumin. After fixation with ethanol, the smeared protoplasts are washed for 20 min in phosphate-buffered saline (PBS), pH 7.0, and then stained with the FITC-labeled antibody for 1 hr. After excess antibody is removed by washing with PBS, the infection percentage is determined under a fluorescence microscope. Under optimal condition, 95% of 3×10^6 protoplasts are infected with TMV RNA encapsulated in liposomes.

Interaction of Liposomes with Protoplasts. As shown above, TMV RNA encapsulated in liposomes was successfully introduced into protoplasts. Then the process of introducing liposomes into protoplasts was examined in an attempt to understand the underlying mechanism. Initially it was thought that this process might involve the fusion of protoplast membranes and liposomes, because the conditions employed used PEG or PVA in combination with high pH–high Ca washing. These same conditions were utilized for the fusion between protoplasts.[3] However, it was not the case as is described below.

The process involving interaction of liposomes with protoplasts was examined by electron microscopy. In this experiment liposomes are prepared from PS or PS and cholesterol (1 : 1) according to the Ca–EDTA

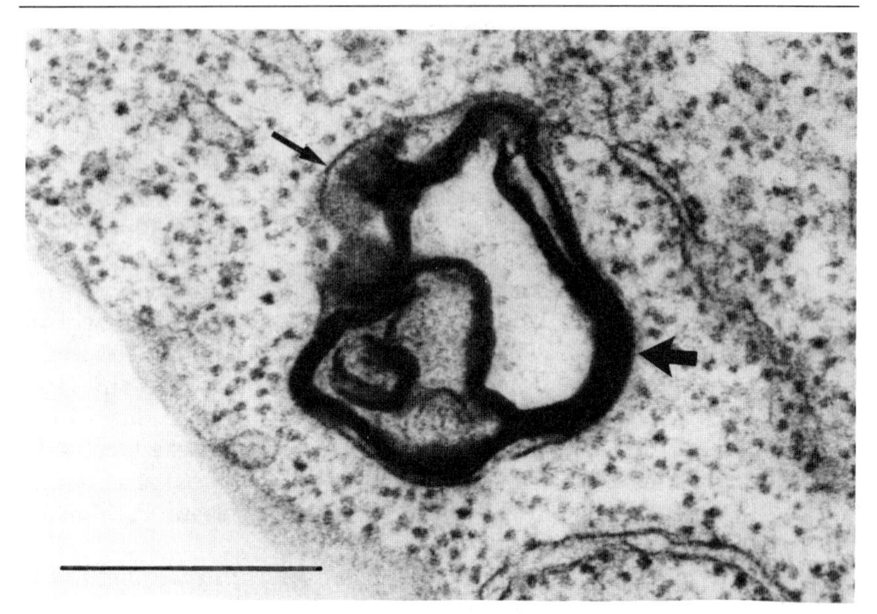

FIG. 1. Electron micrograph of the uptake of liposomes by *Vinca rosea* protoplasts. Large arrow shows the liposome taken up into protoplasts; small arrow shows the plasmalemma which encloses the liposome after introduction into the protoplast. Bar, 0.5 μm.

chelation method and added to the protoplasts from suspension culture cells of *Vinca rosea*.[10] The mixture is treated with PEG or PVA followed by washing with the high pH–high Ca medium. The process is observed using a transmission electron microscope after fixation. Conventional fixation procedure employing glutaraldehyde fixation followed by postfixation with osmium tetroxide cannot be used to observe the interaction between protoplasts and liposomes, as this procedure ruptures liposomes surrounding protoplast surfaces. However, simultaneous fixation with a mixture of glutaraldehyde and osmium tetroxide enabled us to observe the whole process of the uptake of liposomes by protoplasts. The fixative is freshly prepared by adding to 2.5 ml of 0.2 *M* cacodylate buffer, pH 7.2, 0.3 g sucrose, 2.5 ml of 8% glutaraldehyde, and 5.0 ml of 4% osmium tetroxide solution in this sequence in an ice bath. A half-volume of the fixative is added to the protoplast/liposome suspension and the protoplasts are immediately collected by low-speed centrifugation at 100 *g* for 2 min. The protoplasts are suspended in 2 ml of the fresh fixative and

[10] Y. Fukunaga, T. Nagata, I. Takebe, T. Kakehi, and C. Matsui, *Exp. Cell Res.* **144,** 181 (1983).

centrifuged again. This procedure is repeated until the supernatant is no longer a yellowish brown color. The protoplasts are then kept in the fixative for 4 hr in the ice bath. The fixed protoplasts are dehydrated with ethanol and embedded in a low-viscosity epoxy resin. Sections are stained with bismuth subnitrate. Electron microscopy shows that liposomes are taken up very rapidly by protoplasts by the process of endocytosis (Fig. 1).

Concluding Remarks. In this chapter, it has been shown that liposomes of PS and Chol are taken up by plant protoplasts via the process of endocytosis, which is not usually observed in plant cells. The procedure involves treatment with PEG or PVA in combination with the high pH–high Ca washing. However, detailed mechanisms of the interaction between protoplasts and liposomes at molecular level have not yet been resolved. The basic question of why liposomes do not fuse with protoplast membranes, but are taken up endocytotically by protoplasts, has not yet been explained at the molecular level. Furthermore, it has not yet been determined why liposomes of PS and Chol are taken up by protoplasts in preference to other types of liposomes. In order to describe the dynamic nature of plant membranes using liposomes, further comprehensive studies are necessary.

[5] Liposomes as Carriers for the Transfer and Expression of Nucleic Acids into Higher Plant Protoplasts

By MICHEL CABOCHE and PAUL F. LURQUIN

Introduction

Higher plant protoplasts constitute an attractive system for the study of the expression of viral RNA and foreign genes under the control of appropriate promoter regions. Protoplasts, however, are quite refractory to the uptake of nucleic acids which occurs at a significant level only when protoplasts are incubated in the presence of one of several possible "helpers" such as poly(L-ornithine), poly(L-lysine), or polyethylene glycol.

In this chapter, we present an alternative method based on the encapsulation of nucleic acids into large unilamellar liposomes. This technique has been proved to yield high viral RNA-mediated transfection values and, in one case, was shown to be the only one allowing successful transfection of protoplasts with potyviral RNA.[1] In addition, encapsulation of recombinant DNA into such liposomes followed by interaction with protoplasts was shown to enable the recovery of regenerated tobacco plants carrying and expressing the *neo* gene from *Escherichia coli,* that is, transformed to kanamycin resistance.[2]

The earlier literature describing the encapsulation of nucleic acids into a variety of liposome preparations and their interaction with protoplasts has been reviewed.[3,4] An analysis of the published data indicates that nucleic acids trapped in liposomes produced by the reversed-phase evaporation method[5] can be transferred efficiently into plant protoplasts and can be biologically expressed in the cytoplasm and in the nucleus of the recipient cells.[1,2,6–11] In most cases, liposomes were chiefly composed of phosphatidylserine and were, therefore, negatively charged.

The technique described below has been successfully used to transfer and to express tobacco mosaic virus (TMV) RNA and the recombinant plasmid pLGV23 *neo* in tobacco protoplasts. With minor modifications, this method could be applied to a wide variety of plant protoplasts and nucleic acids.

Methods

Isolation of TMV RNA

TMV is purified from the leaves of infected *Nicotiana tabacum* (cv. Samsun) plants by polyethylene glycol (PEG) precipitation.[12] Viral RNA is extracted by proteinase K treatment and phenol–chloroform deproteinization of the virus preparations. After alcohol precipitation, viral RNA is

[1] Z. Xu, C. S. Luciano, S. T. Ballard, R. E. Rhoads, and J. G. Shaw, *Plant Sci. Lett.* **36,** 137 (1984).

[2] M. Caboche and A. Deshayes, *C.R. Seances Acad. Sci., Ser. III* **299,** 663 (1984).

[3] P. F. Lurquin, *in* "Cell Fusion: Gene Transfer and Transformation" (R. F. Beers, Jr., and E. G. Bassett, eds.), p. 227. Raven Press, New York, 1984.

[4] P. F. Lurquin, *in* "Liposome Technology" (G. Gregoriadis, ed.), Vol. 2, p. 187. CRC Press, Boca Raton, Florida, 1984.

[5] F. Szoka and D. Papahadjopoulos, *Proc. Natl. Acad. Sci. U.S.A.* **75,** 4194 (1978).

[6] T. Nagata, K. Okada, I. Takebe, and C. Matsui, *Mol. Gen. Genet.* **184,** 161 (1981).

[7] F. Rollo and R. Hull, *J. Gen. Virol.* **60,** 359 (1982).

[8] R. T. Fraley, S. L. Dellaporta, and D. Papahadjopoulos, *Proc. Natl. Acad. Sci. U.S.A.* **79,** 1859 (1982).

[9] T. Nagata, K. Okada, and I. Takebe, *Plant Cell Rep.* **1,** 250 (1982).

[10] A. A. Christen and P. F. Lurquin, *Plant Cell Rep.* **2,** 43 (1983).

[11] P. Rouze, A. Deshayes, and M. Caboche, *Plant Sci. Lett.* **31,** 55 (1983).

[12] G. V. Gooding and T. T. Heber, *Phytopathol. Notes* **57,** 1285 (1967).

dissolved at a concentration of 1 mg/ml in liposome buffer [5 m*M* Tris–HCl, pH 7.6, 50 m*M* NaCl, 1 m*M* ethylenediaminetetraacetic acid (EDTA), 0.44 *M* mannitol].

Isolation of pLGV23neo DNA

Escherichia coli strain RR1 carrying pLGV23*neo*[13] is lysed by the Triton X-100 procedure described in Kupersztoch-Portnoy *et al.*[14] Plasmid and chromosomal DNA are then separated by CsCl–ethidium bromide density gradient centrifugation. Basically, pLGV23*neo* consists of the coding sequence of the Tn*5* aminoglycoside 3′-phosphotransferase II gene flanked by the *Agrobacterium tumefaciens* T-DNA nopaline synthase promoter and terminator regions cloned in pBR322.

Protoplast Preparation

Protoplasts are prepared from the leaves of a diploid clone, D_8, of *Nicotiana tabacum* (cv. Xanthi) grown in the greenhouse. Leaves are surface sterilized with 5% calcium hypochlorite for 7 min, washed three times with sterile distilled water, and their lower epidermis stripped with forceps. Stripped fragments are floated overnight on To medium containing 0.02% Macerozyme, 0.1% cellulase Onozuka R10, and 0.05% Driselase. To medium contains half-strength Murashige–Skoog macronutrients, Heller's micronutrients, Morel–Wetmore vitamins, 0.45 *M* mannitol, 0.05 *M* sucrose, 5 μ*M* benzyladenine, and 16 μ*M* naphthaleneacetic acid. After digestion protoplasts are washed twice in saline osmoticum [0.3 *M* KCl, 5 m*M* $CaCl_2$, 1 m*M* 2-[*N*-morpholino]ethanesulfonic acid (MES), pH 5.7] and resuspended in fusion medium (0.5 *M* mannitol, 5 m*M* $CaCl_2$, 5 m*M* Tris–HCl, pH 7.6) at a density of 2×10^6 protoplasts/ml.

Encapsulation of TMV RNA into Liposomes

Five micromoles of bovine brain phosphatidylserine (stored under argon at −80°, Sigma Chemicals) and 5 μmol of cholesterol are dried under vacuum and redissolved in 0.8 ml of diethyl ether and 200 μl of the TMV RNA preparation is added. The two phases are mixed under argon by sonication for 30 sec in a bath-type sonicator (Bransonic 220). Ether is removed by evaporation under reduced pressure and the resulting lipo-

[13] L. Herrera-Estrella, M. de Block, E. Messens, J. P. Hernalsteens, M. Van Montagu, and J. Schell, *EMBO J.* **2,** 987 (1983).

[14] Y. M. Kupersztoch-Portnoy, M. A. Lovett, and D. R. Helinski, *Biochemistry* **13,** 5484 (1974).

somes are stored under argon. Liposomes can be separated from nonencapsulated material by flotation onto an isosmotic Ficoll step gradient[8] made in liposome buffer as above. Encapsulation efficiencies range from 20 to 40%.

Encapsulation of pLGV23neo into Liposomes

Five micromoles of phosphatidylserine (Sigma Chemicals) and 2 μmol of cholesterol are dried under vacuum and redissolved in 0.8 ml of diethyl ether. Two hundred μg of plasmid in 200 μl of liposome buffer, containing 0.44 *M* mannitol, 50 m*M* NaCl, and 1 m*M* EDTA in 5 m*M* Tris–HCl buffer, pH 7.6, is then added. The two phases are mixed under argon by sonication for 30 sec in a bath-type sonicator (Bransonic 220). Ether is then removed by evaporation under reduced pressure at 30° and the resulting liposomes are stored under argon after dilution in 1 ml of liposome buffer. Liposomes are separated from the nonencapsulated material by flotation onto an isosmotic Ficoll step gradient made in the liposome buffer. The encapsulation efficiencies vary from 25 to 40%.

Liposome-Mediated Transfection

With TMV RNA. Protoplasts are washed in fusion medium (0.5 *M* mannitol, 5 m*M* $CaCl_2$, 5 m*M* Tris–HCl, pH 7.6) and resuspended at a final concentration of 2×10^6 protoplasts/ml. One million are mixed by hand shaking with liposomes and 30 sec later 5 ml of 20% PEG 6000 (w/v) (Touzart et Matignon, France) dissolved in fusion medium is added. After mixing and incubation for 15 min at room temperature, the viscosity of the preparations is decreased by adding 20 ml of fusion medium and protoplasts are sedimented by centrifugation (100 *g*, 5 min). Protoplasts are washed once more, plated in 10 ml of To medium, and incubated at 28° in the dark.

With pLGV23neo DNA. Fifty microliters of liposome preparation containing 3 μg of encapsulated pLGV23*neo* DNA is mixed with 1 ml of protoplast suspension at room temperature. Five minutes later 5 ml of fusion medium containing 22% polyethylene glycol 6000 (w/v) (Touzart et Matignon, France) is mixed with protoplasts. After incubation for 20 min, 20 ml of saline medium (0.3 *M* KCl, 5 m*M* $CaCl_2$) is added and protoplasts are sedimented at 100 *g* for 5 min. Protoplasts are plated at a density of 5×10^4/ml in To medium.

Detection of TMV RNA and pLGV23neo DNA Expression

Immunofluorescence Assay for TMV Production. Protoplasts are washed after incubation in 0.3 *M* KCl, 20 m*M* $CaCl_2$, and 0.25% bovine serum albumin (BSA). They are then fixed by adding 10 vol of acetone/ethanol (2 : 1) at 4° and then spread on glass slides and dried. Slides are washed with PBS–Tween (25 m*M* potassium phosphate buffer, pH 7.2, 0.15 *M* NaCl, 0.2 mg/ml sodium azide, 0.5 mg/ml Tween 20) and incubated for 1 hr at 37°, with rabbit anti-TMV serum diluted 1/200 in PBS containing 0.5% BSA. After incubation slides are washed three times for 5 min with PBS. The procedure is repeated in the dark, using fluorescein isothiocyanate (FITC)-coupled sheep anti-rabbit immunoglobulin G (IgG) (Institut Pasteur Production) diluted 1/100 in PBS + 0.5% BSA. Slides are mounted in glycerol and observed under the microscope using Leitz fluorescence equipment (515-nm filter). The percentage of fluorescent cells is determined after counting 500–1000 cells.

ELISA Test for TMV Production. Suspensions of protoplasts (1 ml) are concentrated 10 times by centrifugation and resuspended in 50 m*M* potassium phosphate buffer (pH 7.0) in Eppendorf tubes. Protoplasts are then broken by three freeze–thawing cycles, clarified by centrifugation, and the virus content measured by an indirect double-antibody sandwich method. Typically, polystyrene microtiter plates (Linbro ISFB 96, Flow) are coated at 4° overnight with goat anti-TMV immunoglobulins diluted 1/2000 in 0.05 *M* sodium carbonate (pH 9.6) (100 μl/well). After three rinsings, purified TMV standards and protoplast supernatants properly diluted (1/10 to 1/100) are added and incubated for 3 hr at 37°. After rinsing, rabbit anti-TMV immunoglobulins (1/4000) are distributed for a 3-hr incubation at 37°. After three washings conjugate (alkaline phosphatase swine anti-rabbit IgG) is added for 3 hr at 37° and the plates are rinsed three times. Substrate is added (100 μl *p*-nitrophenylphosphate at 1 mg/ml in 10% diethanolamine buffer, pH 9.8) and absorbance (405 nm) is measured at different times with an automatic densitometer (micro-ELISA autoreader, Dynatech). The TMV concentration in the protoplast cultures is calculated from these data and calibration curves obtained with TMV standards (concentration range: 1 ng to 10^4 ng/ml).

Selection of Kanamycin-Resistant Tobacco Cells. Protoplasts treated with liposomes containing pLGV23*neo* are incubated in the dark at 28°. One week later protoplast-derived cells are collected by centrifugation and resuspended in C medium containing 3 m*M* MES, pH 5.9,[15] at a density of 5×10^3 cells/ml. Cultures are incubated in the presence of 70 μg/ml of filter-sterilized kanamycin at 28° under light. Resistant colonies are regenerated in the absence of kanamycin as described in Muller and Caboche.[16]

Assay of Aminoglycoside 3'-Phosphotransferase II [APH(3')II] Activ-

[15] J. P. Bourgin, Y. Chupeau, and C. Missonier, *Physiol. Plant.* **45,** 288 (1979).

[16] J. F. Muller and M. Caboche, *Physiol. Plant.* **57,** 35 (1983).

ity in Plants Regenerated from Kanamycin-Resistant Cells. Two grams of leaf fragments is crushed at 4° in a mortar with a pestle in 2 ml of extraction buffer (10 μg/ml leupeptin, 30 m*M* 2-mercaptoethanol, 20 m*M* EDTA in 0.2 *M* Tris–HCl buffer, pH 7.5) together with 0.2 g Polyclar. After two successive centrifugations in Eppendorf tubes at 12,000 *g* for 15 min, supernatants are used for electrophoresis in a 10% nondenaturing polyacrylamide gel and APH(3′)II activity is revealed according to the method of Reiss *et al.*[17]

Southern Blot Analysis. Plant DNA is isolated by a phenol extraction procedure,[18] purified by two CsCl gradient centrifugations, and dialyzed for 48 hr against 10 m*M* Tris–HCl buffer, pH 8.0, 1 m*M* EDTA. Restriction endonucleases are used according to the instructions of the suppliers to digest plant DNA. Ten micrograms of DNA is digested with 40 U of *Bam*HI or 60 U *Hin*dIII, ethanol precipitated, and resuspended in 30 μl of 10 m*M* Tris–HCl, pH 7.8. Electrophoresis is carried out in 1% high gelling temperature agarose gels. After electrophoresis, DNA is depurinated in 0.25 *M* HCl for 15 min at room temperature, denatured in 1.5 *M* NaCl–0.5 *M* NaOH, neutralized in 3 *M* sodium acetate, pH 5.5, and then transferred to Biodyne A nylon membrane (Pall) with 6× SSC (0.9 *M* sodium chloride and 0.09 *M* sodium citrate) (Southern, 1975). After baking at 80° for 1 hr in a vacuum oven, membranes are incubated in the hybridization buffer (5× Denhardt's reagent, 5× SSC, 50 m*M* sodium phosphate, pH 6.5, 0.1% SDS, 50% formamide, 200 μg/ml of calf thymus DNA). Afterward, membranes are hybridized in the same buffer with nick-translated DNA ($2-4 \times 10^8$ cpm/μg) for 16–70 hr.[19] For this, the 4.1-kb *Hin*dIII fragment of pLGV23*neo* carrying the *neo* gene is recovered directly from a 1% low gelling temperature agarose gel[20] and subsequently nick translated.

Membranes are washed in 2× SSC, 0.1% SDS three times for 5 min at room temperature and then in 0.2× SSC, 0.1% SDS two times for 15 min at 50°. Blots are exposed to Ilford X-ray film with intensifying screen at −70°.

Conclusions

The encapsulation of TMV RNA into large, negatively charged unilamellar liposomes, as described above, leads to the transfection of about

[17] B. Reiss, R. Sprengel, H. Will, and H. Schaller, *Gene* **30,** 217 (1984).

[18] N. Federoff, J. Mauvais, and D. Chaleff, *Mol. Appl. Genet.* **2,** 11 (1983).

[19] P. W. Rigby, M. Diechmann, C. Rhodes, and P. Berg, *J. Mol. Biol.* **113,** 237 (1977).

[20] T. Maniatis, E. F. Fritsch, and J. Sambrook, "Molecular Cloning." Cold Spring Harbor Lab., Cold Spring Harbor, New York, 1982.

70% of the treated tobacco mesophyll protoplasts at a multiplicity of 3×10^5 RNA molecules per protoplast.

DNA-mediated transformation of tobacco mesophyll protoplasts could also be demonstrated using the recombinant plasmid pLGV23*neo* encapsulated into similar liposomes. Transformation frequencies calculated from the number of kanamycin-resistant colonies were of the order of 4×10^{-5} transformant per viable protoplast. Plants regenerated from these colonies did express the *neo* gene as evidenced by the presence of APH(3′)II activity and contained several integrated copies of this gene as shown by Southern blot analysis.

Acknowledgments

This work was supported in part by a grant from the State of Washington High Technology Center (P.F.L.).

[6] Interspecific Transfer of Partial Nuclear Genomic Information by Protoplast Fusion

By OTTO SCHIEDER

The technique of protoplast fusion offers great potential to somatic cell genetics and crop improvement, because the genetic information of species which cannot be crossed by sexual methods can be combined. It has been shown that not only complete nuclear genomes of different plant species can be combined by the fusion technique,[1–3] but also the cytoplasmic genetic information can be transferred from one species into another one.[4,5] However, the transfer of cytoplasmic genetic information is of limited importance for crop improvement because most of the desirable genetic information is located in the nucleus. A transfer of limited nuclear

[1] C. T. Harms, *Experientia, Suppl.* **46,** 69 (1983).

[2] Y. Y. Gleba and K. M. Sytnik, "Protoplast Fusion." Springer-Verlag, Berlin and New York, 1984.

[3] O. Schieder and H. Kohn, *in* "Cell Culture and Somatic Cell Genetics of Plants" (I. K. Vasil, ed.), Vol. 3, p. 569. Academic Press, New York, 1986.

[4] E. Galun and D. Aviv, *in* "Handbook of Plant Cell Culture" (D. A. Evans, W. R. Sharp, P. V. Amirato, and Y. Yamada, eds.), p. 358. Macmillan, New York, 1983.

[5] E. Galun and D. Aviv, *in* "Efficiency in Plant Breeding" (W. Lange, A. C. Teven, and N. G. Hagenboom, eds.), p. 228. Pudoc, Wageningen, 1984.

genomic information is more important. Recently several transformation systems using cloned genes have been developed.[6–9] However, the limitation of this method is the difficulty in isolating and cloning most of the desirable genes, e.g., genes coding for disease resistances. Even greater is the difficulty when the resistant phenotype is polygenically inherited. An alternative method for overcoming this problem might be the uptake of isolated chromosomes in protoplasts[10] or the production of asymmetric somatic hybrids possessing only a part of the nuclear genomic information of one of the two partners combined by protoplast fusion. This chapter will present some details about the production of asymmetric somatic hybrids induced through fusion of heavily irradiated "donor protoplasts" with nonirradiated "recipient protoplasts."[11–16]

Transfer of Nuclear Genomic Controlled Traits

The principal method for the construction of asymmetric somatic hybrid cell lines or plants is the use of protoplasts from a donor plant pretreated with X rays and subsequently fused with the protoplasts of a recipient plant. This principal method has been used successfully also for the transmission of organelle-controlled traits.[17]

[6] J. Schell, L. Herrera-Estrella, P. Zambryski, M. de Block, H. Joos, L. Willmitzer, P. Eckes, S. Rosahl, and M. Van Montagu, *in* "The Impact of Gene Transfer Techniques in Eukaryotic Cell Biology" (J. S. Schell and P. Starlinger, eds.), p. 73. Springer-Verlag, Berlin and New York, 1984.

[7] J. Paszkowski, R. D. Shillito, M. Saul, V. Mandak, T. Hohn, B. Hohn, and I. Potrykus, *EMBO J.* **3,** 2717 (1984).

[8] I. Potrykus, M. W. Saul, J. Petruska, J. Paszkowski, and R. D. Shillito, *Mol. Gen. Genet.* **199,** 183 (1985).

[9] H. Lörz, B. Baker, and J. Schell, *Mol. Gen. Genet.* **199,** 178 (1985).

[10] D. Dudits and T. Praznovszky, *in* "Biotechnology in Plant Science: Relevance to Agriculture in the Eighties" (M. Zaitlin, P. Day, and A. Hollaender, eds.), p. 115. Academic Press, Orlando, 1985.

[11] D. Dudits, O. Fejér, G. Hadlaczky, C. Konc, G. Lázár, and G. Horváth, *Mol. Gen. Genet.* **179,** 283 (1980).

[12] P. P. Gupta, M. Gupta, and O. Schieder, *Mol. Gen. Genet.* **188,** 378 (1982).

[13] P. P. Gupta, O. Schieder, and M. Gupta, *Mol. Gen. Genet.* **197,** 30 (1984).

[14] K. Itho and Y. Futsuhara, *Jpn. J. Genet.* **58,** 545 (1983).

[15] O. Schieder, P. P. Gupta, G. Krumbiegel-Schroeren, T. Hein, and A. Steffen, *in* "Plant Tissue and Cell Culture: Application to Crop Improvement" (F. J. Novák, L. Havel, and J. Dolezel, eds.), p. 371. Czech. Acad. Sci., Prague, 1984.

[16] O. Schieder, *in* "Biotechnology in Plant Science: Relevance to Agriculture in the Eighties" (M. Zaitlin, P. Day, and A. Hollaender, eds.), p. 77. Academic Press, Orlando, 1985.

[17] A. Zelcer, D. Aviv, and E. Galun, *Z. Pflanzenphysiol.* **90,** 397 (1978).

Protoplast Source

So far only a few experiments have been undertaken to transfer nuclear genomic traits from a donor into a recipient plant. In most of the experiments nuclear coded metabolic deficiencies have been used in order to correct them through a transfer via fusion with irradiated donor protoplasts.[11-13] Protoplasts can be isolated from both the donor and the recipient either from mesophyll or from cell suspensions. A prerequisite is that at least the recipient protoplasts can be regenerated to callus or to plants. The isolation and preparation of the protoplasts can be undertaken as described in detail for *Datura*.[18]

Mitotic Inactivation of Donor Protoplasts

In general, X rays have been used for the mitotic inactivation of the donor protoplasts. The principle is the fragmentation of the nuclear genome to avoid the fusion of complete genomes of both the donor and the recipient. Irradiation with γ-rays might also be useful but this technique has not yet been used. Inactivation with chemicals such as iodoacetate, that are useful in transfer experiments of organelle genomes,[19] might not be advantageous because with this treatment no fragmentation of the nuclear genome can be achieved. After isolation and preparation of the donor protoplasts they have to be suspended in 0.6 *M* mannitol and exposed to X rays in a shallow layer. The dose rate for total mitotic inactivation of the protoplasts can differ depending on the species used as donor. For *Datura innoxia,* for example, the application of 5 krad of X rays leads to a total mitotic inactivation of its mesophyll protoplasts,[12] whereas the mesophyll protoplasts of *Physalis minima* and the cell suspension protoplasts of *Nicotiana paniculata* require for total inactivation at least 15 krad.[13,16] It is essential that before starting an experiment the dose rate necessary for a total mitotic inactivation of the donor protoplasts has been determined. Differences in sensitivity against irradiation in different plant species are well known.

Fusion of Protoplasts

The fusion procedure of protoplasts can be carried out by chemical treatment, as in regular somatic hybridization experiments where symmetric somatic hybrids have been produced.[1-3] However, the new fusion

[18] O. Schieder, *in* "Cell Culture and Somatic Cell Genetics of Plants" (I. K. Vasil, ed.), Vol. 1, p. 350. Academic Press, New York, 1984.

[19] V. A. Sidorov, L. Menczel, F. Nagy, and P. Maliga, *Planta* **152,** 341 (1981).

technique using electrical pulse treatment may also be of practical value in future asymmetric hybridization experiments.[20,21]

Culture of Protoplasts

After fusion the protoplasts have to be resuspended in a protoplast regeneration medium suitable for the respective recipient plant species. Numerous regeneration media have already been developed and detailed descriptions for several plant species have been published.[22]

Selection Procedures

As already mentioned, in most of the experiments undertaken so far, metabolic deficiencies have been used in order to correct them via interspecific protoplast fusion. Selection of the asymmetric somatic hybrids was based on their wild-type phenotype. In two experiments the nuclear-coded albinism was corrected by fusion of wild-type protoplasts inactivated with X rays. The albinism in a carrot mutant could be corrected after fusion with inactivated wild-type protoplasts of petrosilium.[11] A similar experiment using an albino mutant of *Datura innoxia* as the recipient and wild-type of *Physalis minima* as the donor was carried out by Gupta *et al.*[13] Obviously, the selection was based on the green color of the asymmetric somatic hybrids. In other experiments the nitrate reductase deficiency in tobacco mutants was partially corrected after fusion with inactivated donor protoplasts of either *Physalis minima* and *Datura innoxia* or *Hyoscyamus muticus*. Selection here was based on the ability of the asymmetric somatic hybrids to develop on a medium containing only NO_3^- ions as the sole nitrogen source.[12,13] Itho[14] fused X-ray-inactivated protoplasts of *Nicotiana langsdorffii* with protoplasts of *Nicotiana glauca*. Selection of asymmetric somatic hybrids is based on their hormone-independent growth behavior, a phenomenon well known in sexual and symmetric hybrids of these two species. When no markers are available which can serve for the selection of asymmetric somatic hybrids the t-DNA of *Agrobacterium tumefaciens* can be incorporated into the donor plant material.[6] Selection of asymmetric somatic hybrids is then based on their hormone-independent growth behavior. With such a method asymmetric somatic hybrid cell lines have been produced between the nitrate reductase-deficient cell line cnx-68 of tobacco as the recipient and a tumorous cell line of *Nicotiana paniculata*.[16] In these experiments besides

[20] U. Zimmermann and P. Scheurich, *Planta* **151,** 26 (1981).

[21] H. Kohn, R. Schieder, and O. Schieder, *Plant Sci.* **38,** 121 (1985).

[22] I. K. Vasil, ed., "Cell Culture and Somatic Cell Genetics of Plants," Vol. 1. Academic Press, New York, 1984.

the T-DNA other nuclear genomic DNA of the donor was simultaneously transferred by this method.[23,24] For transformation of the donor plant material with *Agrobacterium tumefaciens* leaf disks or stem segments of aseptic grown shoots have to be incubated with 100 μl of the late log phase of the bacteria suspension at 27°.[25] After 2 days of incubation wash the leaf disks or stem segments with the respective protoplast regeneration medium containing 0.05% of the antibiotic carbenicillin to avoid overgrowing the bacteria. Since tumors developing after incubation with *Agrobacterium* normally are mixed up with untransformed cells, a leaf disk or stem segment cloning[25] immediately after incubation with the bacteria via protoplasts is necessary for getting cell lines from which all the cells contain the T-DNA. For such a cloning, protoplast isolation, preparation, and culture can be carried out as described elsewhere.[18,22] However, the protoplast regeneration media must also contain carbenicillin (0.025%).[25] When the protoplasts have developed to macroscopic cell colonies transfer them onto agar medium lacking phytohormones but still containing carbenicillin (0.025%). Growing colonies can be picked and used for the establishment of a cell suspension[26] which can serve as the source for the donor protoplasts. Mitotic inactivation of the donor protoplasts with X rays, their fusion with the recipient protoplasts, and culture can be carried out as already described. The selection of the asymmetric somatic hybrid cell lines can be undertaken on hormone-free agar media when the protoplasts have developed into visible colonies.[16] However, the usage of wild-type T-DNA of *Agrobacterium tumefaciens* for the selection has the disadvantage that only tumorous asymmetric somatic cell lines can be obtained. For obtaining nontumorous asymmetric somatic hybrid cell lines and finally plants it has been proposed using T-DNA constructions which are disarmed from all the tumor genes but contain the genetic information coding for the neomycin phosphotransferase (strain pGLV*neo*2103). This would lead to cells resistant to neomycin or kanamycin.[27,28] Figure 1 indicates how such a T-DNA construction can perhaps be used for asymmetric hybridization experiments where

[23] E. Müller-Gensert and O. Schieder, *in* "Genetic Manipulation in Plant Breeding" (W. Horn, C. J. Jensen, W. Odenbach, and O. Schieder, eds.), p. 697. de Gruyter, Berlin, 1986.

[24] E. Müller-Gensert and O. Schieder, *Mol. Gen. Genet.*, in press (1987).

[25] A. Steffen and O. Schieder, *Theor. Appl. Genet.* **72,** 135 (1986).

[26] P. King, *in* "Cell Culture and Somatic Cell Genetics of Plants" (I. K. Vasil, ed.), Vol. 1, p. 130. Academic Press, New York, 1984.

[27] O. Schieder, T. Hein, and H. Kohn, *in* "Genetic Manipulation in Plant Breeding" (W. Horn, C. J. Jensen, W. Odenbach, and O. Schieder, eds.), p. 641. de Gruyter, Berlin, 1986.

[28] R. Hain, P. Stabel, A. P. Czernilofsky, H. H. Steinbiss, L. Herrera-Estrella, and J. Schell, *Mol. Gen. Genet.* **199,** 161 (1985).

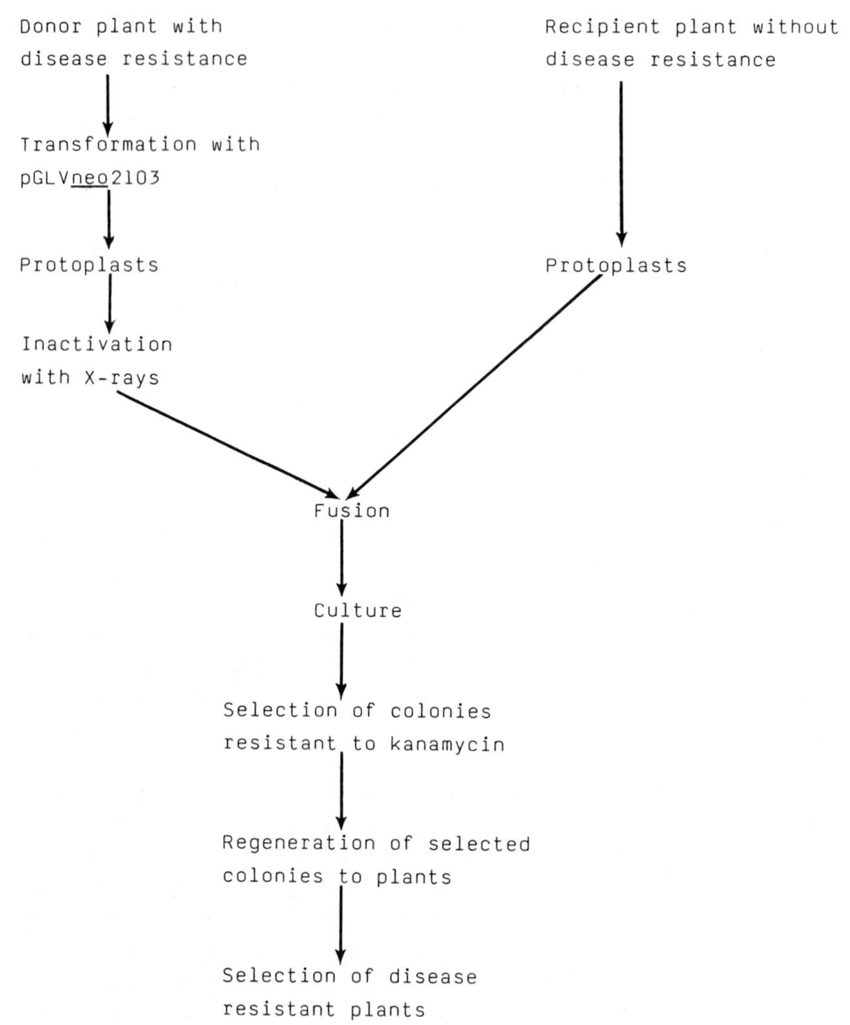

FIG. 1. A scheme indicating how the strain pGLV*neo*2103 of *Agrobacterium tumefaciens*, disarmed from all the tumor genes, can be used for a simultaneous transfer of nuclear genomic information coding for disease resistance.

nuclear genomic information for disease resistances will be transferred from a donor into a recipient plant simultaneously with the T-DNA.

Verification of Somatic Asymmetry

It was concluded in experiments to correct nitrate reductase deficiency in tobacco through X-ray-inactivated wild-type protoplasts of *Phy-*

salis minima and *Datura innoxia* that a transfer of nuclear genomic DNA of the donor is responsible[12] for this correction. No direct proof exists for this predication. Famelaer *et al.*[29] treated nitrate reductase-deficient protoplasts of *Nicotiana plumbaginifolia* with the total DNA of the wild type and the mutant itself and recovered corrected phenotypes from both treatments. They concluded that the mutagenic effects of the donor DNA is the most plausible mechanism. This mutagenic effect may also occur after fusion with inactivated donor protoplasts. Cytological observations that demonstrate the presence of only a few chromosomes of the donor in the selected asymmetric somatic hybrids[13] are better.

The study of the electrophoretic pattern of proteins (isoenzymes) in the selected presumed asymmetric somatic hybrids in comparison to the recipient and the donor plant material can serve as an indication of the transfer of nuclear genomic information.[11–14,30] However, the most convincing method for demonstrating an incomplete transfer of nuclear genomic DNA of the donor into a recipient is the usage of repetitive and species-specific nuclear DNA sequences of the donor in combination with Southern blotting.[31] This method, which worked for identifying symmetric somatic hybrids between *Hyoscyamus muticus* and *Nicotiana tabacum*,[32] is also very useful for confirming asymmetric somatic hybridity.[23,24,33]

For isolation and cloning of repetitive and species-specific nuclear DNA sequences follow the method of Saul and Potrykus.[32] Total DNA of the donor and the recipient plant material as well as that of the presumed asymmetric somatic hybrids can be prepared either according to Bedbrook[34] or to Murray and Thompson.[35] After Southern blotting the differences in the hybridization strength between the restricted total DNA of the donor and the restricted total DNA of the presumed asymmetric somatic hybrids can be used to estimate the amount of transferred donor nuclear DNA.[23,33]

[29] Y. Famelaer, M. Heindrickx, I. Negrutiu, and M. Jacobs, *in* "Genetic Engineering of Plants and Microorganisms Important for Agriculture" (E. Magnien and D. de Nettancourt, eds.), p. 172. Martinus Nijhoff/Dr. W. Junk Publishers, Dordrecht, Netherlands, 1985.

[30] L. R. Wetter, *in* "Cell Culture and Somatic Cell Genetics of Plants" (I. K. Vasil, ed.), Vol. 1, p. 651. Academic Press, New York, 1984.

[31] E. M. Southern, *J. Mol. Biol.* **98,** 503 (1975).

[32] W. Saul and I. Potrykus, *Plant Cell Rep.* **3,** 65 (1984).

[33] J. Imamura, "Annual Report," p. 51. Friedrich-Miescher-Institute, Basel, 1984.

[34] J. Bedbrook, *Plant Mol. Biol. Newsl.* **2,** 24 (1981).

[35] H. G. Murray and W. F. Thompson, *Nucleic Acids Res.* **8,** 4321 (1980).

Section II

Vacuoles and Tonoplasts

[7] Isolation of Mature Vacuoles of Higher Plants: General Principles, Criteria for Purity and Integrity

By GEORGE J. WAGNER

Interest in the structure and function of mature vacuoles of higher plants has intensified in the present decade, with research being focused on the use of vacuoles to affect compartmentation analysis—determine the vacuolar or extravacuolar nature of metabolites[1-5]—and on the definition of primary and secondary transport mechanisms of the tonoplast.[6,7] Prior to 1975 there were no described procedures for preparing substantial quantities of this, the largest of mature plant cell organelles. After a brief introduction, this chapter will focus on recently introduced methodology for preparing mature vacuoles in quantities sufficient for biochemical characterization and studies of solute transport.

Introduction

The means for preparing large quantities of vacuoles from lower plants,[8-10] meristematic higher plant tissues,[8] lutoids (vacuoles) from specialized latex cells of *Hevea*,[11,12] and vacuolar sap from giant algal cells[13]

[1] F. Marty, D. Branton, and R. A. Leigh, *in* "The Biochemistry of Plants" (P. K. Stumpf and E. E. Conn, eds.), Vol. 1, p. 625. Academic Press, New York, 1980.
[2] G. J. Wagner, *Recent Adv. Phytochem.* **16,** 1 (1981).
[3] T. Boller, *Physiol. Veg.* **20,** 247 (1982).
[4] G. Wagner, *in* "Isolation of Membranes and Organelles from Plant Cells" (J. L. Hall and A. L. Moore, eds.), p. 83. Academic Press, New York, 1983.
[5] C. A. Ryan and M. Walker-Simmons, this series, Vol. 96, p. 580.
[6] R. A. Leigh, *Physiol. Plant.* **57,** 390 (1983).
[7] B. P. Marin, ed., "Biochemistry and Function of Vacuolar Adenosine-Triphosphatase in Fungi and Plants," Springer-Verlag, Berlin and New York, 1985.
[8] P. Matile, "The Lytic Compartment of Plant Cells," Cell Biol. Monogr., Vol. 1. Springer-Verlag, Berlin and New York, 1975.
[9] E. J. Bowman and B. J. Bowman, *in* "Biochemistry and Function of Vacuolar Adenosine-Triphosphatase in Fungi and Plants" (B. P. Marin, ed.), p. 141. Springer-Verlag, Berlin and New York, 1985.
[10] Y. Ohsumi, E. Uchida, and Y. Anraku, *in* "Biochemistry and Function of Vacuolar Adenosine-Triphosphatase in Fungi and Plants" (B. P. Marin, ed.), p. 131. Springer-Verlag, Berlin and New York, 1985.
[11] S. Pujarniscle, *Physiol. Veg.* **6,** 27 (1968).
[12] J. D'Auzac, H. Crétin, B. Marin, and C. Lioret, *Physiol. Veg.* **20,** 311 (1982).
[13] K. Sakano and M. Tazawa, *Plant Physiol.* **78,** 673 (1985).

were available prior to 1970. These techniques will not be discussed here and the reader is referred to the cited literature for this information. However, it is important to point out at the onset that methods development and physiological studies of vacuoles from these sources have laid the foundation for more recent investigations of mature higher plant vacuoles. Recently, extensive study of transport mechanisms of *Hevea* lutoids have been made. The reader is referred to recent reviews on this subject.[7,12] An introduction to recent studies of yeast and *Neurospora* vacuoles is found in articles by Bowman and Bowman[9] and Ohsumi *et al.*[10] Methods for preparing seed protein bodies (dehydrated vacuoles) from cotyledon homogenates and protoplasts were recently discussed and described[14,15] as were procedures for preparing proton-pumping vesicles thought to originate from vacuoles.[16,17] Finally, procedures for preparing vacuoles from mature and meristematic plant tissue, *Hevea* latex, yeast, *Neurospora,* and proton-pumping vesicles have been discussed and compared.[18]

There is little doubt that recent studies of plant vacuoles prepared from both higher and lower plants, together with earlier work, will lead to a substantially better understanding of the roles of the mature vacuole in osmotic based tissue support and movement, ion balance and storage, metabolite storage and sequestration, senescence and lytic functions, and intracellular mixing through transvacuolar strands. Most of these functions were recognized by early cytologists including deVries, Went, Darwin, and others.[4,19] An understanding of the underlying mechanisms in certain of these functions is just beginning to emerge.

Higher plant vacuoles were first isolated for biochemical characterization in 1968 by Matile.[8,20] These structures, which were obtained from meristematic tissue, were shown to be enriched in acid hydrolases. Later, particles having some of the same characteristics (density ~1.1 g/cm^3, certain acid hydrolases) were prepared using similar methods from roots and cotyledons of pea, shoots of potato, tomato fruits, and other tissues.[8,18] These isolates were generally considered to be derived from rapidly dividing cells which contain only small vacuoles. However, at

[14] J. Pernollet, *Physiol. Veg.* **20,** 259 (1982).

[15] W. Van Der Wilden and M. J. Chrispeels, *Plant Physiol.* **71,** 82 (1983).

[16] H. Sze, *Physiol. Plant.* **61,** 683 (1984).

[17] H. Sze, *Annu. Rev. Plant Physiol.* **36,** 175 (1985).

[18] G. Wagner, *in* "Modern Methods of Plant Analysis" (H. F. Linskens and J. F. Jackson, eds.), Vol. 1, p. 105. Springer-Verlag, Berlin and New York, 1985.

[19] A. Guilliermond, "The Cytoplasm of the Plant Cell." Chronica Botanica, Waltham, Massachusetts, 1941.

[20] P. Matile, *Annu. Rev. Plant Physiol.* **29,** 193 (1978).

least in the case of tomato, mature fruit was also said to be a source.[21] In contrast to meristematic cell vacuoles, the central vacuole of mature higher and lower plant cells usually occupies 80% or more of the intracellular volume. It is clear that mature vacuoles are multifunctional in nature.[18,19] Primarily the lytic role(s) and the ontogeny of meristematic vacuoles have been studied.[8]

Preparation of Mature Vacuoles of Higher Plants

These organelles have been prepared from lysed protoplasts of various tissues using (1) osmotic lysis,[22] (2) polybase-induced lysis,[23] and (3) lysis due to shear under isotonic conditions.[24] In a fourth basic approach—one which does not utilize protoplasts—vacuoles are released directly by slicing root tissue into plasmolyzing medium.[25] References cited above represent the report which first introduced each basic approach. Practically every application of these approaches has included modifications of original procedures. A survey of the literature shows that in about 70 papers, less than 6% have utilized the original methods unchanged. It is difficult to ascertain if modifications have improved the quality or yield of vacuoles in any given case. However, it seems clear that many investigators consider modifications advantageous.

In order to provide specifics here, one prominant method for each of the four basic approaches listed above will be described in detail. These were selected because they have been successfully used by several laboratories with several different tissues and, in the opinion of this author, they represent generally useful protocols. Clear distinctions which might suggest that one approach is more or less suitable for monocots vs dicots or roots vs leaves, etc., are not obvious.

Protoplast Isolation

Protoplasts have proved to be the best source of vacuoles from the greatest number of tissues and are the starting material for approaches 1–3. Generally useful procedures for obtaining protoplasts are as follows. Tissues are often surface sterilized by brief immersion in or by painting with 70% (v/v) ethanol and are then dissected to allow infiltration of aqueous solutions containing wall-digesting enzymes. In certain cases,

[21] E. Heftmann, *Cytobios* **3,** 129 (1971).
[22] G. J. Wagner and H. W. Siegelman, *Science* **190,** 1298 (1975).
[23] C. Buser and P. Matile, *Z. Pflanzenphysiol.* **82,** 462 (1977).
[24] H. Lorz, C. T. Harms, and I. Potrykus, *Biochem. Physiol. Pflanz.* **169,** 617 (1976).
[25] R. A. Leigh and D. Branton, *Plant Physiol.* **58,** 656 (1976).

surface sterilization is reported to interfere with protoplast isolation.[26] Addition of antibiotics may have undesirable effects on protoplast viability.[26] Many workers make no effort to sterilize tissue and apparently assume that any contaminating bacteria will be diluted during purification of protoplasts and vacuoles obtained from them. Tissue dissection may be achieved by stripping epidermis, slicing tissue into thin (1 mm) strips, chopping it into segments, cutting roots lengthwise,[26] brushing with a stiff brush, or rubbing with carborundum.[27] As when preparing plant cuttings for uptake of solutions through the cut surface, there may be an advantage to preventing air embolisms at cut surfaces by cutting tissue while submerged under water. Access of enzyme solution may also be enhanced by gentle vacuum infiltration of dissected tissue through repeated, short evacuation of submerged tissue.[28]

A generally useful incubation medium contains 2% (w/v) cellulase from *Trichoderma viride* (Cellulysin, Calbiochem Co., or Onozuka RS or R10, Yakult Biochemical Co., Ltd., Nishinomiya, Japan) and 0.1% (w/v) Pectolyase Y-23 (Seishin Pharmaceutical Co., Ltd., Chiba-ken, Japan) in 0.4 to 0.7 *M* mannitol or sorbitol containing 0.2 to 1 m*M* $CaCl_2$, 5 m*M* MES buffer (solution adjusted to pH 5.5 with HCl). Pectinase from *Rhizopus* or *Aspergillus* species (Macerase, Calbiochem Co.; Macerozyme R10, Yakult Biochemical Co., Ltd.; Pectinase No. P-5146, Sigma Chemical Co.) is often included in incubation medium at the 0.5% (w/v) level, but it is the pectolyase Y23 product which appears to be most useful, even with otherwise intractable tissues. The optimum level of osmoticum is that which just exceeds incipient plasmolysis of protoplasm within tissue. Premecz *et al.*[29] have shown that osmotic stress of cells induced by the required plasmolysis of cells—with nonpermeating plasmolitica prior to and during protoplast isolation—is principally responsible for the large increase in ribonuclease which occurs in tobacco leaf protoplasts. These workers conclude that the extent of plasmolysis is more important than the concentration of osmoticum in causing this stress response. It seems prudent to minimally plasmolyze tissue for protoplast isolation. Betaine has been effectively used as osmoticum where sugar analysis of protoplasts or vacuoles is to be made.[30]

The above medium releases protoplasts in good yield from leaves and various aerial tissues and cultured cells of various monocots and dicots in

[26] W. Lin, *Plant Physiol.* **66,** 550 (1980).
[27] R. M. Sindhu and S. S. Cohen, *Plant Physiol.* **76,** 219 (1984).
[28] R. Kringstad, W. H. Kenyon, and C. C. Black, *Plant Physiol.* **66,** 379 (1980).
[29] G. Premecz, T. Olah, A. Gulyas, A. Nyitrai, and G. Palfi, *Plant Sci. Lett.* **9,** 195 (1977).
[30] F. Keller and A. Wiemken, *Plant Cell Rep.* **1,** 274 (1982).

less than 1 hr and from corn roots[26] in 1.5 to 2 hr at 22–30°. Most investigators shake preparations during enzyme digestion, but if penetration of enzyme is efficient, shaking is not essential for obtaining high yields. Shaking is said to have deleterious effects on the photosynthetic capacity of isolated protoplasts.[31] A 150-W tungsten lamp mounted 25 cm above digesting tissue is said to maintain photosynthetic capacity of protoplasts during digestion and perhaps increase the efficiency of digestion.[31] Though not widely employed (or at least described) use of illumination would seem to be prudent for helping to maintain the physiological state of cells of light-grown tissues during protoplast preparation. Changing the enzyme solution during digestion to provide fresh enzyme and remove components from damaged cells may be advantageous and is often done.

Perhaps the most detailed analysis of isolation conditions which yield metabolically competent isolates comes from efforts to obtain photosynthetically competent protoplasts.[31] Therefore, similar conditions for protoplast isolation may apply where, i.e., storage functions of vacuoles are studied by incubation of protoplasts with precursors or ions followed by isolation of vacuoles to determine compartmentation.

After tissue digestion, released protoplasts may be filtered through several layers of cheesecloth, washed several times in isolated medium less wall digesting enzymes (sedimentation at 100 *g*, 2-min resuspension in fresh medium by gentle mixing, and resedimentation), and used directly. Alternatively, they may be purified on Ficoll 400 or metrizamide gradients. In the case of Ficoll the general procedure of Boller and Kende[32] is widely used. Once washed protoplasts are gently mixed with 1 vol of purified[32] 15–20% (w/v) Ficoll 400 (Pharmacia) containing mannitol or sorbitol at the concentration used in isolation medium, and 5 m*M* MES/HCl, pH 7, 0.2 m*M* $CaCl_2$. This solution is overlayered with 8, 6, and 0% (w/v) Ficoll zones (10 ml), each containing mannitol and buffers as in isolation medium. After centrifugation at 300–1500 *g* for 10–20 min at 4°, purified protoplasts are gently collected from the 0–6% interface and diluted into isolation medium at pH 7 less enzymes and collected from this suspension by sedimentation at $\leq$100 *g* for 1–5 min. It may be necessary to vary Ficoll concentrations in the gradient to achieve efficient purification. In the case of protoplasts having high density due to the presence of storage carbohydrates or pigments, other gradient media may be more suitable.[33]

[31] R. C. Leegood and D. A. Walker, *in* "Isolation of Membranes and Organelles from Plant Cells" (J. L. Hall and A. L. Moore, eds.), p. 185. Academic Press, New York, 1983.

[32] T. Boller and H. Kende, *Plant Physiol.* **63,** 1123 (1979).

[33] W. Wagner, F. Keller, and A. Wiemken, *Z. Pflanzenphysiol.* **112,** 359 (1983).

In a number of cases, protoplast purification on Ficoll was made at 10,000 *g* for 2 hr at 4° to achieve high purity. A compromise between purity and yield-condition is made by adjusting the time and centrifugation speed. Complete removal of debris from protoplasts destined for vacuole isolation is not necessary for most applications. However, Boller and Kende[32] have argued for the need to remove debris having bound isolation enzyme where compartmentation analysis of hydrolases is to be made. Thorough washing may also be effective in removing residual isolation enzyme.[34]

Apparently any tissue of any plant which yields essentially naked protoplasts (those with the majority of cell wall removed—perfect spheres) will serve as source material for isolation of vacuoles. However, the condition, purity, and yield may vary greatly with the quality and status of protoplasts from which vacuoles are released. Tissues and isolation media which require long periods (<20 hr) for digestion often yield fragile vacuoles which are difficult to purify and recover.[4] In such cases it is advisable to test various commercial wall-digesting enzymes (including pectolyase Y23) in order to minimize isolation time. Vacuoles having highly acidic sap are often difficult to recover, presumably because it is difficult to maintain basic pH in the vacinity of emerging vacuoles.

Only a few attempts have been made (using the same tissue) to directly compare various basic approaches for isolating vacuoles from protoplasts. Even these are flawed and somewhat contradictory. Aerts and Schram[35] compared the suitability of the three basic approaches for preparing vacuoles from protoplasts and concluded that the polybase procedure was superior in terms of yield and purity. However, the osmotic lysis procedure used was modified by making slow addition of KPO_4—results in low yields of impure vacuoles (unpublished observation). The "mechanical" isolation procedure utilized isotonic medium but did not include raising the pH, which generally accompanies such procedures. Therefore, basic aspects of both approaches 1 and 2 were modified in this study. In contrast to the conclusions of Aerts and Schram, the polybase and osmotic lysis approaches were deemed equally suitable for vacuole isolation from leaves of *Vicia* but only the latter released guard cell vacuoles.[36] Glund *et al.*[36a] concluded that an osmotic lysis procedure based on KPO_4 was unsuitable for preparation of vacuoles from suspension cells of tomato while a procedure employing shear lysis in isotonic medium together with pH adjustment from pH 6 to 8 was useful. Earlier studies

[34] G. J. Wagner, P. Mulready, and J. Cutt, *Plant Physiol.* **68,** 1081 (1981).

[35] J. M. F. G. Aerts and A. W. Schram, *Z. Naturforsch., C: Biosci.* **40C,** 189 (1985).

[36] H. Schnable and C. Kottmeier, *Planta* **161,** 27 (1984).

[36a] K. Glund, A. Tewes, S. Abel, V. Leinhos, R. Walther, and H. Reinbothe, *Z. Pflanzenphysiol.* **113,** 151 (1984).

suggested that the polybase procedure was not optimal for several tissues[32] (see Wagner[4]).

It would not be surprising to find that subtle differences in surface charge or plasmalemma biochemical characteristics (innate or modified by wall-digesting enzymes) which affect protoplast lysis are expressed by different species or varieties within the same species. Griffing *et al.*[37] reported that surface charge of protoplasts varies from species to species and with tissue type.

Osmotic Lysis of Protoplasts to Prepare Vacuoles

To date, there have been about 40 published papers which describe the use of the osmotic lysis approach for preparing mature plant vacuoles. Protoplasts isolated from leaves, cotyledons, suspension cells, guard cells, roots, petals, stems, and tubers of various plant genus and species have yielded useful amounts of vacuoles by this approach. The procedure involving rapid dilution of protoplasts equilibrated in 0.6 to 0.7 *M* osmoticum, pH 5.5, to 0.2 *M* $KHPO_4/K_2HPO_4$, pH 8,[4,22] or a similar procedure involving dilution of protoplasts in 0.6 *M* mannitol, pH 5.5, to give a suspension containing 0.06 *M* mannitol, 0.1 *M* K_2HPO_4/HCl, pH 8,[32] has been described recently in this series.[5] It continues to be used frequently and gives reasonable yields (about 10 to 30%) of vacuoles which are, in general, at least as free from contamination as those obtained by other methods. So, it will be reiterated here.

This procedure involves rapid (5 sec) dilution of a 1- to 2-ml suspension of protoplasts (washed and Ficoll purified, or washed, see above) in 0.5 to 0.7 *M* mannitol or sorbitol, 5 m*M* MES/HCl, pH 5.5, contained in a round-bottomed glass tube with 10 to 20 ml of 0.1 to 0.2 *M* KH_2PO_4/K_2HPO_4, pH 8, 2 m*M* DTT. Other salts, including KNO_3, $NaNO_3$, $NaVO_3$, and Na_2WO_4, are less effective (but some cause interesting modes of vacuole emergence) while $NaPO_4$ is as effective as the K salt (unpublished). The suspension is gently stirred (about 2 min) using a wooden dowel to effect complete emergence of vacuoles from lysing protoplasts. Aggregating debris is accumulated while the lysate is cooled to 10°. Debris is removed by slow filtration over one to three layers of fine glass wool covering the side of a conical filter or through 50-μm nylon cloth. Lysis medium is removed after sedimentation of vacuoles at 100–300 *g* by gentle resuspension of the vacuole pellet in 0.6–0.7 *M* mannitol, 20 m*M* HEPES/KOH, pH 8, 2 m*M* DTT. To further purify vacuoles, and where low-speed centrifugation will not efficiently sediment vacuoles,

[37] L. R. Griffing, A. J. Cutler, P. D. Shargool, and L. C. Fowke, *Plant Physiol.* **77**, 765 (1985).

lysate is gently mixed with 20% (w/v) Ficoll 400 in 0.6 *M* mannitol to give a suspension containing 5% (w/v) Ficoll, 190 m*M* mannitol, 68 m*M* KPO_4, 0.68 m*M* DTT.[32] Portions of this suspension are overlayered with 10 ml each of 3.75% (w/v) Ficoll followed by layers of 2.5 and 0% Ficoll—all layers containing 190 m*M* mannitol, 68 m*M* K_2HPO_4, and 0.68 m*M* DTT. Gradients are centrifuged at 300 *g* for 20 min and vacuoles are collected from the appropriate interface. As with protoplast purification described above, the density of gradient layers will vary according to the density of vacuoles being prepared.

The linear relationship between osmolality and concentration of commonly used osmotica can be established using the following values determined by freezing point depression. For mannitol and sorbitol in H_2O, 0.2 *M* is 195 mOsm/kg and 0.4 *M* is 410 mOsm/kg. For KPO_4 in H_2O, 0.1 *M* is 240 mOsm/kg and 0.2 *M* is 470 mOsm/kg. Plots drawn using these values can be used to estimate the extent to which osmolality of protoplast bathing medium should be lowered to achieve efficient lysis. Generally reduction is 40 to 70%. The osmotic potential of all components of protoplast isolation medium needs to be considered. For example, 0.1 *M* Na_2EDTA and 0.1 *M* mannitol have the same osmolality.

Major advantages of the osmotic lysis procedure may be its rapidity, simplicity, and the relative lack of contamination of isolated vacuoles with soluble and particulate contaminants. A disadvantage in using KPO_4 is the fact that vacuole sap pH is increased, presumably due to K^+/H^+ exchange. However, anion, other cation, and neutral microsolutes are largely retained in isolated vacuoles.[4] Procedures which do not utilize KPO_4 generally involve lowering the mannitol or sorbitol of protoplast bathing medium from 20 to 80% and raising the pH from ~5.5 to 7.5–8. Adjustment of pH is usually but not always employed in the latter procedures.

Polybase-Induced Lysis of Protoplasts to Prepare Vacuoles

This method has gained in popularity since 1977 when it was introduced by Buser and Matile.[23] At least 15 papers exist which cite its use in different plant systems. Originally devised for preparing yeast vacuoles from spheroplasts, the method has been adapted for use with a variety of tissues including leaves of *Bryophyllum,* sweet clover, bean, *Kalanchoe, Vicia,* petals of petunia, roots of *Beta vulgaris,* and suspension cells of *Catharanthus roseus* and tomato. The method originally described by Boudet *et al.*[38] has been widely adapted and appears to be generally

[38] A. M. Boudet, H. Canut, and G. Alibert, *Plant Physiol.* **68,** 1354 (1981).

useful. Here twice-washed protoplasts (~2×10^6) in 0.7 *M* mannitol, 25 m*M* MES–Tris, pH 6.5, are layered onto a three-step gradient containing the following: first step, 6 ml of above medium with 2.0% (w/v) Ficoll, 4 mg/ml DEAE Dextran (Pharmacia); second step, 2 ml 0.7 *M* mannitol, 25 m*M* HEPES–Tris, pH 8, 5% (w/v) Ficoll, 4 mg/ml Dextran sulfate (Pharmacia); third step, 2 ml 0.7 *M* mannitol, 25 m*M* HEPES–Tris, pH 8, 20% (w/v) Ficoll. The gradient is centrifuged for 30 min at 2000 *g* at 17° and partially purified vacuoles are collected at the interface between the second and third steps. Yield is reported to vary between 3 and 50% based on various reports and levels of contamination with unlysed protoplasts and/or marker activities range from 2 to 15%. This group sometimes uses a modified procedure in which the pH is not changed, DEAE Dextran and Dextran sulfate concentrations are adjusted to 6 mg/ml and 1 mg/ml, respectively, and the bottom gradient zone is 10% Ficoll (w/v).[39] Both the polybase and osmotic lysis procedures appear to yield, on the average, one vacuole per lysed protoplast and undoubtedly the efficiency of protoplast lysis is dependent on the degree to which cell wall is removed during protoplast preparation.

The polybase procedure is as rapid as the osmotic lysis method. That reagents used in these procedures such as DEAE-Dextran,[40] Dextran sulfate, KPO_4, EDTA, PVP, BSA, PEG, etc., can alter the tonoplast cytosol interface or tonoplast is not established. Thom *et al.*[41] have suggested that DEAE Dextran is likely to damage tonoplast. However, while testable, there is no report which describes deleterious effects due to short exposure of tonoplast to this reagent.

Lysis of Protoplasts by Shear under Isotonic Conditions

There have been perhaps 16 papers describing the isolation of vacuoles under isotonic conditions which used modifications of the method described by Lorz *et al.*[24] Many of these involve a pH change during destabilization of protoplasts and maintain stable pH during vacuole purification. In one truly unusual procedure, mannitol concentration is raised from 0.4 to 0.7 *M* during centrifugation to lyse protoplasts.[27] Perhaps this explains the high level of contamination accompanying vacuoles prepared by this means. A widely used adaptation of the procedure of Lorz was first reported by Guy *et al.*[42] Here a concentrated suspension of protoplasts in buffered 0.6 *M* mannitol, pH 5.8, is layered onto a two-step

[39] G. Alibert, A. Carrasco, and A. M. Boudet, *Biochim. Biophys. Acta* **721,** 22 (1982).
[40] H. Gimmler and G. Lotter, *Z. Naturforsch., C: Biosci.* **37C,** 609 (1982).
[41] M. Thom, A. Maretzki, and E. Komor, *Plant Physiol.* **69,** 1315 (1982).
[42] M. Guy, L. Reinhold, and D. Michaeli, *Plant Physiol.* **64,** 61 (1979).

Ficoll gradient composed of 12 and 15% (w/v) steps, each containing 0.6 *M* mannitol, 20 m*M* HEPES, pH 7.7. The gradient is centrifuged at about 200,000 *g* for 2 hr at 4° and the majority of vacuoles are collected from the 12% Ficoll–sample zone interface. Adjustment of Ficoll concentrations may be required depending on the tissue used as a source of vacuoles. When the presence of both soluble and particulate contaminants has been assessed, vacuoles prepared by this general method appear to be more highly contaminated (~10–30% of total extravacuolar marker activity) than those obtained using the osmotic or polybase-induced lysis procedures (~3–10%) or those obtained by mechanically slicing tissue (~3–10%) (see below). Thom *et al.*[41] have argued that vacuoles isolated using the basic method of Lorz *et al.* are not enveloped by plasma membrane as proposed by Lorz *et al.* (vacuoplasts) but that residual plasma membrane (19% that of protoplast) occurs in pieces over the tonoplast. They argued further that the isolation procedure used was chosen—at the expense of purity—to preserve the integrity of the tonoplast so that meaningful studies of transport could be made. Perhaps tonoplast surfaces preserved beneath residual plasmalemma are the most transport-competent surfaces.

Indeed, where transport has been studied successfully for extended periods of time (hours) and in some detail using vacuoles isolated from protoplasts, the isotonic, shear lysis technique has been most often employed. A possible exception to this rule is found in the work of Kaiser and Heber,[43] who studied sucrose transport in barley leaf vacuoles isolated by osmotic lysis. They concluded that sucrose transport was energy independent and carrier mediated. The method of vacuole isolation used in this case was a modification of that of Martinoa *et al.*[44] where protoplasts in 0.4 *M* sorbitol, pH 5.6, are suspended in medium containing 0.4 *M* sorbitol, pH 7.8, 12.5% (w/v) osmotically adjusted Percoll (Pharmacia), and other additives in low concentration. The suspension is pushed through a needle (10 cm long, 0.7-mm diameter) to release vacuoles followed by purification on Percoll gradients. Yield was 25–30% and contamination with soluble markers was said to be ≥5%. Contamination was not assessed using membrane markers. Thom and Maretzki[45] recently reported that sucrose transport in sugarcane vacuoles isolated by isotonic, shear lysis occurs by a group translocation mechanism. Electrogenic proton transport has also been demonstrated in this system.[46]

[43] G. Kaiser and U. Heber, *Planta* **161,** 562 (1984).
[44] E. Martinoa, V. Heck, and A. Wiemken, *Nature* (*London*) **289,** 292 (1981).
[45] M. Thom and A. Maretzki, *Proc. Natl. Acad. Sci. U.S.A.* **82,** 4697 (1985).
[46] M. Thom and E. Komor, *Plant Physiol.* **77,** 329 (1985).

Therefore, if one dares to draw general conclusions from this limited experience, the isotonic, shear lysis approach for preparing vacuoles from protoplasts may be advantageous for transport studies but not ideal for compartmentation analysis where purity is of upmost concern. It is interesting that Thom *et al.*[41] reported carboxypeptidase and acid protease to be partially associated with tonoplast of sugar cane vacuoles isolated by isotonic, shear lysis while others have failed to obtain evidence for such association in vacuoles prepared by osmotic lysis of protoplasts.[4] Either artifactual tonoplast association of protease or preservation of labile membrane association in the former case (which is perturbed by the osmotic or polybase methods) could explain these contradictory results. More study is needed to resolve this point, but it may be an important one in that a number of vacuolar functions may require tonoplast–enzyme associations, associations which may be disrupted during vacuole isolation by certain methods, but not others. It is well known that certain glycolytic enzyme–membrane interactions are sensitive to pH and ionic strength and that membrane binding greatly influences enzymatic activity and possibly pathway function.[47,48]

Particularly in the transport area, more studies are needed where comparison of basic isolation procedures is made. It is the author's view that such efforts should employ the specific methods described here. These have been successfully applied in a number of studies with different tissues and by a number of investigators. Modification of these procedures may negate any attempt at comparison with earlier work because modifications may have substantial effects on the quality of vacuoles isolated. A case in point is the work of Aerts and Schram[35] with regard to the osmotic lysis and "mechanical" isolation procedures which were compared with the polybase method. Basic procedures as described in the literature were not applied for osmotic and mechanical lysis but were substantially modified so as to represent new protocols. It is difficult to assess results, in the sense of methods comparison, from such work.

Isolation of Vacuoles Directly from Tissue

The basic procedure of Leigh and Branton[25,49] for isolating vacuoles from fresh beet root continues to be used with only minor modification. Several investigators have performed slicing of tissue by hand using a

[47] H. Wombacher, *Mol. Cell. Biochem.* **56,** 155 (1983).

[48] J. J. Blum, *in* "Advances in Cell Biology" (J. Schwartz and L. Azair, eds.), p. 510. Van Nostrand-Reinhold, Princeton, New Jersey, 1981.

[49] R. A. Leigh, D. Branton, and F. Marty, *Methodol. Surv. Biochem.* **9,** 69 (1979).

razor blade,[50–52] making unnecessary the construction of the chopping device described by Leigh and Branton.[49] Slices of beet root tissue 5 mm thick are finely sliced by hand with a single-edged razor blade into a shallow dish containing 1 *M* sorbitol, 50 m*M* Tris/HCl, pH 8.0, 25 m*M* 2-mercaptoethanol. A total of about 450 g fresh wt of tissue is so processed to yield 1 liter of homogenate. Tissue pieces are separated by filtration with Miracloth and the homogenate is cooled to 4° and centrifuged at 1300 *g* for 20 min. The pellet is resuspended with 15% (w/v) metrizamide (Nyegaard Co., A/S Oslo, Norway) in 1.2 *M* sorbitol, 10 m*M* Tris/Mes, pH 8.0, 5 m*M* 2-mercaptoethanol, placed in centrifuge tubes, and this suspension is overlaid with 10 and 0% (w/v) metrizamide layers in the above buffer. After centrifugation for 10 min at 430 *g*, vacuoles are recovered from the 10%/0% metrizamide interface, diluted with 2 vol of isolation medium, and the suspension is centrifuged at 650 *g* for 10 min. Yield from this procedure is about 0.2%, but large amounts of tissue can be processed. For red beet, contamination is low as measured using both soluble and particulate marker enzyme activities.[49,50] While sucrose and betanin are retained during isolation, about half of Na^+ and K^+ is lost during initial steps of the procedure.[53] The major drawbacks of the procedure are its limitation to firm root tissue and, because yield is low, its unacceptability for studying the compartmentation of metabolites after radiolabeling tissue. Procedures which result in very low yields of vacuoles may select for specific populations of vacuoles. Miller *et al.*[54] suggested that substitution of Dextran for metrizamide in the isolation procedure of Leigh and Branton results in vacuoles having different properties than those obtained using metrizamide. The implication was made that a different population of vacuoles were obtained in each case. Using pH and solute-reactive vital stains, early cytologists recognized that vacuoles having different composition can occur in a given tissue and even within a single cell.[4,19] Kirkdjian *et al.*[55] recently showed that cells, protoplasts, and vacuoles derived from cultured cell populations have variable vacuolar pH.

[50] H. Barbier and J. Guern, *in* "Plasmalemma and Tonoplast: Their Functions in Plant Cells" (D. Marme, E. Marre, and R. Hertel, eds.), p. 233. Elsevier/North-Holland Biomedical Press, Amsterdam, 1982.

[51] A. B. Bennett, S. D. O'Neill, and R. M. Spanswick, *Plant Physiol.* **74,** 538 (1984).

[52] J. Helminger, T. Rausch, and W. Hilgenberg, *Physiol. Plant.* **58,** 302 (1983).

[53] R. A. Leigh and A. Deri Tomos, *Planta* **159,** 469 (1984).

[54] T. J. Miller, J. J. Brimelow, and P. John, *Planta* **160,** 59 (1984).

[55] A. Kurkdjian, H. Quiauampuix, H. Barbier-Byrgoo, M. Pean, P. Manigault, and J. Guern, *in* "Biochemistry and Function of Vacuolar Adenosine-Triphosphatase in Fungi and Plants" (B. P. Marin, ed.), p. 98. Springer-Verlag, Berlin and New York, 1985.

Criteria for Purity, Integrity, and Stability

As already indicated, different criteria define acceptability and quality–integrity of isolated plant vacuoles. Criteria will differ according to the intended use of the isolated organelles. They will be discussed here in the context of the two principal, current uses of isolated higher plant vacuoles.

Compartmentation Analysis

Two principal criteria which define the quality of vacuole preparations used to determine vacuolar–extravacuolar distribution of metabolites, ions, enzymes, etc., in protoplasts and tissue are the purity of preparations and retention of solutes within vacuoles during and after their isolation. When protoplasts are used, the question of how protoplasts reflect solute compartmentation *in vivo* is another major question.[53] Subcellular localization using protoplasts ignores constituents which may occur in the cell wall, unless tissue is also examined and considered.[32,53]

The extent of contamination of vacuole preparations by extravacuolar cellular membranes is often assessed using marker enzyme[56] and chemical detections including cytochrome *c* oxidase as a marker for inner mitochondrial membrane, chlorophyll as a thylakoid (chloroplast) marker in green tissues, and, in a few cases, glucan synthase II as a plasmalemma marker. Antimycin, an insensitive NADH–cytochrome-*c* reductase frequently used in plant membrane fractionation studies as a marker for smooth and rough endoplasmic reticulum (ER) is inappropriate as an ER marker in assessing vacuole purity because this activity is shown to be enriched in tonoplast.[1,4] This may reflect the common ontogeny of tonoplast and ER.[1] Chrestin *et al.*[57] suggest that in *Hevea* lutoids this activity participates in an NADH-consuming redox system which functions in proton transport out of the lutoid. Such a mechanism, if generally observed in higher plant vacuoles, would suggest an additional role for mature vacuoles, i.e., a role in pH stabilization and regulation of cytosolic metabolism via pH. In any case, a reliable marker is not available for monitoring ER, a prominent plant membrane during vacuole isolation and purification. A possible candidate for such a marker is cinnamic acid 4-hydroxylase *trans*-cinnamate 4-monooxygenase), shown to be ER bound in several systems.[56,58] However, its absence from tonoplast has not been

[56] P. H. Quail, *Annu. Rev. Plant Physiol.* **30,** 425 (1979).

[57] H. Chrestin, X. Gidrol, J. d'Auzac, J. L. Jacob, and B. Marin, *in* "Biochemistry and Function of Vacuolar Adenosine-Triphosphatase in Fungi and Plants" (B. P. Marin, ed.), p. 245. Springer-Verlag, Berlin and New York, 1985.

[58] G. Wagner and G. Hrazdina, *Plant Physiol.* **74,** 901 (1984).

shown directly. While latent UDPase is a useful marker for Golgi,[59] it is seldom applied to studies of isolated vacuoles. Other membranes which might contaminate both intact vacuoles and tonoplast (1.1 g/cm^3) recovered from tissue homogenates are chloroplast, etioplast and mitochondrial outer membranes, and nuclear and peroxisomal membranes. These are seldom monitored. Galactosyltransferase is recognized as a chloroplast (and probably etioplast) envelope marker[56] and was found to be highly enriched in vacuoles isolated from oat protoplasts using an osmotic lysis method.[60] In that study, it was concluded that vacuoles were contaminated with chloroplast outer envelope. Occurrence of this activity in vacuole preparations from other systems prepared using other approaches may prove interesting. The possibility of nuclear contamination, while often ignored, has been assessed by monitoring radiolabeled[39,61] or unlabeled RNA and DNA.[4] Peroxidase has often been used to monitor possible contamination by peroxisomes and fumarase for mitochondria. However, these enzymes are soluble organellar constituents[56] and it is not known if peroxisomes and mitochondria remain intact during vacuole isolation using various approaches. These activities have not been reported to be enriched above other contaminant markers where they have been analyzed. The presence of plasmalemma-enveloping tonoplast (vacuoplast structure)[24] has been monitored using fluorescein diacetate.[62] Limitations of this method have been discussed[4] and reports of its use do not suggest that vacuoplast structures generally occur as a result of any of the basic approaches used to isolate vacuoles.[35,61,63] Thom *et al.*[41] utilized radiolabeled and fluorescent Con A in binding experiments to assess plasmalemma contamination of sugarcane vacuoles. They concluded that while substantial plasmalemma remained with vacuoles isolated using essentially the isotonic, shear lysis method of Lorz, it did not envelope the tonoplast.

Various enzymes have been employed as "soluble" markers for monitoring vacuole preparations including (in order of prevalence of use) glucose-6-phosphate dehydrogenase, malate dehydrogenase, phosphoglucoisomerase (glucose-6-phosphate isomerase), phosphoglucomutase, and others. The first two of these are perhaps the activities of choice since

[59] G. Nagahashi, *in* "Modern Methods of Plant Analysis" (H. F. Linskens and J. F. Jackson, eds.), Vol. 1, p. 66. Springer-Verlag, Berlin and New York, 1985.

[60] B. Verhoek, R. Haas, K. Wrage, M. Linscheid, and E. Heinz, *Z. Naturforsch., C: Biosci.* **38C,** 770 (1983).

[61] K. Glund, A. Tewes, S. Abel, V. Leinhos, R. Walther, and H. Reinbothe, *Z. Pflanzenphysiol.* **113,** 151 (1984).

[62] A. Admon, B. Jacoby, and E. E. Goldschmidt, *Plant Physiol.* **65,** 85 (1980).

[63] R. C. Granstedt and R. C. Huffaker, *Plant Physiol.* **70,** 410 (1982).

they are thought to be at least in large part cytosolic in most plants. Alcohol deoxydrogenase, while not used as a cytosal marker in studies of vacuoles, may be suitable.[56]

An often-used criterion for monitoring contamination is to count the number of residual protoplasts in aliquots of vacuole and protoplast preparations and express protoplast contamination as a percentage of total structures.[4] Where both marker enzymes and protoplast counts have been reported, it is not uncommon to read that contamination levels estimated by these measures agree. Yet, where presented, light micrographs of vacuole presentations—even those which show only a few structures—often show substantial particulate contamination as well as residual protoplasts. In such instances, an underestimation of the degree of contamination by, at least, residual organic matter is suggested.

The list of secondary metabolites, amino acids, sugars, ions, etc., occurring in isolated vacuoles continues to grow.[5,52,64–70] In many cases, the vacuolar content was determined to be like that of protoplasts or, in the cases using the mechanical tissue slicing method, like that of tissue. These results indicate that the majority of various vacuolar neutral, anionic, and cationic microsolutes is retained in organelles isolated by all the basic approaches and suggest that tonoplast is not easily made generally leaky to solute constituents. However, Leigh and Deri Tomos[53] have shown that while sucrose and betanin are completely retained in beet root vacuoles isolated by mechanical slicing of tissue, Na^+ and K^+ are partially lost. Lin *et al.* have shown that petal vacuoles isolated by osmotic lysis in KPO_4 undergo what appears to be K^+/H^+ exchange resulting in the alkalinization of vacuole sap.[4] These results, and other arguments,[71] suggest that porters for inorganic ions occur in tonoplast and that these can facilitate ion exchange under certain conditions. Evidence is growing for the presence of secondary transport porters in tonoplast.[7,17]

Tonoplast can be extended to great lengths without disruption or release of sap pigment.[18,72,73] In this respect, tonoplast is remarkably differ-

[64] E. A. Bray and J. A. D. Zeevaart, *Plant Physiol.* **79,** 719 (1985).

[65] M. Guy and H. Kende, *Planta* **160,** 281 (1984).

[66] H. Hollander-Czytko and N. Amrhein, *Plant Sci. Lett.* **29,** 89 (1983).

[67] V. Sharma and D. Strack, *Planta* **163,** 563 (1985).

[68] P. Ratuboul, G. Alibert, T. Boller, and A. M. Boudet, *Biochim. Biophys. Acta* **816,** 25 (1985).

[69] E. Pahlich, R. Kerres, and H. Jager, *Plant Physiol.* **72,** 590 (1983).

[70] R. Schmitt and H. Sandermann, Jr., *Z. Naturforsch., C: Biosci.* **37C,** 772 (1982).

[71] D. Cornel, C. Grignon, J. P. Rona, and R. Heller, *Physiol. Plant.* **57,** 203 (1983).

[72] G. J. Wagner, *Plant Physiol.* **68,** 499 (1981).

[73] R. K. Salyaev, *in* "Biochemistry and Function of Vacuolar Adenosine-Triphosphatase in Fungi and Plants" (B. P. Marin, ed.), p. 3. Springer-Verlag, Berlin and New York, 1985.

ent from plasmalemma, at least plasmalemma of isolated protoplasts. The extensibility of tonoplast reflects the integrity of membrane of isolated vacuoles and undoubtedly reflects properties important to transvacuolar strand and vacuole dynamics *in vivo*.[18]

It has been suggested that vacuolar sap solutes, particularly ions, may undergo redistribution during protoplast preparation.[53] The extent to which this may occur is seldom addressed. However, the strategy for such analysis is found in the work of Leigh and Deri Tomos,[53] where endogenous pigment (marker enzymes may also serve) is shown to be exclusively localized in the vacuole by a correlative method or by careful comparison of total protoplast and resulting total vacuole sap volumes (see below).[4] The exclusively vacuole sap standard may then be used to determine the percentage of solute in question occurring in the vacuole of tissue (after consideration of a possible cell wall pool) using the following ratio[53]:

$$\text{Percentage vacuolar solute} = \frac{\text{solute/standard in isolated vacuoles}}{\text{solute/standard in tissue homogenate}}$$

Methods for relating protoplasts and vacuoles to determine vacuole/extravacuole distribution of solutes include the use of endogenous pigments,[35] vacuolar enzymes,[3,4] and the distribution of acidophylic substances which equilibrate in acidic compartments—neutral red, radiolabeled alkaloids, benzylamine,[39] etc. Distribution of these is most often determined after counting protoplasts and vacuoles. This is an error-prone procedure because a small error in estimating structure diameter can result in substantial variation in computed volume.[4] For example, an error of 5% in estimating the diameter of a vacuole or protoplast in the light microscope results in an error of 14% in estimating the volume. In vacuole/extravacuole compartmentation analysis, the volume is critical and where light micrographs have been provided, a 5% or greater variation in size of both protoplasts and vacuoles is evident. Therefore, where simple counting of structures is applied, an error of 15% or more in volume estimation can be anticipated and should be considered in drawing conclusions concerning compartmentation.

The enzyme most often used as a vacuole sap marker is α-mannosidose.[3,18] Indeed, in 24 reports using various tissues and methods of vacuole isolation, 19 conclude that 85–130%[35,61,65,66,74–79] of protoplast α-man-

[74] G. Marigo, H. Bouyssou, and M. Belkoura, *Plant Sci. Lett.* **39,** 97 (1985).

[75] T. Boller and U. Vogeli, *Plant Physiol.* **74,** 442 (1984).

[76] F. Keller and P. Matile, *J. Plant Physiol.* **119,** 369 (1985).

[77] J. A. Saunders and J. M. Gillespie, *Plant Physiol.* **76,** 885 (1984).

[78] L. M. V. Jonsson, W. E. Donker-Koopman, P. Uitslager, and A. W. Schram, *Plant Physiol.* **72,** 287 (1983).

nosidose is vacuolar, 3%[36] report 46–82% vacuolar, and 2[60,67] were unable to detect the activity—see Wagner[18] for reports prior to 1982. Vacuolar α-mannosidose appears to be widespread, but each system must be examined individually before assuming the value of this enzyme as a sap marker. Acid protease and *N*-acetylglucosaminidase also appear to be useful vacuole sap markers in many systems. The former activity appears to be common to both meristematic and mature plant vacuoles.[18] Acid phosphatase is a less reliable vacuole marker. In 25 reports, acid phosphatase activity—measured as *p*-nitrophenylphosphatase—was reported to be vacuolar to the following degrees: 85–123%,[36,41,76,75] 50–85%,[43,67,79,80] and ≥50%[61] in 14, 8, and 3 cases, respectively—see Wagner[18] for reports prior to 1982. Phosphodiesterase, RNase, and glycosidase are less studied and, in general, do not appear to be confined to the vacuole.

As noted above, there exists substantial evidence that the bulk of cellular acid protease is localized in mature plant vacuoles.[3,18] However, extravacuolar proteases also occur and the site(s) of preterminal senescence, normal and abnormal protein turnover is little studied. However, Nishimura and Beevers[81] showed storage protein turnover in castor bean endosperm vacuoles and recently Canut *et al.*[82] showed hydrolysis of newly synthesized proteins within vacuoles of protoplasts isolated from cultured *Acer pseudoplantanus* cells. The latter results show that vacuoles are capable of protein hydrolysis. Unfortunately, kinetics of turnover of newly synthesized protein was not compared in cells and protoplasts derived from them, leaving open the possibility that stress during protoplast isolation induced transport of newly synthesized protein into vacuoles and/or hydrolysis of these proteins. This most interesting study shows that the question of the site(s) of normal protein turnover might be resolved using the *Acer* cell system. Recently, chloroplastic proteases and chlorophyll-degrading activities have been recovered from isolated chloroplasts.[83]

There continues to be no reliable tonoplast marker. While vanadate-insensitive, NO_3-inhibited Mg^{2+}-ATPase, pyrophosphatase, and NADH–cytochrome-*c* reductase appear to be associated with tonoplast obtained from a number of systems,[18] and sorbitol oxidase in apple cotyledons,[84] the occurrence of these in other cellular membranes has either been

[79] W. Wagner, F. Keller, and A. Wiemken, *Z. Pflanzenphysiol.* **112,** 359 (1983).

[80] S. Yamaki, *Plant Cell Physiol.* **23,** 881 (1982).

[81] M. Nishimura and H. Beevers, *Nature (London)* **277,** 412 (1979).

[82] H. Canut, G. Alibert, and A. M. Boudet, *Plant Physiol.* **79,** 1090 (1985).

[83] A. M. Nettleton, P. L. Bhalla, and M. J. Dalling, *J. Plant Physiol.* **199,** 35 (1985).

[84] S. Yamaki, *Plant Cell Physiol.* **23,** 891 (1982).

shown or not ruled out. In yeast, α-mannosidase is reported to be bound to tonoplast. This does not appear to be the case in higher plants.[18]

Transport Studies

It is reasonable to suggest that purity of isolated vacuoles is less crucial to transport studies than to compartmentation analysis. However, it is also reasonable to suggest that occurrence of substantial contamination by other intact organelles or membrane vesicles which have similar transport processes, or bacteria, could lead to misleading results. Occurrence of different types of vacuoles in a preparation could also be a confusing factor. Finally, assuming that characteristics of transport measured using a specific isolation procedure are representative of all vacuoles within the tissue being studied may be misleading because only a select population of vacuoles may survive the isolation procedure. This may be especially relevant where low yields are observed.[4] On the other hand, labile transport systems may be lost where isolation conditions are rigorous so as to maximize removal of cellular contaminants.

So, preservation of labile transport systems may be crucial to efficient study of some modes of transport using isolated structures. This has been the assumption made by Thom *et al.*,[41] which was discussed earlier. Another important requirement is that organelle stability be retained over the period required for studying transport.

A number of investigators have examined conditions which effect stability of isolated vacuoles. There is a general consensus on only three factors which appear to contribute to stability; low temperature (4°) storage, maintenance of pH 7 to 8, and minimum mechanical disturbance. Several investigators suggest that addition of low levels (1–5 m*M*) of $MgCl_2$, $CaCl_2$, and serum albumin enhance stability[18,36] while others observe no effect or negative effect with certain of these additives.[85] Knuth *et al.*[85] concluded that 1–5 m*M* EDTA greatly stabilizes cowpea and barley vacuoles. They observed 200 to 300% increased stability in the presence of 1 m*M* EDTA. Also in this study, various antioxidants, reductants ions, and other agents were tested as stabilizing agents in the presence and absence of 1 m*M* EDTA. An outstanding observation was that 10 m*M* $MnCl_2$ greatly destabilized vacuoles in the presence and absence of EDTA while 10 m*M* $MgCl_2$ did not. Calcium chloride (5 m*M*) was also destabilizing. Given that the stability constant[86] (log K_a) of Mn_2EDTA (14.5) is substantially higher than that of Mg_2EDTA (9.6) and also

[85] M. E. Knuth, B. Keith, C. Clark, J. L. Garcia-Martinez, and L. Rappaport, *Plant Cell Physiol.* **24,** 423 (1983).

[86] A. E. Martell and M. Calvin, "Chemistry of Metal Chelate Compounds," p. 613. Prentice-Hall, Englewood Cliffs, New Jersey, 1952.

Ca_2EDTA (11.5), this result may indicate that tonoplast is extremely sensitive to Mn. Further study is needed to clarify this interesting observation. EDTA (1–10 m*M*) has been included by several investigators in isolation medium during vacuole preparation using all of the basic approaches. Yamaki[80] has indicated that EDTA greatly enhances lysis of protoplasts from apple cotyledons.

The pH conferring maximum stability to isolated vacuoles appears to be 7.0 to 8.0. This is close to what is thought to be the cytosolic pH value.[87] The optimum pH for tonoplast Mg^{2+}-ATPase[2,88–91] and proton pump activities[89,92] in various systems—some of which have a vacuolar sap pH as low as 4.5[93]—is pH 6.5 to 7.5. In contrast, sucrose transport in sugarcane vacuoles is optimal at pH 5.5[94] and that of UDPG uptake is pH 6.5.[95] Perhaps sugar translocators are near to tonoplast surface in a microenvironment while tonoplast ATPase[96] protrudes into the cytosol. The reported size of multimeric tonoplast ATPase (400,000 Da) and suggestive ultrastructural evidence[73,97] are consistent with this speculation.

While a better understanding of factors which affect tonoplast stability would be generally useful in both compartmentation analysis and transport studies, several existing isolated vacuole systems—principally those using *Acer*,[55] sugarcane,[94] and *Hevea*[57]—appear to afford sufficient stability to allow transport studies over an extending period of time (hours).

Summary

Four basic approaches have been utilized for isolation of higher plant vacuoles and one specific procedure for each of these is offered here as a broadly tested representative of each of the basic approaches. Criteria for

[87] F. A. Smith and J. A. Raven, *Annu. Rev. Plant Physiol.* **30,** 289 (1979).

[88] R. A. Leigh and R. R. Walker, *in* "Biochemistry and Function of Vacuolar Adenosine-Triphosphatase in Fungi and Plants" (B. P. Marin, ed.), p. 45. Springer-Verlag, Berlin and New York, 1985.

[89] X. Gidrol, B. Marin, H. Chrestin, and J. D. Auzac, *in* "Biochemistry and Function of Vacuolar Adenosine-Triphosphatase in Fungi and Plants" (B. P. Marin, ed.), p. 151. Springer-Verlag, Berlin and New York, 1985.

[90] J. A. C. Smith, E. G. Uribe, E. Ball, S. Heuer, and U. Lüttge, *Eur. J. Biochem.* **141,** 415 (1984).

[91] K. Aoki and K. Nishida, *Physiol. Plant.* **60,** 21 (1984).

[92] D. P. Briskin and R. J. Poole, *J. Plant Physiol.* **119,** 164 (1985).

[93] W. Lin, G. J. Wagner, H. W. Siegelman, and G. Hind, *Biochim. Biophys. Acta* **465,** 110 (1977).

[94] M. Thom, E. Komor, and A. Maretzki, *Plant Physiol.* **69,** 1320 (1982).

[95] M. Thom and A. Maretzki, *Proc. Natl. Acad. Sci. U.S.A.* **82,** 4697 (1985).

[96] S. Mandala and L. Taiz, *Plant Physiol.* **78,** 327 (1985).

[97] F. Marty, *in* "Biochemistry and Function of Vacuolar Adenosine-Triphosphatase in Fungi and Plants" (B. P. Marin, ed.), p. 14. Springer-Verlag, Berlin and New York, 1985.

quality of isolated vacuoles may vary according to the intended use of the organelles. Guidelines and contamination markers for assessing purity of preparations are discussed. The remarkable stability–extensibility of tonoplast has made it possible not only to isolate intact vacuoles (largest isolated vacuoles are about 150 μm in diameter yet tonoplast is about 100 Å thick) but also to manipulate them. Exploitation of isolated vacuoles and tonoplast for fundamental studies, especially transport studies, has just begun.

[8] Isolation of Vacuoles and Tonoplast from Protoplasts

By ALAIN M. BOUDET and GILBERT ALIBERT

General Considerations

The vacuole is the most conspicuous organelle of mature plant cells and plays an important role in the control of a variety of cellular processes. These include the regulation of cell turgidity, the modulation of cytoplasmic component concentrations and cellular functions through cytoplasm/vacuole exchanges across the tonoplast.[1] Most of the present knowledge on the biochemical characteristics of the plant vacuole and its membrane comes from studies on isolated vacuoles. Vacuoles, which are relatively large in size and surrounded by a single membrane, are very fragile organelles. This explains why they were the last cellular organelles to be isolated in a pure state in the mid-1970s and that most of the methods developed include gentle disruption of the protoplasts to release the intact vacuoles.

There is no single or preferable way to isolate vacuoles. The choice of a procedure may depend on the nature of the plant material or on the purpose of the scientist. In this chapter we will describe the polybase-induced lysis of protoplasts originally designed by Dürr[2] for yeast and modified by us for plant vacuoles. In a recent comparison of different experimental procedures potentially suitable for vacuole isolation from *Petunia hybrida* protoplasts, Aerts and Schram[3] conclude that the polybase procedure results in the best vacuole preparation.

[1] A. M. Boudet, G. Alibert, and G. Marigo, *Ann. Pro. Phytochem. Soc. Eur.* **24,** 29 (1984).
[2] M. Dürr, T. Boller, and A. Wiemken, *Arch. Microbiol.* **105,** 319 (1975).
[3] J. M. F. G. Aerts and A. W. Schram, *Z. Naturforsch., C: Biosci.* **40C,** 189 (1985).

When vacuoles are released from protoplasts, whatever the specific procedure used, certain principles must be observed: (1) rapid protoplast isolation procedure to avoid prolonged exposure to wall-degrading enzymes, (2) gentle lysis of the protoplasts in order to avoid rupture of the released vacuoles, (3) efficient lysis in order to limit the occurrence of residual protoplasts which exhibit similar size and density to the vacuoles and are very difficult to discard by centrifugation methods, and (4) rapid vacuole isolation procedure and limitation of the purification steps in order to reduce bursting of the released vacuoles and leakage of solutes through the tonoplast.

Chemicals, Media, and Techniques

Tris, Tris(hydroxymethyl)aminomethane (ICN)
MES [(2-*N*-morpholino)ethanesulfonic acid], HEPES (N^2-hydroxyethylpiperazine-N'-2-ethanesulfonic acid), Dextran sulfate potassium salt (Calbiochem-Behring Corp., La Jolla, CA)
Ficoll 400 and DEAE–Dextran (Pharmacia Fine Chemicals, Uppsala, Sweden)
D-Mannitol, 2-mercaptoethanol, ATP (adenosine 5'-triphosphate), PNP, α-mannoside, and PNP β-galactopyranoside (Sigma)

Media

Medium B: 25 mM Tris–MES buffer, pH 6.5, 0.7 M mannitol
Medium C: 25 mM Tris–MES buffer, pH 8, 0.7 M mannitol

Enzyme Assays

The following activities were estimated using the procedures given in the references: fumarase,[4] G6P dehydrogenase,[5] catalase,[6] α-mannosidase,[7] β-glucosaminidase,[8] ATPase.[9]

Chlorophyll estimation was performed according to Arnon.[10]

[4] T. G. Cooper and H. Beevers, *J. Biol. Chem.* **244,** 3507 (1969).
[5] J. Brulfert, D. Guerrier, and O. Queiroz, *Plant Physiol.* **51,** 220 (1973).
[6] H. Luck, *in* "Methods of Enzymatic Analysis" (H. U. Bergmeyer, ed.), p. 885. Academic Press, New York, 1963.
[7] P. Rataboul, G. Alibert, T. Boller, and A. M. Boudet, *Biochim. Biophys. Acta* **816,** 25 (1985).
[8] T. Boller and H. Kende, *Plant Physiol.* **63,** 1123 (1979).
[9] G. Alibert, A. Carrasco, and B. Citharel, *Physiol. Veg.* **24,** 85 (1985).
[10] D. I. Arnon, *Plant Physiol.* **24,** 1 (1949).

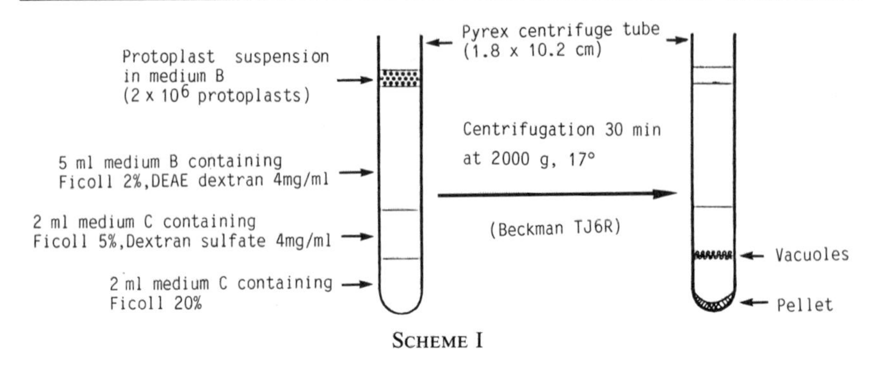

SCHEME I

Example of Vacuole Preparation from Sweet Clover Protoplasts

Procedure

Protoplasts rapidly prepared as in Boudet *et al.*[11] were resuspended in a convenient volume of medium B in order to obtain a final concentration in the range 1.5 to 2 $\times$ 10^6 protoplasts ml^{-1}.

In the procedure described in Scheme I, the lysis of protoplasts and the purification of vacuoles occur during the same centrifugation step.

The vacuoles released during the centrifugation through the DEAE-Dextran layer were collected with a Pasteur pipet at the interface 5/20% Ficoll and counted under a microscope using a Fuchs–Rosenthal hemocytometer. Confirmation that one protoplast yields one vacuole was achieved using the marker enzyme α-mannosidase.[11] The average yield of the preparation procedure ranged between 25 and 35% on the basis of the starting protoplast number.

Characteristics of the Vacuolar Suspension

Purity. The purity of the vacuolar fraction was tested by microscopy and by marker enzyme assays. As can be seen from the data in Table I, contamination of the purified vacuolar fraction was around 5%. This relatively slight level of contamination was mainly due to a few residual protoplasts in the vacuole suspension.

Stability. Once isolated from the plant cell, vaçuoles are rather labile. However, they can be stored in the isolation medium without any losses up to 6 hr at 20° or up to 12 hr at 4°. Other characteristics of these vacuoles have been reported elsewhere.[7,11,12]

[11] A. M. Boudet, H. Canut, and G. Alibert, *Plant Physiol.* **68,** 1354 (1981).

[12] K. Oba, E. E. Conn, H. Canut, and A. M. Boudet, *Plant Physiol.* **68,** 1359 (1981).

TABLE I
ASSESSMENT OF VACUOLE SUSPENSION PURITY BY ASSAY OF EXTRAVACUOLAR MARKERS

Organelles or contaminants	Markers	Contaminant[a] in vacuolar suspension (%)
Mitochondria	Fumarase	5
Peroxisomes	Catalase	4
Cytosol	G6P dehydrogenase	2
Chloroplasts	Chlorophyll	5
Protoplasts	Counting under microscope	5

[a] Expressed taking the levels of contaminants in a similar number of protoplasts as 100%.

Suitability of the Polybase Procedure for Different Plant Materials

Table II gives several examples of the adaptability of the polybase procedure to different plant species.[13–17] The characteristics of the discontinuous gradient and of the centrifugation conditions change slightly from one material to another and are mainly dependent on the vacuole and protoplast densities.

Should it be necessary to adapt the procedure to a new plant material the following preliminary determinations must be performed.

1. Evaluation of the protoplast density after a 10-min centrifugation at 2000 *g* on a preliminary Ficoll gradient (2-ml layers 0, 5, 10, 15, and 20% Ficoll). If the protoplasts do not cross the 5% layer of flotation procedure should be utilized instead. In this case the protoplasts should be resuspended in 20% Ficoll medium and the lysis and the purification steps should be achieved by upward centrifugation through a gradient as described in Aerts and Schram.[3] Alternatively, if the protoplasts do cross the 5% boundary, the Ficoll concentrations of the different layers need to be adjusted to the density of the protoplasts under study
2. Careful determination of the DEAE–Dextran and Dextran sulfate concentrations (usually in the range of 1 to 6 mg/ml) allowing optimal lysis and maximum stability of the vacuoles, respectively

[13] G. Alibert, A. Carrasco, and A. M. Boudet, *Biochim. Biophys. Acta* **721,** 22 (1982).
[14] G. Marigo, H. Bouyssou, and M. Belkoura, *Plant Sci.* **39,** 97 (1985).
[15] Y. Meyer, Y. Chartier, and G. Alibert, *Plant Physiol.* **83,** 713 (1987).
[16] I. Struve, A. Weber, U. Lüttge, E. Ball, and J. A. C. Smith, *J. Plant Physiol.* **117,** 451 (1985).
[17] T. Boller and U. Vogeli, *Plant Physiol.* **74,** 442 (1984).

TABLE II
POLYBASE PROCEDURE FOR VARIOUS PLANT SPECIES[a]

Plant material	Characteristics of gradient layers: Volume of layers (ml)	Gradient constituents	DD (mg/ml)	DS (mg/ml)	pH	Time and speed of centrifugation	Vacuole location in the gradient	Reference
Melilotus alba leaves	5	F, 2%	4	0	6.5	30 min, 1900 *g*	5–20% F	11
	2	F, 5%	0	5	8			
	2	F, 20%	0	0	8			
Acer pseudoplatanus cells	6	F, 2.5%	3	0	6.5	30 min, 1900 *g*	5–10% F	13
	2	F, 5%	0	1	6.5			
	2	F, 10%	0	3	6.5			
Catharanthus roseus cells	6	F, 1.25%	2	0	6.5	30 min, 1900 *g*	2–5% F	14
	2	F, 2%	0	3	6.5			
	2	F, 5%	0	0	6.5			
Nicotiana tabacum leaves	5	F, 2.5%	6	0	6.5	2 min, 500 *g* + 30 min, 1900 *g*	5–20% F	15
	2	F, 5%	0	3	6.5			
	2	F, 20%	0	10	6.5			
Petunia hybrida[b] petals	3	F, 20%	0	0	6.5	45 min, 1600 *g*	0–6% F	3
	3	F, 15%	7	0	6			
	2	F, 10%	0	3	6.5			
	2	F, 6%	0	3	6.5			
	2	F, 0%	0	0	6.5			
Kalanchoe blossfeldiana leaves	4	F, 2%	4	0	6	14 min, 1300 *g*	5–15% F	16
	2	F, 4%	0	0	8			
	1.5	F, 5%	0	0	6			
	2	F, 15%	0	0	6			
Phaseolus vulgaris leaves	2	S, 18% (1 v) M, 9% (5 v)	1.4	0	6.5	30 min, 2000 *g*		
	4	S, 18% (1 v) M, 9% (3 v)	0	1	8		S, 18%–S, 18%, M, 9%	17
	3	S, 18%	0	0	6.5			

[a] F, Ficoll; M, mannitol; S, sucrose; v, volume; DD, DEAE–Dextran; DS, Dextran sulfate.

[b] In the case of *Petunia* petals the protoplasts suspended in a medium containing 20% Ficoll were overlayered with the lighter Ficoll layers; they then moved up to the top of the gradient during centrifugation.

3. Determination of the optimum pH values of the different gradient layers involved in the lysis of protoplasts and the purification of the vacuoles (see Table II for examples)

4. Systematic study of the effects of the centrifugation parameters (e.g., time and speed) on the yield and the purity of the isolated vacuoles

General Comments

As already mentioned the polybase procedure appears, at least for *Petunia* petals, to be superior, as far as purity, yield, and stability are concerned, to other methods of protoplast disruption including mechanical methods and osmotic shock.[3] Other advantages of the method include (1) relative rapidity of the overall procedure (which includes only one centrifugation step). Repeated centrifugation has been suggested to induce the formation of artifactual secondary vesicles[18]; (2) very fast release of the vacuoles from the protoplasts within less than 1 min of the beginning of centrifugation (the 29 additional minutes being necessary for the complete purification of the vacuoles). This characteristic makes the method particularly suitable for performing *in vivo* kinetic studies of solute transfer from the cytoplasm to the vacuoles[7]; and (3) no changes in the osmotic pressure of the medium which could facilitate the leakage of internal solutes from the vacuoles.

However, some inconveniences are also reported: (1) As in the case of other methods starting from protoplasts the preparation scale is reduced. This can be a serious limitation for larger scale biochemical and molecular investigations on vacuolar constituents; and (2) the vacuoles are collected in a medium containing both high Ficoll concentrations and an ionized molecule Dextran sulfate. These organelles (1.5 to 2×10^6 vacuoles ml^{-1}) are difficult to concentrate because of the high viscosity of the Ficoll solution and the fragility of vacuoles during dilution treatments. In addition, the functional properties of the vacuoles may be altered as it is probable that Dextran sulfate binds to the tonoplast and changes its surface charge. These characteristics make the vacuoles isolated by such a procedure not very suitable for *in vitro* uptake studies.

Isolation of Tonoplast from Vacuoles of Sycamore Cells as an Example

Because vacuoles are surrounded by the tonoplast, which is a single membrane, a simple centrifugation step after the rupture of the vacuoles is an easy way to isolate this membrane in a pure state.[9]

[18] F. Sasse, D. Backs-Huseman, and W. Barz, *Z. Naturforsch., C: Biosci.* **34C,** 848 (1979).

Procedure

The vacuole-containing suspension is diluted by 2 vol of 25 m*M* Tris–MES Ficoll buffer, pH 6.5, containing 5 m*M* 2-mercaptoethanol.

The resulting suspension is stored for 10 min in an ice-bath with intermittent shaking every 2 min. Then the medium is centrifuged 10 min at 3000 *g* and the resulting supernatant centrifuged again 20 min at 22,000 *g*. The tonoplast fraction pellet is resuspended in the above dilution buffer and homogenized in a Potter homogenizer by 15 strokes of a Teflon pestle.

Characteristics of the Tonoplast Fraction

Purity. As a specific enzyme marker of the tonoplast is still lacking, the purity and enrichment of the membrane fraction can be only checked by the relative absence of negative markers (Table III).

A very limited contamination by membranes from mitochondrial origin was observed. This could also be the case for membranes from the other cell organelles, including plastids and peroxisomes, which may arise from the contaminating protoplasts. They probably remain intact when the osmolarity of the medium is decreased and are removed by the centrifugation step at 3000 *g*.

Certain contaminating membrane components from the endomembranous system may be detected. They again arise from the disruption of residual protoplasts but they cannot be separated from the tonoplast in the purification procedure. Some membranous adherences surrounding the tonoplast of isolated vacuoles have also been observed. They may reflect the continuity between cytoplasmic membranes.

In any case this tonoplast fraction appears purer than individual mem-

TABLE III
RELATIVE CONTAMINATION OF THE TONOPLAST FRACTION ESTIMATED BY THE DISTRIBUTION OF NEGATIVE MARKER ENZYMES

Membrane sources	Marker enzymes	Percentage recovered activity compared to the starting protoplasts
Mitochondria	Oligomycin-sensitive ATPase	2
Plasmalemma	Glucan synthase I	5
	Vanadate-sensitive ATPase	8
Golgi apparatus	Glucan synthase II	3.5

brane fractions obtained by differential centrifugation of a microsomal preparation.

Other Characteristics. In sycamore cells the tonoplast fraction obtained by such a procedure has a low density (*d* 1.10). It exhibits a lipid/protein ratio of 0.37 and a high phospholipid content (85% of the total lipids). In addition tonoplastic proteins can be reversibly phosphorylated.[19]

Practical Problems and Alternative Procedures

The isolation of the tonoplast from vacuoles is a small-scale procedure for the following reasons: (1) the isolation of vacuoles from protoplasts cannot be extended to a very large number of vacuoles (see above); and (2) the tonoplast represents a small proportion of the total membrane fraction of the cell. Consequently a limited amount of other cell constituents can result in a relatively high contamination of the tonoplast fraction. The purification procedure described above includes a 3000 *g* centrifugation step to remove most of the contaminating cell organelles. A further 22,000 *g* spin was found to be preferable to one at 100,000 *g* as it preferentially sediments the tonoplast fraction, leaving most of the membrane contaminants in the supernatant. Although this step improves the purity (on the basis of vanadate-sensitive ATPase, a specific plasmalemma marker) it does reduce the yield.

However, this isolation procedure will facilitate the study of the basic characteristics of this membrane and particularly the definition of specific markers. In parallel with this approach the separation of membranes from a microsomal fraction through differential centrifugation on sucrose gradient will probably allow the purification of large quantities of tonoplast. This procedure has been used by different authors[20–22] but the lack of unambiguous markers for the tonoplast has limited the strict characterization of the vacuolar membrane. In two cases[23,24] the correlation between the properties of the purified fraction and the tonoplast resulting from isolated vacuoles seems more firmly established. Improvements in separation techniques and a better knowledge of tonoplast markers will surely help to isolate large amounts of pure vacuolar membrane in the future.

[19] C. Teulières, G. Alibert, and R. Ranjeva, *Plant Cell Rep.* **4,** 199 (1985).

[20] A. B. Bennet, S. D. O'Neil, and R. M. Spanswick, *Plant Physiol.* **74,** 538 (1984).

[21] S. D. O'Neil, A. B. Bennet, and R. M. Spanswick, *Plant Physiol.* **72,** 837 (1983).

[22] K. A. Churchill, B. Holaway, and H. Sze, *Plant Physiol.* **73,** 921 (1983).

[23] R. J. Poole, D. P. Briskin, Z. Kratky, and R. M. Johnstone, *Plant Physiol.* **74,** 549 (1984).

[24] J. A. C. Smith, E. G. Uribe, E. Ball, S. Heuer, and U. Lüttge, *Eur. J. Biochem.* **141,** 415 (1984).

[9] Preparation of Tonoplast Vesicles from Isolated Vacuoles

By ROBERT T. LEONARD

Introduction

It is difficult to purify tonoplast vesicles by conventional cell fractionation procedures because a specific biochemical marker for the tonoplast membrane has not been discovered and because the density of tonoplast vesicles in sucrose substantially overlaps that of vesicles from the smooth endoplasmic reticulum and Golgi bodies. The enzymatic activity indicated to be associated with the tonoplast membrane is a nitrate-inhibited, chloride-sensitive adenosinetriphosphatase (ATPase) which functions in H^+ transport into the vacuole, but a similar activity is also found in association with Golgi body[1] and mitochondrial membranes.[2] At present, the most reliable way to obtain a highly purified suspension of tonoplast vesicles is to isolate them from highly purified, intact vacuoles.

The most common procedure for the isolation of intact vacuoles is to release them by gentle lysis of protoplasts followed by separation of vacuoles from cellular debris.[3–10] The recovery of purified vacuoles is usually less than 30% (by number) of the protoplasts, assuming that each protoplast releases only one vacuole. It is important to be able to conveniently obtain large quantities of protoplasts to isolate biochemically useful quantities of vacuoles and tonoplast vesicles. Large quantities of vacuoles can be isolated directly from red beet tissue by a mechanical procedure (slicing the tissues into a suitable buffer) without preparation of protoplasts.[11–13]

[1] A. Chanson and L. Taiz, *Plant Physiol.* **78,** 232 (1985).
[2] H. Sze, *Annu. Rev. Plant. Physiol.* **36,** 175 (1985).
[3] I. Mettler and R. Leonard, *Plant Physiol.* **64,** 1114 (1979).
[4] D. Briskin and R. Leonard, *Plant Physiol.* **66,** 684 (1980).
[5] M. Thom, A. Maretzki, and E. Komor, *Plant Physiol.* **69,** 1315 (1982).
[6] G. Wagner, *in* "Isolation of Membranes and Organelles from Plant Cells" (Hall and Moore, eds.), p. 83. Academic Press, London, 1983.
[7] S. Mandala and L. Taiz, *Plant Physiol.* **78,** 104 (1985).
[8] E. Martinoia, U. Flügge, G. Kaiser, U. Heber, and H. Heldt, *Biochim. Biophys. Acta* **806,** 311 (1985).
[9] M. Dracup and H. Greenway, *Plant, Cell Environ.* **8,** 149 (1985).
[10] W. Kenyon and C. Black, *Plant Physiol.* **82,** 916 (1986).
[11] F. Marty and D. Branton, *J. Cell Biol.* **87,** 72 (1980).
[12] A. Bennett, S. O'Neil, and R. Spanswick, *Plant Physiol.* **74,** 538 (1984).
[13] A. Bennett, R. Leigh, and R. Spanswick, this series, in press.

There are several ways to isolate vacuoles from protoplasts and, in turn, to prepare tonoplast vesicles from the isolated vacuoles.[6] The utility of the various procedures should be evaluated for the plant tissue and species under study. The purity of the vacuole preparation is the overriding factor controlling the successful purification of tonoplast vesicles. Unfortunately, the extent of contamination of cytoplasmic components in the vacuole preparation has been too often overlooked. The procedure described here for isolation of tonoplast vesicles from highly purified intact vacuoles was developed for cultured cells of tobacco,[4] but it can be applied to any situation where it is possible to obtain large quantities (several milliliters of packed volume) of protoplasts.

Isolation of Vacuoles from Protoplasts

A common way to release intact vacuoles from protoplasts is by controlled lysis by lowering the osmotic potential of the external medium.[3,4] The shearing forces associated with high-speed centrifugation into a density barrier[5–7] or with forcing the solution through a syringe needle[8] can also rupture protoplasts and release intact vacuoles. Regardless of the method used, it is essential to determine empirically optimum conditions for breaking the maximum number (preferably 100%) of protoplasts while recovering as many intact vacuoles as possible. For protoplasts from cultured tobacco cells, lysis was achieved by dilution of a concentrated suspension of protoplasts in 10 vol of 0.3 *M* sorbitol, 2 m*M* EDTA–Tris (pH 7.5), and 2.5 m*M* dithiothreitol (DTT). The decrease in osmotic potential and the chelating effects of EDTA result in rapid (10–15 min) lysis of most of the protoplasts. However, this lysis procedure was considerably less effective with protoplasts of corn root cortical cells, which required dilution in 0.1–0.2 *M* KH_2PO_4, 2 m*M* EDTA–Tris (pH 8.0), and 2.5 m*M* DTT for maximum protoplast breakage.

Osmotic lysis is produced by water-driven expansion of the protoplast, leading to stretching of the plasma membrane until it ruptures. At the lowered osmotic potentials required for maximum protoplast lysis water also enters the vacuole, causing dramatic expansion and stretching of the tonoplast membrane. Hence, for maximum stability of the isolated vacuoles, it is essential to remove them from the lysis medium as soon as possible. In addition, the stretched tonoplast membrane of vacuoles obtained by osmotic lysis of protoplasts may not be suitable for research on ion and solute transport.[5]

During protoplast lysis there is considerable aggregation of protoplasm which can be partially removed by gently filtering the lysate through several layers of cheesecloth or another large pore sized material.

Intact vacuoles (including large organelles and unbroken protoplasts) can be collected from the filtered lysate by centrifugation at 500 *g* for 10 min. The force and time required should be based on a determination of percentage of the vacuoles in the filtered lysate which are recovered in the pellet following centrifugation. Large vacuoles from leaf cell protoplasts of certain plant species can be collected by simply allowing them to settle out of the solution, followed by aspiration of the supernatant.[14]

For cultured tobacco cells, the pellet was suspended by very gentle agitation in a small volume of 0.5 *M* sorbitol, 0.25 m*M* EDTA–Tris (pH 7.5) in 1 m*M* DTT (suspension buffer). Very gentle agitation is essential to keep vacuole breakage to a minimum. The higher osmotic potential provided by 0.5 *M* sorbitol is essential to stabilize the vacuoles.

Vacuoles can be separated from organelles and protoplasts by centrifugation in a simple discontinuous density gradient. The composition and exact concentrations for the density gradient should be optimized. For obtaining highly purified vacuoles from cultured tobacco cells, the gradient was formed in a 15-ml conical glass tube and consisted of 5 ml of suspension buffer layered over 3 ml of 3.5% (w/v) Ficoll in suspension buffer.[4] After centrifugation at 1000 *g* for 45 min, vacuoles were collected from the Ficoll interface. Higher concentrations of Ficoll (e.g., 7.5%) and shorter centrifugation times resulted in a greater recovery of vacuoles, but more contamination by unbroken protoplasts.[3]

Determination of Purity of the Vacuole Preparation

The purity of the vacuole preparation can be assessed by a combination of microscopy and marker enzyme assay. For light microscopy, it is desirable to stain the interior of the vacuole by addition of neutral red (final concentration of 0.01%) to the preparation. Neutral red diffuses into the vacuole and is bright pink to red depending on the pH in the vacuole. The number of protoplasts relative to vacuoles can be determined by counting with the aid of a hemacytometer. Typical, but unacceptably high, levels of protoplast contamination range between 10 and 20%, by number. Even at 5% protoplast contamination, over one-third of the protein in the vacuole preparation is from contaminating protoplasts because a protoplast has about 10 times more protein than a vacuole.[6] For example, the vacuole of the cultured tobacco cell accounted for about 7.6% of the 1.3 ng of protein per protoplast.[4] The amount of protein per protoplast will vary depending on the size of the protoplast and the percentage of the protoplast volume occupied by cytoplasm (i.e., the protein concentration of the cytoplasm, including protein-rich mitochondria and plastids, is

[14] C. Buser-Suter, A. Wiemken, and P. Matile, *Plant Physiol.* **69,** 456 (1982).

much greater than that in the vacuole). If the amount of protein in the vacuole preparation expressed per vacuole exceeds 10% of the protein per protoplast, then there is reason to conclude that the vacuole preparation is contaminated by nonvacuolar proteins. For most plant tissues, a vacuole preparation which is suitable for isolation of tonoplast membranes should have less than 1% contamination by protoplasts and the apparent protein content per vacuole should be less than 10% of the total protein per protoplast.

Contamination of cytoplasmic components in the vacuole preparation can be determined by assay of chemical or enzymatic markers.[15-17] The recovery of markers for cytoplasmic components in the vacuole preparation should be less than 1% of the total present in the protoplast. Contamination in excess of this level by organelles could mean that a significant proportion of the membrane in the preparation is not vacuolar in origin. Various markers and the procedures for assay are presented elsewhere in this volume.

Isolation of Tonoplast Vesicles from Broken Vacuoles

The predominant membrane in a highly purified preparation of intact vacuoles is the tonoplast. Hence, membrane vesicles derived from the preparation should be highly enriched in tonoplast vesicles. These vesicles are concentrated by breaking the vacuoles and collecting the tonoplast membrane by centrifugation. The preparation of tonoplast vesicles may contain residual amounts of various vacuole-localized hydrolases such as acid phosphatase, a membrane-associated, nitrate-inhibited ATPase, and ATP-driven H^+ transport activity catalyzed by the nitrate-inhibited ATPase. Assay procedures for these activities are available in this series.[13,18]

Methods used to break the vacuoles include the freezing and thawing of the preparation, osmotic lysis, passage through a hypodermic needle, or sonication. A comparison of the relative effectiveness and suitability of these methods has not been reported and it is possible that a relatively harsh procedure, such as high-intensity sonication, may remove or inactivate components of the tonoplast membrane. For vacuoles from cultured tobacco cells, a 20-sec low-intensity sonication at ice temperature with a probe-type sonicator was effective in breaking the vacuoles.

[15] P. Quail, *Annu. Rev. Plant. Physiol.* **30,** 425 (1979).

[16] D. Moore, C. Cline, R. Coleman, W. Evans, H. Glaumann, D. Headon, E. Reid, G. Siebert, and C. Widnell, *Eur. J. Cell Biol.* **20,** 195 (1979).

[17] G. Nagahashi, *in* "Modern Methods of Plant Analysis" (H. F. Linskens and J. F. Jackson, eds.), p. 68. Springer-Verlag, Heidelberg, 1985.

[18] D. P. Briskin, R. T. Leonard, and T. K. Hodges, this volume [51].

Tonoplast vesicles from lysed vacuoles can be collected by high-speed centrifugation, usually in a density gradient to provide a final purification step. The density of the lysed vacuole preparation is adjusted to the equivalent of 40% (w/w) sucrose with a room-temperature saturated solution of sorbitol in 1 m*M* EDTA–Tris (pH 7.2) and 1 m*M* DTT (gradient buffer). A linear sucrose gradient consisting of 16 ml of 30–40% (w/w) sucrose in gradient buffer is layered over the sonicated vacuole preparation followed by 12 ml of 15% sucrose in gradient buffer. After centrifugation (80,000 to 100,000 *g*) to equilibrium, the tonoplast vesicles band at the 15–30% interface. A gradual transition of density between the lysed vacuole preparation and the 15% sucrose layer facilitates equilibration of tonoplast vesicles in the gradient. Sedimentation of tonoplast vesicles into a linear sucrose gradient can also be used to concentrate and purify tonoplast vesicles. The effective peak density (in sucrose) of tonoplast vesicles from cultured tobacco cells is about 1.12 g/cm^3.[4]

The percentage of vacuolar protein accounted for by the tonoplast vesicles has not been widely reported, but it should be a relatively low number ($<10\%$) because the volume of vacuole sap is large in proportion to the volume of tonoplast membrane. Various hydrolases in the vacuolar sap, even though present at low concentration, account for a high proportion of the total vacuolar protein. Hence, the proportion of total vacuolar protein recovered as tonoplast vesicles should be determined. For most plant cells, the recovery of more than 10% of the vacuolar protein in the tonoplast fraction indicates that the vacuole preparation may be significantly contaminated by nonvacuolar membranes. The purity of the tonoplast vesicles should be examined by assay of markers for cytoplasmic components.

Assuming that a highly purified vacuole preparation is used, then the major band on the density gradient should represent highly purified tonoplast vesicles. Such vesicles should be rich in nitrate-inhibited, chloride-sensitive ATPase activity[11,16]; however, the presence of substrate-specific ATPase activity in the tonoplast preparation has not always been confirmed.[4,10]

Optimum conditions for storage of tonoplast vesicles have not been determined. Initial freezing under liquid nitrogen followed by storage in liquid nitrogen or at $-80°$ in a freezer is recommended.

Comment

Conventional cell fraction of a variety of plant tissues has demonstrated that membrane vesicles recovered at low density (about 1.09 to 1.12 g/cm^3) in a sucrose gradient contain nitrate-inhibited ATPase activity and the associated ATP-driven H^+ transport activity. There is evidence that this region of the gradient contains tonoplast vesicles and/or small intact vacuoles.[2] However, since this region of a sucrose gradient is also rich in vesicles from the endoplasmic reticulum and the Golgi body, it appears that tonoplast vesicles prepared in this manner will usually have a relatively low level of purity.

The preparation of tonoplast vesicles from intact vacuoles can yield preparations of superior purity, but the methods developed for one plant species or tissue may not be readily applied to another species or tissue. The preparation of highly purified tonoplast vesicles is still largely a matter of trial and error, and represents a method that will continue to evolve.

[10] Biochemical and Enzymatic Components of a Vacuolar Membrane: Tonoplast of Lutoids from *Hevea* Latex

By J. D'AUZAC, H. CHRESTIN, and B. MARIN

Latex is collected by tapping *Hevea brasiliensis;* an incision is made in the bark to within 1 or 2 mm of the cambium. The cytoplasmic origin of latex was determined after observing the laticiferous system using electron microscopy.[1,2] Fifty to 60% of the volume of latex collected by tapping consists of particles of rubber 0.03 to 3 μm in diameter, surrounded by a phospholipoprotein layer. Frey-Wyssling particles, discovered by the author of that name in 1929, form 1 to 3% of the volume of latex.[3] They are organelles of the chromoplast type with a double membrane and are rich in carotenoids[1]; polyribosomes functional *in vitro* were also revealed.[4] The latex collected does not contain mitochondria or nuclei since these remain confined in the more viscous parietal cytoplasm which does not flow during tapping.[2]

The most abundant organelles are the lutoids, discovered in 1948.[5] They form 15 to 20% of the volume of fresh latex and are bounded by a single membrane.[6] In 1968, Pujarniscle showed that they possess all the

[1] P. B. Dickenson, *Proc. Natl. Rubber Res. Assoc., Jubilee Conf.*, p. 52 (1968).
[2] C. Hebant and E. de Faÿ, *Z. Pflanzenphysiol.* **97,** 391 (1980).
[3] A. Frey-Wyssling, *Arch. Rubbercult.* **13,** 394 (1929).
[4] M. Coupé and J. d'Auzac, *C.R. Acad. Sci. Paris, Ser. D* **274,** 1031 (1972).
[5] L. N. S. de Haan-Homans, J. W. Van Dalfsen, and G. E. Gils, *Nature (London)* **161,** 177 (1948).
[6] J. B. Gomez and W. A. Southorn, *J. Rubber Res. Inst. Malaya* **21,** 513 (1969).

acid hydrolases typical of animal lysosomes and thus demonstrated their lysosomal nature.[7] The fact that lutoids absorb neutral red[8] and that by direct measurement lutoid pH is different from cytosolic pH (5.5–6.9) reveals the acid pH of lutoid compartment.[9] The vacuolar nature of lutoids was finally shown by the fact that they accumulate certain cytosolic solutes against their concentration gradient. This is the case for mineral ions (P_i, Mg^{2+}, Ca^{2+}, and Cu^{2+}),[10] citrate,[11] and basic amino acids.[12]

As shown in 1942,[13] lutoids also contain proteins with a very high isoelectric point—hevamins A and B,[14] which turned out to be lysozymes.[15] That lutoids contain 80 S ribosomes and rRNA degraded to various extents raises the question of whether they may play an autophagous role.[16,17] It is clear, however, that the lutoids in *Hevea* latex form a polydispersed vacuolar–lysosomal system,[18] doubtless the first to have been so clearly defined in the plant kingdom.

Obtaining Lutoids and Lutoid Tonoplast

Since lutoids are fragile, osmosensitive, single-membrane organelles,[7,19] located in a cytoplasm in which a large proportion of metabolic activity is preserved *in vitro*,[4,20,21] considerable precautions must be taken when latex is collected. In order to obtain undamaged lutoids, latex must be collected in an ice-cold receptacle and taken to the laboratory immediately. The latex produced during the first 5 min of flow is discarded because it contains lutoids which have been damaged by the shearing effect due to decompression which occurs at the end of the latex tubes at tapping.[19] This first fraction of flow might also be contaminated by microorganisms present on the tapping cut.

[7] S. Pujarniscle, *Physiol. Veg.* **6,** 27 (1968).
[8] L. K. Wiersum, *Rev. Gen. Caoutch.* **35,** 276 (1957).
[9] W. A. Southorn and E. Edwin, *J. Rubber Res. Inst. Malaya* **20,** 187 (1968).
[10] D. Ribaillier, J. L. Jacob, and J. d'Auzac, *Physiol. Veg.* **9,** 423 (1971).
[11] J. d'Auzac and C. Lioret, *Physiol. Veg.* **12,** 617 (1974).
[12] J. Brzozowska, P. Hanower, and R. Chezeau, *Experientia* **30,** 894 (1974).
[13] C. P. Roe and R. M. Edwart, *J. Am. Chem. Soc.* **64,** 2628 (1942).
[14] B. L. Archer, *Phytochemistry* **15,** 297 (1976).
[15] S. J. Tata, A. N. Boyce, B. L. Archer, and B. G. Audley, *J. Rubber Res. Inst. Malaysia* **24,** 233 (1976).
[16] B. Marin and P. Trouslot, *Planta* **124,** 31 (1975).
[17] B. Marin, *Planta* **138,** 1 (1978).
[18] J. d'Auzac, H. Chrestin, B. Marin, and C. Lioret, *Physiol. Veg.* **20,** 311 (1982).
[19] W. A. Southorn, *J. Rubber Res. Inst. Malaya* **21,** 494 (1969).
[20] J. d'Auzac and S. Pujarniscle, *C.R. Acad. Sci. Paris* **257,** 299 (1963).
[21] J. L. Jacob, *Physiol. Veg.* **8,** 395 (1970).

Obtaining Lutoids by Low-Speed Centrifugation

Low-speed centrifugation was first used to separate the latex into a white supernatant fraction and a yellow fraction.[22] The latter consists mainly of lutoids colored to varying degrees by Frey-Wyssling particles, whereas the supernatant contains most of the rubber particles, which are lighter.

This technique was also used with refrigerated centrifuges (4°) with centrifugation for 10–15 min at 6000 to 10,000 rpm.[23] The supernatant rubber was removed using a spatula and the sediment was left at the bottom of the tube. The sides of the tube were carefully wiped with filter paper. The yellow fraction was recovered using 5 vol of an isotonic medium containing mannitol (0.5 *M*), triethanolamine–HCl (TEA) buffer (50 m*M*, pH 7.5), and mercaptoethanol (MSH) (5 m*M*). The sediment was suspended by gentle shaking in a container packed in ice. The yellow fraction was washed at least three times by centrifuging and resuspending the sediment. Under these conditions, the very fragile Frey-Wyssling particles were eliminated from the supernatant during the successive washing operations together with the lighter particles of rubber. Lutoids in good condition were obtained.

Obtaining Lutoids by Differential Ultracentrifugation

Ultracentrifugation of latex is a good method for separating the main structured components.[23] Examination of the various phases with the naked eye[24] and phase-contrast microscopy reveal the complexity of the system.[25]

Ultracentrifugation at 70,000–100,000 *g* for 40 min at 4° or at 40,000 *g* for 60 min produces a compact and extremely viscous supernatant rubber phase, an intermediate clear serum which is the cytosol of the latex, and a sediment which consists essentially of lutoids. The extremely fragile Frey-Wyssling particles are beneath the rubber phase and on the sediment, depending on their state of degradation. After careful elimination of the rubber phase and the cytosol, the sediment is put back into suspension by gentle pipetting into an isotonic medium at 4° (mannitol, 0.3 *M*, HEPES–Tris, 50 m*M*, pH 7.0) and then recentrifuging at a lower speed (30,000 *g* for 10 min). This type of washing carried out three times yields lutoids considered to be pure. No characteristic mitochondrial activity can be detected in lutoid sediment washed in this way.

[22] L. N. S. de Haan-Homans, *Inst. Rubber Ind., Trans. Proc.* **25,** 346 (1950).
[23] A. S. Cook and B. C. Seckar, *J. Rubber Res. Inst. Malaya* **14,** 163 (1955).
[24] G. F. J. Moir, *Nature (London)* **184,** 1626 (1959).
[25] G. F. J. Moir, *Nature (London)* **188,** 165 (1960).

Obtaining Lutoids by Ultracentrifugation on a Density Gradient

Centrifugation of latex on a sucrose density gradient was first carried out by Pujarniscle and the acid hydrolase content of lutoids was shown.[7] A linear gradient of 0 to 50% sucrose in the presence of mannitol (0.3 *M*) (124,000 *g* for 2 hr) was employed.

The lutoid fraction (mean specific gravity 1.14) was above the sediment, below the layer of Frey-Wyssling particles, and below the rubber supernatant. Cretin[26] placed 5 ml of latex or 3 ml of a solution containing 50% lutoids in the cytosol on a linear sucrose gradient of 0.6 to 1.8 *M* containing Tris–HEPES (20 m*M*, pH 7), EDTA (0.5 m*M*), and mannitol (0.3 *M*) and on a bed of sucrose (2.3 *M*). Centrifugation was carried out from 70 to 120 min at 85,000 *g* and 8°. Low and Wiemken[27] showed that the centrifugation of latex on a 0 to 10% Ficoll gradient resulted in poor separation of lutoids and Frey-Wyssling particles. However, if the latex is diluted in a ratio of 1 : 12 with a HEPES buffer (pH 7.9), KCl (0.25 *M*), and made heavy by the addition of 5 vol of Ficoll (15%) to the same isotonic medium, flotation (8000 *g* for 15 min) followed by 200,000 *g* for 2 hr, gives very good separation of lutoid marker enzymes (acid phosphatase and β-galactosidase) and Frey-Wyssling particles (polyphenol oxidase).[28]

Obtaining Tonoplast from Lutoid Membrane

Various methods can be used to rupture purified lutoids. Short but repeated ultrasonic treatment of a lutoid suspension cooled to between 0 and 4° can be used. Treatment of lutoids using 0.05–0.1% detergent such as Triton X-100 or X-114[7] causes degradation of the membrane and releases the intravacuolar contents, but it can also contribute to detaching enzymes that are loosely linked or absorbed from the membrane. Treatment in a hypotonic medium (phosphate buffer, 3 m*M*, pH 7.2),[29] accompanied by homogenization in a Teflon–glass homogenizer, produces the same result. This technique nevertheless leads to tonoplast vesicles which must be pelleted by centrifugation and rehomogenized (40,000 *g* for 15 min). This treatment must be repeated several times to prevent tonoplast vesicles from contamination by vacuolar components.

Tonoplast vesicles can also be obtained by putting freeze-dried lutoids in suspension using a Teflon–glass homogenizer. Repetition of this treatment produces, as above, tonoplast vesicles which are relatively free of

[26] H. Cretin (now Chrestin), *J. Membr. Biol.* **65,** 175 (1982).
[27] F. C. Low and A. Wiemken, *Phytochemistry* **23,** 747 (1984).
[28] M. Coupé, S. Pujarniscle, and J. d'Auzac, *Physiol. Veg.* **10,** 459 (1972).
[29] F. Moreau, J. L. Jacob, J. Dupont, and C. Lance, *Biochim. Biophys. Acta* **396,** 116 (1975).

intravacuolar material.[30,31] Whatever the type of treatment used, vesiculated tonoplast fragments are obtained by ultracentrifugation (40,000 *g* for 15 min).

However, experience shows that detergent treatments are usually not sufficient to remove totally the intralutoid enzymes, such as acid phosphatase,[30] that are adsorbed on the tonoplast. Freeze-drying results in almost complete release of an enzyme that is probably tonoplastic—NADH-quinone reductase (EC 1.6.99.2) (see below).

Lutoid Tonoplast: Biochemical Composition

The ease of obtaining lutoids doubtlessly facilitated the analysis of the biochemical composition of a tonoplast. Thin-layer chromatography shows that this tonoplast has an unusual content of phosphatidic acid (82%) and two other phospholipids (18%) which have not been identified.[32] This unusual feature may account for the negative charges carried by the lutoids (and by the particles of rubber) which are necessary to maintain the colloidal stability of the latex. Such high phosphatidic acid content has not been observed in yeast or sugar beet vacuoles.[33]

The fatty acids of phosphatidic acid are $C_{14:0}$, 1.6%, $C_{16:0}$, 20.5%; $C_{18:0}$, 25.3%; $C_{18:2}$, 13.8%; $C_{18:3}$ is not detectable.[32] There are similar quantities of saturated and unsaturated fatty acids, making these membranes relatively rigid.[6,7]

The tonoplast includes two *b*-type cytochromes: cytochrome *b*-563 and *b*-561, but does not include cytochrome *P*-450.[29]

Lutoid Tonoplast: Enzymatic Activities

Mg^{2+}-ATPase: A Tonoplast Proton Pump

Detection and Physiological Role. The fact that the rate of absorption of ^{14}C-labeled citrate by lutoids incubated *in vitro* in an isotonic medium practically doubles in the presence of ATP (5 m*M*)[34] led to searching for an ATPase on the lutoid tonoplast. In fact, it is possible to use freeze-dried lutoids to characterize an Mg^{2+}-dependent ATPase in the presence of 0.1 m*M* ammonium molybdate, which totally inhibits residual lutoidic

[30] J. d'Auzac, *Phytochemistry* **14,** 671 (1975).
[31] J. d'Auzac, *Phytochemistry* **16,** 1881 (1977).
[32] J. Dupont, F. Moreau, C. Lance, and J. L. Jacob, *Phytochemistry* **15,** 1215 (1976).
[33] F. Marty, *in* "The Biochemistry of Plants" (N. E. Tolbert, ed.), Vol. 1, p. 625. Academic Press, New York, 1980.
[34] J. d'Auzac and C. Lioret, *Physiol. Veg.* **12,** 617 (1974).

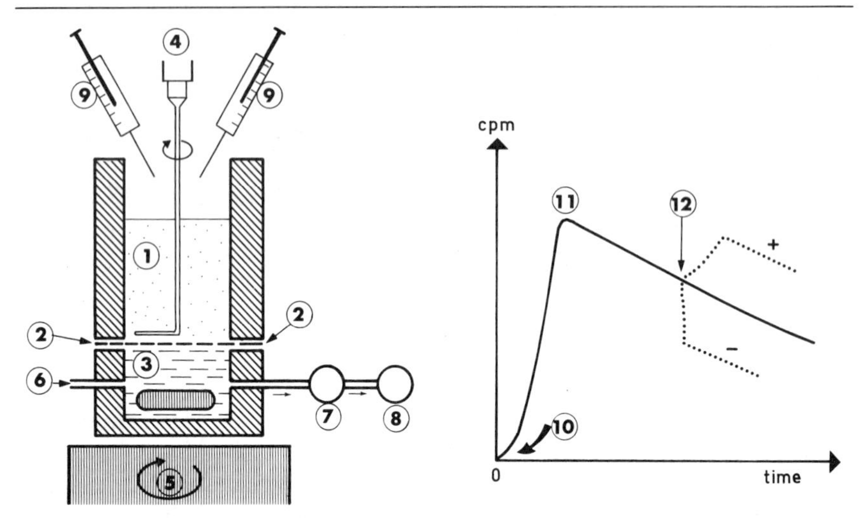

FIG. 1. Continuous flow dialysis. (1) Compartment for incubation containing a lutoid suspension preincubated with a radioactive tracer added at zero time; (2) dialysis membrane; (3) flow dialysis compartment; (4) rotating stirrer (1 revolution per second); (5) magnetic stirrer (5 revolutions per second); (6) buffer solution input; (7) peristaltic pump; (8) continuous liquid scintillation counting; (9) injection of substrates and effectors; (10) appearance of radioactivity in the outflow after addition of labeled lutoids; (11) level of equilibrium of radioactivity efflux; (12) introduction of an effector producing efflux (+) of the radioactivity entrapped in lutoids, or an influx (−) of the radioactivity into lutoids from the suspension medium.

acid phosphatase.[30,35] Partial purification of the enzyme was obtained by three successive washing operations carried out on freeze-dried material (100 mg ml^{-1}) using glycine–NaOH (20 m*M*, pH 7.5) and EDTA (10 m*M*). The membrane fraction was collected by centrifugation (40,000 *g* for 10 min) between each washing operation. This membranous fraction was incubated for 30 min at 20° with $MgCl_2$ (10 m*M*) and sodium deoxycholate (2% of initial freeze-dried material). Incubated material (1.5 ml) was then placed on 11 ml of a medium containing TEA buffer (10 m*M*, pH 7.5), sucrose (5%), $MgCl_2$ (10 m*M*), MSH (5 m*M*). After centrifugation (200,000 *g* for 30 min), the sediment was homogenized (10 ml g^{-1} of freeze-dried material) in a TEA (10 m*M*, pH 7.5), EDTA (10 m*M*), MSH (5 m*M*) buffer. The ratio of ATPase activities measured by enzyme assays of free ADP to phosphatase activities (measured by *p*-nitrophenol phosphate) was greater than 80.[31] ATP is the preferential substrate of the enzymes. K^+ and Na^+ did not have any effect whereas the anions Cl^-, HCO_3^-, malate, and aspartate are activators and NO_3^- strongly inhibit the reaction.[30] Using the flow dialysis technique[26] (Fig. 1), classic centrifugation

[35] J. L. Jacob and N. Sontag, *Biochimie* **56,** 1315 (1974).

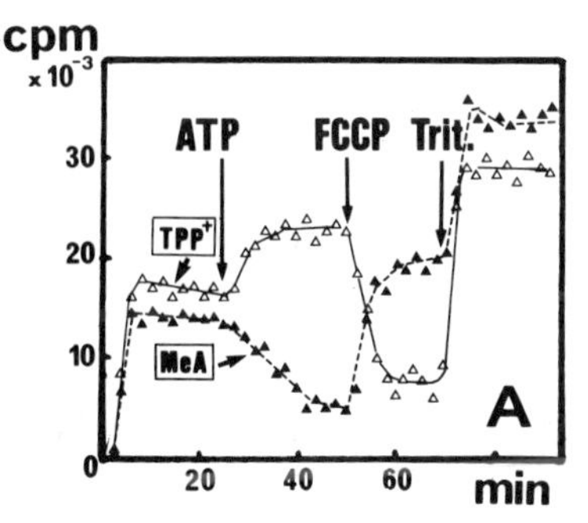

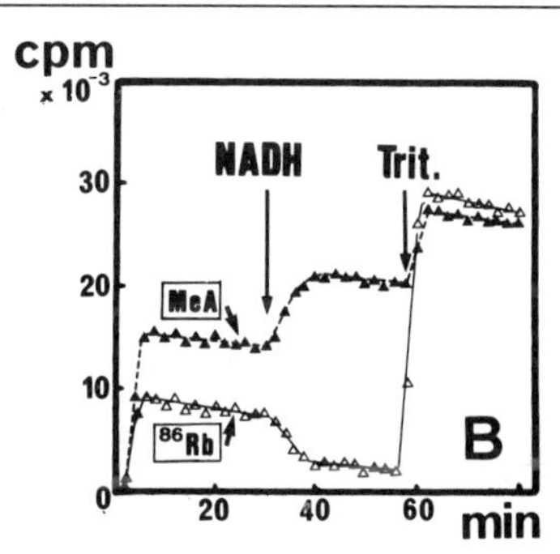

FIG. 2. The existence of two opposite H^+ pumps on the lutoidic tonoplast demonstrated by flow dialysis with fresh lutoids. (A) Tonoplastic H^+-pump ATPase (influx). Lutoids have been incubated 20 min in an isotonic buffered medium (pH 7.15) with a ΔpH probe [[^{14}C]methylamine (MeA)] and a ΔΨ probe [[^{3}H]tetraphenylphosphonium (TPP^+)] and transferred to the compartment of flow cell dialysis (see Fig. 1). After 30 min, Mg ATP (3 m*M*, final concentration) adjusted exactly at the same pH was added in the same compartment. An influx of MeA into lutoids indicates acidification of this compartment while efflux of TPP^+ shows the electrogenicity of the H^+ pump. A further addition of a protonophore (FCCP, 25 μM final concentration) break down the H^+ gradient between lutoids and the incubation medium. A nonionic detergent (Triton X-100, 0.1%) causes a last efflux of the two probes. (B) Tonoplastic NADH-cytochrome *c* H^+ pump (efflux). Lutoids have been incubated as above with ΔpH probe (MeA), a ΔΨ probe (^{86}Rb in KCl, 30 m*M*, + valinomycin, 10 g · ml^{-1}), and oxidized cytochrome *c* (500 μM). In the flow dialysis cell NADH (2 m*M*) was added 30 min after lutoids. The efflux of ΔpH probe and the influx of ΔΨ probe indicates the functioning of an electrogenic efflux H^+ pump for H^+ entrapped into the lutoidic compartment.

and ultrafiltration techniques,[36,37] ΔpH probes such as [^{14}C]methylamine, and ΔΨ probes such as ^{86}Rb, valinomycin,[26] or [^{14}C]tetraphenylphosphonium,[38] it was shown with freshly isolated lutoids that the tonoplastic ATPase activity resulted in an active electrogenic influx of protons to the vacuolar compartment (Fig. 2A).

The proton-pumping ATPase is very probably functional *in vivo*. When freshly isolated intact lutoids are immersed in fresh ultrafiltered latex cytosol (filtration on Amicon PM10 membranes) with exogenous MgATP, they exhibit vacuolar acidification leading to a transtonoplastic pH gradient approaching 2 pH units, membrane depolarization (interior less negative), and a resulting electrochemical gradient of protons ($\Delta\tilde{\mu}H^+$) of nearly 150 mV.[39]

[36] H. Chrestin, Doctorat Thesis, Montpellier II University (1984).

[37] X. Gidrol, Third Cycle Thesis, Marseille Luminy University (1984).

[38] B. Marin, *Planta* **157,** 324 (1983).

[39] X. Gidrol, B. Marin, H. Chrestin, and J. d'Auzac, *in* "Biochemistry and Function of Vacuolar Adenosine-Triphosphatase in Fungi and Plants" (B. P. Marin, ed.), p. 151. Springer-Verlag, Berlin and New York, 1985.

This H^+-pumping ATPase has been exhaustively described using fresh lutoids in artificial and physiological (ultrafiltered) latex cytosol, in aged lutoids,[38] and with vesicles of tonoplastic origin prepared from freeze-dried lutoids.[40–42] Neither the ATPase splitting activity nor the associated proton-pumping activity is sensitive to the specific inhibitors of plasmalemma and mitochondrial ATPase, but they are inhibited by DCCD, DIDS, NEM, quercetin, NO_3^-, and Cu^{2+}.[36,37,43,44] Electrogenic protonophores such as CCCP and FCCP (10 μM) reverse and inhibit H^+-pumping activity and, in contrast, significantly increase the ATP splitting activity, suggesting that the proton-pumping ATPase is at least pH dependent.[45] Furthermore, a constant supply of exogenous ATP is necessary to sustain a high $\Delta\tilde{\mu}H^+$, suggesting that the H^+-pumping ATPase is kinetically dependent on ATP availability.

Finally, it was shown that $\Delta\tilde{\mu}H^+$ generated by the proton-pumping ATPase governs the accumulation of various ions such as citrate, basic amino acids, and Ca^{2+}.[46,47] It is therefore concluded that the H^+-pumping ATPase, located on the lutoidic tonoplast, is not only involved in pH regulation of the cytosol (external medium) but also in the ionic composition of latex cytosol.[39,47,48]

Solubilization, Purification, and Study of Membranous ATPase.[49] *Solubilization.* Several detergents listed in Table I were tested for their ability to solubilize the molybdate-insensitive ATPase from tonoplast membrane. Lutoids were used in the form of freeze-dried powder put back in suspension at a concentration of 2 to 5 mg protein/ml of washing medium containing 25 mM MES, 25 mM HEPES, and 5 mM MSH adjusted to pH 6.0 with Tris–base. They were quickly mixed with this ice-

[40] B. Marin, Doctorat Thesis, Montpellier II University (1981).

[41] B. Marin, M. Marin-Lanza, and E. Komor, *Biochem. J.* **198,** 365 (1981).

[42] H. Cretin (now Chrestin), B. Marin, and J. d'Auzac, *in* "Plasmalemma and Tonoplast: Their Function in the Plant Cell" (D. Marme, E. Marre, and R. Hertel, eds.), p. 201. Elsevier/North-Holland Biomedical Press, Amsterdam, 1982.

[43] B. Marin, *Plant Physiol.* **73,** 973 (1983).

[44] B. P. Marin and X. Gidrol, *Biochem. J.* **226,** 85 (1985).

[45] B. P. Marin, *Biochem. J.* **229,** 459 (1985).

[46] B. Marin, H. Chrestin, and J. d'Auzac, *in* "Plasmalemma and Tonoplast: Their Function in the Plant Cell" (D. Marme, E. Marre, and R. Hertel, eds.), p. 209. Elsevier/North-Holland Biomedical Press, Amsterdam, 1982.

[47] B. Marin and H. Chrestin, *in* "Biochemistry and Function of Vacuolar Adenosine-Triphosphatase in Fungi and Plants" (B. P. Marin, ed.), p. 212. Springer-Verlag, Berlin and New York, 1985.

[48] H. Chrestin, X. Gidrol, J. d'Auzac, J. L. Jacob, and B. Marin, *in* "Biochemistry and Function of Vacuolar Adenosine-Triphosphatase in Fungi and Plants" (B. P. Marin, ed.), p. 245. Springer-Verlag, Berlin and New York, 1985.

[49] B. Marin, J. Preisser, and E. Komor, *Eur. J. Biochem.* **151,** 140 (1985).

TABLE I
EFFICIENCY OF DIFFERENT DETERGENTS AND ORGANIC SOLVENTS TO SOLUBILIZE TONOPLAST ATPase FROM *Hevea* LATEX

Detergent (or organic solvent)	Protein in supernatant (% of initial content)	ATPase activity solubilized (in supernatant)[a] (% of initial activity)
Control	6–8	2–3
5% Lubrol WX (w/w)	25–35	10–12
1% Triton X-100 (v/v)	40–50	25–30
1% Triton X-100 (v/v) + 8 *M* urea	75–85	15–20
1% deoxycholate (w/v)	65–70	40–45
1% cholate (w/v)	35–40	5–10
0.1% SDS (w/w)	60–70	5–10
30 m*M* octylglucoside	44–55	75–80
4 m*M* Zwittergent 3-14	65–70	50–60
5 m*M* CHAPS	60–70	15–25
5 m*M* EDTA	15–20	5–10
1 m*M* KI	35–40	10–15
33.3% dichloromethane (v/v)	10–20	50–60
33.3% chloroform (v/v)	10–20	20–25

[a] Measured in presence of detergent or organic solvent used in the same conditions.

cold buffer and homogenized (50 times in approximately 10 min) in a Teflon–glass homogenizer packed in ice. The addition of protease inhibitor PMSF (phenylmethylsulfonyl fluoride) is recommended in preparations which are particularly rich in protease. The resulting preparation yielded tightly sealed vesicles.[49] The membranes were then pelleted by centrifugation (45,000 *g* for 10 min, 2°) and washed three times using a medium containing 10 m*M* Tris, 1 m*M* EDTA, and 1 m*M* DTT adjusted to pH 7.5 using H_2SO_4. Finally the membranes were resuspended in ATPase solubilization buffer containing 100 m*M* Tris, 1 m*M* DTT, 1 m*M* ATP (in the form of disodium salt), and 1 m*M* $MgSO_4$ adjusted to pH 7.0 using H_2SO_4 and generally incubated at 0° for 30 min with continuous agitation. The resulting suspension was centrifuged (200,000 *g* for 60 min). The ATPase activity measured as release of P_i was estimated for the supernatant and the pellets obtained in this way. Sulfobetaines, such as Zwittergent 3-14, appear to be effective at low concentrations but inactivate ATPase at higher concentrations. Octylglucoside also solubilizes the membrane but only when it is used at 25–30 m*M*. In all cases, however, residual rubber particles were always present in tonoplast fractions.

When solubilization was attempted, a large proportion of the ATPase activity remained linked to the rubber particles which floated at the top of the tube. In light of these difficulties, the most effective procedure for the solubilization of ATPase is to use organic solvents as described initially for mitochondrial membranes.[50]

The suspension of tonoplast membranes obtained in the manner described above was vigorously vortex mixed and 0.5 vol of organic solvent (chloroform or dichloromethane) was rapidly added in drops. Shaking was continued for 10 min at 35–40°. The emulsion was centrifuged in a bench centrifuge (10,000 *g*, 2–5 min, ambient temperature). The upper aqueous layer was removed carefully and diluted 1 : 1 with fresh washing solution. The organic solvent layer was centrifuged (10,000 *g* for 20 min) to yield a clear supernatant and a tightly packed yellowish-brown pellet. The best results were obtained when buffered medium was used containing 1 m*M* ATP, 1 m*M* EDTA, 0.5 m*M* DTT, and 0.5 m*M* PMSF adjusted to pH 7.0–8.0; the supernatant obtained contained 40–45% of the initial ATPase activity.

Purification procedure.[49] *Step 1: Ammonium sulfate precipitation:* The ATPase released as described above was purified by two $(NH_4)_2SO_4$ fractionation steps. First, solid $(NH_4)_2SO_4$ up to 0.176 g ml^{-1} was added slowly to the well-stirred aqueous extract and the mixture was stirred for an additional 15 min in an ice bath before centrifugation (12,000 *g* for 10 min). The supernatant was then carefully removed and the redissolved protein precipitate was treated with ammonium sulfate (0.214 g ml^{-1}) to obtain 60% saturation. After a 15-min incubation and centrifugation, the supernatant was discarded and the yellowish sediment was dissolved in a small volume of washing solution. In some cases the initial solution was centrifugated for 4–5 min in an Eppendorf microfuge to remove insoluble material and a second precipitation operation was carried out using 60% ammonium sulfate. The enzyme generally remained stable and fully active for several weeks when stored at 2° in 60% $(NH_4)_2SO_4$.

Step 2: Sephadex G-200 chromatography: The precipitate from 60% ammonium sulfate-saturated suspension was dissolved in 1 ml of medium containing 100 m*M* Tris, 5 m*M* MSH adjusted to pH 7.0 with H_2SO_4 (generally 0.25–0.40 mg of protein, containing 30–40 U of ATPase). Chromatography was carried out on a Sephadex G-200 column (0.9 × 45 cm) previously equilibrated with Tris–H_2SO_4 buffer. The flow rate was 7.1 ml hr^{-1} and 1.2-ml fractions were collected and analyzed for ATPase activity and protein content. The enzyme generally eluted from the column

[50] R. B. Beechey, S. A. Hubbard, P. E. Linnett, A. D. Mitchell, and E. A. Munn, *Biochem. J.* **148,** 533 (1975).

shortly after the void volume as a single peak for enzymatic activity and protein. This peak contained 80–90% of the loaded activity and 90% of the initial protein was removed.

Properties of the purified enzyme[49]: The purified dichloromethane-solubilized enzyme exhibited a specific activity about 50- to 100-fold greater than the crude membranes and 10- to 20-fold greater than the dichloromethane extract. It could be stored in the cold as a precipitate in ammonium sulfate in the presence of a protecting agent (ethylene glycol or glycerol). In contrast, the membrane-bound ATPase does not require cryoprotective agents.

The molecular mass of the purified ATPase was determined on a Sephacryl S-200 (or S-300) column (2.5 × 96 cm) equilibrated with 0.1 *M* Tris–H_2SO_4 buffer at pH 7.5. The flow rate was 10 ml hr^{-1}. Routinely purified ATPase eluted in five peaks. Only the first, with an apparent mass of 200 kDa, possesses ATPase activity. The other four bands eluted according to molecular mass of approximately 110, 65, 25, and 12 kDa, respectively. These four components corresponded to a slow decomposition of the initial component (with a molecular mass of 200 kDa) due to storage conditions. Mild SDS treatment (1%) led to the decomposition of the 200-kDa protein into these four components.

The ATPase seems to consist of four polypeptides in equal proportions (probably 1 : 1 : 1 : 1). The sum of the molecular mass gave 212 kDa, which is close to the molecular mass of the total ATPase (about 200 kDa).

Isoelectrofocusing of the purified enzyme performed with Servalyt-precotes gels (SERVA), pH 3–10, in aspartic–glutamic acid buffer and arginine/lysine/ethylenediamine buffer for 3 hr at 10° with a final voltage of 1000 V gave only one protein band at pH 5.3.

SDS–polyacrylamide gel electrophoresis of purified ATPase was performed after incubation of proteins for 60 min at 25° in SDS (2%). Electrophoresis was carried out according to Laemmli[51] with gels prepared as an exponential gradient from 10 to 18% or 7.5 to 15%. After 4–5 hr of electrophoresis at 25 mA (8°), the gels were stained with Coomassie brilliant blue K250 and destained in 25% methanol plus 7% acetic acid. After this treatment the solubilized ATPase appeared to consist of two 54-kDa polypeptides (not dissociated by SDS mild treatment) and three polypeptides of 66, 23, and 13 kDa which displayed different sensitivity to specific protein strains as shown by UV analysis.

The principal characteristics of the purified ATPase are listed in Table II. They show that the tonoplastic ATPase of *Hevea* latex is different than the F_1,F_0-type ATPase and E_1,E_2-type ATPase. Consequently

[51] U. K. Laemmli, *Nature (London)* **227,** 680 (1970).

TABLE II
PRINCIPAL CHARACTERISTICS OF THE SOLUBILIZED AND PURIFIED TONOPLASTIC H^+-PUMP ATPase

Parameter	Results observed
pH dependence	Optimum at 7.0
Substrate	
Specificity	ATP > CTP > UTP > GTP
K_m (ATP · Mg^{2+})	0.1–0.4 mM
Divalent cation requirement	Strict with Mg^{2+} (formation of substrate) But also stimulation by Ca^{2+} and inhibition by Cu^{2+}
Monovalent cation effect	Very reduced effect (if any)
Anion effect	According to order of effectiveness: $Cl^- > Br^- > HCO_3^- > SO_4^{2-}$
Inhibitors	Inhibition
Oligomycin	No effect
NaN_3	No effect
Orthovanadate	No effect
Trimethyltin	(I_{50} = 0.15 μM)
DCCD	(I_{50} = 0.25 μM)
DIDS	(I_{50} = 40 μM, K_i = 5–10 μM)
NBD–Cl^-	(K_i = 2.0 μM)
NEM	(I_{50} = 25–30 μM)
Quercetin	(I_{50} = 20–30 μM)
TNP–ATP	(I_{50} = 10–15 μM)
AMP–PNP	(I_{50} = 3–5 μM)
KNO_3	(I_{50} = 50 μM, K_i = 5–20 μM)
Polypeptide composition	
Native form	200 kDa
Catalytic subunit	88 kDa
Other subunits	Function of the mode of solubilization used In the dichloromethane procedure: 2 × 54 kDa (or 110 kDa), 66 kDa, 23 kDa, and 13 kDa

this ATPase seems to belong to a third class of proton-translocating ATPases.[51a]

NADH-Cytochrome c Lutoidic Oxidoreductase

In 1975, Moreau *et al.*[29] demonstrated the activity of an antimycin-insensitive NADH-cytochrome *c* oxidoreductase (EC 1.6.99.3, NADH dehydrogenase) located on the lutoid membrane which can also use ferri-

[51a] B. Marin, X. Gidrol, H. Chrestin, and J. d'Auzac, *Biochimie* **68,** 1263 (1986).

cyanide as an electron acceptor. In 1983, using the flow dialysis technique and the pH and probes described above, it was shown that the addition of NADH plus cytochrome *c* to freshly isolated lutoids leads to an electrogenic transtonoplastic efflux of protons from the lutoids to the external medium and generates a transtonoplastic hyperpolarization (inside more negative) (Fig. 2B).[52] This efflux of H^+ is reversed by FCCP (10–50 μM) while neither KCN (1 mM) and antimycin A (50 μM), known as classic inhibitors of the mitochondrial and bacterial cytochromic chains, nor hydroxamic acids (inhibitors of the alternate pathway), have any effect.

This enzyme also operates in laticiferous cytosol whose protein has been removed by ultrafiltration (Amicon PM10 membrane). It functions *in vivo* in the laticiferous system,[51] although it is unlikely that the electron acceptor is cytochrome *c*. An electrogenic ATPase proton pump operates on the lutoidic tonoplast *in vitro* under almost natural conditions (ultrafiltered cytosol) and acidifies the vacuole contents. An NADH cytochrome *c* reductase electrogenic proton pump handles the efflux of protons from inside the lutoids to the cytosol (external medium). These two pumps have a slightly different optimum pH (6.7 and 7.5, respectively), whereas the physiological pH of latex varies between 6.6 and 7.3.[48] This leads to the consideration that these two tonoplastic enzymes operate as a biophysical pH-stat and at the same time as a biochemical pH-stat using a very active phosphoenolpyruvate carboxylase[53,54] to control the cytosolic pH and the transtonoplastic pH (cytosol–lutoids). These enzymes are well known for their very efficient regulation of the highly pH- and ion-dependent latex metabolism.[55]

Tonoplastic NAD(P)H Reductase: A NAD(P)H-Quinone Reductase

Productivity in *Hevea* latex depends directly on the duration of latex flow after tapping. Coagulation phenomena at the open extremities of the latex vessels are responsible for earlier or later stoppage of flow. The rupture of lutoids leads to coagulation of latex after the release into the cytosol of the destabilizers which they contain (H^+, Mg^{2+}, Ca^{2+}, positively charged proteins, hydrolases, etc.).[56–58] When *Hevea* tapping is

[52] H. Cretin (now Chrestin), *C.R. Acad. Sci. Paris, Ser. 3* **296,** 137 (1983).

[53] J. L. Jacob and J. C. Prevot, *Physiol. Veg.* **17,** 501 (1979).

[54] J. L. Jacob, J. C. Prevot, and J. d'Auzac, *Physiol. Veg.* **21,** 1013 (1983).

[55] H. Chrestin, X. Gidrol, B. Marin, J. L. Jacob, and J. d'Auzac, *Z. Pflanzenphysiol.* **114,** 269 (1984).

[56] W. A. Southorn, *J. Rubber Res. Inst. Malaya* **21,** 494 (1969).

[57] P. Hanower, J. Brzozowska, and C. Lioret, *Physiol. Veg.* **14,** 677 (1976).

[58] J. Brzozowska-Hanower, P. Hanower, and C. Lioret, *Physiol. Veg.* **16,** 231 (1978).

overexploited, the latex flow may stop (bark dryness), leading to coagulation within the laticiferous vessels themselves.[59,60]

In the study of lutoids in latex from partially dry tapping cuts, polarographic analysis shows that the addition of NAD(P)H (0.5 m*M*) causes considerable consumption of the oxygen in the medium; this can be reduced by the addition of superoxide dismutase (SOD), and then almost completely cancelled out by the addition of catalase (10 U ml^{-1} of each enzyme).[61] At the same time there is an inverse evolution in the production of superoxide ion ($O_2^{\cdot-}$) measured by the appearance of formazan monitored at 560 nm after the addition of Nitroblue tetrazolium (NBT) according to Beauchamp and Fridovitch.[62]

The condition of the lutoids is easily checked by determining the ratio of free phosphatase activities to total phosphatase activities after treatment with Triton X-100 (0.1%). *p*-Nitrophenyl phosphate is used as substrate and the bursting index of lutoids is thus determined.[7] When the lutoids obtained from latex of trees whose bark is partially dry are put in the presence of NADH (1 m*M*) and Fe^{3+}/ADP (50/100 μM)[36] and if the following are determined in parallel as a function of the incubation time—bursting of lutoids, release of malondialdehyde (MDA) (lutoids complemented with linoleic acid, 1 m*M*) by assay with thiobarbituric acid,[63] O_2 consumption (polarography), and the disappearance of NADH at 340 nm—there is a joint increase in the bursting of lutoids, in the appearance of MDA at 560 nm, and in the consumption of NADH and O_2.[64]

It is concluded that the various toxic forms of oxygen released lead to perioxidatic degradation of the membranous unsaturated lipids; organelle lysis then occurs, followed by the decompartmentalization of the coagulating factors of the latex.[18,36,61,65]

The following reaction pattern is proposed.

$$\mathrm{NAD(P)H} + 2\mathrm{O}_2 \rightarrow \mathrm{NAD(P)}^+ + 2\mathrm{O}_2^{\cdot-} \qquad [\mathrm{NAD(P)H} - \mathrm{O}_2 \text{ reductase}]$$
$$2\mathrm{O}_2^{\cdot-} + 4\mathrm{H}^+ \rightarrow 2\mathrm{H}_2\mathrm{O}_2 \qquad (\mathrm{SOD})$$
$$\mathrm{H}_2\mathrm{O}_2 \rightarrow \mathrm{H}_2\mathrm{O} + \tfrac{1}{2}\mathrm{O}_2 \qquad (\text{catalase})$$

[59] J. Brzozowska-Hanower, H. Cretin (now Chrestin), P. Hanower, and P. Michel, *Physiol. Veg.* **17,** 851 (1979).

[60] E. de Faÿ and C. Hebant, *C.R. Acad. Sci. Paris, Ser. D* **291,** 865 (1980).

[61] H. Cretin (now Chrestin) and J. Bangratz, *C.R. Acad. Sci. Paris, Ser. 3* **296,** 101 (1983).

[62] C. Beauchamp and I. Fridovitch, *Anal. Biochem.* **11,** 276 (1971).

[63] R. C. Stancliff, M. A. Williams, K. Utsumi, and L. Packer, *Arch. Biochem. Biophys.* **131,** 629 (1969).

[64] W. D. Wosilait and A. Nason, *J. Biol. Chem.* **206,** 255 (1954).

[65] H. Cretin (now Chrestin), J. Bangratz, J. d'Auzac, and J. L. Jacob, *Z. Pflanzenphysiol.* **114,** 261 (1984).

but $O_2^{\cdot-}$ can react nonenzymatically with H_2O_2 by Haber–Weiss-type reactions[66] catalyzed by metal cations which are chelated (or not, such as ADP/Fe^{3+}):

$$H_2O_2 + O_2^{\cdot-} \rightarrow O_2 + OH^- + OH\cdot$$

The free radical OH·, which is particularly reactive, is known to attack the double bonds of membranous ethylenic fatty acids producing, among other things, various molecules including MDA according to the following pattern:

$$-CH=CH-CH_2-CH=CH_2 \xrightarrow{OH\cdot} OHC-CH_2-CHO \quad \text{(MDA)}$$

NAD(P)H–O_2 reductase is localized on the lutoids; nevertheless, if lutoids have been degraded by sonication, the consumption of NADH-dependent O_2 shows a 2-fold increase. Either the access to the enzyme is difficult for substrates or the intralutoidic activators are present.

The K_m of the enzyme has been estimated at 50 μM for NADH and 22–30 μM for O_2. Optimum pH is between 7.0 and 8.0.[36]

NADH–O_2 reductase activity was also observed in a homogenized suspension of freeze-dried lutoids (1 g in 4 ml of TEA, 50 m*M*, pH 7.5 buffer) after addition of NADH (0.35 m*M*). Exactly the same activity was obtained in the clear supernatant (40,000 *g* for 30 min) of the same lutoid suspension. It follows that the enzyme, probably of membranous origin, was solubilized by freeze-drying and/or homogenization in a Teflon–glass homogenizer.

When clear lutoid suspension was desalted on an AcA202 Ultrogel column, no NADH-dependent O_2 consumption was observed using polarography. Enzyme activity is restored if the low-molecular-weight fraction obtained by desalting and which contains the possible substrates for this and other enzymes able to interfere is added to the protein fraction.

The soluble, desalted enzyme was purified by following its diaphorase activity (EC 1.6.99.2) measured by the discoloration of dichlorophenol-indophenol blue (DCPIP, 0.125 m*M*) in the presence of NADH (0.35 m*M*) at 620 nm. A double-beam spectrometer was used to subtract automatically the slow nonenzymatic discoloration of DCPIP by NADH.

Table III shows the various steps of purification: ammonium sulfate precipitation, DEAE–Sepharose CL-6B chromatography, concentration with Amicon PM15 membrane, and molecular sieving on AcA 44 Ultrogel. The molecular weight of the purified enzyme was 101,000 ± 3,000.

The purified enzyme may be determined by various analytical methods. NADH–DCPIP discoloration is the easiest. The disappearance of

[66] F. Haber and J. Weiss, *Proc. R. Soc. London, Ser. A* **147,** 332 (1934).

TABLE III
DIFFERENT STEPS OF NADH-QUINONE REDUCTASE PURIFICATION[a]

Step	Total units (n. kat. DCPIP)	Proteins (mg)	Specific activity (n. kat)	Purification (-fold)	Yield (%)
Lutoidic supernatant	7683	77.4	99	—	—
$(NH_4)_2SO_4$ (35–65%)	6450	61.0	106	1.1	83.8
Desalting: Ultrogel AcA 202	5117	19.8	258	2.6	66.5
DEAE–Sepharose	4083	5.4	756	7.6	53.2
Concentration: Amicon PM15	3783	2.1	1801	18.2	49.3
Molecular sieving: Ultrogel AcA 44	3262	0.7	4667	47.0	42.4

[a] The purification was carried out from a homogenized suspension of freeze-dried lutoids (0.2 g ml^{-1}), the clear supernatant (40,000 *g* × 30 min) was used, and the purification steps were followed by using the discoloration of dichlorophenolindophenol (DCPIP) by reaction with NADH in a double-beam spectrophotometer against a control without enzyme.

NADH was followed at 340 nm in the presence of quinone and duroquinone (respectively), whose absorption spectra do not interfere with the NADH spectrum. The appearance of formazan at 540 nm after addition of NBT (0.15 m*M*) was used for other quinones or quinoid compounds like menadione, 1,4-naphthoquinone, and 1,4-chloranil (5 μM). The latter method determined superoxide ion produced by self-oxidizable semiquinones during NADH-dependent quinone reduction.

After disk gel polyacrylamide electrophoresis according to classic methods, a single enzyme band was revealed by NADH-dependent DCPIP discoloration or by the appearance of a formazan band with NBT, NADH, and the different quinones.

Table IV gives the apparent kinetic contants of the purified enzyme for various quinones or quinoid compounds at a fixed concentration of NADH. It would appear that this tonoplastic enzyme, which reacts like a diaphorase (EC 1.6.99.2) with quinoid compounds such as DCPIP, with numerous quinones, and also with ferricyanide and cytochrome *c* (in the latter case if a quinone is present), is probably a nonspecific NAD(P)H-quinone reductase[67] of the type described by Wosilait and Nason in 1954[64] and more recently by Lind *et al.*[68] The latter authors consider, however,

[67] J. d'Auzac, C. Sanier, and H. Cretin (now Chrestin), *Int. Rubber Conf., 1985, Kuala-Lumpur,* in press.

[68] C. Lind, P. Hochstein, and L. Ernster, *Arch. Biochem. Biophys.* **216,** 178 (1982).

TABLE IV
KINETIC CONSTANTS OF PURIFIED NADH-QUINONE REDUCTASE TO VARIOUS QUINONES[a]

Substrate	K_m (μM)	V_{max} (n. kat. ml^{-1} formazan)
Quinone	130	6
Duroquinone	70	29
Menadione	15	44
1,4-Naphthoquinone	7	47
Chloranil	13	13

[a] The purified enzyme was incubated with NADH (0.35 m*M* final concentration) and Nitroblue tetrazolium (0.15 m*M*). The absorbance was measured at 540 nm in a double-beam spectrophotometer. The control cell contained all the reactives except the enzyme.

that the NAD(P)H–DT diaphorase activity, which reduces quinones to hydroquinones, functions as a system protecting the cell against the formation of self-oxidizable semiquinone-producing superoxide ions. In our case, however, the consumption of oxygen and the formation of formazan clearly show that a proportion yet to be determined of the reducing power of NAD(P)H is used for the incomplete reduction of diquinones to autoxidizable semiquinones with the accompanying formation of superoxide ions (Fig. 3).[67]

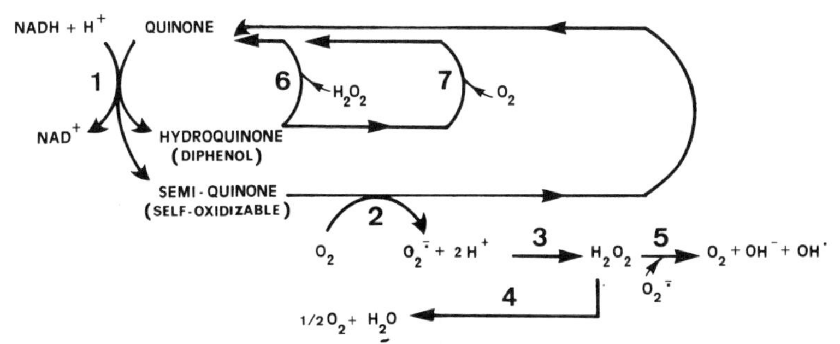

FIG. 3. Hypothetical representation of the functioning of NADH-quinone reductase of lutoids. (1) NADH(P)H-quinone reductase produces in the same time an undetermined ratio of hydroquinones and self-oxidizable semiquinones; (2) the self-oxidizable semiquinones produce oxygen superoxide ion $O_2^{\cdot-}$; (3) $O_2^{\cdot-}$ transformation in H_2O_2 by superoxide dismutase; (4) H_2O_2 decomposition by catalase; (5) a nonenzymatic reaction between $O_2^{\cdot-}$ and H_2O_2 produces the most toxic oxygen form OH·; (6) a peroxidase reacting with hydroquinone (diphenol) and H_2O_2 regenerates quinones; (7) the phenol oxidases with O_2 regenerate quinones from hydroquinones. Peroxidases and phenol oxidases are very abundant in latex.

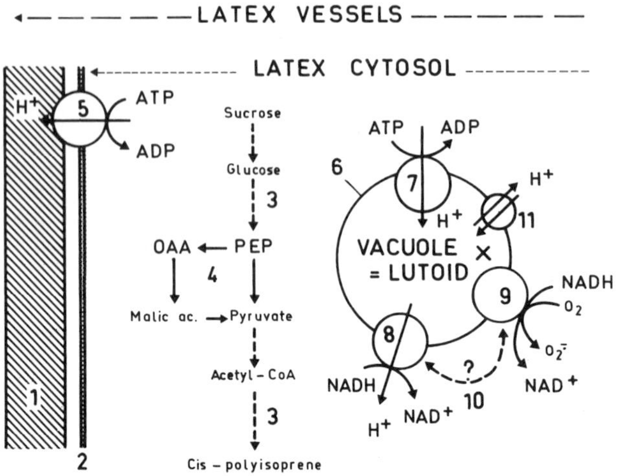

FIG. 4. Schematic representation of the tonoplastic enzymes in laticiferous cell. (1) Latex cell wall; (2) plasmalemma of laticiferous cell; (3) metabolic pathway leading to *cis*-polyisoprene biosynthesis; (4) Davies-type pH stat; (5) plasmalemmic H^+-pump ATPase (?); (6) tonoplast; (7) tonoplastic H^+-pump ATPase (influx); (8) tonoplastic NADH-cytochrome *c* H^+ pump (efflux); (9) NADH-O_2 reductase (NADH-quinone reductase); (10) linking between the two last enzymes; (11) H^+ solutes antiporter into vacuolar compartment.

The ease of obtaining large quantities of lutoid vacuoles from *Hevea* latex and hence of tonoplast was used to obtain better knowledge of a plant tonoplast. A large number of questions still remain unanswered, particularly with regard to the structure of lutoidic ATPase in comparison with other plant ATPases and the nature of the natural receptor of the oxidoreducing, NADH-dependent, tonoplastic proton pump. Finally, there would appear to be a relationship between the activity of the NADH-dependent tonoplastic proton pump (outflow) and the NADH-quinone reductase activity producing toxic forms of oxygen. The former activity is replaced by the latter in artificially aged lutoids or in lutoids suffering from bark dryness (cellular degeneration). This relationship could be the dysfunctioning, the release, or the disappearance of a factor which in the lutoidic tonoplast provides the link between membranous transport of electrons (NADH dependent) and transtonoplastic outflow of protons. This delinking would then make possible transport of electrons to dissolved oxygen.[36]

Figure 4 is a diagrammatic representation of the role of a plant tonoplast in maintaining homeostasis[48] and of cell degeneration,[65] in the special case of laticiferous cells in *Hevea*.

[11] Characterization of Tonoplast Enzyme Activities and Transport

By BENJAMIN JACOBY

Introduction

Mature cells of higher plants are encased by the cell wall and contain two principal membrane-bound compartments: an outer cytoplasmic compartment, which is surrounded by the plasmalemma and includes the cytosol as well as the organelles, and a large inner compartment, the vacuole. The latter occupies about 80–90% of the cell volume, and is encased by another simple membrane, the tonoplast. The vacuole generally functions as a lysosomal compartment.[1]

A priori, no marker enzyme for the tonoplast of higher plant cells was known; hence to characterize this membrane it first had to be obtained from separated and pure vacuoles. The difficulties involved in separation of pure vacuoles and the artifacts encountered were discussed by Alibert and Boudet.[2] One artifact of particular importance in relation to the separation of authentic tonoplasts are vacuoplasts, namely vacuoles which are surrounded by the plasmalemma in addition to the tonoplast and contain very few cytoplasmic remnants between the membranes.[3] They appear in some preparations of vacuoles which are obtained by lysis of protoplasts and can be detected by fluorescein diacetate (FDA) staining.[4]

Pure vacuoles, not containing vacuoplasts, were first obtained from red beet storage tissue[4,5] and from the latex of *Hevea brasiliensis*.[6] Recently vacuoles, which were shown not to stain with FDA, were also obtained from corn coleoptile protoplasts.[7] Authentic tonoplast have been obtained from these vacuoles[7–11] and their properties were compared to those of microsomal fractions separated on density gradients.[7,10,12] Such

[1] P. Matile, *Physiol. Veg.* **20,** 303 (1982).
[2] G. Alibert and A. Boudet, *Physiol. Veg.* **20,** 289 (1982).
[3] H. Lorz, G. T. Harms, and I. Potrykus, *Biochem. Physiol. Pflanz.* **169,** 617 (1976).
[4] A. Admon, B. Jacoby, and E. E. Goldschmidt, *Plant Physiol.* **65,** 55 (1980).
[5] R. A. Leigh and D. Branton, *Plant Physiol.* **56,** 656 (1976).
[6] J. d'Auzac, H. Cretin, B. Marin, and C. Lioret, *Physiol. Veg.* **20,** 311 (1982).
[7] S. Mandala and L. Taiz, *Plant Physiol.* **78,** 104 (1985).
[8] A. Leigh and R. R. Walker, *Planta* **150,** 222 (1980).
[9] A. Admon, B. Jacoby, and E. E. Goldschmidt, *Plant Sci. Lett.* **22,** 89 (1981).
[10] A. B. Bennett, S. D. O'Neil, and R. M. Spanswick, *Plant Physiol.* **74,** 538 (1984).
[11] B. Marin, J. A. C. Smith, and U. Lüttge, *Planta* **153,** 486 (1981).
[12] R. Poole, D. P. Briskin, Z. Kratky, and R. M. Johnstone, *Plant Physiol.* **74,** 549 (1984).

comparison resulted in identification of a tonoplast-enriched light microsomal fraction from microsomes. This fraction is characterized by the presence of an NO_3^--inhibited and vanadate-insensitive Mg^{2+}-ATPase, which serves now as a marker enzyme for the tonoplast.

Separation of Tonoplasts from Beet Vacuoles

The method is based on the observation that vacuoles can be prepared by slicing tissue into an appropriate osmoticum. In suitable, freshly plasmolyzed cells the vacuole has shrunken, but the plasmalemma is still attached to the cell wall at several sites. During slicing the tonoplast-enclosed vacuoles are released into the osmoticum, while the plasmalemma remains attached to the cut cell wall. Beet vacuoles separated in this way are enclosed by a single membrane[13] and do not stain with FDA.[4]

Method

The method for separating large amounts of vacuoles follows Leigh *et al.*[14] with minor modification.[9,10]

Plant Material. Commercial red beets are used. Care should be taken to obtain freshly harvested beets with fresh leafy tops. The detopped beets can be stored up to 6 months in moist vermiculite at 4°.

Collection Medium. Combine 1 *M* sorbitol, 5 m*M* EDTA, 4 m*M* Tris(hydroxymethyl)aminomethane/2-(*N*-morpholino)ethanesulfonic acid (Tris/MES), pH 8.0.

Resuspension Buffer. Combine 1.2 *M* sorbitol, 1 m*M* ethylenediaminetetraacetic acid (EDTA), 2 m*M* dithiothreitol (DTT), 25 m*M* Tris/MES (pH 7.0).

Vacuole Separation. Red beet slices (approximately 5 mm thick) are placed in an ice-cold collection medium (about 500 g in 1.0 liter) and either finely sliced with a razor blade in a shallow dish or in a special apparatus[5] at 17–20°. The sliced tissue is filtered through Miracloth and the filtrate centrifuged at 1300 *g* for 20 min. The pellet is gently resuspended in 20 ml resuspension buffer containing 15% (w/v) metrizamide. The suspension is placed at the bottom of a centrifugation tube and overlaid with 10 ml each of 10% (w/v) and 0% metrizamide in resuspension buffer. Alternatively, to avoid possible metrizamide inhibition,[15] an isotonic sucrose/sorbitol gradient can be used.[9] The tubes are centrifuged at 430 *g* for 10 min and intact vacuoles are collected at the 0/10% metrizamide interface.

[13] R. K. Salyaev, *in* "Membrane Transport in Plants" (W. J. Cram, K. Janacek, R. Rybova, and K. Sigler, eds.), p. 249. Wiley, Chichester, 1984.

[14] R. A. Leigh, T. ap Rees, W. A. Fuller, and J. Banfield, *Biochem. J.* **178,** 539 (1979).

[15] S. Doll, F. Rodeer, and J. Willenbrink, *Planta* **144,** 407 (1979).

Tonoplast Separation. The band containing the vacuoles is diluted 10-fold with 2 m*M* DTT, 5 m*M* Tris/MES (pH 6.5). Membranes of the lysed vacuoles are pelleted by centrifugation at 80,000 *g* for 30 min. The tonoplast pellet is resuspended in the dilution medium containing also 0.25 *M* sucrose.

Separation of Tonoplasts from *H. brasiliensis* Lutoids

Lutoids, which comprise 20% of the *H. brasiliensis* latex volume, constitute a dispersed lysosomal vacuome in a specialized cytoplasm, that is the latex. They are unit-membrane organelles, 1–5 μm in diameter.[6] The latex used for work done on these vacuoles has been obtained from *H. brasiliensis* clones GT, TJ, and RB86 grown in the Institut de Recherches sur le Caoutchoue en Afrique, Abidjan, Ivory Coast.

Method

Latex Collection.[16] The tapping panel is cleaned with 80% ethanol and dried before tapping. The first, bacteria-rich, 50-ml fraction is discarded. The latex is then collected in ice-cold bottles.

Lutoid Separation.[17] The latex is diluted with 1–3 vol of 0.3 *M* mannitol, 2 m*M* mercaptoethanol (ME), 150 m*M* triethanolamine (TEA), pH 7.5 with HCl (TEA–mannitol buffer), and centrifuged at 39,000 *g* for 60 min at 4°. The pellet, which constitutes the crude lutoid fraction, is washed twice with TEA–mannitol buffer. For experimentation with intact lutoids, the pellet is resuspended in 0.3 *M* mannitol, 25 m*M* MES, 25 m*M* 3-(*N*-morpholino)propanesulfonic acid (MOPS), 25 m*M* *N*-(2-hydroxyethylpiperazine)-*N'*-2-ethanesulfonic acid (HEPES), pH 7.5 with TEA.

Tonoplast Separation.[11] To obtain tonoplasts, the washed lutoid pellet is gently homogenized in isotonic buffer at a ratio of 100 ml to 1 g lyophilized lutoids. A glass potter and piston in an ice bath are used. The buffer consists of 0.3 *M* mannitol, 5 m*M* ME, 25 m*M* MES, 25 m*M* MOPS, 25 m*M* HEPES, pH 6.0 with Tris base.

Separation of Tonoplasts from Microsomes

The procedure described is for red beets and essentially follows Bennett *et al.*[10] and Poole *et al.*[12] These authors have shown that the characteristics of the separated membrane fractions are similar to those of tono-

[16] B. Marin and P. Trouslot, *Planta* **124,** 31 (1975).

[17] B. Marin, M. Marin-Lanza, and E. Komor, *Biochem. J.* **198,** 365 (1981).

plasts obtained from vacuoles. A similar method was recently described for separation of tonoplasts from corn coleoptile protoplasts.[7]

Method

Preparation of Microsomes. Red beet storage tissue is cut into a chilled blender containing 0.25 *M* sucrose, 0.1% bovine serum albumin (BSA, fatty acid free), 0.5% poly(vinylpyrolidone) (PVP-40), 4 m*M* DTT, 3 m*M* EDTA, 70 m*M* Tris (pH 8.0 with HCl) at a solution : tissue ratio of 2 : 1, and blended three times for 10 sec each. The homogenate is filtered through four layers of cheesecloth and centrifuged for 10 min at 13,000 *g*. The supernatant is centrifuged again for 30 min at 80,000 *g* and the microsomal pellet is resuspended in suspension medium containing 0.25 *M* sucrose, 2 m*M* DTT, and 5 m*M* Tris/MES (pH 6.5). To the resuspended membranes an equal volume of the suspension medium also containing 500 m*M* KI is added with continuous stirring. The membranes are then repelleted for 30 min at 80,000 *g* and resuspended in suspension medium without KI. The KI treatment reduces contamination with adsorbed phosphatases, but also reduces yield.[12]

Tonoplast Fraction. The microsomal suspension is loaded on the top of a discontinuous 26 and 16% sucrose gradient in 2 m*M* DTT and 5 m*M* Tris/MES (pH 6.5). After centrifugation at 80,000 *g* for 2 hr a tonoplast-enriched fraction is obtained from the 10%/23%[12] or 16%/26%[10] interface (the lower cutoff point decreases contamination with other membranes, but also tonoplast yield[12]). The membranes are diluted with suspension medium without DTT and pelleted for 30 min at 80,000 *g*. The tonoplast pellet is again dispersed in suspension medium without DTT and can be stored for 2 weeks at −90°.

Tonoplast Characteristics

Density

Tonoplast vesicles peak at 1.10–1.13 g ml^{-1} in linear density gradients.

Lipid Composition

Total lipids of yeast tonoplasts were extracted.[18,19] The phosphorus content of the extract was determined[20]; phospholipids were identified by

[18] W. Van der Wilden and P. Matile, *Biochem. Physiol. Pflanz.* **173,** 285 (1978).

[19] R. Letters, *in* "Aspects of Yeasts Metabolism" (A. K. Mills, ed.), p. 303. Blackwell, Oxford, 1968.

[20] G. R. Barlett, *J. Biol. Chem.* **234,** 166 (1959).

two-dimensional thin-layer chromatography[21] and ergosterol was determined according to Kates.[22] The membranes were found to be characterized by a high lipid : protein ratio of 1.42; high phospholipid and low ergosterol contents, 28 and 7.9% of total lipids, respectively; and absence of phosphatidylserine and phosphatidic acid.

Tonoplast-Bound Enzymes and Porters

Enzymes. Two enzymes have been clearly identified as bound to the tonoplast: a Mg^{2+}-ATPase[23] and a pyrophosphatase.[24,25] In addition α-mannosidase was identified as a marker enzyme of yeast tonoplast.[18,26] In higher plant vacuoles α-mannosidase was, however, localized in the soluble fraction and is considered as a marker for the vacuolar sap.[27]

An antimycin-insensitive NADH-cytochrome *c* oxidoreductase, able to function as a proton pump in the presence of exogenous cytochrome *c*, was found in lutoid membranes.[6] In a linear gravity gradient of beet microsomes an NADH-cytochrome *c* oxidoreductase did, however, peak at 11% sucrose equivalent while the tonoplast Mg^{2+}-ATPase peaked at 9–10% sucrose and the enzyme is considered an endoplasmic reticulum marker.[28]

Porters. Various solutes are transported across the tonoplast, apparently aided by porters. Evidence seems to exist for such Ca^{2+}/H^+,[29–31] Na^+/H^+,[32] and sucrose/H^+ [33] antiport; NO_3^-/H^+ symport[34]; and UDP-glucose, sucrose group transport.[35]

[21] K. Hunter and A. H. Rose, *Biochim. Biophys. Acta* **260,** 639 (1972).

[22] M. Kates, *in* "Laboratory Techniques in Biochemistry and Molecular Biology" (T. S. Work and E. Work, eds.), Vol. 3, Part II. Am. Elsevier, New York, 1972.

[23] H. Sze, *Annu. Rev. Plant Physiol.* **36,** 175 (1985).

[24] G. J. Wagner and P. Mulready, *Biochim. Biophys. Acta* **728,** 267 (1983).

[25] P. A. Rea and R. A. Poole, *Plant Physiol.* **77,** 46 (1985).

[26] W. Van der Wilden, P. Matile, M. Schellenberg, J. Meyer, and A. Wieneken, *Z. Naturforsch., C: Biosci.* **23C,** 416 (1973).

[27] T. Boller, *Physiol. Veg.* **20,** 247 (1982).

[28] R. R. Lew and R. M. Spanswick, *Plant Sci. Lett.* **36,** 187 (1984).

[29] F. Rasi-Caldogno, M. I. De Michelis, and M. C. Pugliarello, *in* "Plasmalemma and Tonoplast: Their Function in the Plant Cell" (D. Marme, E. Marre, and R. Hertel, eds.), p. 361. Elsevier North-Holland Biomedical Press, Amsterdam, 1982.

[30] K. S. Schumaker and H. Sze, *J. Biol. Chem.* **261,** 12172 (1986).

[31] E. Blumwald and R. Poole, *Plant Physiol.* **80,** 727 (1986).

[32] E. Blumwald and R. J. Poole, *Plant Physiol.* **78,** 163 (1985).

[33] D. P. Briskin, W. R. Thornley, and R. E. Wyse, *Plant Physiol.* **78,** 871 (1985).

[34] E. Blumwald and R. J. Poole, *Proc. Natl. Acad. Sci. U.S.A.* **82,** 3683 (1985).

[35] A. Maretzki and M. Thom, *Plant Physiol.* **80,** 34 (1986).

Mg^{2+}-ATPase

For a recent survey, cf. Marin.[36]

Method

The colorimetric assay of P_i release follows Hodges and Leonard[37] with modifications. The reaction mixture contains 30 m*M* Tris (pH 8.0 with MES), 50 m*M* KCl, 3 m*M* Tris–ATP, 3 m*M* $MgSO_4$. To inhibit remnants of acid phosphatase as well as mitochondrial and plasmalemma ATPase, 0.1 m*M* $(NH_4)_6Mo_7O_{24}$, 0.5 m*M* NaN_3, and 0.1 m*M* Na_3VO_4, respectively, can be included. ATP hydrolysis by tonoplasts can also be measured with [γ-^{32}P]ATP, or by coupling to NADH oxidation.[38]

Purification[39]

Partial Removal of Other Proteins. The nitrate-sensitive Mg^{2+}-ATPase in the tonoplast-enriched fraction from corn coleoptiles, obtained from a Dextran gradient, was purified about 24 times by a 2-step solubilization and sucrose gradient centrifugation: Sodium deoxycholate (DOC) is added dropwise from a 5% solution in 10 m*M* Tris/MES buffer (pH 7.5) to the tonoplast-enriched fraction while stirring on ice. Final concentrations are about 0.1% protein and 0.15% DOC. The membranes are incubated in DOC for 30 min and centrifuged for 1 hr at 120,000 *g*. Most of the KNO_3-sensitive ATPase remains in the pellet.

ATPase Solubilization. The purified pellet of corn coleoptile tonoplasts is suspended to obtain about 2 mg protein ml^{-1} in 0.25 *M* sorbitol, 1 m*M* EDTA, 1 m*M* DTT, 10 m*M* Tris/MES (pH 7.5); then 200 m*M* octyl-β-D-glucopyranose (OG) in 10 m*M* Tris/MES (pH 7.5) is added to obtain a final concentration of 40 m*M* OG. The suspension is incubated for 30 min and centrifuged for 1 hr at 120,000 *g*. The supernatant contains most of the NO_3^--sensitive Mg^{2+}-ATPase.

The OG supernatant is layered on a linear 10–25% sucrose gradient and centrifuged in a swing-out head at 200,000 *g* for 5 hr at 4° (1 ml of supernatant is layered on 2 ml each of 25 and 10% sucrose). Fractions of 0.5 ml are taken from the top; fractions 6–9 are enriched in the ATPase.

[36] P. B. Marin (ed.), "Biochemistry and Function of Vascular Adenosine-Triphosphatase in Fungi and Plants." Springer, Berlin, 1985.

[37] T. K. Hodges and R. T. Leonard, this series, Vol. 32B, p. 392.

[38] C. J. Griffith, P. A. Rea, E. Blumwald, and R. Poole, *Plant Physiol.* **81,** 120 (1986).

[39] S. Mandala and L. Taiz, *Plant Physiol.* **78,** 327 (1985).

Methods for purification of tonoplast ATPase from *H. brasiliensis* and *Neurospora crassa* were described by Marin *et al.*[40] and Bowman *et al.*[41]

Proton Transport

The tonoplast Mg^{2+}-ATPase functions as a proton pump in sealed, right-side-out, tonoplast vesicles. The initial rate of proton pumping (J_i), indicative of ATPase activity, and the steady state ΔpH, as well as $\Delta\psi$, are measured with fluorescent probes. Alternatively, ΔpH and $\Delta\psi$ can be measured with radioactive probes[11] like [^{14}C]methylamine, [^{3}H]triphenylmethylphosphonium ($TPMP^+$), and $S^{14}CN^-$.

Intravesicular Volume.[11] This is determined with ^3H_2O and ^{14}C-labeled Dextran and calculated as the water-accessible, but Dextran-inaccessible space.

J_i and ΔpH with Quinacrine.[42,43] The reaction medium consists of 1.5 ml 0.25 *M* sorbitol, 25 m*M* MES/bistrispropane (pH 6.5), 5 m*M* $MgSO_4$, 10 μM quinacrine, and 100–200 μg tonoplast vesicle protein. With temperature at 30°, 0.5 *M* Na_2ATP is added to a final concentration of 5 m*M* and the change in fluorescence is followed in a fluorescence spectrophotometer with excitation at 423 nm (slit width, 1–2 nm) and emission at 550 nm (slit width, 2 nm). The ATP-dependent fluorescence quenching should be completely reversible with 6 μM FCCP or 5 μM nigericin.

J_{net} is a function of the rate of fluorescence quenching, dQ/dt [$Q = (F/\Delta F)100$], which is plotted. The initial net proton flux (J_i) is:

$$J_i = (B_i V_i^2/230\,A V_o)dQ/dt \qquad (1)$$

where F is maximal fluorescence; B_i, internal buffer capacity; V_i and V_o, internal and external volumes, and A is vesicle surface area.

$$\Delta pH = \log[Q/(100 - Q)] + \log V_o/V_i \qquad (2)$$

$\Delta\psi$ with Oxonol-V.[44,45] Membrane potential is measured similar to ΔpH, but with oxonol-V [bis(3-phenyl-5-oxoisoxonol)] instead of quinacrine. Oxonol-V is added to a final concentration of 8 μM from a 0.2 *M*

[40] B. Marin, J. Preisser, and E. Komor, *Eur. J. Biochem.* **151,** 131 (1985).

[41] E. J. Bowman, S. Mandala, L. Taiz, and B. J. Bowman, *Proc. Natl. Acad. Sci. U.S.A.* **83,** 45 (1986).

[42] A. B. Bennett and R. M. Spanswick, *J. Membr. Biol.* **71,** 95 (1983).

[43] H. C. Lee, J. G. Forte, and D. Epel, *in* "Intracellular pH: Its Measurement, Regulation and Utilization in Cellular Functions" (R. Naccitelli and D. W. Deaner, eds.), p. 135. Alan R. Liss, Inc., New York, 1982.

[44] R. R. Lew and R. M. Spanswick, *Plant Physiol.* **77,** 352 (1985).

[45] S. Scherman and J. P. Henry, *Biochim. Biophys. Acta* **599,** 150 (1980).

stock solution in 10% (v/v) dimethyl sulfoxide. Fluorescence quenching is recorded with excitation at 580 nm (slit, 2 nm) and emission at 640 nm (slit, 2 nm). For absolute measurement of membrane potential the method has to be calibrated against distribution of radiotracer-labeled permeable ions, $TPMP^+$ and SCN^-.[11]

Characteristics of Tonoplast ATPase

Table I lists the characteristics of ATPase from various plant species.[46–50] The characteristics of the enzyme, in particular, insensitivity to vanadate, suggest that its reaction mechanism does not involve a covalent phosphoenzyme[41,48]; tyroxine and cysteine residues are apparently in the catalytic site.[48]

Porters

The Mg^{2+}-ATPase-dependent proton-motive force (pmf) across the tonoplast can be utilized to energize various porter-aided solute transports. To demonstrate the presence of such porters in the tonoplast, accompanying proton transport should be demonstrated in addition to solute transport. Such flux of an ion should also be reversible in the presence of an appropriate ionophore. Using such criteria, evidence for Ca^{2+}, Na^+, NO_3^-, and sucrose porters in the tonoplast seems to exist.

Measurement with Radiotracers

Reaction Medium. Combine 0.25 *M* sucrose, 25 m*M* MES (pH 7.2 with Tris), 1 m*M* $MgSO_4$, and radiolabeled substrate at desired concentration.

Measurement. To the medium 50 μg ml^{-1} of membrane protein is added and the mixture is incubated at 30°. At various time intervals 1.0 ml is removed and filtered through a 0.45-μm Millipore filter. The precipitate is washed with 6-ml portions of unlabeled reaction mixture at 0°. The Millipore filter with the percipitate is radioassayed. To the mixture Tris–ATP is added to a final concentration of 1 m*M* as well as various promoters, ionophores, and inhibitors as needed. ATP-independent absorption should be subtracted.

[46] A. Bennett, S. D. O'Neil, M. Eilmann, and R. M. Spanswick. *Plant Physiol.* **78,** 495 (1985).

[47] L. Tognoli, *Eur. J. Biochem.* **146,** 581 (1985).

[48] Y. Wang and H. Sze, *J. Biol. Chem.* **260,** 10434 (1985).

[49] B. Marin, *Planta* **157,** 324 (1983).

[50] B. Marin, *Plant Physiol.* **73,** 973 (1983).

TABLE I
CHARACTERISTICS OF TONOPLAST ATPASE

Characteristic	Data	Plant
Molecular weight		
Functional	About 386 kDa, (radiation inactivation)	Corn[39]
Subunits	62; 72 kDa (SDS–PAGE)	Corn[39]
Substrate	MgATP, K_m 0.1–0.2 m*M*	Beet[42]
Free Mg^{2+}	Excess over ATP does not inhibit	Beet[46]
Free ATP	Excess over Mg^{2+} inhibits	Beet[46]
Substrate specificity	ATP > GTP > ITP	*Tulipa*[24]
pH	Optimum 7–8	Beet,[9] *Hevea*[49]
Anion stimulation	$Cl^- > Br^- >$ acetate $> HCO_3^- > I^- > SO_4^{2-}$	Beet,[10] oat[48]
Cation stimulation	Little or none	Beet[10]
Inhibition		
Ca^{2+}	NI[a] (5 m*M*)	Beet,[46] radish[47]
ADP	I_{50}[b] (0.75 m*M*, 0.4 m*M*)	Beet,[46] oat[48]
AMP	I_{50} (0.4 m*M*)	Oat[48]
NO_3^-	I_{50} (8–20 m*M*)	*Hevea*,[49] corn[46]
SCN^-	80% I[c] at 50 m*M*	Corn[46]
DCCD	I_{50} (3 μM, 126 μM)	Corn,[46] *Hevea*[50]
DIDS	I_{50} (45 μM)	*Hevea*[50]
Na_3VO_4	NI (50 μM, 1 m*M*)	Beet,[9] *Hevea*[50]
$(NH_4)_6Mo_7O_{24}$	NI (0.1 m*M*, 1 m*M*)	Beet,[9] *Hevea*[50]
Oligomycin	NI (2 ppm, 20 ppm)	Beet,[9] *Hevea*[50]
NaN_3	NI (10 m*M*)	Beet,[9] *Hevea*[50]
Ouabain	NI (0.15–1 m*M*)	Beet,[9] *Hevea*[50]
PCMS	NI (250 μM)	*Hevea*[50]
TPS	NI (500 μM)	*Hevea*[50]

[a] NI, Not inhibited.
[b] I_{50}, Concentration for 50% inhibition.
[c] I, Inhibition.

Transport. The tracer method was employed to demonstrate Ca^{2+}/H^+ [20,30,31,51] and sucrose/H^+ [33] antiport in tonoplast vesicles from plant and in lutoids. Transport of both solutes was activated by an ATP-dependent pmf (positive inside) and was protonophore sensitive. Addition of the calcium ionophore A23187 released accumulated Ca^{2+}. Transport of Ca^{2+} can also be activated by a ΔpH generated by pH jumps (6.0 to 8.0) or a K^+ gradient in the presence of nigericin.[30]

[51] B. Marin, H. Cretin, and J. d'Auzac, *in* "Plasmalemma and Tonoplast: Their Function in the Plant Cell" (D. Marme, E. Marre, and R. Hartel, eds.), p. 209. Elsevier/North-Holland Biomedical Press, Amsterdam, 1982.

Measurement with Fluorescent Probes

The effect of solutes on fluorescence quenching of appropriate probes for ΔpH and $\Delta\psi$ (see above) is measured. In addition, the rate of Ca^{2+} uptake can also be monitored by the initial rate of chlorotetracycline fluorescence increase.[52]

Na^+/H^+ Antiport. Employing fluorescence quenching, Blumwald and Poole[32] showed in tonoplast vesicles from red beets an electrochemical Na^+/H^+ antiport which was insensitive to K^+, exhibited Michaelis–Menten kinetics, and was inhibited by amiloride. In addition a ψ-mediated Na^+ and K^+ flux, ascribed to conductive pathways, was found. The latter fluxes were abolished when ψ was clamped at zero.

NO_3^-/H^+ Symport. Blumwald and Poole[34] showed that Cl^- as well as NO_3^- dissipated $\Delta\psi$ in tonoplasts vesicles, while only NO_3^- dissipated the ΔpH as well. This indicated ψ-dependent influx, perhaps diffusion of both ions, and NO_3^-/H^+ symport out again.

Pyrophosphatase

Assay[25]

The standard assay procedure is similar to that for the Mg^{2+}-ATPase, but 3 m*M* Tris-PP_i ($P_2O_7^{4-}$) serves as substrate instead of Tris-ATP. Activity is calculated as half the rate of P_i liberation (one PP_i yields two P_i).

Properties

Pyrophosphate hydrolysis by red beet tonoplasts differs from ATP hydrolysis in its insensitivity to NO_3^-, its stimulation by K^+ (much more than by Na^+), and its insensitivity to Cl^-. Pyrophosphatase activity is similar to ATPase in that both are Mg^{2+} dependent, linked to proton transport, and insensitive to molybdate, vanadate, and azide.[25,53,54] The ATPase and PP_iase can be separated by Sephacryl-400 chromatography of tonoplast vesicles solubilized with 2% (w/v) Triton X-100 in 20% (w/v) glycerol, 4 m*M* $MgCl_2$, 1 m*M* EDTA, and 5 m*M* Tris-MES (pH 8.0).[55]

[52] R. Lew, D. P. Briskin, and R. Wyse, *Plant Physiol.* **82,** 47 (1986).

[53] A. Chanson, J. Fichmann, D. Spear, and L. Taiz, *Plant Physiol.* **79,** 159 (1986).

[54] Y. Wang, R. A. Leigh, K. H. Kaestner, and H. Sze, *Plant Physiol.* **81,** 497 (1986).

[55] P. A. Rea and R. Poole, *Plant Physiol.* **81,** 126 (1986).

[12] Preparation of Tonoplast Vesicles: Applications to H^+-Coupled Secondary Transport in Plant Vacuoles

By EDUARDO BLUMWALD, PHILIP A. REA, and RONALD J. POOLE

Although transport has been detected in purified preparations of plasma membrane, secretory vesicles, Golgi, and endoplasmic reticulum of plant cells, most of the transport seen in crude microsomal membrane preparations appears to be attributable to tonoplast vesicles. This is because, on homogenization of plant tissues, tonoplast fragments apparently reseal more readily than plasma membrane. Studies on isolated intact vacuoles have identified two characteristic enzyme activities [an anion-sensitive ATPase and a cation-sensitive inorganic pyrophosphatase (PPase)] which together with the typical density of tonoplast in sucrose gradients ($1.10\ g \cdot cm^3$) are tentatively used as markers for tonoplast isolation. These recent developments have led to a burst of research activity on tonoplast transport.

Tonoplast vesicles have several advantages over intact vacuoles for transport studies: (1) because of their high surface area to volume ratio, they show high rates of transport; (2) they are essentially devoid of endogenous solutes; (3) they rapidly equilibrate with the bulk medium and can therefore be loaded with any desired contents; and (4) if treated with a chaotrope such as KI (see below), they can be stripped of weakly associated peripheral proteins.

Storage tissue of red beet (*Beta vulgaris* L.) appears to be particularly suitable for studies on tonoplast transport. This is because the relative densities of the various cellular membranes allow a better purification of tonoplast from this material, and perhaps also because storage tissue is specialized for vacuolar storage and retrieval of nutrients.

Preparation of Tonoplast Vesicles

Reagents

Homogenization Medium. A suitable homogenization medium consists of 10 m*M* glycerophosphate, 0.65 *M* ethanolamine (adjusted to pH 8.0 with concentrated H_2SO_4 before addition), 0.28 *M* choline chloride, 2 m*M* salicylhydroxamic acid, 0.2% (w/v) BSA (fraction V, essentially fatty acid free), 10% (w/v) insoluble PVP, 0.5 m*M* butylated hydroxytoluene, 1 m*M* nupercaine, 3 m*M* Tris–EDTA, and 26 m*M* $K_2S_2O_5$ and 5 m*M* DTT buffered to pH 8.0 with 70 m*M* Tris–MES buffer. The homogenization medium can be stored for several weeks at −20° without deterioration providing that the $K_2S_2O_5$ and DTT are not added until immediately before use. The pH of the medium should be adjusted before the addition of the insoluble PVP since it strongly binds to glass and impairs electrode performance. Choline and ethanolamine are included in the homogenization medium to diminish membrane degradation by phospholipase D while nupercaine and glycerophosphate are added to inhibit phospholipase A and phosphatidic acid phosphatase activity, respectively.[1] Butylated hydroxytoluene is added to scavenge free radicals and diminish lipid peroxidation. Storage tissue of beet particularly requires large amounts of reductant ($K_2S_2O_5$).

Suspension Medium. The suspension medium consists of 1.1 *M* glycerol, 1 m*M* Tris–EDTA, 0.27 m*M* nupercaine, 0.45 m*M* butylated hydroxytoluene, and 1 m*M* DTT buffered to pH 8.0 with 5 m*M* Tris–MES. Stock solutions of 670 and 230 m*M* nupercaine and butylated hydroxytoluene are prepared in 2-propanol and 0.4 and 2 ml, respectively, are added to 1 liter of the suspension medium with vigorous stirring. The solution is filtered through 0.45-μm Millipore HA filter to remove undissolved nupercaine and butylated hydroxytoluene. Other suspension media may be used, but glycerol and DTT have a marked stabilizing effect on the tonoplast vesicles.

KI and Sucrose Solutions. KI (0.18 *M*) and 10 and 23% (w/w) sucrose are prepared in fresh suspension medium.

Procedure

Plant Material. Both fresh and dormant red or sugar beet (*Beta vulgaris* L.) yield transport-competent tonoplast vesicles. If there is not a ready supply of fresh beets, the roots should be stored in moist vermiculite at 4°. However, fresh beets tend to give the best results.

Homogenization and KI Treatment. Three hundred and thirty grams of beets are thoroughly cooled in ice-cold water, peeled, and cut into 1 × 1 × 3 cm segments and left in ice-cold water until homogenization. All solutions and equipment are precooled. The tissue segments are homogenized at high speed in a precooled Waring blender for 30 sec in 330 ml of homogenization medium and the homogenate is filtered through two layers and then four layers of cheesecloth to remove large cell debris and some of the insoluble PVP. (The relative yield of active tonoplast vesicles decreases with an increase in volume ratio of tissue fresh wt : homogeni-

[1] G. F. E. Scherer and D. J. Morré, *Plant Physiol.* **62,** 933 (1978).

zation medium.) The filtrate is immediately centrifuged at 10,000 *g* for 15 min at 4°. The pellet, containing cell debris and mitochondria, is discarded and the supernatant is centrifuged at 80,000 *g* for 30 min at 4°. After centrifugation, the supernatant is aspirated and the pellet is resuspended with a Teflon pestle homogenizer in suspension medium. The volume is adjusted to 10 ml and 50 ml of 0.18 *M* KI in suspension medium is added with stirring to give a final concentration of 0.15 *M* KI. The membranes are again sedimented at 80,000 *g* for 30 min at 4°. The KI treatment is useful for removing peripheral membrane proteins.

Density Gradient Centrifugation. For the routine preparation of tonoplast vesicles, the KI-treated microsomal pellet is resuspended in 18 ml suspension medium and 3-ml volumes are layered on top of six step gradients of 5 ml 10% (w/w) and 5 ml 23% (w/w) sucrose. After centrifugation in a swing-out rotor at 80,000 *g* for 2 hr at 4°, the membranes at the 10%/23% sucrose interface are removed with a Pasteur pipet. The membranes are diluted 10- to 20-fold with the desired experimental solution, sedimented at 80,000 *g*, and resuspended in 1–2 ml of the same medium to a protein concentration of 0.5–4.0 mg/ml. If plasma membrane as well as tonoplast vesicles are required, the step gradients consist of 2.5 ml of each 10% (w/w), 23% (w/w), 34% (w/w), and 40% (w/w) sucrose. Plasma membrane is collected from the 34%/40% sucrose interface.

Storage. The membranes may be used immediately, or frozen in liquid N_2 and stored at −70°. Freezing and storage have little effect on ATPase activity but increase the passive conductance of the membranes. During experiments, membranes could be kept on ice for 3 to 6 hr before serious deterioration.

Characteristics of Tonoplast Vesicles

The membranes collected from the 10%/23% sucrose have two predominant enzyme activities, an ATPase and a PPase, which are qualitatively similar to those of isolated intact vacuoles. The ATPase[2] is stimulated by halides (Cl^-, Br^-, and I^-), its activity is relatively independent of the nature of the countercation, it has a pH optimum of 8.0, and is inhibited by NO_3^-. The inhibitory action of NO_3^- exhibits pseudocompetitive kinetics with respect to ATP concentration and yields K_i values of 39 and 22 m*M* in the absence and presence of Cl^-, respectively.[3] When expressing the degree of inhibition of a particular membrane preparation by NO_3^-, it is therefore necessary to state the MgATP and Cl^- concentra-

[2] R. J. Poole, D. P. Briskin, Z. Kratky, and R. M. Johnstone, *Plant Physiol.* **74,** 437 (1984).

[3] C. J. Griffith, P. A. Rea, E. Blumwald, and R. J. Poole, *Plant Physiol.* **81,** 120 (1986).

tions employed. Since mitochondrial F_1F_0-ATPases are also subject to inhibition by NO_3^-, it is also imperative to demonstrate that the ATPase activity associated with the tonoplast vesicles is not sensitive to azide, an F_1F_0-ATPase inhibitor. Furthermore, tonoplast ATPase is very sensitive to transmembrane gradients of pH and potential, and is therefore affected by secondary transport of Cl^-, NO_3^-, etc. Direct effects of these ions on the ATPase can be measured after collapsing membrane gradients with ionophores.

The activity of the PPase is stimulated by K^+, relatively independent of the nature of the counteranion, is inhibited by Na^+, and has a broad alkaline pH optimum.[4] In order to accurately assay the PPase activity of a preparation it is important to precisely control the $MgPP_i$ ratio of the reaction system as this enzyme is subject to inhibition by excess Mg^{2+} and free PP_i.[5]

Marker enzyme activities indicate only small amounts of other identifiable membranes in the tonoplast vesicle fraction. Thus, glucan synthase II and vanadate-sensitive ATPase (plasma membrane) are negligible. Cytochrome *c* oxidase activity and azide-sensitive ATPase (mitochondria) are essentially zero. Latent IDPase (Golgi) is very low. Some NADH-cytochrome *c* reductase activity is present, but is also found in intact vacuole preparations.[2]

Applications to H^+-Coupled Transport

For a substrate to be transported across a membrane both a pathway and a driving force are required. Two transport systems serve to build up the proton motive force (PMF) by the direct utilization of metabolic energy (ATP or PP_i). One or both components of the PMF (ΔpH and $\Delta\psi$) serve then as the driving force for various secondary transport processes. Secondary transport mechanisms fall into three basic categories: uniport, symport, and antiport. Ion uniports are systems carrying a net charge driven by $\Delta\psi$, thus tending to collapse $\Delta\psi$. This appears to be the mode of transport of Cl^- and NO_3^- into the vacuole.[6] Symports and antiports are systems carrying both H^+ and substrate, thus they will be driven by ΔpH. It is proposed that the retrieval of NO_3^- from the vacuole is operated by a H^+/NO_3^- symport.[6] In this section we detail methods for measuring pH gradients across the tonoplast membrane and their application for the study of cation/H^+ antiports in this membrane.

[4] P. A. Rea and R. J. Poole, *Plant Physiol.* **77,** 46 (1985).

[5] P. A. Rea and R. J. Poole, *Plant Physiol.* **81,** 126 (1986).

[6] E. Blumwald and R. J. Poole, *Proc. Natl. Acad. Sci. U.S.A.* **82,** 3683 (1985).

ΔpH Measurements

The interior acid pH of tonoplast vesicles is generally determined from the equilibrium distribution of a weak base. At any given pH, the free (B) and protonated forms (BH^+) of a weak base are in equilibrium with the H^+ activity. If B is freely permeable across the membrane and BH^+ is not, BH^+ is trapped in the acidic interior of the vesicle. We have used the fluorescent monoamine acridine orange for the measurement of ΔpH. The method is based on the fluorescence quenching by self-interaction at high internal probe concentrations[7] such that ΔpH across the membrane vesicle is proportional to $\log[Q/(100 - Q)]$, where Q is the percentage of fluorescence quenching.

Tonoplast vesicles (60 μg of membrane protein), 3 m*M* Tris/MES, 250 m*M* sorbitol, and 5 μ*M* acridine orange are added to the sample cell to give a final volume of 2 ml at 25°. Changes in fluorescence are monitored with a Perkin-Elmer spectrofluorimeter model LS-5 at excitation and emission wavelengths of 495 and 540 nm, respectively, and a slit width of 5 nm for both excitation and emission. During the measurement the samples are continuously stirred. For the calibration of the fluorescence quenching, acid-interior pH gradients are imposed by addition of aliquots of concentrated base (e.g., NaOH), and the ΔpH is calculated by measuring the pH in the cuvette before and after addition of base.[6] Such calibrations cannot necessarily be applied directly to measure ΔpH generated by H^+ pumps, because it is likely that only a small proportion of the sealed vesicles participates in ATP-dependent H^+ pumping.[8]

Substrate Loading

Active transport of cations through antiport mechanisms across the tonoplast operates by coupling the flux of cations to the opposite flux of H^+. The measurement of H^+-coupled transport requires, then, prior loading of the vesicles with one of the exchangeable substrates (cation or H^+). By loading the vesicles with H^+ (acid loading), it is possible to follow the recovery of the quenched fluorescence as a function of the addition of cations to the extravesicular medium (cation-dependent H^+ fluxes). A complementary approach is the loading of the vesicles with the cation investigated (cation loading). In these experiments, the vesicles are loaded with the cation and an outwardly directed cation flux is obtained

[7] S. Schuldiner, H. Rottenberg, and M. Avron, *Eur. J. Biochem.* **25,** 64 (1972).

[8] R. J. Poole, R. J. Mehlhorn, and L. Packer, *in* "Biochemistry and Function of Vacuolar Adenosine-Triphosphatase in Fungi and Plants" (B. Marin, ed.), p. 114. Springer-Verlag, Berlin and New York, 1985.

by dilution of the loaded vesicles in a cation-free medium, and the resulting coupled H^+ influx (fluorescence quenching) is measured as described above.

Acid Loading. Vesicles can be loaded with H^+ by two different methods:

1. ATPase dependent ΔpH: a pH gradient (acid inside) can be generated by the activation of the H^+-ATPase. In a routine experiment, tonoplast vesicles (60 μg protein), 30 m*M* Tris/MES, pH 8.0, 250 m*M* sorbitol, 5 μM acridine orange, and 3 m*M* Tris–ATP are added to the sample cell to give a final volume of 2 ml. ATPase activity is initiated by the addition of 3 m*M* $MgSO_4$ and the fluorescence is monitored as described above. After a steady state ΔpH formation is achieved (5 to 10 min), ATP is removed from the system by addition of 2 m*M* D-glucose and 10–20 U of hexokinase (EC 2.7.1.1 from yeast). This method has a serious disadvantage for the measurement of H^+-coupled transport, since there are indications that only a small proportion of the total population of sealed vesicles directly participate in ATP-dependent H^+ pumping.[8] (This could in some cases be an advantage, if it is desired to work with only right-side-out vesicles. The proportion of inside-out vesicles in this preparation has not yet been determined.)
2. Artificial pH gradients (pH jumps): Tonoplast vesicles are loaded with a buffer (say, pH 6.0) of desired ionic composition by suspension and sedimentation at 100,000 *g* for 30 min at 4°. Final membrane pellets are resuspended in the same loading medium to 6 mg protein/ml and incubated at 4° for 2 hr. A preset ΔpH (2 U) is obtained by mixing tonoplast vesicles (30 to 60 μg protein) loaded with buffer at pH 6.0 with a buffer solution at pH 8.0.[9,10]

Cation Loading. For Na^+ loading, for example, tonoplast vesicles are suspended in a buffer containing 150 m*M* sodium gluconate (prepared by titrating solutions of D-gluconic acid lactone with NaOH), 250 m*M* sorbitol, 5 m*M* Tris/MES, pH 8.0, and sedimented, resuspended, and incubated as described for pH jumps.[9]

Cation/H^+ Antiport in Tonoplast

This section deals with the characterization of an $1Na^+/1H^+$ antiport. A similar experimental approach has been used for the characterization of an $1Ca^{2+}/2H^+$ antiport in the same membrane preparation.[10]

[9] E. Blumwald and R. J. Poole, *Plant Physiol.* **78,** 163 (1985).
[10] E. Blumwald and R. J. Poole, *Plant Physiol.* **80,** 727 (1986).

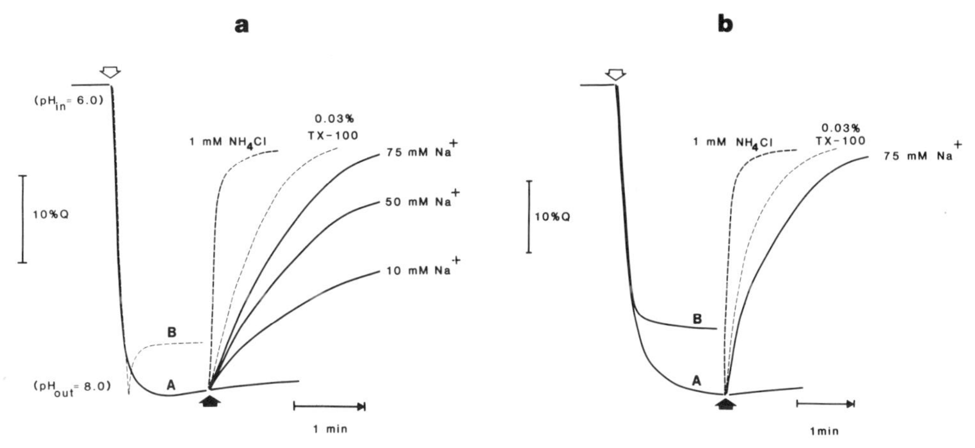

FIG. 1. (a) H^+ gradient-dependent quenching of acridine orange fluorescence, and its reversal by Na^+. At the indicated time (open arrow), a ΔpH of 2 U is obtained by a pH jump in the absence (A) and presence (B) of K^+-valinomycin. At the indicated time (filled arrow), aliquots of Na_2SO_4, NH_4Cl, or Triton X-100 were added and the subsequent recovery of the fluorescence was followed. (b) Quenching of acridine orange fluorescence by dilution of Na^+-loaded tonoplast vesicles. At the indicated time (open arrow), Na^+-loaded vesicles were diluted in a Na^+-free medium in the absence (A) and presence (B) of K^+-valinomycin. At the indicated time (filled arrow), aliquots of Na_2SO_4, NH_4Cl, or Triton X-100 were added and the subsequent recovery of the fluorescence was followed. Methodology described in text.

A $1H^+/1Na^+$ antiport is a carrier that catalyzes the exchange of Na^+ and H^+, i.e., translocates Na^+ and H^+ in opposite directions across the tonoplast. Since it has been shown that the initial $\%Q$/unit time is proportional to the initial H^+ flux,[11,12] the effects of Na^+ on the dissipation of a transmembrane pH gradient, or the development of a pH gradient in response to outwardly oriented Na^+ gradients, can be studied by following the pH-dependent fluorescence quenching of acridine orange.

The effects of Na^+ on the dissipation of a transmembrane pH gradient are measured as follows: 10 μl of tonoplast vesicles loaded with a solution containing 250 mM sorbitol, 1 mM DTT, and 5 mM Tris/MES, pH 6.0, is added to 2.0 ml containing 250 mM sorbitol, 5 mM Tris/MES, pH 8.0, and 5 μM acridine orange. The addition of vesicles loaded at pH 6.0 to a pH 8.0 buffer causes quenching of acridine orange fluoresence (pH jump) (Fig. 1a). The quenched fluorescence is restored upon addition of differ-

[11] J. M. Wolosin and J. G. Forte, *J. Membr. Biol.* **71,** 195 (1983).

[12] A. B. Bennett and R. M. Spanswick, *J. Membr. Biol.* **71,** 95 (1983).

ent Na^+ concentrations. Recovery of fluorescence about the same extent as that obtained with 75 m*M* Na^+ is obtained by collapsing the transmembrane pH gradient with 1 m*M* NH_4Cl or 0.03% Triton X-100.

The dissipation of the preset pH gradient by Na^+ can be due to the action of an Na^+/H^+ antiport and/or to electrically driven H^+ movements through conductive pathways in the isolated tonoplast membranes, i.e., an H^+ diffusion driven by an Na^+ diffusion potential. To distinguish between these pathways, the dissipation of the pH gradient by increasing Na^+ concentrations in the external medium should be measured in the presence of equimolar K^+ concentrations across the membrane and valinomycin, thus in conditions where the membrane potential is zero. For this, 20 m*M* potassium gluconate is added to the loading buffer (pH 6.0), and 20 m*M* potassium gluconate and 3–5 μM valinomycin are added to the dilution buffer (pH 8.0) (trace B).

The development of pH gradients in tonoplast vesicles in response to outwardly oriented Na^+ gradients is measured as follows: 10 μl tonoplast vesicles loaded with a solution containing 150 m*M* sodium gluconate, 100 m*M* sorbitol, 1 m*M* DTT, and 5 m*M* Tris/MES, pH 8.0, is diluted 200-fold in a sodium-free medium containing 2.0 ml of 400 m*M* sorbitol, 5 m*M* Tris/MES, pH 8.0, and 5 μM acridine orange. Dilution of sodium loaded vesicles results in quenching of acridine orange fluorescence (Fig. 1b). The quenched fluorescence is restored upon addition of 75 m*M* Na^+ or by collapsing the transmembrane pH gradient with NH_4Cl or 0.03% Triton X-100. If equimolar K^+ concentrations across the membrane and valinomycin are present during dilution, the resulting trace (trace B) represents the electroneutral Na^+/H^+ antiport activity.

Kinetics are measured as the rate of dissipation of the pH gradient upon addition of Na^+ after a preset ΔpH was obtained by pH jumps. Aliquots from stocks of 0.5 *M* Na_2SO_4 solution, buffered at pH 8.0, are injected during continuous fluorescence recording. The initial rates are determined by drawing the tangents of the recorded traces obtained in the first 10 sec. These rates should be corrected for possible artifacts by subtraction of rates obtained by addition of an impermeable cation (i.e., tetramethylammonium), and are expressed as rate of change in fluorescence (%*Q*) per minute. The electroneutral Na^+-dependent H^+ flux displayed a Michaelis–Menten kinetics relationship to Na^+ concentration with an apparent K_m of 7.5 m*M*.[9]

By subtracting the values of the coupled exchange in the presence of K^+-valinomycin from the values of total Na^+–H^+ exchange in the absence of K^+-valinomycin, the rate of $\Delta\psi$-dependent exchange can be calculated. These values, which should represent passive ion conductances in tonoplast vesicles, must be considered with caution, since they may be a product of the isolation procedure and therefore may not represent the *in vivo* properties of the vacuolar membrane.

Acknowledgments

Supported by the Natural Sciences and Engineering Research Council of Canada and the Department of Education of Quebec.

[13] Purification and Characterization of the Tonoplast H^+-Translocating ATPase

By STEPHEN K. RANDALL and HEVEN SZE

Introduction

An electrogenic H^+-translocating ATPase is associated with the vacuolar membrane (tonoplast) of plant tissues.[1] This enzyme acidifies the vacuolar lumen and provides a proton motive force for active transport of various solutes across the tonoplast. The properties of this ATPase[2,3] differ from those of the mitochondrial (F_1F_0 type)[4] and the plasma membrane (E_1E_2 type) H^+-ATPases.[1] Purification of the tonoplast ATPase is the first step toward understanding its structure, function, and biosynthesis at the molecular level.

Here we describe a purification procedure that takes advantage of the low density of isolated tonoplast vesicles and the large mass of the solubilized tonoplast ATPase. The tonoplast ATPase is relatively abundant in the cell and is estimated to represent 4–8% of the total tonoplast protein. The following protocol includes several improvements not previously reported.[3]

Methods

Purification of Tonoplast-Enriched Vesicles

Solutions

1. Homogenization buffer: 250 m*M* D-mannitol, 6 m*M* EGTA, 0.1% (w/v) bovine serum albumin (BSA), 0.1 m*M* phenylmethylsulfonyl

[1] H. Sze, *Annu. Rev. Plant Physiol.* **36,** 175 (1985).
[2] Y. Wang and H. Sze, *J. Biol. Chem.* **260,** 10434 (1985).
[3] S. K. Randall and H. Sze, *J. Biol. Chem.* **261,** 1364 (1986).
[4] S. K. Randall, Y. Wang, and H. Sze, *Plant Physiol.* **79,** 957 (1985).

fluoride (PMSF), 50 mM HEPES–bistrispropane (HEPES–BTP) at pH 7.4, and 1 mM DTT (added immediately before use)

2. Resuspension buffer: 250 mM D-mannitol, 0.5 mM DTT, 0.1% BSA, 0.1 mM PMSF, 2.5 mM HEPES–BTP at pH 7.2
3. 6% (w/w) Dextran (Sigma Chemical Co., 79,000 Da) made in 2.5 mM HEPES–BTP at pH 7.2 and 250 mM D-mannitol

Oat seeds (*Avena sativa* var. Lang) are germinated over an aerated solution of 0.5 mM $CaSO_4$.[5] Three- to 5-day-old oat roots are excised and washed three times with cold deionized water. All operations are carried out at 0–4°. The roots (50 g fresh wt) are homogenized twice with a mortar and pestle using a medium-to-tissue ratio of 1.5 to 2.0 ml/g. The brei is strained through four layers of cheesecloth, and centrifuged for 5 min at 1000 *g*. To remove mitochondria, the supernatant is centrifuged first at 6000 *g* for 15 min, and then at 12,000 *g* for 10 min. Membranes in the resulting supernatant are then pelleted by centrifugation at 60,000 *g* for 30 min (SW 28, r_{max}). The pellets are gently resuspended in 1 ml resuspension buffer (for some tissues, vesicles must be incubated at 35° for 3–5 min to obtain a uniform suspension), then diluted to 24 ml with cold resuspension buffer. Aliquots of 6 ml are layered onto 10 ml of a 6% (w/w) Dextran cushion. Gradients are centrifuged (slow acceleration and slow brake) for 2 hr at 70,000 *g* (SW 28.1, r_{max}). The 0/6% Dextran interfaces from four tubes are collected, diluted 10- to 20-fold with resuspension buffer, and pelleted at 85,000 *g* for 30 min (SW 28, r_{max}). The pelleted membranes, enriched in tonoplast ATPase,[2,3] are resuspended to a concentration of 4 mg protein/ml. About 1–2 mg vesicle protein is recovered from 50 g of roots. The membrane-bound ATPase may be stored for at least 4 weeks at −76° without loss in activity. The ATPase activity of this preparation is relatively insensitive to orthovanadate and azide, indicating absence or low levels of plasma membrane and mitochondrial ATPases.[1-3]

Solubilization and Purification of the Tonoplast ATPase

Solutions

1. Solubilization buffer: 10% (v/v) Triton X-100, 60% (v/v) glycerol, 2 mM BTP–HEPES at pH 7.4 (Tris can be substituted for BTP), and 0.3 mM ATP
2. Elution buffer: 0.2% (v/v) Triton X-100, 30% (v/v) glycerol, 2 mM BTP–HEPES at pH 7.4, 0.1 mM DTT, 0.15 mM ATP, and 0.1 mg/ml asolectin. Asolectin (Sigma type IV) is partially purified[6] and

[5] T. K. Hodges and R. T. Leonard, this series, Vol. 32, p. 392.

[6] C. I. Ragan, this series, Vol. 63F, p. 715.

stored in $CHCl_3$ under N_2 (at −20°). Before use, it is dried as a thin film under N_2, and then placed *in vacuo* for at least 3 hr. The dried phospholipids are resuspended in elution buffer

One volume of ice-cold solubilization buffer (0.5 ml) is added to an equal volume of tonoplast membranes (0.5 ml) and mixed well. The solution is allowed to stand on ice for 30 min, and is then centrifuged for 1 hr at 105,000 *g* (Beckman Ty 65, r_{max}). The supernatant is considered the crude solubilized ATPase preparation.

The supernatant (1.0 ml) is loaded on a Sepharose CL-6B column (1.5 × 65 cm) equilibrated with the elution buffer at 4°. The solubilized proteins are eluted with a flow rate of 10 ml/hr and fractions of 0.6 ml are collected. The ATPase activity is completely eluted in 5–6 hr. Activity is monitored using a coupled assay (see the next section). The leading portion of the peak ATPase activity (Fig. 1) is collected and combined. The tonoplast ATPase is purified at least 15-fold relative to the microsomal pellet, and 22-fold if it is expressed as NO_3^--sensitive activity (Table I).[2,3]

Speed is critical in this purification step as activity is rapidly lost at 4°.[3] Sepharose CL-6B is preferred over Sepharose 4B[3] because the cross-linked beads permit a faster flow rate. Glycerol and phospholipids are essential for the maintenance of activity during gel filtration. Addition of DTT, ATP, or $MgSO_4$ in the solubilization and elution buffers appears to stabilize the ATPase. However, to study the effect of *N*-ethylmaleimide (NEM) or 7-chloro-4-nitrobenzo-2-oxa-1,3-diazole (NBD-Cl) on the puri-

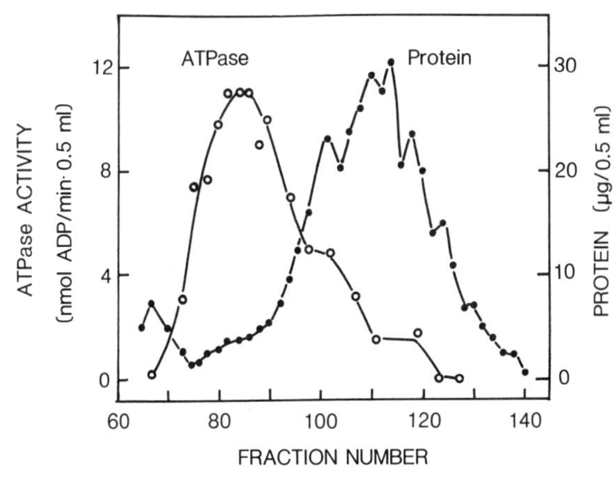

FIG. 1. Partial purification of the solubilized tonoplast ATPase by chromatography on Sepharose CL-6B. Fractions were assayed with the coupled assay in the absence of additional phospholipids. Fractions containing the highest specific activity (No. 72–82) were collected and used to characterize the purified enzyme.

TABLE I
PURIFICATION OF THE TONOPLAST ATPase FROM MICROSOMAL VESICLES OF OAT ROOTS[a]

Fraction	Protein (mg)	Total activity (units)	Specific activity (units/mg)		Purification (-fold)	
			Control	ΔNO_3^-	Control	ΔNO_3^-
Microsomal fraction	10.00	2.37	0.24	0.11	1.0	1.0
0/6% Dextran interface vesicles	2.18	1.10	0.50	0.25	2.1	2.3
Triton X-100 extract	1.72	0.66	0.39	0.24	1.6	2.2
Sepharose CL-6B peak	0.032	0.12	3.75	2.45	15.6	22.3

[a] ATPase activities were measured in the presence of 50 m*M* choline chloride. The Triton-solubilized enzyme was assayed with 0.2 mg/ml phospholipids, and the purified preparation was assayed with 0.03–0.1 mg/ml phospholipids.[3] Fractions equivalent to tubes 72 to 82 of Fig. 1 were combined and referred to as the Sepharose CL-6B peak. Data are the average of two or three experiments. Nitrate sensitivity (ΔNO_3^-) was measured as the difference in activity in the presence and absence of 20 m*M* BTP–NO_3.

fied enzyme, DTT and ATP should be absent from the elution buffer. These changes do not alter the apparent mass of the ATPase. Approximately 50% of ATPase activity is lost under these conditions, and several additional polypeptides are found in this purified ATPase preparation.

Based on recent experiments in our laboratory, several additional modifications of the procedure described above are suggested. Separation of tonoplast proteins by Sepharose CL-6B chromatography at 10–12° (compared to 4°) reduces elution time and increases the activity recovered. Since low-ionic-strength solutions cause dissociation of the peripheral subunits from the membrane sector (see below), we suggest the buffer concentration in the solubilization and elution media be increased to 25 m*M*.

ATPase Assay

ATPase activity can be measured spectrophotometrically by the lactate dehydrogenase–pyruvate kinase coupled assay[7,8] or by determining

[7] M. E. Pullman, H. S. Penefsky, A. Datta, and E. Racker, *J. Biol. Chem.* **235,** 3322 (1960).
[8] L. Josephson and L. C. Cantley, *Biochemistry* **16,** 4572 (1977).

the rate of inorganic phosphate liberation.[5] The coupled assay using an ATP-regenerating system is relatively more sensitive and is preferred since ADP is a competitive inhibitor of the tonoplast ATPase.[2] In the presence of pyruvate kinase, each molecule of ADP released by ATP hydrolysis accepts a phosphate group from phosphoenolpyruvate to form pyruvate and regenerate ATP. With excess lactate dehydrogenase, each mole of pyruvate is rapidly reduced by 1 mol of NADH to form lactate and NAD^+. Formation of NAD^+ is followed by the decrease in absorbance at 340 nm ($\varepsilon_{red-ox} = 6.22$ mM^{-1} cm^{-1}). Using known amounts of ADP, it can be established that 1 μmol of ADP yields 1 μmol of NAD^+. One unit of ATPase activity is defined as the liberation of 1 μmol ADP/min. Specific activity is expressed as units/milligram protein.

The procedure of Josephson and Cantley[8] is used with some modifications.[3] A 0.5-ml reaction mixture contains 30 m*M* HEPES-BTP (pH 7), 50 m*M* potassium iminodiacetate (pH 7), 4.5 m*M* $MgSO_4$, 2 m*M* ATP–BTP (pH 7), 0.66 m*M* phosphoenolpyruvate, 0.2 m*M* NADH, 2.5 U pyruvate kinase, and 6 U lactate dehydrogenase. The phospholipid concentration required for optimal activity depends on the Triton content.[3] Generally, the optimal phospholipid concentration for the crude solubilized and the purified preparation is 0.1 to 0.2 mg/ml and 0.05 to 0.1 mg/ml, respectively. The mixture is equilibrated at 24°, and the reaction is started with 0.2 to 1 μg of purified ATPase. After a constant rate is established, 10 μl of a 2.5 *M* choline chloride solution is added (final concentration, 50 m*M*) to show Cl^- stimulation. The effect of 0.5 m*M* sodium azide or 0.1 m*M* orthovanadate can be tested in the same reaction mixture. To determine NO_3^- sensitivity, 20 μl of a 1.0 *M* BTP–NO_3 (final concentration, 40 m*M*) is added. Other inhibitors should be tested to determine that they do not alter pyruvate kinase or lactate dehydrogenase activity. (Note: Effects of high concentrations of NBD-Cl and NEM cannot be measured with the coupled assay due to their absorbance near 340 nm.)

Analysis of Subunit Composition by SDS–Polyacrylamide Gel Electrophoresis

The purity and subunit composition of the enzyme preparation obtained previously is determined by acrylamide gel electrophoresis using established procedures.[9] Triton and glycerol are removed by precipitating the proteins with TCA followed by 90% cold acetone. Acetone is removed *in vacuo* and the pellet is resuspended in sample buffer[9] containing 1.0 *M* urea[3] and boiled for 2 to 4 min. Immediately before electrophoresis, samples are centrifuged at 12,000 *g* (5 min) to remove aggregates.

[9] U. K. Laemmli, *Nature (London)* **227,** 680 (1970).

TABLE II
PROPERTIES OF MEMBRANE-BOUND AND PURIFIED TONOPLAST ATPase[a]

		ATPase activity (%)	
Inhibitor	Concentration	Membrane bound	Purified
None	—	100	100
Nitrate	5 mM	77	32
	20 mM	45	10
NEM	100 μM	10	20
NBD–Cl	1 μM	87	70
	10 μM	45	35
DCCD	1 μM	85	85
	10 μM	25	55
	100 μM	10	12
Azide	5 mM	100	99
Vanadate	100 μM	85	98

[a] Activities were usually measured with the coupled assay in the presence of 50 mM choline chloride at pH 7.0. Activities in the presence of NEM and NBD–Cl were monitored by P_i release. Since the membrane-bound tonoplast ATPase activity was insensitive to vanadate or azide, the effect of other inhibitors was routinely assayed in the presence of 0.1 mM vanadate and 0.5 mM azide to eliminate ATPase activity from the plasma membrane and mitochondria, respectively.[2,4] Nonspecific acid phosphatases were inhibited with 0.1 mM ammonium molybdate. Specific activities of the membrane-bound and purified tonoplast ATPase were 0.43 and 2.8 U/mg, respectively.

Properties of the Oat Tonoplast ATPase

Substrate Specificity and Stability. The purified enzyme hydrolyzes ATP (K_m = 0.25 mM) preferentially and GTP partially.[3] It is stable for at least 4 weeks at −76° but loses activity with a half-time of 20 hr at 4°.

Activator and Inhibitors. The tonoplast ATPase is insensitive to azide and orthovanadate (Table II). It is stimulated by Cl^- and is inhibited by NO_3^-, 7-chloro-4-nitrobenzo-2-oxa-1,3-diazole (NBD-Cl), *N*-ethyl maleimide (NEM), and *N,N'*-dicyclohexylcarbodiimide (DCCD).[3] Although the tonoplast ATPase has a pH optimum at 7.0 to 7.5, it is most sensitive to NBD-Cl and NEM at pH 8.0.[10]

Physical Properties. The ATPase holoenzyme is estimated to be 400–

[10] S. K. Randall and H. Sze, *J. Biol. Chem.* **262,** in press (1987).

500 kDa, assuming a single Triton X-100 micelle of 90 kDa[11] is associated with the enzyme during gel filtration (Fig. 1). The purified tonoplast ATPase consists of at least three polypeptides of 72, 60, and 16 kDa (Fig. 2). Minor components of 40 kDa and 13 kDa also copurify with ATPase activity, suggesting the holoenzyme is made up of as many as five polypeptides. The 72-kDa subunit may contain the catalytic site based on our studies[10] and those of others.[12,13] Although the 60-kDa subunit contains an ATP- or nucleotide-binding site, its function is unclear.[10,14] The 16-kDa polypeptide binds to dicyclo[^{14}C]hexylcarbodiimide, and is postulated to function in proton conductance.[3] Studies in our laboratory (Shoupeng Lai, unpublished results) indicate that the 72, 60, and 40 kDa polypeptides are peripheral proteins, while the 16 and 13 kDa polypeptides are integral. This suggests that the structure of the tonoplast ATPase is analogous to that of the F_1F_0-ATPases.[15] Immunological data show that not only are the various plant tonoplast ATPases related to each other (Randall, unpublished data) but that the plant tonoplast ATPase is immunologically related to ATPases localized on endocytic vesicles of animal cells.[16] The purification procedure described here may be applicable to a wide variety of vacuolar H^+-ATPases.[1,17]

Comments

Purification of the tonoplast ATPase by gel filtration is simple and rapid, and should be widely applicable to other plant tissues. A similar procedure was used for the purification of the tonoplast ATPase from red beet root.[14]

Triton and glycerol interfere with many enzyme and protein assay procedures, and therefore must be removed. Protein concentrations can be determined by the Lowry procedure,[18] after precipitating twice with trichloroacetic acid[19] to remove glycerol. Sodium dodecyl sulfate (SDS,

[11] A. J. Furth, *Anal. Biochem.* **109,** 207 (1980).

[12] S. Mandala and L. Taiz, *J. Biol. Chem.* **261,** 12850 (1986).

[13] E. J. Bowman, S. Mandala, L. Taiz, and B. J. Bowman, *Proc. Natl. Acad. Sci. U.S.A.* **83,** 48 (1986).

[14] M. F. Manolson, P. A. Rea, and R. J. Poole, *J. Biol. Chem.* **260,** 12273 (1985).

[15] L. M. Amzel and P. L. Pedersen, *Annu. Rev. Biochem.* **52,** 801 (1983).

[16] M. F. Manolson, J. M. Percy, D. K. Apps, X.-S. Xie, D. K. Stone, and R. J. Poole, *in* "Membrane Proteins: Proceedings of the Membrane Protein Symposium (1986)" (S. C. Goheen, L. Hjelmeland, M. McNamee, and R. Gennis, eds.). Bio-Rad, in press.

[17] I. Mellman, R. Fuchs, and A. Helenius, *Annu. Rev. Biochem.* **55,** 663 (1986).

[18] O. H. Lowry, N. J. Rosebrough, N. J. Farr, and R. J. Randall, *J. Biol. Chem.* **193,** 265 (1951).

[19] A. Bensadoun and D. Weinstein, *Anal. Biochem.* **76,** 241 (1970).

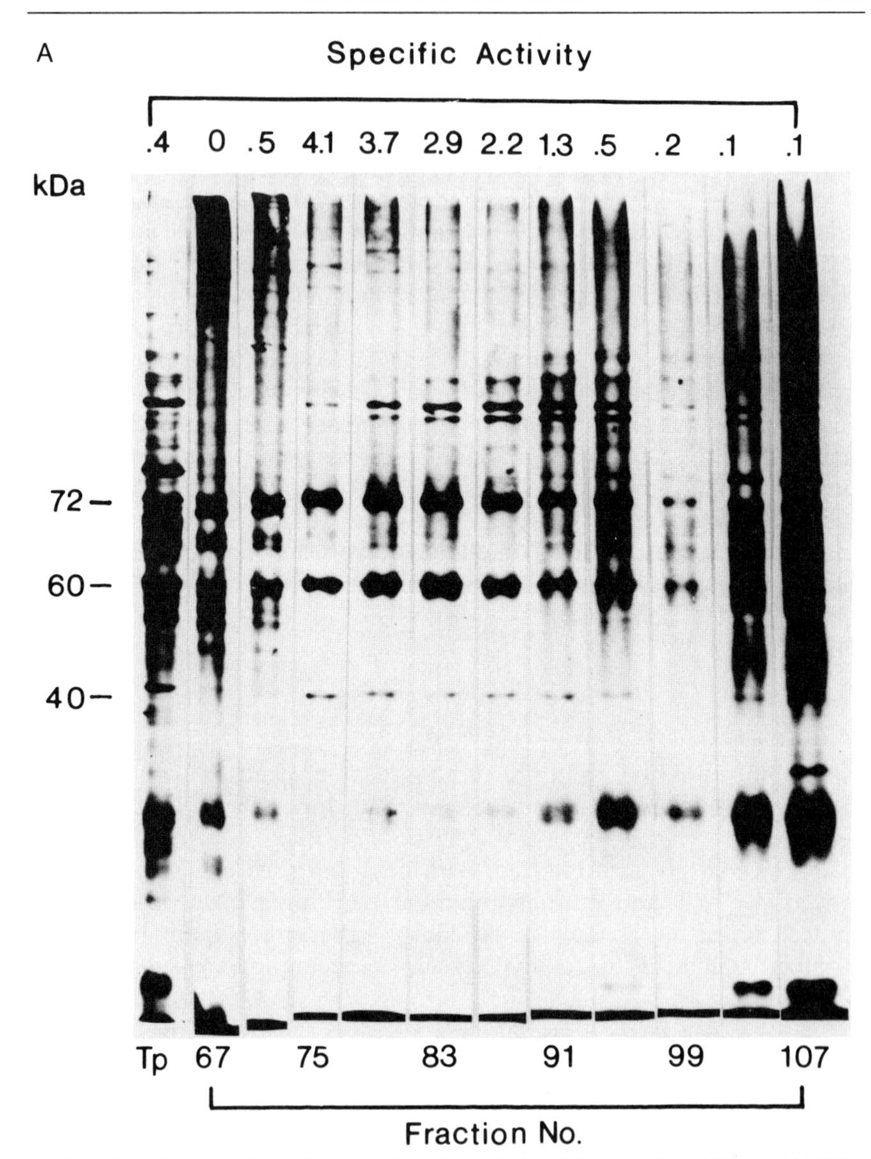

FIG. 2. Polyacrylamide gel electrophoretic analysis of the tonoplast ATPase. (A) Polypeptide profile of tonoplast membranes and purified ATPase. Tp, Tonoplast membranes purified on a 6% Dextran cushion. Tube numbers refer to fractions (#67–107) eluted from the Sepharose CL-6B column shown in Fig. 1. Seventy microliters of each fourth fraction was analyzed. Specific activities (units/milligram) of each fraction are shown. Proteins were resolved on a 10% acrylamide gel and stained with silver, according to T. Marshall, *Anal. Biochem.* **139,** 506 (1984). (B) [^{14}C]DCCD-labeled polypeptides of tonoplast membranes and purified tonoplast ATPase. Triton X-100 was removed from the purified tonoplast ATPase

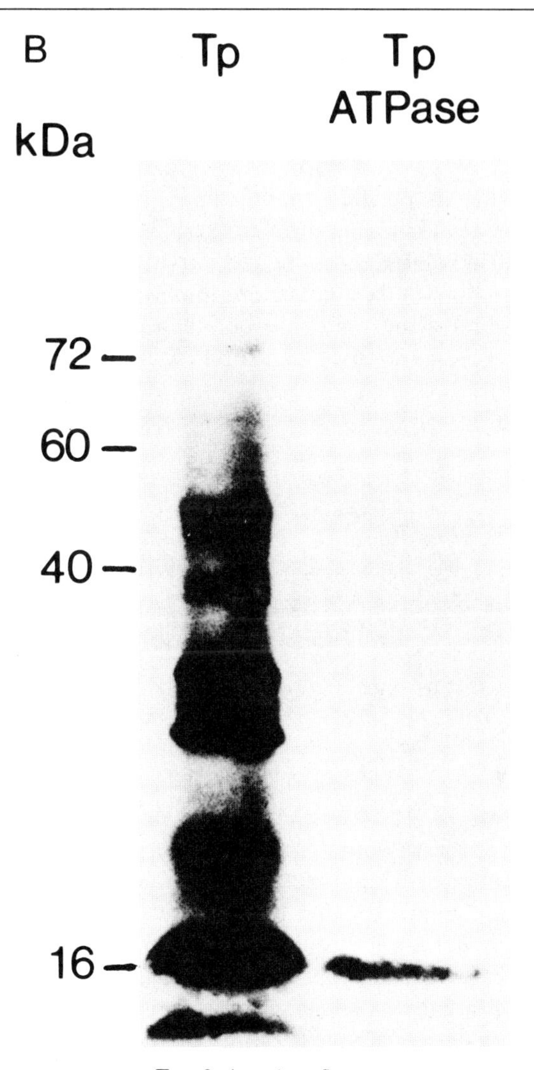

FIG. 2. (*continued*)

by treatment with 20 mg/ml Amberlite XAD-2 for 2 hr at 25°. Membranes (150 μg protein) and purified ATPase (9 μg protein) were treated with 70 μM [^{14}C]DCCD for 2 hr (4°), pelleted with 5% trichloroacetic acid, and washed with 100% cold acetone.[3] After electrophoresis, gels (7.5–15% acrylamide) were washed with water, soaked in Autofluor, dried, and exposed to Kodak XAR-5 film for 2 weeks at −76°. Reprinted from Randall and Sze[3] by permission of the American Society of Biological Chemists, Inc.

1%) is included in the assay to prevent precipitation of Triton X-100. If ATPase activity is assayed by inorganic phosphate release,[5,20] 1% SDS must be included in the acid–molybdate solution. To minimize activity from cation-stimulated ATPases, the tonoplast ATPase from crude membrane fractions is assayed with BTP or Tris salts. This precaution is unnecessary with the purified enzyme.

Octylglucoside (35 m*M*) and Zwittergent 3-14 (6 m*M*) were as effective as Triton X-100 in solubilizing ATPase activity. However, sodium cholate (2–180 m*M*), sodium deoxycholate (2–15 m*M*), CHAPS or CHAPSO (5–12 m*M*), and chloroform or dichloromethane (0.5 vol) were ineffective. Some of these detergents may be useful for reconstitution studies.

[20] C. H. Fiske and Y. Subbarow, *J. Biol. Chem.* **66,** 375 (1925).

[14] High-Performance Liquid Chromatography for Simultaneous Kinetic Measurements of Adenine Nucleotides in Isolated Vacuoles

By M. Hill, A. Dupaix, P. Volfin, A. Kurkdjian, and B. Arrio

Introduction

The vacuoles, i.e., the tonoplast and the cell sap, represent an osmotic system which provides some rigidity to the plant cells. Numerous substances are contained in the vacuoles; some of them are highly concentrated like sugars, alkaloids, and organic acids, which may be representative of the accumulation processes in vacuoles.[1,2] Many data in the literature suggest that transport is energy dependent and linked to proton pumping by a tonoplast ATPase (for a review, see Sze[3]). But, in a complex system like plant vacuoles, the hydrolysis of ATP is catalyzed not only by ATPases but also by other phosphorolytic enzymes which are not involved in the energization process of the membrane. Thus, the estimation of ATPase activities is an important and difficult task, since the results from many usual titration methods are more or less specific for the ATP hydrolysis into ADP. The less specific methods are obviously the

[1] F. Marty, D. Branton, and R. A. Leigh, *in* "Biochemistry of Plants" (P. K. Stumpf and E. E. Conn, eds.), p. 625. Academic Press, New York, 1980.

[2] C. A. Ryan and M. Walker-Simmons, this series, Vol. 96, p. 580.

[3] H. Sze, *Annu. Rev. Plant. Physiol.* **36,** 175 (1985).

titration of phosphate by colorimetric techniques derived from the widely used Fiske and Subbarow method[4–9] or by employing ^{32}P as a tracer.[10,11] The hydrolysis of *p*-nitrophenylphosphate has been described for ATPase activity measurements in some particular membranes free of other activities.[12,13] The most specific is the ATP titration based on the luminescence occurring in the luciferin–luciferase assay[14,15] or the ADP titration by a reaction coupled to an adenylate kinase.[16,17] Whether or not the titration is specific for ATP, the data show different aspects of the adenine nucleotide hydrolysis catalyzed by vacuolar enzymes. It is possible to remove this uncertainty by simultaneously monitoring the evolvement of the adenine nucleotides, in the presence of purified tonoplast vesicles or cell sap, by using reversed-phase (RP) HPLC and rapid isocratic elution. As can be expected, there are no standard conditions: retention time and elution order of nucleotides vary with experimental conditions and columns used. The procedures described in this chapter are compromises between the best separation and the shortest retention times observed on three commercially available columns.

HPLC Technique

Chromatographic Instrumentation

HPLC was carried out with two Gilson model 303 pumps controlled by a gradient manager 702 steered by an Apple IIe computer, a dynamic mixer, and a Rheodyne injection valve 7125 with a 20-μl filling loop. Each sample was manually injected with a 10-μl Hamilton syringe. Nucleotides

[4] C. H. Fiske and Y. Subbarow, *J. Biol. Chem.* **66,** 375 (1925).
[5] S. Mandala and L. Taiz, *Plant Physiol.* **78,** 327 (1985).
[6] B. N. Ames, this series, Vol. 8, p. 115.
[7] I. Struve, A. Weber, U. Lüttge, E. Ball, and J. A. C. Smith, *J. Plant. Physiol.* **117,** 451 (1985).
[8] H. H. Taussky and E. Shorr, *J. Biol. Chem.* **202,** 675 (1953).
[9] B. P. Marin, *Biochem. J.* **229,** 459 (1985).
[10] O. Lindberg and L. Ernster, *Methods Biochem. Anal.* **3,** 1 (1956).
[11] M. Thom and E. Komor, *Eur. J. Biochem.* **138,** 93 (1984).
[12] D. Brethes, J. Chevallier, and J. P. Tenu, *Biochimie* **61,** 109 (1979).
[13] G. Inesi, *Science* **171,** 901 (1971).
[14] A. Hager and W. Biber, *Z. Naturforsch., C: Biosci.* **39C,** 927 (1984).
[15] B. L. Strehler, *in* "Methods of Enzymatic Analysis" (H. U. Bergmeyer, ed.), p. 559. Academic Press, New York, 1965.
[16] H. Adam, *in* "Methods of Enzymatic Analysis" (H. U. Bergmeyer, ed.), p. 573. Academic Press, New York, 1965.
[17] X. Gidrol, B. Marin, H. Chrestin, and J. d'Auzac, *Z. Pflanzenphysiol.* **114,** 279 (1984).

were detected at 259 nm and absorbance values were monitored with a Beckman model 165 variable wavelength detector.

Columns and Mobile Phases. All separations were performed at room temperature. The columns were protected by a guard column Brownlee RP18 (7 μm, 3.2 × 1.5 cm). The eluents were degassed and filtered through a 0.5-μm Millipore filter. The mobile phases and columns used are summarized in Table I. The columns were equilibrated for 15 min with the eluents before the first injection and they were used all day without regeneration. When sample runs were terminated, the columns were flushed by linear gradient elution up to 100% methanol for 60 min; then, if required, returned to initial conditions in 30 min, before the next series.

Chemical and Chromatographic Standards. Methanol, HPLC grade, was obtained from Carlo Erba. Water was deionized and distilled. Sodium dihydrogen phosphate, ammonium heptamolybdate, magnesium chloride,

TABLE I
COLUMNS AND PHASES FOR HPLC

Columns	Eluents	Flow rates (ml/min)	Figures
Zorbax Bio Series: C_8 PEP-RP/1, 5 μm, 6.3 mm × 8 cm (Du Pont)	0.1 *M* NaH_2PO_4, pH 6, 4% (v/v) methanol	1	1a
	0.1 *M* KH_2PO_4, pH 6, 22% (v/v) methanol, 25 m*M* tetra-*n*-butylammonium hydrogen sulfate	1.5	2a
Supelcosil LC-18-DB: 5 μm, 4.6 mm × 15 cm (Supelco)	0.1 *M* NaH_2PO_4, pH 5, 6% (v/v) methanol	1	1b
	0.1 *M* KH_2PO_4, pH 7, 18% (v/v) methanol, 25 m*M* tetra-*n*-butylammonium hydrogen sulfate	2	2b
μBondapak C_{18}: 10 μm, 3.9 mm × 15 cm (Waters Associates)	0.1 *M* NaH_2PO_4, pH 6, 4% (v/v) methanol	2	1c
	0.1 *M* KH_2PO_4, pH 6, 20% (v/v) methanol, 25 m*M* tetra-*n*-butylammonium hydrogen sulfate	1.8	2c

potassium chloride, ammonium acetate, perchloric acid (PCA), tris(hydroxymethyl)aminomethane (Tris), and tetra-*n*-butylammonium hydrogen sulfate, analytical grade, were purchased from Merck. 2-*N*-Morpholinoethanesulfonic acid (MES), adenosine 5′-triphosphate (ATP), adenosine 5′-diphosphate (ADP), and adenosine 5′-monophosphate (AMP) were purchased from Sigma.

Stock Solutions. ATP, ADP, and AMP stock solutions (6 m*M*) were prepared by dissolving the dried material in 50 m*M* Tris–MES buffer, pH 7.5, containing 5 m*M* $MgCl_2$, 50 m*M* KCl, and 2 m*M* ammonium molybdate when specified. Nucleotide concentrations were calculated using the molar absorbance coefficient at 259 nm: 15,400 $M^{-1} \cdot cm^{-1}$.

Preparation of Protoplasts, Vacuoles, and Tonoplast Isolation

Protoplasts are isolated from an *Acer pseudoplatanus* suspension of cultured cells at the exponential phase of growth according to a technique previously described.[18]

Vacuoles are obtained from protoplasts by osmotic shock according to the procedure of Wagner and Siegelman[19] and modified in our laboratory.

Contamination of the vacuole suspensions by whole protoplasts and large protoplast debris has been estimated to be <1% as detected by a light microscope. The vacuole suspension usually contained 10^6 vacuoles/ml.

Protoplasts and vacuoles from *Catharanthus roseus* were prepared according to Renaudin *et al.*[20] with the following modification. Since Nicodenz (Nyegaard, Oslo, Norway) is eluted later than the nucleotides, the delay preceding the next sample injection increased. Therefore, in the two-step gradient, 10% Ficoll is used to replace 7.5% Nicodenz and 10 m*M* HEPES buffer, pH 7.3, containing 0.55 *M* sorbitol and 2.5 m*M* 2-mercaptoethanol, is layered above the 10% Ficoll. After centrifugation at 600 *g* for 10 min vacuoles are collected in the upper layer (10^6 vacuoles/ml). The contamination by protoplasts was <1% as estimated by light microscopy.

Tonoplast Vesicles. The vacuole suspensions are twice diluted in 50 m*M* Tris–MES buffer, pH 7.5, containing 5 m*M* $MgCl_2$ and 50 m*M* KCl. Then the suspensions are homogenized with a small manual potter ho-

[18] A. C. Kurkdjian and H. Barbier-Brygoo, *Anal. Biochem.* **132,** 96 (1983).

[19] G. J. Wagner and H. W. Siegelman, *Science* **190,** 1298 (1975).

[20] J. P. Renaudin, S. C. Brown, H. Barbier-Brygoo, and J. Guern, *Physiol. Plant.* **68,** 695 (1986).

mogenizer (Thomas Co., Philadelphia) in an ice bath. Tonoplast vesicles were recovered by centrifugation at 100,000 *g* and 4° for 5 hr. The pellet was suspended in 700 μl of 50 m*M* Tris–MES buffer, pH 7.5, containing 5 m*M* $MgCl_2$ and 50 m*M* KCl, using the same potter.

Protein Concentrations. The modified method of Bradford was used to estimate the amount of proteins in the various isolated fractions.[21]

Kinetic Assays of Adenine Nucleotidase Activities

In a typical assay, the reaction was started by mixing, under magnetic stirring, 150 μl of pellet or supernatant with 150 μl of the 6 m*M* ATP, ADP, or AMP stock solution. The reactions were performed at 20° in 1.5-ml conical centrifuge tubes. In due time, a 20-μl aliquot was withdrawn and mixed with 20 μl of 10% PCA, to stop the reaction. After centrifugation for 2 min at 15,000 *g* and 20°, 20 μl of supernatant was adjusted to pH 5.5 with 20 μl of 4 *M* ammonium acetate and frozen until HPLC analysis. For the nucleotide separation, 5 μl of the last centrifuged solution was injected. The analysis times were dependent on the elution method: about 3 min with an ion-pair reagent and 4 min without an ion-pair reagent.

Examples and Discussion

An ATP-dependent proton pump has been implicated in the energization processes of the tonoplast. Thus, data on ATP levels and other adenine nucleotides are useful. Two separation methods of adenine nucleotides on RP18 (ODS) columns were performed frequently: (1) in the presence of ion-pair reagents with phosphate buffer/methanol,[22,23] phosphate buffer/acetonitrile,[24,25] and propanol/water as eluents[26]; and (2) in the absence of ion-pair reagents with phosphate buffer/methanol as eluents.[27–29] Both methods were tested in this study. The first method separated complex mixtures of nucleotides and their degradation prod-

[21] S. M. Read and D. H. Northcote, *Anal. Biochem.* **116,** 53 (1981).

[22] D. R. Webster, G. D. Boston, and D. M. Paton, *J. Liq. Chromatogr.* **8**(4), 603 (1985).

[23] O. C. Ingebretsen and A. M. Bakken, *J. Chromatogr.* **242,** 119 (1982).

[24] G. H. R. Rao, J. D. Peller, K. L. Richards, J. McCullough, and J. G. White, *J. Chromatogr.* **229,** 205 (1982).

[25] E. Juengling and H. Kammermeier, *Anal. Biochem.* **102,** 358 (1980).

[26] J. D. Schwenn and H. G. Jender, *J. Chromatogr.* **193,** 285 (1980).

[27] V. Stocchi, L. Cucchiarini, M. Magnani, L. Chiarantini, P. Palma, and G. Crescentini, *Anal. Biochem.* **146,** 118 (1985).

[28] G. Crescentini and V. Stocchi, *J. Chromatogr.* **290,** 393 (1984).

[29] R. A. Hartwick, S. P. Assenza, and P. R. Brown, *J. Chromatogr.* **186,** 647 (1979).

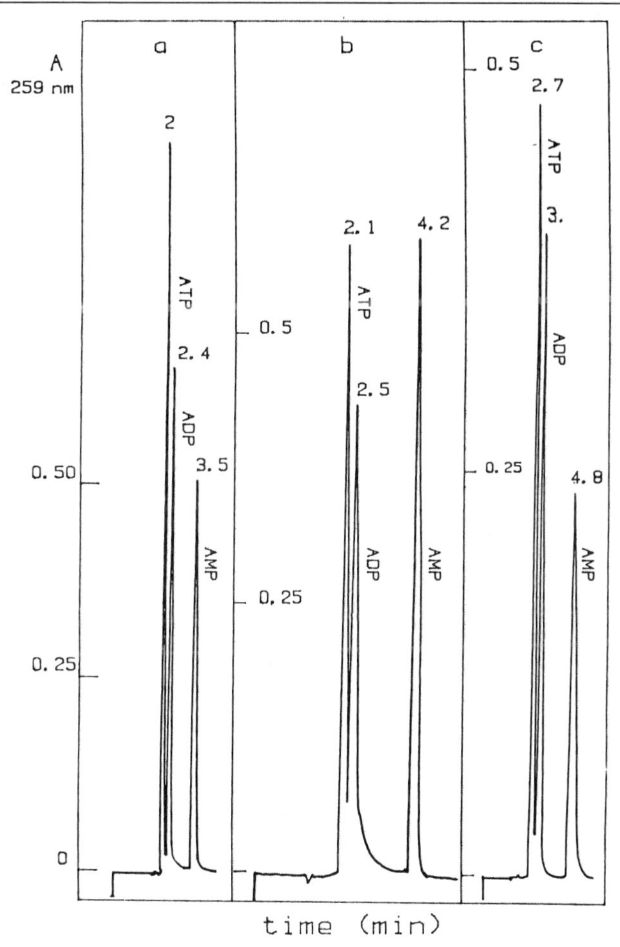

FIG. 1. Reversed-phase high-performance liquid chromatography in the absence of tetrabutylammonium hydrogen sulfate. Separation of a standard mixture of nucleotides: Injection of 10 μl of a solution containing 0.47 m*M* ATP, 0.46 m*M* ADP, 0.45 m*M* AMP. The conditions are detailed in the text. (a) Zorbax Bio Series C_8 PEP.RP/1 column; (b) Supelcosil LC-18-DB column; and μBondapak C_{18} column.

ucts. The elution order in the presence of the ion-pair is AMP, ADP, ATP (Fig. 2a, b, and c). The second method was chosen when the adenine nucleotides were ATP, ADP, and AMP (Fig. 1a–c). Other reversed-phase propylsilyl (C_3) and octylsilyl (C_8) columns were evaluated in terms of resolution, peak shape, selectivity, and reproducibility column to column from the same manufacturer. The three selected columns (Figs. 1 and 2) were adequate to separate numerous samples in the minimum lapse time.

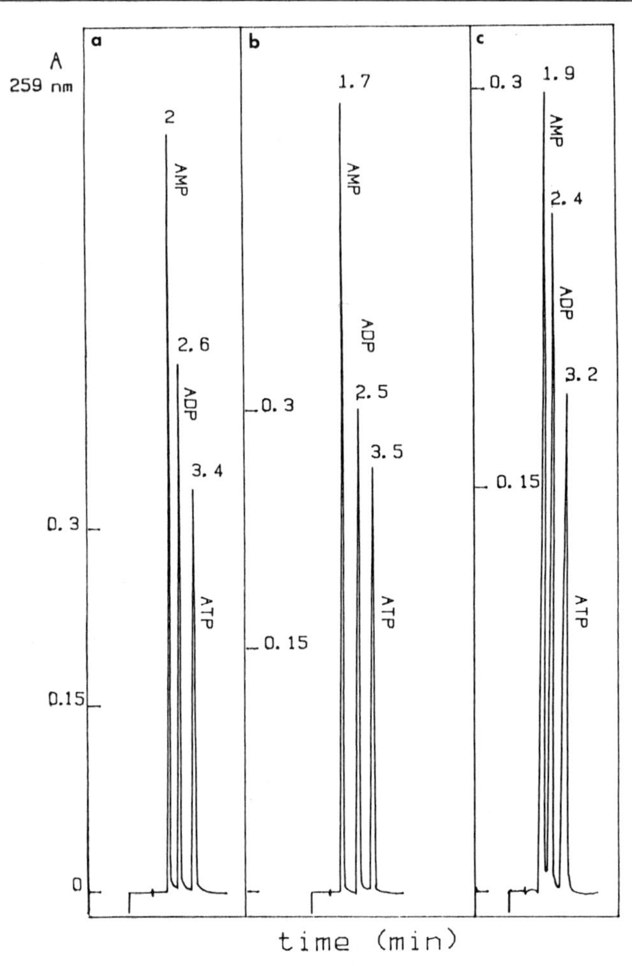

FIG. 2. Reversed-phase high-performance liquid chromatography in the presence of tetrabutylammonium hydrogen sulfate. Separation of a standard mixture of nucleotides: Injection of 10 μl of a solution containing 0.47 m*M* ATP, 0.46 m*M* ADP, 0.45 m*M* AMP. The chromatographic conditions are detailed in the Columns and Mobile Phases section. (a) Zorbax Bio Series C_8 PEP.RP/1 column; (b) Supelcosil LC-18-DB column; (c) μBondapak C_{18} column.

The lowest ATP, ADP, or AMP amount detectable was about 2 pmol. Since the incubation times might be long, the stability of the nucleotides was checked by incubating them for the same time without membranes. The extent of ATP and ADP hydrolysis was estimated to be about 2.5 and 0.5%, respectively, after a 6-hr incubation. AMP remained stable under

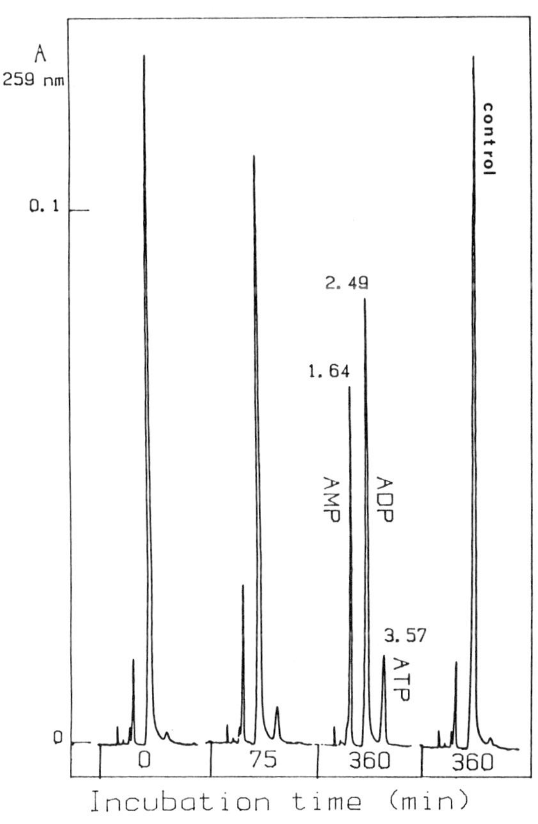

FIG. 3. Sample chromatograms illustrating the hydrolysis of ADP, for various incubation times, in the presence of tonoplast proteins (15 μg) from *Catharanthus roseus*. (ADP is contaminated by 5% AMP and 0.5% ATP.) The chromatographic conditions are the same as in Fig. 2b.

the same conditions. The run-to-run precision of the retention times was 1% within a day and 2% day to day. Routinely, 8 to 10 kinetic assays could be simultaneously performed and 100 samples were analyzed daily.

The chromatograms presented in Fig. 3 are a typical example of adenine nucleotide separation during ADP hydrolysis in the presence of tonoplast vesicles. Kinetic analyses obtained from such chromatograms are presented in Fig. 4. Besides nucleotide hydrolysis, a concomitant ATP synthesis occurs when ADP is the initial substrate. This synthesis may be due to an adenylate kinase-like activity. These phenomena are observed, whatever the cells (*Acer* or *Catharanthus*), in the presence or in the absence of 1 m*M* molybdate.

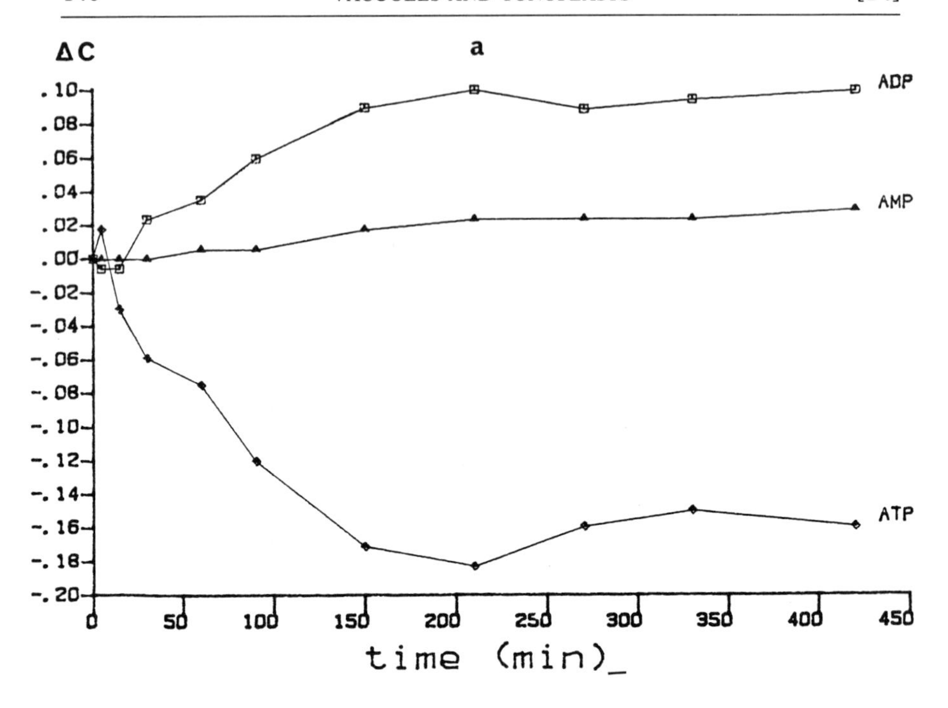
ΔC
a
ADP
AMP
ATP
.10
.08
.06
.04
.02
.00
-.02
-.04
-.06
-.08
-.10
-.12
-.14
-.16
-.18
-.20
0
50
100
150
200
250
300
350
400
450
time (min)

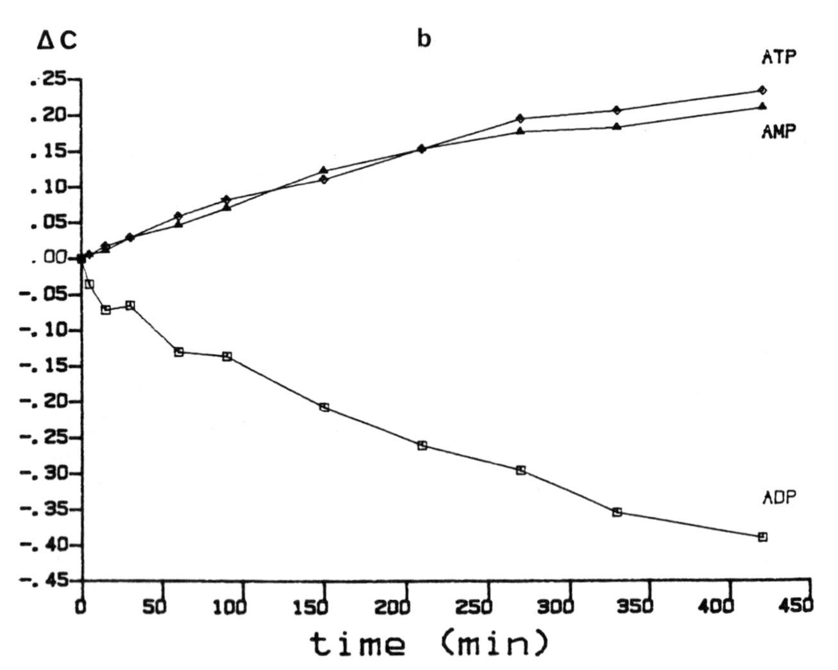
ΔC
b
ATP
AMP
ADP
.25
.20
.15
.10
.05
.00
-.05
-.10
-.15
-.20
-.25
-.30
-.35
-.40
-.45
0
50
100
150
200
250
300
350
400
450
time (min)

Using the RP-HPLC technique for simultaneous kinetic measurements of ATP, ADP, and AMP, several enzymes involved in hydrolysis and synthesis of these nucleotides are revealed. All these observations emphasize the fact that complex kinetics bring different conclusions, whether or not they are monitored by specific techniques. From the data in the literature, the titration of phosphate, usually at a single reaction time, cannot obviously describe the involvement of the nucleotides. This chromatographic technique could be useful in characterizing the adenine nucleotide-related enzymes during tonoplast preparations. Difficulties in this field have been pointed out by Sze[3] and Boller.[30]

[30] T. Boller, *Physiol. Veg.* **20,** 247 (1982).

FIG. 4. Variation of adenine nucleotides concentration ΔC (mol/mg protein) $\times 10^3$ versus incubation time (in minutes). Incubation medium: 0.3 ml of 50 m*M* Tris–MES buffer, pH 7.5, containing 5 m*M* $MgCl_2$, 50 m*M* KCl, 1 m*M* Mo ammonium heptamolybdate, and 11 μg tonoplast proteins from *Acer pseudoplatanus*. (a) ATP, initial concentration 3 m*M*; (b) ADP, initial concentration 3 m*M*.

Section III

Plastids

[15] Isolation of Intact Chloroplasts: General Principles and Criteria of Integrity

By DAVID A. WALKER, ZORAN G. CEROVIC, and SIMON P. ROBINSON

Introduction

The intact chloroplast comprises three distinct entities: (1) The thylakoid membranes, which contain the pigments, components of the electron-transport chain, and the coupling factor for photophosphorylation; (2) the stroma, which contains soluble proteins including those necessary for photosynthetic carbon reduction, plus metabolites and other salts; and (3) the envelope membranes, which enclose the organelle and also contain the metabolite translocators. Many partial reactions of photosynthesis require only one of these entities but it is preferable to isolate intact chloroplasts even if these are subsequently ruptured to obtain one of the components. Isolation of intact chloroplasts is a rapid and relatively simple purification step but it also protects the internal components of the chloroplast from degradative enzymes and substances released when cells are homogenized. Deliberate rupture of intact isolated chloroplasts yields thylakoid membranes which exhibit high rates of electron transport and photophosphorylation and a stromal fraction which has high activities of enzymes of the carbon reduction cycle. For isolation of envelope membranes, isolation of intact chloroplasts as a first step is mandatory.

The isolation of intact chloroplasts has been reviewed previously[1-5] and the basic methodology has not altered significantly since its development in the 1960s. The general principles will be restated in this chapter together with two recent areas of advance, namely the purification of chloroplasts and the isolation of chloroplasts from a wider range of tissues. Those interested in working with a particular species of plant should persevere in attempts to isolate chloroplasts from that species although it should be realized that even with recent advances it is still not possible to isolate chloroplasts from many economically important plants. Where the interest lies in chloroplast properties in general, the use of spinach is still

[1] D. A. Walker, this series, Vol. 23, p. 211.

[2] D. A. Walker, this series, Vol. 69, p. 94.

[3] S. P. Robinson, G. E. Edwards, and D. A. Walker, *Methodol. Surv. Biochem.* **9,** 13 (1979).

[4] R. G. Jensen, *Encycl. Plant Physiol., New Ser.* **2,** 31 (1979).

[5] R. C. Leegood and D. A. Walker, *in* "Isolation of Membranes and Organelles from Plant Cells" (J. L. Hall and A. L. Moore, eds.), p. 185. Academic Press, London, 1983.

preferred. Intact chloroplasts with high rates of photosynthesis are readily isolated from spinach but, more importantly, there is already a vast background literature on the biochemistry of spinach chloroplasts. Where growth of spinach is not possible, peas may be a useful alternative and in some instances, such as studies of protein synthesis or plastid development, the use of a developing and rapidly growing tissue such as young shoots of peas may be preferred. For both spinach and peas, isolation of chloroplasts is best achieved by mechanical grinding of the tissue. Intact chloroplasts capable of high rates of photosynthesis can be prepared from many species, such as cereals, only by first digesting the leaves with cellulase and pectinase to obtain protoplasts and then rupturing these to release the chloroplasts.[3-5]

Plant Material

As stated previously,[1-5] whatever the method employed, growing the plants is of critical importance. The metabolic activity of isolated chloroplasts cannot exceed that of the parent tissue but, in addition, healthy and rapidly growing plants generally have softer leaves which contain smaller amounts of interfering substances (such as phenolics, oxalate, etc.), and thus more readily yield intact chloroplasts with high rates of photosynthesis.

Spinach. It should be noted that the name "spinach" is applied to a great many species other than *Spinacea oleracea* (e.g., "perpetual spinach" is a cultivar of *Beta vulgaris*) but that only *Spinacea oleracea* offers real advantages as a source of mechanically isolated chloroplasts. Many varieties of *Spinacia oleracea* can be used successfully, one of the commonly adopted ones being US National Hybrid 74, from Ferry Morse Seed Company, Mountain View, CA, also sold in the UK and Australia as Yates Hybrid 102. Another variety which can be recommended for water culture is "Virtuosa" from Rijk Zwaan, De Lier, Holland.

It is often erroneously supposed that spinach will not tolerate high temperatures despite the fact that it may have originated in the Middle East. On the contrary, provided that the critical day length (which differs according to variety) is not exceeded by a large factor and it is not subject to water stress or mineral deficiency it will grow extremely well in temperatures up to 40°. The seed contains an endogenous inhibitor of germination so it should be washed for at least 1–2 hr and preferably surface sterilized for 5–10 min in 3% hypochlorite solution to reduce the chance of infection to seedlings. The washed seed is sown in only slightly moistened vermiculite and germinated in the dark for 5–7 days at 20–25°. Vermiculite needs to be kept moist but not overwatered as this reduces germination. Once germinated the seedlings are kept for a further 5–7 days in the light in a growth cabinet or glasshouse before transplanting into hydroponic culture when the cotyledons are fully expanded and the first pair of leaves have just emerged. The composition of the nutrient solution for water culture is not critical but one convenient method is to use two separate stock solutions:

A. 800 m*M* $Ca(NO_3)_2$, 10 m*M* $FeNa(EDTA)_2$
B. 1200 m*M* KNO_3, 200 m*M* KH_2PO_4, 400 m*M* $MgSO_4$, 10 m*M* H_3BO_3, 2 m*M* $MnCl_2$, 0.2 m*M* $ZnSO_4$, 0.1 m*M* $CuSO_4$, 0.02 m*M* Na_2MoO_4

These can be stored indefinitely in a refrigerator. Adding 5 ml (A) plus 5 ml (B) per liter of water gives the following final concentrations: 4 m*M* $Ca(NO_3)_2$, 50 μM $FeNa(EDTA)_2$, 6 m*M* KNO_3, 1 m*M* KH_2PO_4, 2 m*M* $MgSO_4$, 50 μM H_3BO_3, 10 μM $MnCl_2$, 1 μM $ZnSO_4$, 0.5 μM $CuSO_4$, 0.1 μM Na_2MoO_4. A commercially available mixture, Solufeed (ICI Plant Protection Division, UK Department, Surrey, UK) is also a convenient source of most mineral nutrients.

The containers of solution need to be continually aerated (e.g., by an aquarium pump) and kept dark to prevent algal growth. They should hold at least 1 liter of nutrient solution per plant and be kept free of fungal infection. One way of minimizing infection is to use disposable plastic bags to line the containers. Alternatively, seeds may be planted in a mixture of Perlite, Levington (peat) compost and vermiculite in disposable expanded polystyrene drinking cups with perforated floors and watered from above until germination is advanced. The cups are then transferred, in supporting racks, to tanks of water culture solution so that the bottom 1–2 cm of the perforated cups are immersed for 1 or 2 hr a day only. For the rest of the day the solution in the tanks is lowered to a level just below the cups. The developing roots then grow downward into the solution but overwatering of the hypocotyl region is avoided. The diurnal change in level is easily effected by periodic pumping of solution from a lower reservoir to which it is returned via a large overflow (during pumping) and a small overflow. The position of these overflow vents determines the upper and lower levels.

Spinach will tolerate a reasonably wide range of temperature, humidity, and light, but the day length must be restricted to prevent flowering. If virtually total darkness is imposed in the night period (e.g., in a growth cabinet or by using a mechanical shutter) some varieties will tolerate a day length up to 14 hr without flowering. The conditions should be chosen to give rapid growth of the plants. This can be achieved in a glasshouse or in a growth cabinet with fluorescent plus incandescent lights to give an intensity of at least 250 $\mu E \cdot m^{-2} \cdot sec^{-1}$ (PAR). Supplementary artificial light, even of relatively low intensity, may increase growth in winter months in a glasshouse. Best growth is not achieved at day lengths much lower than the critical day length despite the fact that flowering is then completely suppressed. Suitable leaves should be available 2–4 weeks after transplanting the seedlings. As an alternative to water culture, spinach can be grown, though with some difficulty, by nutrient film technique.[2,6] Spinach will also grow in pots of open draining soil or gravel but the growth rates are normally not as high as in water culture.

Peas. Peas have the advantages of much shorter growing periods and less complicated requirements for growth. Seeds can be germinated in trays of moist Vermiculite or Perlite[2,7] or on perforated trays floating on tapwater.[8] Suitable shoots (2–5 cm high) are obtained 9–12 days after planting seed. Given constant environmental conditions, extremely uniform material can be grown on a continuous basis. Most varieties grow well in a glasshouse or in growth cabinets with a 10- to 14-hr photoperiod, temperatures of 20–25°, and light intensity of at least 250 $\mu E \cdot m^{-2} \cdot sec^{-1}$. Because of their large seed size and short growth period peas do not normally require additional nutrients.

Harvesting and Preparation. Ideally, leaves should be harvested just prior to use. Spinach can sometimes be stored successfully in a cold room for days or even weeks but peas undergo rapid postharvest changes which preclude even short-term storage. When leaves cannot be used immediately steps should be taken to minimize water loss, e.g., place leaves in a plastic bag with a small amount of water and chill if possible.

A white ring in the chloroplast pellet indicates the presence of either starch or oxalate. This can be checked with iodine, which stains starch blue-black but does not react with oxalate. If large amounts of starch are present the yield of intact chloroplasts will be reduced, presumably because the rapidly accelerating starch grains rupture the envelope membranes during centrifugation. Excessive starch in leaves harvested at the beginning of the photoperiod is generally indicative of unhealthy or slow growing plants. Starch may be reduced by changing growth conditions; starch accumulation is favored by high day temperatures, low night temperatures, high light, and long photoperiods. If all else fails, leaf starch can be reduced by dark pretreatment of the leaves for 24–48 hr.

Preillumination of the tissue can increase chloroplast yield[2] and decrease the lag phase of photosynthesis. This is preferably achieved without detaching the leaves simply by harvesting the first 1–3 hr of the

[6] A. Cooper, "The ABC of NFT." Grower Books, London, 1979.
[7] S. P. Robinson and J. T. Wiskich, *Plant Physiol.* **58**, 156 (1976).
[8] Z. G. Cerovic and M. Plesnicar, *Biochem. J.* **223**, 543 (1984).

photoperiod and using the leaves immediately. If detached leaves are preilluminated care must be taken to avoid water loss and heating of the leaves. Detached spinach leaves can be floated on cold running water and illuminated with a water bath below the lights to prevent overheating. In some tissues it may be advantageous to chill the leaves by immersing in ice water before grinding.

Isolation Methods

Isolation Media. With tissues such as spinach and peas, a wide variety of media have been used successfully in many laboratories, reflecting the fact that the choice of osmoticum, buffer, and salts is not absolutely critical. Most media have 0.33 *M* sorbitol as osmoticum and have an osmotic potential of −1.0 MPa, which is the osmotic potential of the cell sap of spinach leaves grown hydroponically. For other tissues, it is important to determine the osmotic potential of the cell sap and ensure that the osmotic potential of the medium is matched to that of the leaves. A variety of other sugars, sugar alcohols, and salts can be used to provide osmotic support and recent experiments suggest that the use of 200 m*M* KCl in place of sorbitol can lead to improved rates of CO_2 fixation.[9] Any one of a number of buffers can be employed but pyrophosphate, MES, or HEPES are the most commonly used. For spinach, the most simple effective grinding medium is still that described by Cockburn *et al.*[10] in 1968: 330 m*M* sorbitol, 10 m*M* $Na_4P_2O_7$, 5 m*M* $MgCl_2$, 2 m*M* sodium isoascorbate, adjusted to pH 6.5 with HCl. A wide variety of media have been used to isolate pea chloroplasts[2,7,8] but the above grinding medium is also suitable.

Grinding. The aim is to break as many cells but as few chloroplasts as possible. This is most readily achieved with homogenizers such as the Polytron (Kinematica, Switzerland; Brinkmann, USA) and Ultraturrax (IKA, Janke and Kunkel, West Germany). Waring or kitchen blenders can also be used but generally incur longer blending and give lower yields. Deribbed spinach leaves (30–50 g) are diced into pieces roughly 3 cm square and ground with 200 ml medium for 1–5 sec in a square Perspex vessel (5 × 5 × 20 cm). The grinding medium can be frozen into a slushy consistency to reduce cavitation by the blender and to allow the latent heating of melting to offset heat generated by the mechanical action of the blender.

Filtration and Centrifugation. The brei should be quickly filtered to remove debris and cells. Muslin, Miracloth (Calbiochem, USA) and nylon

[9] S. P. Robinson, *Photosynth. Res.* **10,** 93 (1986).
[10] W. Cockburn, D. A. Walker, and C. W. Baldry, *Plant Physiol.* **43,** 1415 (1968).

mesh (20–50-μm pore size) have all been used successfully. A layer of cotton wool is usually included as its fibers tend to retain thylakoid membranes and nuclei but not intact chloroplasts.

The chloroplasts should then be separated from the filtrate as quickly as possible. It is preferable to use a swing-out rotor and unscratched glass tubes in order to minimize breakage of chloroplasts by contact with the walls of the centrifuge tubes. If the tubes are only half-filled the centrifugal path is reduced and this improves the yield of intact chloroplasts. Normally the filtrate (150–200 ml) is divided among four 100-ml tubes and centrifuged for 1 min at 1500 *g* before being braked as quickly as possible. Alternatively, the centrifuge can be rapidly accelerated to 6000 *g* and immediately braked so that the total centrifugation time is about 90 sec. Longer times or greater forces will increase the yield but will also decrease the percentage of intact chloroplasts and increase contamination by other organelles such as mitochondria and peroxisomes. The chloroplast pellet will also be compacted and more difficult to resuspend.

Resuspension. The supernatant is poured off and any froth adhering to the sides of the tubes wiped out. The upper portion of the pellet generally contains a greater proportion of broken chloroplasts and this can be gently rinsed off with a small quantity of grinding or resuspension medium. Resuspension is best achieved by gently shaking the tubes or by using a soft paintbrush or a rod wrapped with cotton wool. The resuspending medium influences stability of the chloroplasts (see below) and is normally similar to the assay medium but can include BSA. One suitable medium is 330 m*M* sorbitol, 2 m*M* EDTA, 1 m*M* $MgCl_2$, 1 m*M* $MnCl_2$, 50 m*M* HEPES–KOH, pH 7.6, plus 0.2% BSA. The chloroplasts are normally resuspended to give a chlorophyll concentration of 2–4 mg/ml and stored on ice.

Purification. The procedures described above result in a preparation which contains 60–90% intact chloroplasts, depending on the skills of the experimenter, but which is also contaminated, to some extent, by mitochondria, peroxisomes, and other organelles. Further purification of the chloroplasts has often been avoided because the extra time and washing procedure result in loss of photosynthetic activity.[2,3,5] This is probably caused by loss of low-molecular-weight substances from the chloroplasts. Although the chloroplast envelope was considered impermeable to most ions it is now clear that K^+,[11] Mg^{2+},[12] and P_i[13] all diffuse out of isolated chloroplasts. This is hardly surprising since considerable concentration gradients can be established across the chloroplast envelope when the cell

[11] S. P. Robinson and W. J. S. Downton, *Arch. Biochem. Biophys.* **228,** 197 (1984).

[12] R. J. Deshaies, L. E. Fish, and A. T. Jagendorf, *Plant Physiol.* **74,** 956 (1984).

[13] G. Mourioux and R. Douce, *Plant Physiol.* **67,** 470 (1981).

is ruptured and the chloroplast is released into a large volume of grinding medium.

With the introduction of silica sol gradients[13-18] it is possible to purify chloroplasts rapidly and also improve photosynthetic activity. The following method, based on that of Mills and Joy,[16] yields preparations of chloroplasts from spinach or peas which are greater than 95% intact and capable of high rates of CO_2 fixation.[19] The pellets from the first centrifugation are resuspended in 6 ml of resuspension medium and placed into two 50-ml glass centrifuge tubes. Resuspension medium containing 40% (v/v) Percoll (Pharmacia) is layered under the chloroplast suspension by slowly injecting 4 ml into the bottom of each tube using a syringe with a long needle. The tubes are centrifuged at 1700 *g* for 1 min, then brought to rest without braking. Broken chloroplasts form a band at the Percoll layer whereas intact chloroplasts are pelleted. The supernatant, including the Percoll layer, is aspirated and the pellet of intact chloroplasts resuspended. This purification procedure can be achieved in 5–10 min and results in increased percentage of intact chloroplasts, increased rates of CO_2 fixation, and decreased contamination of the preparation by other organelles and by degradative enzymes.[16,20]

Chloroplasts can also be purified on linear gradients of silica sols, resulting in preparations of intact chloroplasts with high rates of CO_2 fixation and with very low contamination by other organelles.[13-18] Such gradients are preferable where purity is important but for routine preparation of chloroplasts the use of a Percoll cushion described above is quick and also cheap since only small quantities of Percoll are required. Purification does reduce yield; in the procedure described only 30–40% of the chlorophyll is pelleted,[20] which is less than expected since the percentage of intact chloroplasts in the initial chloroplast preparation would normally be at least 60%. The purification may select not only for intact chloroplasts but also for those with higher density. It should also be noted that there is one report of Percoll inhibiting photosynthesis,[21] although this has not been the general experience. The inhibition was overcome by dialysis of the Percoll.[21]

[14] J. J. Morgenthaler, C. A. Price, J. M. Robinson, and M. Gibbs, *Plant Physiol.* **54,** 532 (1974).
[15] T. Takabe, M. Nishimura, and T. Akazawa, *Agric. Biol. Chem.* **43,** 2137 (1979).
[16] W. R. Mills and K. W. Joy, *Planta* **148,** 75 (1980).
[17] C. A. Price *et al.,* this volume [16].
[18] K. W. Joy and W. R. Mills, this volume [17].
[19] S. P. Robinson, *Plant Physiol.* **75,** 425 (1984).
[20] S. P. Robinson, *Plant Physiol.* **70,** 645 (1982).
[21] M. Stitt and H. W. Heldt, *Plant Physiol.* **68,** 755 (1981).

Low Ionic Strength Media. Nakatani and Barber[22] found that highly intact chloroplast preparations could be obtained by using media of low ionic strength. The separation of intact and broken chloroplasts during centrifugation is apparently enhanced in low ionic strength media and this may reflect differences in surface charge or the fact that chloroplast thylakoid membranes destack in the absence of Mg^{2+}, resulting in swollen (and therefore less dense) broken chloroplasts. There is no evidence that low-ion media reduce contamination of the chloroplast preparation by other organelles. The possibility of loss of ions such as Mg^{2+} and K^+ from chloroplasts would be increased in low-ion media and this may be the basis of the requirement for high levels of cations for CO_2 fixation in "low-salt" chloroplasts[23,24] although it should be noted that such low-salt preparations are substantially different from those prepared according to the procedures of Cerovic and Plesnicar[8] below. Jensen[4] reported decreased stability of chloroplasts isolated in low-ion media but this may be overcome by the inclusion of cations in the resuspension media.[8] If highly intact chloroplasts are required and the use of silica sol gradients is not appropriate the following media can be used successfully with spinach or pea: grinding medium, 340 m*M* sorbitol, 0.4 m*M* KCl, 0.04 m*M* EDTA, 2 m*M* HEPES–KOH, pH 7.8; resuspension medium, 330 m*M* sorbitol, 10 m*M* KCl, 1 m*M* EDTA, 1 m*M* $MgCl_2$, 1 m*M* $MnCl_2$, 10 m*M* KH_2PO_4, 50 m*M* HEPES–KOH, pH 7.9, 1% BSA. The full advantage of low ionic strength media is only secured if they are used throughout, i.e., as grinding medium followed by resuspension and recentrifugation in the same medium. Even in the hands of the inexperienced and without the more sophisticated blenders, etc., these procedures[8] will normally yield 90% or more intact chloroplasts.

Isolation from Plants Other than Spinach. The mechanical method has almost exclusively been applied to spinach and peas. It is to be hoped that techniques such as silica sol gradient purification and the use of protective agents will extend the usefulness of this technique in the future, but it must be acknowledged that some leaves may never yield active chloroplasts by these procedures because of the presence of interfering substances and degradative enzymes, or the difficulty in disrupting hardened leaves without rupturing chloroplasts. Intact chloroplasts capable of reasonable rates of photosynthesis have recently been isolated from sugarbeet,[25] from the halophyte *Suaeda australis*,[26] and from petunia[27] using

[22] H. Y. Nakatani and J. Barber, *Biochim. Biophys. Acta* **461,** 510 (1977).

[23] W. M. Kaiser, W. Urbach, and H. Gimmler, *Planta* **149,** 170 (1980).

[24] B. Demmig and H. Gimmler, *Plant Physiol.* **73,** 169 (1983).

[25] S. P. Robinson, *Photosynth. Res.* **4,** 281 (1983).

[26] S. P. Robinson and W. J. S. Downton, *Aust. J. Plant Physiol.* **12,** 471 (1985).

[27] S. P. Robinson, unpublished observations (1985).

minor modifications of the methods outlined above. The success in these cases resulted more from the use of soft, rapidly growing leaves from plants grown in water culture than from any improvements in isolation technique. With any new tissue, the initial aim must be to isolate intact chloroplasts but this does not guarantee photosynthetic activity. Preparations of highly intact chloroplasts can be isolated from sunflower but rates of CO_2 fixation are poor despite the fact that both electron transport and carbon cycle activity are equal to those in spinach chloroplasts.[28] Similarly, preparations of greater than 95% intact chloroplasts but with only low rates of CO_2 fixation have also been isolated from lettuce and New Zealand spinach (*Tetragonia tetragoniodes*) leaves by the methods described above.[27] In such instances it seems likely that the low photosynthetic activity may be the result of damage to the chloroplast envelope or of loss of low-molecular-weight compounds. It is of interest that chloroplasts capable of high rates of CO_2 fixation can be isolated from sunflower by the protoplast method.[29] In some tissues, the use of protoplasts as an intermediate step may be the only way to prepare photosynthetically active chloroplasts.

Protoplast Method. The use of protoplasts has greatly extended the range of tissues from which active chloroplasts can be obtained and in some tissues, such as C_3 grasses, C_4 and CAM species, the use of protoplasts is mandatory. The same general considerations apply whether chloroplasts are isolated from protoplasts or by the mechanical disruption of tissue but the use of protoplasts offers two distinct advantages. First, the mechanical force needed to rupture protoplasts is much less than that required to disrupt leaf tissue; hence protoplasts can be disrupted by gentle procedures which greatly reduce the breakage of chloroplasts. Second, any interfering substances present in cell walls, vascular tissue, or specialized cells is removed during preparation of protoplasts.

There are a great variety of methods of protoplast isolation which are discussed in more detail elsewhere.[3–5,30,31] For isolation of chloroplasts it is important that the protoplasts are first purified to remove cells, vacuoles, vascular strands, and also chloroplasts which have been released from broken protoplasts since these are likely to be photosynthetically inactive. Protoplasts can be purified on gradients of Percoll, metrizamide, or more simply by flotation on a step gradient of sucrose and sorbitol.[3–5] The protoplasts are then pelleted and resuspended in a suitable medium for isolating chloroplasts before being ruptured by forcing them through a

[28] M. E. Delaney and D. A. Walker, *Plant Sci. Lett.* **7**, 285 (1976).

[29] G. E. Edwards, S. P. Robinson, N. J. C. Tyler, and D. A. Walker, *Arch. Biochem. Biophys.* **190**, 421 (1978).

[30] M. Nishimura *et al.*, this volume [3].

[31] S. P. Robinson, this volume [18].

syringe needle or through fine (20- to 25-μm pore size) nylon mesh. This breaks all of the protoplasts but leaves chloroplasts and other organelles mostly intact. The chloroplasts can be separated by centrifugation and resuspended and purified in the same way as chloroplasts isolated by mechanical methods.

Storage and Stability. Chloroplasts isolated from spinach or peas by the mechanical method are relatively stable if stored on ice in a concentrated suspension. The rate at which photosynthetic activity is lost varies, depending both on the isolation method and the plant material. Three factors are likely to contribute to loss of photosynthetic activity: rupture of the chloroplast envelope, degradation by hydrolases in the preparation, and loss of low-molecular-weight compounds by diffusion. At least in the first few hours, loss of photosynthetic activity is not generally accompanied by a decrease in the percentage of intact chloroplasts. Mourioux and Douce[13] demonstrated that chloroplast stability was enhanced by purification in Percoll gradients due to the removal of soluble hydrolases present in the initial chloroplast pellet fraction. They also found that diffusion of metabolites such as P_i into the suspending medium contributed to the decline in photosynthesis A, and this was reduced by the use of concentrated chloroplast suspensions and inclusion of P_i in the resuspension medium. Chloroplasts purified on a Percoll gradient and suspended in a medium with 10 m*M* P_i at a chlorophyll concentration of 7–8 mg/ml still retained 80% of their photosynthetic activity 30 hr after isolation.[13] If P_i is included in the resuspension medium it should be diluted to a suitably low concentration in the assay medium (see below). The inclusion of cations in the resuspension medium also improves chloroplast stability,[4,8] possibly by decreasing the loss of cations from the chloroplasts by diffusion. Protective agents such as BSA (1%) and ascorbate or isoascorbate (2 m*M*) added to the resuspension medium can also improve chloroplast stability.

Criteria of Integrity

Chloroplast Intactness. Intact chloroplasts are those which have retained their envelope membranes. Intactness can be determined by any assay in which a substrate or product does not penetrate the chloroplast envelope so that the reaction is latent in intact plastids. The simplest (and most widely) used assay utilizes ferricyanide as an electron acceptor in the Hill reaction. Ferricyanide does not cross the chloroplast envelope and its reduction is therefore only catalyzed by the thylakoid membranes in the preparation. Total activity is then determined after deliberately rupturing the envelopes of the intact chloroplasts by osmotic shock. Ferricyanide reduction can be measured spectrophotometrically or by measuring oxygen evolution with an oxygen electrode in the presence of DL-glyceraldehyde to prevent CO_2-dependent oxygen evolution by the intact chloroplasts. The assay medium is made double strength, i.e., 660 m*M* sorbitol, 4 m*M* EDTA, 2 m*M* $MgCl_2$, 2 m*M* $MnCl_2$, 100 m*M* HEPES–KOH, pH 7.6. For intact chloroplasts, add 0.5 ml double-strength medium, 20 μl of 50 m*M* potassium ferricyanide, 50 μl of 200 m*M* DL-glyceraldehyde, sufficient water to give a total volume of 1 ml, and, lastly, the chloroplasts. After illumination for 1–2 min, add 20 μl of 250 m*M* NH_4Cl to uncouple electron transport. The procedure is repeated but the order of additions altered to measure total activity in osmotically shocked chloroplasts. Add chloroplasts to water (to give at least a 10-fold dilution) and incubate for 10–20 sec before adding double-strength medium and the other reagents. The percentage of intact chloroplasts can be calculated from the two measurements, e.g., if the intact chloroplasts give a rate of 25 and after osmotic shock the rate is 500, the percentage intact is $100(500 - 25)/500 = 95\%$.

The percentage of intact chloroplasts in a preparation can also be determined by phase-contrast light microscopy although this is tedious, as it involves counting large numbers of plastids. Intact chloroplasts are pale yellow-green and refractile with a bright halo apparent around each plastid whereas broken chloroplasts (thylakoid membranes) are dark green, granular, and nonrefractile. The ferricyanide assay may overestimate intactness in circumstances in which electron transport activity of the thylakoids present in a preparation has been inhibited during isolation. The thylakoids of the intact chloroplasts would be protected against inhibition by the envelope membranes. In spinach and pea chloroplast preparations this is not normally the case but it should be considered in species with high levels of interfering substances such as tannins and phenolics. Lilley *et al.*[32] concluded that the envelope membranes of some chloroplasts may rupture and reseal during isolation, resulting in plastids which have intact envelopes but have lost stromal contents and are less photosynthetically active. Chloroplast preparations may also be contaminated by plastids bounded by a cytoplasmic layer[33] but this is normally only a very small proportion of the total. Contamination of chloroplast preparations by other organelles can be determined by the use of marker enzymes.[5]

Chlorophyll is determined by diluting a small aliquot of chloroplasts with 80% acetone followed by filtration or centrifugation to remove insoluble material. Using the extinction coefficient of Bruinsma,[34] absorbance

[32] R. McC. Lilley, M. P. Fitzgerald, K. G. Rienits, and D. A. Walker, *New Phytol.* **75,** 1 (1975).

[33] C. Larsson, C. Collin, and P. A. Albertsson, *Biochim. Biophys. Acta* **245,** 425 (1971).

[34] J. Bruinsma, *Biochim. Biophys. Acta* **52,** 576 (1961).

at 652 nm (1 cm light path) multiplied by the dilution factor and divided by 36 gives the chlorophyll concentration of the original chloroplast suspension in mg/ml. More accurate measurement of chlorophyll can be made by measurement at the peak maxima for chlorophyll *a* and chlorophyll *b*. For extinction coefficients of chlorophylls in various solvents see Lichtenthaler and Wellburn[35] and Inskeep and Bloom.[36] To ensure that the chloroplast thylakoids are light saturated, measure oxygen evolution by osmotically shocked chloroplasts at varying chlorophyll concentrations. Depending on the light source, this will be achieved with 10–50 μg chlorophyll/ml. There are advantages to be derived from working at constant chlorophyll concentration.[2]

Carbon Assimilation. The best indication of chloroplast integrity is to measure the CO_2 assimilation rate and compare this with the parent tissue. Ideally, leaf photosynthesis should be measured at saturating CO_2 to determine maximum photosynthetic capacity in the absence of stomatal limitation and this can be achieved with the leaf disk oxygen electrode.[37] At 20° with saturating light and CO_2, spinach leaves often exhibit rates of 150–250 μmol $\cdot$ mg^{-1} chlorophyll $\cdot$ hr^{-1} and it should be possible to obtain similar rates with isolated chloroplasts. The optimal conditions for measuring CO_2-dependent oxygen evolution in isolated chloroplasts vary, both seasonally and between different laboratories. The two most critical factors are assay medium pH and P_i concentration and the effect of varying these two parameters should be checked. For spinach chloroplasts the following medium is normally suitable: 330 m*M* sorbitol, 2 m*M* EDTA, 1 m*M* $MgCl_2$, 1 m*M* $MnCl_2$, 50 m*M* HEPES–KOH, pH 7.6, 5 m*M* $NaHCO_3$, 0.5 m*M* P_i, and chloroplasts equivalent to 40–100 μg chlorophyll/ml. Addition of inorganic pyrophosphate (5 m*M*) and catalase (1000 U/ml) may give improved rates of CO_2 fixation in some cases.[2,5] With pea chloroplasts, pyrophosphate is inhibitory unless ATP or ADP is also present but the highest rates are normally achieved with both pyrophosphate and ATP.[38,39] Chloroplasts prepared by the protoplast method from wheat or sunflower require the presence of a chelating agent, such as 10 m*M* EDTA, for maximal activity.[29] Addition of cycle intermediates such as 3-phosphoglycerate or dihydroxyacetone phosphate do not normally increase rates of oxygen evolution in spinach chloroplasts but their addition (0.1–1 m*M*) may be worthwhile if low rates are observed.

[35] H. K. Lichtenthaler and A. R. Wellburn, *Biochem. Soc. Trans.* **11,** 591 (1983).
[36] W. P. Inskeep and P. R. Bloom, *Plant Physiol.* **77,** 483 (1985).
[37] T. J. Delieu and D. A. Walker, *Plant Physiol.* **73,** 534 (1983).
[38] S. P. Robinson and J. T. Wiskich, *Plant Physiol.* **59,** 422 (1977).
[39] Z. S. Stankovic and D. A. Walker, *Plant Physiol.* **59,** 428 (1977).

One additional way to express chloroplast integrity is to calculate how much of the reductive potential is utilized. The ratio between the maximum rate of CO_2-dependent oxygen evolution and the rate of uncoupled electron transport (measured with osmotically shocked chloroplasts using ferricyanide) can be such a measure. It is dimensionless and therefore independent of the base for the rate expression and has been termed chloroplast efficacy.[8] In spinach or pea chloroplasts capable of high rates of CO_2 assimilation ratios around 0.5 are obtained. Low ratios, in preparations of high intactness such as with sunflower chloroplasts,[28] would indicate damage to the chloroplast envelope, loss of metabolites, or inhibition of enzymes involved in photophosphorylation and the carbon cycle.

Isolation of Thylakoid Membranes. It is preferable to prepare thylakoid membranes from intact isolated chloroplasts rather than directly from leaf tissue since the envelope membrane protects the thylakoids from degradative enzymes and inhibitory substances during isolation. Osmotic shock yields thylakoid membranes which exhibit high rates of electron transport and photophosphorylation and high ATP/2*e* ratios. Intact chloroplasts are osmotically shocked by diluting at least 10-fold with a buffer containing Mg^{2+} (e.g., 5 m*M* tricine–KOH, 5 m*M* $MgCl_2$, pH 7.9) which prevents swelling and unstacking of the thylakoid grana stacks. The thylakoid membranes can be collected by centrifuging (3000 *g*, 5 min) and resuspension. The supernatant fraction contains stromal proteins which can be added back to the thylakoid membranes together with ferredoxin, NADP, and ADP to form the "reconstituted chloroplast" system.[2,5]

[16] Isolation of Plastids in Density Gradients of Percoll and Other Silica Sols

By CARL A. PRICE, JOHN C. CUSHMAN, LETICIA R. MENDIOLA-MORGENTHALER, and ELLEN M. REARDON

Plastids are a class of organelles which includes chloroplasts, proplastids, etioplasts, amyloplasts, and chromoplasts. Because plastids are relatively large and bound by two or more membranes, they are sensitive to shear, hypo- or hypertonicity, and hydrostatic pressure.

With some tissues and for some purposes, differential centrifugation provides chloroplast preparations of sufficient purity. When, however, it is essential to exclude nuclear material, peroxisomes, mitochondria, and other cytoplasmic contaminants, or thylakoids from broken chloroplasts, a density gradient step is required. Because it is extremely critical to avoid extremes of osmotic potential, gradient materials of low molecular weight, such as sucrose, cannot be used. Because of their low osmotic potentials and relatively high densities, silica sols are the only class of gradient materials that has permitted the isolation of physiologically active chloroplasts and other plastids.[1,2] The most commonly used silica sols are Percoll, which is quite expensive, and Ludox HS and Ludox AM, which are very cheap. Percoll is prepared from Ludox HS such that the particles of silica are bonded to poly(vinylpyrrolidone). The effect of the poly(vinylpyrrolidone) is to mask surface charges on the silica. Ordinary silica sols, such as Ludox HS or Ludox AM, tend to interact with membranes and soluble proteins and also to gel irreversibly at low temperatures or in the presence of moderate concentrations of salt. (A detailed discussion of the properties of silica sols and other gradient materials is presented elsewhere.[3]) Ludox AM can be substituted for Percoll in applications where the plastids may be washed free of the gradient material. In the protocols below, we have indicated the different proportions required for Percoll and Ludox AM.

General Methods

Disruption of Cells and Tissues. For preparing chloroplasts from spinach, Gibbs and Robinson[4] recommend two 5-sec bursts in a Waring blender. Bourque and Zaitlin[5] (cf. also Price[3]) have described a modification of the Waring blender in which a series of double-edged razor blades are mounted on a vertical shaft in place of the standard blades. The strategy in both cases is to subject the chloroplasts to minimum shear.

Because sedimenting starch grains may tear through chloroplast envelopes, it is often difficult to isolate chloroplasts with large starch granules.

[1] J.-J. Morgenthaler, C. A. Price, J. M. Robinson, and M. Gibbs, *Plant Physiol.* **54,** 532 (1974).

[2] J.-J. Morgenthaler, M. P. F. Marsden, and C. A. Price, *Arch. Biochem. Biophys.* **168,** 289 (1975).

[3] C. A. Price, *in* "Isolation of Membranes and Organelles from Plant Cells" (J. L. Hall and A. L. Moore, eds.), p. 1. Academic Press, London, 1983.

[4] M. Gibbs and J. M. Robinson, *in* "Experimental Plant Physiology" (A. San Pietro, ed.), p. 13. Mosby, St. Louis, Missouri, 1974.

[5] D. P. Bourque and D. Zaitlin, unpublished.

Many workers destarch the leaves by placing the plants in the dark for 48 hr prior to harvest.

Nearly all procedures yield a mixture of intact and broken chloroplasts. These can be distinguished by the bright refractility of intact chloroplasts when viewed by phase-contrast microscopy.

Removal or weakening of the cell wall greatly increases the subsequent yield of intact plastids. Nishimura *et al.*[6] obtained nearly 100% intact chloroplasts when spinach leaf protoplasts were disrupted by gentle shear. Spheroplasting has thus far been the *only* means of obtaining intact chloroplasts from *Euglena gracilis*.

Similarly, chloroplasts have been obtained from protoplasts of *Chlamydomonas* by treating the cells with autolysin, a cell wall-digesting enzyme released during the mating of gametes of the alga.[7,8]

Other plastid types are generally more difficult to isolate than chloroplasts, but procedures that are successful with chloroplasts provide a good starting point for the isolation of proplastids, chromoplasts, and amyloplasts.

Differential Centrifugation. In order to minimize exposure of chloroplasts to vacuolar sap, the standard procedure upon disruption of plant tissue is to centrifuge the clarified brei at 4000 *g* in the shortest possible time. Smaller plastids, such as proplastids, may require longer times and/or higher speeds. Amyloplasts are much more fragile than chloroplasts and may be disrupted if the centrifugal field exceeds 150 *g*.

Gradient Centrifugation. Density gradients can be generated by a variety of methods. Silica sol gradients may be generated by sedimenting a homogeneous sol at 90,000 *g* for 30 min (cf. Price[3]). Separations of plastids in density gradients are best achieved by isopycnic sedimentation. The speeds and times required for even small plastids to approach equilibrium in silica sol gradients are remarkably low. Spinach chloroplasts, for example, sediment to within a few percent of their equilibrium density in 15 min at 9000 *g*; *Euglena* proplastids require only 40 min at 13,000 *g*. Longer times or higher speeds will only contaminate the plastid band with smaller particles or damage the plastids through excessive hydrostatic pressure.

Unless one requires a full scan of the density gradient, the plastid band may be recovered by aspirating the less dense zones and removing the plastid zone with a pipet and manual pipet controller.

[6] M. Nishimura, D. Graham, and T. Akazawa, *Plant Physiol.* **58,** 309 (1976).

[7] U. G. Schloesser, H. Sachs, and D. G. Robinson, *Protoplasma* **88,** 51 (1976).

[8] Y. Matsuda, S. Tamaki, and Y. Tsubo, *Plant Cell Physiol.* **19,** 1253 (1978).

Materials

Chlamydomonas Medium A

	Concentration
68.3 g sorbitol	375 m*M*
35 ml 1 *M* HEPES/KOH, pH 7.7	35 m*M*
10 ml 0.2 *M* potassium EDTA	2 m*M*
5 ml 1 *M* $MgCl_2$	5 m*M*
1 ml 1 *M* $MnCl_2$	1 m*M*

Dissolve and dilute to 1000 ml with deionized water.

Chlamydomonas Medium B

	Concentration
68.3 g sorbitol	375 m*M*
35 ml 1 *M* HEPES/KOH, pH 7.7	35 m*M*
10 ml 0.2 *M* potassium EDTA	2 m*M*
1 ml 1 *M* $MgCl_2$	1 m*M*
1 ml 1 *M* $MnCl_2$	1 m*M*

Dissolve and dilute to 1000 ml with deionized water.

Chlamydomonas Medium C

	Concentration
0.1 ml 1 *M* potassium phosphate buffer, pH 6.0	1 m*M* in phosphate
50 mg autolysin	0.5 mg/ml
50 mg bovine serum albumin	0.5 mg/ml
1 ml 1 mg/ml cycloheximide (in ethyl alcohol)	10 μg/ml

Dissolve and dilute to 100 ml with deionized water.

Chlamydomonas Medium D

	Concentration
0.5 ml 1 *M* potassium phosphate buffer, pH 6.5	5 m*M* in phosphate
6 g polyethylene glycol	5% (w/v)
4 ml 1 mg/ml digitonin solution (digitonin solution prepared by heating to 90–100°)	0.004% (w/v)
400 mg bovine serum albumin	4 mg/ml

Dissolve and dilute to 100 ml with deionized water.

Chlamydomonas Medium E

	Concentration
2 ml 1 *M* tricine–NaOH buffer, pH 7.7	20 m*M* in tricine
2.73 g mannitol	150 m*M*
0.1 ml 1 *M* $MgCl_2$	1 m*M*
0.4 ml 0.5 *M* sodium EDTA	2 m*M*

Dissolve and dilute to 100 ml with deionized water.

Chlamydomonas Medium F

	Concentration
10 ml 1 *M* tricine–NaOH buffer, pH 7.7	100 m*M* in tricine
13.67 g mannitol	750 m*M*
0.5 ml 1 *M* $MgCl_2$	5 m*M*
0.5 ml 1 *M* $MnCl_2$	5 m*M*
2 ml 0.5 *M* sodium EDTA	10 m*M*

5× Gradient Mix

	Concentration	Concentration at 1×
136.5 g sorbitol	1.5 *M*	300 m*M*
25 g Ficoll (Pharmacia Fine Chemicals)	5% (w/v)	1% (w/v)
12.5 ml 0.2 *M* HEPES chloride buffer, pH 6.8	15 m*M*	5 m*M*
5 ml 1 mg/ml poly(vinyl sulfate)	10 μg/ml	2 μg/ml
2.19 g NaCl	45 m*M*	15 m*M*
200 ml deionized water		

Dissolve and adjust the pH to 6.8 with 6 *N* NaOH.
Dilute to 500 ml with deionized water.
Store frozen or heat sterilized.

When 5× gradient mix is diluted to 1×, check the pH and readjust to 6.8. Dilute to 500 ml with deionized water. Immediately before use, add 35 μl of 2-mercaptoethanol/100 ml of 1× gradient mix for a final concentration of 5 m*M*.

5× GR (Grind–Resuspension) Mix

	Concentration	Concentration at 1×
4.45 g NaP_2O_7	5 m*M*	1 m*M*
119.15 g HEPES	250 m*M*	50 m*M*
601.2 g sorbitol	1.65 *M*	330 m*M*
40 ml 0.5 *M* Na_2EDTA	10 m*M*	2 m*M*
10 ml 1 *M* $MgCl_2$	5 m*M*	1 m*M*
10 ml 1 *M* $MnCl_2$	5 m*M*	1 m*M*

Dissolve the pyrophosphate in 20 ml boiling water.
Dissolve balance in 1500 ml deionized water and combine.
Adjust pH to 6.8 with 6 *N* NaOH.
Dilute to 2 liters with deionized water.
Divide into 400-ml aliquots and store frozen or heat sterilize.

1× GR Mix

40 ml 5× GR mix
2 ml 0.5 *M* isoascorbate buffer
To 200 ml deionized water

5× HEPES–Sorbitol

	Concentration	Concentration at 1×
5.0 ml 0.2 *M* EDTA	2 m*M*	0.4 m*M*
11.92 g HEPES	100 m*M*	20 m*M*
113.9 g sorbitol	1.25 *M*	250 m*M*
350 ml deionized water		

Dissolve and adjust to pH 7.4 with 6 *N* NaOH.
Dilute to 500 ml with deionized water.

Isoascorbate Buffer

	Concentration
4.4 g isoascorbic (araboascorbic) acid	0.5 *M*
0.595 g HEPES	0.05 *M*

Titrate to pH 7 with 6 *N* NaOH.
Dilute to 50 ml with deionized water.
Store frozen in aliquots of 2 ml.

5× K-P-Sorbitol

	Concentration	Concentration at 1×
400 ml deionized water		
125 ml 1 *M* KH_2PO_4	125 m*M*	50 m*M* in phosphate
125 ml 1 *M* K_2HPO_4	125 m*M*	
346 g sorbitol	1.9 *M*	380 m*M*

Dissolve and dilute to 1000 ml with deionized water.
LCB. See PCB.
LCBF. See PCBF.

PCB or LCB

75 ml Percoll (Pharmacia Fine Chemicals), 100% (v/v) *or* 62.5 ml Ludox AM (E. I. Du Pont), 62.5% (v/v)
(Ludox sols must be purified prior to use to remove biocides and titrated to pH 6.8 with a cation exchange resin.[8] Addition of acids to Ludox may cause irreversible gelation)
6.25 g polyethylene glycol (Carbowax 8000, Union Carbide), 6.25% (w/v)

For PCB, dilute to 100 ml with Percoll.
For LCB, dilute to 100 ml with deionized water.
Heat sterilize.
Immediately prior to use, add 1 g of bovine serum albumin.

If it is desired to have a totally sterile LCB, Ludox AM and polyethylene glycol may be diluted to 90 ml and heat sterilized. The albumin can be dissolved in 10 ml water, filter sterilized, and the two solutions combined.

PCBF or LCBF

100 ml Percoll (Pharmacia Fine Chemicals), 100% (v/v) *or* Ludox AM (E. I. Du Pont) (see above), 100% (v/v)
3 g polyethylene glycol (Carbowax 8000, Union Carbide), 3% (w/v)
1 g bovine serum albumin, 1% (w/v)
1 g Ficoll (Pharmacia Fine Chemicals), 1% (w/v)

60% (v/v) Percoll Mix

	Concentration
6.8 g sorbitol	375 m*M*
3.5 ml 1 *M* HEPES/KOH, pH 7.7	35 m*M*
1.0 ml 0.2 *M* potassium EDTA	2 m*M*

20.4 mg $MgCl_2 \cdot 6H_2O$	5 m*M*
19.8 mg $MnCl_2 \cdot 4H_2O$	1 m*M*
60 ml Percoll	60% (v/v)

Dilute to 100 ml with deionized water.

80% (v/v) Percoll Mix

	Concentration
6.8 g sorbitol	375 m*M*
3.5 ml 1 *M* HEPES/KOH, pH 7.7	35 m*M*
1.0 ml 0.2 *M* potassium EDTA	2 m*M*
20.4 mg $MgCl_2 \cdot 6H_2O$	5 m*M*
19.8 mg $MnCl_2 \cdot 4H_2O$	1 m*M*
80 ml Percoll	80% (v/v)

Dilute to 100 ml with deionized water.

5× Sorbitol–Tricine

	Concentration	Concentration at 1×
150 g sorbitol	1.65 *M*	330 m*M*
22.4 g tricine	0.25 *M*	50 m*M*
300 ml deionized water		

Adjust to pH 8.4 with 6 *N* KOH.
Bring to a final volume of 500 ml with deionized water.
Heat sterilize.

When 5× sorbitol–tricine is diluted to 1×, check the pH and readjust to 8.4.

5× STPF

	Concentration	Concentration at 1×
90.18 g sorbitol	1.65 *M*	0.33 *M*
13.44 g tricine	0.25 *M*	0.05 *M*
45 g polyethylene glycol (Carbowax 8000)	15% (w/v)	3% (w/v)
15 g Ficoll 400	5% (w/v)	1% (wv)
180 ml deionized water		

Measure out components in an Erlenmeyer flask and place on shaker at 37° until dissolved. Adjust to pH 8.4 with 6 *N* KOH. Dilute to 300 ml and heat sterilize. Note: 5× STPF forms two liquid phases at low temperatures.

STPFB

40 ml 5× STPF
2 g bovine serum albumin

Dissolve and dilute to 200 ml with deionized water.

Tums (Euglena digestion mix)

50 ml K-P sorbitol
250 mg trypsin, Sigma type IX (5 mg/ml)
50 mg chymotrypsin, Sigma type II (1 mg/ml)

The solution may be divided into aliquots and stored frozen.

5× Washing Medium

	Concentration	Concentration at 1×
150 g sorbitol	1.65 m*M*	330 m*M*
2.5 ml 1 *M* $MgCl_2$	5 m*M*	1 m*M*
10 ml 0.5 *M* Na_2EDTA	10 m*M*	2 m*M*
22.4 g tricine	250 m*M*	50 m*M*
300 ml deionized water		

Dissolve and adjust to pH 8.4 with 6 *N* KOH.
Dilute to 500 ml with deionized water.

Procedures

Chloroplasts from Spinach.[1,2] Select healthy, intact spinach leaves. For protein synthesis select leaves that are <5 cm long. Float in basin of water and expose to bright lights for 30 min. Blot, remove midribs from large leaves, and weigh out five 60-g batches. Carry out all subsequent operations at 0°. The composition of the required gradients is as follows:

Starting solution	Component	Limiting solution
20 ml	5× GR mix	20 ml
1 ml	Isoascorbate buffer	1 ml
~20 mg	Glutathione	~20 mg
10 ml	LCFB	To 100 ml
or 10 ml	PCFB	To 100 ml
To 100 ml	Deionized water	

The protocol for the isolation is shown in Fig. 1.

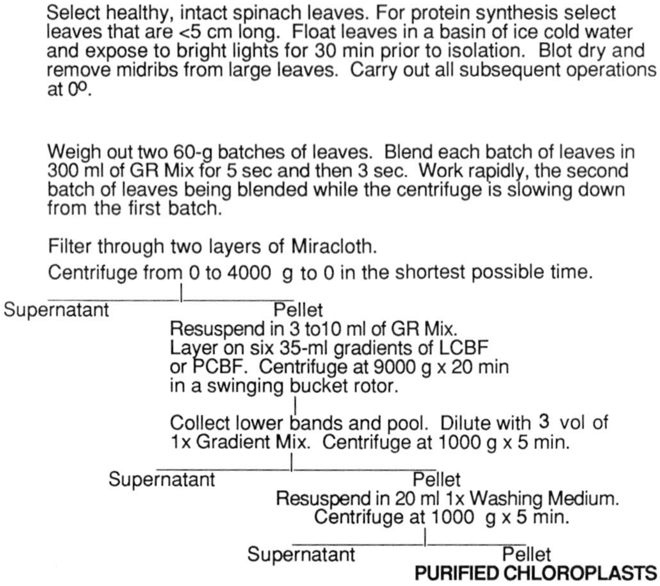

Select healthy, intact spinach leaves. For protein synthesis select leaves that are <5 cm long. Float leaves in a basin of ice cold water and expose to bright lights for 30 min prior to isolation. Blot dry and remove midribs from large leaves. Carry out all subsequent operations at 0°.

Weigh out two 60-g batches of leaves. Blend each batch of leaves in 300 ml of GR Mix for 5 sec and then 3 sec. Work rapidly, the second batch of leaves being blended while the centrifuge is slowing down from the first batch.

Filter through two layers of Miracloth.

Centrifuge from 0 to 4000 g to 0 in the shortest possible time.

Supernatant | Pellet

Resuspend in 3 to10 ml of GR Mix. Layer on six 35-ml gradients of LCBF or PCBF. Centrifuge at 9000 g x 20 min in a swinging bucket rotor.

Collect lower bands and pool. Dilute with 3 vol of 1x Gradient Mix. Centrifuge at 1000 g x 5 min.

Supernatant | Pellet

Resuspend in 20 ml 1x Washing Medium. Centrifuge at 1000 g x 5 min.

Supernatant | Pellet

PURIFIED CHLOROPLASTS

FIG. 1. Isolation of chloroplasts from spinach.[1] In all protocols, *g* values refer to the maximum centrifugal field of the rotor.

Chloroplasts from Chlamydomonas reinhardii. Because *Chlamydomonas* contains one large chloroplast occupying most of the cell and is normally surrounded by a tough cell wall, it has been extremely difficult to obtain intact chloroplasts from this alga. Simple disruption of the cells by mechanical means inevitably results in the disruption of the chloroplasts as well. There are two methods that have been described for isolating intact and functional chloroplasts from *Chlamydomonas*. In method I a cell wall-deficient mutant is disrupted under mild shear; in method II, protoplasts are prepared from normal cells by digestion with autolysin.

Method I.[9,10] This procedure (Fig. 2) takes advantage of the cell wall mutant *cw-15,* which can be broken under low pressures in the Yeda press.

The cultures are grown synchronously in cycles of 10-hr dark/14-hr light and harvested in the middle of the third light period. In general, the cells from 2 liters of culture (~10^7 cells/ml) are employed. After breaking

[9] L. Mendiola-Morgenthaler and A. Boschetti, *in* "Photosynthesis V. Chloroplast Development" (G. Akoyunoglou, ed.), p. 457. Balaban, Philadelphia, Pennsylvania, 1981.

[10] L. Mendiola-Morgenthaler, S. Leu, and A. Boschetti, *Plant Sci.* **38,** 33 (1985).

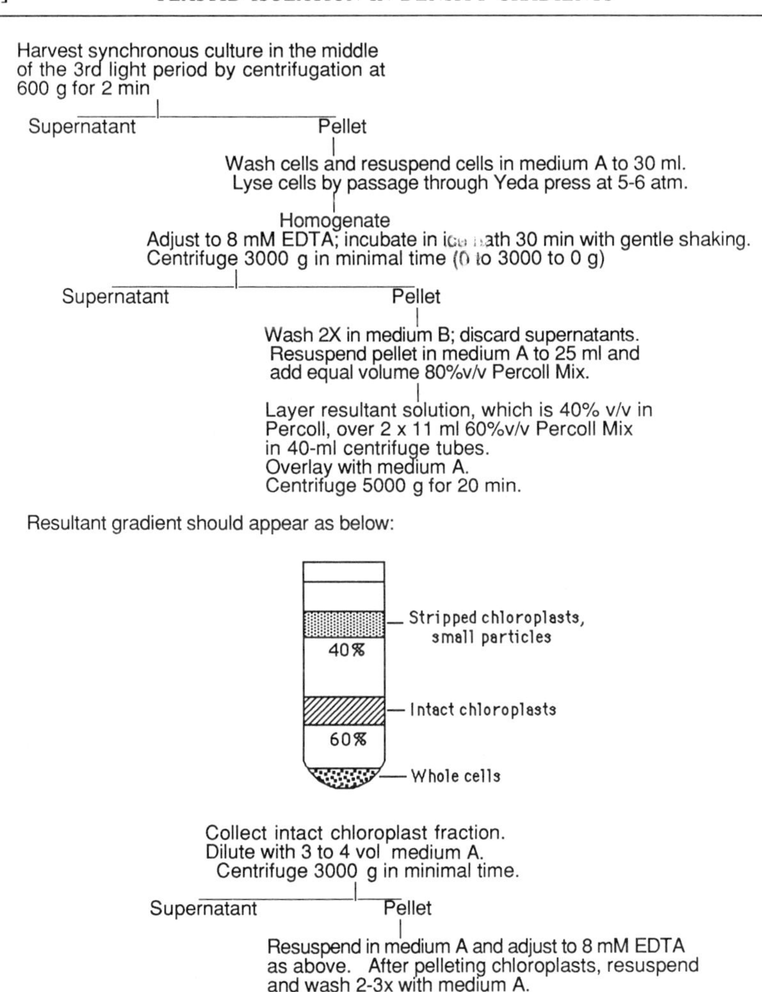

FIG. 2. Isolation of chloroplasts from *Chlamydomonas reinhardii* mutant *cw-15-2* (method I[10,11]). Protocol is based on 2 liters of culture (~10^7 cells/ml). All operations are carried out at 0–2°.

the cells, the homogenates are centrifuged through a two-step gradient consisting of 40% (v/v) and 60% (v/v) Percoll.

About 65% of the cells are broken after passage through the Yeda press. The pressure employed needs to be adjusted to each strain of cell wall mutants. Lower sorbitol concentrations (330 m*M*) may be used with similar results. At still lower concentrations (e.g., 125 m*M*) the yield of broken cells is much lower, although intact chloroplasts are still obtained. For protein synthesis, sterile techniques are used wherever feasible. All of the solutions may be heat sterilized.

The general method may be used for the preparation of chloroplast envelopes.[10a] The final preparation of chloroplasts in this case are washed two to three times with medium B.

The final chloroplast preparations contain mostly 70 S-type ribosomes and few of the cytoplasmic 80 S-type ribosomes, although they are still contaminated with some impurities, such as membranes and mitochondria, which adhere to the chloroplasts. Extensive washings, centrifugation on linear or step gradients of Percoll, or combinations of these do not significantly improve the purity of the plastids. Some of the difficulties in purification may be due to the strong tendency for subcellular particles to clump in the presence of 5 m*M* Mg^{2+}.

Method II.[11] The procedure (Fig. 3) involves the preparation of protoplasts from strain SAG 11-32/b, which possess a normal cell wall. The cell wall-digesting enzyme, autolysin, is prepared from mating gametes of *Chlamydomonas*.[6,7] For optimal yield and intactness of the chloroplasts, cells from fully synchronized cultures are harvested after a 6-hr dark period.

In the method of Klein *et al.*,[11] the protoplasts are subsequently lysed by treatment with digitonin. Intact chloroplasts are then obtained by centrifugation of the lysate through a step gradient of Percoll. The composition of the steps are as follows:

Component/step	40% (v/v)	50% (v/v)
Chlamydomonas medium F	2 ml	2 ml
Percoll	4 ml	5 ml
Bovine serum albumin	10 mg	—
Deionized water	4 ml	3 ml

Klein *et al.*[11] recommend dialyzing the Percoll against 4 vol of distilled water. Treatment with activated charcoal[12] is an alternate procedure for removing residual toxic materials.

The procedure takes about 60 min from cell harvest, starting with 100 ml of algae (30 μg chlorophyll/ml).

In Belknap's version,[13] protoplasts of wild-type or mutant strains are lysed by pressure shock in the Yeda press, instead of by detergent.

[10a] L. Mendiola-Morgenthaler, W. Eichenberger, and A. Boschetti, *Plant Sci.* **41,** 97 (1985).
[11] U. Klein, C. Chen, M. Gibbs, and K. A. Platt-Aloia, *Plant Physiol.* **72,** 481 (1983).
[12] C. Price and E. M. Dowling, *Anal. Biochem.* **82,** 243 (1977).
[13] W. R. Belknap, *Plant Physiol.* **72,** 1130 (1983).

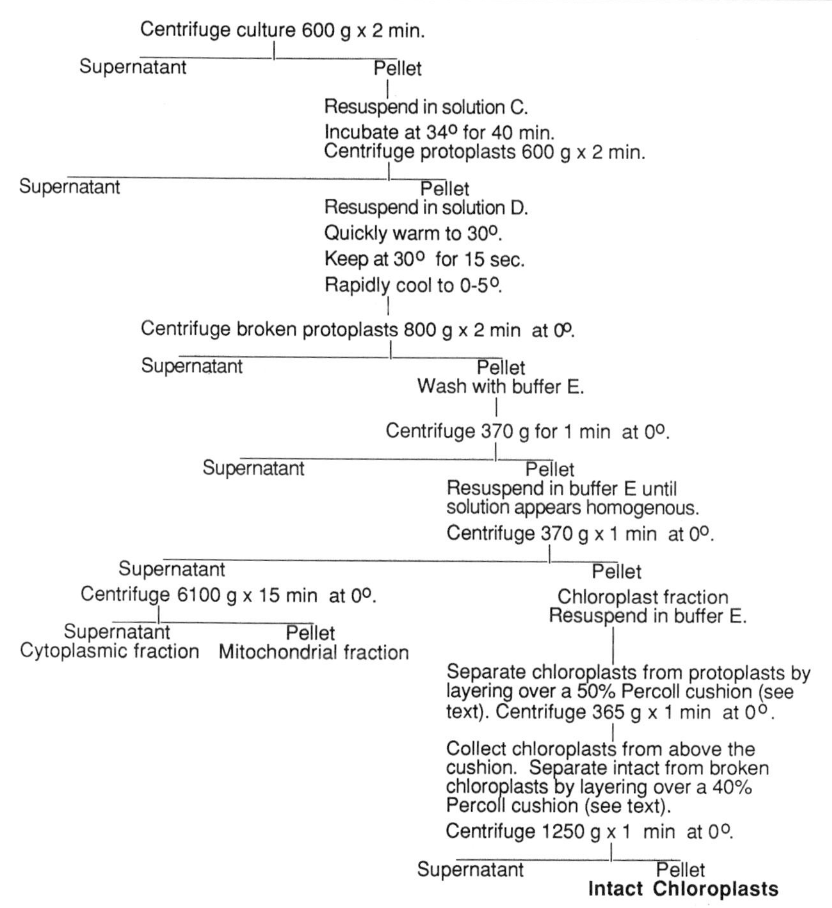

FIG. 3. Isolation of chloroplasts from *Chlamydomonas reinhardii* (method II[12]).

Thus far, none of the available methods yields chloroplasts from *Chlamydomonas* as pure as those from higher plants or *Euglena*.

Chloroplasts and Proplastids from Euglena gracilis.[14–17] Intact chloroplasts or mitochondria can be obtained from *Euglena* only after the cells

[14] W. Ortiz, E. M. Reardon, and C. A. Price, *Plant Physiol.* **66,** 291 (1980).

[15] C. A. Price and E. M. Reardon, *in* "Methods in Chloroplast Molecular Biology" (M. Edelman, N. H. Chua, and R. B. Hallick, eds.), p. 189. Elsevier, Amsterdam, 1982.

[16] M. E. Miller, J. E. Jurgenson, E. M. Reardon, and C. A. Price, *J. Biol. Chem.* **258,** 14478 (1983).

[17] J. C. Cushman and C. A. Price, *Plant Physiol.* **82,** 972 (1986).

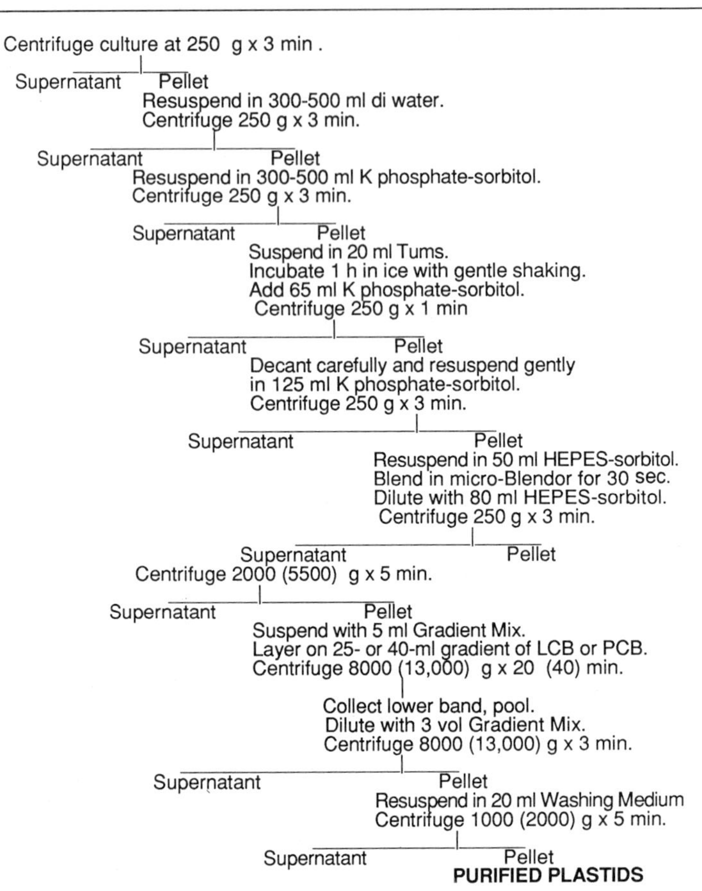

FIG. 4. Isolation of plastids from B_{12}-deficient *Euglena gracilis*.[14-17] Protocol based on 1 liter of culture containing abut 13 g wet wt. Procedure requires about 4 hr and yields ~0.7 mg chlorophyll. Centrifugation fields and times are given for chloroplasts. Where there are differences, fields and times for proplastids are given in parentheses.

have been spheroplasted.[18] Because B_{12}-deficient *Euglena* are exceptionally large, spheroplasts of B_{12}-deficient cells break efficiently under mild shear. For this reason, the original procedure specified growth in a B_{12}-deficient medium.[14-16]

B_{12}-deficient cultures provide the cleanest preparations of plastids

[18] S. Shigeoka, A. Yokota, Y. Nakano, and S. Kitaoka, *Bull. Univ. Osaka Prefect., Ser. B* **32,** 37 (1980).

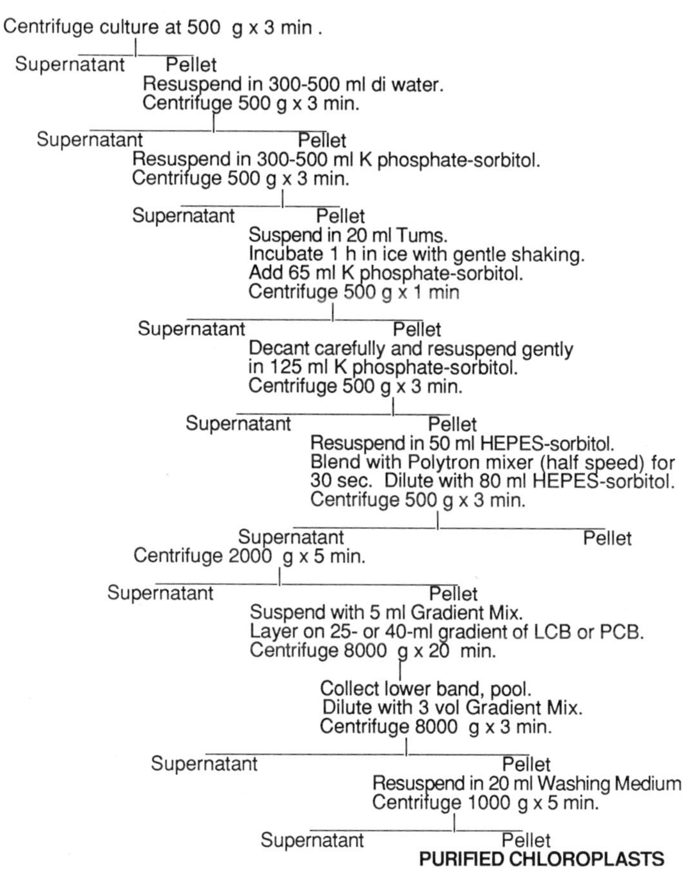

FIG. 5. Isolation of chloroplasts from B_{12}-sufficient *Euglena gracilis*. Protocol based on 1 liter of culture containing about 13 g wet wt. Procedure requires about 4 hr and yields ~0.7 mg chlorophyll (di, deionized).

(protocol of Fig. 4), but we have been able to prepare chloroplasts from B_{12}-sufficient cells with very similar activities in protein synthesis (protocol of Fig. 5). The modifications include the use of increased centrifugal fields for harvesting the cells and the use of a Polytron mixer for disrupting the spheroplasts.

For the preparation of proplastids, *Euglena* cultures must be passaged in the dark for a minimum of 3 weeks. The procedure for proplastids is the same as for chloroplasts, except that centrifugal fields following disruption of the spheroplasts are increased (Fig. 4).

The composition of gradients for isolating plastids from *Euglena* is as follows:

Starting solution	Component	Limiting solution
20 ml	5× gradient mix	20 ml
~20 mg	Glutathione	~20 mg
10 ml	PCB	To 100 ml
or 8 ml	LCB	To 100 ml
To 100 ml	Deionized water	—

Amyloplasts from Pea Epicotyls.[19] Amyloplasts are plastids that are filled with starch. They typically lack chlorophyll, but may have vestigial thylakoids adjacent to the envelope. Because of the ease with which the starch grain can tear through the plastid envelope, amyloplasts are among the most fragile of plant organelles and have been very difficult to purify.

Although intact amyloplasts have been released from potato tubers by careful slicing,[20] the first report of a successful *purification* of amyloplasts is that of Gaynor and Galston.[19] Earlier claims[21] are doubtful, since these authors failed to distinguish between amyloplasts and starch grains.

The protocol shown in Fig. 6 employs step gradients of Urografin. This substance is an iodinated derivative of benzoic acid and is not, therefore, a silica sol. Metrizamide may also be used.

A fraction enriched in amyloplasts is first obtained by differential centrifugation at 120 *g*. Centrifugal fields higher than 150 *g* completely destroyed the amyloplasts. Purification is based on the finding that the banding densities of amyloplasts and starch grains are different in Urografin (or metrizamide) gradients. Most amyloplasts have banding densities in Urografin between 1.41 and 1.43 g/cm^3, whereas starch grains band about 1.44 g/cm^3.

Amyloplasts from Rice Endosperm.[22] During ripening, the seeds of Gramineae typically go through a stage where the endosperm is liquid. This is also a stage where amyloplasts are developing rapidly in size and number, so that such tissue represents an exceptional opportunity for the isolation of intact amyloplasts. Rice is a particularly favorable organism to study, because rice amyloplasts contain a dozen starch grains. It becomes easy, therefore, to distinguish between amyloplasts and starch grains.

Amyloplasts from rice endosperm are too dense to be separated by isopycnic sedimentation in silica sol gradients and were not stable in other

[19] J. J. Gaynor and A. W. Galston, *Plant Cell Physiol.* **24,** 411 (1983).
[20] M. J. Fishwick and A. J. Wright, *Phytochemistry* **19,** 55 (1980).
[21] C. M. Duffus and R. Rosie, *Anal. Biochem.* **65,** 11 (1975).
[22] C. A. Price, E. M. Reardon, and T. Akazawa, unpublished (1982).

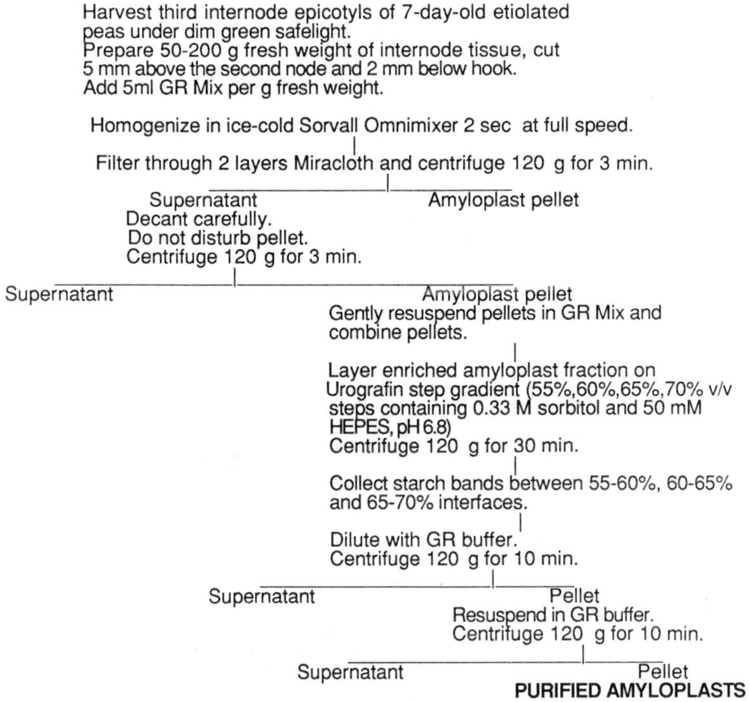

Harvest third internode epicotyls of 7-day-old etiolated peas under dim green safelight.
Prepare 50-200 g fresh weight of internode tissue, cut 5 mm above the second node and 2 mm below hook.
Add 5ml GR Mix per g fresh weight.

Homogenize in ice-cold Sorvall Omnimixer 2 sec at full speed.

Filter through 2 layers Miracloth and centrifuge 120 g for 3 min.

Supernatant — Amyloplast pellet

Supernatant: Decant carefully. Do not disturb pellet. Centrifuge 120 g for 3 min.

Supernatant — Amyloplast pellet

Amyloplast pellet: Gently resuspend pellets in GR Mix and combine pellets.

Layer enriched amyloplast fraction on Urografin step gradient (55%,60%,65%,70% v/v steps containing 0.33 M sorbitol and 50 mM HEPES, pH 6.8)
Centrifuge 120 g for 30 min.

Collect starch bands between 55-60%, 60-65% and 65-70% interfaces.

Dilute with GR buffer.
Centrifuge 120 g for 10 min.

Supernatant — Pellet

Pellet: Resuspend in GR buffer. Centrifuge 120 g for 10 min.

Supernatant — Pellet

PURIFIED AMYLOPLASTS

FIG. 6. Isolation of amyloplasts from *Pisum sativum* by sedimentation in Urografin gradients.[19]

media tested, including metrizamide and cesium chloride. These amyloplasts can, however, be separated from other constituents of the seed by *rate sedimentation* in silica sol gradients.

The procedure, shown in Fig. 7, starts with crushing rice seeds in a mortar and pestle in the presence of STPFB. The suspension is filtered, layered over a gradient whose composition is as follows:

Starting solution	Component	Limiting solution
10 ml	5× STPF	10 ml
0.5 ml	Bovine serum albumin	0.5 g
5 ml	Ludox AM (see above)	To 50 ml
To 50 ml	Deionized water	—

It is important to centrifuge the tubes at no more than 500 rpm (~50 g). The amyloplasts sediment toward the bottom of the gradient. Under these conditions, individual starch grains penetrate only the top part of the gradient. At higher speeds, the amyloplasts disintegrate, releasing their starch grains.

Harvest grains from rice in the "milky" stage, removing the grains from the husks.

All subsequent operations are at 0 to 4°.

Grind the grains with a mortar and pestle with 40 ml of 1 STPFB.
Squeeze the brei through Miracloth.

Place 8-ml portions of the clarified brei in 20-ml centrifuge tubes.
Underlay with 6 ml of 0-80%v/v Ludox AM gradients (see text).
Centrifuge 4 g x 15 min.

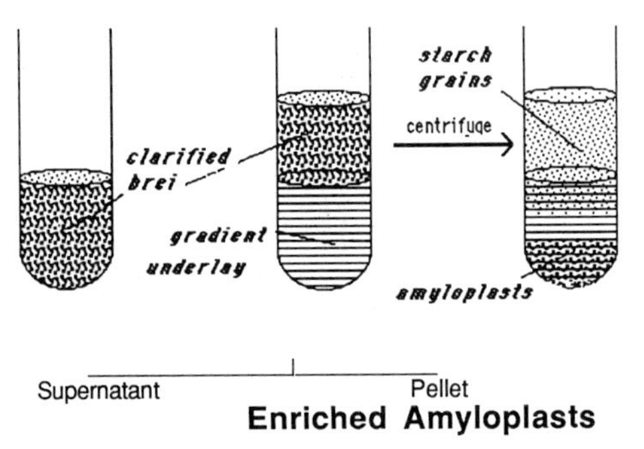

Amyloplasts may be further purified by resuspension in 1x STPFB followed by <u>very gentle</u> centrifugation.

FIG. 7. Isolation of amyloplasts from *Oriza sativa* endosperm by low-speed sedimentation.[22] Procedure is suitable for the preparation of amyloplasts from up to 10 g of rice grains. For larger amounts, see Fig. 9.

Amyloplasts are recovered from the gradient intact and contaminated with only a few free starch grains (Fig. 8). We have not found it possible, however, to wash the amyloplasts free of starch grains without unacceptable losses.

The size and density of rice endosperm amyloplasts are such that they sediment at quite significant rates at 1 *g*. This property can be exploited for their separation on a large scale using a reorienting gradient principle, originally designed for the separation of white blood cells.[23,23a] The extract of crushed endosperm is placed in a wide, flat vessel (Fig. 9) and underlayed with an equal volume of a gradient of the same composition as that shown above. The vessel is then gently turned 90° so that the sample and gradient become thin layers. A large fraction of the amyloplasts sediment

[23] R. G. Miller, *in* "New Techniques in Biophysics and Cell Biology" (R. H. Pain and B. J. Smith, eds.), p. 29. Wiley, New York, 1973.

[23a] J. Burghouts, A. M. Plas, J. Wessels, H. Hillen, J. Steenbergen, and C. Haanen, *in* "Cell Separation Methods" (H. Bloemendal, ed.), p. 27. North-Holland Publ., Amsterdam, 1977.

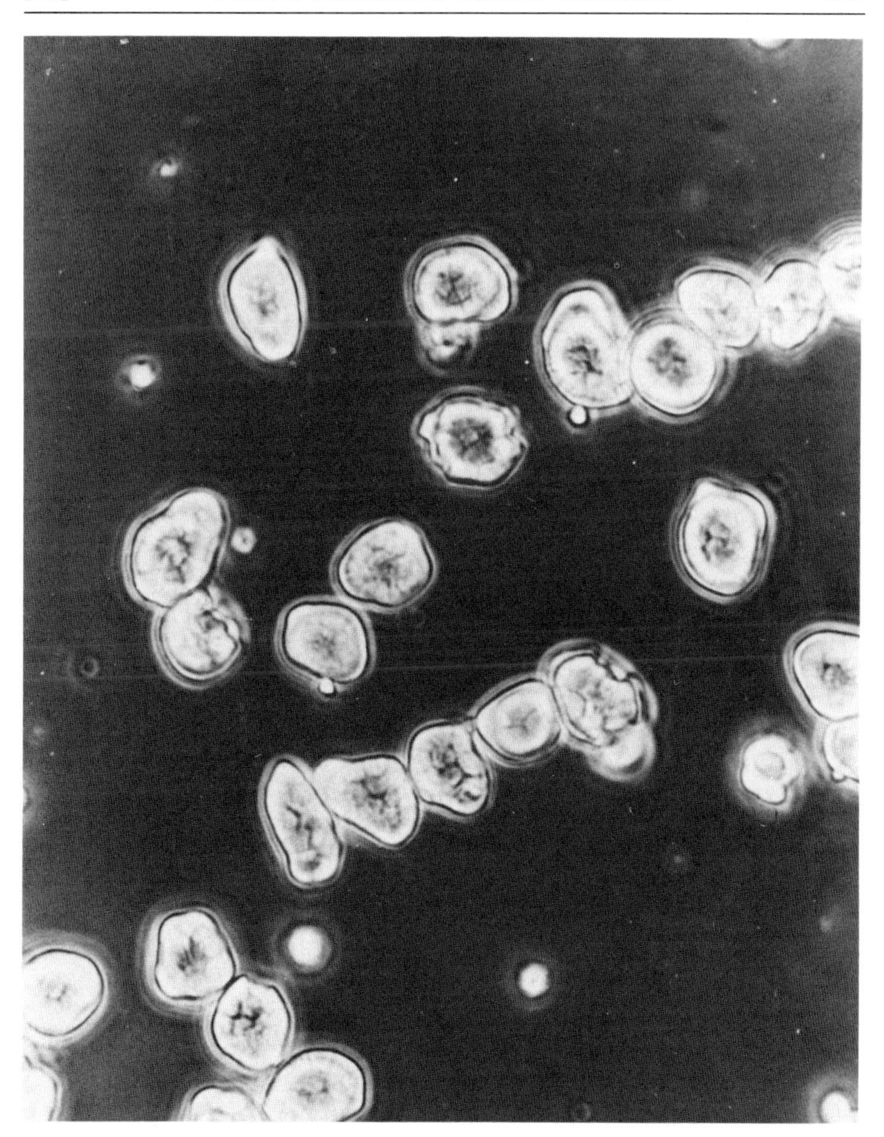

FIG. 8. Photomicrograph of amyloplasts isolated from *Oriza sativa* endosperm.[22] Amyloplasts were isolated by sedimentation through a gradient of Ludox AM in STPFB as described in Fig. 7 and photographed by phase-contrast micrography. Note the large numbers of starch grains visible within individual amyloplasts. The preparation is contaminated by a very few free starch grains.

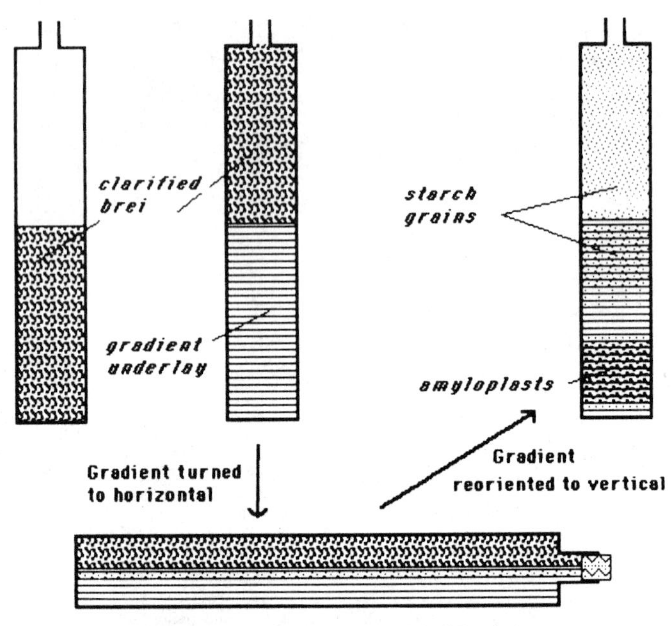

FIG. 9. Isolation of amyloplasts from *Oriza sativa* endosperm by reorienting gradient sedimentation at 1 *g*.[22] This method is suitable for the isolation of amyloplasts from 50 g of rice grains; 100 ml of a clarified brei from 50 g of grains and 200 ml of 1× STPFB, prepared as described in Fig. 7, was placed in a 200-ml Roux bottle in an upright position. The brei was underlaid with 100 ml of a 20–80% (v/v) Ludox gradient in STPFB (see text). The bottle was then stoppered and slowly rotated to a horizontal position. After sedimentation proceeded for 30 min, the bottle was returned to an upright position and the original sample volume removed. The gradient volume was rich in amyloplasts contaminated with free starch grains.

into the gradient in about 40 min. The vessel is then turned upright again, which effectively arrests the process of separation and the gradient layer, enriched in amyloplasts, is removed.

Chromoplasts from Pepper Fruits.[24,25] Camara *et al.*[24] described the extraction of intact chromoplasts from fruits of red pepper, *Capsicum annuum* L., but the preparations were impure. We have modified the conditions so that both chloroplasts and chromoplasts can be separated and added a gradient step to improve purity. The protocol is shown in Fig. 10.

The gradient composition for the separation of chromoplasts is similar

[24] B. Camara, F. Bardat, O. Dogbo, J. Brangeon, and R. Moneger, *Plant Physiol.* **74,** 94 (1983).

[25] N. Hadjeb, I. Gounaris, and C. A. Price, unpublished (1987).

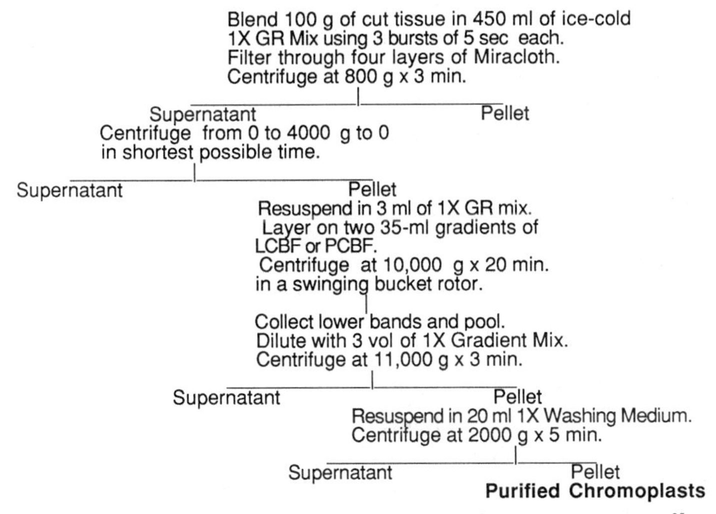

FIG. 10. Isolation of chromoplasts from fruits of *Capsicum annuum*.[25]

to that for spinach chloroplasts, but is of lower density:

Starting solution	Component	Limiting solution
20 ml	5× GR mix	20 ml
1 ml	Isoascorbate buffer	1 ml
17 ml	Glutathione	17 mg
5 ml	LCBF or PCBF	40 ml
To 100 ml	Deionized water	To 100 ml

Bathgate *et al.*[25a] and Iwatsuki *et al.*[25b] separated chromoplasts from tomato fruits on Percoll gradients, but were unable to recover intact chromoplasts from the gradients.

Discussion

Silica sol gradients are generally useful for the isolation of different kinds of plastids from a wide variety of higher plants, including mesophyll and bundle-sheath chloroplasts[26,26a] and heat-bleached chloroplasts,[27] and

[25a] B. Bathgate, M. E. Purton, D. Grierson, and P. W. Goodenough, *Planta* **165,** 197 (1985).
[25b] N. Iwatsuki, R. Moriyama, and T. Asahi, *Plant Cell Physiol.* **25,** 763 (1984).
[26] V. Walbot, *Plant Physiol.* **60,** 102 (1977).
[26a] C. L. D. Jenkins and S. Boag, *Plant Physiol.* **79,** 84 (1985).
[27] R. Hoeinghaus and J. Feierabend, *Protoplasma* **118,** 114 (1983).

from at least three kinds of algae. Separation in silica sol gradients has become the method of choice for the study of plastid composition, for the isolation of plastids other than chloroplasts, and where it is necessary to reisolate plastids from a complex incubation mixture. In the study of the transport of preproteins into chloroplasts,[28] for example, chloroplasts are first incubated with translation products of poly(A)$^+$ RNA and then recovered through a gradient step into order to determine which of the translation products enter the chloroplast.

The first[29] and still a major[14–17,30,30a] use of gradient-purified plastids is in the study of protein synthesis *in organello*. Another major application is in the characterization of plastid envelopes,[10a,31–33] where the purification achieved through the density step is essential to remove contamination by other membranes.

Plastids recovered from silica sol gradients have been used for the isolation of plastid DNA from spinach, tobacco,[34] pepper,[25] and *Cyanophora paradoxas*.[35] Bohnert[34] reports that DNA from chloroplasts purified on Percoll gradients is quite pure, whereas those from Ludox gradients can be contaminated with nuclear DNA. A disadvantage of DNA prepared from silica sol gradients, however, is that it is frequently resistant to cutting with restriction endonucleases.

Gradient-purified plastids have also been used for studies on photosynthesis[36,37] and on the salt[38] and calmodulin[39] contents of chloroplasts.

The components of the gradients can be varied,[30,40] but they invariably include an osmoticum and a buffer and usually include bovine serum

[28] A. R. Grossman, S. G. Bartlett, G. W. Schmidt, J. E. Mullet, and N.-H. Chua, *J. Biol. Chem.* **257,** 1558 (1982).

[29] J.-J. Morgenthaler and L. R. Mendiola-Morgenthaler, *Arch. Biochem. Biophys.* **162,** 51 (1976).

[30] L. E. Fish and A. T. Jagendorf, *Plant Physiol.* **70,** 1107 (1982).

[30a] S. Leu, L. Mendiola-Morgenthaler, and A. Boschetti, *FEBS Lett.* **166,** 23 (1984).

[31] J. Joyard, A. Grossman, S. G. Bartlett, R. Douce, and N.-H. Chua, *J. Biol. Chem.* **257,** 1095 (1982).

[32] K. Cline, J. Andrews, B. Mersey, E. H. Newcomb, and K. Keegstra, *Proc. Natl. Acad. Sci. U.S.A.*, **78,** 3595 (1981).

[33] D. R. McCarty, K. Keegstra, and B. R. Selman, *Plant Physiol.* **76,** 584 (1984).

[34] H. J. Bohnert, personal communication.

[35] H. J. Bohnert, E. J. Crouse, J. Pouyet, H. Mucke, and W. Loeffelhardt, *Eur. J. Biochem.* **126,** 381 (1982).

[36] G. Berkowitz and M. Gibbs, *Plant Physiol.* **70,** 1143 (1982).

[37] P. Schürmann and W. Ortiz, *Planta* **154,** 70 (1982).

[38] S. P. Robinson and W. J. Downton, *Arch. Biochem. Biophys.* **228,** 197 (1984).

[39] H. W. Jarrett, C. J. Brown, C. C. Black, and M. J. Cormier, *J. Biol. Chem.* **257,** 1379 (1982).

[40] G. Mourioux and R. Douce, *Plant Physiol.* **67,** 470 (1981).

albumin, a chelating agent, and a divalent metal ion. In addition to the use of continuous gradients, time may be saved by the use of simple step gradients,[41] as described in some of the protocols here. We have pointed out that the relative position of intact vs stripped chloroplasts can be reversed by changing the concentration of polyethylene glycol in the gradient.[2] In step gradients, therefore, chloroplasts can be recovered as a pellet or in the supernatant solution.

The protocols described here are typically designed to utilize several hundred grams of higher plant tissue or 10 to 100 g of algal cells. The availability of 4-liter, wind-shielded rotors that can be spun at more than 4000 *g* permits one to scale up the protocols about 4-fold. An alternate strategy, which we adapted to the isolation of chloroplasts from kilogram quantities of spinach, is to feed a clarified tissue brei into a continuous flow zonal rotor.[2]

Acknowledgment

This chapter was supported in part by a grant from the Charles and Johanna Busch Memorial Fund and grant no. 85-CRCR-1-1605 from the U.S. Department of Agriculture.

[41] W. R. Mills and K. W. Joy, *Planta* **148,** 75 (1980).

[17] Purification of Chloroplasts Using Silica Sols

By K. W. Joy and W. R. Mills

A supply of pure, intact chloroplasts is essential for many studies on plastids, and for subsequent isolation of chloroplast membranes or stroma. Preparation requires a carefully controlled breakage of plant cells, followed by separation of the chloroplasts from cytoplasm and cell debris to give minimum contamination with other organelles (mitochondria, peroxisomes, nuclei, etc.). The procedure must be rapid, as physiological activities deteriorate quickly after tissue disruption—for example, protein synthesis decreases substantially within 30 to 40 min.

Chloroplast enrichment can be achieved by differential sedimentation of various fractions at different centrifugal forces, but purer preparations require centrifugation through layers or gradients of different densities. Rate of sedimentation of a particle in a centrifugal field is proportional to gravitational force, the difference in density between particle and medium, and the square of the particle diameter, and is inversely proportional to viscosity of the medium. With long centrifugation times and high *g* forces, organelles reach their equilibrium density position in a density gradient, regardless of particle size (isopycnic centrifugation). With shorter centrifugation times organelle size becomes important, the larger particles approaching their equilibrium density more rapidly: this allows separation of organelles with similar density if they are dissimilar in size, for example, chloroplasts and peroxisomes. Traditional density gradient media using low-molecular-weight materials (principally sucrose) have high viscosity and osmotic potential in the denser parts of the medium (e.g., 55%, or 2 *M*, sucrose); the high gravitational forces needed for sedimentation, together with the osmotic effects, may cause some impairment of delicate organelles such as chloroplasts.

Silica sols have two major advantages as media for density gradients, namely low viscosity and negligible osmotic potential, allowing the preparation of less damaging isoosmotic gradients and more rapid sedimentation at lower gravitational forces. The use of silica sols (such as Ludox, Du Pont) in purification of chloroplasts has been suggested for a number of years,[1,2] but was not immediately adopted as a routine method, possibly because of the toxicity of available formulations and the need for purification.[2] Later, a stable, poly(vinylpyrrolidone)-treated preparation with lower toxicity became available (Percoll, Pharmacia Fine Chemicals)[3] and quickly led to a more widespread use in chloroplast purification.[4–6] Use of silica sol media can allow rapid separation of highly intact and physiologically active chloroplasts; however, it should be recognized that the quality of the final preparation will depend on the quality of the initial homogenate; good chloroplasts cannot be resurrected from poor starting material, and appropriate recipes and procedures must be adapted to the plant material and its primary disruption to give optimum results.[7]

A variety of centrifugation procedures can be used with silica sols. The minimum requirement is for a single layer which retains broken chloroplasts and other major contaminants, but allows intact chloroplasts to pass through and collect as a pellet. Several discrete steps can be utilized,

[1] J. W. Lyttleton, *Anal. Biochem.* **38,** 277 (1970).

[2] J. J. Morgenthaler, C. A. Price, J. M. Robinson, and M. Gibbs, *Plant Physiol.* **54,** 532 (1974).

[3] "Percoll Methodology and Applications." Pharmacia Fine Chemicals AB (undated).

[4] T. Takabe, M. Nishimura, and T. Akazawa, *Agric. Biol. Chem.* **43,** 2137 (1979).

[5] C. A. Price, M. Bartolf, W. Ortiz, and E. M. Reardon, *Methodol. Surv. Biochem.* **9,** 25 (1979).

[6] W. R. Mills and K. W. Joy, *Planta* **148,** 75 (1980).

[7] D. A. Walker *et al.*, this volume [15], and this series, Vol. 69, p. 74.

by assembling layers of appropriate density, and this usually produces sharp bands of cellular material at the step boundaries after centrifugation. Alternatively, continuous gradients may be produced, either by the usual mechanical mixing of solutions of different density, or by initial centrifugation of a silica sol medium (before addition of the plant homogenate) at high centrifugal forces (e.g., 30,000 for 15 min) which produces a self-generated gradient caused by sedimentation of the silica particles.[3] Table I summarizes some typical procedures and applications. The general principles of chloroplast extraction, and description of a range of methods, are discussed elsewhere.[7,8]

The procedures described below use Percoll, as it is readily available and relatively free from toxic effects. Other silica sol preparations may be suitable, but should be checked for toxicity and other adverse effects. They may require purification,[2] and addition of extra components such as polyethylene glycol, which can change relative densities of chloroplast fragments and intact chloroplasts.[2,9] In most reports, Percoll does not appear to have had deleterious effects, although an inhibition of "photosynthesis of free chloroplasts"[10] and damage to peroxisomes by prolonged contact with unpurified Percoll[11] has been reported. These effects were alleviated by dialysis or treatment with activated charcoal.

Basic Rapid Method[6]

All operations are carried out at 0–4°.

1. Ten grams or more (up to 60 g can be handled conveniently) of healthy leaf or young shoot material is sliced to small pieces (razor blade or fine scissors) into a homogenization buffer (4 ml/g fresh wt) consisting of the following:

330 m*M* sorbitol
50 m*M* *N*-Tris(hydroxymethyl)methylglycine/KOH, pH 7.9 (Tricine)
2 m*M* EDTA
1 m*M* $MgCl_2$
0.1% bovine serum albumin

Homogenize briefly, preferably using one or two 5-sec bursts in a Polytron tissue homogenizer.

[8] R. C. Leegood and D. A. Walker, *in* "Isolation of Membranes and Organelles from Plant Cells" (J. L. Hall and A. L. Moore, eds.), Chapter 8, p. 185. Academic Press, London, 1983.

[9] J. J. Morgenthaler, M. P. F. Marsden, and C. A. Price, *Arch. Biochem. Biophys.* **168**, 289 (1975).

[10] M. Stitt and H. W. Heldt, *Plant Physiol.* **68**, 755 (1981).

[11] M. R. Schmitt and G. E. Edwards, *Plant Physiol.* **70**, 1213 (1982).

TABLE I
DETAILS OF SOME METHODS FOR CHLOROPLAST PREPARATION USING SILICA SOLS

Reference	Source plant	Gradient[a]	Centrifugation	Intactness	Notes
[b]	Spinach, *Chenopodium*	Ludox, step; densities 1.08 * 1.13, 1.17, 1.22	500 *g*, 15 min	>80%	Zonal rotor also used
[c]	Spinach	Ludox, 10–80%	8000 *g*, 15 min	>75%	CO_2 fixation 103 μmol/mg Chll/hr
[d]	*Euglena*	Ludox, 20–70%	(7000 rpm, 15 min)	85–93%	Capable of protein synthesis
[e]	Spinach	Percoll, 11–90%, step 50% * 90%	8800 *g*, 15 min	90%	Contaminating organelles low
[f]	Pea (leaves or protoplasts)	Percoll step (20%), 40% * 80%	4500 *g*, 4 min or 12,500 *g*, 10 sec	93%	CO_2 fixation 108 μmol/mg Chll/hr; contaminating organelles low, capable of protein synthesis
[g]	*Euglena*	Ludox 5–50% Percoll 10–80%	8600 *g*, 20 min	—[h]	Capable of protein synthesis
[i]	Spinach (protoplasts)	Percoll step 18%, 22% * (50%)	950 *g*, 10 min	—	Starch-containing chloroplasts
[j]	Spinach	Percoll, self-generated gradient from 50%	(10,000 *g*, 100 min)[k] 5000 *g*, 10 min	98%	No detectable contaminating organelles

[l]	*Chlamydomonas*	Percoll (40%)	1250 g, 1 min	90%	Some contamination
[m]	Pea	Percoll, self-generated gradient from 50%	(35,000 g, 30 min)[k] 1000 g, 30 min	—	Used for envelope preparation
[n]	Corn (bundle sheath)	Percoll step 20 or 30% *	700 g, 4 min	>70%	Contaminating organelles low

[a] All gradients contained buffers and various additives (including osmoticum—sucrose, sorbitol, or mannitol, 0.3–0.4 *M*). In step gradients, * indicates position of main band of intact chloroplasts.

[b] J. W. Lyttleton, *Anal. Biochem.* **38,** 277 (1970).

[c] J. J. Morgenthaler, C. A. Price, J. M. Robinson, and M. Gibbs, *Plant Physiol.* **54,** 532 (1974).

[d] J. L. Salisbury, A. C. Vasconcelos, and G. L. Floyd, *Plant Physiol.* **56,** 399 (1975).

[e] T. Takabe, M. Nishimura, and T. Akazawa, *Agric. Biol. Chem.* **43,** 2137 (1979).

[f] W. R. Mills and K. W. Joy, *Planta* **148,** 75 (1980).

[g] W. Ortiz, E. M. Reardon, and C. A. Price, *Plant Physiol.* **66,** 291 (1980).

[h] —, No data given.

[i] M. Stitt and H. W. Heldt, *Plant Physiol.* **68,** 755 (1981).

[j] G. Mourioux and R. Douce, *Plant Physiol.* **67,** 470 (1981).

[k] Initial centrifugation (without chloroplasts) to generate gradient.

[l] U. Klein, C. Chen, M. Gibbs, and K. A. Platt-Aloia, *Plant Physiol.* **72,** 481 (1983).

[m] D. R. McCarty, K. Keegstra, and B. R. Selman, *Plant Physiol.* **76,** 584 (1984).

[n] C. L. D. Jenkins and S. Boag, *Plant Physiol.* **79,** 84 (1985).

2. Drain homogenate through two layers of cheesecloth and two layers of Miracloth filtering material (Calbiochem).

3. In a 50-ml centrifuge tube, place up to 30 ml of filtered homogenate on 14 ml of Percoll medium:

40% (v/v) Percoll
330 m*M* sorbitol
50 m*M* Tricine/KOH, pH 7.9 (pH rechecked after adding Percoll)
0.1% bovine serum albumin

The homogenate may be gently layered onto the Percoll medium, but we find that it is more convenient, with less chance of mixing the layers, to place the homogenate in the tube then add the Percoll medium by dispensing from a syringe. The syringe is fitted with a fine plastic tube or long, wide-bore needle to reach to the base of the centrifuge tube, and sufficient steady pressure is applied to deposit the Percoll medium under the homogenate without disturbance.

4. Centrifuge, timing for 1 min after the system reaches 2500 *g*. A swing-out head is highly preferable, allowing more rapid acceleration and braking, and giving less organelle damage (from wall effects) and mixing of layers than with a fixed angle head.

5. Broken chloroplasts collect at the upper interface; these and other contaminating material are removed by careful aspiration of the upper layer and then the Percoll layer, using for example a small pipet attached to a slow-running water pump. The pellet, which contains largely intact chloroplasts (together with some nuclei) are gently resuspended in homogenization (or other appropriate) medium. Resuspension can be achieved by stirring with a "brush" consisting of the frayed edge of a piece of nylon cloth net, wrapped around the end of a fine glass rod.

This procedure does not give a high recovery of chloroplasts (perhaps 6 to 15% of the original chlorophyll) but allows the production in 6 min or less of chloroplasts which are mostly intact and physiologically active.

Modifications

Starting Preparation

Homogenization. Alternative methods of disruption can be used, and may be necessary if smaller amounts of tissue are to be used. Rapid disruption, rather than complete homogenization, should be the aim. Some workers prefer to chill the homogenization buffer in the freezer for a short while before use, so that a small amount of ice slush is present at the start of homogenization.

Precentrifugation. If the most rapid preparation is not essential, further reduction of contaminating organelles can be achieved by (1) precentrifugation of homogenate (e.g., 500 *g*, 1 min) to remove some debris, and/or (2) sedimentation of chloroplasts (e.g., 4000 *g*, 4 min) followed by gentle resuspension in a small volume of homogenization medium, before placing on the Percoll layer.

Protoplasts. Excellent recovery (>95%) of the total chloroplast population, with very little breakage, can be achieved by gentle rupture of isolated protoplasts,[6,10,12,13] which can be prepared by many methods.[14] Effective rupture is obtained by passing protoplasts (three times) through a nylon mesh with 20-μm pore size (20-μm nylon bolting cloth, e.g., Nitex, available from suppliers of industrial screening and filtering materials), attached to the tip of a syringe.[12] We have found that chloroplasts released from broken (pea) protoplasts are particularly susceptible (in comparison with leaf homogenates) to contamination with other adhering organelles. In general, chloroplast "stickiness" is more pronounced at higher pH and Mg^{2+} concentration, and when the protoplast : buffer ratio is high; protoplast breakage in a medium at pH 6.5 in the presence of bovine serum albumin but lacking magnesium, together with some modification of the centrifugation medium (see below) can alleviate the problem.[6] It should be noted that protoplast isolation can cause major changes to some metabolites,[6] and thus chloroplasts isolated in this way may not be metabolically identical to chloroplasts from fresh material.

Gradient and Centrifugation

Small Volumes. Volumes can be scaled down from the basic method, but it is advisable to maintain a similar depth of the Percoll wash layer (about 3.5 cm), by using smaller diameter centrifuge tubes.

A very useful modification for some studies is to use very small plastic tubes (1.5 ml) in a high-speed (nonrefrigerated) bench microcentrifuge, producing a centrifugal force of 12,500–14,000 *g*. Up to 0.5-ml samples can be layered onto 1 ml of 40% Percoll medium, and the sample is centrifuged for 10 sec, after which the upper layer and Percoll are removed to leave the pelleted chloroplasts. This is particularly useful when a sequence of samples of intact chloroplasts is required from an incubation in which there may be progressive or variable chloroplast breakage.

Step Gradients. One or more additional steps of Percoll medium may be added to the main 40% layer used in the basic method. A bottom

[12] C. K. M. Rathnam and G. E. Edwards, *Plant Cell Physiol.* **17,** 177 (1976).

[13] S. P. Robinson, this volume [18].

[14] A. Ruesink, this series, Vol. 69, p. 69.

"cushion" layer of 80% (or 70%) Percoll (with the same buffer and additions as described above) allows intact chloroplasts to collect at the 40% : 80% interface, and they can be removed after aspiration of the upper layers, or by tube puncture. This step also reduces contamination with nuclei, which pass into the 80% layer.[15]

A more complex step system, for example to determine activity of a range of marker enzymes from ruptured protoplasts, consists of (for example, in a 38-ml tube) 6 ml 80% Percoll, 9 ml 40% Percoll, 6 ml 20% Percoll, all steps containing the following:

330 m*M* sorbitol
50 m*M* 4-(hydroxyethyl)-1-piperazinepropanesulfonic acid, pH 7.5 (EPPS)
0.1% bovine serum albumin

Add 8 ml of sample (e.g., ruptured protoplasts), centrifuge at 4500 *g* for 4 min using a swing-out head. Fractionation of the gradient shows clear resolution of intact chloroplasts from other organelles.[6]

As the densities of contaminating organelles, broken chloroplasts, and intact chloroplasts may vary slightly with source material, and each group has a range of densities, it may be advantageous to adjust the densities of step media to achieve optimum contaminant cut-off and chloroplast recovery in any particular application.

Continuous Gradients. Organelle banding in continuous gradients tends to be more diffuse than in a step gradient, but more subtle separations, particularly of other organelles, is possible. Gradients in the range 10–80% Percoll (with appropriate additions) may be constructed in conventional fashion.[4,16] Alternatively, an initial high-speed centrifugation of the medium can be used to cause sedimentation of the silica sol particles, resulting in self-generation of a density gradient.[3,17,18] In contrast to the centrifugation step used in chloroplast separation, gradient formation in this way cannot be carried out in a swing-out head, and a vertical or steep fixed angle rotor is required.[3] After gradient formation, the sample is added and low-speed centrifugation in a swing-out head is employed as before. See Table I for some details of conventional and self-generated gradients. In general, the procedures are slower than with a step gradient.

Zonal Rotors. There have been a few reports of procedures using zonal rotors for large-scale preparations with silica sols.[1,9] In general, the

[15] D. S. Luthe and R. S. Quatrano, *Plant Physiol.* **65,** 305 (1980).
[16] W. Ortiz, E. M. Reardon, and C. A. Price, *Plant Physiol.* **66,** 291 (1980).
[17] G. Mourioux and R. Douce, *Plant Physiol.* **67,** 470 (1981).
[18] D. R. McCarty, K. Keegstra, and B. R. Selman, *Plant Physiol.* **76,** 584 (1984).

time, cost, and complexity of this system make it less suitable for routine use.

Chloroplast Quality and Applications

When trying a new method or modification, microscopic examination and some estimate of the condition of the chloroplasts should be made. Criteria for chloroplast quality are discussed elsewhere.[7] Chloroplasts isolated on silica sols have been shown to have satisfactory intactness,[4,6,17] physiological function (ability to carry out CO_2 fixation[2,6] and protein synthesis[6,16]), and to have low levels of contamination with other organelles, as determined by the presence of marker enzymes.[4,6,17]

In addition to the general examples already given, silica sols have been used for more specialized preparations, including chloroplasts from mesophyll[19,20] and bundle sheath cells[21,22] of C_4 plants, chloroplast ribosomes,[23] and chloroplast DNA free from nuclear contamination.[24] They have also been used for preparation of chromoplasts[25] and root plastids.[26]

Additional Notes

1. Apparent buoyant densities for chloroplasts and thylakoids in silica sol gradients[4] are lower than commonly reported values derived from sedimentation in sucrose or other high osmotic potential components (which presumably result in removal of water and shrinkage of the organelles). For example, thylakoids and chloroplasts are reported to have buoyant densities of about 1.17 and 1.21, respectively,[27] yet broken chloroplasts do not pass through 40% Percoll/sorbitol ($\rho = \sim 1.11$), and intact chloroplasts are retained by an 80% layer ($\rho = \sim 1.16$).

2. Chloroplasts containing large starch grains can be damaged by the rapid acceleration of the grains during centrifugation (although the problem may be less severe in silica sol gradients, which have been used to prepare chloroplasts which contain starch[10]). A common practice to overcome this problem has been to darken the plants for 24 hr or more prior to harvest. We feel that it is preferable to reduce starch content by decreas-

[19] C. L. D. Jenkins and V. J. Russ, *Plant Sci. Lett.* **35,** 19 (1984).
[20] M. E. Rumpho and G. E. Edwards, *Plant Physiol.* **76,** 711 (1984).
[21] V. Walbot, *Plant Physiol.* **60,** 102 (1977).
[22] C. L. D. Jenkins and S. Boag, *Plant Physiol.* **79,** 84 (1985).
[23] E. I. Minami and A. Watanabe, *Arch. Biochem. Biophys.* **235,** 562 (1984).
[24] P. R. Rhodes and S. D. Kung, *Can. J. Biochem.* **59,** 911 (1981).
[25] N. Iwatsuki, R. Moriyama, and T. Asahi, *Plant Cell Physiol.* **25,** 763 (1984).
[26] Y. Oji, M. Watanabe, N. Wakiuchi, and S. Okamoto, *Planta* **165,** 85 (1985).
[27] P. H. Quail, *Annu. Rev. Plant Physiol.* **30,** 425 (1979).

ing light intensity for a couple of days (e.g., by placing plants on the floor of a growth cabinet or greenhouse) without change of photoperiod, rather than producing the shock of prolonged darkness.

3. Removal of silica sol may be advantageous if there is any possibility of an effect on physiological function, or interaction (e.g., precipitation or gelling of the sol) with added reagents. Residual sol is fairly low when chloroplasts are collected by pelleting, but is greater when they are taken from an interface. Most of the sol can be removed by dilution with suspension buffer, followed by sedimentation and resuspension. Further washes can be given if necessary.

[18] Separation of Chloroplasts and Cytosol from Protoplasts

By SIMON P. ROBINSON

Introduction

The development of techniques for isolating large numbers of metabolically competent protoplasts from leaves has provided a useful method for studying whole cell metabolism in a homogeneous suspension, free of the constraints imposed by the presence of cell walls. It has also greatly extended the range of tissues from which intact chloroplasts can be prepared and allowed isolation of other fragile organelles such as nuclei and vacuoles in an intact state. Because the cell walls have been removed, much gentler procedures can be employed to rupture protoplasts, reducing the damage to subcellular organelles and this is one of their greatest advantages. Mechanical disruption of leaves ruptures more than half of the chloroplasts and although the intact plastids can be readily separated by centrifugation the remaining supernatant is heavily contaminated by proteins and metabolites released from the broken chloroplasts. By contrast, upon gentle rupture of protoplasts, more than 90% of the chloroplasts are kept intact and when these are collected by centrifugation the supernatant provides a source of cytoplasm relatively free of contamination by chloroplasts.

Protoplast Preparation

Methods for isolating leaf protoplasts vary widely, possibly reflecting the diversity of tissues which have been utilized. The tissue needs to be made accessible to the digesting enzymes and with monocots this is normally achieved by cutting the leaves transversely with a sharp razor blade into segments 0.5–1 mm wide. With dicots, the epidermis can be stripped off or abraded by rubbing with fine carborundum powder or with a fine nylon brush. Vacuum infiltration of the enzyme medium may assist penetration in some tissues where large intercellular air spaces are present but should be avoided unless it is necessary since it can result in decreased photosynthetic capacity in the protoplasts. The digestion medium normally contains sufficient solutes to plasmolyze the leaf cells, e.g., 0.5–0.7 *M* sorbitol or mannitol. The pH of the digestion medium is in the range pH 5–6, which is near the optimum for the digesting enzymes. A wide range of enzymes are available but many are relatively crude preparations and most contain a mixture of several different enzyme activities. There may be considerable variation between different suppliers and even between different batches from the same supplier. The choice depends on the leaf tissue but for many plants cellulase Onozuka R10 and Macerozyme R10 give good digestion and yield viable protoplasts. These can be obtained from Yakult Honsha Company, Shingikancho, Nishinomiya, Japan. Similar enzyme preparations are available from Boehringer-Mannheim, West Germany, Calbiochem, La Jolla, CA, and Sigma, St. Louis, MO. For material which is difficult to digest with these two preparations, there are a range of alternative enzymes. More active cellulase preparations have been prepared from selected strains of *Trichoderma* such as cellulase Onozuka RS (Yakult) and purified cellulase (Worthington, Freehold, NJ). Pectolyase Y23 (Seishin Pharmaceutical Company, Noda, Chiba, Japan) is a pectinase preparation from *Aspergillus japonicus* and may be more effective than Macerozyme with some tissues. Driselase (Kyowa Hakko Kogyo, Ohtemachi, Chiyoda-ku, Tokyo, Japan) is a mixture of cellulase, pectinase, and other enzymes useful for difficult-to-digest tissues. Both pectolyase Y23 and Driselase can have deleterious effects on metabolism and should be used at the lowest possible concentrations (generally in the range 0.1–1%).

Protoplasts can be prepared by a two-step procedure[1] in which the tissue is first digested with pectinase to yield isolated cells and the cell walls are then removed by a second digestion with cellulase to yield protoplasts. For many tissues, protoplasts can be more simply isolated by a single digestion using both enzymes to yield protoplasts directly.[2–4]

[1] M. Nishimura and T. Akazawa, *Plant Physiol.* **55,** 712 (1975).

[2] S. C. Huber and G. E. Edwards, *Physiol. Plant.* **35,** 203 (1975).

[3] S. P. Robinson, G. E. Edwards, and D. A. Walker, *Methodol. Surv. Biochem.* **9,** 13 (1979).

[4] G. E. Edwards, S. P. Robinson, N. J. C. Tyler, and D. A. Walker, *Plant Physiol.* **62,** 313 (1978).

Various combinations of incubation time and temperature have been used but digestion for 2–4 hr at 25–30° often gives good yields of metabolically active protoplasts. It may be advantageous to replace the digestion medium at intervals as the enzymes can become inactivated by substances released from broken cells or by binding to cell surfaces. Gentle shaking (40 strokes/min) may improve digestion but is not necessary in many instances. The following method is suitable for many grasses and for some dicots such as spinach and sunflower.[2–4] Leaf slices or pieces (10–15 g) are incubated in a 19-cm-diameter crystallizing dish with 40 ml of digestion medium: 500 m*M* sorbitol, 1 m*M* $CaCl_2$, 5 m*M* MES–KOH, 2% cellulase Onozuka R10, 0.3% Macerozyme R10, pH 5.5. The dish is covered with clear plastic film to minimize evaporation and incubated for 3 hr at 28° without shaking. Weak illumination tends to improve rates of photosynthesis by the protoplasts and can be provided by a 150-W incandescent bulb 25 cm above the dish. After incubation the digestion medium is carefully removed and discarded as it normally contains very few protoplasts. The tissue is then washed three times by shaking gently with 20 ml of wash medium: 500 m*M* sorbitol, 1 m*M* $CaCl_2$, 5 m*M* MES–KOH, pH 6.0. After each wash the tissue is collected by pouring through a tea strainer (0.5- to 1-mm pore size) and the combined washes are then filtered through nylon mesh (100- to 200-μm pore size) to remove vascular tissue and undigested material. The protoplasts are collected by centrifuging for 3 min at 50–100 *g* and the supernatant is aspirated and discarded.

This crude protoplast preparation also contains some cells and chloroplasts and it is important to purify the protoplasts to remove these contaminants. This can be achieved with an aqueous two-phase system of Dextran and polyethylene glycol,[5] using gradients of Percoll[6] or metrizamide,[7] or more simply with solutions of sucrose and sorbitol of different densities.[3,4] The protoplast pellet is gently resuspended in 40 ml of 500 m*M* sucrose, 1 m*M* $CaCl_2$, 5 m*M* MES–KOH, pH 6.0, and this suspension is divided among two 100-ml centrifuge tubes. To each tube add slowly 5 ml of the following: 400 m*M* sucrose, 100 m*M* sorbitol, 1 m*M* $CaCl_2$, 5 m*M* MES–KOH, pH 6.0, and then overlayer this with 5 ml of wash medium to make a three-step gradient. After centrifuging at 300 *g* for 5 min the protoplasts collect as a band at the interface between the top two layers and can be carefully removed with a Pasteur pipet. The protoplasts should be examined with a light microscope to ensure that the preparation is free of cells and chloroplasts. If a large proportion of the

[5] R. Kanai and G. E. Edwards, *Plant Physiol.* **52,** 484 (1973).

[6] M. Nishimura, I. H. Nishimura, and S. P. Robinson, *Plant Sci. Lett.* **37,** 171 (1984).

[7] W. Wirtz, M. Stitt, and H. W. Heldt, *Plant Physiol.* **66,** 187 (1980).

protoplasts are pelleted in this sucrose/sorbitol gradient the density of the lower layers can be increased by adding 5–10% Dextran (15,000–20,000 M_r) or 10–20% Ficoll to make more of the protoplasts float. The purified protoplasts can be concentrated by diluting with 10 ml of wash medium, centrifuging at 100 *g* for 3 min and then resuspending the pellets in a small amount of medium by gently shaking the tubes. The protoplasts are stable for up to 24 hr if stored on ice. Photosynthetic activity of the protoplasts can be determined by measuring CO_2-dependent O_2 evolution with an oxygen electrode, providing rapid stirring is avoided as this will break some of the protoplasts. A suitable medium is as follows: 500 m*M* sorbitol, 1 m*M* $CaCl_2$, 30 m*M* Tricine–KOH, 5 m*M* $NaHCO_3$, with protoplasts equivalent to 50 μg chlorophyll/ml. Protoplasts normally exhibit a relatively broad pH optimum but at more acidic pH values the bicarbonate concentration should be lowered. It should be possible to isolate protoplasts with photosynthetic rates similar to the parent tissue and capable of incorporating ^{14}C into the same spectrum of products as leaf tissue.[8]

Release of Organelles

Protoplasts are generally ruptured by mechanical shear either by forcing the suspension through a fine needle,[9] through a piece of Miracloth,[10] or through narrow-aperture nylon mesh.[4] The last method is the most easily reproducible and is suitable for chloroplast isolation, relying on the fact that chloroplasts (2- to 6-μm diameter) are generally much smaller than the protoplasts, which are normally 30–60 μm in diameter. The nylon mesh should have a pore size of approximately 20 μm and as large an open area as possible, preferably greater than 12%. Suitable mesh types are type N20 HD (Henry Simon, Stockport, UK) or Nybolt P-20 (Schweizerische Seidengazefabrik, Zurich, Switzerland). The needle end of a disposable 1- or 2-ml plastic syringe is cut off and discarded and then a washer or O ring which fits tightly against the outside of the syringe barrel is used to hold a piece of nylon mesh across the open end of the barrel. Washers can be manufactured by slicing sections of the correct diameter from disposable pipet tips. Protoplasts, suspended in a suitable medium at a chlorophyll concentration of about 0.2 mg chlorophyll/ml, are ruptured by sucking them up into the syringe through the nylon mesh and then ejecting them back through the mesh. Avoid sucking air bubbles into the syringe as this causes frothing and ruptures chloroplasts. The

[8] S. P. Robinson and D. A. Walker, *Plant Physiol.* **65,** 902 (1980).

[9] G. Kaiser, E. Martinoia, and A. Wiemken, *Z. Pflanzenphysiol.* **107,** 103 (1982).

[10] M. Nishimura, D. Graham, and T. Akazawa, *Plant Physiol.* **58,** 309 (1976).

suspension should be examined under a light microscope to determine the number of passages through the mesh required to break all of the protoplasts. Depending on the operator, two or three passes should be sufficient. If the chlorophyll concentration is too high or if the protoplast preparation contains particles such as cells or vascular strands, the mesh may become blocked and resist flow of the suspension. This should be avoided as it will generally result in breakage of chloroplasts. The chloroplasts can be collected from the protoplast extract by centrifugation at 500 *g* for 1 min. Alternatively, purification of the chloroplasts can readily be achieved by layering the protoplast extract over 2 ml of the same medium containing 40% (v/v) Percoll and centrifuging for 1 min at 1000 *g*. The pellet contains intact chloroplasts relatively free of contamination by broken chloroplasts and cytoplasmic material. For studies of enzyme localization, the protoplast extract can be layered onto a sucrose density gradient for separation of organelles.[10]

Rapid Fractionation

The method described above allows isolation of intact chloroplasts and other organelles from protoplasts and determination of the subcellular localization of enzymes. The fact that most chloroplasts remain intact following protoplast rupture also provides the potential to measure the subcellular compartmentation of metabolites but this requires faster separation of the organelles. A method of rapid fractionation of protoplasts has been developed[11] based on the fact that protoplasts are ruptured when forced through fine nylon mesh and using centrifugation through silicone oil to separate the chloroplasts from the remainder of the cytoplasm. The procedure utilizes 0.4-ml plastic microcentrifuge tubes and adaptors to hold the protoplasts in the top of these tubes, which can be readily made by cutting the tip from disposable 1-ml syringes. For older models of Beckman microfuge, modified tube holders may be needed to take the tubes with adaptors inserted and these can be fashioned from shim steel.[11] The silicone oil should have a density of 1.04 g/ml and a viscosity of 150 centistokes and can be made by mixing 1 g of AR20 and 9 g of AR200 (obtained from Wacker Chemie, Munich, West Germany).

Add 30 μl of 1 *N* $HClO_4$ to each microfuge tube and then centrifuge briefly to ensure that all the acid is in the bottom of the tube. Next add 100 μl silicone oil followed by 50 μl of 0.4 *M* sorbitol. The adaptors are then inserted into the top of the tube with a piece of 20-μm nylon mesh in between (Fig. 1). The protoplasts (100 μl, equivalent to 5 μg chlorophyll)

[11] S. P. Robinson and D. A. Walker, *Arch. Biochem. Biophys.* **196,** 319 (1979).

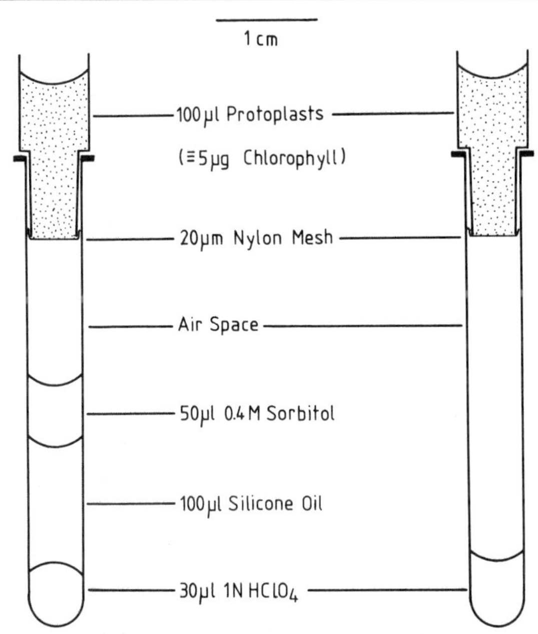

FIG. 1. Diagrammatic representation of the microcentrifuge tubes used for rapid fractionation of protoplasts. The protoplasts are held in adaptors in the top of 0.4-ml microcentrifuge tubes and kept in place above the air space in the tubes by a piece of nylon mesh. When the centrifuge is started the protoplasts pass through the mesh and are ruptured, but 90% of the chloroplasts remain intact and sediment through the silicone oil layer shown in the left-hand tube. The supernatant contains the remainder of the protoplast extract and can be separated from the chloroplasts in the pellet fraction by slicing the tubes through the silicone oil layer. When the silicone oil layer is omitted (right-hand tube) the whole protoplast extract is collected and rapidly quenched.

are added into the adaptors and are kept in place by the nylon mesh. The tubes are then placed in a Beckman microfuge and can be illuminated in the centrifuge from above if necessary. At the end of the desired incubation period the tubes are centrifuged for 20 sec. The protoplasts are ruptured in less than 1 sec as they travel through the mesh and the chloroplasts then sediment through the silicone oil whereas the remainder of the protoplast extract remains in the supernatant layer. The tubes can be cut through the silicone oil layer to separate the two fractions. Within 6–8 sec of starting the centrifuge, 90% of the chloroplasts are separated into the pellet fraction. If the layer of $HClO_4$ is replaced by 0.4 *M* sucrose, the distribution of marker enzymes can be used to assess the cross-contamination of the two fractions.[11] The pellet fraction normally contains 90% of the chloroplasts and only 10–15% of the mitochondria and peroxisomes and less than 5% contamination by the cytosol. The silicone oil and sorbitol layers can be omitted (Fig. 1) to obtain a total protoplast extract which is rapidly quenched by the perchloric acid. Comparison of the sum of metabolites in the pellet and supernatant fractions with that in the total protoplast extract will indicate the effect of any metabolism during the fractionation procedure. Such metabolism can be minimized by acidifying the supernatant fraction with 10 μl of 5 *N* $HClO_4$ immediately after centrifuging. Alternatively, 0.03 *N* HCl can be included in the sorbitol layer, which is sufficient to lower the pH of the supernatant fraction to about 2.5 without rupturing the chloroplasts as they pass through.[12]

If the protoplasts are allowed to photosynthesize in the presence of $^{14}CO_2$ prior to fractionation, the levels of metabolites can be determined by their radioactivity. Combination of protoplast fractionation and separation of the metabolites by HPLC[12] allows rapid quantitation of most photosynthetic metabolites and determination of their subcellular localization. Alternatively, metabolites can be determined by enzymatic analysis.[7] Photosynthetic products are retained within isolated protoplasts[8] and in the first 5–10 min they do not enter the vacuole[9] so that the supernatant fraction reflects metabolites in the cystol. Over longer incubations, metabolites in the vacuolar compartment of protoplasts can be determined separately by isolating the vacuoles.[9] By fractionating protoplasts at various times, kinetic analysis of the metabolite patterns can yield information about pool sizes, flux of metabolites through pools, and their transport between different compartments.[7,12,13] The rapid fractionation technique has also been extended by using an additional layer of silicone oil to yield a fraction enriched in mitochondria.[14] For determination of metabolites with rapid turnover times (such as adenine nucleotides), Lilley *et al.*[15] have developed a technique in which the various fractions are separated by filtration, which is quicker than sedimentation through silicone oil. Using this method, chloroplasts and mitochondria can be separated from the remainder of the protoplast and quenched with perchloric acid in less than 0.1 sec.[15]

[12] C. Giersch, U. Heber, G. Kaiser, D. A. Walker, and S. P. Robinson, *Arch. Biochem. Biophys.* **205,** 246 (1980).

[13] M. Stitt, W. Wirtz, and H. W. Heldt, *Biochim. Biophys. Acta* **593,** 85 (1980).

[14] R. Hampp, *Planta* **150,** 291 (1980).

[15] R. M. Lilley, M. Stitt, G. Mader, and H. W. Heldt, *Plant Physiol.* **70,** 965 (1982).

[19] Use of Thermolysin to Probe the Cytosolic Surface of the Outer Envelope Membrane from Plastids

By JACQUES JOYARD, ALBERT-JEAN DORNE, and ROLAND DOUCE

Proteolytic digestion of isolated cells, organelles, or membranes is an invaluable tool to assess the asymmetric orientation and distribution of proteins across membranes.[1] Such a strategy applied to isolated intact chloroplasts indeed provided information on the specific proteins and corresponding functions of the outer envelope membrane.[2,3] In this chapter, we wish to describe the methodology that has been utilized to probe the cytosolic surface of intact chloroplasts. Thermolysin, a metalloprotease which is active on hydrophobic amino acids (mostly leucine and phenylalanine), has been selected over several proteases to produce a specific digestion of the outer envelope polypeptides facing the cytosol.[2] The methodology described below can be easily applied to other plastids, such as proplastids, amyloplasts, and etioplasts, or to other cell organelles such as mitochondria, provided that they were obtained in a reasonably pure state. Methods are detailed in this volume to purify plastids from cauliflower buds[4] or plant mitochondria.[5]

Isolation and Protease Treatment of Intact Chloroplasts

Plant Material

Improvements in the methods available for chloroplast preparation have made possible the use of a large number of plant species. However, chloroplasts with high rates of CO_2-dependent O_2 evolution are more easily prepared from spinach or peas. Spinach leaves can be obtained in large amounts during most of the year, almost universally, whereas peas can be grown very easily in greenhouses or growth chambers. When only small blenders are available, it may be convenient to cut spinach leaves in small pieces prior to grinding. This is not necessary for peas, which can be

[1] A. H. Etemadi, *Biochim. Biophys. Acta* **604,** 347 (1980).
[2] J. Joyard, A. Billecocq, S. G. Bartlett, M. A. Block, N.-H. Chua, and R. Douce, *J. Biol. Chem.* **258,** 10000 (1983).
[3] A.-J., Dorne, M. A. Block, J. Joyard, and R. Douce, *FEBS Lett.* **145,** 30 (1982).
[4] E.-P. Journet, this volume [23].
[5] R. Douce *et al.*, this volume [37].

homogenized whole, provided that only leaves were cut off from 1- to 2-week-old seedlings.

Preparation of Intact and Purified Chloroplasts

The method used for purification of intact chloroplasts is adapted from those described by Mourioux and Douce[6] and Douce and Joyard.[7] The use of Percoll, a nontoxic silica sol gradient material [colloidal silica coated with poly(vinylpyrrolidone)] developed by Pharmacia has permitted the introduction of rapid purification procedures utilizing isosmotic and low viscosity conditions.[8] The efficiency and adaptability of Percoll gradients are highly suitable for intact chloroplasts purification.[9] However, as demonstrated by Neuburger *et al.*[10] for purification of plant mitochondria, the osmoticum should be carefully selected for the plant species or tissue used.

Solutions

Isolation medium: mannitol (or sucrose), 330 m*M*; tetrasodium pyrophosphate, 50 m*M* (or Tricine–NaOH, 30 m*M*, pH 7.8); 0.1% defatted bovine albumin; adjust to pH 7.8 (at 2°) with HCl when pyrophosphate is used

Washing medium: mannitol (or sucrose), 330 m*M*; MOPS (or Tricine)–NaOH, 10 m*M*, pH 7.8 (at 2°)

Percoll solutions: Percoll, 10, 40, 50, or 80% (v/v), mannitol (or sucrose), 330 m*M*; MOPS (or Tricine)–NaOH, 10 m*M*, pH 7.8 (at 2°)

Materials

Nylon blutex (toile à bluter) 50-μm aperture (Tripette et Renaud, Sailly-Saillisel, Combles, France)

Muslin or cheesecloth, 60 cm large

Motor-driven blender, 1 gal (Waring blender), for spinach

Polytron, probe PT 35K, for pea

[6] G. Mourioux and R. Douce, *Plant Physiol.* **67,** 470 (1981).

[7] R. Douce and J. Joyard, *in* "Methods in Chloroplast Molecular Biology" (M. Edelman, R. B. Hallick, and N.-H. Chua, eds.), p. 239. Elsevier/North-Holland Biomedical Press, Amsterdam, 1982.

[8] "Percoll Methodology and Applications." Pharmacia Fine Chemicals, Uppsala, Sweden (undated).

[9] J. J. Morgenthaler, M. P. F. Mardsen, and C. A. Price, *Arch. Biochem. Biophys.* **168,** 289 (1975).

[10] M. Neuburger, E. P. Journet, R. Bligny, J.-P. Carde, and R. Douce, *Arch. Biochem. Biophys.* **217,** 312 (1982).

Superspeed centrifuge, refrigerated (RC5, Sorvall, or equivalent) with the Sorvall ARC-1 automatic rate controller and the following rotors:

Fixed angle rotor, 8 × 50 ml (SS-34, Sorvall)
Fixed angle rotor, 6 × 500 ml (GS-3, Sorvall)
Swinging bucket rotor, 4 × 150 ml polycarbonate tubes #00289 with adapters #00458 (HS-4 with buckets #00479, Sorvall)
Vertical rotor, 8 × 34 ml (SV-288, formerly SS-90, Sorvall)
Swinging bucket rotor, 4 × 50 ml (HB-4, Sorvall)

Procedure

Place 6- to 8-week-old deribbed spinach leaves in a 1-gal Waring blender (1-kg leaves for 2 liters isolation medium) and homogenize the material at low speed for 3 sec. A more efficient method for pea leaves is the use of a Polytron set at 6; again homogenization should be limited to few seconds. Longer blendings or homogenization improves the yield of recovered chlorophyll but increases the proportion of broken chloroplasts. In addition, all operations are carried out in the cold (0–5°) and are performed as fast as possible.

Filter the homogenate rapidly through six layers of muslin and one layer of nylon blutex. The filtered suspension is then centrifuged in 500-ml bottles and spun to 1500 *g* (maximum) for 10 min in a precooled Sorvall GS-3 rotor. Pour off the supernatant and aspire the remaining medium with a vacuum aspirator pump. Each pellet is resuspended using a plastic spatula, by addition of 3–4 ml of washing medium. Filter the suspension on a nylon blutex to separate the aggregates prior to purification on Percoll gradients. When the chloroplast suspension to be purified corresponds to more than 50 mg chlorophyll, it is recommended to use large swinging bucket rotors (HS-4, Sorvall). Discontinuous (formed by two layers of 50 and 20 ml of 40 and 80% Percoll solutions, respectively) or continuous (10–80% Percoll solutions) gradients can be used.[7] For small amounts of chloroplasts, purification can be achieved very elegantly with a preformed linear gradient in a vertical rotor.[6]

Aliquots of the crude chloroplast suspension are layered on top of the gradients. Spin at 4000 *g* (maximum) for 20 min (HS-4 rotor) or 5000 *g* for 10 min (SS-90 rotor, Fig. 1). In all cases, two green bands are obtained. Discard the upper band by aspiration, and collect carefully the lower band containing intact chloroplasts. Dilute the intact purified chloroplasts with washing medium (10 to 1 vol of chloroplast suspension) and recover them as a pellet after centrifugation (3500 *g* for 90 sec, in 50-ml tubes, with SS-34 rotor, or 3000 *g* for 10 min in 500-ml bottles, with GS-3 rotor). Dis-

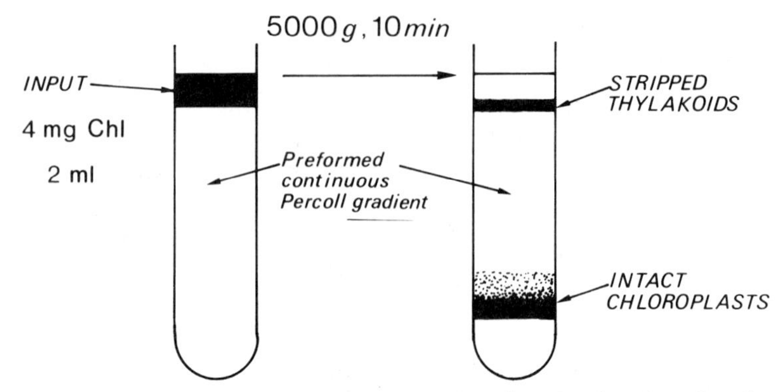

FIG. 1. Schematic representation of the Percoll gradient fractionation of crude chloroplasts to yield intact chloroplasts and "stripped thylakoids"; the rotor used to achieve this fractionation is a vertical rotor (SS-90, Sorvall). Reproduced with permission from Mourioux and Douce.[6]

card supernatant and remove the last drop by aspiration. Resuspend the chloroplast pellet in a minimum volume of washing medium, with gentle agitation.

Chlorophyll Determination

Mix 10 μl of chloroplast suspension in 10 ml acetone–water solution [80%, (v/v)]. Spin the mixture for 10 min at 4500 *g* (in 30-ml Corex tubes, SS-34 rotor). Absorbance of the supernatant is measured at 652 nm. The chlorophyll concentration of the original chloroplast suspension is estimated with the following formula: chlorophyll (mg/ml) = $OD_{652} \times 26$.

Thermolysin Treatment of Isolated Intact Chloroplasts

Solutions

Incubation mixture: mannitol (or sucrose), 330 m*M*; Tricine–NaOH, 10 m*M*, pH 7.8 (at 2°); $CaCl_2$, 1 m*M*; thermolysin (from *Bacillus thermoproteolyticus*, Calbiochem)

Incubation mixture for control: same as above plus EGTA, 10 m*M*

Percoll solution: Percoll, 40% (v/v), mannitol (or sucrose), 330 m*M*; Tricine–NaOH, 10 m*M*, pH 7.8 (at 2°), 5 m*M* ε-aminocaproic acid, 1 m*M* benzamidine–HCl, and 1 m*M* phenylmethylsulfonyl fluoride (PMSF)

Washing solution: same as Percoll solution minus Percoll

Materials

Gyratory shaker (New Brunswick, model G-2)
Superspeed centrifuged (RC5, Sorvall or equivalent) with SS-34 and HB-4 rotors (see above)

Procedure[2]

For reproducibility of data, it is essential to operate under standardized conditions: Chloroplasts (1 mg chlorophyll/ml) are incubated for 1 hr at 4° in 20 ml incubation mixture (such a volume is suggested when envelope membranes are prepared from thermolysin-treated chloroplasts) under constant and gentle agitation. The digestion is terminated by addition of EGTA (10 m*M*). The suspension is then agitated for five additional minutes. Control experiments are carried out under the same conditions except that at the same time, thermolysin, $CaCl_2$, and EGTA are added to the incubation mixture.

After digestion, chloroplasts are recovered by centrifugation through a 40% Percoll solution (5000 *g*, HB-4 rotor, 10 min) containing protease inhibitors in order to remove both thermolysin and broken chloroplasts. The treated, repurified intact chloroplasts are recovered as a pellet and resuspended in washing medium containing protease inhibitors (10 vol for 1 vol of chloroplast suspension) and spin for 90 sec at 3500 *g* (SS-34 rotor). An additional washing is necessary when envelope membranes have to be purified in order to remove all traces of Percoll. As pointed out by Douce and Joyard,[7] traces of Percoll in the chloroplast suspension strongly hamper envelope membrane purification.

Integrity of Thermolysin-Treated Chloroplasts

After thermolysin digestion, the chloroplast preparation contains primarily intact chloroplasts: envelope integrity, determined by following the reduction of ferricyanide before and after an osmotic shock,[11] is maintained above 90%. In addition, centrifugation through the 40% Percoll cushion allows the selection of intact chloroplasts present in the mixture and therefore increases further the integrity to almost 100%. In addition, repurified thermolysin-treated chloroplasts evolve O_2 in the presence of bicarbonate at the same rates as controls (average rates, 80–100 μmol/hr/mg chlorophyll). These measures, together with the analyses of the polypeptide pattern of envelope membranes from thermolysin-treated chloroplasts and studies with specific antibodies raised against envelope

[11] U. Heber and K. A. Santarius, *Z. Naturforsch., B: Anorg. Chem., Org. Chem., Biochem., Biophys., Biol.* **25B,** 718 (1970).

polypeptides, demonstrated that under the above-described experimental conditions thermolysin restricts its action to the outer envelope membrane polypeptides facing the cytosol.[2] Further evidence was provided with membrane fractions enriched in outer and inner envelope membranes by Block *et al.*[12] and confirmed by Cline *et al.*[13]

Analyses of Envelope Polypeptides from Thermolysin-Treated Chloroplasts

Envelope Membrane Purification

Detailed procedures for the preparation of whole envelope membranes are adapted from the original work of Douce *et al.*[14] and the reader is referred to exhaustive descriptions by Douce and Joyard.[7,15] Control or thermolysin-treated chloroplasts are lysed in a hypotonic medium and envelope membranes are purified from the lysate by centrifugation through a step sucrose gradient. The only difference with the previously described procedures is that protease inhibitors (5 m*M* ε-aminocaproic acid, 1 m*M* benzamidine–HCl, and 1 m*M* PMSF) are added to the hypotonic solution and all the different sucrose layers. It is also possible to separate the outer and the inner envelope membranes from thermolysin-treated chloroplasts.[12,13]

Electrophoretic Analyses of Envelope Polypeptides

Method. Polypeptides of the chloroplast envelope are analyzed by sodium dodecyl sulfate (SDS)–polyacrylamide gel electrophoresis as described by Joyard *et al.*[16] Envelope polypeptides were fractionated into two groups on the basis of their solubility in a chloroform/methanol [2 : 1 (v/v)] mixture. Therefore two types of samples (chloroform/methanol extracts and insoluble residues) were analyzed. Electrophoresis was performed at room temperature in slab gels with a 12–18% acrylamide gradient in the presence of 8 *M* urea. The experimental conditions for gel

[12] M. A. Block, A.-J. Dorne, J. Joyard, and R. Douce, *J. Biol. Chem.* **258,** 13273 (1983).

[13] K. Cline, M. Werner-Washburne, J. Andrews, and K. Keegstra, *Plant Physiol.* **75,** 675 (1984).

[14] R. Douce, R. B. Holtz, and A. A. Benson, *J. Biol. Chem.* **248,** 7215 (1973).

[15] R. Douce and J. Joyard, this series, Vol. 69, p. 290.

[16] J. Joyard, A. Grossman, S. G. Bartlett, R. Douce, and N.-H. Chua, *J. Biol. Chem.* **257,** 1095 (1982).

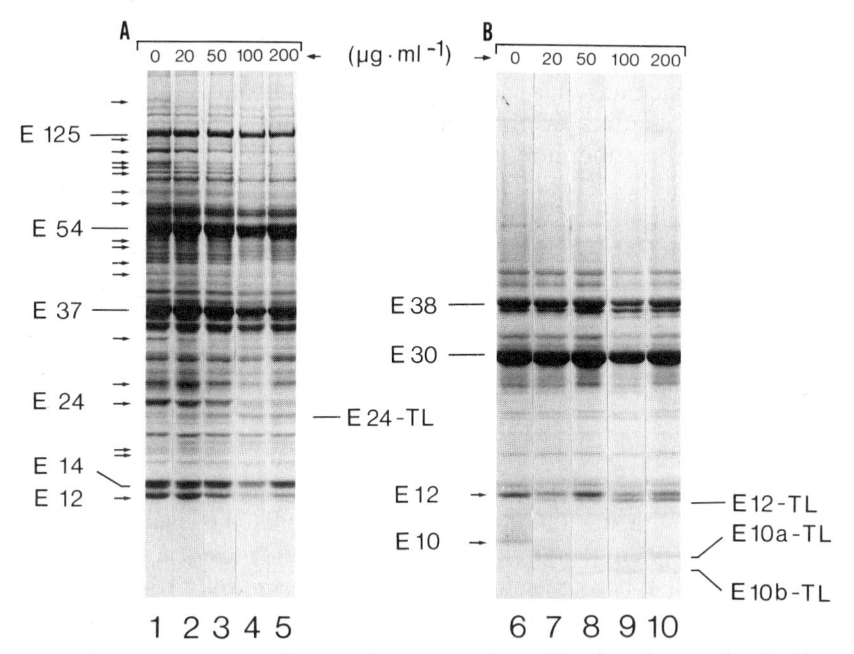

FIG. 2. Electrophoretic analyses of envelope polypeptides from intact spinach chloroplasts treated with thermolysin. SDS–PAGE at room temperature with a 12–18% acrylamide gradient in the presence of 8 *M* urea. Protein load in each slot was equivalent to 100 μg of total envelope protein. Protease treatment of intact chloroplasts was performed with thermolysin concentration of 0 to 200 μg/ml of incubation medium. Lanes 1 to 5: chloroform/methanol [2 : 1 (v/v)]-insoluble fraction of envelope. Lanes 6 to 10: chloroform/methanol [2 : 1 (v/v)]-soluble fraction of envelope. Lanes 1 and 6: control experiment. Lanes 3 to 5 and 7 to 10: treatment with increasing concentrations of thermolysin in the incubation medium. Arrows indicate thermolysin-susceptible envelope polypeptides. E24-TL, E12-TL, E10a-TL, and E10b-TL refer to proteolytic derivatives of E24, E12, and E10. Reproduced with permission from Joyard *et al.*[2]

preparation, sample extraction and solubilization, electrophoresis, and gel staining have been detailed by Chua[17] and Piccioni *et al.*[18]

Comments. Thermolysin concentration in the incubation mixture should be carefully chosen with respect to the outer envelope polypeptide and/or activity to be analyzed. As shown in Fig. 2, at low thermolysin

[17] N.-H. Chua, this series, Vol. 69, p. 434.

[18] R. Piccioni, G. Bellemare, and N.-H. Chua, *in* "Methods in Chloroplast Molecular Biology" (M. Edelman, R. B. Hallick, and N.-H. Chua, eds.), p. 985. Elsevier/North-Holland Biomedical Press, Amsterdam, 1982.

concentration some polypeptides such as E10 are completely susceptible to digestion (Fig. 2, lanes 1, 2, 6, and 7) whereas E24 is digested only at thermolysin concentrations higher than 50 μg/ml. Maximum effects on the polypeptide pattern are obtained when the incubation mixture contains about 200 μg thermolysin/ml. Some additional stained bands are visible in the polypeptide pattern of thermolysin-treated envelope (Fig. 2). Immunochemical studies with specific antibodies raised against E10 and E24 demonstrated that E10a,b-TL and E24-TL (Fig. 2B) share the same antigenic domains as E10 and E24, respectively, and therefore derive from them after cleavage of a small part of the polypeptide by thermolysin.[12]

Use of Other Proteases

The effect of proteolytic digestion of isolated intact chloroplasts on the polypeptide pattern and specific activities of envelope membranes have been analyzed with several other proteases: acetylated trypsin, TPCK (L-1-tosylamido-2-phenylethyl chloromethyl ketone)-trypsin, α-chymotrypsin, papain, and V8 protease from *Staphylococcus aureus* V8. Thermolysin is obviously much more adapted to studies with chloroplasts than all the other proteases. V8 protease and acetylated trypsin have almost negligible effects on the envelope polypeptide pattern and therefore are of limited interest. On the other hand, trypsin, even when used at low concentration, has very deleterious effects on chloroplast integrity and envelope membranes (Fig. 3) and its use must be avoided to analyze outer envelope polypeptides, for example as a mixture with α-chymotrypsin.[13,16,19] Papain and α-chymotrypsin have almost similar effects as thermolysin, but are far less convenient to use.[2] However, these proteases might be very useful for some specific purposes (i.e., when the activity analyzed cannot be destroyed by thermolysin when isolated envelope are digested). Under these conditions, appropriate protease inhibitors and activators should be used.[2]

Applications

Thermolysin treatment of intact chloroplasts is most useful when some specific outer envelope membrane enzymatic activities have to be further localized within the membrane. A good example is provided by the galactolipid : galactolipid galactosyltransferase. Dorne *et al.*[3] have

[19] K. Cline, J. Andrews, B. Mersey, E. H. Newcomb, and K. Keegstra, *Proc. Natl. Acad. Sci. U.S.A.* **78,** 3595 (1981).

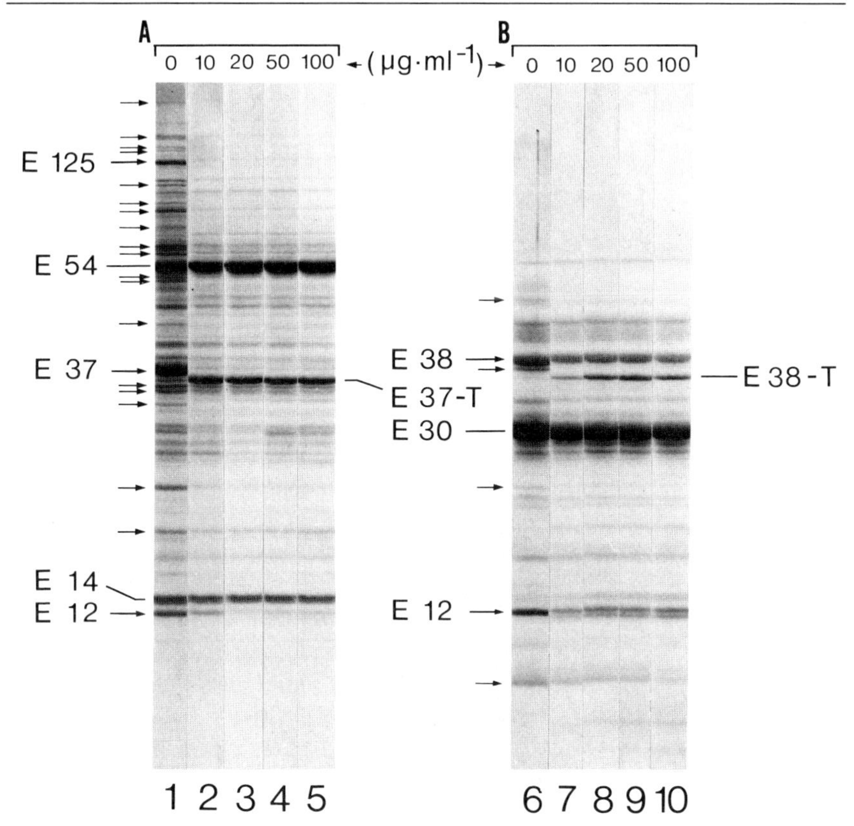

FIG. 3. Electrophoretic analyses of envelope polypeptides from intact spinach chloroplasts treated with trypsin. SDS–PAGE at room temperature with a 12–18% acrylamide gradient in the presence of 8 *M* urea. Protein load in each slot was equivalent to 100 μg of total envelope protein. Protease treatment of intact chloroplasts was performed with trypsin concentration of 0 to 100 μg/ml of incubation medium. Lanes 1 to 5: chloroform/methanol [2 : 1 (v/v)]-insoluble fraction of envelope. Lanes 6 to 10: chloroform/methanol [2 : 1 (v/v)]-soluble fraction of envelope. Lanes 1 and 6: control experiment. Lanes 3 to 5 and 7 to 10: treatment with increasing concentrations of trypsin in the incubation medium. Arrows indicate trypsin-susceptible envelope polypeptides. Reproduced with permission from Joyard *et al.*[2]

shown the thermolysin treatment of isolated intact chloroplasts fully inhibits the intergalactolipid exchange of galactoses and prevents the formation of diacylglycerol at the expenses of monogalactosyldiacylglycerol (MGDG) during the course of envelope membrane preparation. Thus, thermolysin treatment prevented the functioning of the galactolipid : galactolipid galactosyltransferase, which therefore was localized on the

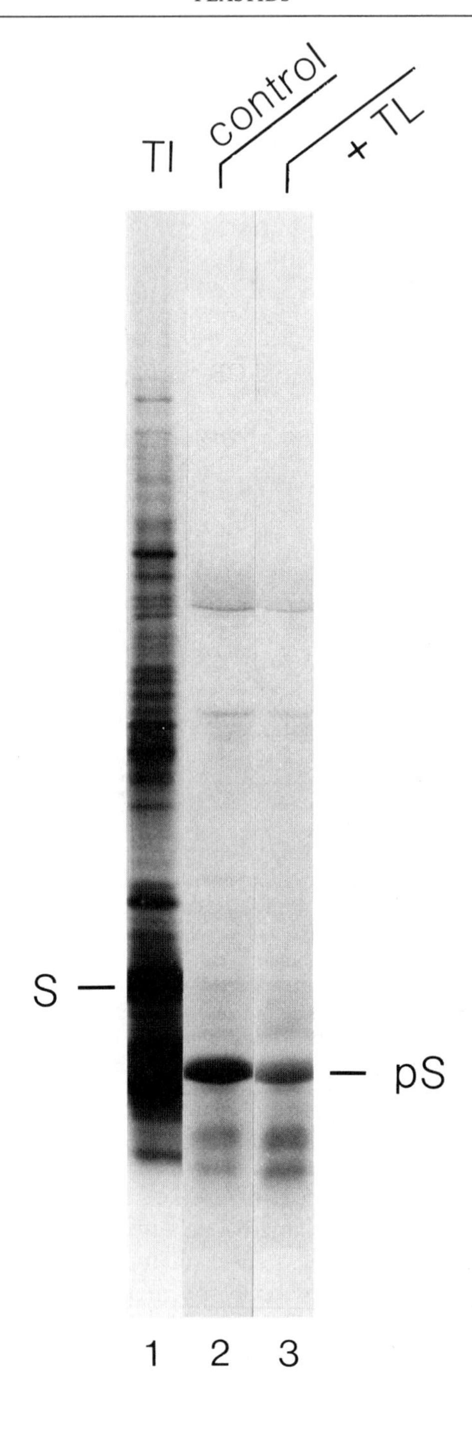
TI
control
+ TL
S
pS
1
2
3

TABLE I
LIPID COMPOSITION OF ENVELOPE MEMBRANES FROM NONTREATED AND THERMOLYSIN-TREATED INTACT SPINACH CHLOROPLASTS[a]

	Nontreated		Thermolysin treated	
Lipids[b]	μg fatty acids/mg protein	(%)	μg fatty acids/mg protein	(%)
MGDG	154	13.5	454	38
DGDG	381	33.5	354	30
TGDG	36	3	0	0
TTGDG	16	1.5	0	0
SL	86	7.5	85	7
PC	172	15	160	13.5
PG	102	9	108	9
PI	34	3	25	2
PE	0	0	0	0
DPG	0	0	0	0
Diacylglycerol	155	13.5	Trace	<0.1
Totals	1136	99.5	1186	99.5

[a] Thermolysin treatment of isolated intact and purified spinach chloroplasts was done as described by Dorne *et al.*[3]

[b] MGDG, Monogalactosyldiacylglycerol; DGDG, digalactosyldiacylglycerol; TGDG, trigalactosyldiacylglycerol; TTGDG, tetragalactosyldiacylglycerol; SL, sulfoquinovosyldiacylglycerol (sulfolipid); PC, phosphatidylcholine; PG, phosphatidylglycerol; PI, phosphatidylinositol; PE, phosphatidylethanolamine; DPG, diphosphatidylglycerol (cardiolipin).

cytosolic face of the outer envelope membrane. Destruction of this enzyme by thermolysin digestion of isolated intact plastids prior to isolation of envelope membranes is necessary when the polar lipid composition of these membranes has to be determined.[3] An example is given in Table I,

FIG. 4. Effect of thermolysin treatment of isolated intact spinach chloroplasts on protein uptake by the organelle. Fluorogram of SDS–polyacrylamide–8 *M* urea gel (12–18%). The experimental conditions for wheat germ extract preparation, mRNA purification, and polypeptide uptake are described by Mullet and Chua.[21] Lane 1: translation products (Tl) of pea mRNA by wheat germ extract in presence of [^{35}S]methionine. The precursor (pS) of the small subunit (S) of ribulose-1,5-bisphosphate carboxylase is one of the major translation products present in the mixture. Lane 2: stromal proteins containing S as the major product posttranslationally imported into the chloroplast. Lane 3: same condition as lane 2 except that spinach chloroplasts were treated by thermolysin (TL) prior to protein uptake. This treatment strongly reduces the capability of isolated intact spinach chloroplasts to import S. Unpublished observations of A. R. Grossman, S. G. Bartlett, J. Joyard, R. Douce, and N. H. Chua (1980).

showing that envelope membranes from thermolysin-treated spinach chloroplasts are devoid of diacylglycerol and of unnatural galactolipids (TGDG and TTGDG) and contain higher amounts of MGDG, a composition which probably reflects more precisely the *in vivo* polar lipid composition of envelope membranes.[3]

Another interesting application of thermolysin treatment of isolated intact chloroplasts is the study of protein transport across the plastid envelope. Chloroplast biogenesis is dependent on the posttranslational uptake of polypeptides that are coded for by nuclear DNA and are translated on cytoribosomes.[20] Thermolysin treatment of isolated intact chloroplasts (prior to incubation with mRNA translation products)[21] prevents in part uptake of the precursor form of cytoplasmically synthesized polypeptides that are destined to function within the chloroplast (Fig. 4). Therefore, such a methodology is most interesting as far as studies of the putative protein receptor of the outer envelope membrane are concerned.

Finally, the lack of effect of thermolysin digestion of intact plastids on a given enzymatic activity does not prove its absence on the cytsolic surface of the plastids but might be due to the lack of specific aminoacids susceptible to proteolytic digestion at the periphery of the protein.

Acknowledgments

Professor Nam-Hai Chua is gratefully acknowledged for his collaboration in the development and optimization of this procedure.

[20] N. H. Chua and G. Schmidt, *J. Cell Biol.* **81,** 461 (1979).
[21] J. E. Mullet and N.-H. Chua, this series, Vol. 97, p. 502.

[20] Characterization of Plastid Polypeptides from the Outer and Inner Envelope Membranes

By JACQUES JOYARD, MARYSE A. BLOCK, JACQUES COVÈS, CLAUDE ALBAN, and ROLAND DOUCE

Sodium dodecyl sulfate–polyacrylamide gel electrophoresis (SDS–PAGE) of chloroplast envelope membranes has shown a rather complex pattern: more than 70 distinct polypeptides can be resolved in 2-dimensional gel electrophoresis.[1] Such a pattern represents a mixture of poly-

[1] J. Joyard, A. R. Grossman, S. G. Bartlett, R. Douce, and N.-H. Chua, *J. Biol. Chem.* **257,** 1095 (1982).

peptides from both the outer and the inner envelope membranes. In this chapter, we describe the methods used in our laboratory to characterize polypeptides from the outer or inner envelope membranes with monospecific antibodies, using isolated intact chloroplasts[2] and membrane fractions enriched in outer or inner envelope membranes.[3]

Preparation of Antibodies to Envelope Polypeptides[1]

Purification of Envelope Membrane Antigens

Although methods are available for the purification of outer and inner envelope membranes[3,4] (see also below), the use of whole envelope membranes prepared according to Douce *et al.*[5] is far more convenient: the yield of envelope recovery is much higher (about 10 mg protein from 2 kg spinach leaves) than in any other procedure (for experimental details, see Douce and Joyard[6,7]). To simplify the polypeptide pattern of the gels obtained after SDS–PAGE, it is suggested to fractionate envelope proteins, prior to polypeptide separation, on the basis of their solubility in a chloroform/methanol mixture [2 : 1 (v/v)].[1,8] Polypeptides of the two fractions obtained (chloroform/methanol extracts and residues) are then separated in two cycles of preparative SDS–PAGE, as described by Piccioni *et al.*[8] and Chua *et al.*[9] The desired bands are excised and pooled, and eluted by electrodialysis.[8] We have selected the polypeptides to be used for raising antibodies on the basis of their sensitivity to mild proteolytic digestion of intact chloroplasts by thermolysin.[2,10] Such a procedure allows the selection of polypeptides localized on the cytosolic side of the outer envelope membrane.[2]

[2] J. Joyard, A. Billecocq, S. G. Bartlett, M. A. Block, N.-H. Chua, and R. Douce, *J. Biol. Chem.* **258,** 10000 (1983).

[3] M. A. Block, A.-J. Dorne, J. Joyard, and R. Douce, *J. Biol. Chem.* **258,** 13273 (1983).

[4] K. Cline, J. Andrews, B. Mersey, E. H. Newcomb, and K. Keegstra, *Proc. Natl. Acad. Sci. U.S.A.* **78,** 3595 (1981).

[5] R. Douce, R. B. Holtz, and A. A. Benson, *J. Biol. Chem.* **248,** 7215 (1973).

[6] R. Douce and J. Joyard, *in* "Methods in Chloroplast Molecular Biology" (M. Edelman, R. B. Hallick, and N.-H. Chua, eds.), p. 239. Elsevier/North-Holland Biomedical Press, Amsterdam, 1982.

[7] R. Douce and J. Joyard, this series, Vol. 69, p. 290.

[8] R. Piccioni, G. Bellemare, and N.-H. Chua, *in* "Methods in Chloroplast Molecular Biology" (M. Edelman, R. B. Hallick, and N.-H. Chua, eds.), p. 985. Elsevier/North Holland Biomedical Press, Amsterdam, 1982.

[9] N.-H. Chua, S. G. Bartlett, and M. Weiss, *in* "Methods in Chloroplast Molecular Biology" (M. Edelman, R. B. Hallick, and N.-H. Chua, eds.), p. 1063. Elsevier/North-Holland Biomedical Press, Amsterdam, 1982.

[10] J. Joyard, A.-J. Dorne, and R. Douce, this volume [19].

Prior to injection of the polypeptides into rabbits for raising antibodies, check the purity of the eluted SDS–polypeptide complex.

Preparation and Characterization of Rabbit Antibodies to Envelope Polypeptides

Detailed procedures for rabbit immunization, sera preparation and screening, and IgG purification have been described by Chua *et al.*[9] We also use a fast and reliable method for IgG purification from rabbit antisera by chromatography on DEAE–Trisacryl M (IBF, Villeneuve la Garenne, France) developed by Saint-Blancard *et al.*[11] We have characterized antibodies to envelope polypeptides E10, E24, E30, and E37 by crossed-immunoelectrophoresis,[1,12] but immunoblotting is also a very convenient method.[3]

Immunochemical Studies of Intact Chloroplasts[2]

It is assumed that intact and purified chloroplasts have been prepared, preferably by using Percoll purification procedures, as described in this volume by Joyard *et al.*[10] It is most important to use chloroplast preparations with more than 95% intact organelles and mostly devoid of extraplastidial contaminants.

Solutions

Suspension medium: sucrose (or sorbitol), 330 m*M*; HEPES–NaOH (or tetrasodium pyrophosphate–HCl), 10 m*M*, pH 8.0; intact and purified chloroplasts (0.1 mg chlorophyll/ml)

Incubation medium: sucrose (or sorbitol), 330 m*M*; HEPES–NaOH, 10 m*M*, pH 8.0; containing when necessary antibodies to envelope polypeptides (diluted from 1/2.5 to 1/160)

Sheep antibodies (normal or fluorescein labeled) raised against rabbit IgG (Institut Pasteur Production, Paris, France)

Materials

Glass plates (9 × 13 cm), with 20 round enameled circles to delimit incubation cuvettes (for agglutination assays)

Gyratory shaker (New Brunswick, model G-2, or equivalent)

[11] J. Saint-Blancard, J. Fourcart, F. Limonne, P. Girot, and E. Boschetti, *Ann. Pharm. Fr.* **39,** 403 (1981).

[12] N.-H. Chua and F. Blomberg, *J. Biol. Chem.* **254,** 215 (1979).

Superspeed centrifuge, refrigerated (RC5 Sorvall, or equivalent) with all the equipment to prepare chloroplasts.[11] In addition, microtubes, 7 × 50 mm (ref. 003103), with their specific adaptors (ref. 00408) to SS-34 rotor are recommended for immunofluorescence experiments

Fluorescence and phase-contrast microscope (Zeiss photomicroscope, or equivalent) with HBO 200 lamp; excitation filters (Zeiss LP 455 and KP 500). Chlorophyll fluorescence was blocked using a Zeiss 65 filter

Procedure

Agglutination Assay. Incubate 20 μl of the chloroplast suspension in one cuvette of the glass plate with 50 μl of the incubation medium. It is recommended to place the plate in a Petri dish on a wet Whatman paper to prevent evaporation of the incubation mixture. Shake the glass plates carefully for 5 min on a gyratory shaker set at 50 rpm, and examine the suspension under phase-contrast microscopy to follow agglutination. Under phase-contrast microscopy, intact chloroplasts appear highly reflective and surrounded by a bright halo whereas broken chloroplasts appear dark and granular (Fig. 1). If antigenic domains of the polypeptides analyzed are accessible to the corresponding antibodies, incubation of isolated intact chloroplasts with these antibodies will lead to agglutination. For example, strong agglutination of intact chloroplasts (but not of broken ones) was obtained with anti-E10 and anti-E24 (antibodies raised against outer envelope polypeptides), but not with anti-E30 and anti-E37 (antibodies raised against inner envelope polypeptides) (Fig. 1).

However, the binding of an antibody to the corresponding antigen may produce only a weak agglutination or even no agglutination at all. To control whether antibody–antigen complexes have been formed, add to the samples in which no direct agglutination was observed 20 μl of incubation medium containing sheep antibodies to rabbit IgG. Carefully agitate the glass plates on the gyratory shaker set at 50 rpm for five additional minutes prior to observation under phase-contrast microscopy. The lack of agglutination under these conditions allows us to conclude that the antigen analyzed was not accessible from the cytosolic side of isolated intact chloroplasts.

Immunofluorescence Assay. Centrifuge 100 μl of the chloroplast suspension at 3000 *g* for 2 min in microtubes. Add 100 μl of the antibody-containing incubation mixture to the pellet. Gently resuspend the pellet and incubate for 30 min at 4°. Lightly agitate the tube to prevent chloroplast sedimentation. After incubation, spin the tube at 3000 *g* for 2 min and wash the pellet two times with 0.4 ml of the incubation mixture

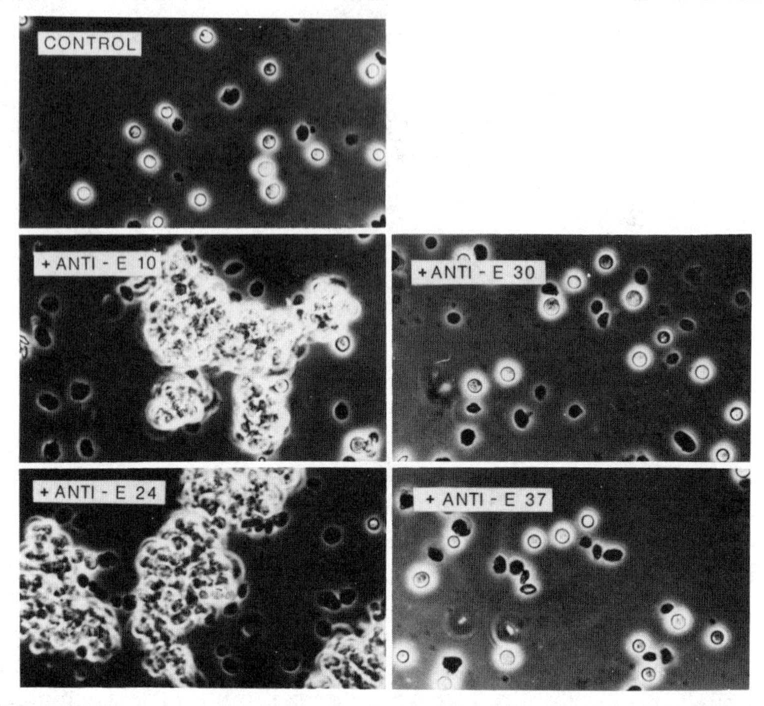

FIG. 1. Phase-contrast microscopy of spinach chloroplasts incubated in the presence of antibodies raised against envelope polypeptides E10, E24, E30, and E37. Chloroplasts were incubated in the presence of antibodies diluted to one-fourth. A control experiment was performed in the presence of nonspecific rabbit IgG. Note that only incubation of chloroplasts with anti-E10 and anti-E24 led to a strong agglutination of intact chloroplasts (which are highly reflective and present a bright halo). Note also that broken chloroplasts (which are dark and granular under phase-contrast microscopy) did not react with antibodies to E10 and E24. Reproduced with permission from Joyard *et al.*[2]

devoid of antibodies. Add 100 μl of incubation medium containing fluorescein-labeled sheep antibodies to the washed chloroplast pellet. Resuspend the pellet and incubate for 30 min at 4°. Eliminate unreacted antibodies by two successive washing/centrifugation steps, as above. To the pellet obtained after the last wash, add 100 μl incubation medium devoid of antibodies. Examine the suspension under phase-contrast and fluorescence microscope.

Comments

It is necessary to run simultaneously control experiments with preimmune serum (see Fig. 1).

A critical step, especially for immunofluorescence assays, is to prevent breakage of the plastid envelope during incubation, centrifugation, and pellet resuspension. Vortex should be avoided to resuspend the pellets. In preliminary experiments, we have used glutaraldehyde-fixed chloroplasts [0.25% glutaraldehyde in suspension medium (v/v)]. We do not recommend such a treatment since it does not increase the percentage of intact chloroplasts in the incubation mixture and induces some artifactual agglutination or fluorescence.

Applications

The technique described above, when associated with mild proteolytic digestion of intact chloroplasts by thermolysin, is a very useful tool to characterize envelope polypeptides located on the cytosolic side of intact chloroplasts.[2] It has also been employed to control the integrity of the outer envelope membrane after phospholipase C treatment of isolated intact chloroplasts.[13] It has also been used to demonstrate the presence, in the outer leaflet of the outer envelope membrane, of the characteristic plastid glycerolipids such as galactolipids[14,15] and sulfolipid.[16]

Immunochemical Analyses of Membrane Fractions Enriched in Outer and Inner Envelope Membranes

To characterize carefully membrane fractions derived from the plastid envelope and assumed to be enriched in outer or inner membrane, the use of monospecific antibodies is far more superior to any other techniques.

Preparation of Membrane Fractions Enriched in Outer and Inner Envelope Membranes from Spinach Chloroplasts[3]

Large amounts of intact and purified spinach chloroplasts (150–200 mg chlorophyll) are necessary to prepare reasonable amounts of purified outer and inner envelope membrane fractions. A procedure to prepare these chloroplasts has been described by Douce and Joyard[7] and in this volume.[10]

Solutions (See Fig. 2)

Materials

Yeda press (Linca Scientific Instruments, Tel Aviv, Israel) supplied with N_2

[13] A.-J. Dorne, J. Joyard, M. A. Block, and R. Douce, *J. Cell Biol.* **100,** 1690 (1985).
[14] A. Billecocq, *Biochem. Biophys. Acta* **352,** 245 (1974).
[15] A. Billecocq, R. Douce, and M. Faure, *C.R. Hebd. Seances Acad. Sci.* **275,** 1135 (1972).
[16] A. Billecocq, *Ann. Immunol. (Paris)* **126C,** 337 (1975).

Intact and purified chloroplasts suspended in hypertonic medium : 0.6M mannitol, 4mM $MgCl_2$, 10mM tricine-NaOH pH 7.8 for 10 min

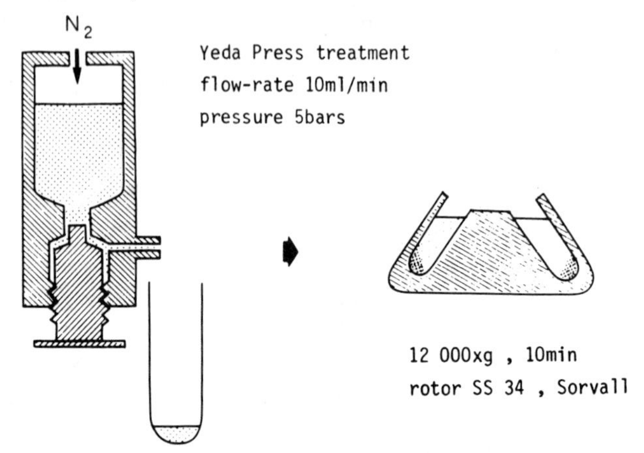

Supernatant diluted to 80ml with 10mM tricine-NaOH pH 7.8, 4mM $MgCl_2$,
13ml of the mixture layered on top of a discontinuous sucrose gradient made of 3 layers (8ml each) containing 1M ; 0.65M ; 0.4M sucrose , 10mM tricine-NaOH pH 7.8 and 4mM $MgCl_2$.

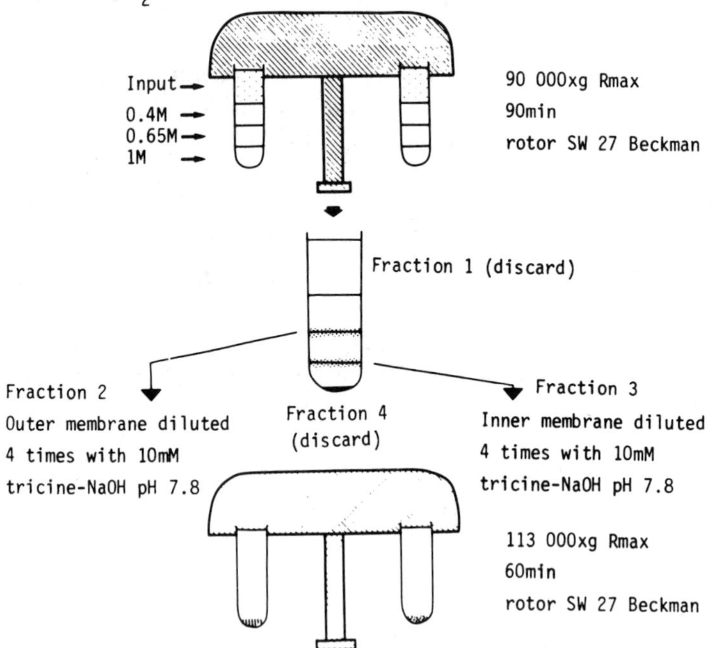

Pellets of outer membrane (∿500μg protein) and inner membrane (∿1mg protein)

FIG. 2. Purification of membrane fractions enriched in outer and inner envelope membranes from spinach chloroplasts. Reproduced with permission from Block *et al.*[3]

Superspeed centrifuge, refrigerated (RC5, Sorvall, or equivalent) with all the equipment to prepare chloroplasts in large amounts[7,10]
Ultracentrifuge (L2 65B, Beckman, or equivalent) with SW-27 rotor

Procedure (See Fig. 2). Incubate the intact and purified chloroplasts for 10 min in 60 ml (final volume) hypertonic medium, to shrink the chloroplasts and induce the formation of large empty spaces in between the two envelope membranes. Load 30 ml of the chloroplast suspension into the chamber of the precooled (0–4°) Yeda press. Increase pressure in the chamber using N_2 until 5 bars (applied pressure) is reached. Open the needle valve gently and slowly so that the entire chloroplast suspension is extruded through the orifice at a flow rate of about 10 ml/min.

Figure 2 describes the series of differential and density gradient centrifugations used to fractionate the Yeda press-treated chloroplasts. Two membrane fractions are obtained: a light fraction (d = 1.08 g/cm^3) or fraction 2 (at the interface 0.45 M/0.6 M sucrose) enriched in outer envelope membrane and a heavy fraction (d = 1.12 g/cm^3) or fraction 3 (at the interface 0.6 M/1 M sucrose) enriched in inner envelope membrane.

Analyses of the Polypeptides from the Envelope Membrane Fractions by SDS–PAGE

Gradient gels (7.5–15% linear polyacrylamide gradients), when run at room temperature, give best results for the outer envelope membrane fraction.[3] Gels containing 8 M urea, or electrophoresis performed at 4° (in presence of lithium dodecyl sulfate, LDS), should be avoided. The outer envelope membrane contains very large amounts of glycerolipids (2.5–3 mg lipids/mg protein). SDS(LDS)–lipid micelles which run at the front of the electrophoretogram strongly distort the polypeptide pattern in the low-molecular-weight region, thus leading to poor resolution. In addition, it is recommended that outer envelope polypeptides (corresponding to 0.5–1 mg protein) be extracted (at 0–4°) with 8–10 ml chloroform/methanol [2 : 1 (v/v)] mixture prior to analysis. After a 5-min incubation, remove the tube from the ice bucket and dry the solvent under a stream of N_2. Extract the lipids from the residue by addition of 8–10 ml diethyl ether (at 0–4°) without shaking to avoid resuspension of the residue. After a few seconds, remove diethyl ether from the tube, and prepare the sample for electrophoresis. The general conditions for gel preparation, sample extraction, and solubilization, electrophoresis, and gel staining have been described by Chua[17] and Piccioni *et al.*[8] It is important that the diethyl ether extraction does not remove some hydrophobic polypeptides. An

[17] N.-H. Chua, this series, Vol. 69, p. 434.

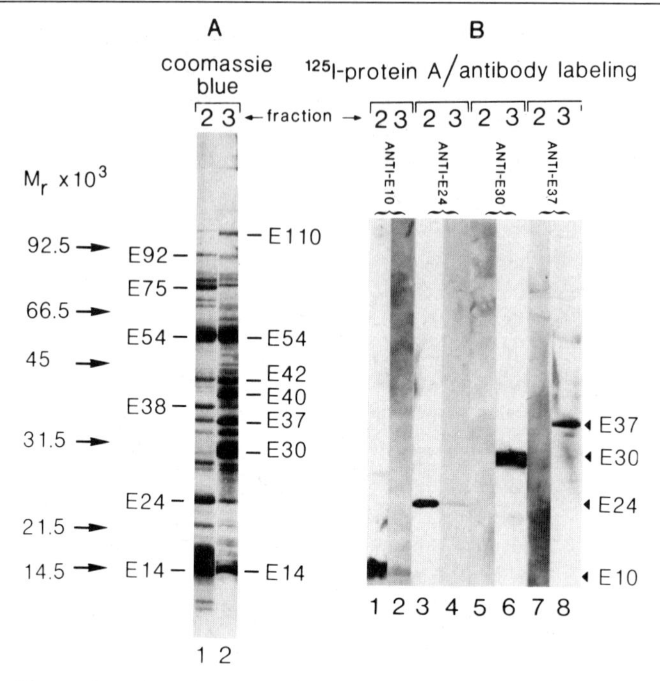

FIG. 3. Electrophoretic and immunochemical characterization of membrane fractions enriched in outer and inner envelope membranes from spinach chloroplasts. (A) Coomassie blue staining of polypeptides from fractions 2 and 3 separated by SDS–polyacrylamide gel electrophoresis at room temperature with a 7.5 to 15% acrylamide gradient. Lane 1: fraction 2; lane 2: fraction 3. (B) Autoradiography of antigen–antibody ^{125}I-labeled protein A complexes after electrophoretic transfer of polypeptides from polyacrylamide gels to nitrocellulose sheets. Antibodies to envelope polypeptides E10 (lanes 1 and 2), E24 (lanes 3 and 4), E30 (lanes 5 and 6), and E37 (lanes 7 and 8) were used. Lanes 1, 3, 5, and 7: fraction 2; lanes 2, 4, 6, and 8: fraction 3. Fractions 2 and 3 were prepared as described in Fig. 2. Reproduced with permission from Block *et al.*[3]

example of the separation of polypeptides from fractions 2 and 3 is shown in Fig. 3A.

Characterization of the Two Envelope Membrane Fractions by Immunoblotting

Solutions

Transfer medium: Tris, 25 m*M*; glycine, 192 m*M*; methanol, 20% (v/v); pH 8.3

Medium A: Tris–HCl, 10 m*M*, pH 7.4; NaCl, 9‰ (w/v); bovine serum albumin (BSA), 30 g/liter

^{125}I-Labeled protein A (New England Nuclear, immunology grade) diluted (10^6 cpm/ml) in medium A devoid of BSA

Materials

Gel transfer apparatus with power supply (Trans-Blot Cell, Bio-Rad, or equivalent)
Nitrocellulose sheets
Transparent plastic sheets, 100 μm thick
Welding apparatus for plastic sheets (Soud-Sac, Calor, France)
X-Ray films (X-OMAT, XAR5, Kodak, or equivalent)
Gyratory shaker (New Brunswick, model G-2, or equivalent

Procedure. The procedure recommended for the use of the Trans-Blot cell is simple and provides reliable results for envelope polypeptides. Electrophoretic transfer from polyacrylamide gel is performed at room temperature, the power supply being set at 60 V, 0.22 A. Transfer is complete after 3 hr electrophoresis and almost none of envelope polypeptides remained in the gel after this period of time. Wear plastic gloves. Cut the stained nitrocellulose sheet in 0.8–1 cm large bands (corresponding to one slot of the polyacrylamide gel), place it between two clean plastic sheets, and seal the plastic sheets around the nitrocellulose band with a welding apparatus, leaving one of the small sides open to fill the small bag (1.5 × 15–18 cm) thus formed. The procedure used for immunoblotting is adapted from Burnette.[18] Fill the plastic bag containing the nitrocellulose band with about 8–10 ml medium A and seal the last side of plastic bag. Incubate for 1 hr at 37° or overnight at 4°, preferably under agitation on a gyratory shaker set at 50 rpm. Remove medium A and replace it by 2–3 ml medium A containing antibodies to envelope polypeptides (about 0.6 mg IgG/ml). Incubate the sealed bag overnight at 4° on a gyratory shaker (50 rpm). After incubation, remove excess antibodies by a series of washings (at least six) with 10 ml medium A (devoid of BSA) each time, for 30 min at room temperature. Incubate washed nitrocellulose sheets for 1 hr at room temperature, under agitation as above, with 2–3 ml medium A (devoid of BSA) and containing ^{125}I-labeled protein A. After incubation remove excess ^{125}I-labeled protein A by a series of washing, as above. Dry the washed nitrocellulose sheets on Whatman #3 paper in an oven at 60°. Reveal specific binding of antibodies to envelope polypeptides by autoradiography.

Application of this procedure to membrane fractions enriched in outer and inner envelope membranes from spinach chloroplasts and antibodies

[18] W. N. Burnette, *Anal. Biochem.* **112,** 195 (1981).

raised against four different envelope polypeptides (E10, E24, E30, and E37) is shown in Fig. 3.[3]

Estimation of the Level of Cross-Contamination of the Two Envelope Membrane Fractions

The procedure developed by Chua and Blomberg[12] for crossed immunoelectrophoresis can easily be adapted for a quantitative analysis of envelope membrane fractions by immunoelectrophoresis. A composite gel, formed of four different agarose gels (cathodal, intermediate, antibody, and anodal gels), is prepared on a thick glass plate as described in detail by Chua *et al.*[9] The antibody concentration in the corresponding gel is about 0.7 mg IgG/ml.

Solutions

TAE buffer: Tris, 80 m*M*; sodium acetate, 40 m*M*; ethylenediaminetetraacetic acid, disodium salt, 1 m*M*; pH adjusted to 8.6 (25°) with glacial acetic acid

Solubilization medium: SDS, 0.2% (w/v); sodium deoxycholate, 1% (w/v); in TAE buffer

Washing solution: NaCl, 0.15 *M*

Materials

Horizontal electrophoresis cell and power supply (LKB multiphor unit or equivalent)

Procedure. Punch 3.5-mm-diameter holes in the cathodal gel, with a well-to-well spacing of 1 cm. Extract envelope membrane fractions with 8–10 ml chloroform/methanol mixture [2 : 1 (v/v)] followed by diethyl ether extraction, as described above. Solubilize the protein residue in solubilization medium (final protein concentration: 3–5 mg/ml). Load *the same amount of protein in each well* (50 to 80 μg protein) and run electrophoresis for 12–15 hr, at 15°, under a constant voltage of 100 V. After immunoelectrophoresis, remove the agarose gel from the glass plate, soak the gel at room temperature in washing solution for 1 day, and then in several changes of distilled water for an additional 3–4 days. Immunoprecipitates can be stained with Coomassie blue.[8] Measure the area under the rockets formed by antibody–antigen complexes. This area is proportional to the amount of antigen present in the original protein residue. An application to the quantitative determination of the level of cross-contamination of membrane fractions enriched in outer and inner envelope membranes is given in Fig. 4 with antibodies against outer envelope polypeptide E24 and inner envelope polypeptide E37.[3]

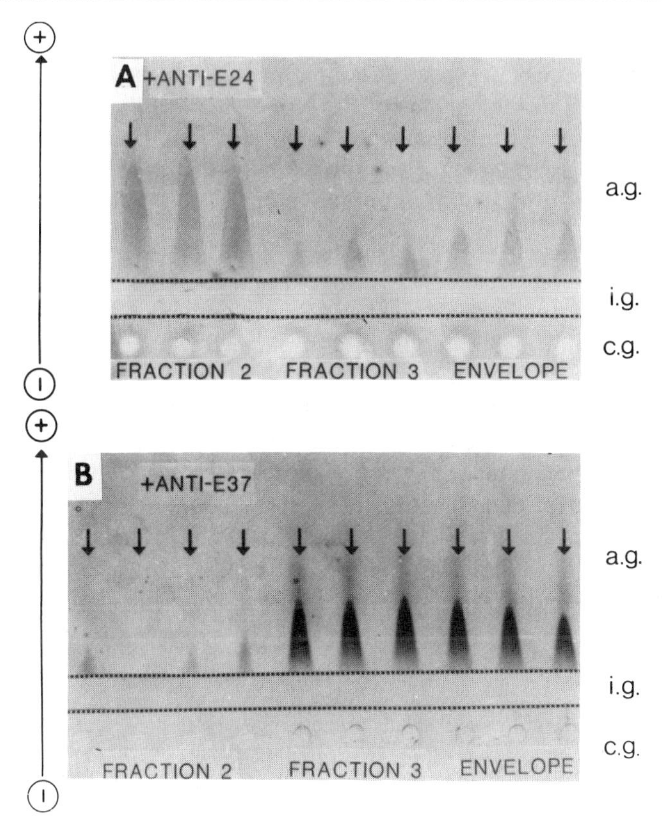

FIG. 4. Rocket immunoelectrophoresis of membrane fractions enriched in outer and inner envelope membranes and of whole envelope membranes in agarose gels containing antibodies raised against envelope polypeptides. The antigen loads were the same in each well (60 μg of protein). The composite gels used were made as described by Chua and Blomberg[12] and consisted of four different gels: cathodal (c.g.) intermediate (i.g.), antibody (a.g.), and anodal (not presented). The antibody gels contained 0.7 mg of IgG (to E24 or to E37) per milliliter. Each sample of membrane fraction was from a different experiment. The areas under the rockets were as follows: (A) (antibody to E24): fraction 2, 0.96 $\pm$ 0.08 cm^2; fraction 3, 0.21 $\pm$ 0.04 cm^2. (B) (antibody to E37): fraction 2, 0.07 $\pm$ 0.05 cm^2; fraction 3, 0.52 $\pm$ 0.06 cm^2. Since the different areas under the rockets are proportional to the amount of antigens in each sample, we calculated (from each set of sample) the percentage of outer envelope proteins in fractions 2 and 3. Fraction 2 contained 91.6 $\pm$ 8.0% of outer envelope membrane proteins whereas fraction 3 contained 19.5 $\pm$ 3.5% of outer envelope membrane proteins. Fractions 2 and 3 were prepared as described in Fig. 2; whole envelope was purified from swollen chloroplasts as described by Douce *et al.*[5] Reproduced with permission from Block *et al.*[3]

Immunodots

We have successfully used the Bio-Dot microfiltration apparatus (Bio-Rad) for quantitative analyses of envelope membrane polypeptides. The procedure described in the Bio-Rad instruction manual works very well with envelope membranes, in the range of 5–10 μg to a few nanograms envelope polypeptides with antibodies diluted to 1/500 to 1/1000.

Comments

The procedures described above—polyacrylamide gel electrophoresis, immunoblotting, immunoelectrophoresis, or immunodots—are widely used for protein characterization. The specific problems related to the application of these methods to envelope membranes are due to the very high glycerolipid content of the outer envelope membrane.[3] Most of these glycerolipids should be removed prior to solubilization of the envelope polypeptides. The chloroform/methanol–diethyl ether procedure described above is a convenient but critical step. Treatment should be done very carefully and should be controlled to prevent any specific removal of hydrophobic polypeptides by this treatment.

Acknowledgments

Professor Nam-Hai Chua is gratefully acknowledged for his collaboration in the preparation of antibodies to E10, E24, E30, and E37. Our thanks are also owed to Dr. Agnès Billecocq for her collaboration in the optimization of the procedure for immunochemical studies on isolated, intact chloroplasts.

[21] Isolation of Amyloplasts from Suspension Cultures of Soybean

By TOM AP REES and FRASER D. MACDONALD

Insufficient is known of the structure and properties of nonphotosynthetic plastids for us to be able to define precisely the stage of differentiation at which they may be called amyloplasts. In this chapter we define amyloplasts as nonphotosynthetic plastids in which starch accumulation is the dominant activity. As starch is widespread in nonphotosynthetic cells and is made only in plastids, it follows that amyloplasts are widely distributed. However, the most obviously developed amyloplasts are found in the major starch storing tissues such as the endosperm of cereals and the tubers of potatoes.

Amyloplasts vary in size according to their stage of development and to the plant species. In maize roots the differentiation of amyloplasts involves a 25- to 50-fold increase in plastid volume and results in a mature amyloplast that has a diameter of 2.17 μm.[1] In barley endosperm amyloplasts of up to 35 μm in diameter are found.[2] Amyloplasts are bounded by an envelope that consists of two membranes,[1,3] but, compared to chloroplasts, the internal membrane systems are poorly developed. Recent studies of amyloplast structure suggest the existence of two types of membrane.[1] First is the amyloplast envelope and a series of vesicles and tubules apparently in continuity with the envelope. Second is a separate series of "thylakoid-like" vesicles within the stroma of the amyloplast. In mature amyloplasts by far the greater part of the plastid is occupied by starch granules. The number and size of the latter vary. Amyloplasts from barley endosperm contain a single starch granule,[3] those from maize roots may contain up to 8 granules,[1] whereas those from the endosperm of oat have been reported to contain up to 100.[3]

Despite the fact that the bulk of mankind's food is formed within amyloplasts, there have been very few reports of attempts to develop methods for the isolation of amyloplasts. The ultimate aim of any such study must be the preparation, in reasonable yield, of intact, functional, pure amyloplasts. At the time of writing this has not been reported. If we consider the enormous amount of work that was required to achieve this aim for chloroplasts, it is perhaps not surprising that the more difficult problem of isolating functional amyloplasts has not been solved. The prime difficulty is the extreme fragility of amyloplasts, which almost certainly results from their very high ratio of starch to stroma. The method that we describe below will give appreciable yields of pure amyloplasts, but the method is limited in that, as yet, we have no proof that these purified amyloplasts have retained all the physiological and biochemical properties that they possessed in the cell. Notwithstanding this limitation, the procedure can be used to characterize the basic composition of the amyloplasts, an essential step in understanding their function. The method that we describe for amyloplast isolation involves sucrose density gradient centrifugation of lysates of protoplasts made from suspension cultures of soybean cells.

[1] P. W. Barlow, C. R. Hawes, and J. C. Horne, *Planta* **160,** 363 (1984).
[2] J. M. Williams and C. M. Duffus, *Plant Physiol.* **59,** 189 (1977).
[3] M. S. Buttrose, *J. Ultrastruct. Res.* **4,** 231 (1960).

Growth of Suspension Cultures

The cultures of soybean cells, derived originally from cotyledons of germinating soybeans (*Glycine max* [L.] Merr. cv. Acme), were obtained from Dr. Miller in 1970 and have been maintained in Cambridge ever since.[4] The culture is best maintained as a callus grown at 25° in the dark on the medium shown in Table I[5] but solidified with 1% agar; subculturing should be done at monthly intervals. To obtain a suspension culture add 10 g fresh weight of callus to a 500-ml Erlenmeyer flask that contains 100 ml of the medium shown in Table I, then incubate the flask for 7–14 days on an orbital shaker at 100 rpm at 25° in the dark. The resulting cell suspension (30 ml) is then used to inoculate fresh medium (100 ml) in a similar flask that is then incubated as described above to give a suspension culture. The latter should be subcultured every 7 days. The suspension cultures are intolerant of hypoxia and require constant agitation to ensure adequate aeration. It is prudent to maintain callus cultures on solid medium so as to be able to replace suspension cultures if the latter are lost.

Preparation of Protoplasts

Reagents

Sorbitol medium: 0.35 *M* sorbitol, 4 m*M* $CaH_4(PO_4)_2$, 10 m*M* MES (pH 5.7)

Digestion medium: 0.35 *M* sorbitol, 4 m*M* $CaH_4(PO_4)_2$, 10 m*M* MES (pH 5.7), 270 U pectinase (polygalacturonase, EC 3.2.1.15, from Sigma Chemical Co.), and 5% (w/v) Driselase (Uniscience, Cambridge)

Method

Use cultures that are 4 to 5 days old. Working at room temperature (20°), take 100 ml of cell suspension (4 to 5 g fresh weight of cells) and collect the cells by gentle suction onto glass fiber filter paper (Whatman GF/C) in a Büchner funnel. Rinse the cells in the funnel with 10 ml sorbitol medium and then use a further 10 ml of the same medium to wash the cells off the paper into a 50-ml beaker. Incubate the resulting cell

[4] C. O. Miller, *in* "Modern Methods of Plant Analysis" (H. F. Linskens and M. V. Tracey, eds.), Vol. 6, p. 194. Springer-Verlag, Berlin and New York, 1963.

[5] C. O. Miller, *Ann. N.Y. Acad. Sci.* **144,** 251 (1967).

TABLE I
MEDIUM[a] FOR GROWTH OF SOYBEAN CULTURES

Component	(mg/liter)	Component	(mg/liter)
Sucrose	30,000	$ZnSO_4 \cdot 7H_2O$	3.8
myo-Inositol	100	H_3BO_3	1.6
KNO_3	1,000	$Cu(NO_3)_2 \cdot 3H_2O$	0.35
NH_4NO_3	1,000	$(NH_4)_6Mo_7O_{24} \cdot 4H_2O$	0.1
$Ca(NO_3)_2 \cdot 4H_2O$	500	KI	0.8
KH_2PO_4	300	Nicotinic acid	0.5
$MgSO_4 \cdot 7H_2O$	71.5	Pyridoxine–HCl	0.1
KCl	65	Thiamin–HCl	0.1
$MnSO_4 \cdot 4H_2O$	14	Naphthalene-1-acetic acid	2.0
Ethylenediaminetetraacetic acid, ferric monosodium salt	13.2	Kinetin	0.5

[a] Macronutrients are added as solids, micronutrients are added from stock solutions. The latter are kept at −20° except that those for KI, naphthalene-1-acetic acid, and kinetin are always freshly prepared. Naphthalene-1-acetic acid and kinetin are dissolved by boiling for 1 hr in distilled water. The pH of the medium is not adjusted; sterilization is by autoclaving for 20 min at 121° at 104 kN m^{-2}. Adapted from Miller.[5]

suspension for 20 min without shaking. Next, collect the cells onto a glass fiber filter paper as described above and wash them off the paper with 15 ml digestion medium. Divide the resulting suspension into thirds and incubate each lot of 5 ml at 25° in a 25-ml beaker for 3 hr on a reciprocal shaker at 60 strokes/min for the first hour and 10 strokes/min for the last 2 hr. Protoplasts appear after about 1 hr; after 3 hr, 40–60% of the cells are converted to protoplasts.

The purification of the protoplasts is done at room temperature (ca. 20°). First, filter by gravity the suspension of cells in digestion medium through stainless steel gauze (mesh size 43 μm). In this and all subsequent transfers of protoplasts use a Pasteur pipet with a wide mouth (2–3 mm). To recover protoplasts from the above filtrate centrifuge at 100 *g* (r_{av} 11 cm) for 1 min. Gently resuspend the protoplasts in 10 ml sorbitol medium with a fine paint brush. Centrifuge as above and gently resuspend the protoplasts in 10 ml 42% (w/w) Percoll (Pharmacia, Ltd.) in sorbitol medium. Now carefully overlay the protoplast suspension with 5 ml sorbitol medium and centrifuge at 200 *g* (r_{av} 11 cm) for 5 min. Intact protoplasts float up to the interface and should be removed carefully with a Pasteur pipet and added to 15 ml sorbitol medium.

Fractionation of Protoplasts

Lysis

Sediment the purified protoplasts by centrifugation for 1 min at 100 *g* (r_{av} 11 cm) and then resuspend the 0.5 to 1.0 ml packed volume of protoplasts in 1.0 ml homogenization medium [50 m*M* Tricine, pH 7.4, 10 m*M* KCl, 1 m*M* $MgCl_2$, 1 m*M* EDTA, 0.1% bovine serum albumin, 21% (w/w) sucrose] at 4°. This suspension is loaded via the open end into the barrel of a 2-ml disposable syringe in which a piece of Miracloth (Calibiochem, San Diego, CA) has been put between the barrel and the needle (0.7 × 40 mm). The protoplast suspension is now forced out through the cloth and needle. Repetition of this process twice, to give three passages through the needle in all, generally gives almost complete lysis of the protoplasts. At this stage it is important to check with a microscope that lysis is almost complete; repeat the process until it is. The lysate is finally filtered by gravity through two layers of Miracloth and made up to 15 ml with ice-cold homogenization medium to give the unfractionated lysate. Lysis is carried out at room temperature.

Fractionation

A portion (8 ml) of the unfractionated lysate is placed carefully onto a 27-ml linear gradient of 16–60% (w/w) sucrose supported by 2 ml 60% (w/w) sucrose. It is important to ensure that the density of the unfractionated lysate is the same as that of the solution at the top of the gradient.[6] The sucrose in the gradient is dissolved in homogenization medium. The lysate is at 4° and the gradient is precooled to 4°. Gradients should be prepared and cooled within 8 hr of and 2 hr prior to use. The gradient plus lysate are centrifuged at 4° for 1 hr in an SW-27 rotor, Beckman L5-50 centrifuge, at 21,000 rpm (90,000 *g* at r_{max} 16.1 cm). Fractionate the gradient from the bottom. It is convenient to take 1.0-ml fractions except for the gradient pellet, which should be resuspended in 2.0 ml homogenization medium containing 21% (w/w) sucrose, and for the top 8 ml of the gradient, equivalent to the volume of lysate added, which is best treated as a single fraction.

Distribution of Marker Enzymes

To assess the effectiveness of the fractionation the following markers may be used to locate the amyloplasts and possible contaminants in the gradient fractions. The best indication of the presence of amyloplasts is

[6] M. Emes and M. W. Fowler, *Planta* **144**, 249 (1979).

TABLE II
ACTIVITIES OF MARKER ENZYMES IN AMYLOPLAST PREPARATIONS[a]

Enzyme	Activity in unfractionated lysate (nmol/min/ml lysate)	Percentage of activity in unfractionated lysate that was recovered in:	
		Amyloplasts	Sum of all gradient fractions
Starch synthase	0.8 ± 0.1	23.4 ± 1.7	102 ± 10
ADPglucose pyrophosphorylase	3.4 ± 1.4	28.9 ± 6.5	120 ± 16
Nitrite reductase	20.9 ± 4.4	23.1 ± 1.7	107 ± 6
Alcohol dehydrogenase	260 ± 35	0.2 ± 0.1	97 ± 12
Cytochrome-*c* oxidase	191 ± 36	0.6 ± 0.2	53 ± 4
Cytochrome-*c* reductase	71 ± 11	1.8 ± 0.3	110 ± 3
Catalase	14,300 ± 300	0.3 ± 0.2	69 ± 5

[a] Protoplast lysates were fractionated as in Fig. 1. For each enzyme the activities in fractions 1 and 2 are summed to give activity in amyloplasts. Values are means ± SE from at least five lysates. From Macdonald and ap Rees.[8]

coincidence between the distribution of starch and the enzymes starch synthase (EC 2.4.1.21), ADPglucose pyrophosphorylase (EC 2.7.7.27, glucose-1-phosphate adenylyltransferase), and nitrite reductase (EC 1.6.6.4).[7] For contaminants the following markers may be used: cytosol, alcohol dehydrogenase (EC 1.1.1.1), phosphoenolpyruvate carboxylase (EC 4.1.1.31), pyrophosphate–fructose-6-phosphate 1-phosphotransferase (EC 2.7.1.90)[7]; mitochondria, cytochrome-*c* oxidase (EC 1.9.3.1); endomembrane systems, cytochrome-*c* reductase (NADH dehydrogenase, EC 1.6.99.3); microbodies, catalase (EC 1.11.1.6), hydroxypyruvate reductase (EC 1.1.1.81).[8]

A typical fractionation is shown in Fig. 1 and the results from a number of such fractionations are averaged and shown in Table II. Apart from cytochrome oxidase, the recoveries of the enzymes from the gradient are satisfactory. The modest recovery of cytochrome-*c* oxidase is not due to the presence of an inhibitor in the amyloplast fraction,[8] but is probably due to the difficulty in obtaining a linear rate when the standard assay for this enzyme is used with unfractionated lysates. Appreciable and comparable proportions of the amyloplast markers, starch synthase, nitrite reductase, and ADPglucose pyrophosphorylase, are recovered in the gradi-

[7] F. D. Macdonald and J. Priess, *Planta* **167,** 240 (1986).
[8] F. D. Macdonald and T. ap Rees, *Biochim. Biophys. Acta* **755,** 81 (1983).

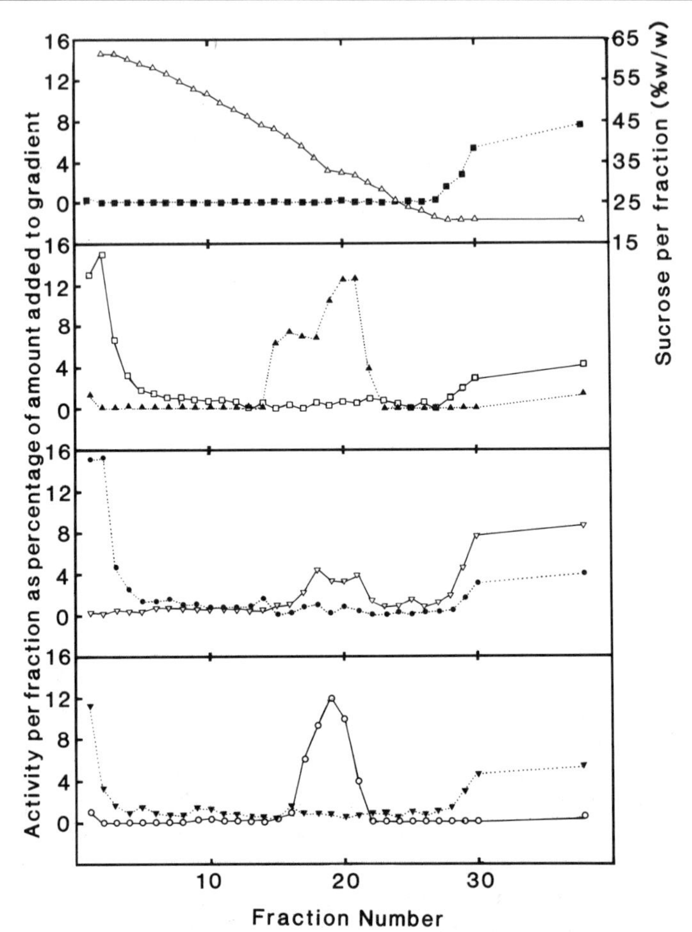

FIG. 1. Sucrose density gradient fractionation of lysates of soybean protoplasts. The gradient pellet was resuspended in 2 ml homogenization medium and is fraction 1: the next 29 ml of the gradient was divided into 1-ml fractions (numbers 2–30), and the top 8 ml of the gradient was taken as a single fraction, numbered 38. Total activity is plotted for each fraction except fraction 38, where activity/ml of fraction is given. Each fraction was assayed for: sucrose (△—△), alcohol dehydrogenase (■---■), nitrite reductase (□—□), catalase (▲---▲), starch synthase (●---●), cytochrome-*c* reductase (▽—▽), cytochrome-*c* oxidase (○—○), α-glucan phosphorylase (▼---▼).

ent pellet and the fraction immediately above it. These fractions also contained 78% of the starch added to the gradient. These fractions are essentially free of contamination by any of the other marker enzymes assayed. Thus the gradient pellet and the fraction immediately above it may be regarded, collectively, as forming the amyloplast fraction. The fact that this fraction contains more of the starch than the stromal enzymes of the amyloplast is almost certainly due to the presence in the fraction of starch released from broken amyloplasts. Evidence that the presence of the stromal enzymes in the amyloplast fraction is not due to selective adsorption of these enzymes onto starch grains is provided by demonstrations that the activities of such enzymes in the unfractionated lysate are both latent, and protected from the action of trypsin.[8]

Comments

Particular attention should be given to the following three points. First, the age of the suspension cultures used to make amyloplasts has a critical effect on the yield of amyloplasts. If the cultures are too young and contain too little starch the yield is reduced and much of the plastid markers are spread as a diffuse band between fractions 6 and 17.[9] The latter probably represents plastids that have not fully differentiated into amyloplasts. Second, rupture of the protoplasts must be done carefully so as to ensure that all the protoplasts are broken with the minimum damage to amyloplasts. Unbroken protoplasts are generally recovered in the diffuse band (fractions 6–17) and can be detected by the presence of all the marker enzymes.[9] Third, the integrity of protoplast and amyloplast membranes is severely reduced by detergents, thus all glassware and solutions should be scrupulously prepared to be free of detergent.

The above procedure gives a respectable yield of uncontaminated amyloplasts, 23–29%, which compares favorably with yields of chloroplasts from leaves. Protection experiments carried out on the isolated amyloplasts from the gradient indicated that 30% of the purified amyloplasts were intact.[10] The extent to which such amyloplasts are functional has not yet been established.

Finally we stress that the above method is not the only one available for the isolation of amyloplasts. Echeverria *et al.* have described a promising method for the isolation of amyloplasts from maize endosperm.[11] In the latter, two of the potential disadvantages of our method, relatively high centrifugal forces and high concentrations of sucrose as osmoticum, are avoided by allowing the amyloplasts to settle under gravity through a gradient of Ficoll. Yields were 5–10% and estimates of intactness of the isolated amyloplasts ranged from 27 to 93% depending upon the methods used to assess it. Journet and Douce have recently reported the successful

[9] P. Gross and T. ap Rees, *Planta* **167,** 140 (1986).
[10] F. D. Macdonald, Ph.D. Thesis, University of Cambridge (1981).
[11] E. Echeverria, C. Boyer, K.-C. Liu, and J. Shannon, *Plant Physiol.* **77,** 513 (1985).

purification of small plastids from cauliflower buds.[12] The yield was low but intactness was very high, 90%. Although these cauliflower plastids contained starch, they were also active in fatty acid synthesis. So the extent to which they represent amyloplasts is not yet known. In terms of yield of intact amyloplasts the three methods do not differ greatly, 30% of 23–29%, 27–90% of 5–10%, 90% of 1%. The crucial comparison will be the extent to which the isolated plastids are functional. No data on this question have been published yet.

[12] E. P. Journet and R. Douce, *Plant Physiol.* **79,** 458 (1985).

[22] Isolation of Amyloplasts from Developing Endosperm of Maize (*Zea mays* L.)

By JACK C. SHANNON, EDGARDO ECHEVERRIA, and CHARLES BOYER

Introduction

Amyloplasts are colorless organelles specialized for the synthesis and accumulation of reserve starch. They develop from proplastids, as do chloroplasts, and are bounded by a double membrane. Under certain conditions amyloplasts develop into chloroplasts and vice versa.[1] Amyloplasts may contain single or multiple starch granules, depending on the species and the tissue within which the amyloplasts occur.[2] Normal maize endosperm amyloplasts generally contain a single starch granule.[2] However, the number and shape of starch granules in amyloplasts vary in different maize endosperm mutants. The reader is referred to the review by Shannon and Garwood[2] for a more complete discussion of the genetics and physiology of starch development.

Procedures for the successful isolation of relatively pure, intact plastids from castor bean endosperm were reported several years ago[3,4] and these have been very useful in demonstrating that plastids contain the

[1] J. T. O. Kirk and R. A. E. Tilney-Basset, "The Plastids: Their Chemistry, Structure, Growth and Inheritance." Elsevier/North-Holland Biomedical Press, New York, 1978.

[2] J. C. Shannon and D. L. Garwood, *in* "Starch: Chemistry and Technology" (R. L. Whistler, J. N. BeMiller, and E. F. Paschall, eds.), p. 26. Academic Press, New York, 1984.

[3] P. D. Simcox, E. E. Reid, D. T. Canvin, and D. T. Dennis, *Plant Physiol.* **59,** 1128 (1977).

[4] M. Nishimura and H. Beevers, *Plant Physiol.* **62,** 40 (1978).

enzymes of glycolysis and gluconeogenesis.[3–5] The glycolytic enzymes, triose-phosphate isomerase (EC 5.3.1.1), glyceraldehyde-3-phosphate dehydrogenase (EC 1.2.1.12), phosphoglycerate kinase (EC 2.7.2.3), enolase (EC 4.2.1.11), and phosphoglycerate mutase (EC 5.4.2.1), in plastids are isozymes of the corresponding enzymes located in the cytosol.[6] Although useful quantities of plastids can be isolated following a very brief homogenization of the tissue,[3] Nishimura and Beevers[4] reported that much higher yields of intact plastids were recovered from castor bean endosperm when protoplasts were used as the starting material.

Amyloplasts, with starch granule inclusions, are even more fragile than castor bean endosperm plastids and extra care must be taken during cell rupture and purification. Duffus and Rosie[7] and Williams and Duffus[8] isolated starch granules from potato tubers and immature barely endosperm, respectively. Although these researchers called their preparations amyloplasts, they presented no evidence that the starch granules were indeed surrounded by intact membranes. Rijven[9] isolated enzymatically active starch granules from developing wheat endosperms. This isolation procedure yielded a granule preparation containing the presumed amyloplast enzymes, starch phosphorylase and soluble and bound starch synthase, and a cytosol enzyme, sucrose synthase. These enzymes apparently are insoluble in the isolation medium which contained 300 m*M* polyethylene glycol 1000 and cosedimented with the starch granules. Thus their association with the starch granules was not related to the presence of an intact amyloplast membrane.

Recently, amyloplasts containing several small starch granules have been isolated from etiolated pea epicotyls,[10] maize coleoptiles,[11] and cauliflower buds.[12] Gaynor and Galston[10] stressed the importance of using low-gravity centrifugation (less than 150 *g*), but Journet and Douce[12] used centrifugation forces up to 40,000 *g* in the purification of cauliflower bud amyloplasts which contained medium-size starch granules. Journet and Douce[12] state that plastids with larger starch granules are ruptured when the starch granules are thrown through the plastid envelope membrane during blending and/or centrifugation. Sack *et al.*[11] prepared a crude low-gravity pellet from maize coleoptiles. Although the pellet was composed

[5] M. Nishimura and H. Beevers, *Plant Physiol.* **64,** 31 (1979).
[6] J. A. Miernyk and D. T. Dennis, *Plant Physiol.* **69,** 825 (1982).
[7] C. M. Duffus and R. Rosie, *Anal. Biochem.* **65,** 11 (1975).
[8] J. M. Williams and C. M. Duffus, *Plant Physiol.* **59,** 189 (1977).
[9] A. H. G. C. Rijven, *Plant Physiol.* **75,** 323 (1984).
[10] J. J. Gaynor and A. W. Galston, *Plant Cell Physiol.* **24,** 411 (1983).
[11] F. D. Sack, D. A. Priestley, and A. C. Leopold, *Planta* **157,** 511 (1983).
[12] E.-P. Journet and R. Douce, *Plant Physiol.* **79,** 458 (1985).

largely of free starch granules, examination by electron microscopy and by fluorescence microscopy following incubation with fluorescein isothiocyanate bound to cationized ferritin showed that about 5% of the particles were intact amyloplasts. Macdonald and ap Rees[13] obtained amyloplasts from lysed protoplasts derived from soybean suspension cultures. Although these researchers demonstrated amyloplast intactness in the protoplast lysates, such a demonstration was not possible following centrifugation (60,000 *g*) through a sucrose density gradient. However, their purified amyloplast preparations were shown to be relatively free of contamination by cytosol or by other organelles likely to be involved in carbohydrate metabolism. Amyloplasts from both cauliflower bud[12] and soybean suspension cultures[13] contained glycolytic and gluconeogenic enzymes.

Amyloplasts containing a single large starch granule, such as potato and maize, are very prone to rupture during even low-gravity centrifugation. However, if one is able to release intact amyloplasts from the cell, then advantage can be taken of the starch granule mass to separate the amyloplasts from the cytosol and other organelles by settling through a density gradient at 1 *g*. Fishwick and Wright[14] took advantage of the heavy starch granules to purify potato amyloplasts by settling (1 *g*) through a sucrose density gradient. Although care was taken to minimize amyloplast membrane rupture, yield of intact amyloplasts present in the preparations rarely exceeded 16% on a starch weight basis. We have also used a 1 *g* settling procedure to isolate intact amyloplasts from developing maize endosperm[15] and the details of our procedure follow.

Aqueous Isolation of Amyloplasts from Developing Maize Endosperm

Probably the most critical stage in the isolation of intact amyloplasts from maize endosperm is the release of the amyloplasts from the cells. The most satisfactory procedure, to date, is to first produce protoplasts from which the amyloplasts are isolated.

Protoplast Formation

Maize ears are harvested 14 to 17 days after pollination. Remove the pericarp and embryo, cut the endosperms transversely into three sections with a sharp razor blade, and collect them at room temperature in distilled water until all sections are prepared (about 30 min). Rinse the slices with distilled water to remove cellular contents from the cut cells.

[13] F. D. Macdonald and T. ap Rees, *Biochim. Biophys. Acta* **755,** 81 (1983).
[14] M. J. Fishwick and A. J. Wright, *Phytochemistry* **19,** 55 (1980).
[15] E. Echeverria, C. Boyer, K.-C. Liu, and J. Shannon, *Plant Physiol.* **77,** 513 (1985).

About 1.5 g of the endosperm sections is placed in a 50-ml Erlenmeyer flask with 10 ml of incubation buffer containing 1% Cellulysin (Calbiochem) and 0.02% pectolyase Y-23 (Seishin Pharmaceutical Co., Ltd.). The incubation buffer contains 0.7 *M* sorbitol, 20 m*M* $CaCl_2$, 0.5 m*M* DTE, 0.05% BSA, 25 m*M* MES, pH 5.6, and 10 m*M* arginine. Vacuum infiltrate the tissue and incubate at 30° for 4 to 5 hr. Incubation without shaking is critical because essentially all protoplasts are ruptured by even slow swirling. Following incubation, protoplasts are released from the endosperm slices by very carefully rotating the flask. The digest is then filtered through a 140-μm nylon mesh to eliminate large debris and undigested tissue.

Trichoderma reesei-derived cellulases, available from Worthington Diagnostics (CEL), can be substituted for Cellulysin but the quantity of enzyme used should be varied depending on the specific activity of the cellulase preparation. Cellulase Y-C from Seishin Pharmaceutical Co., Ltd., effectively produces protoplasts from maize endosperm, but completely inactivates branching enzyme, an amyloplast marker enzyme, and fructose-1,6-bisphosphate aldolase. The purified CEL cellulase was least detrimental to these enzymes. We have not determined the suitability of cellulases from other sources for the production of useful protoplasts. Various batches of Cellulysin have been shown to inhibit branching enzyme differently. Thus, the effect of the cellulases used on enzymes of interest should be carefully determined. Although we have not made an extensive survey of all available pectinases, pectolyase Y-23 added to the Cellulysin made the difference between the production of usable quantities of protoplasts from maize endosperm and release of no protoplasts. Although usable quantities of protoplasts are produced by the above procedure, many cells do rupture and a relatively low proportion of the total number of endosperm cells are recovered as protoplasts. In fact, a much higher number of protoplasts were recovered when the digestion was carried out for 5 days at 4° in a medium containing 0.7 *M* sucrose and 0.2 m*M* $CaCl_2$ in place of the sorbitol and 20 m*M* $CaCl_2$, respectively[15]. However, tissue digestion at 30° in the presence of sucrose is much slower than that in sorbitol. In the preferred incubation medium, the $CaCl_2$ concentration was increased from 0.2 to 20 m*M* to reduce premature protoplast rupture.

Protoplast Purification

The filtered protoplast preparation is layered on a discontinuous Ficoll (M_r 400,000; Pharmacia Co., Uppsala, Sweden) density gradient consisting of 1 ml each of 20% (w/v), 15, and 12.5%, and 2 ml of 10% Ficoll. The Ficoll is in a buffer solution containing 0.7 *M* sorbitol, 10 m*M* arginine, 10 m*M* DTE, 25 m*M* MES, pH 5.6, 5 m*M* $CaCl_2$, and Ficoll at the appropriate concentrations. Intact protoplasts settle to the bottom of the gradient within 1 hr at 4°. Initially,[15] we used a gradient containing 30, 27, 21, 15, and 10% Ficoll in the same buffer solution as above except sucrose was substituted for sorbitol. Protoplasts collected in the 27% Ficoll layer. Both gradients give equal recovery of protoplasts. Therefore, we prefer the simpler gradient.

Amyloplast Isolation

The upper three Ficoll layers are removed and the combined 15 and 20% Ficoll layers and pellet are slowly diluted with an equal volume of the above buffer solution less sorbitol and Ficoll and this mixture is further diluted with an equal volume of the same buffer without Ficoll but containing 0.35 *M* sucrose. The suspension is passed through a nylon mesh (10- or 20-μm opening) positioned at the lower end of a 50-ml disposable plastic syringe. The protoplasts are ruptured by one or two passages through the mesh. Small protoplasts may escape through the 20-μm mesh. The resulting suspension is layered on a discontinuous Ficoll density gradient composed of 1.0 ml of 30% (w/v), 1.0 ml of 15%, 0.5 ml of 12.5%, and 1.0 ml of 10% Ficoll. The Ficoll solutions contain the same components as those used above except 0.35 *M* sucrose is substituted for the sorbitol. The amyloplasts and starch granules are allowed to settle at 1 *g* in the cold (4°) for approximately 3 hr depending on the size of the starch granules.

With this procedure, we found that starch granules and the activities of amyloplast marker enzymes (branching enzyme, nitrite reductase, and soluble starch synthase) were distributed throughout the gradient.[15] Although the 15% Ficoll layer contained only about 10% of the plastid marker enzyme present in the protoplast, it contained the highest proportion of intact amyloplasts and was free of contamination by other cellular components.[15] With the above procedure, we believe that the most serious amyloplast membrane damage occurs during protoplast rupture. Although castor bean endosperm protoplasts were ruptured with little damage to plastids by passage through nylon mesh,[4] the presence of large starch granules in the maize amyloplasts creates a special problem. For example, during maize endosperm protoplast rupture the starch granules quickly settle and cake on the surface of the nylon mesh. Once this occurs, the pressure necessary to force the protoplasts through the mesh also ruptures the amyloplast membrane. Thus, it is important to rupture only a small quantity of protoplast at a time and frequently move to new positions on the nylon mesh. We are continuing to try and improve this step in the procedure.

Determination of Amyloplast Intactness. Latency: Latency studies can be used to determine the percentage of enzyme activity in the preparation which is located within intact amyloplasts. Latency experiments are based on the assumption that the substrate for the plastid marker enzyme does not cross the amyloplast envelope.[12,13,15] Consequently, activity of enzymes within the amyloplasts will not be measured so long as the membrane remains intact, but will be measured once it is ruptured.

Duplicate aliquots of the amyloplast suspension from the 15% Ficoll layer are added to a standard enzyme reaction mixture modified to contain 0.35 *M* sucrose. One tube also contains 0.1% (v/v) Triton X-100. The Triton X-100 causes rupture of the amyloplast envelope and gives a measure of the total enzyme activity. Only enzyme activity exterior to the membrane is measured in the tube without Triton X-100. It is important to exercise extreme care when pipetting the amyloplast suspension and when mixing the enzyme reaction to avoid rupture of the amyloplasts envelope. The enzyme activity of the ruptured preparation (total) minus that of the intact preparation is called latent activity. If it is assumed that the activity of a plastid marker enzyme, such as branching enzyme,[15] is the same for all amyloplasts in the protoplasts, then the number of intact amyloplasts in the 15% Ficoll fraction can be estimated. For example, the number of intact amyloplasts in the 15% Ficoll layer equals the latent activity of branching enzyme (or any other specific amyloplast enzyme) times the total activity of branching enzyme in the ruptured protoplasts divided by the total number of starch granules in the ruptured protoplasts. This calculation is based on the fact that amyloplasts from normal maize endosperm generally contain one starch granule per amyloplast. It follows then that percentage intactness in the 15% Ficoll fraction equals the number of intact amyloplasts in this fraction divided by the total number of starch granules in the 15% Ficoll fraction times 100.

The above procedure assumes that the activity of branching enzyme per amyloplast is the same in all amyloplasts from the tissue. However, this has not been proven. Thus, we suggest below a second method (fluorescence microscopy) for estimating the percentage of amyloplast intactness using fluorescence microscopy.

Fluorescence microscopy: Sack *et al.*[11] demonstrated that cationized ferritin (CF) binds to the negatively charged amyloplast membranes. When CF is tagged with fluorescein isothiocyanate (FITC), membrane-bound plastids can be distinguished from free starch granules by their brighter fluorescence when viewed with fluorescence microscopy.

Fluorescence microscopy can be used to determine the percentage of starch granules in the 15% Ficoll layer which are bounded by an intact membrane. Samples for fluorescence microscopy are collected in a gradient in which the 15% Ficoll layer contains 2.5% glutaraldehyde. After an overnight fixation at 4°, the fixation solution is removed, the amyloplasts thoroughly washed with water, and spread over glass slides. When dry, two drops of acetone are added to fix the amyloplasts to the glass slides. Cover the dry amyloplasts with 0.5 ml of FITC-CF and place the slides in a humid chamber for 30 min. [FITC-CF is prepared by combining 0.5 ml of CF (25 mg/ml, Sigma) with 1.5 ml of amyloplast isolation medium containing 1.5 mg FITC and stirring the mixture 30 min at room temperature before use.] Very carefully rinse off the unbound FITC-CF with distilled water and dry the slides on a hot plate at 40°. Examine the slides with a fluorescence microscope. With the above procedure, Echeverria *et al.*[15] estimated that 45 to 60% of the starch granules in the 15% Ficoll layer were surrounded by an intact membrane. Although we did not attempt to expose the amyloplasts to the FITC-CF prior to adding them to the slides, such a procedure should also be satisfactory.

Protection experiments: Protection experiments have been used to estimate plastid intactness.[13] The rationale is that enzymes within the plastid are protected from inactivation by trypsin. In our studies, trypsin treatment was shown to increase the rate of maize amyloplast membrane rupture and thus resulted in a low estimate of percentage intactness.[15] Since this procedure was not satisfactory for measuring intactness of maize amyloplasts it will not be discussed further. The reader is referred to the paper by Macdonald and ap Rees[13] for protection experiment methods.

Evidence of Purity. Purity of the amyloplast preparation can be assessed by assaying for enzymes specific to the cytosol, mitochondria, or glyoxysomes such as sucrose synthase, isocitrate dehydrogenase, and catalase, respectively. Echeverria *et al.*[15] reported activity for each of these enzymes in the ruptured protoplasts, and in the upper layers of the gradient, but there were no detectable activities of any in the 15% Ficoll layer.

Enzyme Composition. Echeverria *et al.*[16] showed that purified amyloplasts collecting in the 15% Ficoll layer contained all enzymes necessary to convert dihydroxyacetone phosphate (DHAP) to starch. DHAP readily crosses the maize amyloplast membrane and is converted to starch.[17] Although the definitive studies are incomplete, preliminary results indicate that a triose-P/P_i transporter is responsible for moving DHAP across the maize amyloplast membrane.

[16] E. Echeverria, C. D. Boyer, P. A. Thomas, and J. C. Shannon, unpublished observations.
[17] E. Echeverria, C. D. Boyer, and J. C. Shannon, unpublished observations.

Nonaqueous Isolation of Maize Amyloplasts

Endosperm slices are prepared as above and quickly frozen in Freon-12 (CCl_2F_2) chilled to its freezing point (−156°) in liquid nitrogen. The slices are freeze dried at 5–10 μm Hg vacuum for 48 hr. The tissue must be maintained below −25° while the vacuum is being drawn down. Store the freeze-dried tissue below −25° in a desiccator over P_2O_5 until used.

Pulverize the tissue 30 sec in a Spex mixer-mill (Spex Industries, Inc., Metuchen, NJ) and add 0.5 g to a 50-ml plastic centrifuge tube containing 15 ml of cold (0–4°) glycerol. Homogenize the tissue with a Polytron for four 10-sec pulses. To reduce heat build-up during homogenization hold the centrifuge tube in an ethanol–ice bath. All solutions used for the nonaqueous isolations should be dried over silica gel beads (Sigma, type II) for at least 24 hr before use.

Centrifuge the homogenate at 500 *g* for 10 min at 0 to 4° and vacuum filter the supernatant through two layers of Miracloth. Suspend the 500 *g* pellet in 10 ml of glycerol and homogenize 10 sec with the Polytron and filter. Rinse the filter with 5 ml of glycerol. Continually stir during filtration to keep the starch granules suspended.

Collect the combined filtrate (about 30 ml) in a 50-ml plastic centrifuge tube containing 10 ml of 3-chloro-1,2-propanediol and centrifuge at 25,000 *g* for 60 min at 0 to 4° using a Sorvall HB-4 rotor. Decant the supernatant, and wipe the tube walls with a lint-free tissue to remove residual supernatant. Suspend the starch granule pellet in 10 ml of glycerol and inject 3 ml of 3-chloro-1,2-propanediol under the suspension using a syringe and cannula. Centrifuge as above and discard the supernatant. The resulting pellet contains starch granules with associated metabolites.

Liu and Shannon[18] concluded that starch granules prepared as above were relatively free of cytosol and nuclear contamination based on the low quantity of RNA and DNA recovered in the starch granule preparation. Based on ultrastructural observations of freeze-dried endosperm tissue, they suggested that the amyloplast stroma and membrane become fixed to the starch granule surface during freeze-drying. Thus, the metabolites associated with the isolated starch granules represent the metabolites of the amyloplasts or at least those metabolites near the starch granules *in situ*. Starch granules prepared as above were shown to contain neutral sugars, inorganic phosphate, hexose and triose phosphates, organic acids, adenosine and uridine nucleotides, sugar nucleotides, and amino acids.[19] Unfortunately, 3-chloro-1,2-propanediol inactivates at

[18] T.-T. Y. Liu and J. C. Shannon, *Plant Physiol.* **67,** 518 (1981).

[19] T.-T. Y. Liu and J. C. Shannon, *Plant Physiol.* **67,** 525 (1981).

least some of the starch synthetic enzymes and we were unable to determine enzyme compartmentation in these nonaqueously isolated starch granule preparations. As noted earlier we failed to detect hexokinase, sucrose synthase, and invertase in aqueously isolated maize amyloplasts.[16] Therefore, we now question whether the glucose, fructose, and sucrose measured in the nonaqueously isolated starch granule preparation[19] were located within the amyloplasts *in situ*, or whether they became fixed to the starch granules during freeze drying. Additional studies are needed to answer this question.

Acknowledgment

Contribution No. 86, Department of Horticulture, The Pennsylvania State University. Authorized for publication as paper No. 7371 in the journal series of the Pennsylvania Agricultural Experiment Station.

[23] Isolation of Plastids from Buds of Cauliflower (*Brassica oleracea* L.)

By ETIENNE-PASCAL JOURNET

Plastids exist in a number of different forms, with different functions, but they are all, to some extent, interconvertible. Generally, all the plastids in a given plant cell will be of the same type, which is determined by the tissue in which the cell is located.[1] The most commonly studied plastid is the chloroplast. Unfortunately, knowledge about higher plant nongreen plastids has lagged behind that concerning chloroplasts, because serious problems are encountered while preparing intact, active nongreen plastids free from contaminating membranes and mitochondria. Such purification is necessary for the accurate investigation of membrane lipids, nucleic acids, protein, and enzymatic capacities. The aqueous methods reported to date have included linear or discontinuous sucrose gradient centrifugation of both mechanically prepared crude plastids[2] and those from broken protoplasts.[3,4] All of these methods have been partially successful, but

[1] J. T. O. Kirk and R. A. E. Tilney-Bassett, *in* "The Plastids: Their Chemistry, Structure, Growth and Inheritance" (J. T. O. Kirk and R. A. E. Tilney-Bassett, eds.), 2nd ed., p. 219. Elsevier/North-Holland Biomedical Press, Amsterdam, 1978.

[2] J. M. Miflin and H. Beevers, *Plant Physiol.* **53,** 870 (1974).

[3] F. D. Macdonald and T. ap Rees, *Biochim. Biophys. Acta* **755,** 81 (1983).

[4] M. Nishimura and H. Beevers, *Plant Physiol.* **62,** 40 (1978).

also suffer from one drawback: dilution to isosmolar conditions must proceed very slowly to avoid envelope rupture. Furthermore, in the case of amyloplasts, a critical point in the isolation procedure is that starch granules exhibit a high sedimentation coefficient and tear through the plastid envelope at supraoptimal *g* forces and during longer blending. Recently, Echeverria *et al.*[5] reported a procedure using maize endosperm protoplasts and allowing amyloplasts to settle at 1 *g* through a Ficoll layer; these authors obtained a fraction containing pure and 70% intact amyloplasts.

We sought to develop a technique suitable for the large-scale preparation of intact active and pure plastids from cauliflower buds. This tissue is an interesting source for slightly differentiated nongreen plastids. The method outlined below employs mechanical blending of the tissue and purification of plastids on a self-generating isosmolar density gradient of Percoll.

Preparation of Intact and Purified Plastids

Plant Material

Cauliflower buds can be obtained from local markets, in large amounts and during most of the year. The best plastid preparations are yielded by buds obtained in an early stage of development. This stage is morphologically characterized by a compact structure of the inflorescence, showing little free space around the axial stem, and by a smooth, light beige-colored surface. In deep winter season, cauliflower buds are quiescent and most plastids contain large starch grains, being thus potentially very fragile. This situation leads to low yields in the recovery of intact, purified plastids. In order to overcome this problem, cauliflower buds should be stored for 48 hr in a dark cold room (4–8°), with little ventilation to avoid fading of the tissue. Dark storage induces starch degradation and a subsequent decrease of the plastidial starch content is systematically observed.

Discard the bracts, wash in cold deionized water, and drain the buds. Cut and collect the superficial part of the inflorescences (5 mm thick), and discard all underlying stem tissue containing large vacuolated cells.

Other Materials

Nylon blutex (toile à bluter), 50-μm aperture (Tripette et Renaud, Sailly-Saillisel, Combles, France)
Motor-driven blender, 1-gal (Waring blender)

[5] E. Echeverria, C. Boyer, K.-C. Liu, and J. Shannon, *Plant Physiol.* **77**, 513 (1985).

Superspeed centrifuge, refrigerated (RC5, Sorvall, or equivalent) with the Sorvall ARC-1 automatic rate controller and the following rotors:
- Fixed angle rotor, 6 × 500 ml (GS-3, Sorvall)
- Fixed angle rotor, 8 × 40 ml (SS-34, Sorvall)
- Swinging bucket rotor, 4 × 50 ml (HB-4, Sorvall)

Solutions

Isolation medium: 0.3 *M* mannitol (or sucrose); 20 m*M* pyrophosphate–HCl buffer (pH 7.6); 1 m*M* EDTA; 0.2% (w/v) defatted BSA (fraction V powder, Sigma); 4 m*M* cysteine; add cysteine just before the blending step

Washing medium: 0.3 *M* sucrose; 10 m*M* phosphate–KOH buffer (or MOPS–KOH buffer), pH 7.2; 1 m*M* EDTA; 0.1% defatted BSA

Percoll solutions: 35% (v/v) Percoll; 0.3 *M* sucrose; 10 m*M* phosphate–KOH buffer (or MOPS–KOH buffer), pH 7.2; 1 m*M* EDTA; 0.1% defatted BSA. The use of Percoll, a nontoxic silica sol gradient material [colloidal silica with poly(vinylpyrrolidone)] developed by Pharmacia, has permitted the introduction of rapid purification procedures maintaining isosmolar and low viscosity conditions.[6] The efficiency and adaptability of Percoll density gradients are quite suitable for intact nongreen plastid purification.[7]

Procedure

Temperature of the successive suspending media must be maintained at 0 to 4° throughout plastid isolation procedure. Place the top of inflorescences in a 1-gal Waring blender (1 kg of cut material for 2 liters of isolation medium) and grind the material at low speed for 3 sec. Longer blending improves the yield of broken cells, but increases the proportion of damaged plastids. Filter the brei rapidly through one layer of nylon blutex (50-μm aperture). In order to increase the yield of plastids, the material retained by the nylon blutex can be homogenized a second time in 1 liter of fresh isolation medium. The filtered suspension is then placed in 500-ml bottles and centrifuged for 20 min at 3000 *g* in a precooled Sorvall GS-3 rotor. Pour off the supernatant and suck off the remaining medium with a vacuum pump. Each pellet is resuspended using a plastic spatula, by addition of 10 ml of washing medium. Filter the suspension on a nylon blutex to separate the aggregates prior to the second differential centrifugation cycle.

Pour the filtered suspension in 40-ml tubes and centrifuge for 5 min at

[6] "Percoll Methodologies and Applications." Pharmacia Fine Chemicals AB (undated).

[7] E.-P. Journet and R. Douce, *Plant Physiol.* **79,** 458 (1985).

200 *g* in a precooled Sorvall SS-34 rotor. In this step, free starch granules sediment together with a sticky fraction of unknown origin. Slowly pour off the supernatant in new 40-ml tubes and spin again for 10 min at 3000 *g* (SS-34 rotor) to yield the crude plastid fraction. The main contaminant of plastids prepared from cauliflower by differential centrifugation is mitochondria.

The total pellet is carefully resuspended in washing medium (final volume ca. 12 ml) with a spatula or by mild vortex mixing. Aliquots (3-ml sample, 20–30 mg protein) are layered over 36 ml of 35% (v/v) Percoll solution, in Sorvall SS-34 angle head rotor tubes, and centrifuged for 30 min at 40,000 *g*. Since rapid acceleration or deceleration generate adverse effects on gradient stability, the use of an automatic rate controller is strongly advocated. Under these conditions, centrifugation of Percoll in an angle-head rotor results in spontaneous formation of a continuous (but nonlinear), smooth density gradient.[6] The plastids containing medium starch grains are found in a broad, yellow (sometimes greenish) band near the bottom of the tube (fraction 3), while mitochondria and lighter plastids (i.e., almost devoid of starch) remain near the top of the tube (fraction 2), just below a sticky layer (fraction 1). Discard the upper part of the gradient and carefully collect fraction 3. In order to lower the Percoll concentration, dilute this fraction with washing medium (4 vol to 1 vol of fraction 3) and centrifuge for 10 min at 3500 *g* (SS-34 rotor) or at 5000 *g* (HB-4 rotor; such a swinging bucket rotor allows organelles to sediment directly onto the bottom of the tube without rubbing against the tube walls).

In order to remove the last contaminating mitochondria, purification procedure is repeated once. Resuspend pelleted fraction 3 in washing medium (final volume 4 ml), layer the suspension on 36 ml of 35% (v/v) Percoll solution, in a Sorvall HB-4 swing-out rotor tube, and centrifuge for 30 min at 25,000 *g*. Centrifugation in a swing-out rotor does not result in spontaneous formation of a density gradient, owing to the long path length and inequal *g* forces along the tube. The plastids are found in a yellow band near the bottom of the tube while mitochondria stay at the top of the Percoll cushion. Carefully remove the plastids as described above and recover them as a pellet after diluting with washing medium (10 vol to 1 vol of plastid suspension) and centrifuging for 10 min at 5000 *g* (HB-4 rotor). Resuspend the purified plastid pellet in a minimum volume of washing medium, with gentle agitation (final protein concentration 40–60 $mg \cdot ml^{-1}$) and store it in an ice bath.

Comments

In spite of numerous attempts, we could not determine any single-step purification method routinely giving a plastid fraction in a reasonably pure state (i.e., less than 5% mitochondrial protein) without extensive loss of material. On the contrary, the two-step method outlined above satisfactorily meets that goal.

The final yield of intact plastids, however, is very low. On a carotenoid basis, this yield is roughly 0.5 to 1% of the total amount of plastids present in the initial cauliflower bud tissue. Starting from 1 kg of cut tissue, the protein content of the purified plastid fraction amounts to approximately 10–15 mg. Two reasons to explain this fact include (1) the necessary gentle grinding of the tissue leaves most cells unbroken; and (2) the sizes of starch granules present in cauliflower bud plastids show wide disparity, reflected in the diversity of plastid apparent densities. In fact, the first pellets obtained after tissue disruption contain variable but always conspicuous amounts of free starch granules and intact plastids containing large starch grains. This observation demonstrates that most of the plastids containing larger starch granules burst during blending and/or centrifugation. Furthermore, variable amounts of the plastid markers [carotenoid, latent phosphogluconate dehydrogenase (EC 1.1.1.44)] appear as a diffuse band along with mitochondria near the top of the first Percoll gradient. This is due to intact plastids that contain smaller, but variable amounts of starch. Since these plastids share very similar characteristics with mitochondria (average diameter and apparent density), comigration is not due to membrane–membrane interactions and it is very difficult to find conditions suitable for an effective separation of cauliflower bud mitochondria and lighter plastids, either by differential or isopycnic centrifugation. Lighter plastids and mitochondria present in fraction 2 must therefore be discarded together. Consequently, the purified plastids obtained by our method contain middle-size starch granules (from 1 to 10 μmol starch, expressed as equivalent glucose per milligram plastid protein).

Plastids containing bigger starch granules should be considered and treated as amyloplasts during their isolation procedure: mild blending of the tissue[8] or use of broken protoplasts releasing intact amyloplasts that can be collected by low gravity settling.[5] Unfortunately, maceration of cauliflower bud tissue in the presence of cell wall-degrading enzymes did not yield significant amounts of viable protoplasts.

Determination of Plastid Purity

Assays for marker enzymes[9] specific to the cytosol (alcohol dehydrogenase and phosphoenolpyruvate carboxylase), mitochondria (fumarase),

[8] J. J. Gaynor and A. W. Galston, *Plant Cell Physiol.* **24,** 411 (1983).
[9] P. H. Quail, *Annu. Rev. Plant Physiol.* **30,** 425 (1979).

microbodies (catalase), and plastids (nitrite reductase) support the evidence that sedimentation of cauliflower bud plastids through two successive Percoll gradients almost completely eliminates cytosolic and mitochondrial contamination.[7] Calculated from the lower limit of detection of alcohol dehydrogenase, the possible cytosolic contamination cannot exceed a value of 0.2% on a protein basis. The first step of purification reduces the mitochondrial contamination from 20 to 5% (on a protein basis), while the second step yields contamination values as low as 0–2.5%. The only extraplastidial marker which does show noticeable activity is catalase. It is possible, however, that catalase present in the plastid fraction is due to a nonspecific adsorption of the enzyme to the plastid envelope.[7]

Assessing Plastid Intactness

The intactness of plastid preparations can be evaluated by measurement of phosphogluconate dehydrogenase (EC 1.1.1.44) latency in a quick and reliable spectrophotometric assay.[7] As a matter of fact, this enzyme shows high specific activity in cauliflower bud plastids (220 nmol · min^{-1} per milligram protein).[7] The reaction medium was optimized to obtain the maximal phosphogluconate dehydrogenase activity at 25°; in these conditions, the activity is linear with respect to time for at least 5 min and is proportional to the amount of organelle protein in the range of 50 to 250 $\mu g \cdot ml^{-1}$. Plastids (ca. 50 μg protein) are added to a spectrophotometer cuvette containing 0.5 ml of the following isosmotic reaction mixture: 300 m*M* sucrose, 25 m*M* Tricine–NaOH buffer, pH 8.2, 5 m*M* $MgCl_2$, 500 μM NADP. It is important to disperse the plastid sample into the reaction medium with gentle spatula strokes in order to avoid envelope rupture. The reaction is initiated by addition of 2 m*M* phosphogluconate and the change in absorbance at 340 nm is recorded. After 2 min, 0.025% (v/v) Triton X-100 is added by vigorous mixing so that the plastid envelope is quasi-instantaneously ruptured. The enzyme activity of the ruptured preparation (total) minus that of the intact preparation is called "latent activity."[3] Therefore, the proportion of intact plastids in a given preparation can be rapidly estimated from the percent of latency of phosphogluconate dehydrogenase, i.e., from the relative rates of NADP reduction by untreated and Triton X-100-shocked plastids. Note that 0.025% Triton X-100 causes an 8% inhibition of phosphogluconate dehydrogenase activity and the rates recorded with detergent-treated plastids have to be corrected accordingly.

The use of plastid enzyme latency to assess plastid intactness lies upon the assumption that the substrates and products of the corresponding reaction do not rapidly cross the plastid envelope.[3] In fact, in 50 experiments, the latency of phosphogluconate dehydrogenase thus measured on freshly prepared cauliflower bud plastids ranges from 85 to 98% with a mean value of 90%. Since high values of latency (99%) can be observed in the best preparations, it is clear that some or all of the substrates and products of phosphogluconate dehydrogenase (NADP, gluconate 6-P; NADPH, ribulose 5-P) cannot readily penetrate the stromal space. Therefore, the latency of this activity appears to be a good criterion for purified plastid intactness. We have observed that cauliflower bud plastids purified on Percoll gradients remain fully intact after 24-hr aging in ice-cold washing medium, provided that 1 m*M* $MgCl_2$ is added externally.

In the case of a crude plastid fraction, the presence of contaminating phosphogluconate dehydrogenase activity (cytosolic isozyme) can lead to underestimated values of plastid intactness. It is advantageous here to use the latency of a typical plastidial enzyme, such as glucose-1-phosphate adenylyltransferase (ADP-glucose pyrophosphorylase)[10] (also see Echeverria *et al.*[5]), to estimate plastid intactness, in a slightly more complicated assay. In the case of purified plastids, we always observed that the latency value for glucose-1-phosphate adenylyltransferase (ADP-glucose pyrophosphorylase) is identical with that for phosphogluconate dehydrogenase.

Applications

The method outlined above, when followed carefully, will produce plastids containing middle-size starch grains with a high degree of intactness. They approach ultimate purity as judged by electron microscopy and analysis using marker enzymes.[7] Percoll-purified cauliflower bud plastids are thus suitable for a large range of studies. For example, the first exhaustive description of the enzymatic capacities of a nongreen plastid type for basic carbon metabolism (glycolysis, pentose-P pathway, starch metabolism), nitrite assimilation, and lipid synthesis (long-chain fatty acid and galactolipid synthesis) could be drawn up by using purified cauliflower bud plastids.[7] The maximal activities recorded for the enzymes involved in these pathways showed that cauliflower bud plastids are recovered in an excellent physiological state. Cauliflower bud plastids are also suitable for envelope phospholipid and pigment analysis, native plastid–DNA isolation, metabolite transport studies, and should help to increase knowledge about these biochemically very flexible organelles.

[10] H. P. Ghosh and J. Preiss, *J. Biol. Chem.* **241,** 4491 (1966).

[24] Isolation of Membranous Chromoplasts from Daffodil Flowers

By BODO LIEDVOGEL

Introduction

Chromoplasts are the yellow- to red-colored plastids of many flower petals and fruits. Their fine structure represents the most heterogeneous category of plastids.[1] Most of the chromoplasts investigated so far by electron microscopic methods belong to the globulous, tubulous, membranous, or crystallinous type as determined by the carotenoid-bearing fine structural element prevailing in their final state of development.

Of these types the membranous chromoplast is the rarest one. In its typical form, it has so far been found only in flowers—in particular in the cells of the bright yellow corona ("Nebenkrone") of the daffodil *Narcissus pseudonarcissus* L.[2-4] The corona represents the aroma-producing organ of the flower which probably attracts insects.[5] It exhibits the highest numbers of chromoplasts per cell whose plastids are all uniform in their final state of differentiation.[6]

Inside the double chromoplast envelope, the periphery of the plastid body is made up of numerous (up to 20), concentric, closely appressed double membranes, the thickness of a single membrane sheet being about 6 nm. The build-up of these extremely lipid-rich internal membranes probably originates from invaginations of the inner sheet of plastidal envelopes that are frequently observed by electron microscopy of ultrathin sections from early developmental stages.[4] Inside the membrane convolution a relatively dense stroma matrix material is embedded, sometimes crossed by a few flat or tubular membranes with numerous ribosome-like particles as well as some plastoglobules. A few less dense regions probably contain the chromoplasts DNA.[3]

The multimembrane coat which surrounds the plastidal stroma pro-

[1] P. Sitte, H. Falk, and B. Liedvogel, *in* "Pigments in Plants" (F.-C. Czygan, ed.), p. 117. Fischer, Stuttgart, 1980.

[2] H. H. Mollenhauer and C. Kogut, *J. Microsc. (Paris)* **7,** 1045 (1968).

[3] K. V. Kowallik and R. G. Herrmann, *Protoplasma* **74,** 1 (1972).

[4] B. Liedvogel, P. Sitte, and H. Falk, *Cytobiologie* **12,** 155 (1976).

[5] S. Vogel, *Abh. Akad. Wiss. Lit. (Mainz), Math.-Nat. Kl.* **10** (1962).

[6] Besides of the daffodil, membranous chromoplasts have been found in some close relatives like *Corbularia* [J. F. Mesquita, *Rev. Biol. (Lisbon)* **10,** 127 (1976)] and *Sternbergia* (H. Falk, personal communication).

vides excellent mechanical as well as osmotic stability to these organelles and, additionally, outstanding preservation of the stroma enzymes during the isolation procedure.

Isolation Procedure for Daffodil Chromoplasts

Daffodil flowers from *Narcissus pseudonarcissus* L. (var. "Golden Harvest") are used. In the very early investigations wild-type plant material was also used which was collected in the southern Vosges (France) near Gérardmer.[4,7,8] The cultivated forms of *N. pseudonarcissus* grow taller and possess larger and more deeply colored flowers but the general chemical composition of the chromoplasts originating from both plant materials remains more or less unchanged.

For chromoplast isolation the coronas ("Nebenkronen") were used exclusively. It has been shown by electron microscopy that the only plastid types present in all tissues of this part of the plant are chromoplasts. Because petals contain some chloroplasts in their basal veins, they have not been used in our investigations.

Only freshly bloomed flowers or those just in the state of opening should be used because chromoplasts from fully expanded flowers are less active with respect to their lipid biosynthetic capabilities.[9] Furthermore, with older plant material the chromoplasts sometimes exhibited a tendency to aggregate during the very early steps of isolation; therefore, they should be excluded from the beginning.

The coronas were cut off at their bases and the stamens carefully removed. For smaller preparations (up to 100 coronas) the homogenization step was done with a knife homogenizer (rotating knife; Bühler, Tübingen, FRG) in portions containing about 10 flowers. The coronas were first cut into small pieces with razor blades in a small amount of ice-cold isolation medium. For the routinely performed large-scale preparations (up to 1000 coronas/day) a rebuilt kitchen mixer equipped with two pairs of razor blades was used for the homogenization of the tissue.[10]

The isolation medium consists of 0.47 *M* sucrose, 5 m*M* $MgCl_2$, 0.2% (w/v) poly(vinylpyrrolidone) K 90, M_r ca. 360,000 (Roth KG, Karlsruhe, FRG), and 67 m*M* phosphate buffer (Soerensen type buffer; solution A: KH_2PO_4; solution B: $Na_2HPO_4 \cdot 2H_2O$; approximate mixture for pH 7.5: 1 vol A : 5.7 vol B). For homogenization a tissue-to-medium ratio of 1 : 3 to 1 : 4 (w/v) was used. Homogenization was carried out by four short

[7] B. Liedvogel and P. Sitte, *Naturwissenschaften* **61,** 131 (1974).

[8] H. Falk, B. Liedvogel, and P. Sitte, *Z. Naturforsch., C: Biosci.* **29C,** 541 (1974).

[9] H. Kleinig and B. Liedvogel, *Eur. J. Biochem.* **83,** 499 (1978).

[10] C. G. Kannangara, S. P. Gough, B. Hansen, J. N. Rasmussen, and D. J. Simpson, *Carlsberg Res. Commun.* **42,** 431 (1977).

strokes (5 sec each; in the case of the Bühler homogenizer at half-maximum speed, ca. 35,000 rpm). The tissue brei was filtered through four layers of nylon cloth (40-μm mesh), and cellular debris, nuclei, and cell wall material were removed by a low-speed centrifugation step (15 min, 1000 *g*, 4°). Chromoplasts were sedimented from the supernatant by a subsequent centrifugation step (20 min, 16,500 *g*). This crude chromoplast pellet also contained mitochondria, which have to be removed from the plastids during a sucrose gradient centrifugation step.

The pellets were gently resuspended with the aid of a rather loose-fitting Dounce glass homogenizer in 67 m*M* Soerensen buffer, pH 7.5, containing 50% (w/v) sucrose and 5 m*M* $MgCl_2$. The resuspended organelles were placed in centrifugation tubes to form a layer at the bottom (ca. 8 ml, if 34-ml swing-out tubes were used, or ca. 14 ml in the case of 60-ml tubes). The organelle suspension was carefully overlaid with equal volumes of 40, 30, and 15% (w/v) sucrose in the same buffer to form a discontinuous gradient. It was centrifuged for 1 hr in a 3 × 34 ml and 3 × 60 ml swinging bucket rotor (Weinkauf, Brandau, FRG), respectively, at 50,000 *g*. The chromoplasts, concentrated by flotation between 40 and 30%, and between 30 and 15% sucrose, respectively, were carefully removed with a Pasteur pipet.

It should be mentioned that by conventional centrifugation, i.e., migration of the organelles from top to bottom of the gradient, such a high degree of purity of the fraction could not be achieved; the necessary elimination of mitochondrial contaminants was not very successful by this method. Therefore, flotation-like centrifugation of the gradient was an indispensable requirement in order to obtain pure chromoplast fractions.

The combined fractions from the two interphases were slowly diluted with 67 m*M* phosphate buffer, pH 7.5, 5 m*M* $MgCl_2$ to a final concentration of 15% sucrose, as measured from the refractive index by an Abbe refractometer. A pure chromoplast pellet was obtained by a final centrifugation step (20 min, 16,500 *g*). All steps of the isolation procedure were carried out in the cold (4°).

The chromoplast fraction was used immediately after isolation when investigating acyl- and galactosyltransferases or the fatty acid-synthesizing machinery of plastids. For studies on carotenogenic enzymes,[11] the chromoplasts were diluted in the appropriate buffer, treated in a French pressure cell, and then stored at −70° until use. (There was no substantial loss with respect to the enzymatic activity for several months.)

In this context it should be remarked that it is also possible to isolate—in addition to the chromoplasts—the mitochondria of the corona cells. These organelles are found in the 40%/50% sucrose interphase of the

[11] P. Beyer, this volume [36].

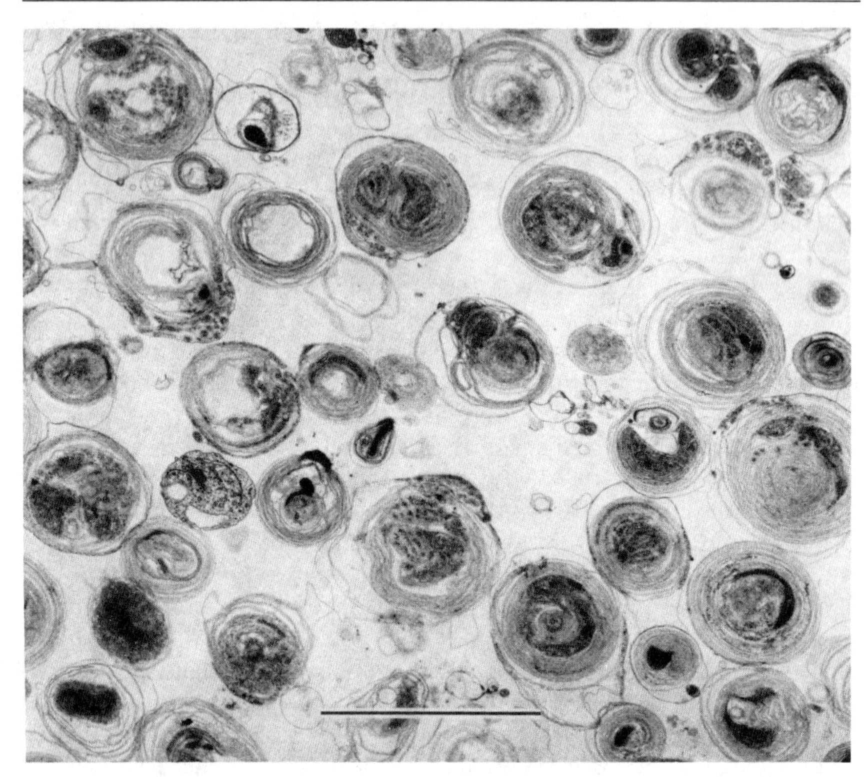

FIG. 1. Ultrathin section of isolated chromoplasts of *Narcissus pseudonarcissus*. Survey of the fraction (×6000; bar, 5 μm). The electron micrograph was kindly provided by Dr. H. Falk.

gradient, but still contaminated with chromoplasts. The method for mitochondria purification is given elsewhere.[12]

Basing on the procedure presented above, a modification has been described which allows the simultaneous isolation of chromoplasts, as well as mitochondria, from daffodil coronas.[13]

Criteria for the Purity of the Chromoplast Fraction

The isolation procedure described above resulted in a clean fraction, rich in unbroken chromoplasts (Fig. 1). The morphometric evaluation of a number of electron micrographs of several different preparations is summarized in Table I. More than 92% and probably as much as 97% of the

[12] F. Lütke-Brinkhaus, B. Liedvogel, and H. Kleinig, *Eur. J. Biochem.* **141,** 537 (1984).

[13] P. Hansmann and P. Sitte, *Z. Naturforsch., C: Biosci.* **39C,** 758 (1984).

TABLE I
CHARACTERIZATION OF *Narcissus pseudonarcissus* CORONA CHROMOPLAST FRACTION BY PARTICLE COUNTING ON ULTRATHIN SECTIONS

Particle type	Total counted particles (%)
Chromoplasts	92.3
Mitochondria	0.6
Microbodies	0.5
Nuclei, and nuclear envelope fragments	—
Small plasma droplets with inclusions	0.8
Bacterial cells	0.6
Other particles, not identifiable[a]	5.2

[a] None of these particles is mitochondria or bacteria, which are easily identified in sections. Part of this category is likely to be unusually formed or degraded chromoplasts.

TABLE II
ENZYMATIC PROPERTIES OF ISOLATED DAFFODIL CHROMOPLASTS

A. Activities associated with chromoplast membranes
  1. Reactions concerned with glycerolipid metabolism
    Glycerol-3-phosphate acyltransferase[14]
    Acylglycerol-3-phosphate acyltransferase[14]
    Phosphatidate phosphatase[14]
    Galactosyltransferases (galactosylating diacylglycerol and monogalactosyldiacylglycerol)[14–16]
    Acylation of phospholipids from acyl-CoA[14,16]
  2. Further acyl and glycosyl transfer reactions
    Acylation of monogalactosyldiacylglycerol[16]
    Acylation of sterol glycoside[16]
    Glycosylation of sterols[15,16]
  3. Acyl-CoA synthetase[14] and acyl-CoA hydrolase[9,14,16]
  4. Prenyl lipid metabolism
    Carotenogenic enzymes (including phytoene synthase and prenyltransferases)[11,17–19]
    Chlorophyll synthase[20]
  5. Membrane translocators (phosphate and adenylate translocator)[21]
B. Soluble chromoplast enzymes
  1. Complete enzyme complement for long-chain fatty acid synthesis from acetate, including acetyl-CoA carboxylase[9,14,22,23]
  2. Acetyl-CoA synthetase[24]
  3. Glycolytic enzymes (starting from dihydroxyacetone phosphate)[22]
  4. Pyruvate dehydrogenase complex[23,25]

particles counted on those micrographs were identified as chromoplasts. Less than 1% of the particles were mitochondria, microbodies, or bacterial cells. Contaminating droplets with cytoplasmic inclusions did not exceed 1%, and nuclei or nuclear envelope material were totally absent. Support for the latter statement came from investigations on the chromoplast DNA—linear DNA strands exceeding a contour length of about 45 μm, which is the size of a complete plastid DNA circle, were never found.[8]

In accordance with the high purity of the chromoplast fraction determined by electron microscopy, marker enzymes typically associated with plant mitochondria and microbodies could not be detected.

The membranous chromoplast of the daffodil has proved to be an outstanding example of a nonphotosynthesizing, energy-heterotrophic plastid exhibiting high activities for several metabolic pathways. In particular, lipid synthetic capabilities were investigated. Table II presents a survey of the rather extensively studied enzymatic properties of daffodil chromoplasts.[14-25]

[14] B. Liedvogel and H. Kleinig, *Planta* **144,** 467 (1979).
[15] B. Liedvogel and H. Kleinig, *Planta* **129,** 19 (1976).
[16] B. Liedvogel and H. Kleinig, *Planta* **133,** 249 (1977).
[17] P. Beyer, K. Kreuz, and H. Kleinig, *Planta* **150,** 435 (1980).
[18] K. Kreuz, P. Beyer, and H. Kleinig, *Planta* **154,** 66 (1982).
[19] P. Beyer, G. Weiss, and H. Kleinig, *Eur. J. Biochem.* **153,** 341 (1985).
[20] K. Kreuz and H. Kleinig, *Plant Cell Rep.* **1,** 40 (1981).
[21] B. Liedvogel and H. Kleinig, *Planta* **150,** 170 (1980).
[22] H. Kleinig and B. Liedvogel, *Planta* **150,** 166 (1980).
[23] B. Liedvogel and H. Kleinig, *in* "Biogenesis and Function of Plant Lipids" (P. Mazliak, P. Benveniste, C. Costes, and R. Douce, eds.), p. 107. Elsevier/North-Holland Biomedical Press, Amsterdam, 1980.
[24] B. Liedvogel, *Anal. Biochem.* **148,** 182 (1985).
[25] B. Liedvogel, *Z. Naturforsch., C: Biosci.* **40C,** 182 (1985).

[25] Structure and Function of the Inner Membrane Systems in Etioplasts

By SATORU MURAKAMI

In etiolated plants grown in total darkness the proplastids develop to etioplasts, which are devoid of chlorophyll (Chl) and contain substantial amounts of protochlorophyllide (Pchld) and carotenoids. Photosynthetic electron carriers such as cytochromes, plastocyanin, and ferredoxin, and H^+-ATPase (CF_0-CF_1 complex) are accumulated in etioplasts in the dark, but photochemical reaction centers and light-harvesting chlorophyll protein complexes are not formed.[1] Etioplasts have a unique inner membrane system: a paracrystalline three-dimensional tubular lattice called a prolamellar body (PLB) and a prothylakoid (PT) extending into stroma from the periphery of the PLB.[2]

Etioplasts are transformed into the chloroplasts (etiochloroplasts) by exposure to light for about 1 day. Upon illumination Pchld accumulated in the dark is converted instantaneously into chlorophyllide (Chld) and additional Chl is synthesized progressively after a short lag period. This greening process is accompanied by changes in the structure of the inner membrane system, leading to the formation of the photosynthetic machinery. Because of these events etioplasts have been the object of studies on chlorophyll formation and on the development of the structure and photosynthetic functions of the chloroplasts.

Plant Materials

Various kinds of plant material have been used for developmental and biochemical studies of etioplasts. The most widely used material is the first leaves of monocots such as oat,[3–7] barley,[8–10] wheat,[11–14] and corn.[15,16] They are a good source of etioplasts, but it should be mentioned that their etioplasts manifest a sequential differentiation from the base to tip along the leaf axis. The second group is the first leaves of dicots, including bean[5]

[1] N. R. Baker, *in* "Chloroplast Biogenesis" (N. R. Baker and J. Barber, eds.), p. 207. Elsevier, Amsterdam, 1984.

[2] S. Murakami, N. Yamada, M. Nagano, and M. Osumi, *Protoplasma* **128,** 147 (1985).

[3] C. Lütz, *in* "Chloroplast Development" (G. Akoyunoglou and J. M. Argyroudi-Akoyunoglou, eds.), p. 481. Elsevier/North-Holland Biomedical Press, Amsterdam, 1978.

[4] C. Lütz, U. Röper, N. S. Beer, and T. Griffiths, *Eur. J. Biochem.* **118,** 347 (1981).

[5] R. P. Oliver and W. T. Griffiths, *Plant Physiol.* **70,** 1019 (1982).

[6] J. Kesselmeier and B. Urban, *Protoplasma* **114,** 133 (1983).

[7] U. Röper, P. Bergweiler, and C. Lütz, *Z. Pflanzenphysiol.* **112,** 89 (1983).

[8] W. T. Griffiths, *Biochem. J.* **146,** 17 (1975).

[9] W. T. Griffiths, *Biochem. J.* **174,** 681 (1978).

[10] K. Apel, H.-T. Santel, T. E. Redlinger, and H. Falk, *Eur. J. Biochem.* **111,** 251 (1980).

[11] M. Ryberg and C. Sundqvist, *Physiol. Plant.* **56,** 125 (1982).

[12] M. Ryberg and C. Sundqvist, *Physiol. Plant.* **56,** 133 (1982).

[13] E. Selstam and A. S. Sandelius, *Plant Physiol.* **76,** 1036 (1984).

[14] K. Dehesh and M. Ryberg, *Planta* **164,** 396 (1985).

[15] W. T. Griffiths, *Biochem. J.* **152,** 623 (1975).

[16] S. Kohn and S. Klein, *Planta* **132,** 169 (1976).

and pea.[17] They contain uniform etioplasts. The third group is the cotyledons of dicots such as cucumber[18,19] and squash.[20,21] In some plants storage substances such as lipid, protein, and starch in the cotyledons may interfere with the isolation of intact etioplasts. However, if this defect is overcome by starvation of the ingredients and careful isolation, the cotyledons provide sufficient amounts of pure and uniform intact etioplasts for biochemical study.

Isolation of Intact Etioplasts

About 20 g of etiolated leaves (e.g., squash cotyledons[21]) is soaked in 100 ml of isolation medium and homogenized twice, for 6 sec each, in a Waring blender at maximum speed. The isolation medium contains 1 m*M* $MgCl_2$, 1 m*M* Na_2EDTA, 5 m*M* 2-mercaptoethanol, 0.5 *M* sucrose, and 40 m*M* Tricine–NaOH (pH 7.7). Other isolation media, e.g., those used by Griffiths[8] and Ryberg and Sundqvist,[11] can also be used. A Polytron-type homogenizer is recommended especially for laminal leaves from monocots.[22]

The homogenate is filtered through four layers of nylon mesh (25 μm) and centrifuged at 2000 *g* for 2 min. The pellet containing crude etioplasts is resuspended in the isolation medium, then centrifuged at 100 *g* for 1 min to remove cell debris and aggregates of cellular components with nuclear materials.

The supernatant is laid on a discontinuous gradient of 10 ml of 40% and 5 ml of 70% Percoll solution, then centrifuged in a swing rotor at 6500 *g* for 10 min. The Percoll solution contains 5% (w/v) polyethylene glycol 6000, 1% (w/v) Ficoll, 1 m*M* $MgCl_2$, 1 m*M* Na_2EDTA, 5 m*M* 2-mercaptoethanol, and 40 m*M* Tricine–NaOH (pH 7.7). The intact etioplasts located at the boundary between the 40 and 70% Percoll are collected with a syringe, diluted with the isolation medium, and pelleted at 4500 *g* for 10 min. The resulting pellet is resuspended and washed again by centrifugation under the same condition.

The organelles found in this fraction are mostly etioplasts; few mitochondria and microbodies are found. About 94% of these etioplasts are apparently intact.

[17] K. W. Henningsen, S. W. Thorne, and N. K. Boardman, *Plant Physiol.* **53,** 419 (1974).
[18] A. Tanaka and H. Tsuji, *Biochim. Biophys. Acta* **680,** 265 (1982).
[19] A. Tanaka and H. Tsuji, *Plant Cell Physiol.* **24,** 101 (1983).
[20] M. Ikeuchi and S. Murakami, *Plant Cell Physiol.* **23,** 1089 (1982).
[21] M. Ikeuchi and S. Murakami, *Plant Cell Physiol.* **24,** 71 (1983).
[22] S. Murakami, M. Miyao, and M. Ikeuchi, *Plant Cell Rep.* **2,** 148 (1983).

Separation of PLBs and PTs

Percoll Density Gradient Centrifugation. The intact etioplasts are ruptured osmotically by suspending them in a sucrose-free hypotonic medium (5 m*M* $MgCl_2$, 5 m*M* 2-mercaptoethanol, and 40 m*M* Tricine–NaOH, pH 7.7). Deoxyribonuclease I (Sigma Chemical Co., electrophoretically pure) is added to a final concentration of 100 μg/ml to prevent the formation of aggregates of the membranes which makes separation difficult.[21] After incubation for 30 min at 25°, the broken etioplasts are gently ground in a glass homogenizer with a Teflon pestle and centrifuged at 4500 *g* for 5 min. Almost all the PLBs are pelleted, and most of the PTs remain in the supernatant.

The pellet is resuspended in the same medium and centrifuged again at 4500 *g* for 5 min to remove PT remnants. The supernatants obtained by these centrifugations are combined and centrifuged at 30,000 *g* for 30 min, the PTs being collected in the pellet. The resulting pellets (purified PTs) are resuspended in the isolation or other appropriate medium and used for biochemical studies.

The crude PLB pellet is resuspended in the hypotonic medium, then laid on a 5–50% Percoll density gradient. Percoll solution contains 1 m*M* $MgCl_2$, 5 m*M* 2-mercaptoethanol, 5% polyethylene glycol 6000, and 40 m*M* Tricine–NaOH (pH 7.7). The gradient is centrifuged in a swing rotor at 51,000 *g* for 1 hr. Yellow bands are obtained. The upper broad band in 24% Percoll contains PTs, PLBs and their aggregates. The purified PLBs in the lower band in 34% Percoll are collected and pelleted by centrifugation at 4500 *g* for 5 min.

All these procedures are performed at 4° in the dark or under a dim green safe light.

The purified PLB fraction contains mostly paracrystalline PLBs and only a small amount of vesicles. The PT fraction is rich in vesicular structures, whose outer surface is densely covered with the CF_1 particles, identified as the PTs. Larger, smooth-surfaced vesicles, some of which have double membranes and are regarded as envelopes, are also found. Neither PLBs nor their fragments are contained in the PT fraction. Percoll density gradient centrifugation was shown to be a useful method for isolation of PLBs and PTs from squash etioplasts by Ikeuchi and Murakami.[21]

Sucrose Density Gradient Centrifugation. Intact etioplasts suspended in a sucrose-free medium (1 m*M* $MgCl_2$, 1 m*M* EDTA, 20 m*M* TES, 10 m*M* HEPES, adjusted to pH 7.2 with KOH) are ground in a glass homogenizer, 10 strokes up and down with a Teflon pestle, and centrifuged at 7700 *g* for 15 min to remove envelopes and stroma.[11] The pellet containing etioplast inner membranes is resuspended in a medium containing 50% sucrose and subjected to brief sonication (3×, 5 sec each) to break the connection between the tubules of PLBs and PTs.[3]

The sonicated membranes in 50% sucrose medium are placed at the bottom of a continuous sucrose density gradient [10–50% (w/w)] by using a syringe with a translucent plastic tube adapted. After centrifugation at 25,400 *g* for 120 min, two bands are obtained; the upper band at a density of ca. 1.17 g ml^{-1} contains purified PLBs and the lower band at a density of ca. 1.21 g ml^{-1} contains PTs. The bands are removed with a syringe, diluted with a sucrose-free medium, and centrifuged at 42,500 *g* for 90–120 min. The resulting pellets are resuspended in the isolation (or other) medium and used for experiments.

The bottom-loaded sucrose density gradient centrifugation method mentioned above was successfully used by Ryberg and Sundqvist[11] for wheat etioplasts.

Top-loaded sucrose density gradient centrifugation can also be used for separation of the PLBs and PTs.[3] Total etioplast inner membranes which have been obtained from intact etioplasts by osmotic lysis and centrifugation in a hypotonic medium are briefly sonicated and loaded on top of a continuous sucrose density gradient (10–60%). After centrifugation at 72,000 *g* for 3–4 hr two bands are obtained. The upper and lower bands are rich in PTs and PLBs. The PT band obtained by this method is broader and more diffuse than that obtained by the bottom-loaded method.[11]

Simple Method for Isolation of PLBs

Etiolated leaves or cotyledons are homogenized twice in a Waring blender at maximum speed for 6 sec each. The grinding medium contains 0.5 *M* sucrose, 1 m*M* $MgCl_2$, 1 m*M* EDTA, 5 m*M* 2-mercaptoethanol, and 40 m*M* Tricine–NaOH (pH 7.7). The homogenate is filtered through four layers of nylon mesh (25 μm) and centrifuged at 2000 *g* for 2 min. The etioplast membranes (PLBs plus PTs) in the supernatant are pelleted by centrifugation at 4500 *g* for 5 min, suspended in a hypotonic medium containing 1 m*M* $MgCl_2$, 5 m*M* 2-mercaptoethanol, and 40 m*M* Tricine–NaOH (pH 7.7), and ground in a glass homogenizer. The PLBs are purified by linear sucrose density gradient centrifugation at 30,000 *g* for 30 min. The gradient contains 0.9–1.5 *M* sucrose, 1 m*M* $MgCl_2$, 5 m*M* 2-mercaptoethanol, and 40 m*M* Tricine–NaOH (pH 7.7). The yellow band including the purified PLBs is collected with a syringe and pelleted at 30,000 *g* for 30 min.

Characterization of PLBs and PTs

The absorption spectrum of squash etioplast inner membrane systems which are composed of large PLBs and associated PTs shows a peak at 648–650 nm and a shoulder at 640 nm. The purified PLBs from which the PTs are detached show a similar absorption spectrum with a peak at 648 nm and a shoulder at 638–639 nm.[11,21] The absorption spectrum of the PTs has a broad peak at 633–635 nm and a shoulder at 647 nm. The content of acyl lipids and pigments on a protein basis is higher in the PLBs than in the PTs.[13,23,24] Three galactolipids, monogalactosyldiacylglycerol (MGDG), digalactosyldiacylglycerol (DGDG), and sulfoquinovosyldiacylglycerol (SQDG), are the dominant lipids of both the PLBs and PTs, in this order. The molar ratio of MGDG to DGDG is higher in the PLBs (1.6–1.8) than in the PTs (1.1–1.2). Steroidal saponins are not detected in the PLBs or in the PTs.[25,26] Phosphatidylglycerol and phosphatidylcholine, which make up about 10% of the total acyl lipids, are the major phospholipids.

Highly purified PLBs contain a large amount of a unique protein with a mobility corresponding to 36,000 Da.[11,21] The 36-kDa protein in the PLBs was identified as NADPH-protochlorophyllide oxidoreductase (Pchld reductase, EC 1.6.99.-).[20] The amount of this protein is more than 70% of the total PLB protein. The PLB has a much higher activity of Pchld reductase (7 nmol Chld formed $flash^{-1} \cdot mg^{-1}$ protein) than the PTs (0.1–0.3 nmol Chld formed $flash^{-1} \cdot mg^{-1}$ protein). It is well documented that more than 95% of the Pchld reductase activity and Pchld content in the etioplasts are concentrated in the PLBs, whereas Pchl is distributed equally in both the PLBs and PTs.[21]

[23] M. Ryberg, A. S. Sandelius, and E. Selstam, *Physiol. Plant.* **57,** 555 (1983).

[24] C. Lütz, U. Nordmann, H. Tönissen, P. Bergweiler, and U. Röper, *Adv. Photosynth. Res.* **4,** 627 (1984).

[25] S. Murakami and M. Ikeuchi, *in* "Cell Function and Differentiation" (G. Akoyunoglou *et al.,* eds.), Part B, p. 13. Alan R. Liss, Inc., New York, 1982.

[26] S. Murakami, M. Ikeuchi, and M. Miyao, *Plant Cell Physiol.* **24,** 581 (1983).

[26] Isolation Procedures for Inside-Out Thylakoid Vesicles

By HANS-ERIK ÅKERLUND and BERTIL ANDERSSON

Introduction

To understand the function of a biological membrane like that of chloroplast thylakoids, it is important to understand the arrangement of its different protein and lipid components. Preparations that have proved to be particularly suited for such studies are those consisting of membrane vesicles which are turned inside out. Inside-out vesicles from the thylakoid membrane were first obtained from spinach chloroplasts by a combination of mechanical fragmentation and separation by aqueous two-phase partition.[1,2] By the same or very similar procedures inside-out thylakoid vesicles have now also been obtained from other plant sources such as pea,[3] barley,[4] mangrove (*Avicennia marina*),[5] lettuce,[6] and the photosynthetic bacteria *Rhodopseudomonas viridis*.[7] Since the isolation procedure does not involve the use of detergents the inside-out thylakoids have a preserved membrane structure and are ideally suited for structure–function studies. They have been used extensively in studies on thylakoid membrane topography[8–11] and for the identification of proteins associated with the oxygen evolution process.[12,13] An important finding was that the inside-out vesicles are only formed from appressed thylakoid membranes

[1] B. Andersson, H.-E. Åkerlund, and P.-Å. Albertsson, *FEBS Lett.* **77,** 141 (1977).
[2] B. Andersson and H.-E. Åkerlund, *Biochim. Biophys. Acta* **503,** 462 (1978).
[3] R. W. Mansfield, H. Y. Nakatani, J. Barber, S. Mauro, and R. Lannoye, *FEBS Lett.* **137,** 133 (1982).
[4] L. E. A. Henry, J. Dalgård-Mikkelsen, and B. Lindberg-Møller, *Carlsberg Res. Commun.* **48,** 131 (1983).
[5] B. Andersson, C. Critchley, I. J. Ryrie, C. Jansson, C. Larsson, and J. M. Anderson, *FEBS Lett.* **168,** 113 (1984).
[6] M. D. Unitt and J. L. Harwood, *Biochem. Soc. Trans.* **10,** 249 (1982).
[7] F. A. Jay and M. Lambillotte, *Eur. J. Cell Biol.* **37,** 7 (1985).
[8] C. Jansson, B. Andersson, and H.-E. Åkerlund, *FEBS Lett.* **105,** 177 (1979).
[9] H.-E. Åkerlund and C. Jansson, *FEBS Lett.* **124,** 229 (1981).
[10] W. Haehnel, R. J. Berzborn, and B. Andersson, *Biochim. Biophys. Acta* **637,** 389 (1981).
[11] B. Andersson, J. M. Anderson, and I. J. Ryrie, *Eur. J. Biochem.* **123,** 465 (1982).
[12] H.-E. Åkerlund, C. Jansson, and B. Andersson, *Biochim. Biophys. Acta* **681,** 1 (1982).
[13] U. Ljungberg, H.-E. Åkerlund, and B. Andersson, *FEBS Lett.* **175,** 225 (1984).

while right-side-out vesicles derive from nonappressed membranes.[14] As a result, the usefulness of inside-out vesicles could be extended to include studies on the lateral organization of the thylakoid membrane.[15,16] The use of inside-out thylakoids in studies on structure and function of the thylakoid membrane has recently been reviewed by Andersson *et al.*[17] The isolation of inside-out thylakoid vesicles can be divided principally into two steps: fragmentation and subsequent separation. The ionic conditions imposed during fragmentation have a profound influence on both the yield and the properties of the inside-out vesicles.[14]

Three different fragmentation procedures will be described. The first and original procedure[1,2] gives rise to inside-out vesicles originating from the appressed thylakoid region, highly enriched in photosystem II, but relatively low in yield. The second procedure[18] gives inside-out vesicles in a higher yield but somewhat lower photosystem II purity. The inside-out vesicles obtained by these two procedures are particularly suitable for studies on the lateral distribution of thylakoid components and for studies on the organization and function of photosystem II and its oxygen-evolving system. The third procedure[14] gives inside-out vesicles with the same relative amounts of photosystem I and II as whole thylakoids and is therefore especially suited for studies on transverse arrangements of all components of the thylakoid membrane.

The key step for the isolation of the inside-out vesicles from the right-side-out vesicles is aqueous polymer two-phase partition.[19–22] This method separates membrane particles according to differences in surface properties such as charge and hydrophobicity. It is therefore ideally suited for separation of membrane vesicles which only differ in sidedness. This separation step is essentially the same for all three preparation procedures.

[14] B. Andersson, C. Sundby, and P.-Å. Albertsson, *Biochim. Biophys. Acta* **559,** 391 (1980).
[15] B. Andersson and J. M. Anderson, *Biochim. Biophys. Acta* **593,** 427 (1980).
[16] U. K. Larsson, B. Jergil, and B. Andersson, *Eur. J. Biochem.* **136,** 25 (1983).
[17] B. Andersson, C. Sundby, H.-E. Åkerlund, and P.-Å. Albertsson, *Physiol. Plant.* **65,** 322 (1985).
[18] H.-E. Åkerlund and B. Andersson, *Biochim. Biophys. Acta* **725,** 34 (1983).
[19] P.-Å. Albertsson, this series, Vol. 31, p. 761.
[20] P.-Å. Albertsson, B. Andersson, C. Larsson, and H.-E. Åkerlund, *Methods Biochem. Anal.* **28,** 115 (1982).
[21] P.-Å. Albertsson, "Partition of Cell Particles and Macromolecules," 3rd ed. Wiley, New York, 1986.
[22] H. Walter, D. E. Brooks, and D. Fisher, eds., "Partitioning in Aqueous Two-Phase Systems; Theory, Methods, Uses and Applications to Biotechnology." Academic Press, Orlando, Florida, 1985.

Generation of Inside-Out Thylakoid Vesicles

Fragmentation Procedure I: Inside-Out Vesicles Highly Enriched in Photosystem II

Intact and stacked thylakoids, isolated by conventional procedures,[18] are suspended in 50 m*M* sodium phosphate, pH 7.4, 150 m*M* NaCl, at a concentration of 0.2–1 mg chlorophyll/ml, and passed dropwise twice through a Yeda press (Yeda, Rehovot, Israel) at a nitrogen gas pressure of 10 MPa. The press homogenate is centrifuged at 40,000 *g* for 30 min. The supernatant which contains stroma lamellae membranes highly enriched in photosystem I is carefully removed. The sediment, containing 90–95% of the starting material, is drained carefully to reduce carryover of excess salt and is resuspended in 10 m*M* sodium phosphate, pH 7.4, 5 m*M* NaCl, and 100 m*M* sucrose to a concentration of 1 mg chlorophyll/ml and then passed twice more through the Yeda press.

Fragmentation Procedure II: High Yield of Inside-Out Vesicles Enriched in Photosystem II

Intact thylakoids are suspended in 5 m*M* $MgCl_2$, 10 m*M* sodium phosphate, pH 7.4, 5 m*M* NaCl, 100 m*M* sucrose at a concentration of 4 mg chlorophyll/ml and passed twice through a Yeda press at 10 MPa. The homogenate is made 5 m*M* with respect to sodium EDTA (pH 7.4, 0.1 *M* stock) and passed twice more through the Yeda press. If the medium used for isolation of the intact thylakoids contains divalent cations it is necessary to wash the thylakoids in a medium composed of 50 m*M* sodium phosphate, pH 7.4, 10 m*M* NaCl, 50 m*M* sucrose prior to the resuspension for Yeda press treatment. This is done to avoid carryover effects which may alter the balance between Mg and EDTA.

Fragmentation Procedure III: Inside-Out Vesicles with Photosystems I and II

Intact thylakoids are suspended in 10 m*M* Tricine, pH 7.4, 100 m*M* sucrose and incubated in this medium for 1.5 hr to allow complete destacking and lateral randomization of membrane components. The pH is adjusted to 4.7, using 0.01–0.1 *M* HCl, to reduce electrostatic repulsion and create artificial membrane pairing. The material is then passed twice through the Yeda press at 10 MPa, followed immediately by readjustment of the pH back to 7.4 using 0.1 *M* NaOH. The homogenate is centrifuged at 40,000 *g* for 30 min. The sediment is resuspended in 10 m*M* sodium phosphate, pH 7.4, 5 mM NaCl, 100 mM sucrose and passed twice more through the Yeda press.

Remarks

The procedures described here were developed for thylakoids from spinach but should be suitable for most other species too. All work is performed at 0–4° unless otherwise stated. The first pair of Yeda press treatments in each procedure is the key step for the formation of the inside-out thylakoids. The idea behind the second pair of press treatments is to convert material that was only partly fragmented in the first treatment into right-sided vesicles. This improves the following separation steps. Alternatives to the Yeda press for thylakoid fragmentation is the use of a French pressure cell[4] or sonication. However, these methods appear less reproducible and care must be taken to avoid extensive fragmentation.

Separation by Aqueous Two-Phase Partition

The aqueous two-phase partition method and its applications have been described in detail[19–22] and the basic properties of the method will be described only briefly here.

When two water-soluble polymers such as Dextran and polyethylene glycol are mixed with water above certain concentrations, a two-phase system is formed. Biological material added to such a system will partition between the two phases. To which phase material partitions is determined by its surface properties and the method can therefore be used for separation. The partition can be controlled through changes in (1) polymer concentrations, (2) type and concentration of salt, and (3) the temperature at which the partition is performed. These are the most critical factors in the procedure described below and it is therefore important to know how they influence the partition. Increasing the concentrations of both Dextran and polyethylene glycol generally favors the collection of particles at the interface or in the lower phase. Conversely, lowering of the polymer concentrations will favor the collection of material in the upper phase. Different ions have different affinities for the two phases and inclusion of salt in the phase system therefore creates a Donnan type of electric potential between the two phases. This interfacial potential is determined by the dominating salt and is usually different for different salts.[23] For the phase system applied to isolate inside-out thylakoids, an

[23] G. Johansson, *Acta Chem. Scand., Ser. B* **B28,** 873 (1974).

increase in NaCl concentration or a decrease in sodium phosphate concentration will favor the partition of the material to the lower phase. A change in pH will also have an influence since both the ratios between different ionic species and the membrane charge will change. An increase in the temperature at which the partition is performed generally favor collection of the material in the upper phase. A complication is that the available polymer batches vary in their molecular weight distributions. The phase composition given below may therefore need some small adjustment in each particular case. Thus, if the partition is too low, a small decrease in concentration of both polymers is usually sufficient to obtain a successful separation. The opposite holds if the partition is too high.

Polymer Stock Solutions

Polyethylene glycol 3350 (earlier named polyethylene glycol 4000), 40% (w/w): Dissolve 400 g polyethylene glycol (obtainable as Carbowax 3350 from Union Carbide, NY) in 600 g distilled water.

Dextran T500, 20% (w/w): 220 g Dextran powder (obtainable from Pharmacia Fine Chemicals, Uppsala, Sweden) is layered on 780 g of water. This is then heated on a boiling water bath with occasional gentle stirring until the Dextran has dissolved. Since the Dextran powder contains some water the concentration of this solution will not be 22% but approximately 21% (w/w). The exact concentration of the solution is determined polarimetrically as follows. Approximately 5 g of the solution is diluted to 25 ml with water and the optical rotation (a) at 589 nm is measured (specific rotation = $+199° \times$ ml/g $\times$ dm). The concentration C is then calculated from the following formula,

$$C = aV100/199ml$$

where V is the volume in milliliters (25 ml in this case), l is the optical path length in decimeters, and m is the amount in grams of the original Dextran solution used for the determination. The stock solution can then be adjusted to 20% with water. If a polarimeter is not available the approximate value of 21% given above can be used. However, in this case care should be taken to minimize the dissipation of water vapor during the heating. A frozen Dextran solution is stable for at least 1 year.

Separation Procedure

Thylakoid material from any of the three fragmentation procedures is first subjected to centrifugation at 1000 g for 10 min to remove starch grains and unfragmented material. From the supernatant, 5 ml (from procedures I or III) or 1 ml (from procedure II) is taken and added to a preformed phase mixture to give a 25-g phase system with the final composition:

5.7% (w/w) Dextran T500
5.7% (w/w) polyethylene glycol 3350
10 m*M* sodium phosphate, pH 7.4
5 m*M* NaCl
20 m*M* sucrose

The temperature of the phase system is kept at 4° to obtain reproducible results. The phase system is mixed by 50 inversions and allowed to settle. The phase settling is facilitated by centrifugation at 1500 *g* for 3 min. The inside-out vesicles will partition predominantly to the lower phase while the right-side-out material prefers the upper phase. The amount of material found in the lower phase and interface should be roughly 20, 60, and 15% for the three procedures, respectively. If this is not the case, adjust the polymer concentrations as outlined earlier. To improve the purity of the inside-out thylakoids the upper phase is removed, 13 ml of new pure upper phase is added to the lower phase, and the mixing and settling procedure is repeated. This washing step is usually repeated twice. After the last partition step the material at the interface is removed together with the upper phase. The purity of the inside-out thylakoids can be further improved at the expense of yield by repeated extraction of the lower phase with fresh upper phase. Lowering the concentration of both polymers, to 5.55% for example, will have a similar effect.[24]

The right-side-out material partitioning to the upper phase can be purified in a manner corresponding to that for the inside-out material. After the first partition step the upper phase is removed and transferred to 9 ml of a new, pure lower phase. The mixing and settling procedure is then repeated. This washing step is usually repeated twice.

The pure phases required for the washing steps are obtained as follows. A large phase system with the same composition as that above is prepared in a separatory funnel. The phase system is adjusted to 4°, mixed, and allowed to settle. The two phases are collected separately and the interface is discarded. If only inside-out thylakoids are to be isolated, an alternative way to obtain pure upper phase which consumes less Dextran is to make a phase system of 1.5% (w/w) Dextran, 7.5% (w/w) polyethylene glycol, 10 m*M* sodium phosphate, pH 7.4, 5 m*M* NaCl, and 20 m*M* sucrose. This phase system has a large upper phase with the same composition as the former one.

[24] B. Andersson, *Adv. Photosynth. Res., Proc. Int. Congr. Photosynth., 6th, 1983,* Vol. 3, p. 223 (1984).

Removal of Polymers

The inside-out vesicles in the lower phase can be used directly for activity measurements, but for most studies the polymers have to be removed and the thylakoid material concentrated. This may be done in three different ways. (1) The lower phase is diluted at least 3-fold with buffer solution, to reduce the viscosity, and centrifuged at 100,000 *g* for 60 min to sediment the thylakoids; (2) 15 ml of pure upper phase adjusted to room temperature is added. In this way a phase system at 10–15° is created in which, after mixing and settling, the thylakoid material is collected in the upper phase. The upper phase is removed, diluted at least 2-fold, and centrifuged as above; (3) 2–5 ml of an upper phase in which 25% of the polyethylene glycol is exchanged for the positively charged trimethylaminopolyethylene glycol (obtained from Aqueous Affinity, Arlöv, Sweden) is added. After mixing and settling the thylakoid material is collected in the small upper phase, which is removed, diluted at least 2-fold, and centrifuged as above. The advantage with methods (2) and (3) is that the thylakoid material is obtained in a smaller volume, which facilitates the centrifugation step. Furthermore, the sedimentation of the thylakoid material can be obtained already at 40,000 *g* for 30 min if 10 m*M* $MgCl_2$ is included in the medium.[25]

Properties of the Isolated Inside-Out Thylakoids

In order to determine whether the isolation has been successful the following criteria can be used. As an indicator of sidedness of the isolated vesicles, the direction of pH changes associated with the photosynthetic electron transport reactions is determined. The membranes are first removed from buffer by two consecutive centrifugations at 100,000 *g* for 30 min using 40 m*M* KCl for resuspensions. The light-induced pH change is measured in a thermostatted vessel (20°) using a combined glass electrode and with a medium composition of 40 m*M* KCl, 0.4 m*M* phenyl-*p*-benzoquinone, and thylakoid material corresponding to 50–100 μM chlorophyll/ml. The initial pH is adjusted to 6.5. Light of sufficient intensity is obtained if two projectors are placed one on each side of the reaction vessel. For inside-out thylakoids a light-induced pH decrease in the external medium is expected. A limitation of this method is that it only indicates whether right-side-out or inside-out thylakoids are dominating in the preparation. A quantitative determination can be made by freeze-fracture electron microscopy.[26]

[25] P.-Å. Albertsson, *FEBS Lett.* **149,** 186 (1982).

[26] B. Andersson, D. J. Simpson, and G. Høyer-Hansen, *Carlsberg Res. Commun.* **43,** 77 (1978).

Since the inside-out vesicles obtained by the first two methods described should be enriched in photosystem II, this can be used as a criterion for a successful isolation of these vesicles. The most convenient assay for this enrichment is to measure the chlorophyll *a*/*b* ratios. For the first two procedures the inside-out vesicles should have chlorophyll *a*/*b* ratios of 0.8–0.9 U lower than the original thylakoids, i.e., 2.3–2.4 if the original thylakoids have a value of 3.2. For the third procedure the chlorophyll *a*/*b* ratio should not be more than 0.1 U lower than the original thylakoids.

Acknowledgments

We wish to thank Prof. P.-Å. Albertsson and Dr. C. Sundby for stimulating collaboration and discussions. The work has been supported by the Swedish Natural Science Research Council.

[27] Characterization of Chloroplast Cytochromes

By DEREK S. BENDALL and STEPHEN A. ROLFE

The cytochrome components of thylakoid membranes occur mainly in the cytochrome *bf* complex (*f* and *b*-563) and photosystem II (b-559$_{HP}$). Cytochromes *b*-563 and *f* are analogous to the cytochromes *b* and *c* (c_1) of cytochrome *bc* complexes from other sources. Thus there are two hemes *b* per heme *f*, but both hemes *b* reside on the same polypeptide. Nevertheless, the two *b*-563 components have been shown to have different characteristic redox potentials and slightly different optical spectra in the α-band region.

The above components have been relatively well characterized; they have been purified and their genes identified and sequenced. By contrast, little is known about cytochromes b-559$_{LP}$ and *b*-560.[1] The former is a regular component of chloroplasts, occurring in about the same concentration as cytochrome *f*; it is probably distinct from b-559$_{HP}$ because it occurs in freshly isolated chloroplasts and tends to fractionate with the *bf* complex (although more highly purified preparations of the complex lack it), but neither the polypeptide nor the gene have been identified. Cytochrome *b*-560 was first observed in a mutant of *Chlamydomonas reinhardii* lacking the cytochrome *bf* complex, but there is evidence for its occur-

[1] D. S. Bendall, *Biochim. Biophys. Acta* **683,** 119 (1982).

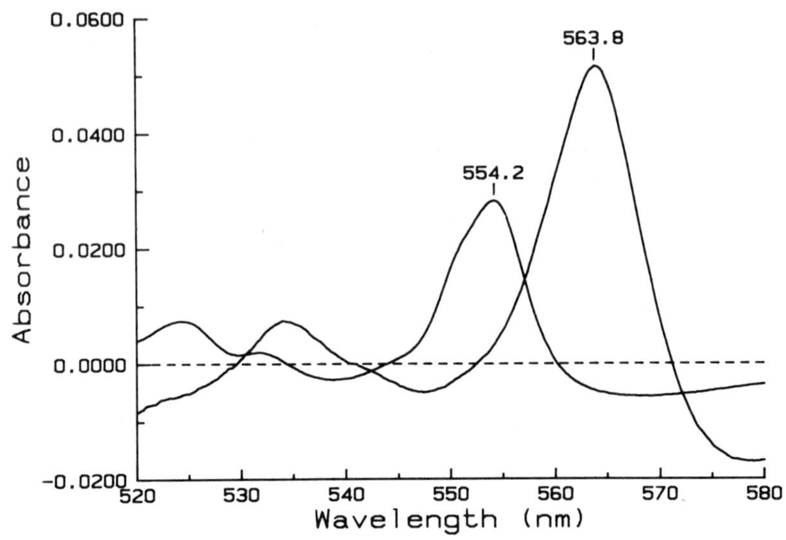

FIG. 1. Difference spectra (reduced minus oxidized) of cytochromes *f* and *b*-563 in the cytochrome *bf* complex purified from lettuce chloroplasts. The baseline was drawn through isosbestic points at 560 (*f*) and 571 (*b*-563); a small error has probably been introduced at shorter wavelengths as a result of bleaching of yellow pigments by the reducing agents. Spectra were measured on a sample in 50 m*M* MES, pH 6.5, with the following redox reagents: cytochrome *f*, 1 m*M* $K_3Fe(CN)_6$, and 1 m*M* hydroquinone; cytochrome *b*-563, 1 m*M* hydroquinone and a few crystals of $Na_2S_2O_4$ in the presence of catalase.

rence in wild-type membranes and also in higher plant chloroplasts; nevertheless, it appears to be absent from thylakoid membranes of a blue-green alga, *Phormidium laminosum*,[2] and thus may not be an essential intermediate in photosynthetic electron transport.

The characteristic properties of chloroplast membrane cytochromes are listed below. Details of analytical procedures and the detection of cytochromes on polyacrylamide gels are then given. The soluble algal *c*-type cytochromes that are functional replacements of plastocyanin have not been included.

Optical Spectra

Spectra of the main chloroplast components, cytochromes *f*, *b*-563, and *b*-559, in the α-band region, are shown in Figs. 1–3. The spectra in Fig. 2 were measured at 77 K and demonstrate that whereas cytochrome *b*-563_{LP} has a single peak *b*-563_{HP} has a satellite peak on the short-wave

[2] S. A. Rolfe and D. S. Bendall, unpublished work.

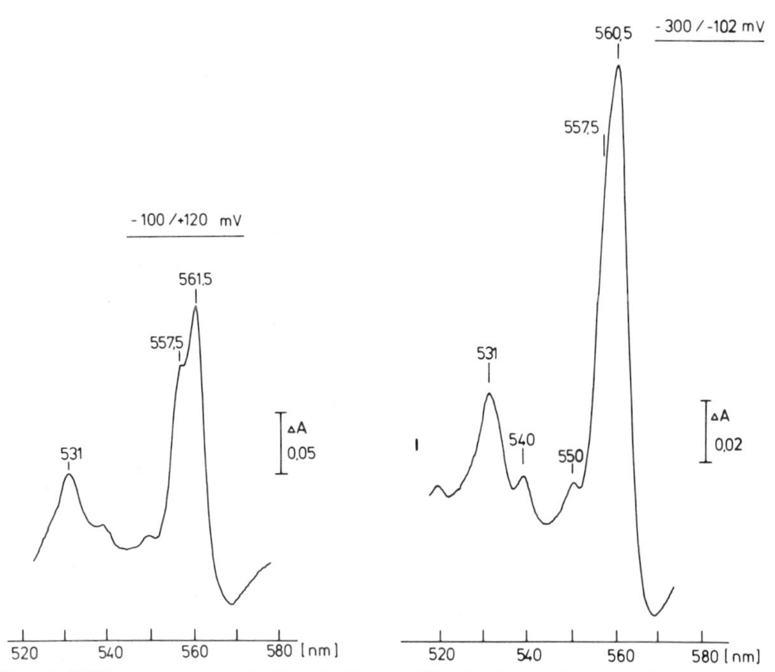

FIG. 2. Difference spectra (reduced minus oxidized) of cytochrome *b*-563 at 77 K. Cytochrome *b*-563_{HP} (left) has a more pronounced satellite band at 557.5 nm than cytochrome *b*-563_{LP} (right). Reprinted with permission from E. C. Hurt and G. Hauska, *FEBS Lett.* **153,** 413 (1983).

side, although some of this may be contributed by contaminating cytochrome *b*-559_{LP}. Isosbestic points are recorded in Table I. Extinction coefficients for cytochrome α-bands and other components that absorb in the green region are listed in Table II. When an appropriate extinction coefficient is not available for a cytochrome component the assumption of $\varepsilon = 20\ \text{m}M^{-1} \cdot \text{cm}^{-1}$ seems to provide a fair approximation. A major problem is the occurrence of several components with overlapping spectra, and it is thus necessary to take measurements at as least as many wavelengths as the number of components thought to be involved, and to calculate from a set of simultaneous equations the contributions of the individual components. Table III presents a set of extinction coefficients for such multicomponent analysis which may be useful in some circumstances. A slightly different method of analysis has been described by Joliot and Joliot.[3,4]

[3] P. Joliot and A. Joliot, *Biochim. Biophys. Acta* **765,** 210 (1984).

[4] P. Joliot and A. Joliot, *Biochim. Biophys. Acta* **806,** 398 (1985).

TABLE I
ISOSBESTIC POINTS OF CHLOROPLAST CYTOCHROMES

Cytochrome	Wavelength (nm)
f	$543\frac{1}{4}$, $560\frac{1}{4}$
b-563[a]	548, $570\frac{1}{2}$(b-563_{HP}), 572(b-563_{LP})
b-559[b]	538, 548, 570[c]

[a] R. D. Clark and G. Hind, *Proc. Natl. Acad. Sci. U.S.A.* **80,** 6249 (1983).
[b] G. T. Babcock, W. R. Widger, W. A. Cramer, W. A. Oertling, and J. G. Metz, *Biochemistry* **24,** 3638 (1985).
[c] W. A. Cramer, personal communication.

TABLE II
EXTINCTION COEFFICIENTS OF CHLOROPLASTS CYTOCHROMES AND RELATED COMPONENTS

Component	Wavelength (nm)	Reduced–oxidized ($mM^{-1} \cdot cm^{-1}$)	Ref.
Cytochrome *f*	554	19.3[a]	*b*
		19.7[a]	*c*
		20.8[d]	*e*
		20.3[d]	*f*
Cytochrome *b*-559	559	17.5	*g*
Cytochrome *b*-563	563	18.2[h]	
P-700	703	−64	*i*
Plastocyanin	595	−4.9[j]	*k, l*
C-550	540	1.2	*m*
	550	−1.0	*m*

[a] Parsley; [b] D. S. Bendall, *in* "CRC Handbook of Biosolar Resources" (A. Mitsui and C. C. Black, eds.), Part I, p. 153. CRC Press, Boca Raton, Florida, 1982; [c] G. Forti, M. L. Bertolè, and G. Zanetti, *Biochim. Biophys. Acta* **109,** 33 (1965); [d] spinach; [e] J. Singh and A. R. Wasserman, *J. Biol. Chem.* **246,** 3532 (1971); [f] N. Nelson and J. Neumann, *J. Biol. Chem.* **247,** 1817 (1972); [g] W. R. Widger, W. A. Cramer, M. Hermodson, D. Meyer, and M. Gullifor, *J. Biol. Chem.* **259,** 3870 (1984), and W. A. Cramer, personal communication; [h] calculated from difference spectra of the purified cytochrome *bf* complex, assuming $\varepsilon_{mM} = 19.3$ for cytochrome *f* and $b/f = 2$; [i] T. Hiyama and B. Ke, *Biochim. Biophys. Acta* **267,** 160 (1972); [j] spinach; slightly different values have been reported for other species (see Ref. *l*); [k] S. Katoh, I. Shiratori, and A. Takamiya, *J. Biochem.* (*Tokyo*) **51,** 32 (1962); [l] D. Boulter, B. G. Haslett, D. Peacock, J. A. M. Ramshaw, and M. D. Scawen, *Int. Rev. Biochem.* **13,** 1 (1977); [m] H. J. Van Gorkom, *Biochim. Biophys. Acta* **347,** 439 (1974).

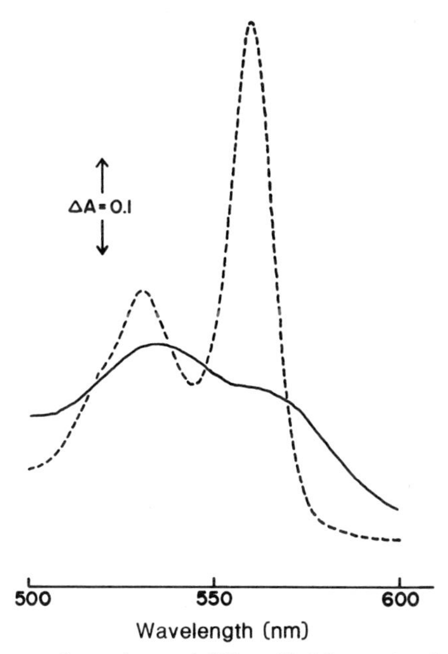

FIG. 3. Absolute spectra of cytochrome *b*-559 purified from spinach chloroplasts. Continuous line, oxidized; broken line, reduced. Reprinted with permission from G. T. Babcock, W. R. Widger, W. A. Cramer, W. A. Oertling, and J. G. Metz, *Biochemistry* **24,** 3638 (1985).

TABLE III
EXTINCTION COEFFICIENTS (REDUCED–OXIDIZED) FOR MULTICOMPONENT ANALYSIS[a]

Component	(mM^{-1} cm^{-1}) at given wavelength (nm)			
	542	554	563	575
Cytochrome *f*	−1.4	19.3[b]	−2.9	−4.4
Cytochrome *b*-563	0.3	1.8	18.2[b]	−1.9
P-700[c]	−4.2	−3.6	−2.8	−0.3
Plastocyanin[d]	−1.7	−2.6	−3.3	−4.2

[a] Values relative to those at absorbance peaks were measured by P. R. Rich (personal communication).

[b] See Table II.

[c] Values assume $\varepsilon = -64$ $mM^{-1} \cdot cm^{-1}$ at 703 nm; T. Hiyama and B. Ke, *Biochim. Biophys. Acta* **267,** 160 (1972).

[d] Values assume $\varepsilon = -4.8$ $mM^{-1} \cdot cm^{-1}$ at 595 nm.

TABLE IV
EPR SPECTRA OF CHLOROPLAST CYTOCHROMES

Component	g_x	g_y	g_z	Ref.
f			3.5–3.51	*a, b, c, d, e, f*
	1.6	2.07	3.48[g]	*h*
b-559$_{HP}$	1.36	2.16	3.08	*a, b, c, i, m*
b-559[n]	1.55	2.26	2.93	*k*
b-559[o]		2.26	2.9	*a*
b-559[p]		2.26	2.94	*c*
b-559$_{LP}$		2.26	2.9–2.95	*b, d*
b-563			3.4–3.7	*e, f, l*

[a] R. Malkin and T. Vänngård, *FEBS Lett.* **111,** 228 (1980); [b] M. S. Crowder, R. C. Prince, and A. Bearden, *FEBS Lett.* **144,** 204 (1982); [c] J. Bergström and T. Vänngård, *Biochim. Biophys. Acta* **682,** 452 (1982); [d] P. R. Rich, P. Heathcote, M. C. W. Evans, and D. S. Bendall, *FEBS Lett.* **116,** 51 (1980); [e] J. C. Salerno, J. W. McGill, and G. C. Gerstle, *FEBS Lett* **162,** 257 (1983); [f] J. Bergström, L.-E. Andréasson, and T. Vänngård, *FEBS Lett.* **164,** 71 (1983); [g] purified spinach protein; [h] J. N. Siedow, L. E. Vickery, and G. Palmer, *Arch. Biochem. Biophys.* **203,** 101 (1980); [i] J. H. A. Nugent and M. C. W. Evans, *FEBS Lett.* **112,** 1 (1980); [j] purified protein from spinach in a low-protein form; [k] G. T. Babcock, W. R. Widger, W. A. Cramer, W. A. Oertling, and J. G. Metz, *Biochemistry* **24,** 3638 (1985); [l] J. W. McGill and J. C. Salerno, *Biophys. J.* **47,** 241a (1985); [m] D. F. Ghanotakis, C. F. Yocum, and G. T. Babcock, *Photosyn. Res.* **9,** 125 (1986); [n] purified protein from spinach in a low-potential form; [o] photosystem II subchloroplast fragments; [p] Tris-washed chloroplasts.

EPR Spectra

All the chloroplast cytochromes, with the exception of *b*-560, have been shown to exhibit EPR spectra characteristic of low-spin heme. Of the three anisotropic *g* values, g_z is the most prominent, and g_x has been observed so far only with purified cytochromes *f* and *b*-559. The spectra are characteristic of the ferric state and disappear on reduction. Liquid helium temperatures are necessary. A list of reported *g* values is given in Table IV. The low-spin signal near $g = 3.5$ from cytochrome *b*-563 is only observed at a potential at which cytochrome *f* is reduced (Fig. 4). Denatured cytochrome *b*-563 gives signals at $g \approx 2.9$ and around $g \approx 6$.[5]

[5] J. C. Salerno, J. W. McGill, and G. C. Gerstle, *FEBS Lett.* **162,** 257 (1983).

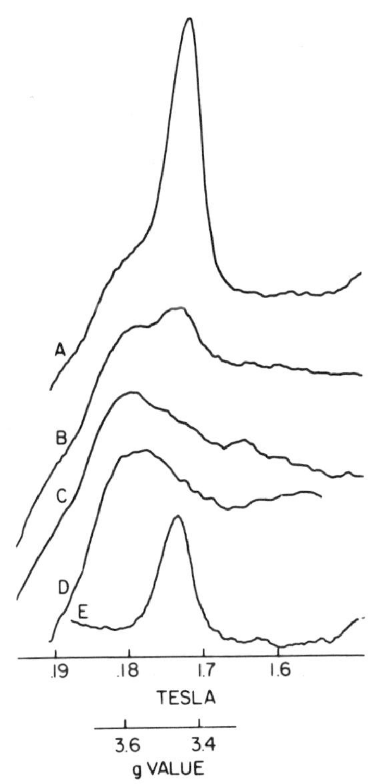

FIG. 4. EPR spectra of cytochromes *f* and *b*-563 at 13 K in cytochrome *bf* complex purified from spinach. (A) In the presence of 0.5 m*M* $K_3Fe(CN)_6$; (B) complex as isolated; (C) in the presence of 1 m*M* ascorbate; (D) as in (C), with the addition of 0.5 m*M* antimycin A; (E) (A) minus (C). Reprinted with permission from J. C. Salerno, J. W. McGill, and G. C. Gerstle, *FEBS Lett.* **162,** 257 (1983).

Oxidation–Reduction Potentials

Differences in midpoint redox potentials provide an important method for discriminating between cytochrome components, and values reported for those present in chloroplasts and cytochrome *bf* preparations are collected in Table V. Despite the range of values found for cytochromes *f* and *b*-559_{HP}, these two components have very similar and unusually high midpoints. Cytochrome *b*-559_{LP} has a potential about 300 mV lower, but various treatments, including aging, convert *b*-559_{HP} into forms with lower potentials which may render distinction from the native *b*-559_{LP} difficult. Cytochrome *b*-559_{LP} can be extracted from chloroplasts with digitonin, however, whereas the high-potential component and its modified forms remain associated with photosystem II.

TABLE V
REDOX POTENTIALS OF CHLOROPLAST CYTOCHROMES

Component	Source	E_m (mV)	pH	pH dependence	Ref.
f	Chloroplasts	335–390	7–8.2		*a–e*
	bf preparation	330–375	5–8.5	−60 mV/pH, above pH 9	*f–h*
	Purified protein	365, 370	6–8	−60 mV/pH, above pH 8.4	*i–k*
b-559$_{HP}$	Chloroplasts	325–395	6–8.2		*a, c, e, l, m–o*
b-559$_{LP}$	Chloroplasts	20–77	7–7.8		*e, g, l*
	bf preparation	85	6.2–7.8		*g*
b-563	Chloroplasts	−30[p]	7		*q*
	Chloroplasts	−110	7		*g*
	Chloroplasts	−147	7		*e*
	Chloroplasts	5	7		*r*
	bf preparation	−90	7	−15 mV/pH, pH 5–11	*g*
	bf preparation	−100	7		*f*
b-563$_{HP}$	*bf* preparation	−40	6.5	−18 mV/pH, pH < 7.5	*h*
	bf preparation	−3	5.6		*r*
	bf preparation	−30[s]	7.5		*t*
	bf preparation	70	7.3		*u*
b-563$_{LP}$	*bf* preparation	−172	6.5	−18 mV/pH, pH < 7.5	*h*
	bf preparation	−146	5.6		*r*
	bf preparation	−150	7.5		*t*
	bf preparation	−50	7.3		*u*

[a] D. S. Bendall, *Biochem. J.* **109,** 46P (1968); [b] H. N. Fan and W. A. Cramer, *Biochim. Biophys. Acta* **216,** 200 (1970); [c] D. B. Knaff and D. I. Arnon, *Biochim. Biophys. Acta* **226,** 400 (1971); [d] R. Malkin, D. B. Knaff, and A. J. Bearden, *Biochim. Biophys. Acta* **305,** 675 (1973); [e] F. A. L. Peters, J. E. Van Wielink, H. W. Wong Fong Sang, S. De Vries, and R. Kraayenhof, *Biochim. Biophys. Acta* **722,** 460 (1983); [f] N. Nelson and J. Neumann, *J. Biol. Chem.* **247,** 1817 (1972); [g] P. R. Rich and D. S. Bendall, *Biochim. Biophys. Acta* **591,** 153 (1980); [h] E. Hurt and G. Hauska, *J. Bioenerg. Biomembr.* **14,** 405 (1982); [i] H. E. Davenport and R. Hill, *Proc. R. Soc. London, Ser. B* **139,** 327 (1952); [j] E. Matsuzaki, Y. Kamimura, T. Yamasaki, and E. Yakushiji, *Plant Cell Physiol.* **16,** 237 (1975); [k] J. C. Gray, *Eur. J. Biochem.* **82,** 133 (1978); [l] P. Horton and E. Croze, *Biochim. Biophys. Acta* **462,** 86 (1977); [m] N. K. Boardman, J. M. Anderson, and R. G. Hiller, *Biochim. Biophys. Acta* **234,** 126 (1971); [n] P. Horton, J. Whitmarsh, and W. A. Cramer, *Arch. Biochem. Biophys.* **176,** 519 (1976); [o] D. B. Knaff, *FEBS Lett.* **60,** 331 (1975); [p] midpoints were reported to vary between 0 and −80 mV with different preparations; [q] M. E. Girvin and W. A. Cramer, *Biochim. Biophys. Acta* **767,** 29 (1984); [r] E. Hurt and G. Hauska, *FEBS Lett.* **153,** 413 (1983); [s] an anomalous slope corresponding to $n = 2$ was reported; [t] R. D. Clark and G. Hind, *Proc. Natl. Acad. Sci. U.S.A.* **80,** 6249 (1983); [u] J. C. Salerno, J. W. McGill, and G. C. Gerstle, *FEBS Lett.* **162,** 257 (1983).

Results with cytochrome *b*-563 have been listed more fully in Table V because of the wide variation in the reported values. The cause of the variation is not understood, but we tend to favor the lower values, partly because menadiol, which provides a quick and easy method of achieving an E_h in the range 0 to −20 mV at pH 6, reduces very little cytochrome *b*-563.[6] Despite the disagreement over absolute values, a consistent difference of 120–130 mV has been found between b-563_{HP} and b-563_{LP} whenever the distinction has been made. The presence of the two forms has not yet been clearly demonstrated in chloroplasts, as distinct from *bf* preparations.

The technique of redox potentiometry has been fully described elsewhere.[7,8]

Molecular Weights and Behavior on Polyacrylamide Gels

Definitive molecular weights for the major components are now available from DNA sequences, and these are listed in Table VI. The cytochrome *f* gene from three species (pea, wheat, and spinach) has been sequenced and in each case codes for a mature protein of 285 amino acid residues and a molecular weight of 31,000–32,000. In some cases cytochrome *f* has been shown to run on SDS–polyacrylamide gels with an apparent molecular weight of about 31,000,[9–11] but in other cases higher values have been obtained, the extreme being 37,300 for the pea protein.[12–14] The reason for these variations is not well understood but the mobility seems to depend on sample preparation as well as gel system.[9] Cytochrome *f* from charlock (*Sinapis arvensis*) runs as a polypeptide of 28.5 kDa,[14] and the protein from several other crucifers behaves similarly, but this is likely to be due to proteolysis.[15]

Cytochrome *f* and other *c*-type cytochromes can be specifically detected on polyacrylamide gels by a staining reaction based on the peroxidatic activity of the heme group covalently bound to the denatured proteins. A sensitive method in which a blue stain is developed in the

[6] P. R. Rich and D. S. Bendall, *Biochim. Biophys. Acta* **591,** 153 (1980).

[7] P. L. Dutton, *Biochim. Biophys. Acta* **226,** 63 (1971).

[8] P. L. Dutton, this series, Vol. 54, p. 411.

[9] D. L. Willey, C. J. Howe, A. D. Auffret, C. M. Bowman, T. A. Dyer, and J. C. Gray, *Mol. Gen. Genet.* **194,** 416 (1984).

[10] K.-H. Süss, *FEBS Lett.* **70,** 191 (1976).

[11] J. Singh and A. R. Wasserman, *J. Biol. Chem.* **246,** 3532 (1971).

[12] E. Hurt and G. Hauska, *Eur. J. Biochem.* **117,** 591 (1981).

[13] A. Doherty and J. C. Gray, *Eur. J. Biochem.* **98,** 87 (1979).

[14] J. C. Gray, *Eur. J. Biochem.* **82,** 133 (1978).

[15] J. C. Gray, personal communication.

TABLE VI
MOLECULAR WEIGHTS OF CHLOROPLAST CYTOCHROMES CALCULATED FROM DNA SEQUENCES

Component	Source	Polypeptide	Hemoprotein	Reference
f	Pea	31,096	31,712[a]	*b*
	Wheat	31,302	31,918	*c*
	Spinach	31,323	31,939	*d*
b-563	Spinach	23,385	24,618	*e, f*
b-559_{HP}	Spinach	9255; 4307 (α, β subunits)	14,179[g]	*h, i*
	Wheat	9313; 4367	14,297	*j*
	Oenothera	9183; 4293	14,093	*k*
	Tobacco	9265; 4367	14,249	*k*

[a] One heme for *f* and *b*-559_{HP}, two for *b*-563; [b] D. L. Willey, A. D. Auffret, and J. C. Gray, *Cell (Cambridge, Mass.)* **36,** 555 (1984); [c] D. L. Willey, C. J. Howe, A. D. Auffret, C. M. Bowman, T. A. Dyer, and J. C. Gray, *Mol. Gen. Genet.* **194,** 416 (1984); [d] J. Alt and R. G. Herrmann, *Curr. Genet.* **8,** 551 (1984); [e] W. Heinemeyer, J. Alt, and R. G. Herrmann, *Curr. Genet* **8,** 543 (1984); [f] W. R. Widger, W. A. Cramer, R. G. Herrmann, and A. Trebst, *Proc. Natl. Acad. Sci. U.S.A.* **81,** 674 (1984); [g] assuming a heterodimer structure (αβ, one heme); [h] R. G. Herrmann, J. Alt, B. Schiller, W. R. Widger, and W. A. Cramer, *FEBS Lett.* **176,** 239 (1984); [i] W. R. Widger, W. A. Cramer, M. Hermodson, and R. G. Herrmann, *FEBS Lett.* **191,** 186 (1985); [j] S. M. Hird, D. L. Willey, T. A. Dyer, and J. C. Gray, *Mol. Gen Genet.* (in press); [k] N. Carrillo, P. Seyer, A. Tyagi, and R. G. Herrmann, *Curr. Genet.* (in press).

presence of 2,3,5,6-tetramethylbenzidine and H_2O_2 has been described.[16] More surprisingly, in view of the noncovalent heme–protein linkage, cytochrome *b*-563 can also be detected (although the method of Thomas *et al.*[16] was originally described for cytochrome P-450). Cytochrome *b*-559, on the other hand, is more labile, and such staining has only been observed with lithium dodecyl sulfate gels run at 4° and pH 7.[17]

Cytochrome *b*-563 has been identified as a band running at 23.5 kDa on SDS–polyacrylamide gels of the purified spinach *bf* complex.[12,18] The relative staining of this polypeptide and cytochrome *f* with Coomassie blue and amido black suggested a 1 : 1 stoichiometry, and therefore that both *b*-563 hemes reside on the one polypeptide.[18] This is supported by sequence homology with mitochondrial cytochrome *b* and by the presence of four histidine residues that provide putative heme ligands in ap-

[16] P. E. Thomas, D. Ryan, and W. Levin, *Anal. Biochem.* **75,** 168 (1976).

[17] W. R. Widger, W. A. Cramer, M. Hermodson, D. Meyer, and M. Gullifor, *J. Biol. Chem.* **259,** 3870 (1984).

[18] E. Hurt and G. Hauska, *J. Bioenerg. Biomembr.* **14,** 405 (1982).

propriate positions according to a model based on a hydropathy plot of the sequence.[19]

Purified preparations of cytochrome *b*-559 give a main band at 9–10 kDa on SDS–urea or lithium dodecyl sulfate polyacrylamide gels together with more weakly staining bands at 6, 8, and 10 kDa.[17,20] The 10-kDa band has been reported to respond to the peroxidase stain, and antibodies to it react with the 10-kDa band observed in photosystem 2 preparations.[17] Thus the 10-kDa polypeptide is derived from cytochrome b-559$_{HP}$. Optical and resonance Raman spectra have shown that both axial heme ligands are histidine residues,[21] but the 10-kDa polypeptide contains only one histidine.[22] The other is probably provided by the 6-kDa polypeptide, which occurs in approximately 1 : 1 stoichiometry with the 9- to 10-kDa component.[23] The gene for the smaller polypeptide has been located immediately downstream from that for the larger component, and codes for a polypeptide of actual M_r = 4307 (Table VI).[22]

Estimation of Cytochromes *in Situ*

The methods given below are modified from those previously described,[24] and as far as possible are based on spectra and extinction coefficients measured for α-bands of purified components. Significant inaccuracy (at least 10%) may still be present because the necessary information is incomplete or uncertain, and in the case of chloroplast preparations interfering spectral changes may not have been fully allowed for. Measurements must be made with a sensitive spectrophotometer suitable for use with turbid samples; the commonest arrangement is a large end-window photomultiplier placed as close as possible to the sample cuvette. A computer-linked system is preferable, in which case use of a single-beam instrument is possible although not essential, otherwise a split-beam arrangement with sample and reference cuvettes is necessary. The most sensitive instrument is probably the one described by Joliot *et al.*,[3,25]

[19] W. R. Widger, W. A. Cramer, R. G. Herrmann, and A. Trebst, *Proc. Natl. Acad. Sci. U.S.A.* **81,** 674 (1984).

[20] J. G. Metz, G. Ulmer, T. M. Bricker, and D. Miles, *Biochim. Biophys. Acta* **725,** 203 (1983).

[21] G. T. Babcock, W. R. Widger, W. A. Cramer, W. A. Oertling, and J. G. Metz, *Biochemistry* **24,** 3638 (1985).

[22] R. G. Herrmann, J. Alt, B. Schiller, W. R. Widger, and W. A. Cramer, *FEBS Lett.* **176,** 239 (1984).

[23] W. R. Widger, W. A. Cramer, M. Hermodson, and R. G. Herrmann, *FEBS Lett.* **191,** 186 (1985).

[24] D. S. Bendall, H. E. Davenport, and R. Hill, this series, Vol. 23, p. 327.

[25] P. Joliot, D. Béal, and B. Frilley, *J. Chem. Phys.* **77,** 209 (1980).

which can be used for wavelength scanning as well as kinetic measurements. Whatever the instrument, results will be obtained as difference spectra (reduced minus oxidized).

Cytochromes f and b-559$_{HP}$

These components have almost identical midpoint potentials (Table V) and will normally be observed together in chemical difference spectra.

1. Thylakoid membrane preparations are diluted into a medium containing 0.3 *M* sorbitol and 50 m*M* phosphate (or MES) buffer, pH 6.5. Normally a chlorophyll concentration of 150 μg/ml is appropriate when using a 10-mm cuvette. A slightly acid pH is important as at higher values the spectra are unstable owing to autoxidation of the hydroquinone added to the reference cuvette. The difference spectrum is recorded between 520 and 580 nm 3 min after addition of 1 m*M* hydroquinone to the sample cuvette and 1 min after addition of 1 m*M* $K_3Fe(CN)_6$ to the reference cuvette. Volumes of additions to each cuvette should be identical. A baseline is drawn between 538 and 570 nm (isosbestic points for cytochrome *b*-559) and absorbance differences above the baseline determined at 552 (H_{552}) and 560 (H_{560}) nm. The concentrations of cytochromes *b*-559$_{HP}$ and *f* are calculated from the following pair of simultaneous equations:

$$c_b = 60.1H_{560} - 10.5H_{552} \text{ nmol/ml}$$

$$c_f = 58.3H_{552} - 16.7H_{560} \text{ nmol/ml}$$

In deriving the above equations the extinction coefficients given in Table VII were assumed.

TABLE VII
EXTINCTION COEFFICIENTS FOR CALCULATION OF CONCENTRATIONS OF CYTOCHROMES *f* AND *b*-559$_{HP}$

	ε_{mM}	
Wavelength (nm)	Cytochrome *b*-559$_{HP}$	Cytochrome *f*
538	0	−1.5
552	5	15.5[a]
560	17.5	0
570	0	−3.9

[a] Based on ε_{mM} = 19.3 at 554 nm.

The difference spectrum obtained is likely to be distorted, particularly at the wavelength extremes, by bleaching of a small amount of chlorophyll in the reference cuvette by hydroquinone. The above calculation assumes that the effect is linear between 538 and 570 nm. Similarly, linearity is assumed for the small contributions of P700 and plastocyanin to the spectrum.

2. Cytochrome *f* can be observed independently by taking advantage of the fact that E_m of cytochrome b-559$_{HP}$ is lowered about 300 mV by exposure to certain detergents.[24] Reaction mixtures are prepared as under (1) except that they include in addition 1% (w/v) Triton X-100 (with some samples, e.g., preparations from *Chlamydomonas* and cyanobacteria, 2% Triton X-100 may be necessary). The concentration of cytochrome *f* is determined from the peak height at 554 nm above a baseline drawn through the spectrum at 543 and 560 nm (isosbestics), with the assumption of $\varepsilon_{mM} = 19.3$. Care should be taken that the spectrum is measured not much more than 1 min after addition of ferricyanide because in the presence of Triton X-100 ferricyanide causes a progressive bleaching of chlorophyll and other absorbance changes in the region of 554 nm.

Cytochromes b-559$_{LP}$ and b-563

Despite the variations in the reported midpoint potentials for cytochrome b-563 and the reluctance of some authors to accept the separate identity of cytochrome b-559$_{LP}$ in intact chloroplasts, method (1) below provides a quick and simple assay for b-559$_{LP}$ which can be applied to freshly isolated intact preparations (although it remains true that a low-potential form of the component associated with photosystem 2 would also be detected, there is no clear evidence for this). Method (2) is a slight modification of an earlier method which relies on the fact that in chloroplasts reduction of cytochrome b-563 by dithionite takes a few minutes but reduction of b-559$_{LP}$ takes a few seconds (or less).[24]

1. The method depends on the use of menadiol (2-methyl-1,4-naphthoquinol) as a selective reductant of cytochrome b-559$_{LP}$.

Preparation and Use of Menadiol.[26] Dissolve 0.5 g menadione in 50 ml diethyl ether and reduce it by shaking with a fresh solution of sodium dithionite in 0.1 *M* phosphate buffer, pH 7. Remove the aqueous layer and repeat the process with a fresh solution of dithionite. Wash the upper phase several times with a saturated solution of NaCl and then pour it through a column of anhydrous sodium sulfate in a sintered glass funnel, allowing it to flow directly into a round-bottomed flask flushed with N_2.

[26] P. R. Rich, personal communication.

Immediately evaporate the solution to dryness in a rotary evaporator. The product should be white with no more than a tinge of purple color. It can be stored at room temperature in a well-stoppered bottle flushed with N_2 and kept in the dark, but will deteriorate after a few weeks.

For use, prepare a solution of 0.5 *M* menadiol in ethanol containing 10 m*M* HCl. This solution can be kept at $-20°$ and used as long as it remains colorless. To use menadiol as a reducing agent the reaction mixture need not be anaerobic provided it is buffered at pH 6.0 to minimize autoxidation.

Estimation of Cytochromes. For estimation of both components three cuvettes are required, each containing chloroplasts diluted into a medium containing 0.3 *M* sorbitol, 50 m*M* MES, pH 6.0, and catalase (0.5 μl crystalline suspension in water/ml; this may be omitted when dithionite is not to be used) at a final chlorophyll concentration of 150 μg/ml (10-mm cuvettes). As reductants 1 m*M* methylhydroquinone, 1 m*M* menadiol, and a few crystals of sodium dithionite (or for each milliliter of suspension 10 μl of a freshly prepared solution, 50 mg/ml, in buffer) are added, one reagent to each cuvette. Difference spectra should be recorded 3 min after addition of quinols and at least 5 min after addition of dithionite. The concentration of cytochrome b-559_{LP} is calculated from the height of the peak above a baseline drawn through assumed isosbestic points of 548 and 568 nm in the menadiol minus methylhydroquinone spectrum. An extinction coefficient of 20 $mM^{-1} \cdot cm^{-1}$ is assumed. Similarly, the concentration of *b*-563 is calculated from the peak height above a baseline through 548 and 571 nm in the dithionite minus menadiol spectrum using $\varepsilon_{mM} = 18.2$. As before, the spectra are influenced by effects of the reagents on bulk chlorophyll which to a first approximation may be assumed to be linear between the isosbestics, but in addition there will be a small uncorrected contribution from C550. Cytochrome *b*-560 contributes to the *b*-563 spectrum but a satisfactory correction is not yet available.

2. Thylakoid membranes are diluted into a medium containing 0.3 *M* sorbitol, 50 m*M* phosphate or MES buffer, pH 6.5, 1 m*M* hydroquinone, and catalase (0.5 μl crystalline suspension/ml) to give a chlorophyll concentration of 150 μg/ml and the suspension is divided between two cuvettes. One is used as the reference and to the other is added a freshly prepared solution of sodium dithionite in buffer (50 mg/ml; 10 μl/ml suspension) at zero time. The spectrum is scanned rapidly and repeatedly between 530 and 580 nm, but at increasing time intervals, over a period of 7–10 min. Reduction of cytochrome *b*-563 is pseudo-first-order with $t \approx 80$ sec, depending on dithionite concentration. Baselines are drawn through each spectrum between 548 and 571 nm (isosbestics for cytochrome *b*-563) and peak heights in absorbance units (*H*) measured at 559

and 563 nm. A plot of $\log(H_{max} - H)$ against time for each wavelength gives a pair of parallel lines that can be extrapolated to zero time. The concentrations of the two cytochromes (c_{559} and c_{563}) are given by the following equations:

$$c_{559} = 50[H_{max} - \text{antilog}(Y_{559})] \text{ nmol/ml}$$

$$c_{563} = 54.9 \text{ antilog}(Y_{563}) \text{ nmol/ml}$$

where Y_{559} and Y_{563} are intercepts and ε_{mM} has been taken to be 20 and 18.2, respectively.

Cytochrome *b*-560 has not been included explicitly in the above calculation. It is reduced slowly by dithionite at about the same speed as *b*-563.

Cytochrome b-560

A satisfactory method for estimation of cytochrome *b*-560 in thylakoid membranes has not yet been devised. Its E_m lies within the range for cytochrome *b*-563 and it is also reduced slowly by dithionite. It may be distinguished from *b*-563, however, by its resistance to extraction by a mixture of sodium cholate and octyl glucoside.[27]

Thylakoid membranes are resuspended with a glass homogenizer to give a chlorophyll concentration of 2 mg/ml in a medium containing 20 m*M* octyl-β-D-glucopyranoside, 0.33% (w/v) sodium cholate, 0.4 *M* sucrose, 0.4 *M* $(NH_4)_2SO_4$, and 20 m*M* Tricine–NaOH buffer, pH 8.0. The suspension is stirred for 1 hr at 4° in darkness and then centrifuged for 1 hr at 100,000 *g*. The supernatant contains most of the cytochrome *bf* complex. The depleted membranes are resuspended in 50 m*M* MES, pH 6.0, and assayed for cytochrome *b*-560 by taking a dithionite minus menadiol difference spectrum as described above for cytochrome *b*-563.

Acknowledgments

Work from the authors' laboratory was supported by the Science and Engineering Research Council. We thank Dr. P. R. Rich, Dr. J. C. Gray, and Dr. W. A. Cramer for unpublished information, and Mary Taylor for a sample of cytochrome *bf* complex purified from lettuce chloroplasts.

[27] M. Sanguansermsri, S. A. Rolfe, and D. S. Bendall, unpublished work.

[28] Cell-Free Reconstitution of Protein Transport into Chloroplasts

By MICHAEL L. MISHKIND, KAREN L. GREER, and GREGORY W. SCHMIDT

Reconstitution of protein transport from the cytoplasm to the chloroplast with isolated components has revealed some of the fundamental aspects of chloroplast biogenesis.[1,2] Foremost of these is that the cytoplasmically synthesized proteins of the organelle are synthesized as soluble, higher molecular weight precursors that are then imported into plastids by a posttranslational mechanism.[3,4] This process is ATP dependent[5,6] and is mediated by an amino acid extension at the N-terminus of all precursors so far examined. The extension, termed a transit or leader sequence,[7] is both necessary[8] and sufficient[9] for directing a polypeptide to the chloroplast.

The use of cell-free systems that reconstitute critical steps in membrane transport and precursor maturation follows in the methodological tradition of workers in the area of protein secretion.[10] In recent years, this approach has led to a rather detailed understanding of the molecular mechanisms for the early steps of the secretory pathway.[11] In contrast, studies on protein import into chloroplasts have remained, for the most part, at the phenomenological level. Since the initial demonstrations of the posttranslational import pathway,[3,4] the technique has been used mainly to verify that most, if not all, chloroplast proteins are transported by a posttranslational route.[12–17] Insight into interactions between precur-

[1] A. Cashmore, L. Szabo, M. Timko, A. Kausch, G. Van den Broeck, P. Schreier, H. Bohnert, L. Herra-Estrella, M. Van Montagu, and J. Schell, *Bio/Technology* **3,** 803 (1985).

[2] G. W. Schmidt and M. L. Mishkind, *Annu. Rev. Biochem.* **55,** 879 (1986).

[3] P. E. Highfield and R. J. Ellis, *Nature (London)* **271,** 420 (1978).

[4] N.-H. Chua and G. W. Schmidt, *Proc. Natl. Acad. Sci. U.S.A.* **75,** 6110 (1978).

[5] A. Grossman, S. Bartlett, and N.-H. Chua, *Nature (London)* **285,** 625 (1980).

[6] K. Cline, M. Werner-Washburne, T. H. Lubben, and K. Keegstra, *J. Biol. Chem.* **260,** 3961 (1985).

[7] N.-H. Chua and G. W. Schmidt, *J. Cell Biol.* **81,** 461 (1979).

[8] M. L. Mishkind, S. R. Wessler, and G. W. Schmidt, *J. Cell Biol.* **100,** 226 (1985).

[9] G. Van den Broeck, M. P. Timko, A. P. Kausch, A. R. Cashmore, M. Van Montagu, and L. Herrera-Estrella, *Nature (London)* **313,** 358 (1985).

[10] G. Blobel and B. Dobberstein, *J. Cell Biol.* **67,** 835 (1975).

[11] P. Walter, R. Gilmore, and G. Blobel, *Cell (Cambridge, Mass.)* **38,** 5 (1984).

[12] G. W. Schmidt, S. G. Bartlett, A. R. Grossman, A. R. Cashmore, and N.-H. Chua, *J. Cell Biol.* **91,** 468 (1981).

METHODS IN ENZYMOLOGY, VOL. 148

sors and chloroplast envelope membranes at the molecular level have been limited to a small number of reports on the energetics[5,6] and the functional determinants in the transit sequence for protein import.[8,9] One intent of this chapter is to extend the application of this technique in the study of chloroplast biogenesis.

Cell-free reconstitution of protein transport eventually should enable various components of the process to be isolated, manipulated, and assayed for functional competence. In addition to its use in such traditional cell fractionation analyses, the technique is amenable to contemporary approaches that involve *in vitro* mutagenesis of chloroplast protein precursors and their use in resolving the molecular basis of transit sequence function. Finally, by combining elements from highly divergent organisms, evolutionarily conserved functional determinants for the transport process can be perceived.[8]

The diversity of components that can be combined to generate a transport system is striking. Cell-free translation products of mRNAs from *Chlamydomonas reinhardii*[8] and etiolated soybean hypocotyls,[17] for example, contain protein precursors that are posttranslationally transported into chloroplasts isolated from pea leaves. The method thus serves as an efficient assay for nuclear encoded chloroplast proteins. It is especially useful in identifying plastid proteins that are either of low abundance or present in tissues that contain plastids that are difficult to isolate in an intact form. Nuclear encoded chloroplast heat shock proteins, for example, were initially detected using the *in vitro* transport system.[16,17] When employed with hybrid selected mRNAs[6] or mRNAs generated in the SP6 transcription system,[18] the method can serve to screen genomic or cDNA libraries for sequences that encode chloroplast proteins.[19]

Cell-Free Translation

Protein precursors for transport into isolated chloroplasts can be generated by *in vitro* and *in vivo* methods. Although the wheat germ cell-free

[13] A. R. Grossman, S. G. Bartlett, G. W. Schmidt, J. E. Mullet, and N.-H. Chua, *J. Biol. Chem.* **257,** 1558 (1982).

[14] F. van der Mark, W. van den Briel, and H. G. Huisman, *Biochem. J.* **214,** 943 (1983).

[15] G. Meyer and K. Kloppstech, *Eur. J. Biochem.* **138,** 201 (1984).

[16] K. Kloppstech, G. Meyer, G. Schuster, and I. Ohad, *EMBO J.* **4,** 1901 (1985).

[17] E. Vierling, M. L. Mishkind, G. W. Schmidt, and J. L. Key, *Proc. Natl. Acad. Sci. U.S.A.* **83,** 361 (1986).

[18] P. A. Krieg and D. A. Melton, *Nucleic Acids Res.* **12,** 7057 (1984).

[19] R. Broglie, G. Bellemare, S. G. Bartlett, N.-H. Chua, and A. R. Cashmore, *Proc. Natl. Acad. Sci. U.S.A.* **78,** 7304 (1978).

translation system has been preferred by workers in the chloroplast field, reticulocyte lysates also produce proteins capable of transport into isolated chloroplasts. The expression in *Escherichia coli* of cloned DNAs that encode chloroplast proteins also has been used to generate precursors for transport.[9]

For many plant species published protocols are available for the purification of high-molecular-weight translatable mRNA.[20] These procedures take into account pertinent obstructions to the isolation of active plant mRNA such as the abundant ribonucleases and phenolics present in plant tissue and the tendency of carbohydrates that inhibit cell-free translation systems to copurify with nucleic acids. When an organism proves to be especially intractable to RNA isolation, a method must be devised empirically, usually by combining components of available procedures.[21]

Total RNA fractions generally are adequate for *in vitro* synthesis of chloroplast precursors. Most often, however, it is useful to enrich for the mRNAs of interest by purifying polyadenylated RNA on oligo(dT)-cellulose[22] or poly(U)-Sepharose.[23] Alternatively, mRNAs that encode specific polypeptides can be obtained by hybrid selection using cDNAs specific for cytoplasmic mRNAs that encode chloroplast proteins.[6,19,24] Procedures in which the cloned DNAs are immobilized on nitrocellulose filters can be used for obtaining sufficient mRNA for a small number of transport assays. Large-scale preparation of mRNA by hybrid selection requires that the cloned DNAs be covalently attached to insoluble supports.[25] When properly maintained, these supports can be used repeatedly.

Recently, mRNAs produced by *in vitro* transcription of cDNAs cloned into the genome of the phage SP6[18] have been used to generate chloroplast precursor proteins.[1] This system can be used to produce large amounts of a single species of translatable mRNA.

Wheat Germ Translation System

Although optimization of the efficiency of the wheat germ translation system has been attempted by many groups, we find that a variation of the method of Roberts and Paterson[26] reproducibly yields a system capable of

[20] A. R. Cashmore, *in* "Methods in Chloroplast Molecular Biology" (M. Edelman, R. B. Hallick, and N.-H. Chua, eds.), p. 387. Elsevier/North-Holland Biomedical Press, Amsterdam, 1982.

[21] M. M. Kirk and D. L. Kirk, *Cell (Cambridge, Mass.)* **41,** 419 (1985).

[22] T. Maniatis, E. F. Fritsch, and J. Sambrook, "Molecular Cloning." Cold Spring Harbor Lab., Cold Spring Harbor, New York, 1982.

[23] M. Adesnik and J. E. Darnell, *J. Mol. Biol.* **67,** 397 (1972).

[24] J. S. Gantt and J. L. Key, *Mol. Gen. Genet.* **202,** 186 (1986).

[25] H. Bunemann, P. Westhoff, and R. G. Hermann, *Nucleic Acids Res.* **10,** 7163 (1982).

[26] B. E. Roberts and B. M. Paterson, *Proc. Natl. Acad. Sci. U.S.A.* **70,** 2330 (1973).

high levels of synthesis of proteins of a wide molecular weight range. Since detailed accounts of the wheat germ cell-free system and its preparation are available,[27] only a brief description of a reliable protocol is offered here. We stress, however, that care must be taken to avoid contamination of the extract with ribonucleases. All solutions should be prepared in sterile water; stock solutions of buffers and salts should be autoclaved prior to use; 2-mercaptoethanol and dithiothreitol are added after the autoclaved solutions have cooled. Glassware, pipets, and other materials should be sterilized by autoclaving, or, preferably, by baking at 150–200° overnight.

Fresh wheat germ (obtainable from General Mills, Vallejo, CA, or the Pillsbury Company, Minneapolis, MN) is defatted by shaking 10 g of wheat germ in 500 ml of cyclohexane. Alternatively, a fraction enriched in whole embryos can be produced by flotation in mixtures of organic solvents. Unless indicated otherwise, all steps in the subsequent extraction procedure are performed at 4°.

After filtering and air drying, 6 g of the defatted germ is ground in a chilled mortar and pestle with 6 g of sand and 28 ml of grinding buffer:

100 m*M* Potassium acetate
1 m*M* Magnesium acetate
2 m*M* Calcium chloride
6 m*M* 2-Mercaptoethanol
20 m*M* *N*-2-Hydroxyethylpiperazine-*N*′-2-ethanesulfonic acid (HEPES) adjusted to pH 7.5 with KOH

The extract is centrifuged at 30,000 *g* for 10 min in a 30-ml Corex tube. The supernatant is removed from the tube, taking care to avoid the pellet and floating lipid (a syringe fitted with an 18-gauge needle is useful for this purpose). The supernatant (15 ml) is adjusted to:

2.5 m*M* Magnesium acetate (37.5 μl of 1 *M* stock)
1 m*M* Adenosine 5′-triphosphate (150 μl of 100 m*M* stock adjusted to pH 7 with 1 *M* potassium hydroxide; pH indicator paper should be used for the titration)
20 μ*M* Guanosine 5′-triphosphate (7.5 μl of 40 m*M* stock)
2 m*M* Dithiothreitol (DTT) (30 μl of 1 *M* stock)
8 m*M* Creatine phosphate (150 μl of 0.8 *M* stock)
40 μg/ml Creatine phosphokinase [150 μl of 4 mg/ml stock; 50–100 U/mg; prepared in 0.1 *M* HEPES, pH 7.5, 10% (v/v) glycerol]

The indicated stock solutions can be stored at −20° for months without loss of activity.

[27] A. H. Erickson and G. Blobel, this series, Vol. 96, p. 38.

After a 10-min incubation at 30°, the extract is depleted of amino acids by passage through a 50 × 2 cm Sephadex column (G-25 or G-50, coarse or fine grade) that has been equilibrated in:

120 m*M* Potassium chloride (or acetate)
5 m*M* Magnesium acetate
20 m*M* HEPES–KOH, pH 7.5
6 m*M* 2-Mercaptoethanol

The column is run by gravity flow at 1.5–2.0 ml/min. When the column effluent begins to appear turbid, 1- to 2-ml fractions are collected. The most turbid of these are pooled, dispensed into small aliquots (usually 50–200 μl), and frozen in liquid nitrogen. The column also can be monitored by collecting drops into a solution of 10% (w/v) trichloroacetic acid. Fraction collecting begins when the column effluent yields a flocculant white precipitate upon contact with the acid. When stored in either liquid nitrogen or at −80°, the extract remains active for at least 1 year.

Prior to use, the extract can be depleted of endogenous mRNA by treatment with Ca^{2+}-dependent nuclease ("micrococcal") from *Staphylococcus aureus*. Stock solutions of the enzyme are prepared at 2000 U/ml in sterile distilled water and stored at −80°. An aliquot of the wheat germ extract (50 μl) is mixed with 1 μl of nuclease stock solution and 1 μl of 0.1 *M* calcium chloride. After incubation at room temperature for 5 min, 2 μl of 0.1 *M* EGTA is added to chelate calcium and stop the reaction.

Translations can be conveniently performed by adding RNA, labeled amino acid(s), and the wheat germ extract to the cocktail described below. The cocktail contains unlabeled amino acids, buffer, components of an ATP regenerating system, and other necessary factors. Since creatine phosphokinase is added to the wheat germ extract at an earlier step, it is not necessary to provide additional enzyme prior to performing the translation. Aliquots of cocktail can be stored for several months at −80° and periodically thawed and refrozen without loss of activity.

For approximately 1 ml of translation cocktail, combine the indicated volumes of the following stock solutions:

72 μl 1 *M* HEPES–KOH, pH 7.5
240 μl 25 m*M* Magnesium acetate
228 μl 1 *M* Potassium acetate
30 μl 100 m*M* ATP, pH 7.0
30 μl 40 m*M* GTP
30 μl 0.8 *M* Creatine phosphate
6 μl 1 *M* DTT
240 μl 0.5 m*M* Spermine, pH 7.0

150 μl Amino acid mixture containing 0.5 mM of each of the 19 amino acids not supplied in labeled form

Depending on the requirements of the experiment and the purity of the mRNA, translation mixtures of 10 to 100 μl are usually sufficient. Each mixture contains 0.33 vol of cocktail, 0.2 vol of wheat germ extract, and 0.47 vol of water, labeled amino acid(s), and RNA. Typically, incorporation of label into protein occurs maximally at concentrations of polyadenylated RNA of 40 μg or total RNA of 400 μg/ml of translation mixture. Since the amount of radioactivity incorporated into protein is proportional to the amount of isotope added,[13] labeled amino acids are included at levels appropriate to the experiment and the budget. If necessary, labeled amino acids can be concentrated by lyophilization. The most frequently used labeled amino acid, [^{35}S]methionine, can be included up to 750 μCi/ml of translation, but for most purposes, 150 to 500 μCi/ml suffices.

Wheat germ systems can be further modified by including ribonuclease and protease inhibitors and by varying the salt concentrations to enhance the translation of particular mRNAs. Generally, this is not essential.

Reticulocyte Lysates

In contrast to earlier reports,[28] we find that chloroplast protein precursors synthesized in rabbit reticulocyte lysates can be used for transport into isolated chloroplasts. The efficiency of transport does not differ markedly from precursors synthesized in the wheat germ system and the profile of imported proteins is also quite similar for the two systems. Furthermore, addition of wheat germ extract to transport mixtures containing reticulocyte lysate products does not cause major quantitative or qualitative changes in the pattern of imported polypeptides.

Differences in import between wheat germ extracts and reticulocyte lysates are consequences of the translational properties of the cell-free systems. Wheat germ extracts preferentially translate mRNAs for lower molecular weight proteins whereas reticulocyte lysate translation products are skewed toward the higher molecular weight range. These patterns are reflected in the profiles of imported proteins in transport assays reconstituted with translation products of total polyadenylated RNA from *Chlamydomonas*. Among the proteins imported from wheat germ extracts, those in the lower molecular weight range predominate; higher

[28] S. G. Bartlett, A. R. Grossman, and N.-H. Chua, *in* "Methods in Chloroplast Molecular Biology" (M. Edelman, R. B. Hallick, and N.-H. Chua, eds.), p. 1081. Elsevier/North-Holland Biomedical Press, Amsterdam, 1982.

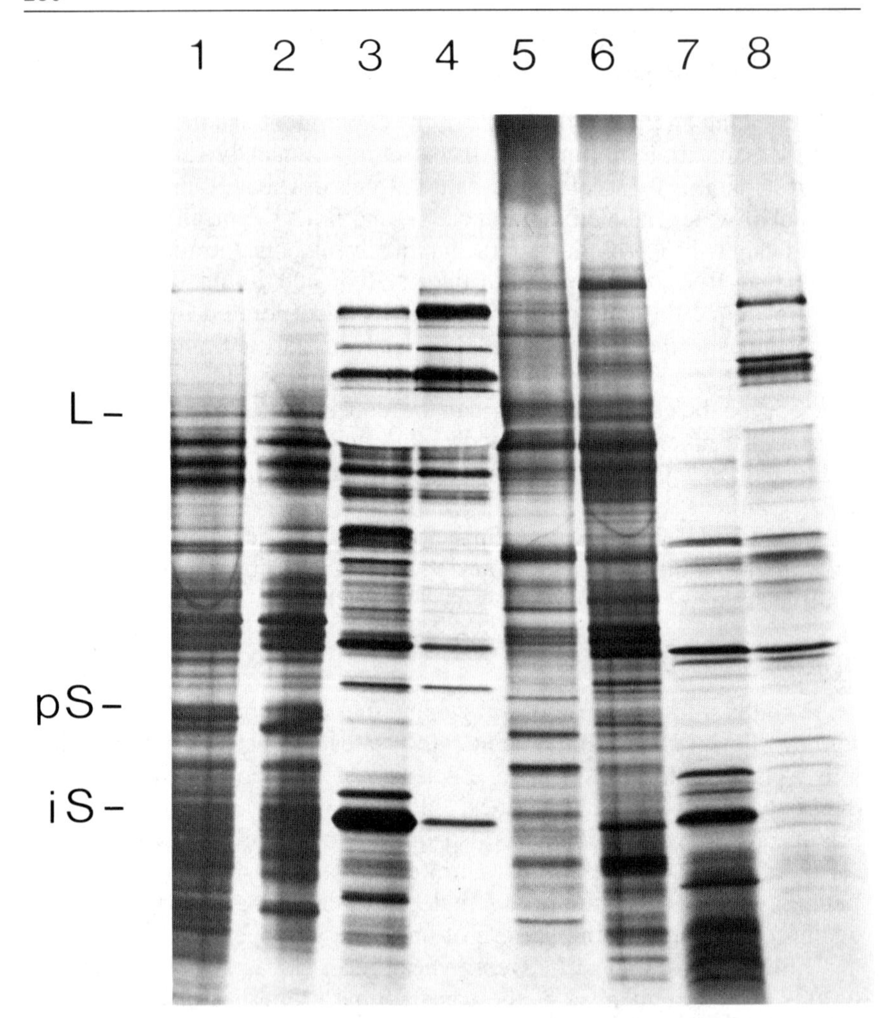

FIG. 1. Transport into isolated chloroplasts of *in vitro* translation products of the wheat germ and reticulocyte lysate cell-free systems. Translation products labeled with [^{35}S]methionine of *Chlamydomonas* mRNA generated in the wheat germ (lane 1) and reticulocyte lysate (lane 6) systems were incubated with isolated pea chloroplasts. Afterward, the chloroplasts were reisolated, treated with trypsin and chymotrypsin to remove polypeptides associated with the outer surface of the envelope, and then were lysed in the presence of protease inhibitors. The chloroplast lysates were separated into soluble and membrane fractions as described in the text. Soluble transported polypeptides generated in the wheat germ system (lane 3) are enriched in lower molecular weight proteins whereas those produced in reticulocyte lysates (lane 4) are enriched in higher molecular weight proteins. Similar differences are found between the *in vitro* transported membrane polypeptides generated in the wheat germ (lane 7) and reticulocyte (lane 8) systems. Wheat germ extract (lane 2) and reticulocyte lysate (lane 5) polypeptides that remained outside the chloroplasts were

molecular weight proteins are more abundant in a transport assay reconstituted with reticulocyte lysate (Fig. 1). Thus reticulocyte lysates offer a viable method for generating chloroplast precursor proteins for transport, especially those of higher molecular weight (>50,000).

Although published procedures are available for producing reticulocyte lysates,[29] commercial preparations are utilized most often. Lysates can be obtained from several suppliers including New England Nuclear and Amersham; the translation and transport products shown in Fig. 1 were generated in a lysate kit from Bethesda Research Laboratories. Variability in the composition of lysates, which reputedly can cause chloroplasts to lyse,[28] may explain the previously reported failure to reconstitute transport with the reticulocyte lysate system.

Chloroplast Isolation

Reconstitution of protein transport into chloroplasts became feasible with the advent of methods for purifying the organelle in an intact, physiologically active form. Although low levels of polypeptides imported *in vitro* were recovered from plastids isolated on sucrose gradients,[3] the introduction of colloidal silica sols as a medium for chloroplast purification[30] resulted in more efficient cell-free systems.[4,13] The low viscosity and low osmotic character of silica sol density gradients allows for the isolation of photosynthetically competent chloroplasts with intact envelope membranes. The medium most frequently used is Percoll, a purified, poly(vinylpyrrolidone)-coated silica sol produced by Pharmacia. It is available directly from the manufacturer or from Sigma.

The method outlined below allows intact chloroplasts from several vascular plants, including spinach, pea, romaine lettuce, and barley, to be isolated. For certain species, notably maize[31] and soybean,[17] osmolarity and gradient densities must be individually optimized. Specialized meth-

[29] R. J. Jackson and T. Hunt, this series, Vol. 96, p. 50.

[30] J.-J. Morgenthaler, M. P. F. Marsden, and C. A. Price, *Arch. Biochem. Biophys.* **168,** 289 (1975).

[31] V. Walbot and D. A. Hoisington, *in* "Methods in Chloroplast Molecular Biology" (M. Edelman, R. B. Hallick, and N.-H. Chua, eds.), p. 211. Elsevier/North-Holland Biomedical Press, Amsterdam, 1982.

obtained by brief centrifugation as described in the text. The radioactive polypeptides were resolved by SDS–PAGE and detected by fluorography. L, Large subunit of RuBPCase (present as an unlabeled band of 55 kDa in lanes 3 and 4); pS, precursor to the small subunit of RuBPCase (lanes 2 and 5); iS, an intermediate form of the algal small subunit precursor produced by one of two proteases in chloroplasts that cleave transit sequences (see Ref. 8).

ods are also required for isolating plastids from tissues such as bundle sheaths of C_4 plants[31] and dark-grown leaves[32] or from algae.[33,34] A useful review on chloroplast isolation has recently appeared.[33]

Although direct comparisons have not been published, the highest yields and most actively transporting chloroplasts are obtained from young expanding leaves. Possibly, immature leaves contain the most rapidly developing and, hence, most actively protein-transporting plastid populations. Since plant growth rates vary widely with light and temperature regimes, physiological and chronological ages are not strictly correlated. Generally, however, pea shoots or young plants 7 to 15 days old are suitable. Mullet and Chua have discussed the quantitative aspects of chloroplast isolation.[35]

Production of a Crude Chloroplast Preparation

Before they are purified by gradient centrifugation, a suspension enriched in intact chloroplasts is prepared. To avoid the introduction of protease-secreting bacterial and fungal contaminants, all solutions should be prepared in sterile water. Concentrated stock solutions of salts and buffers should be autoclaved prior to storage at 4°. Allowance should be made for the decrease in pH of 1 *M* HEPES–KOH stocks that occurs upon dilution. At 25°, a 50 m*M* HEPES–KOH solution prepared from 1 *M* HEPES–KOH, pH 8.0, has a pH of 7.7.

For convenience in this and later steps, a 5× grinding medium (GR) is prepared as follows:

1.65 *M* Sorbitol
10 m*M* EDTA
5 m*M* Magnesium chloride
5 m*M* Manganese chloride
250 m*M* HEPES–KOH, pH 8.0

The solution, sterilized by filtration (0.45-μm pore size), is stable for several months at 4°. Some batches of sorbitol produce yellow solutions which can be decolorized with powdered charcoal followed by filtration (Whatman #50 paper followed by a 0.45-μm pore filter). The impurities do not affect chloroplast integrity as assayed by transport, however.

[32] R. M. Leech and B. M. Leese, *in* "Methods in Chloroplast Molecular Biology" (M. Edelman, R. B. Hallick, and N.-H. Chua, eds.), p. 221. Elsevier/North-Holland Biomedical Press, Amsterdam, 1982.

[33] C. A. Price and E. M. Reardon, *in* "Methods in Chloroplast Molecular Biology" (M. Edelman, R. B. Hallick, and N.-H. Chua, eds.), p. 189. Elsevier/North-Holland Biomedical Press, Amsterdam, 1982.

[34] L. Mendiola-Morgenthaler, S. Leu, and A. Boschetti, *Plant Sci.* **38,** 33 (1985).

[35] J. E. Mullet and N.-H. Chua, this series, Vol. 97, p. 502.

Dilution of the 5× stock to single strength and addition of ascorbic acid to a final concentration of 5 m*M* completes preparation of the solution for use. All steps in the chloroplast isolation, unless indicated, should be performed at 4° or with solutions kept on ice.

Leaves or whole shoots are harvested into a cooled container and macerated into small pieces. A razor blade attached to a hemostat serves as an efficient chopper. GR is added (2 to 4 ml/g of tissue) and the tissue is rapidly homogenized; a compromise between maximal disruption of cells and retention of large numbers of chloroplasts in an intact form is sought. A Polytron (Brinkmann Instruments) is useful for this purpose; blenders also accomplish the task but result in lower yields. Cellular debris is removed by filtration; four layers of Miracloth[35] (available from Calbiochem), six layers of muslin followed by a layer of nylon,[36] or eight layers of cheesecloth with a layer of cotton between the top two layers[37] have been used.

The chloroplasts in the filtered homogenate are pelleted by centrifugation: a fixed angle rotor is accelerated to 6000 *g* and then immediately stopped by hand braking. Alternately, chloroplasts are pelleted by centrifugation at 3000 *g* for 3 min, also followed by hand braking. A more refined procedure is to underlay the homogenate with about one-quarter volume of a 40% Percoll solution and centrifuge 2 min at 4000 *g*.[38] Intact plastids gently pellet while the broken plastids collect at the interface between the buffer and the Percoll. Since this step removes most of the broken chloroplasts, smaller linear gradients are required for the subsequent final purification.

Percoll is prepared for use by adding 30 mg of polyethylene glycol 8000 (identical to PEG 6000, but recently redesignated by Sigma to PEG 8000), 10 mg bovine serum albumin, and 10 mg Ficoll (type 400) per milliliter of Percoll. The resultant "PBF" Percoll serves as a stock for both gradients and underlay solutions. It should be freshly prepared for each use. The 40% Percoll underlay consists of 0.2 vol 5× GR, 0.4 vol of PBF Percoll, and 0.4 vol of water.

Gradient Purification of Intact Chloroplasts

The crude chloroplast pellet is resuspended by gentle pipetting in GR. The volume and concentration of chloroplast suspension loaded onto the

[36] R. Douce and J. Joyard, *in* "Methods in Chloroplast Molecular Biology" (M. Edelman, R. B. Hallick, and N.-H. Chua, eds.), p. 239. Elsevier/North-Holland Biomedical Press, Amsterdam, 1982.

[37] K. Cline, J. Andrews, B. Mersey, E. H. Newcomb, and K. Keegstra, *Proc. Natl. Acad. Sci. U.S.A.* **78,** 3595 (1981).

[38] W. R. Mills and K. W. Joy, *Planta* **148,** 75 (1980).

gradients is not critical. However, the greatest separation between broken and intact plastids is achieved when no more than 10% of the gradient volume is applied to linear gradients. When step gradients are employed, chloroplast suspensions corresponding to 30 to 40% of the gradient volume can be used. In either case, the major consideration is that the suspension should not be so dense that it is difficult to layer on the top of the gradient. Percoll gradients have been produced by several methods. A linear 10–80% gradient can be generated with starting solutions of 10 and 80% Percoll. Alternatively, gradients can be formed by centrifuging 50% Percoll solutions at 43,000 *g* for 30 min or 10,000 *g* for 100 min prior to loading the sample. Discontinuous gradients composed of steps of 0.3 vol of 80% and 0.7 vol of 40% Percoll have also been employed. Gradient solutions are prepared by combining appropriate volumes of PBF Percoll (see previous section) and 0.2 vol of 5× GR. Water is added to the final volume where appropriate. Immediately prior to use, the gradient solutions are brought to 5 m*M* ascorbic acid and 0.5 m*M* glutathione by adding the reagents in solid form. Gradients are stable and serviceable if stored overnight at 4°.

Centrifugation times and velocities appropriate for optimal separation of broken from intact chloroplasts vary with the physiological state of the organelles. Values ranging between 5000 and 15,000 *g* for 10 to 15 min have been used. For most purposes maximal separation is not required. Heavily loaded gradients in which the bands of broken and intact organelles almost fuse still provide sufficiently purified intact chloroplasts if the lower region is used. During centrifugation in a swinging bucket rotor, lysed chloroplasts and membrane fragments collect in an upper band whereas the intact organelles reach equilibrium toward the bottom of the tube. After removal of the upper band by aspiration, the lower band is transferred by pipet to a chilled centrifuge tube and diluted at least 3-fold with GR. Recentrifugation up to 4000 *g* followed immediately by braking or at 1500 *g* for 6 min pellets the intact chloroplasts. Although plastids at this point are usually adequately free of Percoll, they can be washed once in GR by centrifugation prior to suspension in a small volume of GR. The washing and final suspension can also be performed in "transport buffer," i.e., GR that lacks salts, EDTA, and ascorbic acid (see below).

Uptake of Translation Products by Isolated Chloroplasts

The initial reports of cell-free transport systems[3,4] demonstrated the import of only the most abundant soluble polypeptide synthesized in the cytoplasm, the small subunit of ribulose-1,5-bisphosphate carboxylase (RuBPCase). Subsequent work to optimize several incubation parameters[13] led to conditions that enable import and maturation of hundreds of distinct soluble and membrane polypeptides to be detected, several of which have been identified. *In vitro* transport procedures, as defined by Grossman *et al.*,[13] Bartlett *et al.*,[28] and Mullet and Chua,[35] have become standard.

Relative to rates of import that occur *in vivo*, however, the cell-free reconstituted system is still far from optimal. Isolated chloroplasts sluggishly import small numbers of molecules over many minutes. Protein transport into chloroplasts in living cells occurs with such efficiency that cytoplasmic precursors to plastid polypeptides have not yet been demonstrated directly; to date, the only evidence for their existence is from cell-free reconstitution studies. Insights into the mechanisms that control rates of transport both *in vivo* and *in vitro* are important goals of current research.

Factors Influencing in Vitro Transport

Energy, time, chlorophyll concentration, and pH are among the factors that have been shown to affect the extent of protein transport by chloroplast preparations. Analysis of the energy requirements indicated that transport is dependent on stromal ATP concentration rather than envelope membrane potential.[5,6] ATP levels in chloroplasts most likely vary over a wide range as a function of both the species and physiological state. Preparations with high endogenous ATP levels will not be dependent on either photophosphorylation or exogenously supplied ATP for transport. Although light-dependent transport is more easily demonstrated in pea than spinach chloroplasts, the reported extent of light-stimulated transport into pea chloroplasts is rather modest.[5]

The rate of polypeptide uptake into isolated chloroplasts has not been rigorously investigated. Although published time courses show nearly linear uptake rates during 1 hr, transport incubation times of 30 min or less have been used recently without an apparent reduction in the level of imported product (K. Keegstra, personal communication).

The number of chloroplasts, expressed on a chlorophyll or chloroplast number basis, that should be included in the transport mixture depends on the nature of the subsequent analysis of the imported polypeptides. Systems reconstituted with *in vitro* translation products are clearly situations of chloroplast excess; the total amount of radioactivity recovered from chloroplasts incubated with a constant level of *in vitro* translation products increases only 50% over a 6-fold range of chlorophyll concentrations (0.17–1.0 mg/ml), whereas the specific radioactivity of chloroplast proteins recovered per milligram chlorophyll is 6-fold greater at 0.17 mg/ml than at 1.0 mg/ml.[13] Since analyses of imported polypeptides often rely upon one-dimensional polyacrylamide gel electrophoresis (PAGE) and fluorography, the specific activity of the imported polypeptides is an important consideration and is discussed in more detail below.

The influence of pH on protein transport into chloroplasts has been addressed in the literature only by Grossman *et al.*[13] They found that radioactivity in the soluble fraction from plastids incubated with translation products of total polyadenylated RNA is enhanced about 2-fold as pH is increased from 6.5 to 8.5. However, recovery of *in vitro* transported membrane polypeptides remains relatively constant over this range. Following the recommendation of that report, the standard buffer for transport incubations has become 50 m*M* HEPES–KOH, pH 8.0.

Other components of the transport system are included (or omitted) more as a matter of previous practice and convenience than as a result of rigorous demonstration of necessity. The protocol of Morgenthaler *et al.*[30] is used to isolate intact chloroplasts, i.e., magnesium and manganese salts and ascorbic acid are included in the grinding buffer to maintain maximal protein synthetic and photosynthetic rates. When transporting proteins into the isolated organelles, however, the procedure reverts to a simple buffer containing only HEPES, sorbitol, and whatever is carried along with the cell-free translation products.[13,28,35] Clearly, analysis of the requirements for protein transport into isolated chloroplasts merits a more detailed approach, especially in light of the observation of Cline *et al.*[6] that EDTA (10 m*M*) inhibits the import of the chlorophyll *a*/*b*-binding polypeptides but has no effect on the import of the RuBPCase small subunit.

Incubation of Isolated Chloroplasts with in Vitro Translation Products

To determine chlorophyll concentration in the suspension of purified chloroplasts, a small volume of chloroplasts (5–10 μl) is extracted in 1 ml of 80% (v/v) acetone. Precipitated protein is removed by brief centrifugation, and the optical density at 652 nm of the chlorophyll solution is measured. After the chlorophyll determination (mg/ml extinction coefficient of 34.5), the chloroplasts are diluted with GR (or 50 m*M* HEPES, pH 8.0, containing 0.33 *M* sorbitol) noting that the suspension will be diluted another 6-fold in the transport incubation mixture. When GR is used to dilute the chloroplasts, the amount of EDTA carried over to the transport mixture (final concentration of 0.33 m*M*) is well below the level that inhibits transport of the chlorophyll *a*/*b*-binding polypeptides.

Transport mixtures of 300 μl total volume, containing *in vitro* translation products generated by leaf polyadenylated RNA, yield sufficient amounts of imported protein for efficient autoradiographic analysis of abundant polypeptides such as the RuBPCase small subunit and the chlorophyll *a/b*-binding polypeptides. Larger mixtures can be used as required. Incubations are performed in siliconized (e.g., Sigmacote, from Sigma) disposable 13 × 100-mm borosilicate culture tubes or other appropriate containers. The silicone coating minimizes chloroplast lysis as well as losses caused by sticking of the organelles to the sides of the tube. Tubes can be capped to avoid loss of fluid by evaporation.

Since the transport of proteins into chloroplasts occurs posttranslationally, ribosomes can be removed from the translations by centrifugation (140,000 *g* for 15–60 min) prior to the transport incubation. In the case of wheat germ translations, this step removes the nascent polypeptide chains that tend to accumulate. Since ribosomes do not seem to interfere with the import process, however, use of postribosomal supernatants is not required.

The translation mixture is adjusted to a final volume of 250 μl containing 50 m*M* HEPES–KOH (the HEPES present in the translation need not be accounted for), pH 8.0, 0.33 *M* sorbitol, and 10 m*M* methionine. As suggested by Bartlett *et al.*,[28] this is accomplished conveniently by adding the following to 100 μl of translation mixture: 12.5 μl of 1 *M* HEPES–KOH, pH 8.0, 40 μl of 2 *M* sorbitol, 10 μl of 250 m*M* methionine, and 85 μl of water. Appropriately diluted chloroplasts (50 μl at 1–6 mg/ml of chlorophyll) are added. The mixture is incubated under illumination at 26° and continuously agitated so that the chloroplasts remain in suspension—a rotary shaker set at 200 rpm is sufficient if the tubes are supported in a slanted position.

Inhibitors of Transport

Inhibitors added to the transport incubation mixtures can be useful in analysis of the protein import process. So far, use of inhibitors has focused on energy requirements. Uncouplers of photophosphorylation such as salicylanilide XIII (2 μ*M*),[5] SF6847 (2 μ*M*),[5] and nigericin[6] (400 n*M*) inhibit the import of both stromal and membrane proteins. Exogenously supplied ATP reverses this block, suggesting that these inhibitors act by depleting the chloroplast of ATP rather than by collapsing pH and electrochemical gradients. Although the uncouplers block transport across the envelope membranes, at least in the case of nigericin, the binding of protein precursors to the surface of the chloroplast is not affected[6] (see below). These results are in apparent contrast to mitochondrial protein import, where transport of most polypeptides is blocked by uncouplers, even in the presence of ATP.[2]

Polyclonal antibodies to chloroplast precursor proteins specifically inhibit the import of the antigen to which they are directed. Antibodies to the RuBPCase small subunit of *Chlamydomonas,* for example, block both import and envelope binding of the precursor polypeptide (M. L. Mishkind and G. W. Schmidt, unpublished). Presumably, the immunoglobulins sterically hinder the interaction of the precursor with transport receptors present at the chloroplast's surface. Furthermore, bulky immunoglobulin molecules may interfere with the movement of the polypeptide across lipid bilayers. More selective probes, such as monoclonal antibodies directed to the carboxy terminus of precursor proteins, are likely to block import but not binding to the envelope receptors. Under saturating conditions possible with the large amounts of precursor polypeptides that can be produced by expression of cloned genes, antibody arrest of transport might serve to block selected classes of receptors and/or import pathways, thus permitting analysis of various modes by which proteins enter chloroplasts.

The study of protein import into plastids will benefit from more extensive use of selective inhibitors including proteases, membrane modifying reagents, and antibodies to putative receptors for precursor proteins.

Analysis of Chloroplasts after Transport

Recovery of Chloroplast and Supernatant Fractions

The following procedure is useful in determining the kinetics of transport and whether a particular translation product is in fact a chloroplast precursor. Transport mixtures can be rapidly divided into chloroplast and supernatant fractions by brief centrifugation. An aliquot (20 μl or more) of the transport mixture is transferred to a microfuge tube, centrifuged for 15 sec at 12,000 *g*, and the supernatant (containing the polypeptides that remain outside the chloroplast) is removed for analysis by PAGE and fluorography. The pelleted chloroplasts are washed once in GR and prepared for analysis as described below. In this manner, the disappearance of precursor polypeptides from translation products (supernatant) and the concomitant appearance of mature products within chloroplasts can be efficiently monitored.[8] When resolved by two-dimensional gel electrophoresis, the removal by transport of low abundance precursors from translation mixtures can be followed.

Proteolytic Treatment of Intact Chloroplasts

To verify that a particular polypeptide recovered in a chloroplast-derived fraction has been transported across the envelope and not bound to the surface of the organelle, the intact chloroplasts are treated with proteases prior to lysis and fractionation. Digestion can be performed directly in the transport mixture or the chloroplasts can be washed (in GR or HEPES–sorbitol) prior to adding the proteases. For washing, the chloroplasts are diluted in 5 ml of either buffer, centrifuged to 4000 *g* followed by immediate braking, and resuspended in 0.5 ml of fresh buffer. Stock solutions of trypsin and chymotrypsin are each prepared at 5 mg/ml in GR and added to the chloroplasts to a final concentration of 160 μg/ml. Proteinase K[39] (100 μg/ml) and thermolysin[6] (100 μg/ml) have also been used for this purpose. The plastids are kept on ice for 30 min after which the protease activity is inhibited by adding 5 ml of a buffer containing 50 m*M* HEPES–KOH, pH 8.0, 0.33 *M* sorbitol, 1.0 m*M* phenylmethylsulfonyl fluoride (PMSF), 1.0 m*M* benzamidine (BAM), and 5 m*M* ε-amino caproic acid (ACA). PMSF can be prepared as a 40 m*M* stock solution in isopropanol and stored at 4° and a stock solution containing both 200 m*M* ACA and 40 m*M* BAM can be autoclaved and stored at 4°. Thermolysin is inhibited by adjusting the chloroplast suspension to 10 m*M* EDTA. Intact chloroplasts are then reisolated by underlaying the suspension with about 2 ml of the 40% Percoll solution described above, centrifuging for 2 min at 4000 *g*, and removing the supernatant. Bartlett *et al.*[28] indicate that inclusion of PEG in the Percoll solution is unnecessary when purifying protease-treated chloroplasts. For convenience, however, the Percoll underlay used initially to isolate the chloroplasts is adequate for this step.

Protease treatment weakens the envelope membranes and thereby reduces the yield of intact chloroplasts upon reisolation. After verifying that the polypeptide of interest has in fact been transported into the organelle, the protease digestion step can be omitted in order to increase the yield of reisolated intact chloroplasts.

Chloroplast Subfractionation

Reisolated chloroplasts are washed once by centrifugation in the appropriate buffer containing protease inhibitors as described above. The chloroplasts then are lysed in a small volume (0.3–1.0 ml) by vigorous resuspension of the final pellet in water or 10 m*M* HEPES or Tris (pH 7.5–8.5) containing the protease inhibitors at concentrations present in the wash buffer. Highly purified chloroplast components, such as the RuBPCase holoenzyme and envelope or thylakoid membranes, can be obtained from the reisolated chloroplasts by following standard fractionation protocols scaled to small samples. The chloroplasts should be lysed, however, in a buffer compatible with the purification procedure.

[39] J. Pfisterer, P. Lachmann, and K. Kloppstech, *Eur. J. Biochem.* **126**, 143 (1982).

Chloroplast membranes can be rapidly separated from the soluble proteins in the lysate by adjusting the suspension to 0.25 *M* NaCl (stock NaCl at 4 *M* is convenient for this purpose) and centrifuging for 5 min at 12,000 *g*. The pellet, containing envelope and thylakoid membranes, can be dissolved in sample buffer for PAGE. The proteins in the supernatant can be concentrated by precipitation [by adjusting the supernatant either to 10% (w/v) trichloroacetic acid or to 90% (v/v) acetone] prior to being dissolved in gel sample buffer [60 m*M* Tris–HCl, pH 8.5, or 60 m*M* Tris–base for acid precipitates, 60 m*M* DTT, 5 m*M* ACA, 1 m*M* BAM, 2% (w/v) sodium or lithium dodecyl sulfate, 15% (w/v) sucrose]. Pellets can be resuspended by sonication or vortexing—a freeze–thaw cycle facilitates resuspension.

High levels of endogenous unlabeled chloroplast proteins tend to distort migration patterns of polypetides. This is especially true of those proteins whose molecular weights are similar to the large (55K) and small (16K) subunits of RuBPCase and the chlorophyll *a*/*b*-binding proteins (25K–28K), all of which occur in prodigious amounts in chloroplasts. However, samples that contain relatively high levels of unlabeled protein can be analyzed without distortion by two-dimensional gel electrophoresis.

Anomalous migration patterns on one-dimensional gels can be reduced by limiting the amount of chloroplasts used in the transport mixtures to 0.17–0.34 mg chlorophyll/ml. In the case of proteins whose mRNAs are of minor abundance, however, it may be critical to maximize yields of imported translation products by employing larger numbers of chloroplasts in the assay (0.6–1.0 mg chlorophyll/ml of transport mixture). Procedures are available, however, to avoid distortion caused by the most abundant polypeptides. RuBPCase holoenzyme can be removed from the soluble fraction of chloroplast lysates by differential centrifugation. The holoenzyme pellets through a 15% sucrose cushion (prepared in 25 m*M* Tris–HCl, pH 7.5, 200 m*M* NaCl, 5 m*M* DTT) upon centrifugation at 140,000 *g* overnight; polypeptides not present in high-molecular-weight complexes are recovered from the supernatant. When membrane polypeptides are to be analyzed, chloroplasts lacking the chlorophyll *a*/*b*-binding proteins may be helpful. For example, the inner leaves of romaine lettuce are deficient in the chlorophyll *a*/*b*-binding proteins.[12] Alternatively, plants grown with intermittent illumination (e.g., cycles of 2 hr dark and 1 min light) contain the full complement of reaction center proteins but lack the chlorophyll *a*/*b*-binding polypeptides.[40] Despite the absence of the light-harvesting pigment–protein complexes, chloroplasts

[40] A. C. Cuming and J. Bennett, *Eur. J. Biochem.* **118,** 71 (1981).

from intermittently illuminated leaves import and assemble the chlorophyll *a*/*b*-binding polypeptides at efficiencies similar to those from plants grown in continuous light (M. L. Mishkind and G. W. Schmidt, unpublished).

The fractionation schemes outlined primarily concern separating the chloroplast into soluble and membrane components and provide limited information about the disposition of imported proteins in the organelle. Several procedures, however, resolve certain structural details of the location of imported proteins. Since the crude membrane fraction obtained by NaCl aggregation contains the envelopes as well as the thylakoids, side-by-side comparison of this fraction from chloroplasts that have been digested with proteases with those that have not will distinguish polypetides bound to the external surface of the organelle from polypeptides in protected compartments. Use of this procedure has demonstrated, for example, that nigericin leads to increased binding of precursors to putative receptors on the surface of the envelope.[6]

Joyard *et al.*[41] and Cline *et al.*[42] have shown that trypsin and chymotrypsin digest polypeptides of both the inner and outer envelope membranes. These enzymes presumably also digest proteins present in the space between the envelope membranes. In contrast, thermolysin gains access only to envelope polypeptides of the outer lipid bilayer. Thus, soluble polypeptides that are sensitive to trypsin and chymotrypsin but resistant to thermolysin are possible occupants of the interenvelope membrane space.

Ultrastructural Autoradiography

Ultrastructural autoradiography offers another means of locating polypeptides within subchloroplast compartments. This procedure will be especially useful in conjunction with transport experiments performed with translation products of hybrid-selected[6] or SP6-generated[1] mRNAs whereby the location of individual polypeptides can be determined. Tritium is the standard isotope for autoradiography at the ultrastructural level, but ^{35}S is also compatible with autoradiographic emulsions, but with a 2-fold decline in sensitivity with respect to tritium.[43] Sensitivity can be significantly enhanced, however, by a recently introduced fluorographic technique for ultrastructural autoradiography.[44] Resolution, defined as

[41] J. Joyard, A. Billecocq, S. G. Bartlett, M. A. Block, N.-H. Chua, and R. Douce, *J. Biol. Chem.* **258,** 10000 (1983).

[42] K. Cline, M. Werner-Washburne, J. Andrews, and K. Keegstra, *Plant Physiol.* **75,** 675 (1984).

[43] M. M. Salpeter, *Methods Cell Physiol.* **2,** 239 (1966).

[44] M. Gregg and H. M. Reznik-Schuller, *J. Microsc.* (*Oxford*) **135,** 115 (1984).

the half-distance (*HD*), or the distance from a line source of radioactive isotope within which 50% of the silver grains are located, is considerably better for ^{3}H ($HD = 50$ nm) than for ^{14}C ($HD = 230$ nm).[45,46] The resolution of ^{35}S is presumably similar to that of ^{14}C. Standard [^{35}S]methionine-labeled cell-free translation products thus can be used for certain localization studies. In terms of the fine structure of the chloroplast, internal and external sites can be most easily distinguished with ^{35}S-labeled proteins whereas localization of imported proteins in the stroma, thylakoid membrane, and thylakoid lumen would require ^{3}H-labeled proteins and careful statistical analysis of the silver grain distribution. Salpeter *et al.* discuss statistical methods for analysis of autoradiographs.[47] Potentially, if chloroplasts are plasmolyzed so as to separate the inner and outer envelope membranes, compounds within the intermembrane space can be visualized.

Chloroplasts are prepared for autoradiography by standard fixation and embedding procedures. When manipulating isolated chloroplasts, however, care must be taken to avoid loss of the outer envelope membrane. Osmoticum is provided until after postfixation in osmium tetroxide. Chloroplasts are pelleted by brief centrifugation and gently resuspended. Fixation and embedding are conveniently performed in 1.5-ml microfuge tubes.

Reisolated, intact chloroplasts can be plasmolyzed by incubation in buffer containing 0.6 *M* sorbitol for 10 min at room temperature. The plastids should be suspended in a buffer (e.g., 0.1 *M* phosphate, pH 7.4) compatible with aldehyde fixatives. Subsequent steps are performed at 4°. An equal volume of twice-strength fixative [4% (w/v) glutaraldehyde containing 2% (w/v) paraformaldehyde or another electron microscopy fixative], prepared in the osmotically adjusted buffer, is added to the chloroplast suspension. After incubation (1 hr to overnight) the plastids are washed four times by brief centrifugation and then resuspended in 2% (w/v) osmium tetroxide prepared in phosphate buffer. After 2 hr the plastids are washed, dehydrated in an ethanol or acetone series, and embedded in epoxy resin (e.g., Spurrs, Epon-Araldite). Silver–gold sections are cut on an ultramicrotome, collected on Formvar-coated nickel grids, and checked in the electron microscope to verify the adequacy of the preservation. Sections which are to be coated with emulsion are not exposed to the electron beam so as to prevent possible destruction of the signal by high-energy electrons.

[45] M. M. Salpeter, L. Bachmann, and E. E. Salpeter, *J. Cell Biol.* **41,** 1 (1969).
[46] M. M. Salpeter, F. A. McHenry, and E. E. Salpeter, *J. Cell Biol.* **76,** 127 (1978).
[47] M. M. Salpeter, C. D. Smith, and J. A. Matthews-Bellinger, *J. Electron Microsc. Tech.* **1,** 63 (1984).

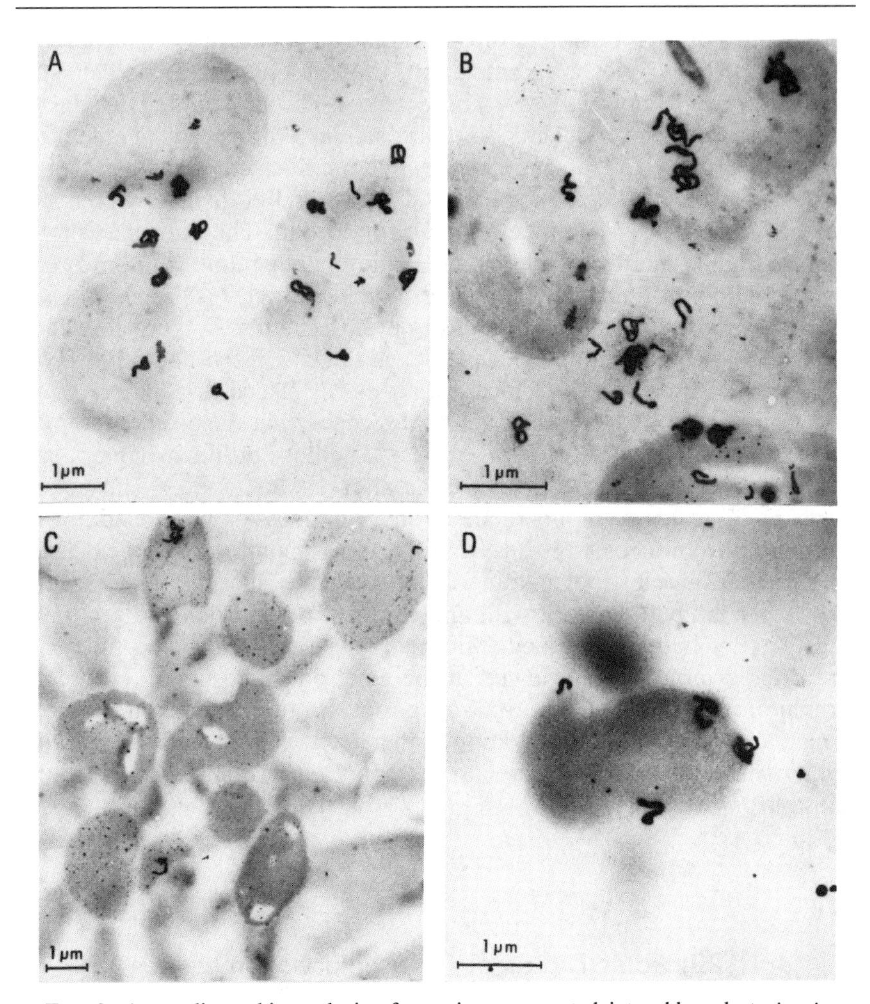

FIG. 2. Autoradiographic analysis of proteins transported into chloroplasts *in vitro*. Isolated chloroplasts from pea seedlings were incubated with [^{35}S]methionine-labeled translation products of mRNA purified from *Chlamydomonas reinhardii*. Following transport, the isolated chloroplasts were fixed, embedded, sectioned, and processed for EM autoradiography as described in the text. Sample micrographs taken after development of exposed photographic emulsion illustrate isolated chloroplasts which were treated with chymotrypsin and trypsin (160 μg/ml) (A and C) or without proteases (B and D). In addition, transport in (C) and (D) was performed in the presence of the ionophore nigericin (0.4 μM). Exposed silver grains appear in the vicinity of the labeled polypeptides but the grains actually may be somewhat separated from the source due to the lateral movement of the emission particles before contacting the emulsion.

Detailed discussions of autoradiographic technique are available (e.g., Ref. 43). Briefly, grids are coated with Ilford L4 photographic emulsion (available from Polysciences). The emulsion can be manipulated in the presence of a dark red (Kodak #2) 25-W safelight for periods of at least 10 min but exposure to the safelight should be minimized. Emulsion (5 g) is weighed into a glass container, diluted with 10 ml of distilled water, and melted by heating in a 45° water bath for 15 min. The emulsion should then be chilled in an ice bath for 10–15 min; chilling time is optimized so that a thin film forms across a copper wire loop after it is dipped into the emulsion and the excess is allowed to drain. The film is then allowed to fall across grids which have been attached to glass microscope slides with double-stick tape. Coated slides are air dried, placed in a slide box enclosed in a light tight box containing desiccant, and exposed at 0–4°.

Exposure time varies with the specific activity of the specimen, so a general procedure is to develop grids after 1 week and on subsequent dates as indicated by appearance of silver grains. As mentioned above, fluorography reduces exposure times (from 3 months to 1 week in the published example), but commercially available fluorographic reagents are rather difficult to work with and can destroy sections and grid coatings.[44] After being developed (Microdol-X, Kodak) and fixed, the grids are dried overnight and viewed in the electron microscope. Autoradiographic localization of *in vitro* transported proteins is shown in Fig. 2 which also presents a visualization of the effects of protease and inhibitor treatments on the import of precursors from *Chlamydomonas* into chloroplasts from pea seedlings.

[29] Polar Lipids of Chloroplast Membranes

By DAVID J. CHAPMAN and JAMES BARBER

The polar lipids of the membranes of higher plant and algal chloroplasts share little in common with the lipids which constitute the membranes of other cytoplasmic structures of eukaryotic cells, including the functionally related mitochondria. They also differ from the polar lipids of the photosynthetic bacteria but, in line with the theory of endosymbiosis, they do compare closely with the membrane lipids of the oxygen-evolving cyanobacteria. The special feature of the chloroplast membranes is that they contain high levels of electroneutral glycolipids and relatively low levels of phospholipids. The outer envelope membrane probably contains

50% galactolipid and 30% phosphatidylcholine (PC) while the inner envelope and thylakoid membranes contain very low levels of PC, less than 5%, and higher levels of glycolipids: about 50% monogalactosyldiacylglycerol (MGDG; also called diacylgalactosylglycerol) and 30% digalactosyldiacylglycerol (DGDG; also called diacylgalabiosylglycerol). All three chloroplast membranes contain low proportions (each less than 15%) of the acidic glycolipid, sulfoquinovosyldiacylglycerol (SQDG, also called diacylsulfoquinovosylglycerol) and the electrically charged phospholipid, phosphatidylglycerol (PG). The acyl chains associated with these lipid classes also have their specific features, there being a large proportion of trienoic acids and the presence in PG of a very unusual trans isomer, *trans*-Δ_3-hexadecenoic acid. The uniqueness of the chloroplast polar lipids means that standard methods for lipid analyses need to be modified in order that reliable information can be gained rapidly. Although such information on composition has been forthcoming for many years and there was a review of techniques in "Methods in Enzymology" in 1971,[1] interest in the chloroplast polar lipids has increased recently, with a shift much more toward the importance of their structure and the understanding of their functional role. For example, our modern view of the mechanisms of light interception, electron flow, and photophosphorylation relies on viewing the thylakoid membrane as a lipoprotein system having organizational and dynamic properties as envisaged in the fluid-mosaic model of Singer and Nicolson.[2] Now, new questions are being asked about the importance of fluidity in this membrane,[3,4] the heterogeneity of acyl lipids along and across the bilayer,[4-6] and the significance of the effect of electrical neutrality of charges on function and organization of different lipid classes.[6,7] Moreover, there is a need to know about specific protein–lipid interactions and how these relationships control functional activities.[7-9] All these considerations mean that we can now expect far more investigators of photosynthetic mechanisms to turn their attention to lipid analyses. Therefore, in this chapter we wish to summarize procedures which

[1] C. F. Allen and P. Good, this series, Vol. 23, p. 523.

[2] S. L. Singer and G. L. Nicolson, *Science* **175,** 720 (1972).

[3] P. A. Millner and J. Barber, *FEBS Lett.* **169,** 1 (1984).

[4] J. Barber, *in* "Photosynthetic Mechanisms and the Environment" (J. Barber and N. R. Baker, eds.), p. 91. Elsevier, Amsterdam, 1985.

[5] P. J. Quinn and W. P. Williams, *in* "Photosynthetic Mechanisms and the Environment" (J. Barber and N. R. Baker, eds.), p. 1. Elsevier, Amsterdam, 1985.

[6] C. Sundby and C. Larsson, *Biochim. Biophys. Acta* **813,** 61 (1985).

[7] J. Barber and K. Gounaris, *Photosynth. Res.* **9,** 239 (1986).

[8] U. Pick, K. Gounaris, M. Weiss, and J. Barber, *Biochim. Biophys. Acta* **808,** 415 (1985).

[9] K. Gounaris and J. Barber, *FEBS Lett.* **188,** 68 (1985).

are commonly in use in our laboratory and also give reference, where appropriate, to established techniques used by others.

Isolation of Membranes

Methodology

The chloroplasts of higher plants and algae contain three different membranes, two closely associated boundary membranes (inner and outer), known together as the envelope, and the thylakoid membrane. The envelope acts as a diffusion barrier between the chloroplast stroma and the cytosol, has important roles in the synthesis of chloroplast lipids and proteins, and contains the molecular machinery which governs the import and export of metabolites by this organelle. The thylakoid membrane is the site of light interception, electron transport and phosphorylation, and its isolation is readily accomplished by leaf maceration and differential centrifugation.[10] However, isolation of the envelope membranes, uncontaminated by other membranes, is more difficult. The usual starting point is to isolate intact chloroplasts possessing the complete envelope system. To do this a variety of methods have been reported and include gentle maceration of leaves followed by differential centrifugation under strictly controlled ionic conditions,[11] density gradient procedures,[12,13] and the disruption of isolated protoplasts.[14] Douce and Joyard[15,16] have given a comprehensive description of the isolation of envelopes from intact chloroplasts, and have also summarized techniques for separating inner and outer membranes. Even more than with the isolation of thylakoids, the procedures for envelope isolation have been restricted to only a few plant species, predominantly spinach and pea, and, of course, the yields are significantly lower than those of thylakoids.

The thylakoid membrane is differentiated into two regions: appressed lamellae of the grana, and the nonappressed regions of the stromal and end-granal lamellae. For many years mechanical fragmentation[17,18] and

[10] S. G. Reeves and D. O. Hall, this series, Vol. 69, p. 85.

[11] H. Y. Nakatani and J. Barber, *Biochim. Biophys. Acta* **461,** 510 (1977).

[12] R. M. Leech, *Biochim. Biophys. Acta* **79,** 637 (1964).

[13] W. R. Mills and K. W. Joy, *Planta* **148,** 75 (1980).

[14] R. C. Leegood and D. A. Walker, *in* "Isolation of Membranes and Organelles from Plant Cells" (J. L. Hall and A. L. Moore, eds.), p. 185. Academic Press, London, 1983.

[15] R. Douce and J. Joyard, this series, Vol. 69, p. 29.

[16] R. Douce and J. Joyard, *in* "Chloroplast Biogenesis" (N. R. Baker and J. Barber, eds.), p. 71. Elsevier, Amsterdam, 1984.

[17] P. V. Sane, D. J. Goodchild, and R. B. Park, *Biochim. Biophys. Acta* **216,** 162 (1970).

[18] G. Jacobi, this series, Vol. 23, p. 289.

detergent treatments,[19,20] followed by differential centrifugation, have been used to separate granal and stromal membranes. More recently these procedures have been modified to isolate appressed and nonappressed membranes. In one approach differential centrifugation has been replaced by phase-partition methodology,[21] while in another, differential centrifugation is maintained, but the detergent treatment modified.[22–24] Although these fragments are of interest for compositional and functional studies, their structure is too complex for identifying specific protein–lipid interactions. For this, further detergent solubilization is required. Indeed, all the major multipeptide complexes of the thylakoid membrane have been isolated and, in some cases, already found to have specific lipid classes associated with them.[8,9]

Pitfalls

Variation. There are numerous reports of both genotypic and phenotypic variation in lipid composition of thylakoids and variable results of lipid analyses must be expected if membranes are not isolated from a consistent source of leaf tissue. A major genotypic difference to be aware of is the occurrence of high amounts of hexadecatrieneoic acid in some species (e.g., *Spinacea oleracea,* spinach; *Lycopersicon esculentum,* tomato), but not in others (e.g., *Pisum sativum,* pea; *Triticum aestivum,* wheat; *Vicia faba,* broad or faba bean). In addition, there are less obvious genotypic differences, such as variations which seem to be related to the ability to withstand temperature stress,[25] which may even occur between different varieties within the same species.[26] Even with a genetically consistent source of material, or taking the precaution of harvesting leaves from a population of plants rather than genetically distinct individuals, there can be variations in the thylakoid lipid composition resulting from changes in plant growth conditions.[5,27,28] Besides the obvious need to stabilize the growth environment as far as possible, conclusions need to

[19] N. K. Boardman, this series, Vol. 23, p. 268.

[20] L. P. Vernon and E. R. Shaw, this series, Vol. 23, p. 289.

[21] H.-E. Akerland and B. Andersson, *Biochim. Biophys. Acta* **725,** 34 (1983).

[22] D. A. Berthold, G. T. Babcock, and C. F. Yocum, *FEBS Lett.* **134,** 231 (1981).

[23] T. Kuwabara and N. Murata, *Plant Cell Physiol.* **23,** 533 (1982).

[24] Y. Yamamoto, T. Uda, H. Shinkai, and M. Nishimura, *Biochim. Biophys. Acta* **679,** 347 (1982).

[25] N. Murata, *Plant Cell Physiol.* **24,** 81 (1983).

[26] H. A. Norman, C. McMillan, and G. A. Thompson, Jr., *Plant Cell Physiol.* **25,** 1437 (1984).

[27] D. J. Chapman, J. De-Felice, and J. Barber, *Plant Physiol.* **72,** 225 (1983).

[28] D. J. Chapman, J. De-Felice, and J. Barber, *Photosynth. Res.* **8,** 257 (1986).

be based on replication in source material and not just on repetition of analytical procedures.

Contamination. The differences in lipid composition of the envelope and thylakoid membranes, compared to the other membranes of plant cells, emphasizes the need to minimize cross-contamination during membrane isolation procedures. Also the contamination of envelope membranes by thylakoids may occur but fortunately can be detected and allowed for because the latter contain chlorophyll as a convenient marker. The contamination of isolated thylakoids by envelope membranes can, however, be more problematic and may not be satisfactorily dealt with by measuring specific activities of marker enzymes (rate of substrate utilization per milligram membrane protein) because of the lower protein-to-lipid ratio of the envelope. Simply to ignore envelope contamination because of its relatively low level could be even more dangerous especially when isolated thylakoids are further fragmented and separated into different fractions, one of which could preferentially contain the envelope contaminant.

A major source of lipid contamination in chloroplast membrane preparations can be the plastoglobuli, which contain mainly neutral lipids and quinones,[29] and possibly some polar acyl lipids.[30] Removal of plastoglobuli from isolated thylakoids can be achieved by sonication or treatment in a high-pressure cell.[29] For example, we have found that two slow passages through a Yeda press (Linca Scientific Instruments, Tel Aviv, Israel) at 1500 psi and 4°, followed by sedimentation at 50,000 *g* for 1 hr, is sufficient to deplete the plastoglobuli content of thylakoids as monitored by reduction in the level of plastoquinone-9.[31]

Clearly, cross-contamination of membranes can be a problem and the use of immunological methods to monitor such contamination should be seriously considered.

Chemical and Enzymatic Breakdown. Membrane isolation procedures must be chosen to limit the likelihood of lipid degradation before their extraction from the membrane. For example, it is well known that the mechanical disruption of cells can be followed by the loss of lipids through lipase activity and the breakdown of fatty acids via oxidative processes. Such degradations are often selective to particular head groups and fatty acids.[32,33] A number of measures can be taken to reduce this

[29] D. Steinmuller and M. Tevini, *Planta* **163,** 201 (1985).

[30] J. L. Bailey and A. G. Whyborn, *Biochim. Biophys. Acta* **78,** 163 (1963).

[31] D. J. Chapman, J. De-Felice, and J. Barber, *Planta* **166,** 280 (1985).

[32] P. S. Sastry and M. Kates, *Biochemistry* **3,** 1280 (1964).

[33] T. Galliard, this series, Vol. 31, p. 520.

problem with chloroplast membranes but none is totally effective. It is vital to be rapid in removing the isolated chloroplasts from the crude homogenate of leaf tissue. Even then lipase activity can still be detected with the separated sample.[32] In order to reduce the effect of this residual enzymatic activity the lipids should be immediately extracted or else the membrane preparation stored at liquid nitrogen temperature. Extraction should be conducted at low temperatures and we also take the additional precaution against lipid breakdown by using suspension media with low divalent cation content having 5 m*M* EDTA or EGTA and a pH above 6.0 (usually pH 6.5). With these conditions we attempt to minimize lipase and lipoxygenase activities. We also avoid high light intensities which can speed the breakdown of phospho-[34] and galactolipids.[35] Unfortunately there is no general compound to totally inhibit enzymatic breakdown of chloroplast lipids, although when extracting lipids from leaf tissue an initial homogenization in 2-propanol can be used with some success.[36] Nupercaine was reported to inhibit both lipoxygenase and lipolytic acyl hydrolase activities but in some circumstances there is evidence of significant stimulation of activities by this compound.[37] With other compounds, such as diisopropyl phosphofluoridate, there is partial inhibition but only at concentrations as high as 5 m*M*.[38] Attempts to avoid the oxidation of fatty acids can be made by the inclusion of antioxidants, such as 5–20 m*M* sodium ascorbate, particularly in the media used for the initial maceration of leaf tissue. Bovine serum albumin (BSA) can partially protect against lipase activity[33] and is commonly included in media used for isolating chloroplasts. However, although there is some evidence of a direct effect of BSA on lipases it is more likely that its mode of action is to bind free fatty acids released by acyl hydrolase activity and thus prevent further disruption of the membrane.

Selection of a particular plant species for isolating chloroplast membranes can also have important consequences for lipid degradation. For example, many *Phaseolus* species should be avoided and even commonly used species like *Spinacea oleracea* are not completely devoid of the

[34] S. Hoshina, T. Kaji, and K. Nishida, *Plant Cell Physiol.* **16,** 465 (1975).

[35] A. S. Sandelius, L. Axelsson, B. Klockane, H. Ryberg, and K-O. Widell, *in* "Photosynthesis V. Chloroplast Development" (G. Akoyunoglou, ed.), p. 305. Balaban, Philadelphia, Pennsylvania, 1981.

[36] B. W. Nichols, *Biochim. Biophys. Acta* **70,** 417 (1963).

[37] P. J. Walcott, J. R. Kenrick, and D. G. Bishop, *in* "Biochemistry and Metabolism of Plant Lipids" (J. F. G. M. Wintermans and P. J. C. Kuiper, eds.), p. 297. Elsevier, Amsterdam, 1982.

[38] T. Galliard, *Biochem. J.* **121,** 379 (1971).

problem.[32] Much of our own work has used *Pisum sativum* and with this experimental material we have not detected significant breakdown of lipids in thylakoids in a 2-hr period after isolation, even at room temperature. However, we recommend that lipase activity should always be tested when adopting new isolation procedures or when changing to another source of leaf tissue. A further check of lipid breakdown during isolation is to compare the lipid analyses of thylakoids and leaf tissue. If little or no breakdown occurs it is expected that, as a proportion of total lipid, the galactolipids, sulfolipid, *trans*-hexadecenoic acid, and linolenic acid will all be higher in the thylakoid fraction. Furthermore, when subfractionating thylakoids or envelopes, a composition similar to that derived by analysis of the intact membrane system should be obtained by calculating an average of the fractions based on the composition of each fraction weighted to allow for the total lipid recovered. Indeed, Henry *et al.*[39] carried out a careful accounting procedure of this type and demonstrated that considerable breakdown of specific lipids occurred during detergent fractionation of barley thylakoids and the subsequent phase partitioning of photosystem II membrane fragments.

Monoacylglycerols, phosphatidic acid, free fatty acids, and fatty acid peroxidation products occur only in trace amounts in most plant membranes so that their appearance in lipid extracts indicates that lipid degradation has occurred. However, in the case of galactolipids, loss of these lipids can occur without the formation of easily detectable lyso derivatives.[32,34,40]

Continued Synthesis and Interconversion. Envelopes and thylakoid membranes contain many of the enzymes involved in lipid synthesis and a significant proportion of the synthetic pathways can continue to function in chloroplasts when isolated intact.[41] Therefore some consideration must be given to the possibility of continued activity of some of these enzymes during membrane purification or incubation, particularly if essential cofactors are present. This seems to happen during the isolation of protoplasts, there being alterations in lipid metabolism and composition.[42,43] A good example of an enzyme activity which could be a problem is the formation of DGDG from MGDG by a galactosyltransferase. Conditions

[39] L. E. A. Henry, J. D. Mikkelsen, and B. L. Moller, *Carlsberg Res. Commun.* **48,** 131 (1983).

[40] T. Galliard, *Phytochemistry* **9,** 1725 (1970).

[41] P. G. Roughan and C. R. Slack, *Annu. Rev. Plant Physiol.* **33,** 97 (1982).

[42] M. S. Webb and J. P. Williams, *Plant Cell Physiol.* **25,** 1541 (1984).

[43] M. S. Webb and J. P. Williams, *Plant Cell Physiol.* **25,** 1551 (1984).

controlling maximum activity are given by Heemskerk *et al.*[44] and use of a suspension buffer containing no divalent cations should reduce the rate of interconversion.

Unrepresentative Samples. It is not usually possible to gain full recovery of a specific membrane during its purification and the danger of obtaining an unrepresentative sample is exemplified by chloroplast isolation because it is known that there are heterogeneous populations of chloroplasts within leaves.[45] The problem can be compounded at each stage of purification and this has been a concern in some of our own work where recoveries of chloroplast envelopes were much lower than estimates of envelope/thylakoid ratios based on ultrastructural studies.[46] In the case of purification by phase-partition methods, there could be a further difficulty arising from the reported specificity of this selection process for particular lipid polar head groups.[47]

Artifactual Lipid–Protein Interactions. There is a serious possibility that artificial specific lipid–protein interactions occur during membrane fractionation, particularly when using detergent treatments which can disrupt the *in vivo* relationship. A possible approach to the assessment of this problem is the use of double-isotope radioactive labeling experiments. Heinz and Siefermann-Harms[48] used ^{14}C to biosynthetically label galactolipids in thylakoids before conducting detergent fractionation and electrophoretic separation of pigment–protein complexes in the presence of ^{3}H-labeled galactolipid. They concluded that 10 to 20% of total thylakoid galactolipid was associated with the complexes but most of this was a result of interactions formed during the isolation procedure.

Lipid Isolation

Solvents

Analar-grade solvents can be used for the analyses of the main lipid classes but redistillation, perhaps of lower grade solvents, is recommended for the most accurate analyses of low quantities of specific lipids. Routine procedures used in many laboratories also include deaeration by

[44] J. W. M. Heemskerk, G. Bogemann, and J. F. G. M. Wintermans, *Biochim. Biophys. Acta* **754,** 181 (1983).

[45] I. Terashima and I. Inoue, *Plant Cell Physiol.* **26,** 781 (1985).

[46] J. Forde and M. W. Steer, *J. Exp. Bot.* **27,** 1137 (1976).

[47] E. Eriksson and P.-A. Albertsson, *Biochim. Biophys Acta* **507,** 425 (1978).

[48] E. Heinz and D. Siefermann-Harms, *FEBS Lett.* **124,** 105 (1981).

bubbling N_2 through the solvent and addition of antioxidants such as 2,6-di-*tert*-butyl-*p*-cresol or butylated hydroxytoluene at levels between 50 and 100 mg/liter.

Extraction

Membrane preparations are extracted in chloroform : methanol [2 : 1 (v : v)] and for thylakoids (measured by chlorophyll content) we use a ratio of solvent to chlorophyll of approximately 20 ml to 1 mg. Insoluble material is removed by passage through a solvent-resistant filter (e.g., Whatman GF/C) which must be carried out rapidly and is facilitated by applying suction. The filter is then washed repeatedly with about 15 ml additions of the solvent mixture until, in the case of thylakoids, all chlorophyll is extracted. We usually then give one final 15-ml wash with chloroform alone. For the extraction we use a hand homogenizer combined with vortex mixing.

Removal of Water and Nonlipid Contaminants

Separation of the hydrophobic lipid components from any hydrophilic contaminants in the extract can be rapidly and efficiently achieved by using Dextran gel and the procedure of Williams and Merrilees.[49] This is done by transferring the contaminated solvent extract to a 250-ml flask to which dry Dextran gel (Sephadex G-25–medium) is added ready for rotary film evaporation. The water and nonlipid contaminants enter the gel leaving lipids at the gel and glass surfaces. The solvent is removed during the rotation of the flask in a water bath at about 30° with a reduced pressure supplied by a vacuum system. When all the solvent is removed, chloroform is added, evaporated, and this cycle repeated once more. For thylakoids equivalent to 1 mg chlorophyll we use about 0.5 g of Dextran gel and approximately 15 ml of chloroform for each addition. Although air can be used to release the vacuum in the evaporator, we recommend the use of N_2. A final gel–chloroform slurry is made and poured onto a glass-sinter which retains the Sephadex and allows repeated washings with chloroform until all the chlorophyll is removed (note that often some carotenoids remain behind). The combined washings are taken to dryness by rotary evaporation and redissolved in a small volume of chloroform.

It is worth mentioning a column procedure which is a useful alternative when large amounts of lipid are needed and recovery of gel is important.[50]

[49] J. P. Williams and P. A. Merrilees, *Lipids* **5**, 367 (1970).
[50] R. E. Wuthier, *J. Lipid Res.* **7**, 558 (1966).

Alternative Procedure for Isolation and Purification

Previous chapters in this series[1,51] describe commonly used procedures of extraction based on the methods of Folch *et al.*[52] and Bligh and Dyer.[53] These involve homogenization in organic solvent followed by the formation of a two-phase system by addition of methanol and water. The lipids partition into the lower chloroform phase. This approach, however, may be unsatisfactory because of traces of methanol and water in the chloroform leading to degradation of unsaturated fatty acids, and could have the additional problem of incomplete partitioning between the two phases of the more polar lipids if the composition of the two phases is not carefully controlled.[54]

Storage

Immediately after redissolving the dried lipid extract in chloroform, a glass vial is almost filled with sample and the remaining air replaced with N_2. The vial is then made gas tight with a screw cap having a solvent-insoluble seal and ideally the sample analyzed as soon as possible. We routinely store our samples at $-15°$ for up to 7 days and for longer periods it is advisable to drop the storage temperature to $-40°$ and add antioxidants. No traces of water or methanol should be present and chlorophyll, if exposed to strong light, could sensitize photooxidation of lipids containing unsaturated fatty acids.[55]

Fatty Acid Analysis

After preparation of lipid extracts the most common analytical approach is to determine relative and total amounts of the fatty acids, either in the total lipid extract or in fractions produced by the separation of lipid classes (see Fig. 1). Gas–liquid chromatography (GLC) is the most widely used method and gives fast, accurate, and reproducible results.

Preparation of Derivatives of Fatty Acids for GLC

The fatty acids must be cleaved from the rest of the lipid molecules and, for the fatty acids in the molecular weight range present in the lipids

[51] N. S. Radin, this series, Vol. 14, p. 245.

[52] J. Folch, M. Lees, and G. H. S. Stanley, *J. Biol. Chem.* **226,** 497 (1957).

[53] E. G. Bligh and W. J. Dyer, *Can. J. Biochem.* **37,** 911 (1959).

[54] W. W. Christie, "Lipid Analysis." Pergamon, Oxford, 1982.

[55] R. Yamauchi, M. Kojima, M. Isogai, K. Kato, and Y. Ueno, *Agric. Biol. Chem.* **46,** 2815 (1982).

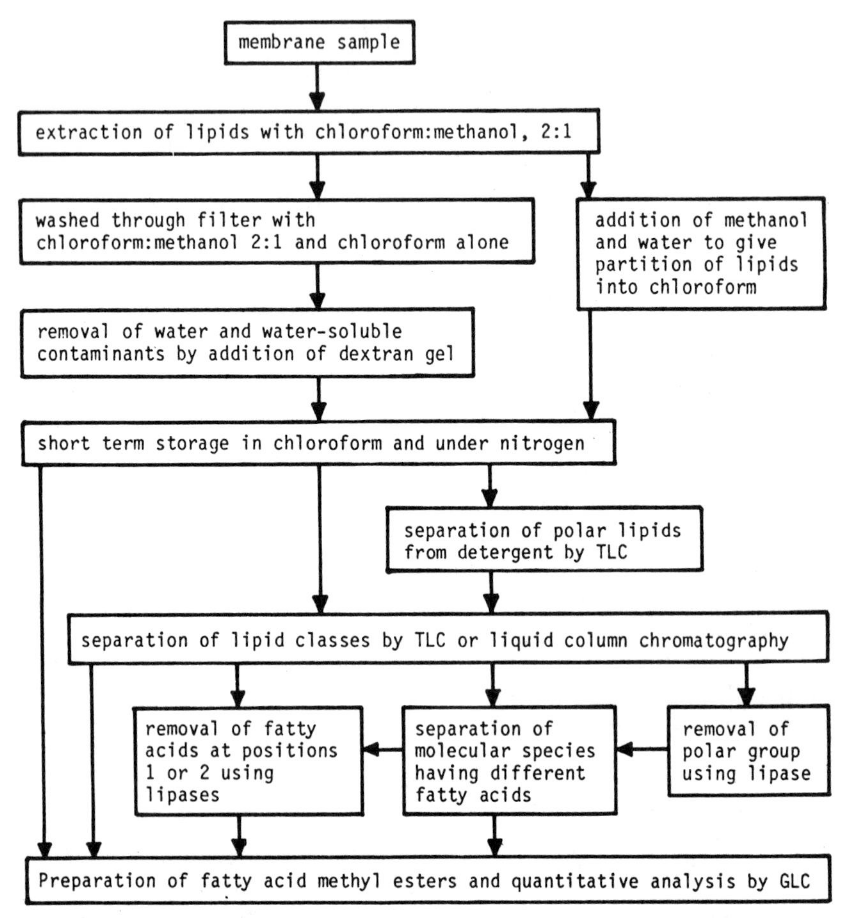

FIG. 1. Combinations of methods commonly used for the analysis of polar lipids of chloroplast membranes.

of chloroplast membranes, it is necessary to convert them to the more volatile fatty acid methyl esters (FAMEs). Separate hydrolysis and methylation reactions are not necessary as this can be achieved in one step by a variety of reagents, e.g., 0.5 *M* sodium methoxide in anhydrous methanol, 5% (w/v) anhydrous hydrogen chloride in methanol, or 1 to 2% (v/v) sulfuric acid in methanol. Methods for their preparation and use are fully discussed elsewhere.[54] We have, however, preferred to use the commercially available reagent, boron trifluoride in methanol (10–15%). Although it is a convenient and rapid approach, the reagent does have a limited shelf life and can give rise to minor spurious contamination peaks in GLC traces. To avoid these problems it should be stored at 4° and in such sizes of containers that each is opened for only a limited number of analyses. For the methylation reaction, screw-top glass tubes with Teflon-lined (PTFE) caps are reuseable as long as they are thoroughly washed with acid, rinsed, and dried. A sample of the lipid extract in chloroform, up to 1 mg lipid, is completely dried down in the tube under a stream of N_2 before addition of about 1 ml of reagent (excess). The sample is capped, after a N_2 flush, and heated to 80° for 3 min. The caps are then firmly tightened and heating continued for a further 25 min. The tubes are allowed to cool to room temperature and additions made as follows: a known amount of pentadecenoic acid methyl ester (as a standard), 1 ml of hexane, and finally 0.1 ml of water. Although a greater sensitivity can be achieved, we usually aim to analyze between 50 and 300 μg of a lipid sample which, in the case of thylakoids, is about 20 to 100 μg chlorophyll. For these levels we add about 50 nmol of pentadecenoic acid as our internal standard. The methanol/hexane solvent system is vortex mixed and centrifuged at 800 *g* to aid formation of an upper hexane phase which contains the FAMEs. This is carefully removed, to be combined with a further hexane washing of the methanol phase, and then stored under N_2 in an almost full, securely capped vial. There is a danger of breakdown of unsaturated fatty acids at this stage and GLC analysis should be carried out as soon as possible, and certainly within 24 hr. Prior to analysis the sample is taken to dryness under a stream of N_2 and redissolved in a small volume of hexane. The exact volume of hexane used for the resuspension (usually between 50 and 200 μl) and for injection into the gas chromatogram (1 to 5 μl) will depend on the amount of lipid and the performance characteristics of the particular injection system, column, and detector.

GLC Separation of FAMEs

There is a wide variety of GLC techniques and equipment suitable for the analysis of FAMEs and this particularly applies to the central component: the chromatography column. The spectrum of fatty acids in photosynthetic membranes is comparatively simple and in routine analyses standard packed columns are more desirable than the support-coated (SCOT) and wall-coated (WCOT) open tubular capillary columns because they are generally cheaper and easy to prepare and maintain. Also, although the latter columns have very high resolving power this is not normally needed. The important factor which determines the success of the separation is the liquid phase which coats the packing material. Analysis of FAMEs on the basis of molecular weight, and further resolution due to the number of double bonds, is commonly achieved on polyester liquid phases such as diethylene glycol succinate (DEGS) and EGSS-X (a copolymer of a methyl silicone and polymeric ethylene glycol succinate). A range of more recently developed stationary phases, which are now widely used, are the alkylpolysiloxanes; e.g., Silar 10C, SP2330, and OV275. We use 10% Silar 10C (also known as Reoplex 400) on Chromosorb W-HP (100–120 mesh) with N_2 carrier gas flow rates of 20 to 30 ml/min and an isothermal column oven temperature set at between 180 and 210°. An alternative column, useful for checking results, is 10% EGSS-X on Gas-Chrom Q used at about 20 ml/min and 180°. Exact running conditions vary between individual columns and alter with column age. Silar 10C has the advantage of giving a good separation of hexadecatrienoic acid methyl ester from other FAMEs and can resolve cis and trans isomers of hexadecenoic acid, although chromatography times are about 25% longer than when using EGSS-X. Typical GLC traces for analyses of FAMEs are usually published by commercial suppliers of packing materials and the exact relative retention times should be determined with standards.

GLC Detection and Estimation of FAMEs

The most widely used GLC detector system for FAMEs is the hydrogen flame ionization detector. The response is almost exactly proportional to the number of carbon atoms and determination of response factors is determined by calibration of the column with repeated injections of equimolar standards of FAMEs. The presence of an internal standard of pentadecenoic acid in the experimental sample provides a means of readily calculating amounts of FAMEs in the original sample by accounting for losses during the hexane extraction and injection procedure:

$$\text{Moles FAME} = \left(\frac{\text{FAME detector response}}{\text{internal standard detector response}}\right) \times \text{moles internal standard}$$

High-Pressure or High-Performance Liquid Chromatography

Although GLC is the most widely used method for analysis of fatty acids, it is also possible to achieve good resolution, reproducibility, and quantitative results by HPLC. Wood and Lee[56] have published a method in which phenacyl or naphthacyl derivatives are prepared and separated on an octadecyl reversed-phase column using solvent mixtures of acetonitrile : water. In this form the fatty acids are detected and quantities measured by spectrophotometry at 242 nm.

[56] R. Wood and T. Lee, *J. Chromatogr.* **254,** 237 (1983).

Lipid Class Analysis

Thin-layer chromatography (TLC) is preferred to liquid column methods for quantitative analysis of lipid classes because it is more rapid and can give more reproducible results. However, with current technological developments, high-performance liquid chromatography (HPLC) is set to supersede TLC as an even more efficient and reliable technique for lipid class analyses. The disadvantage of HPLC is the initial cost of purchasing suitable equipment and maintaining it. Thus it is likely that TLC will always be a useful alternative approach.

Thin-Layer Chromatography

Chromatographic Separation. The presence of both glyco- and phospholipids in extracts of chloroplast membranes presents a special problem for successful lipid class analyses by TLC. This can be overcome by two-dimensional methods of the type described in a previous chapter in this series.[1] However, one-dimensional TLC has the advantage of being more rapid, subjecting fatty acids to less risk of oxidation and allowing the loading of larger samples. A successful one-dimensional method is that adopted by Khan and Williams.[57] TLC plates are coated with silica gel G suspended in 0.15 *M* $(NH_4)SO_4$ and, after drying at room temperature for 5 min, are dried further at 100° for 4 hr. When drying and cooling are complete, lipid samples are loaded as narrow bands. For extracts of thylakoids a 150-mm-wide band containing 1 mg chlorophyll (corresponding to about 3 mg lipids) is suitable for loading on a 0.4-mm-thick layer of silica gel. The developing solvent is acetone : benzene : water at a 91 : 30 : 8 (v/v) ratio and the solvent front will reach a position close to the top of a standard 200-mm plate in under 1 hr.

Detection. After the run, the TLC plates are dried in nitrogen and sprayed to detect lipids. Some useful spraying agents and their lipid specificities[58–65] are given in Table I. When lipids are to be recovered for subsequent GLC analysis of fatty acids, the plates are viewed under UV light after spraying with a nondestructive fluorescent dye such as 0.2 mg ml^{-1}

[57] M.-U. Khan and J. P. Williams, *J. Chromatogr.* **140,** 179 (1977).
[58] A. N. Siakotos and G. Ronser, *J. Am. Oil Chem. Soc.* **42,** 913 (1965).
[59] L. Svennerholm, *J. Neurochem.* **1,** 42 (1956).
[60] H. Jatzkewitz and E. Mehl, *Hoppe-Seyler's Z. Physiol. Chem.* **320,** 251 (1960).
[61] N. Shaw, *Biochim. Biophys. Acta* **164,** 435 (1968).
[62] H. Jatzkewitz and E. Mehl, *Hoppe-Seyler's Z. Physiol. Chem.* **320,** 251 (1960).
[63] V. E. Vaskovsky and E. Y. Kostetsky, *J. Lipid Res.* **9,** 396 (1968).
[64] J. Sherma and S. Bennett, *J. Liq. Chromatogr.* **6,** 1193 (1983).
[65] E. K. Ryu and M. MacCoss, *J. Lipid Res.* **20,** 561 (1979).

TABLE I
SPRAY REAGENTS USED FOR SPECIFIC DETECTION OF GLYCO- AND PHOSPHOLIPIDS ON TLC PLATES

Reagents	Specificity	Reference
α-Naphthol solution	Glycolipids	58
Orcinol–sulfuric acid	Glycolipids	59
Diphenylamine	Glycolipids	60
Periodate–Schiff reagent	Glycolipids	61,62
Phosphomolybdic acid	Phospholipids	63–65

of 2′,7′-dichlorofluorescein in methanol or 0.1 mg ml^{-1} rhodamine 6G in 95% ethanol. Lipids are detected as areas of different fluorescence intensity which can be compared with the positions of authentic standards run in tracks adjacent to the unknown samples.

Recovery. Areas of the silica gel containing lipids are marked with a needle and then scraped from the glass plate into screw-top tubes. As long as the silica gel has dried, the presence of the detection dye does not normally interfere with subsequent analyses of FAMEs. A word of warning, however, is that the period of exposure of the loaded TLC plates to air should be minimized to reduce the risk of fatty acid oxidation, particularly of the unsaturated fatty acids. Even with care in manipulating the silica gel, there is usually variation in the recovery of lipid and more than one TLC analysis of each sample is recommended.

Direct Quantitative Analysis of TLC Plates. Two approaches can be used to obtain quantities of lipid classes without GLC analyses or elution from the silica gel. In the first, lipids are visualized with a spray reagent, such as sulfuric acid, and the plates then scanned in a spectrophotometric densitometer[66] or photographed prior to analyses of negatives.[67] The second approach involves chromatography on a tubular thin-layer system followed by passage of the chromatogram through a furnace so that the products of pyrolysis are taken into a flame ionization detector by an inert carrier gas.[68] A variation on this method uses silica-covered quartz rods which are passed directly through the flame of an ionization detector.[69] It is generally felt that good reproducibility of results can be difficult with these methods and information on fatty acid compositions is sacrificed.

[66] O. S. Privett, *in* "Quantitative Thin-layer Chromatography" (J. C. Touchstone, ed.), p. 57. Wiley (Interscience), New York, 1973.

[67] A. Rawyler and R. A. Siegenthaler, *J. Biochem. Biophys. Methods* **2,** 271 (1980).

[68] H. V. Mangold and K. D. Mukherjee, *J. Chromatogr. Sci.* **13,** 398 (1975).

[69] F. B. Padley, *J. Chromatogr.* **39,** 37 (1969).

However, their speed facilitates replication and analysis of large numbers of samples.

Analyses of Extracts Containing Nonionic Detergents. Detergent solubilization of thylakoids is commonly used for isolating membrane fragments or individual lipoprotein complexes. In these cases it is inevitable that detergent will be present in subsequent lipid extracts and this can interfere with TLC separations and identification of lipid classes, particularly with the one-dimensional method described above. We have overcome this sort of problem involving Triton X-100 by a TLC step prior to lipid class separation. Unlike neutral lipids and this type of nonionic detergent, the polar acyl lipids remain within 10 mm of the loading point on silica gel G plates when developed in nonpolar solvents, such as a mixture of hexane : diethyl ether : acetic acid, 90 : 10 : 1 (v/v), or petroleum ether : diethyl ether : acetic acid, 80 : 20 : 10 (v/v). The acyl lipids are recovered from the adsorbent by washing with chloroform : methanol, 1 : 1 (v/v).

Liquid Column Chromatography

Low-Pressure Liquid Chromatography. Column chromatography can be time consuming and generally is not used for routine lipid analyses of large numbers of samples. However, when crude subdivisions or bulk preparations are needed, several methods are applicable, including ion-exchange chromatography[70] and controlled elution from columns of silicic acid. In the latter case advantage is taken of the different solubilities of neutral lipids, glycolipids, and phospholipids in solvents such as chloroform, acetone, and methanol.[71] Gounaris *et al.*[72] have used acid-washed Florisil (magnesium silicate) rather than silicic acid to separate the main lipid classes of photosynthetic membranes because it has a greater adsorbtive capacity for the acidic lipids. The sequence of lipid elution is as follows: neutral lipids and pigments by chloroform (10× bed volume), MGDG by chloroform : acetone [1 : 1 (v/v); 5× bed volume], DGDG and some SQDG by acetone (10× bed volume), remaining SQDG and some PG and PC by chloroform : methanol [1 : 1 (v/v); 10× bed volume). Final purification of individual lipid classes is then usually achieved by TLC.

High-Pressure or High-Performance Liquid Chromatography. There is an increasing interest in the liquid column methods of HPLC. The main attractions of this analytical approach are, in addition to the highly repro-

[70] N. Murata, N. Sato, N. Takahashi, and Y. Hamazaki, *Plant Cell Physiol.* **23,** 1071 (1982).

[71] M. L. Vorbeck and G. V. Marinetti, *J. Lipid Res.* **6,** 3 (1965).

[72] K. Gounaris, D. A. Mannock, A. Sen, A. P. R. Brain, W. P. Williams, and P. J. Quinn, *Biochim. Biophys. Acta* **732,** 229 (1983).

ducible results that it gives, the full recoveries of lipid without oxidation of fatty acids and the ability to fully automate the analyses of large numbers of samples. Silicic acid columns appear to be the best choice for separation of different lipid classes. Several methods have been published for separation of phospholipids,[73,73a] but inclusion of glycolipids and pigments in extracts of photosynthetic membranes make the problem more complex. The method of Marion *et al.*[74] separates all the major lipid classes with a single-pump HPLC system which uses a constant (isocratic) solvent composition [acetonitrile : methanol : sulfuric acid, 135 : 5 : 0.2 (v/v); 0.7 ml min^{-1}] for elution from a column of 10-μm particle size Lichrospher Si 100 (Merck, GFR). In this system SQDG is found at the solvent front and may coelute with a minor quantity of phospholipid such as phosphatidylinositol. The use of a modified solvent composition at the beginning of the analysis should overcome this problem although this will need a dual pump gradient elution HPLC system. Spectrophotometric detection at UV wavelengths between 195 and 220 nm is suitable for monitoring the separation products but the response is dependent on the number of double bonds and therefore there is a need for a careful calibration as well as consideration of the effect of differences in fatty acid unsaturation between samples. If a more accurate quantitative analysis is required then it is better to subject main eluate fractions to GLC analysis of FAMEs.

HPLC methods can be adapted for preparation of large amounts of pure lipid classes by loading more samples onto bigger columns as well as using facilities available for repeated automatic injection and fraction collection. The silicic acid column technique of Marion *et al.*[74] is suitable for this larger scale separation, as could be the column chromatography approach of Gounaris *et al.*[72] if modified for HPLC. The latter alternative, however, would require the introduction of automatic solvent switching valves and doing without UV spectrophotometric detection because of the use of UV absorbing solvents. An alternative approach, which we have used for preparative separation of thylakoid lipids, involves a reversed-phase column (Spherisorb RP-18, 10 μm) with a solvent of methanol : water, 93 : 7 (v/v) changing to methanol : ethanol, 1 : 1 (v/v) to elute neutral lipids and pigments, and run at 2.5 ml min^{-1} flow rate.[75] This

[73] G. M. Patton, J. M. Fasulo, and S. J. Robins, *J. Lipid Res.* **23,** 190 (1982).

[73a] A. Tremolieres, *Physiol. Veg.* **24,** 539 (1986).

[74] D. Marion, G. Gandemer, and R. Douillard, *in* "Structure, Function and Metabolism of Plant Lipids" (P.-A. Siegenthaler and W. Eichenberger, eds.), p. 139. Elsevier, Amsterdam, 1984.

[75] P. A. Millner, R. A. C. Mitchell, D. J. Chapman, and J. Barber, *Photosynth. Res.* **5,** 63 (1984).

method ensures good separation of polar lipids from chlorophylls but not a clear division between lipid classes. Further purification is carried out by TLC on 0.6-mm-thick silica plates.

Analysis of Molecular Species

The lipid class fractions separated as described in the previous section are in fact a mixture of different molecular species, each with its own fatty acids attached in a particular order to the carbons, 1 and 2, of the glycerol backbone. Therefore a complete lipid analysis requires a separation of these molecular species and determinations of the type and positional distribution of the acyl chains.

Silver Nitrate TLC

TLC on silica gel plates impregnated with $AgNO_3$ will separate the molecular species of the glycolipids into fractions containing lipids with different numbers of double bonds.[76–78] However, there is a variation in the exact choice of ratio of silica gel to $AgNO_3$ [often 4 : 1 (w/w) or less], the period of heat activation (usually overnight at 100°), and the precise composition of the solvents used for developing the plates. Good results will be achieved with practice and by fine tuning of the technique to suit particular circumstances. The usual solvent system is a mixture of chloroform : methanol : water and it may be advisable to use different ratios for each galactolipid, for example, 65 : 25 : 4 (v/v) for MGDG and 60 : 30 : 4 (v/v) for DGDG.[78] Molecular species separation of SQDG has been possible using the former solvent system[77] and improved resolution obtained by developing plates at −15° rather than at room temperature.[79] The position of lipids can be revealed with nondestructive spray reagents (section on TLC of Lipid Classes) and the bands scraped from the plates for GLC analysis of FAMEs. This approach can give almost a complete description of molecular species composition because it not only identifies and quantifies bands where just a single molecular species is present but with multicomponent bands it can give relative amounts, provided that each lipid type contains different fatty acids, e.g., the tetraenoic band (four double bonds) will contain dienoic : dienoic (2 : 2) and anenoic : trienoic (1 : 3) species.

[76] M. Nishihara, K. Yokota, and M. Kito, *Biochim. Biophys. Acta* **617,** 12 (1980).
[77] H. P. Siebertz, E. Heinz, J. Joyard, and R. Douce, *Eur. J. Biochem.* **108,** 177 (1980).
[78] J. P. Williams, *Biochim. Biophys. Acta* **618,** 461 (1980).
[79] R. G. Nichols and J. L. Harwood, *Phytochemistry* **18,** 1151 (1979).

In the case of PG, chromatography on $AgNO_3$ plates gives poor results, probably because the interplay between silver ions and double bonds is masked by a stronger interaction with the negatively charged phosphate-containing group. This problem experienced with phospholipids can be overcome by a prior digestion with phospholipase C, which removes the polar groups and leaves the more easily separated neutral diacylglycerols.[80] The chromatographic resolution is in fact increased by derivitizing the diacylglycerols with acetic anhydride and pyridine.[81] This procedure also protects against possible changes in the positional distribution of fatty acids during the analysis. It is also possible to use direct acetylation of galactolipids[82] and sulfolipid[76] to increase the resolution of molecular species of these lipids by $AgNO_3$ chromatography. Suitable methods for phospholipase C digestion, extraction of diacylglycerols, and acetylation can be based on those described by Kito.[81] A 1.25-ml suspension of PG (about 2 mg ml^{-1}) is suspended by sonication in 10 m*M* Tris/HCl buffer, pH 7.4, and shaken with 0.5 mg phospholipase C of *Bacillus cereus* plus 2 ml of ether. After incubation at 30° for 100 min, the diacylglycerols are extracted from the incubation mixture four times with 1.5 ml of chloroform : methanol (2 : 1, v/v). The ether and chloroform layers are washed with water, then combined before being dried under N_2. Monoacetyldiacylglycerols are formed by incubation at 30° for 24 hr in acetic anhydride : pyridine (2 : 1, v/v). The chloroform layer is washed three times with 30 ml of water and then dried under N_2.

GLC of Lipid Derivatives

Molecular species separation and the determination of their quantities can be carried out in one step by GLC. Galactolipids have been analyzed as trimethylsilyl ether derivatives of the intact lipid molecules[83,84] but better results are probably achieved by acetylation of diacylglycerols (as described above) produced by hydrolysis with H_2SO_4.[76,85] In fact $AgNO_3$ TLC is usually preferred for studies of galactolipids while the main use of GLC has been for the analysis of phosphatidylglycerol molecular species. The acetylated diacylglycerols can be separated according to fatty acid chain lengths and numbers of double bonds on a GLC column, e.g., 2%

[80] F. Haverkate and L. L. M. Van Deenen, *Biochim. Biophys. Acta* **106,** 78 (1965).

[81] M. Kito, *Eur. J. Biochem.* **54,** 55 (1975).

[82] R. O. Arunga and W. R. Morrison, *Lipids* **6,** 768 (1971).

[83] G. Auling, E. Heinz, and A. P. Tulloch, *Hoppe-Seyler's Z. Physiol. Chem.* **352,** 905 (1971).

[84] J. P. Williams, G. R. Watson, M. Khan, S. Leung, A. Kuksis, O. Stachnyk, and J. J. Myher, *Anal. Biochem.* **66,** 110 (1975).

[85] M. Noda and N. Fujiwara, *Biochim. Biophys. Acta* **137,** 199 (1967).

Silar 10C on Gaschrom Q run at 255° and 40 ml min^{-1}.[86] Alternatively, trimethylsilyl derivatives are analyzed by a method based on that of Myher and Kuksis[87] and modified for chloroplast phospholipids by Lynch and Thompson.[88,89] In this case the diacylglycerols are incubated at 35° for 30–60 min in pyridine : hexamethyldisilazone : trimethylchlorosilane (9 : 3 : 1, v/v/v; Sylon HTP from Supelco, Bellefonte, PA). The preparation is dried and the derivatives taken up in petroleum ether at 30–60° for GLC using split injection (split ratio set at 10 : 1 to 20 : 1) onto a 10 × 0.25 mm internal diameter open tubular glass column coated with SP2330 (Supelco). A N_2 head pressure of 0.5 kg cm^{-2} is maintained and the column temperature programmed to hold 200° for 5 min before increasing to 250° at 10° min^{-1}. Injector and detector temperatures are set at 270 and 300°, respectively.

HPLC

The separation of molecular species on reversed-phase HPLC columns (C_{18}-alkylated silicas, 5- or 10-μm particle size) has been successfully developed for galactolipids[90,91] and phospholipids.[26,73,92] There is no need for derivatization and an analysis is typically completed within 60 min. MGDG and DGDG molecular species are separated with a solvent mixture of methanol : water, 95 : 5 (v/v)[90,91] and PG with 30 m*M* choline chloride in methanol : water : acetonitrile, 70 : 5 : 25 (v/v).[93] However, although there are various alternatives for the detection of eluted lipids (see HPLC of Lipid Classes) the identification and quantification is best done by collection of fractions and GLC analysis of FAMEs.

Determination of Acyl Chain Position

Lipase. The acyl chain at position one of the glycerolipids (MGDG, DGDG, SQDG, and PG) can be determined by its removal with a stereospecific enzyme, such as pancreatic lipase (EC 3.1.1.3, triacylglycerollipase), followed by GLC analysis of the released fatty acid and lysolipid

[86] N. Murata, *Plant Cell Physiol.* **24,** 81 (1983).
[87] J. J. Myher and A. Kuksis, *Can. J. Biochem.* **60,** 638 (1982).
[88] D. V. Lynch and G. A. Thompson, Jr., *Plant Physiol.* **74,** 193 (1984).
[89] D. V. Lynch and G. A. Thompson, Jr., *Plant Physiol.* **74,** 198 (1984).
[90] R. Yamanchi, M. Kojima, M. Isogai, K. Kato, and Y. Ueno, *Agric. Biol. Chem.* **46,** 2847 (1982).
[91] D. V. Lynch, R. E. Gunderson, and G. A. Thompson, Jr., *Plant Physiol.* **72,** 903 (1983).
[92] G. M. Patton, J. M. Fasulo, and S. J. Robins, *J. Lipid Res.* **23,** 190 (1982).
[93] H. A. Norman, C. McMillan, and G. A. Thompson, Jr., *Plant Cell Physiol.* **25,** 1437 (1984).

product. Other sources of this specific hydrolytic activity are the molds *Rhizopus arrhizus* and *R. delemar*. These enzyme preparations may contain contaminants so it is usual to compare the original lipid with the monoacylglycerolipid product rather than rely on the analysis of the fatty acid fraction. With PG it is possible to hydrolyze the ester bond at position two rather than position one using snake venom phospholipase A_2 (EC 3.1.1.4) from *Crotalus adamanteus* or *Ophiophagus hannah*. Full details of specific incubation conditions which have been used successfully with the position one lipases or phospholipase A_2, are given in the literature for MGDG and DGDG,[94-96] SQDG,[95,97] and PG.[80,95] The incubation medium usually consists of a buffer at the pH optimum of the enzyme (pH 7.0 to 8.0, e.g., Tris/HCl), a solvent or detergent (e.g., 0.5% Triton X-100) to assist dispersal of the lipid substrate, and 5 to 10 m*M* $CaCl_2$ to increase activity. The precise ratio of lipid substrate to lipase is dependent on the activity of the enzyme preparation available ranging from 2 : 1 to 20 : 1 (w/w) and the reaction time is usually within 60 min at 20°. Precautions against fatty acid degradation can include flushing with N_2 and inclusion of bovine serum albumin (1 mg/10 mg lipid). The reaction can be stopped by acidification to pH 4.0, or below, with acetic or sulfuric acid. A lipid fraction is obtained by the extraction into chloroform : methanol prior to TLC to separate the free fatty acids from the monoacylglycerolipid components. A typical solvent mixture is hexane : diethyl ether : formic acid, 60 : 40 : 2 (v/v). The fatty acid remaining attached to the lipid can then be determined by GLC analysis of a derivative of this molecule[83,84] or a preparation of FAMEs (see previous section).

GLC–Mass Spectrometry. The equipment is not always readily available although this can be a powerful technique for lipid molecular species analyses. This is because it can determine the positional distribution of fatty acids at the same time as separating molecular species. The diacylglycerol moiety must be obtained by phospholipase C treatment (as in the section on GLC derivatives) and is then analyzed as acetyl[81,86,98] or *tert*-butyldimethysilyl[99] derivatives. Besides the details given in these references, a description of the technique of mass spectrometry and identification of spectra of lipids has been presented in other chapters in this

[94] M. Noda and N. Fujiwara, *Biochim. Biophys. Acta* **13,** 199 (1967).

[95] W. E. Fischer, E. Heinz, and M. Zeus, *Hoppe-Seyler's Z. Physiol. Chem.* **354,** 1115 (1973).

[96] R. Safford and B. W. Nichols, *Biochim. Biophys. Acta* **210,** 57 (1970).

[97] A. P. Tulloch, E. Heinz, and W. Fischer, *Hoppe-Seyler's Z. Physiol. Chem.* **354,** 879 (1973).

[98] K. Hasegawa and T. Suzuki, *Lipids* **10,** 667 (1975).

[99] J. J. Myher, A. Kuksis, L. Marai, and S. K. F. Yeung, *Anal. Chem.* **50,** 557 (1978).

series.[100,101] Compared with the usual resolution of components obtained by GLC, the use of GLC-mass spectrometry can often give poor separations unless high temperatures and temperature gradients are employed. In addition to this, the results of a complete molecular species analysis can be very complex and it is often preferable to carry out a preliminary separation by $AgNO_3$ TLC, or to compare analyses of samples before and after hydrogenation so as to assist with peak identifications.

Complete Structural Analyses

This chapter has reviewed commonly used methods for identifying and quantifying polar lipids, and their component fatty acids, which occur in chloroplast membranes. Although reference has been made briefly to mass spectrometry, it is beyond the scope of this chapter to deal in any depth with complete structural analyses of lipids. For those interested in this aspect we recommend specialized texts such as that by D. Chapman.[102]

Quantification and Expression of Results

GLC

After determination of fatty acid content in a lipid sample by GLC, as described previously, a simple calculation can be made to give the total weight, or moles of lipid; this relies on the knowledge that the main polar lipid classes of chloroplast membranes have two fatty acids per molecule (thus: total lipid = fatty acid content × 0.5). Interconversion between weight and moles can be made using 827 as an approximate average molecular weight of thylakoid lipids.

Radioactive Labeling in Vivo

The detection of relatively low amounts of specific lipids is made easier if a radioactive isotope is incorporated prior to membrane preparation and lipid analyses. This approach can also be used to determine absolute amounts if the organism is grown so that the sole source of an element (C, P, S) is a radioactive isotope at a known and constant specific radioactivity [for ^{14}C: moles lipid = Ci in isolated lipid × (specific radioac-

[100] J. A. McCloskey, this series, Vol. 14, p. 382.

[101] R. A. Hites, this series, Vol. 35, p. 348.

[102] D. Chapman, "The Structure of Lipids by Spectroscopic and X-ray Techniques." Methuen, London, 1965.

tivity as Ci/mol $CO_2)^{-1}$ × (number carbon atoms in the lipid)$^{-1}$], where Ci is the radioactive disintegration rate as curies, 3.7×10^{10} disintegrations/sec. The chloroplast membrane must be isolated from the organism and both the growth and fractionation procedures are most easily carried out with algal cultures. However, it should be noted that to use this approach it must be assumed that there is no significant sublethal radiation damage and that biosynthetic pathways do not discriminate between radioactive and nonradioactive molecules. Indeed, negligible discrimination has been reported between ^{12}C and ^{14}C in the photoassimilation of CO_2.[103] An example of the use of this method is the estimation of small amounts of SQDG tightly bound to the coupling factor complex isolated from chloroplast thylakoids of *Dunaliella salina*.[8]

Other Methods of Quantification

The choice of alternative methods of quantitative lipid analysis can depend on the equipment available for routine measurements of the large number of samples which are often produced. The polar group of chloroplast membrane lipids provides the opportunity to use standard quantitative analyses of galactose, sulfur, and phosphorus,[104] which depend on spectrophotometric measurements. Other methods of sample digestion for phosphorus determination[105–107] are worth considering because they avoid the potentially hazardous use of perchloric acid. Flame ionization detection after chromatography requires specialized equipment as does scanning densitometry and photographic analysis of TLC plates after charring with acid (see section on TLC analysis of lipid classes). Approaches which also involve analysis after acid digestion of the lipid are spectrophotometric analysis at 375 nm[108] and measurements in a scintillation counter after suspension of the charred sample in an emulsifier–scintillator mixture.[109]

Although derivatization can add a time-consuming step to an analysis it can also be an answer to the problems of quantitative detection during HPLC. Phenacyl and naphthacyl derivatives of fatty acids[56] or benzoyl derivatives of galactolipids[110] are examples of UV-absorbing compounds. Derivatization can often be worthwhile if it increases sensitivity of detec-

[103] E. W. Yemm and R. G. S. Bidwell, *Plant Physiol.* **44,** 1328 (1969).
[104] J. C. Dittmer and M. A. Wells, this series, Vol. 14, p. 482.
[105] C. G. Duck-Chong, *Lipids* **14,** 492 (1979).
[106] E. K. Ryu and M. MacCoss, *J. Lipid Res.* **20,** 561 (1979).
[107] J. C. M. Stewart, *Anal. Biochem.* **104,** 10 (1980).
[108] J. B. Marsh and D. J. Weinstein, *J. Lipid Res.* **7,** 574 (1966).
[109] J. H. Shand and R. C. Noble, *Anal. Biochem.* **101,** 427 (1980).
[110] K. Shimomura and Y. Kishimoto, *Biochim. Biophys. Acta* **794,** 162 (1984).

tion and in this respect radioactive reagents could be very useful. This can be done with [^{14}C]acetic anhydride using a procedure such as that given in the section on molecular species analysis by silver nitrate TLC.[111] Limited use of immunological procedures has been made for detection of specific lipid classes[112,113] and with future improvements this approach could be more widely used for both detection and quantification.

Expression of Lipid Quantities

Amounts of a lipid class or fatty acid in a sample can be expressed as an absolute amount and this is usually as a weight or number of moles. However, to express this value in a way which allows comparison with lipids from another membrane preparation it must be given as a quantity of lipid in relation to a standard amount of a second component of the membrane samples, i.e., a ratio. It must be borne in mind that, particularly when comparing samples, the value of a ratio is dependent on the amount of both the lipid and the second component in the sample. For chloroplast thylakoids it is convenient to use chlorophyll as the second component because it is fully extracted into the lipid sample and amounts of total chlorophyll, or chlorophylls *a* and *b* alone, are easily measured. However, the lipid : chlorophyll ratio is not relevant for studies of envelope membranes because chlorophyll is not present. Even with thylakoids, using chlorophyll has a disadvantage in not being a direct indication of the total amount of membrane. These problems are solved in part by measuring protein content but this is not entirely satisfactory. One reason is that measurement of membrane protein can be slow or inaccurate, particularly when detergents have been used in membrane fractionation. Another difficulty is in being sure how much protein is a true membrane component because the presence of extrinsic proteins can depend on the method of membrane preparation. The same problems are relevant when use of membrane dry weight is considered and no ideal solution is available. Another approach has been to express results as lipid per plastid and this has proved useful for developmental studies.[114] A great deal of interest in lipid studies concerns the properties of the lipid matrix rather than the membrane as a whole. Therefore a relevant additional method of expression for individual lipids is the ratio of a lipid to the total lipid content of the membrane, i.e., the lipid as a proportion, or

[111] M. W. Banschbach, R. L. Geison, and J. F. O'Brien, *Anal. Biochem.* **59,** 617 (1974).
[112] A. Radunz, *Z. Naturforsch., C: Biosci.* **35C,** 1024 (1980).
[113] K. Gounaris, M. Lambillote, J. Barber, K. Muehlethaler, and F. Jay, *in* "Structure, Function and Metabolism of Plant Lipids" (P.-A. Siegenthaler and W. Eichenberger, eds.), p. 485. Elsevier, Amsterdam, 1984.
[114] B. M. Leese and R. M. Leech, *Plant Physiol.* **57,** 789 (1976).

percentage, of the total. When lipid compositions are expressed in this way it is helpful if the lipid : chlorophyll ratio is given for the arithmetic total of the lipid types from the sample. Of course, a distinction has to be drawn between this figure and the amount of total lipid determined before chromatographic separations of lipid classes or molecular species were begun. Any difference between the two values will indicate losses or errors during the chromatography.

Expression of Fatty Acid Content

The amounts of fatty acids in a lipid extract are usually calculated as moles, or weight, using the internal standard method for analysis of FAMEs (see the section, GLC Detection and Estimation of FAMEs). After addition of these values the total amount of fatty acid in a sample is usually expressed on the basis of chlorophyll or protein in the preparation from which lipid was extracted. The individual fatty acids are given as percentages of this total and in some cases a single index of fatty acid unsaturation is used to supplement or even replace the complete data on composition. These indices make comparison of samples easier and can often be more readily related to physical properties of the lipids than the detailed results. The most commonly used index is simply the average number of double bonds in a lipid molecule (two fatty acids) and, in some cases, average number in a fatty acid. In chloroplast membranes the range of values will be 0 to 6 and 0 to 3, respectively.

Average number of double bonds per lipid molecule =

$$2\left[\frac{(\%\text{ monoenoic acids}) + (2 \times \%\text{ dienoic acids}) + (3 \times \%\text{ trienoic acids})}{100}\right]$$

In some studies a distinction needs to be made between those fatty acid with and those without double bonds. The relevant index is the total anenoic (saturated) fatty acids as a percentage of the total fatty acids. In some cases a ratio of unsaturated to saturated fatty acids is calculated, but this should be used with caution because it can give deceptively large differences between samples when fatty acid percentage compositions are in fact only slightly different. This is usually the case with most chloroplast lipids because of their very high proportions (over 80%) of fatty acids which are unsaturated.

Unidentified Lipids

In this chapter we have concentrated on techniques which can be used to determine the identities and quantities of polar lipids which are known to be located in chloroplast membranes. With greater lipid depletion during preparation of membrane fractions, the less common lipids can become relatively more abundant,[8] and it is likely that lipids will be found which previously have not been recognized as components of chloroplast membranes. Suitable methods for identification and quantification of a wide range of different lipids will be found in several general texts, including those by Christie,[54] Kates,[115] Sastry,[116] and Volumes 14 and 35 of this series.

Acknowledgments

We are grateful to our colleague K. Gounaris for advice and careful reading of the manuscript and the Agricultural and Food Research Council for financial support. D.J.C. wishes to thank J. P. Williams and M. Khan for introducing him to the techniques developed by them for analysis of plant lipids.

[115] M. Kates, "Techniques of Lipidology." North-Holland Publ., Amsterdam, 1972.
[116] P. S. Sastry, *Adv. Lipid Res.* **12,** 251 (1974).

[30] Isolation and Reconstitution of Thylakoid Lipids

By SALLIE G. SPRAGUE and L. ANDREW STAEHELIN

Introduction

Many methods are currently available for reconstituting a variety of lipids into bilayer vesicles.[1] All the methods allow roughly cylindrical lipids, e.g., phosphatidylcholine (PC) and digalactosyldiglyceride (DG), to form bilayer liposomes.[2] In contrast, molecules which are more "cone shaped," e.g., phosphatidylethanolamine (PE) or monogalactosyldiglyceride (MG), form hexagonal II or inverted micellar phases in aqueous media.[3] PC has been used for reconstitution of many proteins because it so readily forms liposomes. For mammalian systems where PC often constitutes the major or a significant portion of the membrane lipids this practice is convenient. For thylakoids, where MG and DG constitute 51 and 26%, respectively, and PC only 3% of the membrane lipid,[4] we need

[1] F. Szoka and D. Papahadjopoulos, *Annu. Rev. Biophys. Bioeng.* **9,** 467 (1980).
[2] P. R. Cullis and B. deKruijff, *Biochim. Biophys. Acta* **559,** 399 (1979).
[3] S. M. Gruner, P. R. Cullis, M. J. Hope, and C. P. S. Tilcock, *Annu. Rev. Biophys. Biophys. Chem.* **14,** 211 (1985).
[4] R. Douce and J. Joyard, *Adv. Bot. Res.* **7,** 1 (1979).

to consider more seriously the price of this convenience. It has been reported that (1) significant membrane destruction occurs after the addition of phospholipids to thylakoids,[5,6] (2) MG promotes and PC inhibits energy transfer from light-harvesting complexes (LHC) to photosystems I and II (PSI, PSII) in isolated thylakoid fragments,[7] and (3) isolated LHC and PSII cores adopt "more native" configurations in DG liposomes than either complex does in PC liposomes.[8] To further investigate the structure and function of reconstituted thylakoid proteins we felt it was necessary to develop a rapid and reproducible method for obtaining bilayer vesicles from thylakoid lipids. Simple hydration methods have been used to show that pure MG forms hexagonal II phases and DG forms bilayers,[9] and that a 2:1 mixture of MG and DG forms a complex mixture of hexagonal II tubes and inverted micelles.[10–14] However, the reliability of such methods has been questioned.[15,16] We describe here several methods by which various combinations of thylakoid lipids can be reconstituted into bilayer and mixed bilayer and hexagonal II phases. For ease in preparation and reproducibility of results we have found the reversed-phase evaporation method[15] (RPEV) to be by far the most efficient in producing pure bilayer liposomes from mixtures of thylakoid lipids with $\geq$30% MG.[16]

Thylakoid Isolation

Commercial spinach leaves were washed and destemmed prior to homogenization in a Waring blender in cold buffer (0.05 *M* NaH_2PO_4, 0.1 *M* KCl, 0.3 *M* sucrose, pH 7.2). The homogenate was filtered through 8 and 12 layers of cheesecloth and centrifuged at 300 *g* for 2 min at 4° to remove unbroken cells and large debris. Crude thylakoids were collected from the

[5] C. O. Siegel, A. E. Jordan, and K. R. Miller, *J. Cell Biol.* **91,** 113 (1981).

[6] P. A. Millner, J. P. Grouzis, D. J. Chapman, and J. Barber, *Biochim. Biophys. Acta* **722,** 331 (1983).

[7] D. Siefermann-Harms, J. W. Ross, K. H. Kaneshiro, and H. Yamamoto, *FEBS Lett.* **149,** 191 (1982).

[8] S. G. Sprague, E. L. Camm, B. R. Green, and L. A. Staehelin, *J. Cell Biol.* **91,** 113 (1981).

[9] G. G. Shipley, J. P. Green, and B. W. Nichols, *Biochim. Biophys. Acta* **311,** 531 (1973).

[10] P. J. Quinn and W. P. Williams, *Biochim. Biophys. Acta* **559,** 399 (1979).

[11] A. Sen, A. P. R. Brain, P. J. Quinn, and W. P. Williams, *Biochim. Biophys. Acta* **737,** 223 (1983).

[12] A. Sen, W. P. Williams, A. P. R. Brain, M. J. Dickens, and P. J. Quinn, *Nature (London)* **293,** 488 (1981).

[13] A. Sen, W. P. Williams, A. P. R. Brain, and P. J. Quinn, *Biochim. Biophys. Acta* **685,** 297 (1982).

[14] A. Sen, W. P. Williams, and P. J. Quinn, *Biochim. Biophys. Acta* **663,** 380 (1981).

[15] S. G. Sprague and L. A. Staehelin, *Plant Physiol.* **75,** 502 (1984).

[16] S. G. Sprague and L. A. Staehelin, *Biochim. Biophys. Acta* **777,** 306 (1984).

300 *g* supernatant by centrifugation at 1000 *g* for 10 min (4°) and washed once in the same buffer. For ease of subsequent lipid extraction thylakoids from 300–400 g of prepared leaves were resuspended in 30 ml of buffer.

Lipid Extraction

Freshly prepared thylakoids in buffer (30 ml = 1 vol) were extracted[17] with 10 vol of cold chloroform/methanol (1 : 2) and filtered through four layers of cheesecloth. The residue was washed with 3.3 vol of chloroform. Five volumes of aqueous 0.1 *M* NaCl was added to the combined organic phase to give chloroform/methanol/0.1 *M* NaCl = 1/1/0.9.[17] NaCl was used to increase the yield of sulfolipid (SL) in the organic phase. The organic phase was collected and the aqueous NaCl phase washed twice with 1 to 2 vol of chloroform. The combined organic fractions were concentrated by rotary evaporation (Buchler Instruments, NJ) to 40–60 mg lipid/ml and stored under N_2 at −20°.

Purification of Lipid Classes

The concentrated extract was separated by column chromatography on silicic acid (SIL-LC, Sigma Biochemicals, St. Louis, MO) according to a procedure modified from that of Rouser *et al.*[18] For preparative work we used a 5.5-cm (o.d.) glass column containing 60–70 g silicic acid. This gives a column height of 6–8 cm and allows separation of 0.8 to 0.9 g of lipid in a single run (silicic acid/lipid ≥75).[18] The columns were gravity fed and flow rates were usually 3–5 ml/min. Alternatively, 25 g of silicic acid was used to make an analytical column 2.5 cm (o.d.) × 17 cm. Solvents were bubbled with N_2 before use. Elution of the column was as follows: Wash with 2 column volumes of chloroform, load sample. Elute pigments with 10 vol of chloroform. Elute MG with 5 vol of chloroform : acetone (1 : 1). Elute DG and SL with 40 vol of acetone. DG comes off at 10–20 vol and SL 30–40 vol. Elute remaining PL with 10 vol of methanol.

A total nonpigmented lipid fraction was obtained by eluting the bulk of the pigments with 10 vol of chloroform and then eluting all the remaining lipids with chloroform : methanol (1 : 1, v/v). Similarly, an MG-depleted sample was obtained by removing the pigments with chloroform and MG with chloroform : acetone before eluting the remaining lipids with chloro-

[17] C. F. Allen and N. Good, this series, Vol. 23, p. 523.

[18] G. Rouser, G. Kritchevsky, and A. Yamamoto, *in* "Lipid Chromatographic Analysis" (G. W. Marinetti, ed.), p. 713. Dekker, New York, 1976.

form : methanol. Samples were concentrated by rotary evaporation and stored under N_2 at $-20°$.

Analysis of Lipids

Routine identification of lipids was carried out by thin-layer chromatography (TLC) on silica gel G with inorganic binder (Rediplates, Fisher Scientific). The plates were always prerun in the appropriate solvent to wash a yellowish colored contaminant to the top of the plate. Activation was for 1 hr at 120°. For two-dimensional separation we used the solvent system of Allen and Good,[17] which employs a mixture of chloroform/methanol/water (65 : 25 : 4) in the first direction and chloroform/methanol/concentrated ammonium hydroxide (65 : 35 : 5) in the second. Iodine vapor was used to locate nonpigmented material. The iodinated plates could be wrapped in plastic film and photocopied for a permanent record. Initial identification of galactolipids was made using the α-naphthol/sulfuric acid spray reagent[19] for sugars. We observed pinkish spots for MG and DG, a purple spot for SL, and yellowish spots for PLs. DG standard was obtained from PL Biochemicals (Milwaukee, WI) or Sigma (St. Louis). Phospholipids were stained using ammonium molybdate/stannous chloride spray reagents.[19]

X-Ray energy spectrometry in a Kevex 5500 series spectrometer attached to a Cambridge S4 scanning electron microscope was used to confirm the presence of sulfur (SL), phosphorus (PL), and magnesium (chlorophyll) in column fractions (Fig. 1). An aliquot (20–60 μl) of a concentrated column fraction in chloroform was applied to a carbon support stub, air dried, and mounted in the specimen chamber of the scanning electron microscope. The stage was oriented 55° to the electron beam.[16] Since this method can be used only to detect atoms larger than sodium ($Z = 10$), its utility is limited to finding the right fractions within a specific group after you have characterized the elution profile by TLC. It proved useful for locating SL after we knew that the DG would be off the column. The major advantage is the speed with which one can analyze a single sample (~5 min). The method is useful only for qualitative analysis.

Electron Microscopy

Samples of reconstituted lipids were routinely dip frozen in (1) liquid nitrogen-cooled Freon 12 after glycerination to 35% (v/v) or (2) liquid

[19] M. Kates, "Techniques in Lipidology." Am. Elsevier, New York, 1972.

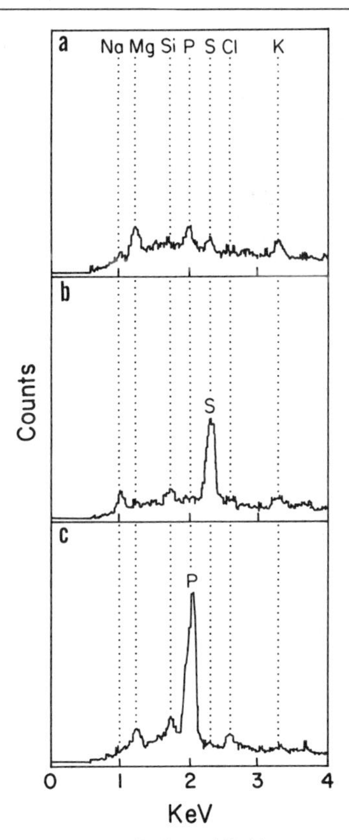

FIG. 1. X-Ray energy spectrograms of (a) total lipid extract applied to column, (b) pooled fractions containing sulfur (SL), and (c) pooled fractions containing P (PL). Data were collected until 1000 counts were obtained at (a) 404 sec, (b) 223 sec, and (c) 279 sec. Atoms responsible for the observed peaks are marked along the *x* axis.

nitrogen-cooled propane (nonglycerinated samples). Alternatively samples were jet frozen with and without cryoprotectant.[16] The best micrographs were obtained from glycerinated samples. Qualitatively the same structures were observed in glycerinated and nonglycerinated samples.[16] Freeze fracture and replication of samples was carried out in a Balzers 360M freeze-etch device equipped with electron guns. Standard fracture and replication was done at −110°, etching was done at −103°. Replicas were cleaned in commercial bleach, distilled water, and chloroform/methanol (1 : 1, v/v). Samples were observed in a JEOL 100C (60 or 80 keV) or a Hitachi 300 (75 keV) electron microscope.

Reconstitution of Thylakoid Lipids

Mixtures and individual classes of thylakoid lipids were transferred to an aqueous medium by (1) a simple hydration procedure in distilled water, buffer (0.05 *M* sodium phosphate, pH 7.2, 0.1 *M* KCl), or 0.1 *M* NaCl solution, (2) solubilization in Triton X-100 followed by detergent removal with Biobeads, or (3) RPEV from Freon 11 as outlined in Fig. 2. All samples were dried under a stream of N_2 to remove chloroform. Aqueous samples were sealed with an N_2 atmosphere during heating and sonication.

Simple Hydration

Aqueous solutions were added directly to the dried lipid film. Heating times were varied from 5 to 10 min at 30° to 60 min at 80°. Sonication was in a bath for 15–45 min at room temperature. Typically 2–10 mg of lipid was dried and hydrated in 1 ml aqueous solution. MG was selectively

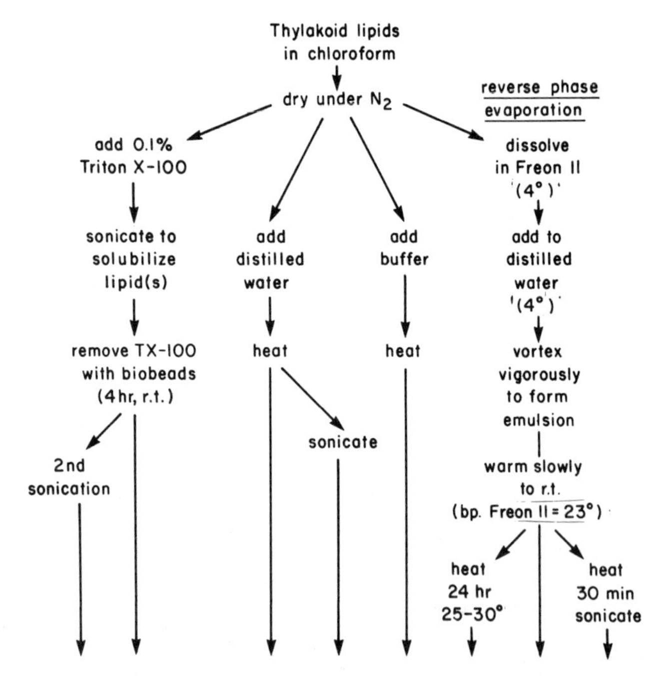

FIG. 2. Methods for formation of lipid structures. Details in text. r.t., Room temperature. From Sprague and Staehelin.[15]

retained on the glass[16,20] after extensive heating and sonication, even in the total nonpigmented lipid sample.

Detergent Solubilization

Lipids (2, 10, 20 mg) were solubilized in 1 ml of 0.1% Triton X-100 (Sigma) and sonicated with two or three 30-sec bursts with a Branson microprobe. Power was set just below that which caused cavitation around the probe. Samples were maintained in a water jacket (room temperature) to prevent excessive heating. Biobeads (Bio-Rad, Richmond, CA) were prepared according to the manufacturer's directions. Samples were incubated with the beads for 4 hr at room temperature with constant stirring. The samples were removed from the beads with a Pasteur pipet after centrifugation for 10 min at maximum speed in a table top centrifuge (New Brunswick). Samples containing more than 40% MG formed large clumps during detergent removal. These clumps could be used for freeze-fracture electron microscopy but were not suitable for reconstitution of proteins. The second sonication was for 10–45 min in a bath sonicator and served to disperse some of the smaller clumps, or to reduce the size of bilayer vesicles.[16]

Reversed-Phase Evaporation

To accommodate larger quantities of lipid as well as to demonstrate the suitability of Freon 11 (trichlorofluoromethane, bp 23°; Matheson, East Rutherford, NJ) as solvent for reversed-phase evaporation[15] we modified the method of Cafiso *et al.*[21] as follows. MG and DG were soluble to 40–50 mg/ml, the highest concentrations we tested, in Freon 11. For routine preparation of liposomes, 10 mg of the nonpigmented lipids was dissolved in 1 ml of ice-cold Freon 11. Ice-cold distilled water (2 ml) was added and an emulsion was formed by vortexing. The sample was warmed slowly (5–10 min) to room temperature, vortexing continuously to maintain the emulsion. The rate of warming can be varied by the positioning of the experimenter's fingers along the glass test tube. Holding more than the very top of the tube warmed the sample too quickly (2 min), resulting in vigorous boiling that spattered lipid over the inside of the tube (recovery was then difficult in some cases). Any remaining traces of dis-

[20] S. S. Sprague and L. A. Staehelin, *in* "Biosynthesis and Function of Plant Lipids" (W. W. Thomson, J. B. Mudd, and M. Gibbs, eds.), p. 144. Am. Soc. Plant Physiol., Rockville, Maryland, 1983.

[21] D. S. Cafiso, H. R. Petty, and H. M. McConnell, *Biochim. Biophys. Acta* **649**, 129 (1981).

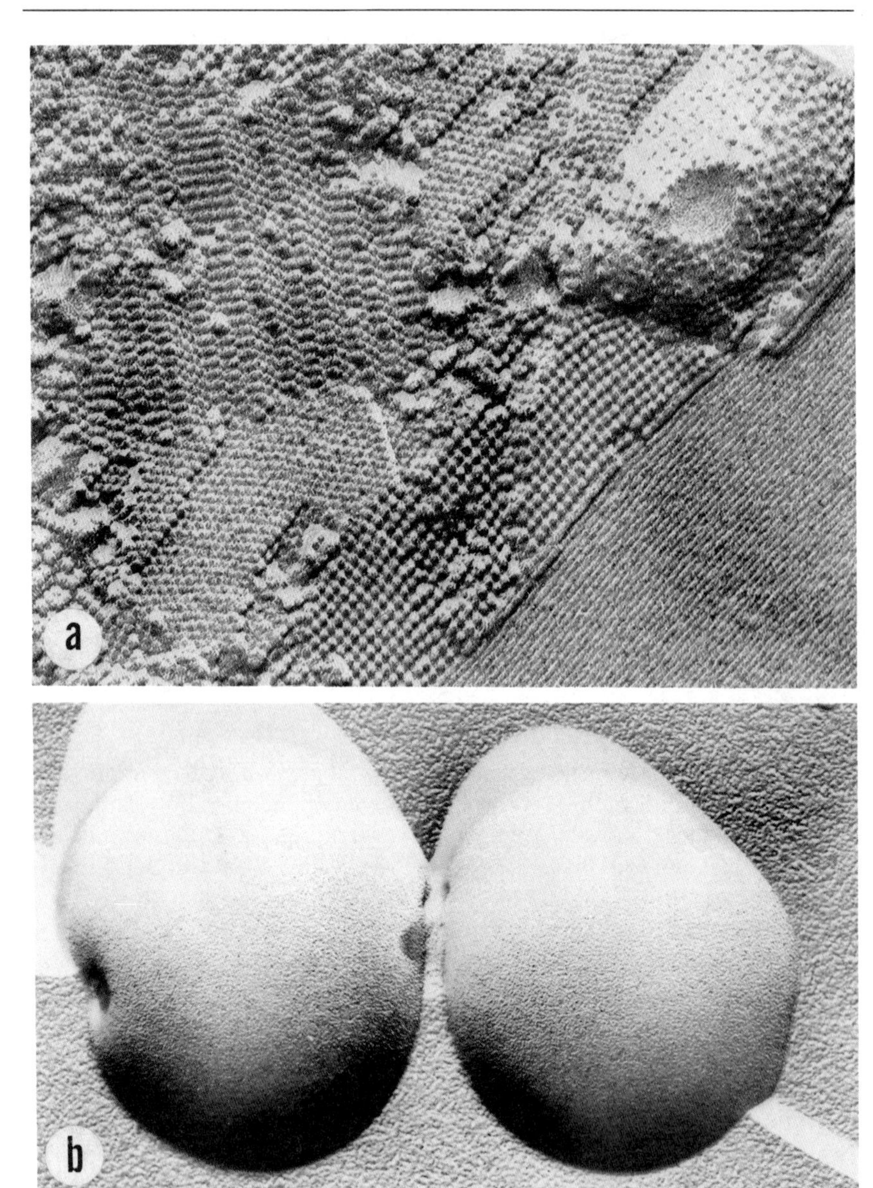

FIG. 3. Freeze-fracture electron micrographs showing (a) hexagonal II spheres and tubes in a 50/50 mixture of MG/DG after reconstitution by the detergent solubilization method and (b) primarily bilayer structures seen after reconstitution of the same mixture by RPEV. Magnifications: (a) ×125,000, (b) ×65,000.

solved Freon 11 were removed by placing the sample tube in hot tap water for 5 min or under house vacuum for 5–15 min. To make more concentrated liposome solutions (10 mg/ml), we repeated the above procedure using the freshly prepared and cooled liposomes (2 ml) instead of the distilled water. This repetitive method produces liposomes with fewer nonbilayer (e.g., hexagonal II) structures than observed when 20 mg of lipid is used in the original 1 ml of Freon 11. Changing the ratio of Freon to water (2 : 1 in our experiments) did not alter the results as long as the emulsion was maintained (by vortexing) during warming.

General Comments

All methods are suitable for making bilayer liposomes from DG. None of them will force MG into anything other than hexagonal II forms.[16] Obtaining sufficient amounts of MG in the aqueous phase for electron microscopy was much simpler with reversed-phase evaporation. For binary mixtures of MG and DG, the same structures were observed, qualitatively, after Triton X-100 solubilization and removal and reversed-phase evaporation. Quantitatively it appeared that reversed-phase evaporation forced (allowed?) more MG into bilayers. At 50/50 MG/DG (by weight) Triton-prepared samples were composed of extensive hexagonal II tubes and particles (Fig. 3a) whereas RPEV-prepared samples (Fig. 3b) contained primarily bilayer vesicles with "fusion pores" and only few "lipidic particles."[16,20] However, even when RPEV is employed it appears that the full compliment of thylakoid lipids is required to obtain pure bilayers with an MG : DG ratio of 2 : 1, the same ratio found in native membranes (51% MG : 26% DG : 7% SL : 9% phosphatidylglycerol : 3% PC).

[31] Long-Chain Fatty Acid Synthesis and Utilization by Isolated Chloroplasts

By GRATTAN ROUGHAN

Introduction

A considerable measure of the progress in our understanding of plant lipid metabolism over the last decade may be attributed to detailed studies on the light-dependent incorporation of [1-^{14}C]acetate into the lipids of isolated chloroplasts. The fatty acid synthase in leaves of higher plants comes conveniently packaged within chloroplasts,[1] which are relatively easily isolated from a number of plants, and rates of [1-^{14}C]acetate incorporation into long-chain fatty acids by chloroplasts isolated from spinach (*Spinacia oleracea*)[2-4] and pea (*Pisum sativum*)[5,6] are now routinely reported in the range 1–2 μmol/hr/mg of chlorophyll. This is equivalent to the estimated rates of fatty acid synthesis *in vivo*.[7,8] However, there are relatively few plants from which chloroplasts retaining light-dependent, biosynthetic activities may be isolated, and success with a particular plant species in one laboratory does not guarantee a similar success in another laboratory. Subtle differences in the way in which plants are grown may affect the biosynthetic capabilities of chloroplasts *in vitro* in ways that are not yet understood. Spinach may be less sensitive than other plants (e.g., peas) to minor environmental differences and remains the species most likely to yield highly active chloroplasts at any given location.[9]

Preparation of Plant Material

Chloroplasts isolated from rapidly expanding leaves of young (6–8 weeks), well-grown plants are most likely to exhibit high rates of fatty acid synthesis from [1-^{14}C]acetate. The current water status of a plant may have a profound effect upon biosynthetic activities of chloroplasts isolated from that plant.[10] Too little water or too much unaerated water around the roots produce similar symptoms in whole plants, namely wilted leaves and starch accumulation, and it seems likely that chloroplasts from stressed plants will have suboptimal activities long before these obvious symptoms appear. Therefore, to avoid the possibility of even transient water stress, it is recommended that plants be grown in aerated solution culture. Containers for solution (or water) culture of plants may be made from domestic bowls measuring 30 × 30 × 12 cm

[1] J. B. Ohlrogge, D. N. Kuhn, and P. K. Stumpf, *Proc. Natl. Acad. Sci. U.S.A.* **76,** 1194 (1979).

[2] P. G. Roughan, R. Holland, and C. R. Slack, *Biochem. J.* **184,** 193 (1979).

[3] D. J. Murphy and D. A. Walker, *Planta* **156,** 84 (1982).

[4] J. Soll and P. G. Roughan, *FEBS Lett.* **146,** 189 (1982).

[5] E. Heinz and P. G. Roughan, *Plant Physiol.* **72,** 273 (1983).

[6] S. E. Gardiner, E. Heinz, and P. G. Roughan, *Plant Physiol.* **74,** 890 (1984).

[7] D. J. Murphy and R. M. Leech, *Plant Physiol.* **68,** 762 (1981).

[8] J. Browse, P. G. Roughan, and C. R. Slack, *Biochem. J.* **196,** 347 (1981).

[9] R. C. Leegood and D. A. Walker, *in* "Isolation of Membranes and Organelles from Plant Cells" (J. L. Hall and A. L. Moore, eds.), p. 185. Academic Press, London, 1983.

[10] J. S. Boyer, *Phytopathology* **63,** 466 (1973).

($l \times w \times h$) and fashioned in opaque plastic. Squares of a rigid, black plastic, 36 × 36 cm and about 3 mm thick, serve as covers for the bowls and provide support for the plants. They are drilled with arrays of 13 holes (7-mm diameter) evenly spaced within 24 × 24 cm, giving 3 rows of 3 holes and 2 rows of 2 holes, while an additional hole is drilled nearer to the edge of each cover to accommodate a plastic tube carrying air. The bowls hold 10 liters of nutrient solution (modified Hoaglands at half-strength) which is aerated using equipment available for domestic aquariums, and which is completely replaced at least weekly. Common-sense hygiene should be observed.

Seeds are germinated at 20–25° in vermiculite (it may be necessary to experiment with different sources and different grades of vermiculite to ensure a high germination percentage) under natural or artificial lighting and, after 2–3 weeks, the seedlings are transferred to solution culture taking care to minimize damage to roots and hypocotyls in the process. The roots are passed through the holes in the bowl covers (one seedling per hole) into the nutrient solution and the seedlings are held in place by small wads of plastic foam wedged loosely into the holes. After a further 2–3 weeks (amaranthus, *Amaranthus lividus*) or 3–4 weeks (spinach) at 20–25° in 10-hr days with light intensities of 500–600 μmol/m^2/sec, rapidly expanding leaves suitable for chloroplast isolation are produced in quantity. Spinach plants grown in <10-hr day will not flower nor will the young leaves contain starch—unless aeration of the nutrient solution is inadequate. Safflower (*Carmathis tinctoris*), quinoa (*Chenopodium quinoa*), and nightshade (*Solanum americanum*) have also yielded highly active chloroplasts when cultured in this system.[6] The most active chloroplasts seem to be isolated from plants grown in completely controlled environments.[2]

Pea seedlings are easier to raise and may be grown in natural lighting at all times of the year. However, the activities of pea chloroplasts isolated in different laboratories appear to vary considerably. Pea seeds (60 g) are imbibed overnight in running tap water (>10°) and then planted 2 cm deep into a bed of vermiculite or pumice/peat at least 5 cm in depth and which is kept thoroughly damp until the seedlings are about 5 cm tall (9–12 days). Alternatively, the imbibed peas are spread onto a 22 × 22 cm stainless steel mesh positioned over water in a plastic bowl, the whole enclosed in a plastic bag and incubated in the dark at 20–25° for 3–4 days. The emerging radicles will, with a little help, pass through the mesh and into the water whereupon the plastic bag is removed and the system transferred to the light for a further six to eight 10-hr days. The water is aerated over this latter period.

Chloroplast Isolation

Rates of lipid synthesis from [1-^{14}C]acetate are probably enhanced when chloroplasts are transferred intact from leaf cells to incubation mixtures as rapidly as feasible. Mechanical disruption of cells is recommended and procedures based on low ionic strength media[11,12] will, with practice, permit the rapid (5–7 min) and economical isolation of chloroplasts of 80–90% intactness from most of the plant species cited above. The following procedure is recommended for the preparation of chloroplasts, equivalent to 1.5–3 mg of chlorophyll, from spinach and peas. Leaves or shoots are harvested between 30 min and 2 hr after the beginning of the light cycle. About 20 g of lamina from spinach leaves, which are less than one-half fully expanded, is sliced into 2- to 3-mm-wide strips using a new razor blade and is plunged into 100 ml of semifrozen buffer. Pea shoots (20 g) are cut off immediately below the first pair of leaves, which are not yet fully unfolded, transferred to the semifrozen buffer, and chopped into smaller pieces with scissors. The homogenizing buffer consists of 0.33 *M* sorbitol, 2 m*M* HEPES–KOH, pH 7.8, 0.4 m*M* KCl, and 0.04 m*M* EDTA and is conveniently made up as required by combining 6 g of sorbitol, 0.8 ml of 0.25 *M* HEPES–KOH, pH 7.8, 40 μl of 1 *M* KCl, 40 μl of 0.1 *M* Na_3EDTA, and water to 100 ml. It is partially frozen in a glass vessel with a 5 × 5 cm cross section and a capacity of about 200 ml, e.g., a Dijon mustard jar or equivalent from the local supermarket. The leaf tissue is homogenized in this slush using two 3-sec bursts at three-quarter to full speed of the PTA 20S probe on a Polytron PT 10/35 homogenizer and the homogenate is filtered through two layers of Miracloth. The filtrate is divided between two 50-ml polycarbonate tubes and centrifuged at 4000 rpm for 30 sec in the HB-4 rotor in a Sorval RC-2B, superspeed centrifuge. The rotor is hand braked so that total time in the centrifuge is <1 min. Broken chloroplasts are gently rinsed from the surface of each pellet using 5 ml of fresh homogenizing buffer before the rest of each pellet is resuspended very gently in 2 ml of the same buffer using a 5-ml pipet with an outlet >1 mm in diameter to minimize the effects of shear forces. Another 15 ml of buffer is mixed into each tube and chloroplasts are repelleted as above. The supernatants are decanted and both tubes kept inverted while the interior walls are wiped clean with a tissue clasped in forceps. One pellet is resuspended very gently in exactly 1 ml of homogenizing buffer using a 2-ml pipet with an outlet >1 mm in diameter and is added to the other pellet which is resuspended in like manner. The volume of the chloroplast pellet is now measured in the 2-ml pipet and

[11] H. Y. Nakatani and J. Barber, *Biochim. Biophys. Acta* **461,** 510 (1977).

[12] Z. G. Cerović and M. Plesnicar, *Biochem. J.* **223,** 543 (1984).

additional buffer is added so that 1 ml of buffer is used to resuspend each 0.1 ml of chloroplast pellet. This will give close to 1 mg of chlorophyll/ml of resuspension for most chloroplast types. The yield of chloroplasts will depend primarily upon the efficiency of the homogenization.

Variations to This Procedure. The homogenizing buffer may be made to pH 6.5 using 2 m*M* MES–KOH in place of HEPES–KOH—but the washing/resuspending buffer should still be pH 7.8; BSA (0.1–0.2%, w/v) may be added to the original homogenizing buffer; the original filtrate may be underlain with 5 ml of 40% Percoll in washing buffer and the initial centrifugation extended to 60 sec.[13] This latter modification, in particular, may increase the proportion of intact organelles in the preparation and adds <2 min to the procedure. Chloroplasts prepared from amaranthus must be purified by density gradient centrifugation in Percoll[5] or by sedimentation through a cushion of Percoll[14] before exhibiting high biosynthetic activities. The latter procedure is probably more suitable for preparing chloroplasts in larger quantities.

Fatty Acid Synthesis from [1-^{14}C]Acetate

The labeled substrate is added to incubation media designed originally to support high rates of CO_2-dependent O_2 evolution by spinach chloroplasts.[15] A concentrated incubation medium is made up at least weekly and consists of 1.2 g of sorbitol, 2.5 ml of 0.25 *M* HEPES–NaOH, pH 8.0, 0.5 ml of 0.5 *M* $KHCO_3$, 0.5 ml of 0.1 *M* Na_3EDTA, 0.25 ml each of 0.1 *M* $MgCl_2$, 0.1 *M* $MnCl_2$, and 0.05 *M* K_2HPO_4, and water to make 10 ml. Basal reaction mixtures contain 100 μl of this concentrate, 90 μl of water, 10 μl of 5 m*M* Na-[1-^{14}C]acetate (0.4 μCi) and, to start the reaction, 50 μl of chloroplasts equivalent to 50 μg of chlorophyll. The specific radioactivity of the [1-^{14}C]acetate may be increased from 8 to about 60 Ci/mol when detailed analyses of weakly labeled constituents are envisaged. Reactions are carried out in 20 × 125 mm culture tubes which can be sealed with Teflon-lined screw caps. They are shaken in an illuminated water bath (e.g., Braun model VL166 with lighting system; Gilson model IGRP-14) at 25° with just sufficient vigor to ensure thorough mixing of the reactants. Light intensities at the level of the reactions should be at least 150 μmol/m^2/sec. Additions are made to the basal mixture by substituting 10 μl of, e.g., 12.5 m*M* CoA, 25 m*M* ATP, 10 m*M* *sn*-glycerol 3-phosphate, 5 m*M* UDPgalactose, 25 m*M* CTP, 0.2% (w/v) Triton X-100 for an equivalent

[13] W. R. Mills and K. W. Joy, *Planta* **148,** 75 (1980).

[14] C. L. D. Jenkins and J. J. Russ, *Plant Sci. Lett.* **35,** 19 (1984).

[15] T. Delieu and D. A. Walker, *New Phytol.* **71,** 201 (1972).

TABLE I
EFFECTS OF VARIOUS ADDITIONS TO INCUBATION MEDIA UPON THE LIPID CLASSES SYNTHESIZED FROM [1-^{14}C]ACETATE BY CHLOROPLASTS ISOLATED FROM SPINACH AND PEA[3,5,6,16–18]

Additions to basal medium	Isolated chloroplasts; lipid class predominantly labeled[a]	
	Spinach	Pea
None or CoA[b]	UFA	UFA
CoA + ATP	UFA, acyl-CoA	UFA, acyl-CoA
G3P + Triton X-100	DAG	UFA, PA
G3P + Triton X-100 + UDPgalactose	DGG	UFA, PA
G3P + Mg^{2+}	DAG, PA	UFA, PA
G3P + Mg^{2+} + CTP	DAG, PA, PG	UFA, PA

[a] DAG, 1,2-Diacylglycerol; DGG, diacylgalactosylglycerol; G3P, *sn*-glycerol 3-phosphate; PA, phosphatidic acid; PG, phosphatidylglycerol; UFA, unesterified fatty acids.
[b] Typical concentrations for these additions are given in the text.

amount of water (above). The phosphorylated compounds should be neutralized before they are added to the medium so that the final pH in the reactions is always between 7.8 and 7.9. Incubation mixtures are constituted before leaves are harvested so that freshly prepared chloroplasts may be added to start the reactions with a minimum of delay. Chlorophyll analysis is performed at some convenient time after reactions have been initiated. Reaction volumes may be increased to 0.5 ml in the 20 × 125 mm tubes and 25-ml Erlenmeyer flasks should be used for reaction volumes ≥1 ml. Acetate incorporation into long-chain fatty acids should be linear from time zero but with some chloroplasts (e.g., those from safflower), there may be a delay of 2–4 min before linearity is established. Rates of fatty acid synthesis *in vitro* are enhanced by the presence of CoA (0.5 m*M*) and Triton X-100 (0.13 m*M*) in incubation media but high rates are not maintained for longer than about 20 min. The lipid classes synthesized from [1-^{14}C]acetate by isolated chloroplasts are influenced by additions made to incubation media as shown in Table I.[16–18] Acetate is the preferred substrate for fatty acid synthesis in isolated spinach chloro-

[16] P. G. Roughan, J. B. Mudd, T. T. McManus, and C. R. Slack, *Biochem. J.* **184,** 571 (1979).
[17] P. G. Roughan, *Biochim. Biophys. Acta* **835,** 527 (1985).
[18] S. A. Sparace and J. B. Mudd, *Plant Physiol.* **70,** 1260 (1982).

plasts; label from $^{14}CO_2$ or from [2-^{14}C]pyruvate is incorporated at much lower rates and from [2-^{14}C]malonate not at all.[19]

Analysis of Reaction Products

The following analyses are performed using tubes which can be sealed with Teflon-lined, screw-on caps.

The simplest method of measuring amounts of [1-^{14}C]acetate incorporated into fatty acids is to stop reactions by adding 1 ml of 10% KOH (w/v) in methanol and heating to 80° for 60 min in the capped tubes. The cooled saponification mixture is acidified with 1 ml of 1.8 *M* H_2SO_4 and fatty acids are essentially quantitatively transferred into 5 ml of light petroleum (bp 40–60°) with one thorough extraction. Following brief centrifugation (acceleration to 900 *g*, then decelerated), portions (0.5–1.0 ml) of the extract are concentrated and applied as 1-cm streaks within 2-cm lanes scored onto thin layers of silica gel G. Fatty acids are separated by developing chromatograms to a height of 10 cm using petroleum ether/diethyl ether/glacial acetic acid (80 : 20 : 1, by volume) and are detected by staining with I_2 vapor. Following evaporation of the I_2, the zones are scraped into vials and radioactivity determined in detergent-stabilized, scintillation counting emulsions containing $\geq$10% (v/v) water.

Much more information is obtained, however, when the different lipid classes are recovered intact from reactions. To a 0.25-ml reaction is added 2.5 ml of chloroform/methanol (1 : 1, v/v), followed by 0.9 ml of 0.2 *M* H_3PO_4 in 2 *M* KCl with thorough mixing. After a brief centrifugation, the lower chloroform layer is recovered with the aid of a 5-ml, glass syringe and a 3.5-cm, 19-gauge needle tipped with 2 cm of Teflon tubing. Two milliliters of petroleum ether (bp 40–60°) is used to rinse the syringe and to complete the recovery of chloroform from the original partition. These extracts are combined, dried at 30–40° under a stream of nitrogen, and the lipid residue is finally dissolved in 0.5 ml of chloroform. The aqueous methanol phase is retained for further analysis (below). This chloroform solution should not be used directly for the determination of "lipid-soluble" ^{14}C as [1-^{14}C]acetate may be converted to nonlipid, chloroform-soluble material particularly when ATP, CoA, and DTT are present in reaction mixtures.[20] Instead, unesterified fatty acids, 1,2-diacylglycerols, and polar lipids are conveniently separated from total lipids on thin layers of 5% (w/w) H_3BO_3 in silica gel G in a single development of 17–18 cm using 4% (v/v) acetone in chloroform.[6] This system clearly separates 1,2-diacylglycerols from pigments which could otherwise interfere with the

[19] P. G. Roughan, R. Holland, C. R. Slack, and J. B. Mudd, *Biochem. J.* **184,** 565 (1979).
[20] G. B. Stokes and P. K. Stumpf, *Arch. Biochem. Biophys.* **162,** 638 (1974).

determination of radioactivity. Ten 1-cm streaks, each representing 10–20% of a whole reaction, may be applied within 2-cm lanes scored onto a 20 × 20 cm layer. Reference oleic acid (10 μg) and 1,2-diacylglycerol (10 μg, from egg phosphatidylcholine) should be cochromatographed with each sample so that appropriate bands may be detected using I_2 vapor and, along with the origins representing polar lipids, scraped up for scintillation counting. Alternatively, autoradiograms may be prepared from developed chromatograms in 3–4 days and may be used as templates to locate the zones to be counted. 1,2-Diacylglycerols migrate in front of unesterified fatty acids and their separation is enhanced on layers which have not been heat activated immediately prior to use. The aqueous methanol phase may contain labeled fatty acids, as acyl-CoA. These are recovered by adding 0.25 ml of 40% (w/v) KOH and heating to 80° for 60 min. Unlabeled oleic acid (20 μg) is added followed by 1 ml of 1.8 *M* H_2SO_4 and the fatty acids are extracted and chromatographed as above. Total incorporation of [1-^{14}C]acetate into long-chain fatty acids is determined from the sum of the radioactivities in unesterified fatty acids, 1,2-diacylglycerols, polar lipids, and acyl thioesters.

Individual components of the polar lipid fraction are separated for counting by two-dimensional thin-layer chromatography on silica gel without $CaSO_4$ binder. Chloroplast lipids equivalent to 25–50% of the reaction, along with total lipids from 25 mg of spinach leaf (to provide markers for the different glycerolipids), 25 μg of phosphatidic acid, and 25 μg of lysophosphatidic acid (from phosphatidic acid by the *Rhizopus* lipase) are all applied on a 1-cm, oblique streak in one corner of a 200 × 200 × 0.25 mm layer and the chromatogram is developed in the first direction with chloroform/methanol/concentrated ammonia (e.g., 65 : 25 : 3, by volume). After drying 5 min in air and 30–60 min *in vacuo*, the chromatogram is developed in the second direction using chloroform/methanol/glacial acetic acid/water (e.g., 85 : 15 : 10 : 3, by volume). Separated glycerolipids are stained with I_2 vapor prior to radioactivity measurement by scintillation counting. Chromatograms may also be developed without the unlabeled reference compounds and autoradiograms prepared to locate separated lipids for scintillation counting. In practising the two-dimensional technique, total lipids from 25 to 100 mg of leaf are chromatographed, then separated components are visualized by spraying with 40% (w/v) H_2SO_4 followed by charring at 120° for 10–20 min (pink = sterols; dull red = glycolipids; light brown = phospholipids) and are identified by comparison with a model separation.[21]

[21] W. W. Christie, "Lipid Analysis," 2nd ed., pp. 119 and 162. Pergamon, Oxford, 1982.

Analysis of Newly Synthesized Fatty Acids

Oleic and palmitic acids normally account for >90% of the label incorporated from [1-^{14}C]acetate; the remainder is distributed between stearic and linoleic acids. Argentation thin-layer chromographic separation of fatty acids methyl esters into saturated, monoenoic, dienoic, and trienoic fractions therefore provides a rapid analysis of multiple samples, which is adequate for most purposes and which is more sensitive than radio gas–liquid chromatography for the more weakly labeled compounds. For total fatty acids, 0.1–0.25 ml of the chloroform solution and 0.2 μmol of egg yolk phosphatidylcholine is dried in a 16 × 125 mm tube, 1 ml of 2.5% (w/v) H_2SO_4 in methanol is added, and the mixture is heated for 60 min in an 80° water bath. When cool, 5 ml of petroleum ether (bp 40–60°) and 1 ml of water are added and the tube shaken vigorously until an emulsion forms. The upper phase containing fatty acid methyl esters is recovered following brief centrifugation and the solvent removed under a flow of nitrogen. The residues are redissolved in 100 μl of petroleum ether (bp 60–80°) and transferred quantitatively to 1-cm streaks within 2-cm lanes scored onto thin layers of silica gel G containing 5% (w/w) $AgNO_3$ and which had been stored over a saturated solution of $CaCl_2$. Chromatograms are developed with benzene, sprayed lightly with 0.1% (w/v) dichlorofluorescein in methanol, and viewed under UV light. The four bands, saturated, monoenoic, dienoic, and trienoic esters, in order of increasing mobilities, are scraped into vials for scintillation counting. When a more detailed analysis is required, these bands are transferred instead to 16 × 125 mm tubes and the fatty acid methyl esters eluted from the adsorbent by adding 2.5 ml of acetone, 2.5 ml of petroleum ether (bp 40–60°), and 2.5 ml of water, shaking vigorously and centrifuging. The upper layers are recovered and the lower layers reextracted with another 2.5 ml of petroleum ether. The combined extracts are dried under nitrogen and the residues transferred in petroleum ether to 1-cm streaks within 2-cm lanes scored onto thin layers of silica gel G impregnated with liquid paraffin [predeveloped with 5% (w/v) liquid paraffin in petroleum ether]. Methyl oleate and methyl eicosenoate (10 μg each) are added to the streak of saturated fatty acid methyl esters, chromatograms are developed using acetonitrile/methanol/water (6 : 3 : 1, by volume), and the separated bands are detected in I_2 vapor. Labeled palmitate and stearate cochromatograph with unlabeled oleate and eicosenoate, respectively. The least mobile band corresponds to stearate and each discrete band above stearate represents either a progressive decrease in acyl chain length by two carbons or a progressive increase by one in the number of double bonds in a given chain length. Hence myristate, palmitoleate, and linoleate all migrate together.

Glycerolipids separated by one- or two-dimensional thin-layer chromatography are transmethylated in the presence of the adsorbent and 0.2 μmol of phosphatidylcholine by adding 1 ml of 0.5 *M* $NaOCH_3$ in methanol containing 5% (v/v) dimethoxypropane (as a water scavenger) and standing at room temperature for 20 min. Petroleum ether (5 ml, bp 40–60°) is mixed in followed by 2 ml of water and shaking until an emulsion forms. Fatty acid methyl esters are recovered in the upper phase after a brief centrifugation and are analyzed as above. The unesterified fatty acid fraction separated by thin-layer chromatography is also derivatized in the presence of the adsorbent and 0.2 μmol of phosphatidylcholine by adding 1 ml of 2.5% (w/v) H_2SO_4 in methanol and heating to 80° for 60 min in capped tubes. The resulting fatty acid methyl esters are recovered and analyzed as for glycerolipids. Chloroplasts isolated from well-grown spinach and nightshade will desaturate both the palmitate and oleate moieties of diacylgalactosylglycerol[5,18] and spinach chloroplasts will desaturate the oleate moieties of phosphatidylglycerol.[17,18]

Larger amounts of the different lipid classes are recovered by streaking total lipids equivalent to 100 μg of chlorophyll, along with 40 μg each of oleic acid and 1,2-diacylglycerol, across 4 cm on layers of 5% (w/w) H_3BO_3 in silica gel G and developing chromatograms with 4% (v/v) acetone in chloroform. The origins are scraped into 16 × 125 mm tubes for elution of polar lipids (below) before the unesterified fatty acid and 1,2-diacylglycerol fractions are located by lightly staining with I_2. These fractions are separately eluted from the adsorbent as for fatty acid methyl esters (above) after evaporation of the I_2. Polar lipids are eluted from the adsorbent by adding in sequence 2.5 ml of methanol, 0.5 ml of 0.2 *M* H_3PO_4 in 2 *M* KCl, and 2.5 ml of chloroform, mixing, and centrifuging. The supernatant is decanted, the adsorbent reextracted, and the combined extracts are then washed with a further 3.3 ml of H_3PO_4 in KCl. The lower layers are recovered after centrifugation, are concentrated at 40° under a stream of nitrogen, and excess H_3BO_3 is removed by adding 0.5 ml of methanol and reevaporating. Individual polar glycerolipids, along with 10 μg each of phosphatidic and lysophosphatidic acids, are now separated by two-dimensional thin-layer chromatography and may be eluted from the adsorbent using the same procedure as used for total polar lipids.

Molecular species of 1,2-diacylglycerols, diacylgalactosylglycerol, and phosphatidic acid (after conversion to dimethyl phosphatidate with diazomethane) are separated on layers of 5% (w/w) $AgNO_3$ in silica gel G. Suitable solvent systems are 4% (v/v) methanol in chloroform, chloroform/methanol/acetone (16 : 3 : 1, by volume), and 8% (v/v) methanol in chloroform, respectively. Separate bands are detected by autoradiography and radioactivity is determined by scintillation counting.

Positional distributions of newly synthesized fatty acids within glycerolipids is determined using specific lipases. Radioactive glycolipids, along with 0.5–1.0 μmol of appropriate, unlabeled carrier, are dissolved in 0.5–1.0 ml of washed diethyl ether and continuously shaken at 22–25° with 0.1 ml (monogalactosyldiacylglycerol, 1,2-diacylglycerol) or 0.2 ml (digalactosyldiacylglycerol, sulfoquinovosyldiacylglycerol) of a *Rhizopus* lipase solution—20 mg/ml of the lipase from *Rhizopus sp.* (58 U/mg, Serva) in 0.1 *M* K-borate, pH 7.4. Monogalactosyldiacylglycerol and 1,2-diacylglycerol are incubated for 15–20 min, and digalactosyldiacylglycerol and sulfoquinovosyldiacylglycerol for 30 min, before the ether is removed under nitrogen and the mixtures lyophilized. Phospholipids are treated in the same way except that the lipase solution is 0.1 ml of *Ophiophagus hannah* venom (3 mg/ml, Sigma) in 0.1 *M* K-borate, pH 7.4, containing 4 m*M* $CaCl_2$, and incubations were for 60 min. Hydrolysis products are dissolved in chloroform/methanol (3 : 1, v/v) and are separated on thin layers of silica gel G.[21]

[32] Phosphatidylglycerol Synthesis in Chloroplast Membranes

By J. BRIAN MUDD, JAEN E. ANDREWS, and S. A. SPARACE

$$\text{Acyl-ACP} + \text{1-acyl-}sn\text{-glycerol-3-P} \rightarrow \text{PA} + \text{ACP} \quad (1)$$
$$\text{PA} + \text{CTP} \rightarrow \text{CDP-DG} + \text{PP}_i \quad (2)$$
$$\text{CDP-DG} + \text{glycerol-3-P} \rightarrow \text{PGP} + \text{CMP} \quad (3)$$
$$\text{PGP} \rightarrow \text{PG} + \text{P}_i \quad (4)$$

EC 2.3.1.51 1-acylglycerol-3-phosphate acyltransferase (acyl-ACP)
EC 2.7.7.41 phosphatidate cytidylyltransferase
EC 2.7.8.5 CDPDiacylglycerol–glycerol-3-phosphate 3-phosphatidyltransferase
EC 3.1.3.27 phosphatidylglycerolphosphatase

Phosphatidylglycerols were first reported by Benson and Maruo.[1] Their report concerned the isolation and characterization of the compound from green algae and higher plants. Since that time phosphatidylglycerol (PG) has been isolated from all plant material and also from animals and bacteria. The PG from plant material has a unique characteristic in the presence of *trans*-3-hexadecenoic acid at the *sn*-2 position of the molecule.[2] Although PG is found in the mitochondria and the endoplasmic reticulum of higher plants, it is only the PG of the plastids which contains the *trans*-3-hexadecenoic acid.

The unusual fatty acid composition of the PG of the plastid indicated that its biosynthetic origin may be different from the PG found in the mitochondria and the endoplasmic reticulum, but early attempts to detect the synthesis of PG in plastid preparations were not successful.[3] The deduction that the PG of the plastid could by synthesized in the mitochondria or endoplasmic reticulum and then transported to the plastid was given support by the discovery of phospholipid exchange proteins which could transfer PG from other subcellular organelles to the plastid.[4]

It has turned out that the reaction conditions which were optimum for

[1] A. A. Benson and B. Maruo, *Biochim. Biophys. Acta* **27,** 189 (1958).

[2] F. Haverkate and L. L. M. van Deenen, *Biochim. Biophys. Acta* **106,** 78 (1965).

[3] M. O. Marshall and M. Kates, *Biochim. Biophys. Acta* **260,** 558 (1972).

[4] J.-C. Kader, D. Douady, M. Julienne, M. Grosbois, F. Querbette, and C. Vergnolle, *in* "Biochemistry and Metabolism of Plant Lipids" (J. F. G. M. Wintermans and P. J. C. Kuiper, eds.), p. 107. Elsevier/North-Holland Biomedical Press, Amsterdam, 1982.

the assay of PG synthesis in microsomal fractions from plant leaves[3] are actually inhibitory for the synthesis of PG by plastids.

Phosphatidylglycerol Synthesis by Intact Chloroplasts

The synthesis of PG by isolated chloroplasts was first described by Mudd and DeZacks[5] and later refined by Sparace and Mudd.[6]

Chloroplast Isolation

Chloroplasts were isolated by homogenizing the leaves in 0.33 *M* sorbitol, 0.2 m*M* $MgCl_2$, and 20 m*M* MES brought to pH 6.5 with Tris, according to the directions of Nakatani and Barber.[7] After the initial preparation of chloroplasts, they were purified by centrifugation on a step gradient of Percoll according to the directions of Mills and Joy.[8] The gradient was established by layering into a centrifuge tube 2 ml 90% Percoll, 4 ml 40% Percoll, and 4 ml 20% Percoll. In all cases the molarity of sorbitol was maintained at 0.3 *M*. The chloroplast preparation was layered on the Percoll gradient and then centrifuged for 20 min at 3500 *g*. The intact chloroplasts were recovered from the 40–90% interface. The chloroplasts can be separated from the Percoll by an additional centrifugation from resuspension buffer.

Reaction Mixture

The basic reaction mixture for studying the synthesis of PG was composed of sorbitol, 0.3 *M*; Tricine–NaOH, pH 7.9, 33 m*M*; $MgCl_2$, 2 m*M*; ATP, 1.5 m*M*; sodium acetate, 0.15 m*M*; $KHCO_3$, 10 m*M*; KH_2PO_4, 0.2 m*M*; CoA, 0.2 m*M*; DTT, 0.5 m*M*; *sn*-glycerol 3-phosphate, 0.2 m*M*, and chloroplast suspension equivalent to 30–180 μg chlorophyll; in a final volume of 1.0 ml. Chlorophyll was determined by the method of Bruinsma.[9] Depending on the type of experiment either the acetate or the glycerol 3-phosphate was radioactive. The reaction mixture was shaken and illuminated for the standard reaction time of 30 min.

The reaction was stopped by the addition of 3.0 ml of chloroform/methanol/acetic acid (50/100/5 by volume). To the homogeneous mixture were added 1 ml chloroform and 1 ml water and the samples centrifuged in

[5] J. B. Mudd and R. DeZacks, *Arch. Biochem. Biophys.* **209,** 584 (1981).
[6] S. A. Sparace and J. B. Mudd, *Plant Physiol.* **70,** 1260 (1982).
[7] H. Y. Nakatani and J. Barber, *Biochim. Biophys. Acta* **461,** 510 (1977).
[8] W. R. Mills and K. W. Joy, *Planta* **148,** 75 (1980).
[9] J. Bruinsma, *Biochim. Biophys. Acta* **52,** 576 (1961).

a benchtop centrifuge. The chloroform layer was recovered and washed twice with water/methanol/chloroform (48/47/3 by volume). The chloroform solution of radioactive lipids was evaporated to dryness and then redissolved in 0.5 ml chloroform. Two 25-μl aliquots were taken for the determination of radioactivity by scintillation counting, and the remainder was used for product analysis.

Product Analysis

For product analysis the chloroform solutions were streaked in 2-cm bands on the start line of a silica gel thin layer plate. The plates were developed first with acetone/acetic acid/water (100/2/1 by volume), after which the plates were dried for 2 hr under a stream of nitrogen. The plates were then developed again the same direction using the solvent chloroform/methanol/ammonia/water (65/35/2/2 by volume), permitting this second solvent to travel half as far as the first solvent. This system gave good separation of all the compounds that can be labeled from either acetate or glycerol 3-phosphate: diacylglycerol, monoacylglycerol, free fatty acid, phosphatidic acid, lysophosphatidic acid, and phosphatidylglycerol (DG, MG, FFA, PA, LPA, and PG). The radioactive lipids can be detected by autoradiography and the specific areas can then be scraped for more accurate determination of radioactivity by scintillation counting, or the recovered compounds can be used in degradation studies.

Characteristics of the System

The synthesis of PG from acetate or G-3-P by isolated chloroplasts was stimulated by illumination of the chloroplasts. Breakage of the chloroplasts by osmotic shock eliminated the synthesis of PG. Measurement of product distribution as a function of time of incubation did not indicate any product–precursor relationships; the proportion of PG was unchanged during sampling times from 15 to 120 min. In contrast to systems previously studied for the synthesis of PG, the chloroplast system was not stimulated by detergents, indeed synthesis of PG was specifically inhibited by the presence of low (0.0033%) concentrations of Triton X-100.

The requirements in the standard reaction mixture can be demonstrated by omission of single components. The absence of either bicarbonate or acetate lowered the incorporation of radioactive G-3-P markedly, indicating the need for concomitant synthesis of fatty acids when PG was being synthesized. The omission of ATP and CoA lowered the overall incorporation into PG although the proportion of the total product found as PG is not changed. These effects of ATP and CoA are difficult to understand, and have led some authors to suggest that the synthesis of PG depends on the formation of acyl-CoA's in the reaction mixture.

The synthesis of PG by the intact chloroplasts did not reveal expected intermediates such as CDPdiacylglycerol (CDP-DG) and phosphatidylglycerophosphate (PGP). The proportion of PG labeled from radioactive G-3-P was increased slightly by the addition of CTP to the reaction mixture. Roughan[10] has now demonstrated with pea chloroplasts a clear dependence on CTP for the synthesis of PG.

Degradation of Biosynthesized PG

Sparace and Mudd[6] have analyzed the PG synthesized from either radioactive acetate or radioactive G-3-P. The radioactive PG was isolated by preparative thin-layer chromatography. It was then subjected to enzymatic digestion.

The reaction mixture for digestion with phospholipase A_2 consisted of 100 m*M* Tris–Cl, pH 7.5; 10 m*M* $CaCl_2$; 1 mg phospholipase A_2 from cobra venom, and 2.5 μg Triton X-100 in a final volume of 2.0 ml. The Triton X-100 was added as a chloroform solution of the lipid to be digested. The chloroform was evaporated and the remaining components of the reaction mixture were added. The reaction was allowed to proceed for 30 min.

The reaction mixture for digestion with phospholipase C consisted of 100 m*M* Tris–Cl, pH 7.5; 10 m*M* $CaCl_2$; 10 U of phospholipase C; and 1 ml diethyl ether in a final volume of 3.0 ml. The ether and other components of the reaction mixture were added to a dried sample of the lipid to be digested. The reaction was allowed to proceed for 4 hr.

The reaction mixture for digestion with phospholipase D consisted of 40 m*M* acetate buffer, pH 5.5; 40 m*M* $CaCl_2$; 0.35 m*M* SDS; and 1 ml of the phospholipase D preparation,[11] in a final volume of 2.0 ml. The reaction was allowed to proceed for 4 hr.

The reaction mixture for digestion by *Rhizopus arrhizus* lipase was exactly as described by Fischer *et al.*[12]

The chloroform-soluble products released from the *Rhizopus arrhizus* and phospholipase A_2 digestions (lyso-PG and FFA) were separated by thin-layer chromatography using the double-solvent system described above. The chloroform-soluble products from the phospholipase C and D

[10] P. G. Roughan, *Biochim. Biophys. Acta* **835,** 527 (1985).

[11] R. H. Quarles and R. M. C. Dawson, *Biochem. J.* **112,** 787 (1969).

[12] W. Fischer, E. Heinz, and M. Zeus, *Hoppe-Seyler's Z. Physiol. Chem.* **354,** 1115 (1973).

digestions (DG and PA, respectively) were separated by thin-layer chromatography on silica gel plates developed with hexane : diethyl ether : acetic acid (80 : 20 : 2) and chloroform : methanol : NH_4OH : water (65 : 35 : 2 : 2), respectively. Water-soluble products of the digestions, glycerol phosphate and glycerol, were separated by thin-layer chromatography on silica gel developed with 2-propanol : NH_4OH : water (6 : 3 : 1). Products were recovered from the plates and converted to the fatty acid methyl esters by the method of Beare-Rogers.[13] The fatty acid methyl esters were analyzed by radio-gas chromatography using a stream splitter which supplied 5% of the effluent stream to the flame ionization detector and 95% to a Packard model 895 gas proportional counter.

Digestion by the lipolytic enzymes demonstrated with the PG labeled from radioactive acetate that there were equal amounts of radioactivity in the *sn*-1 and *sn*-2 positions. The radiolabeled fatty acid at the *sn*-2 position was primarily 16 : 0. At the *sn*-1 position the radiolabeled fatty acids were primarily 18 : 1 and 18 : 2. Digestion by the lipolytic enzymes of the PG labeled from radioactive G-3-P demonstrated that the glycerol backbone and the glycerol head group were equally labeled. These results demonstrate that the PG synthesized by the isolated chloroplasts is formed *de novo* rather than by exchange reactions for either the head group or the fatty acid moieties.

Phosphatidylglycerol Synthesis by Chloroplast Envelopes

The results obtained with intact spinach chloroplasts showed the synthesis of PG to require illumination and intact chloroplasts. With that system it was not possible to determine the single enzymatic steps in the synthesis. Spinach belongs to the group of plants known as "16 : 3" plants because the fatty acid 16 : 3 is a significant proportion of the fatty acids in MGDG at the *sn*-2 position. This composition is a consequence of the use of DG moieties synthesized in the plastid for MGDG synthesis. The DG is formed from PA in the plastid by the action of PA phosphatase. Since PA is the putative precursor of PG (via CDP-DG) and is susceptible to the PA phosphatase in the chloroplasts, spinach is not a preferred choice for the study of individual steps of PG synthesis. A better choice would be a so-called "18 : 3" plant where the PA phosphatase is less active.[14]

These considerations led to the choice of pea chloroplasts for the study of individual enzymatic steps of PG synthesis.

[13] J. L. Beare-Rogers, *Can. J. Biochem.* **47,** 257 (1969).

[14] S. E. Gardiner and P. G. Roughan, *Biochem. J.* **210,** 949 (1983).

Membrane Preparation

Pea chloroplasts were prepared by the methods described above. The chloroplasts were fractionated by suspending in 0.6 *M* sucrose and broken by slow freezing and thawing. The suspension was then diluted with buffer to 0.36 *M* sucrose and 5 ml pipetted over a sucrose density gradient composed of 1.3 *M*, 2 ml; 1.2 *M*, 1 ml; 1.1 *M*, 4 ml; 0.5 *M*, 5 ml. The tubes were centrifuged at 116,000 *g* for 90 min. The envelopes were removed from the 0.5/1.1 *M* interface with a Pasteur pipet, diluted with 10 m*M* Tricine–NaOH buffer, pH 7.5, and collected by centrifugation at 90,000 *g* for 60 min. They were then resuspended either in buffer or in 0.2 *M* sucrose.

Inner and outer envelope membranes were prepared according to the directions of Cline *et al.*[15] The chloroplasts were broken by freeze–thaw lysis and diluted to 0.2 *M* sucrose. Thylakoids were removed by centrifugation at 4500 *g* for 15 min. Envelopes were recovered from the supernatant by centrifugation at 45,000 *g* for 45 min. The pellet was resuspended in 0.2 *M* sucrose and layered on a linear sucrose density gradient (0.5–1.2 *M* sucrose). After centrifugation the outer envelope fraction was recovered from the region $p = 1.08$ g/cm^3, and the inner envelope fraction was recovered from the lower portion of the heavy membrane band ($p = 1.125$ g/cm^3). The membranes were pelleted by centrifugation at 90,000 *g* for 90 min and resuspended in 0.2 *M* sucrose.

PA Synthesis

PA was synthesized from 1-oleyl-G-3-P and [^{14}C]palmitoyl-ACP according to the directions of Frentzen *et al.*[16]

CDP-DG Synthesis

The methods described by McCaman and Finnerty[17] and Ganong *et al.*[18] were followed for the assay of CDP-DG synthesis. The reaction mixture consisted of 250 m*M* MOPS–KOH, pH 7.4; 1 m*M* PA; 1 m*M* [^{14}C]CTP; 30 m*M* $MgCl_2$; and enzyme in a final volume of 150 μl. At the end of the reaction period, 2.3 ml of chloroform/methanol, 1/1, was added. CDP-DG was added as carrier before the addition of 1 ml of 1 *M*

[15] K. Cline, J. Andrews, B. Mersey, E. H. Newcomb, and K. Keegstra, *Proc. Natl. Acad. Sci. U.S.A.* **78,** 3595 (1981).

[16] M. Frentzen, E. Heinz, T. A. McKeon, and P. K. Stumpf, *Eur. J. Biochem.* **129,** 629 (1983).

[17] R. E. McCaman and W. R. Finnerty, *J. Biol. Chem.* **243,** 5074 (1968).

[18] B. R. Ganong, J. M. Leonard, and C. R. H. Raetz, *J. Biol. Chem.* **255,** 1623 (1980).

KCl in 0.2 *M* H_3PO_4. The chloroform layer was removed for assay of radioactivity and product analysis by thin-layer chromatography.

PG Synthesis

The synthesis of PG was measured in a reaction mixture similar to that described by Marshall and Kates[3]: 250 m*M* Tris–HCl, pH 7.3; 0.1 m*M* CDP-DG; 1 m*M* [^{14}C]G-3-P; 30 m*M* $MgCl_2$; 0.15% Triton X-100; and enzyme in a final volume of 150 μl. At the end of the reaction period, the radioactive products were extracted by the method of Bligh and Dyer.[19] Reaction products were separated by thin-layer chromatography and detected by autoradiography. Individual compounds were scraped from the plate for assay of radioactivity.

Characteristics of the Syntheses Catalyzed by the Envelope Preparations

When the chloroplasts were broken and the components separated into envelope, thylakoid, and stroma fractions, the distribution of protein and enzyme activity could be assayed. Whereas 74% of the protein of the chloroplasts was recovered in the stroma fraction, there was no activity of PG synthesis or CDP-DG synthesis. The envelope fraction contained only 0.84% of the recovered protein but had 27.8% of the recovered activity for CDP-DG synthesis and 32.6% of the activity for PG synthesis. The thylakoid fraction therefore contains significant amounts of the two enzymes, but this is considered to be attributable to envelopes which adhere to the thylakoids during the preparation and is typical for envelope enzymes. The specific activity of PG synthesis was 12 times higher in the envelope as in the thylakoid. The specific activity of CDP-DG synthesis was 15 times higher in the envelope as in the thylakoid.[20]

In the report of Mudd and DeZacks[5] it was demonstrated that the synthesis of PG by intact chloroplasts was inhibited by low concentrations of Triton X-100. On the other hand Roughan claimed that Triton X-100 had little effect on PG synthesis.[10] Use of the envelope fraction helps to understand this apparent discrepancy since it was demonstrated that CDP-DG synthesis was almost 100% inhibited by 0.1% Triton X-100 while the synthesis of PG from CDG-DG was stimulated 2.5-fold by 0.05% Triton X-100.[21] Thus the synthesis of PG from G-3-P will demonstrate opposing effects of Triton X-100 on two different steps of the synthesis.

[19] E. G. Bligh and W. J. Dyer, *Can. J. Biochem. Physiol.* **37,** 911 (1959).

[20] J. Andrews and J. B. Mudd, *Plant Physiol.* **79,** 259 (1985).

[21] J. Andrews and J. B. Mudd, *in* "Structure, Function and Metabolism of Plant Lipids" (P.-A. Siegenthaler and W. Eichenberger eds.), p. 131. Elsevier, Amsterdam, 1984.

Recalculation of Roughan's data shows that in the presence of Triton X-100 the proportion of PA going on to PG is very much less in reactions containing Triton X-100, exactly as would be expected when the synthesis of CDP-DG is inhibited.

The complete synthesis of PG can be demonstrated in envelopes by the judicious use of specific substrates and inhibitors. If only palmitoyl-ACP and LPA are added to the envelopes, the reaction product is entirely PA. If PA is allowed to accumulate and then CTP and magnesium ions added, the major product accumulating is CDP-DG. If the CDP-DG is allowed to accumulate and then G-3-P added to the reaction mixture then the major product is PG. If in the latter case mercury ions are added at the same time as the G-3-P, the PGP accumulates because of the inhibition of the PGP phosphatase.

Activities Associated with Inner and Outer Envelopes

When the inner and outer envelope membranes were separated, the resolution was not complete. Between the fractions of inner and outer membrane fractions is a region in which the membranes are mixed. Each of these fractions contained about one-third of the recovered protein. The outer envelope fraction contained less than 1% of the CDP-DG synthesis activity and 1.4% of the PG synthetic activity. The inner envelope fraction contained 58 and 55% of the CDP-DG synthetic activity and PG synthetic activity, respectively. The specific activity of CDP-DG synthesis was 68 times higher in the inner envelope fraction than in the outer. The specific activity of PG synthesis was 32 times higher in the inner envelope fraction than in the outer.[20]

Conclusions

The methods described above, when taken with the results of other laboratories, permit the detailed interpretation of PG synthesis in the chloroplast.

G-3-P is acylated using acyl-ACP in the stroma.[22] This reaction is specific for the *sn*-1 position, and in the cases so far described is very specific for oleyl-ACP.[16] The second acylation takes place in the inner envelope membrane and is specific for palmitoyl-ACP. The product, 1-oleoyl-2-palmitoyl-*sn*-G-3-P, is the precursor of PG. Fatty acid analysis of PG from isolated chloroplasts reflects this pathway of synthesis.

[22] J. Joyard and R. Douce, *Biochim. Biophys. Acta* **486,** 273 (1977).

[33] Galactolipid Biosynthesis in Chloroplast Membranes

By NORA W. LEM and JOHN P. WILLIAMS

The plant galactolipids, monogalactosyldiacylglycerol (MGDG) and digalactosyldiacylglycerol (DGDG), are synthesized by reactions which require galactosyltransferases. (1) UDPgalactose : 1,2-diacylglycerol galactosyltransferase (EC 2.4.1.46, 1,2-diacylglycerol 3-β-galactosyltransferase) catalyzes the following reaction:

$$\text{UDPgalactose} + \text{1,2-diacylglycerol} \rightarrow \text{MGDG} + \text{UDP}$$

(2) UDPgalactose : MGDG galactosyltransferase catalyzes the following reaction:

$$\text{UDPgalactose} + \text{MGDG} \rightarrow \text{DGDG} + \text{UDP}$$

DGDG is also synthesized by galactolipid : galactolipid galactosyltransferase (galactosidase):

$$\text{MGDG} + \text{MGDG} \rightarrow \text{DGDG} + \text{diacylglycerol}$$

These reactions have been observed in cell-free preparations from, for example, pea seedlings, in the form of Triton extracts and acetone powders,[1] isolated spinach chloroplasts,[2] and isolated and purified spinach chloroplast envelopes.[3] Recently, the inner and outer membranes of chloroplast envelopes have been isolated[4,5] and evidence has been obtained for the localization of UDPgalactose : diacylglycerol galactosyltransferase in both the outer[6] and inner[5] membranes of the envelope. Galactolipid : galactolipid galactosyltransferase may be localized in the (outer surface of the) outer membrane of the envelope,[5,6] or on the inner envelope.[7] The results of experiments[8] in which *in vivo* feeding and subsequent isolation of membranes from leaf tissue is used show that MGDG and DGDG may

[1] M. Siebertz and E. Heinz, *Hoppe-Seyler's Z. Physiol. Chem.* **358,** 27 (1977).
[2] A. Ongun and J. B. Mudd, *J. Biol. Chem.* **243,** 1558 (1968).
[3] R. Douce, *Science* **183,** 852 (1974).
[4] K. Cline, J. Andrews, B. Mersey, E. H. Newcomb, and K. Keegstra, *Proc. Natl. Acad. Sci. U.S.A.* **78,** 3595 (1981).
[5] A.-J. Dorne, M. A. Block, J. Joyard, and R. Douce, *FEBS Lett.* **145,** 30 (1982).
[6] K. Cline and K. Keegstra, *Plant Physiol.* **71,** 366 (1983).
[7] J. W. M. Heemskerk, G. Bogemann, T. J. M. Peeters, and J. F. G. M. Wintermans, *in* "Structure, Function and Metabolism of Plant Lipids" (P.-A. Siegenthaler and W. Eichenberger, eds.), p. 119. Elsevier, Amsterdam, 1984.
[8] J. P. Williams, E. E. Simpson, and D. J. Champman, *Plant Physiol.* **63,** 669 (1979).

also be synthesized in the thylakoid membranes. The inability to detect galactolipid synthesis *in vitro* in thylakoid membranes could be due to their lack of diacylglycerol substrate (see below). These experiments are also difficult to evaluate as it is impossible at present to isolate reliably pure thylakoid and envelope fractions. The nonphotosynthetic chromoplast inner membranes from the corona of the daffodil, *Narcissus pseudonarcissus*,[9] have galactolipid-synthesizing ability. The reader is referred to these references for descriptions of the methods of preparation of plant cell fractions and membranes.

Procedure

The assays for MGDG and DGDG biosynthetic activity are based upon the ability of the enzyme to react with (water-soluble) UDP[^{14}C]galactose and membrane-bound (organic solvent-soluble) endogenous diacylglycerol to form the (organic solvent-soluble) galactolipids. The course of the reaction may be monitored by liquid scintillation counting or by liquid scintillation counting of individual galactolipids separated by thin-layer chromatography (TLC). The procedure described below follows the assay developed by Joyard and Douce.[10]

Materials and Reagents

$MgCl_2$, 100 m*M*
Tricine–NaOH buffer, 10 m*M*, pH 7.2
UDP[^{14}C]galactose, 2 μCi/μmol, 10 m*M*
Methanol–chloroform (MeOH–$CHCl_3$) (2 : 1, v/v)
Plant material of interest, resuspended in 10 m*M* Tricine–NaOH, pH 7.2, 300 m*M* sucrose
Scintillation fluid (e.g., Aquasol from New England Nuclear)
Silica gel 60 precoated plates (Merck)
Chloroform–methanol–water (65 : 25 : 4, v/v)

The following are added to small screw-capped test tubes:

2 μmol Tricine–NaOH buffer
0.2 μmol $MgCl_2$
enzyme (50–100 μg plant protein)

The reaction is started by the addition of 0.1 μmol UDP[^{14}C]galactose, and the incubation mixture (0.2–0.4 ml) is allowed to stand at room temperature for various periods of time (1–30 min). The reaction is stopped

[9] Liedvogel and H. Kleinig, *Planta* **129,** 19 (1976).
[10] J. Joyard and R. Douce, *Biochim. Biophys. Acta* **424,** 125 (1976).

by the addition of 750 μl MeOH–$CHCl_3$ (2 : 1); the reaction tubes are flushed with N_2, and allowed to stand at room temperature for 20–30 min prior to the addition of 250 μl $CHCl_3$ and 250 μl H_2O. The tubes are flushed with N_2, vortexed, and centrifuged to facilitate the separation of two phases, a lower organic phase and an upper aqueous phase. The upper phase is taken off with a Pasteur pipet and the lower phase is washed two times with small amounts of water. After the second addition of water, the organic (lower) phase is taken off with a Pasteur pipet and put into a clean, screw-capped test tube. The organic phase is dried under N_2 and dissolved in a known volume of $CHCl_3$ (1 ml).

Aliquots of known volume (100–200 μl) are injected into scintillation vials with 10 ml of scintillation fluid and counted in a liquid scintillation counter. The amount of radioactivity, corresponding to total galactolipid biosynthesis, is expressed as picomoles or nanomoles of galactose incorporated into lipid per milligram protein per minute.

The individual galactolipids are separated by TLC. Aliquots of the $CHCl_3$ mixture are spotted onto a precoated silica gel plate and developed with the solvent system, chloroform–methanol–water, (65 : 25 : 4, v/v). MGDG and DGDG are identified by cochromatography with authentic standards or by R_f values. (The R_f value for MGDG is 0.91, for DGDG is 0.51, for trigalactolipid is 0.21, and for tetragalactolipid is 0.05.[5]) The radioactive areas corresponding to the radioactive galactolipids are scraped into scintillation vials with 10 ml scintillation fluid and counted in a liquid scintillation counter. The unusual galactolipids, tri- and tetragalactolipid, produced in these *in vitro* assays, are artifacts of the activity of the galactosidase, which is stimulated during the preparation of the plant fractions.[2,5] Further analysis of MGDG and DGDG, such as molecular species separation, is possible but will depend upon the amount of radioactivity incorporated into the galactolipids. If they are sufficiently radioactive, the galactolipids may be spiked with "cold" MGDG or DGDG and separated into their constituent molecular species by argentation-TLC.[11]

Characteristics of Galactolipid Synthesis

Kinetic analyses of the two galactosyltransferase enzymes show that the pH optimum for the first galactosyltransferase (forming MGDG) is >pH 8.0[10] while the second galactosylating enzyme has its maximal activity at pH 6.0–6.5.[1,10] The pH maximum for galactolipid synthesis may range from pH 7.4 to 9.0.[1,6] The K_m UDPgalactose for UDPgalactose : diacylglycerol galactosyltransferase is 30–45 μM for pea

[11] N. W. Lem and J. P. Williams, *Phytochemistry* **17**, 1893 (1978).

and spinach envelope[6] but is also reported to be approximately 90 μM for spinach envelope membranes[10]; the V_{max} will vary depending on the tissue preparation.

The diacylglycerol substrate for diacylglycerol galactosyltransferase is endogenously available, and attempts to boost or prolong the activity of the enzyme by adding a dispersion of diacylglycerol to the reaction mixture have been unsuccessful for two reasons: (1) the insolubility of the diacylglycerols in aqueous medium and (2) the lack of information about the nature of the molecular species of the diacylglycerol substrates of the enzyme—i.e., the degree of unsaturation of the fatty acids esterified to the glycerol backbone.

The first difficulty appears to be circumvented in an experiment which was designed to measure the activity of the UDPglucose : diacylglycerol galactosyltransferase (which forms glucolipid, the precursor of MGDG, in the cyanobacterium, *Anabaena variabilis*) by directly impregnating lyophilized membranes with diacylglycerol before performing the assay.[12] Perhaps this procedure may be useful for higher plant membranes.

Information about the diacylglycerol molecular species required for MGDG (and DGDG) biosynthesis has been obtained largely from *in vivo* feeding experiments using $^{14}CO_2$. There are at least two different groups of plants defined by the lack of ("18 : 3 plants") or by the presence of large amounts of 16 : 3 fatty acid ("16 : 3 plants") esterified to MGDG and DGDG. Data from *in vivo* experimens using the 18 : 3 plant, *Vicia faba*, show that the diacylglycerol substrates for MGDG synthesis contain a high proportion of 18 : 2 and 18 : 3 fatty acids, i.e., they are highly unsaturated and consist mainly of 18 : 3/18 : 3 and 18 : 2/18 : 3 molecular species.[13] However, in 16 : 3 plants (*Nicotiana, Lycopersicon, Spinacea, Brassica*), the predominant fatty acids esterified to the diacylglycerol precursor of MGDG are 16 : 0 and 18 : 1, and therefore less highly unsaturated. The plant species substrate diacylglycerol differences may be important in those experiments where attempts have been made to increase the amount of diacylglycerol substrate available for MGDG biosynthesis in plant cell preparations (see above).

The efficiency of DGDG biosynthesis in *Pisum sativum* (an 18 : 3 plant) preparations varies with the degree of fatty acid unsaturation of the MGDG molecular species added exogenously.[1] Maximum galactosylation to form DGDG was achieved with MGDG 18 : 3/18 : 3 substrate. This result agrees well with *in vivo* experiments using *Vicia faba*.[11]

The galactolipid transferase enzyme shows preference for diacyl-

[12] N. Sato and N. Murata, *Plant Cell Physiol.* **23,** 1115 (1982).

[13] N. W. Lem and J. P. Williams, *Biochem. J.* **209,** 513 (1983).

glycerol and MGDG substrates containing specific fatty acids. Thus, the extent of the observable galactosylation reactions in any tissue preparation depends upon the molecular species of diacylglycerol and MGDG available in the preparation. The type and amount of substrate available in the plant preparation may be altered by both biotic and abiotic factors.

Factors which may alter the rate of galactolipid biosynthesis and the nature (molecular species) of the galactolipids formed include (1) the temperature of growth of plant tissue,[14] (2) herbicide treatment of plant material,[13] (3) addition of drugs to the assay mixture,[15] etc. The light intensity during growth of the plant appears to have little effect on galactolipid metabolism.[11]

[14] J. P. Williams, M. U. Khan, and K. Mitchell, *in* "Structure, Function and Metabolism of Plant Lipids" (P.-A. Siegenthaler and W. Eichenberger, eds.), p. 123. Elsevier, Amsterdam, 1984.

[15] N. W. Lem and P. K. Stumpf, *Plant Physiol.* **75**, 700 (1984).

[34] Chlorophylls and Carotenoids: Pigments of Photosynthetic Biomembranes

By HARTMUT K. LICHTENTHALER

Introduction

The photosynthetic pigments chlorophylls and carotenoids, together with sterols, prenylquinones, and prenols, belong to the group of isoprenoid plant lipids, which in 1976 were named prenyl lipids by Goodwin[1] and Lichtenthaler.[2] Carotenoids as tetraterpenoids are simple or pure prenyl lipids, the carbon skeleton of which is made up solely of isoprenoid units. Chlorophyll *a* and *b* are mixed prenyllipids: they possess an isoprenoid phytyl chain which is bound to a nonisoprenoid porphyrin ring system. The possession of the phytyl side chain, which is esterified to the carboxyl group of the ring, gives the chlorophylls their lipid character. Because of their biogenetic relationship (isopentenoid pathway), chlorophylls and carotenoids are also referred to as prenyl pigments.

[1] T. W. Goodwin, *in* "Lipids and Lipid Polymers in Higher Plants" (M. Tevini and H. K. Lichtenthaler, eds.), p. 29. Springer-Verlag, Berlin and New York, 1977.

[2] H. K. Lichtenthaler, *in* "Lipids and Lipid Polymers in Higher Plants" (M. Tevini and H. K. Lichtenthaler, eds.), p. 231. Springer-Verlag, Berlin and New York, 1977.

Chlorophylls

The chlorophylls of higher plants, ferns, mosses, and green algae, as well as of the prokaryotic organism Prochloron, consist of chlorophyll *a* as the major pigment and of chlorophyll *b* as an accessory pigment. Both chlorophylls are genuine components of the photosynthetic membranes and occur in a ratio (*a*/*b*) of approximately 3 to 1.[2–5] Growth conditions and environmental factors can modify this *a*/*b* ratio. High-light and sun-exposed plants (high-light chloroplasts) exhibit *a*/*b* ratios of 3.2 to 4, whereas shade plants (low-light chloroplasts) possess lower values for the *a*/*b* ratio (e.g., 2.5 to 2.9).[4–6] In older literature minor amounts of an additional chlorophyll (*a*′) were found.[7] This is a C-10 epimer of chlorophyll *a* and is now regarded in the form of the *a*′ dimer as identical with the reaction center chlorophyll *P*-700 of the photosynthetic photosystem I.[8]

Carotenoids

The group of primary plant carotenoids can be divided (1) into the oxygen-free carotenes and (2) into the xanthophylls, which contain oxygen in different forms, such as one or several hydroxy or epoxy groups. There are two types of isomers on each oxidation level, the α-carotene derivatives (one ε-ionone and one β-ionone ring) and the β-carotene derivatives (2 β-ionone rings). Introduction of hydroxy and epoxy functions into α-carotene (ionone rings) gives rise to lutein and lutein epoxide. β-Carotene is the precursor of the xanthophylls zeaxanthin, violaxanthin, and antheraxanthin. The carotenoids of green, photosynthetically active plant tissue, which are needed for photosynthetic function, are classified as primary carotenoids, whereas those of red fruits and flowers have been termed secondary carotenoids.

The carotenoids of functional chloroplasts include β-carotene, lutein, violaxanthin, and neoxanthin as major and regular components of the photochemically active thylakoids of chloroplasts of higher plants and

[3] R. Willstätter and A. Stoll, "Untersuchungen über Chlorophyll." Springer-Verlag, Berlin, 1913.
[4] A. Seybold and K. Egle, *Planta* **26,** 491 (1970).
[5] H. K. Lichtenthaler, C. Buschmann, M. Döll, H.-J. Fietz, T. Bach, U. Kozel, and U. Rahmsdorf, *Photosynth. Res.* **2,** 115 (1981).
[6] H. K. Lichtenthaler, G. Kuhn, U. Prenzel, C. Buschmann, and D. Meier, *Z. Naturforsch., C: Biosci.* **37C,** 464 (1982).
[7] H. H. Strain, B. T. Cope, and W. A. Svec, this series, Vol. 23, p. 452.
[8] T. Watanabe, M. Kobayashi, A. Hongu, M. Nakazato, T. Hiyama, and N. Murata, *FEBS Lett.* **191,** 252 (1985).

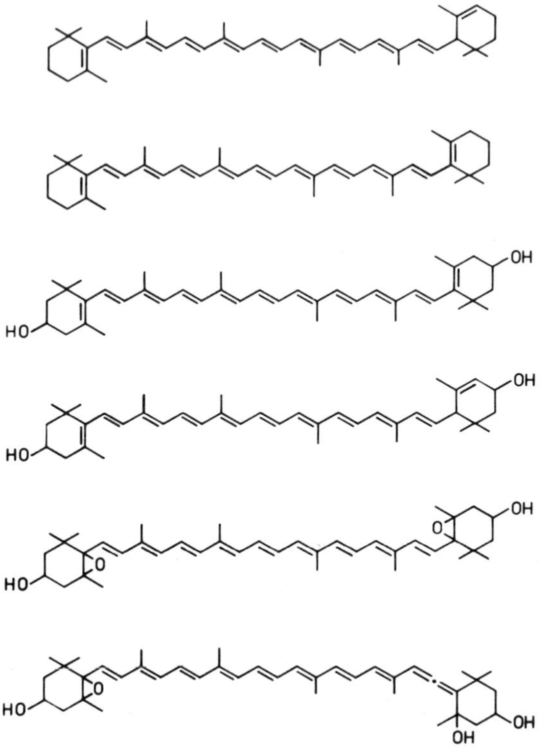

α-Carotene

β-Carotene

Zeaxanthin

Lutein

Violaxanthin

Neoxanthin

FIG. 1. Chemical structure of the main chloroplast carotenoids of higher plants.

green algae (Fig. 1). Depending on growth conditions and stress factors, the percentage composition of carotenoids (weight %) can vary within the ranges: β-carotene 25–40% lutein, 40–57%, violaxanthin, 9–20%, and neoxanthin, 5–13%. The xanthophylls antheraxanthin, luteinepoxid, and zeaxanthin are regular but minor carotenoid components. In green algae α-carotene shows up as an additional carotene to β-carotene. Under high-light conditions the xanthophyll violaxanthin (two epoxy, two hydroxy groups) can by deepoxidized via antheraxanthin to zeaxanthin, the corresponding deepoxy derivative, which exhibits a chromatographic mobility similar to lutein. Though this xanthophyll cycle is well established,[9–11] its

[9] A. Hager, *in* "Pigments in Plants" (F. C. Czygan, ed.), p. 57. Fischer, Stuttgart, 1980.

[10] D. Siefermann-Harms, *in* "Lipids and Lipid Polymers in Higher Plants" (M. Tevini and H. K. Lichtenthaler, eds.), p. 218. Springer-Verlag, Berlin and New York, 1977.

[11] K. Grumbach, *Z. Naturforsch., C: Biosci.* **38C,** 393 (1983).

complete physiological role is not yet known. Lutein and neoxanthin apparently only function as accessory pigments in the photosynthetic light absorption. β-Carotene may partially serve as a light-absorbing pigment; its main function, however, seems to be the protection of chlorophyll *a* from photooxidation in or near the reaction center.

The main paths of biosynthesis of chlorophylls and carotenoids are established. Details concerning chlorophylls may be found in the reviews.[12,13] An open question seems to be whether chlorophyll *b* (—CHO instead of —CH_3 in the second pyrrole ring of the porphyrin ring) derives via oxidation of chlorophyll *a* or by prenylation of a special precursor (chlorophyllide *b*?), which has not yet been detected. Basic information on carotenoid biosynthesis can be found in Goodwin[1] and Davies.[14] Though certain aspects of compartmentalization are not yet fully clear, there is agreement that at least the final steps of carotenoid biosynthesis (including dimerization, desaturation, and cyclization) proceed within the chloroplasts.

The aim of this chapter on chlorophylls and carotenoids is to give practical directions toward their quantitative isolation and determination in extracts from leaves, chloroplasts, thylakoid particles, and pigment proteins. Hence the main emphasis is placed on the spectral characteristics and absorption coefficients of chlorophylls, pheophytins, and carotenoids, which are the basis for establishing equations to quantitatively determine them. Many data in this field were based on rather old observations and old instrumentation and needed revision. Therefore the specific absorption coefficients of the pigments were reevaluated and in part also first established for the purpose of this chapter. This was achieved by using a two-beam spectrophotometer of the new generation, which allows programmed automatic recording and printing out of the proper wavelengths and absorbancy values.

Chlorophyll–Carotenoid Proteins

Much of the present research work on photosynthetic pigments concerns isolation and characterization of pigment proteins. The main features of their isolation and pigment content will therefore be treated here.

[12] H. A. W. Schneider, *Ber. Dtsch. Bot. Ges.* **88,** 83 (1975).

[13] W. Rüdiger and J. Benz, *in* "Chloroplast Biogenesis" (R. J. Ellis, ed.), p. 225. Cambridge Univ. Press, London and New York, 1984.

[14] B. H. Davies, *in* "Chemistry and Biochemistry of Plant Pigments" (T. W. Goodwin, ed.), 2nd ed., Vol. 2, p. 38. Academic Press, New York, 1976.

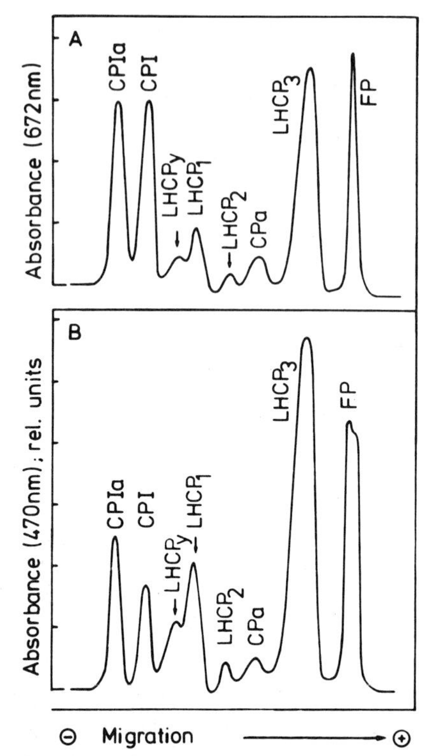

FIG. 2. Densitometer scans of the pigment proteins of *Raphanus* chloroplasts separated by SDS–PAGE [(A) Scan at 672 nm and (B) scan at 470 nm]. As compared to 672 nm, absorbance of the light-harvesting proteins (LHCPs) increases at 470 nm due to their high content in xanthophylls and chlorophyll *b*. In contrast, the relative absorbance of the chlorophyll *a* proteins CPI, CPIa, and CPa, which contain only little chlorophyll *b*, decreases at 470 nm. With respect to the free pigment zone (FP), the scan at 672 nm shows only the peak of free chlorophylls; this peak is broadened by the free carotenoids, when the scan is performed at 470 nm.

Except for violaxanthin, which to some extent is bound to the chloroplast envelope,[15] all carotenoids and chlorophylls are bound to the thylakoids, the photochemically active photosynthetic biomembranes.[16] Within the thylakoids the pigments are associated with several chlorophyll–carotenoid proteins which possess differential chlorophyll and carotenoid composition.[16–24] By polyacrylamide gel electrophoresis (PAGE) of digitonin or sodium dodecyl sulfate-treated chloroplasts or thylakoid preparations

[15] H. K. Lichtenthaler, U. Prenzel, R. Douce, and J. Joyard, *Biochim. Biophys. Acta* **641,** 99 (1981).

[16] J. P. Markwell, J. P. Thornber, and R. T. Boggs, *Proc. Natl. Acad. Sci. U.S.A.* **76,** 1223 (1979).

TABLE I
PRENYL PIGMENT RATIOS AND PERCENTAGE COMPOSITION OF CAROTENOIDS IN PIGMENT PROTEINS

Protein	Chloroplasts	CPI_a	CPI	CP_a	$LCHP_{1-3}$	FP
Pigment ratios						
Chlorophyll *a*/*b*	3.2	4–7	9–21	3–8	1.1–1.3	2.6
$a + b/x + c^b$	4	5–6	7–9	3–5	3–5	3
a/c^b	12	12	9	4–8	60–180	14
Percentage carotenoid composition[c]						
β-Carotene	**30**	**56**	**70**	**75**	1–2	24
Lutein	**45**	24	15	13	**56–65**	37
Neoxanthin	6	9	4	5	**21–29**	10
Violaxanthin	**11**	5	6	4	4–5	**20**
Others	8	6	5	3	6	9

[a] Isolated by SDS–PAGE from radish chloroplasts. (After Lichtenthaler *et al.*[23,24])
[b] $a + b/x + c$ = weight ratio chlorophylls to carotenoids; a/c = weight ratio chlorophyll *a*/β-carotene.
[c] The enrichment of particular carotenoids in a pigment protein as compared to whole chloroplasts is indicated in boldface.

(Fig. 2), one obtains the following: (1) the P700-containing chlorophyll *a*/β-carotene proteins CPI and CPIa of photosystem I, (2) CPa, the presumable reaction center of photosystem II, which is also a chlorophyll *a*/β-carotene protein, and (3) several forms of LHCPs ($LHCP_1$, $LHCP_2$, $LHCP_3$, $LHCP_y$, etc.), the light-harvesting chlorophyll *a*/*b*–xanthophyll proteins.[16–24]

The xanthophylls lutein and neoxanthin are preferentially bound together with chlorophyll *b* and some chlorophyll *a* to the LHC–proteins, which contain β-carotene only in traces. β-Carotene is, however, the main carotenoid component of the chlorophyll *a*–proteins CPI, CPIa, and CPa. The exact composition of the different pigment proteins has been established for *Raphanus* chloroplasts (Table I). Violaxanthin is present

[17] J. P. Thornber, *Annu. Rev. Plant Physiol.* **26,** 127 (1975).
[18] J. M. Anderson, J. C. Waldron, and S. W. Thorne, *FEBS Lett.* **92,** 227 (1978).
[19] O. Machold, D. J. Simpson, and B. L. Müller, *Carlsberg Res. Commun.* **44,** 235 (1979).
[20] R. Bassi, O. Machold, and D. Simpson, *Carlsberg Res. Commun.* **50,** 145 (1985).
[21] J. Argyroudi-Akoyunoglou and G. Akoyunoglou, *FEBS Lett.* **104,** 78 (1979).
[22] H. K. Lichtenthaler, G. Burkard, G. Kuhn, and U. Prenzel, *Z. Naturforsch., C: Biosci.* **36C,** 421 (1981).
[23] H. K. Lichtenthaler, U. Prenzel, and G. Kuhn, *Z. Naturforsch., C: Biosci.* **37C,** 10 (1982).
[24] U. Prenzel and H. K. Lichtenthaler, *in* "Biochemistry and Metabolism of Plant Lipids" (J. F. G. M. Wintermans and P. C. Kuiper, eds.), p. 565. Elsevier/North-Holland Biomedical Press, Amsterdam, 1982.

in the various chlorophyll–carotenoid proteins only in minor amounts, which does not account for the total violaxanthin present in the thylakoids. Violaxanthin, however, is found in a rather high proportion in the free pigment fraction.[24] From this it is assumed that this apparently not protein-bound pool of violaxanthin is identical with those 70 to 80% of chloroplast violaxanthin, which can be transformed at high light intensities into zeaxanthin.[9,10]

A similar carotenoid distribution among different pigment proteins has also been found using the isoelectric focusing technique as a method for pigment protein isolation.[25–27]

The individual levels of the pigment proteins can vary depending on the developmental stage and on the environmental conditions. At low-light growth conditions and in the shade a higher proportion of LHCPs are found in comparison to CPI + CPIa.[6,28,29] This fully explains the lower values for the ratios chlorophyll a/b and the higher values for the ratios chlorophyll a/β-carotene (a/c) and $a + b/x + c$ in low-light plants.

There are different methods for the isolation of pigment particles and pigment proteins, such as the PAGE technique[16–22] or isoelectric focusing,[26,27] including the classical differential centrifugation of partially detergent-digested chloroplast or thylakoid preparations. The latter often precedes the gel electrophoretic methods. Details of the procedures may be found in the original older literature[16,18–27] and the most recent results in the current journals of plant physiology and plant biochemistry.

Mild procedures (low detergent concentrations, short incubation times, and low temperatures) allow the detection of larger pigment protein aggregates (e.g., CPIa and LHCPs 1 and 2), whereas at higher concentrations primarily the oligomeric forms (e.g., CPI and $LHCP_3$) show up. By choosing the right conditions it is also possible to detect several transitional stages between the larger aggregates and the oligomeric forms. Very pure and well-characterized pigment proteins are those of Machold *et al.*[19,20] A recent overview on thylakoid proteins including pigment proteins and molecular weights is found in Critchley.[30]

The general problem with the isolation of pigment proteins is that the free chlorophyll portion often exceeds 15 or even 20% of the total thylakoid chlorophyll applied and that it should be well beyond 10% to obtain representative results for the pigment composition of a thylakoid. An-

[25] D. Siefermann-Harms, *Biochim. Biophys. Acta* **811,** 1 (1985).

[26] D. Siefermann-Harms and N. Ninnemann, *FEBS Lett.* **104,** 71 (1979).

[27] T. Omata, N. Murata, and K. Satoh, *Biochim. Biophys. Acta* **765,** 403 (1984).

[28] H. K. Lichtenthaler, G. Kuhn, U. Prenzel, and D. Meier, *Physiol. Plant.* **56,** 183 (1982).

[29] T.-Y. Leong and J. M. Anderson, *Biochim. Biophys. Acta* **723,** 391 (1983).

[30] C. Critchley, *Biochim. Biophys. Acta* **811,** 33 (1985).

other problem which is not yet fully resolved is the quantitative extraction of chlorophylls and carotenoids from the pigment proteins on the polyacrylamide gels. Repetitive extraction of the material seems to be the method of choice.

Isolation and Chromatography of Pigments

Extraction

All prenyl pigments, chlorophylls and carotenoids, are fat-soluble compounds that can be extracted from water-containing living plant tissue by organic solvents (grinding with quartz sand or use of a homogenizer) such as acetone, methanol, or ethanol, which can take up water. Though aqueous solutions of these solvents are also suitable (and may sometimes be preferable) for extraction, their water content should not exceed 5 or 10%. The widely used method to extract isolated chloroplasts by 80% aqueous acetone[31] does not fully extract the less polar pigment chlorophyll *a* and the polyene β-carotene. An addition step of extraction with 100% acetone is needed to guarantee complete extraction. By addition of petroleum hydrocarbon (or hexane) and water the pigments are transferred to the water-free epiphase, and concentrated in a rotary evaporator to a certain standard volume. In a water-free organic solvent, pigments can be stored longer than in aqueous solution.

Freeze-dried plant material can be extracted directly with hexane or petroleum hydrocarbon; this, however, will mainly remove β-carotene and some chlorophyll *a*. Addition of small amounts of a polar solvent (acetone) to the extracting hydrocarbon solvent will fully extract the chlorophylls and carotenoids. Since carotenoids and chlorophylls are extremely light sensitive and easily become photobleached, the extraction and concentration procedures should be performed in dim light. For correct pigment values the quantitative determination of chlorophylls and total carotenoids in the whole leaf extract should be carried out immediately after preparation of the extract. If storage of the extract is required, it should be done in concentrated form and under nitrogen.

Chromatographic Procedures

Several procedures have been developed for the separation of the photosynthetic pigments including column (CC), paper (PC), and thin-layer chromatography (TLC) and also high-pressure liquid chromatogra-

[31] D. I. Arnon, *Plant Physiol.* **24,** 1 (1949).

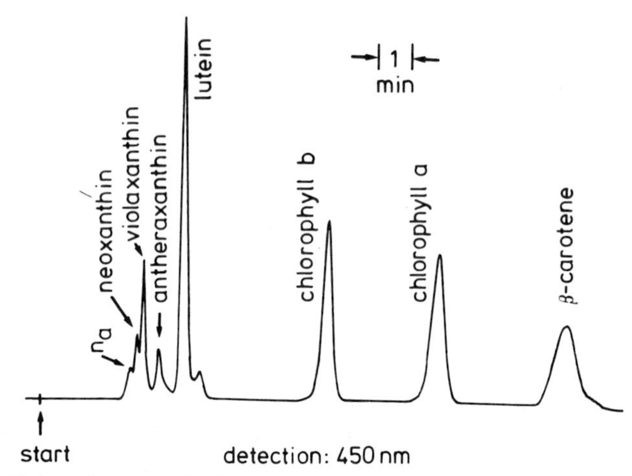

FIG. 3. High-performance liquid chromatography of a leaf pigment extract (reversed-phase HPLC). Column: 25 cm, Nucleosil 5 C-8; solvent: 9% water in methanol; flow rate: 1.3 ml/min. n_a, Neoxanthin *a*.

phy (HPLC). Basic and detailed information on these methods is found in a recent extensive chromatography review by Lichtenthaler.[32]

The separation of a plant pigment extract on $CaCO_3$ and MgO columns into green (chlorophylls) and yellow pigments (carotenoids) by Tswett can be regarded as the beginning of chromatography.[33] Today TLC is the preferred method for the isolation of the photosynthetic pigments of higher plants. Chromatography on precoated silica gel plates using 70 ml petroleum hydrocarbon (bp 40–60°), 30 ml dioxane, and 10 ml 2-propanol as developing solvent results in the separation of the two chlorophylls and the major carotenoids.[34] The chromatographic sequence (R_f values) is β-carotene (0.85), a trace of pheophytin (0.67), chlorophyll *a* (0.6), chlorophyll *b* (0.5), lutein + zeaxanthin (0.4), antheraxanthin (0.34), violaxanthin (0.3), and neoxanthin (0.2). The extract should be applied to the plate in a water-free solvent. Elution of the pigments is performed with ethanol or acetone. A very suitable TLC method for carotenoids, which also separates α- and β-carotene derivatives, e.g., lutein and zeaxanthin, is given by Hager and Meyer-Bertenrath.[35]

[32] H. K. Lichtenthaler, *in* "CRC Handbook of Chromatography: Lipids" (H. K. Mangold, ed.), Vol. 2, p. 115. CRC Press, Boca Raton, Florida, 1984.

[33] M. Tswett, *Ber. Dtsch. Bot. Ges.* **24,** 384 (1906).

[34] H. K. Lichtenthaler, D. Meier, G. Retzlaff, and R. Hamm, *Z. Naturforsch., C: Biosci.* **37C,** 889 (1982).

[35] A. Hager and T. Meyer-Bertenrath, *Planta* **76,** 149 (1967).

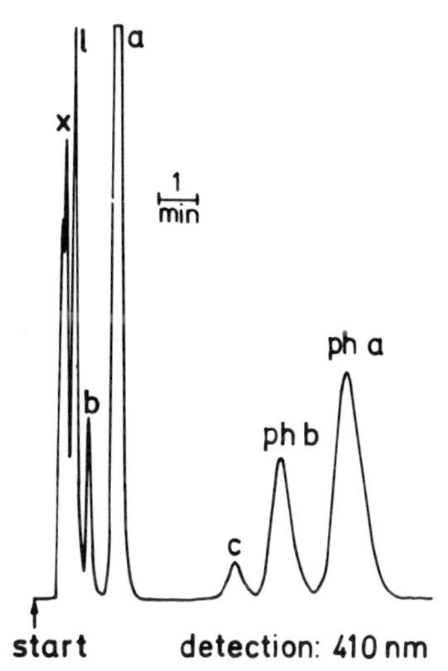

FIG. 4. Reversed-phase HPLC of leaf pigments including pheophytins. Column: 12.5 cm, Nucleosil 7 C-18; solvent: methanol; flow rate: 1.8 ml/min.

An example of the HPLC of prenyl pigments is shown in Fig. 3. Adsorption and reversed-phase HPLC have been applied.[36–41] For further examples and details see Table 3 in Lichtenthaler.[32] HPLC also allows the separation of pheophytin *a* and *b* (Fig. 4).

The Chlorophylls

Spectral Characteristics of Chlorophylls

The absorption maxima of chlorophylls are found in the red and blue regions of the visible spectrum (Figs. 5 and 6). The position for λ_{max} of chlorophyll *b* (e.g., 641.8 and 452.2 nm in dried diethyl ether) lies between those of chlorophyll *a* (660 and 428 nm in dried diethyl ether) in all sol-

[36] K. Eskins, C. R. Scholfield, and H. J. Dutton, *J. Chromatogr.* **135,** 217 (1977).

[37] H. Stransky, *Z. Naturforsch., C: Biosci.* **33C,** 836 (1978).

[38] U. Prenzel and H. K. Lichtenthaler, *in* "Advances in the Biochemistry and Physiology of Plant Lipids" (L.-A. Appelqvist and C. Liljenberg, eds.), p. 319. Elsevier/North-Holland Biomedical Press, Amsterdam, 1979.

[39] T. Braumann and L. Grimme, *J. Chromatogr.* **170,** 264 (1979).

[40] M. Tevini, W. Iwanzik, and U. Thoma, *Planta* **153,** 388 (1981).

[41] T. Watanabe, M. Nakazato, H. Mazaki, A. Hongu, M. Konno, S. Saitoh, and K. Honda, *Biochim. Biophys. Acta* **807,** 110 (1985).

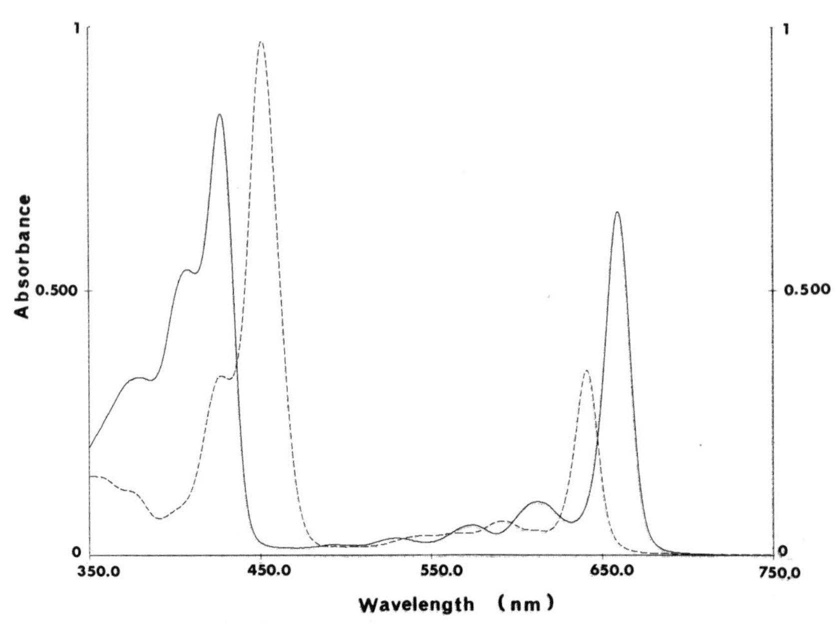

FIG. 5. Absorption spectrum of freshly isolated chlorophyll *a* (solid line) and chlorophyll *b* (broken line) in diethyl ether (pure solvent). The positions of the absorption maxima are given in Table II.

vents assayed (see Table II). The shortest wavelength for λ_{max} in the blue and red is found for both chlorophylls in water-free diethyl ether. The two major absorption maxima in the red and blue of both chlorophylls shift to longer wavelengths with increasing polarity and/or water content of the solvent. The shift at λ_{max} red is less pronounced than that at λ_{max} in the blue region and it is larger for chlorophyll *b* than for chlorophyll *a*. The wavelengths of the much smaller, minor absorption maxima also undergo this shift.

The specific absorption coefficients are highest in dried (water-free) diethyl ether and decrease via diethyl ether (pure solvent), water-saturated diethyl ether, to acetone (100 → 80%), ethanol (95%), and methanol (100 → 90%) (Table II). The absorption in the red and blue maxima is highest in freshly isolated chlorophylls and also decreases in older solutions, particularly in those which are light exposed. Quantitative chlorophyll determinations should therefore be carried out immediately after extraction. In the case of chlorophyll *a* the decrease in absorbance in the blue is stronger than in the red region, which is documented by a continu-

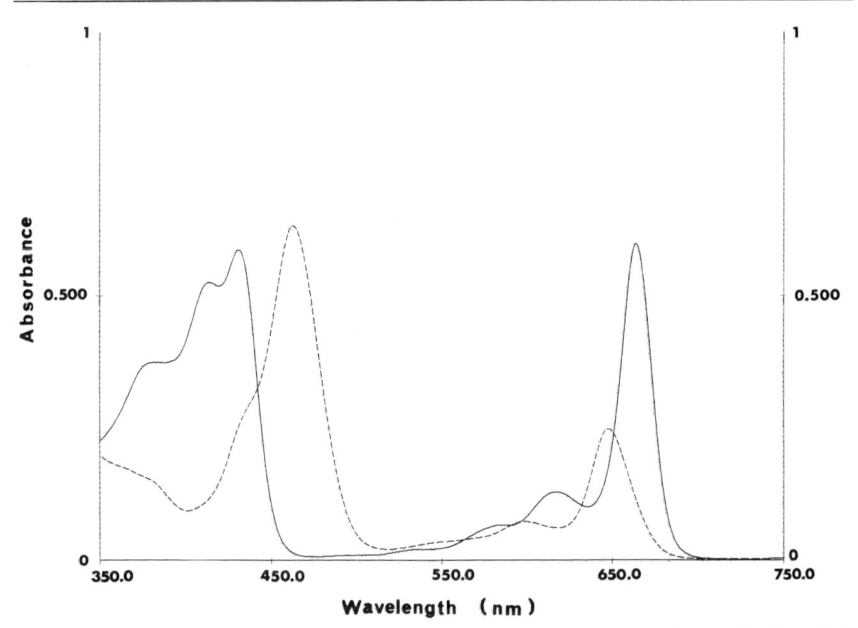

FIG. 6. Absorption spectrum of freshly isolated chlorophyll *a* (solid line) and chlorophyll *b* (broken line) in aqueous ethanol (95%). The wavelengths of the absorption maxima are given in Table II.

ous decrease of the values for the ratio of absorption coefficients of the blue and the red maximum from 1.28 to 0.96 (Table IV). For chlorophyll *b* the decrease is about the same in the blue and in the red region, as seen from the ratio of the absorption coefficients, which remains about the same with values around 2.7 (Table IV).

With increasing polarity and water content of the solvent, the absorption maxima in the blue and red not only become smaller, but at the same time broader. Parallel to the broadening of the red maximum of chlorophyll *a* there occurs a considerable increase in the absorption coefficients of the minor maxima at about 614 to 618 nm and 576 to 582 nm. Due to the shift in the wavelengths, the specific absorption coefficients of chlorophyll *a* and *b* at 470 nm, the reference point of simultaneous carotenoid determination in a whole leaf pigment extract, rise continuously with increasing polarity and water content of the solvent. In ethanol (95%) and in methanol (100 and 90%), the two minor absorption maxima (in the range of 576 to 582 nm and 531 to 537 nm) do not show up (see also Fig. 6).

TABLE II

SPECIFIC ABSORPTION COEFFICIENTS FOR CHLOROPHYLL *a* AND *b* REDETERMINED[a]

Diethyl ether (water free)		Diethyl ether (pure solvent)		Diethyl ether (H_2O saturated)		Acetone (100%)	
(nm)	(sp. coeff.)	(nm)	(sp. coeff.)	(nm)	(sp. coeff.)	(nm)	(sp. coeff.)
				Chlorophyll *a*			
660	101.9	**660.6**	101[c]	**661.6**	98.46	**661.6**	92.45
—		—		—		—	
641.8	15.20	642.2	15.0	643.2	15.31	644.8	19.25
614.4	15.7	*614.8*	15.88	*615.8*	16.07	*616.0*	17.25
576.2	8.7	*576.6*	8.80	*578.2*	8.86	*579.0*	9.92
531.0	4.67	*531.6*	4.48	*532.6*	4.19	*534.2*	4.34
470	1.3	470	1.43	470	1.38	470	1.90
460	1.5	460	1.71	460	1.79	460	2.44
450	3.6	450	3.87	450	5.47	450	7.40
428.0	130.4	**428.4**	127.21	**429.4**	120.96	**429.6**	112.36
408.8	83.15	*409.4*	82.83	*410.0*	81.22	*410.6*	83.18
				Chlorophyll *b*			
660	4.7	660.6	6.0	661.6	7.20	661.6	9.38
641.8	62.3	**642.2**	62.0[c]	**643.2**	58.29	**644.8**	51.64
—							
593.2	11.57	*593.6*	12.35	*594.2*	12.02	*595.8*	12.11
470	33.12	470	35.87	470	48.05	470	63.14
460	122.17	460	121.70	460	128.88	460	133.86
450	162.3	450	158.51	450	139.70	450	122.08
452.2	171.33	**452.6**	167.87	**453.8**	156.69	**455.8**	145.36
428.2	60.13	*428.4*	62.09	*430.2*	58.57	—	—

[a] By use of a UV-260 Shimadzu UV-visible recording spectrometer, which allows auto-absorption maxima of the pigments. sp. coeff., Specific absorption coefficient.

[b] Wavelengths (nm) of the two major absorption maxima in the red and blue spec-The solvents used for the determination of the absorption coefficients were of purest stadt, FRG, No. 929; diethyl ether (pure solvent), diethyl ether extra pure DAB_7 qual-residue analysis, Merck No. 12; ethanol (95%), ethanol for fluorometry, Merck No. water to 95%; methanol, methanol z.A. (analytical grade), Merck No. 6009, or metha-

[c] Accepted from Smith and Benitez[49] (for a light path of 1 cm).

FOR ORGANIC SOLVENTS OF DIFFERENT POLARITY AND WATER CONTENT[b]

Acetone, 80% (v/v) (aqueous sol.)		Ethanol, 95% (v/v) (aqueous sol.)		Methanol (100%)		Methanol, 90% (v/v) (aqueous sol.)	
(nm)	(sp. coeff.)	(nm)	(sp. coeff.)	(nm)	(sp. coeff.)	(nm)	(sp. coeff.)
663.2	86.3	**664.2**	84.60	**665.2**	79.24	**665.2**	78.95
646.8	20.49	648.6	25.06	652.4	35.52	652.4	35.36
618.2	18.15	*617.8*	19.14	*617.6*	18.68	*618.2*	18.84
582.4	10.35	—	—	—	—	—	—
537.0	4.10	—	—	—	—	—	—
470	1.82	470	2.13	470	1.63	470	1.91
460	2.87	460	4.32	460	6.11	460	5.94
450	11.62	450	18.42	450	28.80	450	26.37
431.2	95.82	**431.6**	83.89	**431.8**	77.05	**431.8**	75.97
412.6	78.76	*414.4*	76.60	—	—	*417.6*	—
663.2	11.20	664.2	16.0	665.2	21.28	665.2	19.84
646.8	49.18	**648.6**	41.2	**652.4**	38.87	**652.4**	35.97
597.6	12.80	*600.6*	11.81	*602.6*	12.74	*602.8*	11.78
470	85.02	470	97.64	470	104.96	470	95.15
460	134.14	460	103.32	460	87.48	460	82.03
450	94.89	450	69.95	450	63.07	450	58.83
459.0	134.50	**463.6**	107.45	**469.2**	105.36	**469.2**	99.51
—	—	—	—	—	—	—	—

matic recording and printing of wavelength and absorbance in the major and minor

tral region are in boldface, those of the minor absorption maxima (italics).
obtainable quality: Diethyl ether (water free), diethyl ether dried GR, Merck, Darm-
ity, Merck No. 926; acetone (100%), acetone extra pure, Merck No. 13, and acetone for
973, or ethanol absolute, DAB_8 quality, Roth Karlsruhe, FRG, No. 5054 diluted with
nol dried, Merck No. 6012.

Quantitative Determination of Chlorophylls

There exist various equations from different authors for different solvents for the determination of chlorophylls in leaf pigment extracts (see reviews of Szestak[42] and Holden[43]). The disadvantage of these equations is that the pigment values obtained in one solvent are not comparable with those of another. The discrepancies are particularly large in the values for the ratio of chlorophyll *a*/*b*. The cause of this is that the equations are based on rather early determinations of the specific absorption coefficients at times where the resolution of spectrophotometers was not as good as today. The widely used equations given by Arnon[31] for 80% acetone are still based on the specific absorption coefficients of Mackinney.[44] Though the equations of Arnon may still be used for estimating the total chlorophyll $a + b$ content, they can never be applied to determine the ratio of chlorophyll *a*/*b* of a pigment extract. Several attempts have been made to correct and modify the equations for acetone (100 and 80%), e.g., by Vernon,[45] Ziegler and Egle,[46] and Wintermans and de Mots.[47] Yet there remained differences in the wavelength position of the red peak maxima and in the relative absorption of chlorophyll *a* at λ_{max} of chlorophyll *b* and vice versa. The same holds true for the determination of chlorophylls in diethyl ether solutions. The original absorption coefficients of Comar and Zscheile,[48] were redetermined by Smith and Benitez,[49] and their data were confirmed by Falk,[50] who applied a different technique of magnesium determination. The equations given by Smith and Benitez[49] for diethyl ether,

$$C_a = 10.1A_{662} - 1.01A_{644}$$
$$C_b = 16.4A_{644} - 2.57A_{662}$$

were later slightly modified by Ziegler and Egle[46]:

$$C_a = 10.05A_{662} - 0.89A_{644}$$
$$C_b = 16.37A_{644} - 2.68A_{662}$$

[42] Z. Szestak, *in* "Plant Photosynthetic Production: Manual of Methods" (Z. Šesták, J. Catsky, and P. G. Jarvis, eds.), p. 672. Dr. W. Junk Publ., The Hague, 1971.

[43] M. Holden, *in* "Chemistry and Biochemistry of Plant Pigments" (T. W. Goodwin, ed.), 2nd ed., Vol. 2, p. 1. Academic Press, New York, 1976.

[44] G. Mackinney, *J. Biol. Chem.* **140,** 315 (1941).

[45] L. P. Vernon, *Anal. Chem.* **32,** 1144 (1960).

[46] R. Ziegler and K. Egle, *Beitr. Biol. Pflanz.* **41,** 11 (1965).

[47] J. F. G. M. Wintermans and A. de Mots, *Biochim. Biophys. Acta* **109,** 448 (1965).

[48] C. L. Comar and F. P. Zscheile, *Plant Physiol.* **17,** 198 (1942).

[49] J. H. C. Smith and A. Benitez, *in* "Modern Methods of Plant Analysis" (K. Paech and M. V. Tracey, eds.), Vol. 4, p. 142. Springer-Verlag, Berlin and New York, 1955.

[50] H. Falk, *Planta* **51,** 49 (1958).

More recently, in establishing equations for the simultaneous determination of total carotenoids, together with chlorophyll *a* and *b*, of leaf extracts in different solvents[51] we reinvestigated and confirmed the absorption coefficients of Smith and Benitez and adjusted the equations for a Unicam SP 8000 spectrophotometer:

$$C_a = 10.05A_{662} - 0.766A_{644}$$
$$C_b = 16.37A_{644} - 3.14A_{662}$$

The differences found by authors for the same solvent are mainly due to (1) the resolution properties of the spectrophotometers applied in the red spectral region (which will determine the position of λ_{max} of the red absorption maxima of chlorophyll *a*/*b* and the relative absorption at λ_{max} of the other chlorophyll), (2) the accuracy with which a particular wavelength can be chosen for the absorbancy readings, and (3) the purity of the solvents: slight differences in the content of water or organic material of the solvent will largely influence the refractive index as well as height and broadness of the maxima.

By applying a spectrophotometer of the new generation, which exhibits automatic baseline correction and can be programmed for peak picking, automatic readings, and printing out of absorbancy readings at certain wavelengths, the absorption coefficients were again redetermined using pure solvents of different sources (cf. footnote of Table II). These new spectrophotometers guarantee a better selection of wavelengths (or parts of wavelengths) and absorbancy readings. The position of λ_{max} for chlorophyll *a* and *b* in diethyl ether, as determined by repetitive recording of spectra (in a UV-260 Shimadzu spectrometer) of freshly isolated chlorophyll from several plants (with five aliquots per isolation) are at shorter wavelengths (660.6 and 642.2 nm) than so far reported using other instruments (662 and 644 nm). The new equations for diethyl ether (pure solvent) for the determination of chlorophylls *a* and *b* in a UV-260 spectrophotometer are as follows:

$$C_a = 10.05A_{660.6} - 0.97A_{642.2}$$
$$C_b = 16.36A_{642.2} - 2.43A_{660.6}$$

Based on the redetermined specific absorption coefficients in various other solvents (Table II), equations are given for the simultaneous determination of chlorophyll *a* and *b* in a pigment extract solution (Table III). The validity of the equations was cross-checked by dissolving the same amount of plant extract in each of the solvents. The advantage and progress of the new equations is that the pigment values obtained in one

[51] H. K. Lichtenthaler and A. R. Wellburn, *Biochem. Soc. Trans.* **603,** 591 (1983).

TABLE III

EQUATIONS FOR THE DETERMINATIONS OF THE CONCENTRATIONS OF CHLOROPHYLL *a* (C_a), CHLOROPHYLL *b* (C_b), OF TOTAL CHLOROPHYLLS (C_{a+b}) AND OF TOTAL CAROTENOIDS (C_{x+c}) IN LEAF PIGMENT EXTRACTS FOR SOLVENTS OF DIFFERENT POLARITY AND WATER CONTENT[a]

Solvent	C_a	C_b	C_{a+b}	C_{x+c}
Diethyl ether (pure solvent):	$= 10.05A_{660.6} - 0.97A_{642.2}$	$= 16.36A_{642.2} - 2.43A_{660.6}$	$= 7.62A_{660.6} + 15.39A_{642.2}$	$= \frac{1000A_{470} - 1.43C_a - 35.87C_b}{205}$
Diethyl ether (water free):	$= 9.93A_{660} - 0.75A_{641.8}$	$= 16.23A_{641.8} - 2.42A_{660}$	$= 7.51A_{660} + 15.48A_{641.8}$	$= \frac{1000A_{470} - 1.30C_a - 33.12C_b}{213}$
Diethyl ether (water saturated):	$= 10.36A_{661.6} - 1.28A_{643.2}$	$= 17.49A_{643.2} - 2.72A_{661.6}$	$= 7.64A_{661.6} + 16.21A_{643.2}$	$= \frac{1000A_{470} - 1.38C_a - 48.05C_b}{211}$
Ethanol, 95% (v/v):	$= 13.36A_{664.2} - 5.19A_{648.6}$	$= 27.43A_{648.6} - 8.12A_{664.2}$	$= 5.24A_{664.2} + 22.24A_{648.6}$	$= \frac{1000A_{470} - 2.13C_a - 97.64C_b}{209}$
Acetone, 100% (pure solvent):	$= 11.24A_{661.6} - 2.04A_{644.8}$	$= 20.13A_{644.8} - 4.19A_{661.6}$	$= 7.05A_{661.6} + 18.09A_{644.8}$	$= \frac{1000A_{470} - 1.90C_a - 63.14C_b}{214}$
Acetone, 80% (v/v):	$= 12.25A_{663.2} - 2.79A_{646.8}$	$= 21.50A_{646.8} - 5.10A_{663.2}$	$= 7.15A_{663.2} + 18.71A_{646.8}$	$= \frac{1000A_{470} - 1.82C_a - 85.02C_b}{198}$
Methanol, 100% (pure solvent):	$= 16.72A_{665.2} - 9.16A_{652.4}$	$= 34.09A_{652.4} - 15.28A_{665.2}$	$= 1.44A_{665.2} - 24.93A_{652.4}$	$= \frac{1000A_{470} - 1.63C_a - 104.96C_b}{221}$
Methanol, 90% (v/v):	$= 16.82A_{665.2} - 9.28A_{652.4}$	$= 36.92A_{652.4} - 16.54A_{665.2}$	$= 0.28A_{665.2} + 27.64A_{652.4}$	$= \frac{1000A_{470} - 1.91C_a - 95.15C_b}{225}$

[a] The equations are based on the redetermined specific absorption coefficients listed in Table II. The pigment concentrations obtained by inserting the measured absorbance values are micrograms per milliliter plant extract solution.

solvent can be compared directly with those of another solvent, including the ratio of chlorophyll *a*/*b* and the total amount of leaf carotenoids.

Spectral Characteristics of Carotenoids

General Remarks

All chloroplast carotenoids exhibit a typical absorption spectrum which is characterized by three absorption maxima (violaxanthin, neox-

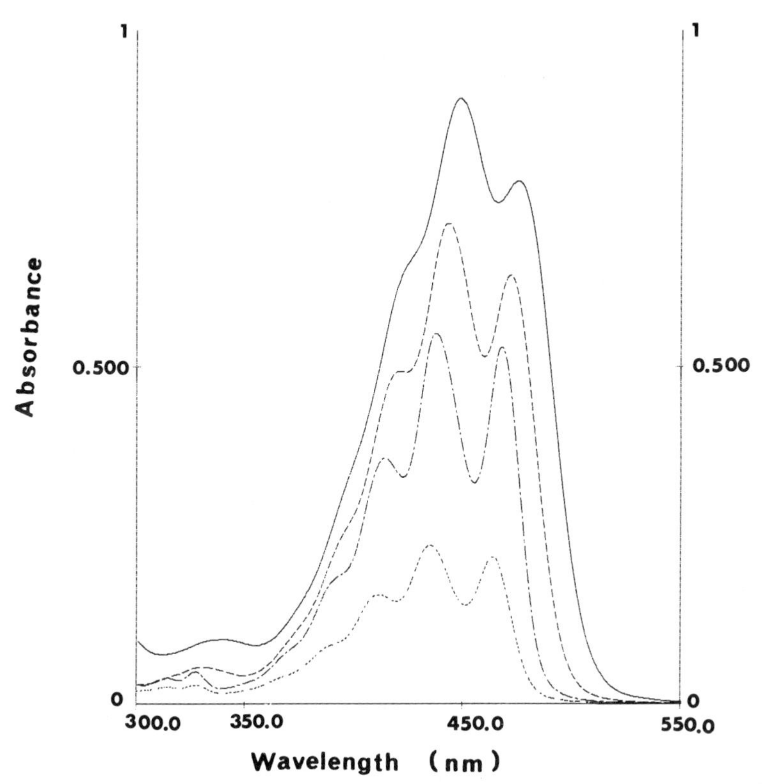

FIG. 7. Absorption spectrum of the major leaf carotenoids in diethyl ether. The wavelengths at the absorption maxima are given in Table V. β-Carotene (—), lutein (– – –), violaxanthin (·–·), and neoxanthin (····).

anthin) or two maxima with one shoulder (lutein and β-carotene) in the blue spectral region (Fig. 7). The wavelength position of the maxima of oxygen-free carotenoids such as β-carotene is oriented to longer wavelengths than those of the oxygen-bearing xanthophylls. With increasing amounts of hydrophilic groups in the tetraterpenoid carbon skeleton of carotenoids the maxima (and shoulder) are shifted to shorter wavelengths. The absorption maxima of the pure polyene β-carotene are 475.4 and 449.2 nm in methanol, whereas those of neoxanthin (with three hydroxy groups and one epoxy group) are 464 and 435.6 nm (Table V).

The identification of pigments can also be performed with the new method of photoacoustic spectroscopy (PAS method).[52,53] One obtains

[52] C. Buschmann, H. Prehn, and H. K. Lichtenthaler, *Photosynth. Res.* **5,** 29 (1984).
[53] E. Nagel, C. Buschmann, and H. K. Lichtenthaler, *Physiol. Plant.* **70,** in press (1987).

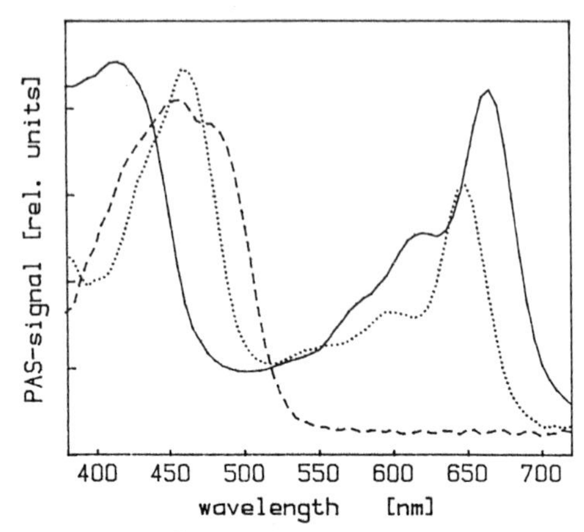

FIG. 8. Photoacoustic spectra[53] of leaf pigments as measured on a silica gel plate after HPTLC of leaf pigments.[34] Chlorophyll *a* (—), chlorophyll *b* (····), β-carotene (– – – –). (Spectra taken by E. Nagel, Karlsruhe.)

photoacoustic spectra which resemble absorption spectra but which are distinctly different from pure absorption spectra. An example for photoacoustic spectra of β-carotene as well as chlorophyll *a* and *b* is shown in Fig. 8.

The wavelength position of the absorption maxima of carotenoids also depends on the type of solvent and its water content. The shortest wavelengths are found for pure methanol. There is an increasing shift of the maxima to longer wavelengths from methanol via diethyl ether, ethanol (95%), acetone (100%), to acetone (80%) (Table V). In solvents such as methanol (90%) or water-saturated diethyl ether, the maxima are shifted by 0.4 to 0.6 nm to longer wavelengths as compared to pure water-free solvent. At a higher water content [e.g., acetone (80%) versus acetone (100%)] the shift can be 1 nm or more (Table V). In the series methanol, diethyl ether, ethanol, and acetone the shift of the maxima to longer wavelengths does not follow the polarity of the solvents. The series of shift for carotenoids is also different from that of chlorophylls, where the absorption maxima shift to longer wavelengths in the series diethyl ether, acetone, ethanol, methanol (cf. Table II).

Specific Absorption Coefficients

The molecular weights of carotenoids differ due to the differential oxygen content (β-carotene, 536.85, lutein, 568.85, violaxanthin, 600.85, neoxanthin, 600.85). The specific absorption coefficients for the individual carotenoids found in literature vary considerably. Most of them lie in the range of 2400 to almost 2600 for a 1% solution and a light path of 1 cm ($A_{1\,cm}^{1\%}$ values). To avoid the problem of the proper coefficient, it was recommended to apply for the sum of leaf carotenoids[54] and for all individual carotenoids[55] an $A_{1\,cm}^{1\%}$ of 2500 at their main absorption maximum. This is still the best approach today. The specific absorption coefficients also vary depending on the type of solvent and its water content. Per definition the $A_{1\,cm}^{1\%}$ of 2500 for the individual carotenoids is determined here with diethyl ether (pure solvent) as a reference basis.

Determination of Total Carotenoids (x + c)

The sum of carotenoids (xanthophylls + β-carotene; $x + c$) can easily be determined in a total leaf or chloroplast pigment extract together with chlorophylls by measuring the absorbance not only in the absorption maxima of chlorophyll *a* and *b*, but additionally at a wavelength where carotenoids show good absorption, e.g., at 470 nm (Fig. 7). The absorbance of the pigment extract measured at 470 nm is mainly due to light absorption by carotenoids; a smaller amount comes from chlorophyll *b*, whereas chlorophyll *a* contributes very little to the absorption at this wavelength. The concentration of total carotenoids (C_{x+c}) can therefore be determined by deduction of the relative absorption of chlorophyll *a* and *b* from the absorbancy read at 470 nm followed by division by the absorption coefficient of total carotenoids at 470 nm. Equations for determination of C_{x+c} in different solvents are shown in Table III. Though the method is quite elegant and reliable, certain precautions have to be taken.

First of all, this method requires a correct determination of the chlorophyll *b* concentration beforehand. If the chlorophyll *b* level were overestimated, the resulting values for total carotenoids would be too low and vice versa. The equation for C_{x+c} given in Table III is based on measuring the relative absorbance of the chlorophylls at 470 nm with respect to the absorption at λ_{max} in the red in a UV-260 Shimadzu spectrometer (cf. Table IV). The equations presented here are certainly also valid for other modern two-beam spectrophotometers. Depending on the technical setup of other, often older, two-beam spectrophotometers, the wavelengths of λ_{max} of chlorophyll *a* and *b* may deviate from the values in Table II by 0.8 to 1.5 nm. This may then cause considerable differences in the absorbancy readings, particularly at λ_{max} of chlorophyll *b* at 470 nm, which will affect the determination of total carotenoids. The spectrophotometer applied in a laboratory should therefore be checked with freshly isolated

[54] T. W. Goodwin, *in* "Modern Methods of Plant Analysis" (K. Paech and M. V. Tracey, eds.), Vol. 3, p. 272. Springer-Verlag, Berlin and New York, 1955.

[55] H. K. Lichtenthaler, *Z. Pflanzenphysiol.* **53,** 388 (1965).

TABLE IV

RATIOS OF THE SPECIFIC ABSORPTION COEFFICIENTS AT λ_{max} IN THE BLUE TO λ_{max} IN THE RED FOR CHLOROPHYLL *a* AND *b* AND AVERAGE ABSORPTION COEFFICIENTS $A^{1\%}_{1cm}$ AT 470 nm FOR TOTAL LEAF CAROTENOIDS (XANTHOPHYLLS PLUS CAROTENES, $x + c$) IN SOLVENTS OF DIFFERENT POLARITY AND WATER CONTENT

Solvent	Ratio of the specific coefficients at λ_{max} blue to λ_{max} red: Chlorophyll *a*	Chlorophyll *b*	Total carotenoids $(x + c)$: $A^{1\%}_{1\,cm}$ at 470 nm[a]
Diethyl ether (water free)	1.28	2.75	2130
Diethyl ether (pure solvent)	1.26	2.71	2050[b]
Diethyl ether (water saturated)	1.23	2.69	2110
Acetone (pure solvent)	1.22	2.81	2140
Acetone, 80% (v/v) (aqueous solution)	1.11	2.73	1980
Ethanol, 95% (v/v) (aqueous solution)	0.99	2.61	2100
Methanol (pure solvent)	0.97	2.71	2210
Methanol, 90% (v/v) (aqueous solution)	0.96	2.66	2250

[a] The mean values listed are based on the relative carotenoid composition of green leaf tissue from several plants (e.g., radish, spinach, tobacco, maize, oat, and wheat).

[b] This value consists of the relative absorption of the individual leaf carotenoids (β-carotene, lutein, violaxanthin, and neoxanthin) at 470 nm as compared to the absorption maximum. An absorption coefficient $A^{1\%}_{1\,cm}$ of 2500 in diethyl ether at the main absorption maximum (between 436 and 450 nm) was assumed for all individual leaf carotenoids.

chlorophyll *a* and *b* solutions, and the wavelength deviation at λ_{max} and the shift at 470 nm be accounted for (see adjustment below). Further requirements are clear solutions and a correct measurement of the baseline.

The recording of the full absorption spectrum of the pigment extract solution between 750 and 450 nm is necessary to be able to judge from the shape of the spectrum whether one measures a clear pigment solution. In fully clear pigment extract solutions of normal green leaves the absorbency readings at 750 nm should be zero, and in the range of 550 to 510 nm they should be very low and amount to far less than 10% of the absorbancy readings at the red absorption maximum of the extract (in the range of 660–665 nm). If this is not the case, a centrifugation step (caution: use closed flasks to avoid evaporation of the solvent) has to be applied.

Strictly sticking to another rule is essential for correct pigment values: after the recording of the full spectrum the absorbancy readings need to be performed by setting the proper wavelengths (absorption maxima of chlorophyll a and b as well as 470 nm) on the spectrometer and by marking the height of the absorbancy by a line with the help of the spectrometer pen. The measurement of the absorbancy on the recording paper, after the spectrum had been recorded, by trying to visually determine the proper wavelengths, is to be avoided. It results in errors up to 20 or 25% of the chlorophyll b and total carotenoid amounts as well as of the ratio of chlorophyll a/b.

The determination of the concentration of $x + c$ in the total pigment extract is the fastest and therefore best method and should always be applied as a reference before other methods, HPLC or TLC, which allow determination of individual carotenoids, are applied. TLC with subsequent elution and spectrophotometric determination of separated carotenoids always yields lower values (for β-carotene about -5% and for xanthophylls up to -13%). The sum of carotenoids determined by evaluation of the individual carotenoid peaks obtained by HPLC often does not add up to the $x + c$ content obtained by the direct measurement of the leaf extracts or exceeds it to a large extent, depending on where the baseline is drawn on the HPLC scans. It is therefore recommended to proof the values for individual carotenoids obtained by HPLC with the TLC technique[34] mentioned above (see Chapter 2).

The total sum of carotenoids can also be determined in a pigment extract together with pheophytin *a* and *b*, provided that the absorbancy readings are performed within 10 min after pheophytinization (e.g., with HCl). A longer standing in acid solution will progressively destroy carotenoids. The $x + c$ concentrations obtained using this method (see equations in Table VII) represent a close but only a first approximation. Since the spectra of pheophytins change with the acid content, one should stick to the standard conditions (see The Pheophytins *a* and *b*, p. 373).

$A^{1\%}_{1\,cm}$ of the Sum of Carotenoids

The $A^{1\%}_{1\,\mathrm{cm}}$ values at 470 nm given in Table IV and applied in the equations for C_{x+c} in Table III are calculated from the relative absorbance of the individual carotenoids as compared to their absorbance at the main absorption maximum. For the latter a common absorption coefficient of

TABLE V

WAVELENGTH POSITION (λ, nm) OF THE ABSORPTION MAXIMA OF THE MAJOR PRIMARY CAROTENOIDS OF CHLOROPLASTS IN SOLVENTS OF DIFFERENT POLARITY AND WATER CONTENT[a]

Carotenoid	Methanol (pure solvent)	Diethyl ether (pure solvent)	Diethyl ether (H_2O saturated)	Ethanol, 95% (v/v) (aqueous sol.)	Acetone (pure solvent)	Acetone, 80% (v/v) (aqueous sol.)
β-Carotene	475.4	476.4	476.8	477.4	478.9	480.0
(pure hydrocarbon)	**449.2**	**449.8**	**449.8**	**451.8**	**453.2**	**454.0**
Lutein	472.0	472.8	473.2	473.6	475.4	476.2
(two hydroxy groups)	**443.6**	**444.4**	**444.8**	**445.6**	**447.4**	**448.0**
Violaxanthin	468.2	469.0	469.2	469.0	470.6	471.2
(two epoxy, two	**438.4**	**438.8**	**439.6**	**442.4**	**442.6**	**442.8**
hydroxy groups)	415.8	416.0	417.2	418.0	419.4	420.0
Neoxanthin	464.0	465.0	465.6	465.8	467.0	468.0
(one epoxy, three	**435.6**	**436.6**	**436.8**	**437.2**	**438.4**	**439.0**
hydroxy groups)	412.6	413.2	413.6	415.2	415.6	416.6

[a] The main absorption maximum lying in the range of 435.6 to 454.6 nm is in boldface.

2500 in diethyl ether was assumed for the individual carotenoids.[55] In view of the different wavelength position of the two (or three) carotenoid maxima and the nearby depressions (cf. Fig. 7), which cancel each other out in a total carotenoid solution, it is clear that the $A_{1\,cm}^{1\%}$ value of the sum of chloroplast carotenoids is lower than 2500. The $A_{1\,cm}^{1\%}$ value in diethyl ether of total carotenoids as isolated from green leaf extracts of seven different plants lies in the range of 2400 to 2450 depending on the carotenoid composition of the plant that is the relative proportions of xanthophylls to β-carotene. The same arguments (partial compensation of the second carotenoid maximum near 470 nm of one carotenoid with a nearby depression of another carotenoid) hold true for the $A_{1\,cm}^{1\%}\,x + c$ value at 470 nm. The values given in Table IV represent mean values based on the carotenoid composition of green leaves of seven plants.

The Pheophytins *a* and *b*

General Remarks

Chlorophylls *a* and *b* are easily transformed by weak acids to their magnesium-free derivatives, the pheophytins *a* and *b*. Their molecular weight (pheophytin *a*, 871.18; pheophytin *b*, 885.16) is 22.3 weight units smaller than that of the corresponding chlorophylls. Small amounts of pheophytin *a* have always been found in plant pigment extracts and seem already to be present in the intact leaf. In fact, pheophytin *a* is a functional component of the reaction center of photosystem II (RC II), where it plays a role in the charge separation and electron transfer from *P*-680.[56,57]

Chlorophyll *a* is more sensitive to pheophytinization than chlorophyll *b*, as can be seen from thin-layer chromatography (TLC) of pigment extracts on silica gel plates. Silicic acid, as a weak acid, will transform part of the chlorophyll *a* into pheophytin *a*, whereas chlorophyll *b* is little affected. Both pheophytins are formed under the influence of the endogenous organic acids during pigment extract preparation, while grinding the leaf tissue in the presence of an extracting solvent with quartz sand or by maceration in a homogenizer. Addition of some magnesium oxide or calcium carbonate will neutralize the endogenous organic acids. This is particularly needed in the case of plants which are especially known to store organic acids in their vacuoles. Pheophytinization is much enhanced with increasing temperature. Homogenization of the leaf tissue in a homoge-

[56] V. V. Klimov, A. V. Klevanik, V. A. Shuvalov, and A. A. Krasnovsky, *FEBS Lett.* **82,** 183 (1977).

[57] T. Omata, N. Murata, and K. Satoh, *Biochim. Biophys. Acta* **765,** 403 (1984).

nizer should therefore be kept very short (20 to 30 sec) and lie far below 1 min. It is also recommended to perform the extraction with solvents which have been stored in the cold room (e.g., at 3 to 5°).

Once a pigment extract has been prepared, the chlorophyll content should be determined immediately thereafter. The level of pheophytin *a* and *b* will increase with increasing storage time of the extract. This is particularly the case in organic solvents which are miscible with water such as acetone, methanol, or ethanol. In cases where the storage of a pigment extract is required, it should only be done in a water-free organic solvent, i.e., petroleum hydrocarbon (e.g., bp 40 to 70°) to minimize pheophytinization. This can be accomplished in a separatory funnel by transferring the pigments of the aqueous extraction solvent, adding petroleum hydrocarbon and water, into the organic hydrocarbon epiphase. The aqueous acetone (methanol or ethanol) hypophase should contain at least 50% water to guarantee a quantitative transfer of the more polar pigments (chlorophyll *b* and xanthophylls) into the water-free epiphase also.

Transformation of Chlorophylls into Pheophytins by HCl and Quantitative Determination

There are many cases in present scientific research (e.g., senescent tissue, leaves from plants fumigated with acid gases such as SO_2 or NO_2, plants under heat or light stress, extracts from herbicide-treated chloroplast particles, or from chlorophyll proteins after gel electrophoresis), where pigment extracts may contain pheophytins together with chlorophylls, and one would like to determine the total amount of pigment in terms of chlorophyll equivalents. This is possible by quantitatively transforming the remaining chlorophylls into pheophytins under defined conditions (see below). The total amount of pigments in terms of chlorophyll *a* and *b* equivalents is then determined by two-wavelength spectrometry of the resulting pheophytins at the red absorption maxima of pheophytin *a* and *b*. This method goes back to Willstätter and Stoll.[3]

Equations for the quantitative determination of pheophytin *a* and *b* are given by Bauer[58] for diethyl ether (transformation with HCl) on the basis of the absorption coefficients for chlorophyll *a* and *b* of Smith and Benitez.[49]

Diethyl ether[59]:

$$C_{\mathrm{ph}\ a} = 16.7A_{667} - 3.20A_{665}$$
$$C_{\mathrm{ph}\ b} = 25.3A_{655} - 8.10A_{667}$$
$$C_{\mathrm{ph}\ a+b} = 8.60A_{667} + 22.10A_{655}$$

[58] A. Bauer, *Planta* **51,** 84 (1958).

[59] $C_{\mathrm{ph}\ a,\,b,\,a+b}$, Concentration of pheophytins *a*, *b*, *a* + *b* in μg/ml extract solution.

Equations in 80% acetone (transformation with oxalic acid) were determined by Vernon[45]:

$$C_{\text{ph }a} = 20.15A_{663} - 5.87A_{655}$$
$$C_{\text{ph }b} = 31.90A_{655} - 13.40A_{666}$$
$$C_{\text{ph }a+b} = 6.75A_{663} + 26.03A_{655}$$

Though these equations in both solvent systems provide good values for the total amounts of pheophytin *a* + *b* (or chlorophyll *a* + *b*), the relative levels of pheophytin *a* and *b* are not accurate, which results in values for the ratio of pheophytin *a* to *b* which are too low.

By use of a modern spectrophotometer (UV-260 Shimadzu), which allows automatic recording and printing out of the proper wavelengths and absorbance values at the major and minor absorption maxima, we redetermined specific absorption coefficients of pheophytin *a* and *b* in diethyl ether and 80% acetone and also established the coefficients for several other solvents (Table VI). This was performed by transforming chlorophyll *a* and *b*, which were freshly isolated by TLC,[34] into their pheophytins. The mean values shown in Table VI are based on at least five chlorophyll isolations per solvent and on the new specific absorption coefficients for chlorophylls shown in Table II. The repetitive recording was carried out at different chlorophyll (pheophytin) concentrations (2 to 15 μg/ml). The differences from the older values of Bauer[58] and Vernon[45] are mainly due to a more accurate determination of the proper wavelengths of the main absorption maxima with the new generation of spectrophotometers and to the fact that the absorbance readings, e.g., of pheophytin *a* at λ_{max} of pheophytin *b*, can be performed automatically.

Determination of Pheophytin a and b

On the basis of the new specific absorption coefficients for pheophytins (Table VI), equations were established which allow the determination of pheophytin *a* and *b* of plant pigment extracts in different solvents (Table VII). The equations for the different solvents were cross-checked using pigment extracts from seven different plants. For routine analysis we highly recommend the solvents diethyl ether (equations given in Table VII) and 80% acetone (equations in Table VII). In these solvents the spectral characteristics of pheophytins are almost independent of the acid content of the solution.

Pheophytinization under Defined Conditions

The transformation of chlorophylls in 5 ml of a leaf pigment extract solution (range of 5 to 20 μg *a* + *b*/ml) to their pheophytins proceeds

TABLE VI

SPECIFIC ABSORPTION COEFFICIENTS (sp. coeff.) FOR PHEOPHYTIN *a* AND *b* Newly Determined[a] FOR ORGANIC SOLVENTS OF DIFFERENT POLARITY AND WATER CONTENT (IN THE PRESENCE OF HYDROCHLORIC ACID)[b]

Diethyl ether (pure solvent)		Diethyl ether (H_2O saturated)		Acetone, 80% (v/v) (aqueous sol.)		Acetone (pure solvent)		Ethanol, 95% (v/v) (aqueous sol.)		Methanol (100%) (pure solvent)		Methanol, 90% (v/v) (aqueous sol.)	
(nm)	(sp. coeff.)	(nm)	(sp. coeff.)	(nm)	(sp. coeff.)	(nm)	(sp. coeff.)	(nm)	(sp. coeff.)	(nm)	(sp. coeff.)	(nm)	(sp. coeff.)
						Pheophytin *a*							
666.6	60.1	*666.6*	58.04	*665.4*	51.88	*653.4*	48.51	*662.2*	42.8	*654.2*	50.4	*655.2*	47.2
654.2	20.65	654.2	18.2	653.4	24.0	652.6	48.02	654.2	35.0	647.6	36.1	653.4	45.83
609.0	8.94	**609**	9.0	**608.2**	9.79	**601**	9.19	**607**	10.66	**602**	9.88	**604**	9.7
561.0	3.49	**561**	2.96	—	—	**566**	8.63	**563.7**	7.06	**566.8**	9.20	**567.4**	—
533.4	11.45	**533.4**	11.35	**535.8**	11.75	**529**	9.56	**536.2**	11.17	**532.6**	9.56	**533.6**	9.5
505.2	13.04	**505.2**	13.22	**506.2**	12.11	—	—	**507.1**	9.24	—	—	—	—
471	4.69	**472.6**	7.22	**473.4**	4.53	—	—	—	—	—	—	—	—
470	4.39	470	7.09	470	4.28	470	2.02	470	4.32	470	1.53	470	2.66
408.8	126.51	*408.8*	120.67	*411*	122.43	*418.2*	206.35	*417.4*	142.58	*417.4*	206.14	*418.2*	181.87
						Pheophytin *b*							
666.6	8.76	666.6	5.65	665.4	8.79	653.4	33.19	662.2	17.9	654.2	33.2	655.2	31.4
654.2	43.62	*654.2*	32.7	*653.4*	28.96	*652.6*	34.22	*654.2*	32.6	*647.6*	43.5	*653.4*	32.75
599.2	9.51	**599.2**	6.61	**598.4**	7.68	**597**	10.96	**599.4**	9.29	**596.6**	11.0	**598.4**	9.10
523.6	14.44	**523.6**	10.05	**527**	10.83	**527.2**	13.62	**528.4**	11.93	**534.5**	9.34	**528.6**	11.99
470	6.44	470	4.51	470	4.78	470	5.97	470	7.30	470	1.57	470	6.38
433.2	218.87	*433.2*	164.32	*436*	146.82	*433.8*	174.52	*437.4*	140.5	*420*	213.76	*436.2*	129.56

[a] Application of a "UV-260 Shimadzu" UV-visible recording spectrometer and by using the redetermined absorption coefficients of chlorophyll *a* and *b* (see Table II) as reference basis.

[b] Wavelengths (nm) of the two major absorption maxima in the red and blue spectral region are underlined (italics) as are those of the minor absorption maxima (boldface). Freshly isolated solutions of purified chlorophyll *a* and *b* were transformed into pheophytins by addition of one drop of 25% aqueous HCl solution (Merck No. 316 analytical grade) per 5 ml of chlorophyll solution. The coefficients given are those determined in the presence of this amount of hydrochloric acid. For diethyl ether and 80% aqueous acetone the specific coefficients and the wavelengths of the major and minor maxima shown here are very similar to those found in acid-free solution. For the other solvents the spectral characteristics of acid-free pheophytin are different (see text).

TABLE VII

EQUATIONS FOR THE DETERMINATION OF THE CONCENTRATIONS OF PHEOPHYTIN a ($C_{ph\,a}$), PHEOPHYTIN b ($C_{ph\,b}$), TOTAL PHEOPHYTINS ($C_{ph\,a+b}$), AND OF TOTAL CAROTENOIDS (C_{x+c}) IN LEAF PIGMENT EXTRACTS, IN WHICH THE CHLOROPHYLLS a AND b HAD BEEN COMPLETELY TRANSFORMED TO PHEOPHYTINS[a]

Diethyl ether (pure solvent)[b]

$$C_{ph\,a} = 17.87A_{666.6} - 3.58A_{654.2}$$
$$C_{ph\,b} = 24.62A_{654.2} - 8.46A_{666.6}$$
$$C_{ph\,a+b} = 9.41A_{666.6} + 21.03A_{654.2}$$
$$C_{x+c} = \frac{1000A_{470} - 4.39A_{666.6} - 6.44A_{654.2}}{174}$$

Diethyl ether (water saturated)[c]

$$C_{ph\,a} = 18.22A_{666.6} - 3.15A_{654.0}$$
$$C_{ph\,b} = 32.33A_{654} - 10.14A_{666.6}$$
$$C_{ph\,a+b} = 7.08A_{666.6} + 29.18A_{654}$$
$$C_{x+c} = \frac{1000A_{470} - 7.09A_{666.6} - 4.51A_{654}}{135}$$

Ethanol, 95% (v/v)

$$C_{ph\,a} = 42.41A_{666.2} - 23.28A_{654}$$
$$C_{ph\,b} = 55.67A_{654} - 45.53A_{662.2}$$
$$C_{ph\,a+b} = 32.39A_{654} - 3.12A_{662.2}$$
$$C_{x+c} = \frac{1000A_{470} - 4.32A_{662.2} - 7.30A_{654}}{160}$$

Acetone (pure solvent)

$$C_{ph\,a} = 516.7A_{653.4} - 501.2A_{652.6}$$
$$C_{ph\,b} = 732.5A_{652.6} - 725.1A_{653.4}$$
$$C_{ph\,a+b} = 321.3A_{652.6} - 208.4A_{653.4}$$
$$C_{x+c} = \frac{1000A_{470} - 2.02A_{653.4} - 5.97A_{652.6}}{148}$$

$$C_{ph\,a+b} = 22.3A_{653}$$

Acetone, 80% (aqueous, v/v)

$$C_{ph\,a} = 22.42A_{665.4} - 6.81A_{653.4}$$
$$C_{ph\,b} = 40.17A_{653.4} - 18.58A_{665.4}$$
$$C_{ph\,a+b} = 3.84A_{665.4} + 33.36A_{653.4}$$
$$C_{x+c} = \frac{1000A_{470} - 4.28A_{665.4} - 4.78A_{653.4}}{164}$$

Methanol (pure solvent)

$$C_{ph\,a} = 43.77A_{654.2} - 33.40A_{647.6}$$
$$C_{ph\,b} = 50.71A_{647.6} - 36.32A_{654.2}$$
$$C_{ph\,a+b} = 7.45A_{654.2} + 17.31A_{647.6}$$
$$C_{x+c} = \frac{1000A_{470} - 1.53A_{654.2} - 1.57A_{647.6}}{160}$$

Methanol, 90% (aqueous, v/v)

$$C_{ph\,a} = 306.8A_{655.2} - 294.2A_{653.4}$$
$$C_{ph\,b} = 442.2A_{653.4} - 429.4A_{655.2}$$
$$C_{ph\,a+b} = 148A_{653.4} - 122.6A_{655.2}$$
$$C_{x+c} = \frac{1000A_{470} - 2.66A_{655.2} - 6.38A_{6.53.4}}{126}$$

$$C_{ph\,a+b} = 23.1A_{654}$$

[a] The transformation of chlorophylls into their pheophytins was performed in all solvent systems by adding one drop of a 25% aqueous hydrochloric acid solution (HCl) to 5 ml of the total leaf pigment extract.

[b] The equations given for solvents of different polarity and water content are based on the newly determined specific absorption coefficients listed in Table VI. The concentrations obtained by inserting the appropriate measured extinction (absorbance) values (at λ_{max} of pheophytin a, pheophytin b, and at 470 nm) are micrograms pigment per milliliter extract solution. The amounts of pheophytin a and b are expressed as micrograms chlorophyll a and b equivalents. By multiplication with the factors 0.975 and 0.9754 one obtains the values, in micrograms, of pheophytin a and b, respectively.

[c] Pigment extracts in water-saturated diethyl ether become turbid upon addition of one drop of HCl solution (+ shaking) and need to be centrifuged (5 min) before the absorbancy readings.

instantaneously after addition of one drop of a 25% HCl solution (mild shaking) and is terminated after 30 to 60 sec for most solvents. Only extracts in diethyl ether solutions do not fully absorb the droplet of HCl solution and become slightly turbid, but become clear after a few minutes. Only in the case of water-saturated diethyl ether is a short centrifugation (laboratory centrifuge) needed to clear the solution. In order to avoid evaporation of diethyl ether with subsequent absorbancy readings, which are too high, the tubes should be closed. One further precaution: since the specific absorption coefficients and the wavelength position of the absorption maxima of pheophytins in the solvents acetone (100%), methanol (100 and 90%), and ethanol (95%) are dependent on the acid content of the pigment solution (see spectral characteristics below) one should stick to the defined conditions for pheophytinization described here, when one would like to use the equations listed in Table VII.

Total Carotenoid Content in Pheophytinized Leaf Extracts

The total carotenoid content of the pheophytinized leaf extracts (C_{x+c}) can be determined in a first approximation by inserting the measured absorbance at 470 nm—in addition to that at λ_{max} of pheophytin *a* and *b*—into the equations given in Table VII for different solvents. This requires prompt readings of the absorbancy after pheophytinization before a slow decomposition of carotenoids takes place. The acid and pH dependency of the pheophytin spectrum also affects the absorbancy readings of a leaf extract at 470 nm. When the conditions of pheophytinization mentioned above are adhered to, one obtains good and fully reproducible carotenoid amounts which correspond to those found in the pigment extracts containing chlorophyll.

Spectral Characteristics of Pheophytins

As compared to chlorophylls, the red absorption maxima of isolated pheophytins in acid-free solution are shifted to longer wavelengths with a distinct decrease of the specific absorption coefficients (Table VI; Fig. 9). The maxima in the blue spectral region, however, are shifted to shorter wavelengths, whereby the specific coefficient values of the blue maxima are either little affected or become larger by comparison to those of chlorophylls. Due to these facts the ratio of the specific absorption coefficients at λ_{max} in the blue to λ_{max} in the red are much higher (Table VIII) than those of chlorophylls (Table IV). These general statements hold true for the acid-free pheophytins in all tested solvent systems. They are also valid for diethyl ether and 80% acetone in the presence of some HCl (one drop 25% HCl/5 ml extract solution).

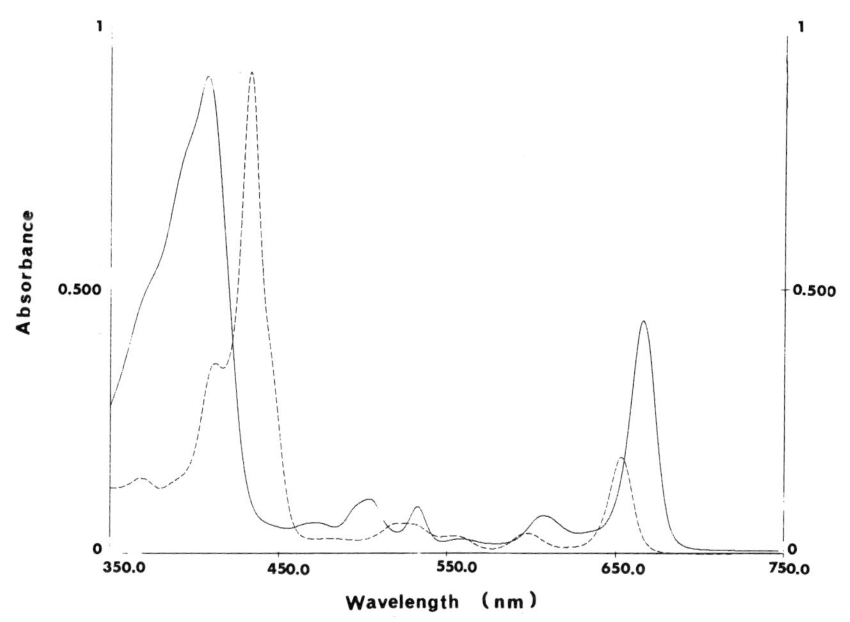

FIG. 9. Absorption spectra of freshly isolated pheophytin *a* (solid line) and pheophytin *b* (broken line) in diethyl ether. The positions of the absorption maxima are given in Table VI.

In acetone (100%), methanol (100 and 90%), and ethanol (95%) the red absorption maximum of chlorophyll *a*, however, shifts to shorter wavelengths upon pheophytinization with HCl (see Tables II and VI). The blue maximum only shifts to ca. 418 nm with a considerable increase in the absorption in the place of 411 nm in acid-free solution. By increasing addition of water there is a change in the spectral characteristics of pheophytin *a* with the appearance of the proper spectra of the acid-free pheophytin. Correspondingly the coefficient ratio (λ_{max} blue to λ_{max} red) is much higher for pheophytin *a* in acid solution than under neutral conditions (see Table VIII). The differential spectral behavior of pheophytin *a* in acid and neutral solution is a particular property and has nothing to do with incomplete pheophytinization. By addition of one drop of HCl to a neutral solution of isolated pheophytin *a*, one changes the absorption spectrum to that of pheophytin *a* in the presence of acids (see Fig. 10).

For pheophytin *b* a shift to shorter wavelength only occurs in 100% methanol. In the other solvents one obtains even in the presence of some HCl spectra which are fairly similar in wavelengths and absorption values to those of the acid-free solvent solution. This indicates that the spectrum of pheophytin *b*—with the exception of pure methanol solutions—is less sensitive to changes in an acid milieu than that of pheophytin *a*.

TABLE VIII
RATIOS OF THE SPECIFIC ABSORPTION COEFFICIENTS[a] AT λ_{max} IN THE BLUE TO λ_{max} IN THE RED SPECTRAL REGION FOR PHEOPHYTIN *a* AND *b*

Solvent	Ratio of the specific coefficients at λ_{max} blue to λ_{max} red[b]	
	Pheophytin *a*	Pheophytin *b*
Diethyl ether (pure solvent)	2.10	5.02
Diethyl ether (water saturated)	2.08	5.03
Acetone, 80% (v/v) (aqueous solution)	2.36	5.07
Acetone (pure solvent)	4.25 (2.15)	5.10 (4.91)
Ethanol, 95% (v/v) (aqueous solution)	3.33 (2.09)	4.31 (4.99)
Methanol (pure solvent)	4.09 (2.08)	4.91 (4.72)
Methanol, 90% (aqueous solution)	3.85 (2.12)	4.03 (4.57)

[a] The coefficients were determined in the presence of some HCl, which was applied to transform the chlorophylls into their pheophytins (see Table 6).

[b] Values in parantheses represent the ratios of coefficients (at λ_{max} blue to red) in the acid-free solvent.

Adjustment of Equations to Other Spectrophotometers

In order to check whether the equations for pigment determination (Table III) can be used with the absorbancy readings of a particular spectrophotometer present in a laboratory, one should isolate fresh chlorophyll *a* and *b* (e.g., by TLC, see Lichtenthaler *et al.*[34]) and determine (1) the wavelength position of their maximum absorption in the red spectral region and (2) their relative absorption at λ_{max} of the other chlorophyll as well as that at 470 nm. This is best performed at concentrations with absorbancy values between 0.4 to 0.8/1 cm light path.

When the specific absorption coefficient at λ_{max} red of chlorophyll *a* and *b* in a particular solvent is accepted from Table II—which can be done in most cases even though the position of λ_{max} may be different by 1 or 1.5 nm from the wavelengths given in Table II—one can recalculate the equations (C_a, C_b, C_{x+c}) for one's own instrument by use of the equations

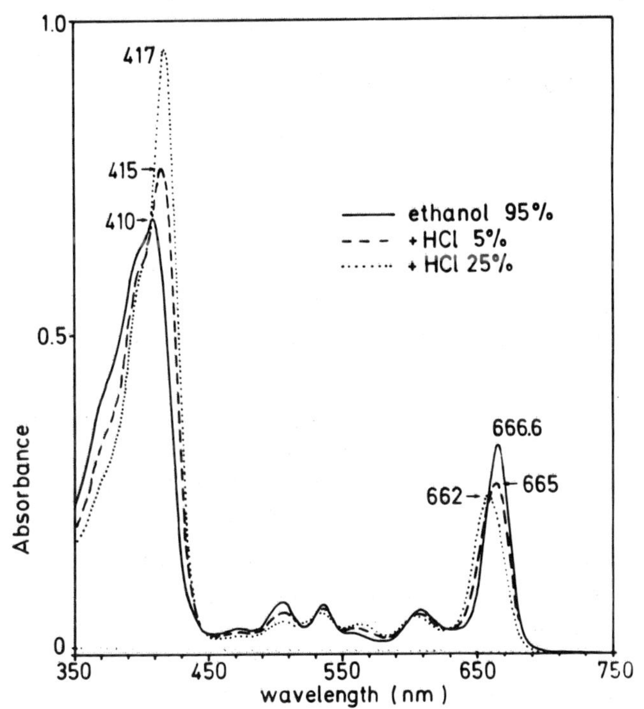

FIG. 10. Absorption spectrum of pheophytin *a* in aqueous (95%) ethanol (solid line) and after addition of one droplet (~0.005 ml) of a 5% (– – –) and of one droplet of a 25% HCl solution (····).

given below. The absorbance of a pigment extract solution (light path 1 cm) at λ_{max} of chlorophyll *a* ($A_{max\,a}$), at λ_{max} of chlorophyll *b* ($A_{max\,b}$), and at 470 nm is defined by the following three equations:

$$A_{max\,a} = aC_a + b'C_b \quad (1)$$
$$A_{max\,b} = bC_b + a'C_a \quad (2)$$
$$A_{470\,nm} = a''C_a + b''C_b + kC_{x+c} \quad (3)$$

a, Specific absorption coefficient of chlorophyll (chl) *a* at λ_{max} (mg/liter); *a'*, specific absorption coefficient of chl *a* at λ_{max} of chl. *b* (mg/liter); *a''*, specific absorption coefficient of chl *a* at 470 nm (mg/liter); *b*, specific absorption coefficient of chl *b* at λ_{max} (mg/liter); *b'*, specific absorption coefficient of chl *b* at λ_{max} of chl. *a* (mg/liter); *b''*, specific absorption coefficient of chl *b* at 470 nm (mg/liter); *k*, specific absorption coefficient of total carotenoids *x* + *c* at 470 nm (mg/liter) (= $A^{1\%}_{1\,cm}$ for $x + c \times 0.1$);

C_a, C_b, concentration of chl a and b; C_{x+c}, concentration of total carotenoids.

Transformation of Eqs. (1) and (2) results in the equations

$$C_a = [(b/z)A_{\max a}] - [(b'/z)A_{\max b}] \quad (4)$$
$$C_b = [(a/z)A_{\max b}] - [(a'/z)A_{\max a}] \quad (5)$$

where

$$z = ab - a'b'$$

When the specific absorption coefficients a and b (= 100%) are taken from Table II the specific coefficients a', a'', b', and b'' can be calculated as follows:

$a' = aE_a/100$ $\quad E_a$ = % absorbance of chl a at λ_{max} chl b
$a'' = aE_{470}/100$ $\quad E_{470}$ = % absorbance of chl a at 470 nm
$b' = bE_b/100$ $\quad E_b$ = % absorbance of chl b at λ_{max} chl a
$b'' = bE_{470}/100$ $\quad E_{470}$ = % absorbance of chl b at 470 nm

By inserting the recalculated values for a' and b' in Eqs. (4) and (5) one obtains the new equations for C_a and C_b, which are adjusted to the particular spectrophotometer.

The equation to determine the concentration of total carotenoids C_{x+c} (μg/ml) is then obtained using the redetermined a'' and b'' values:

$$C_{x+c} = (1000A_{470} - a''C_a - b''C_b/0.1\ A_{1\,\mathrm{cm}}^{1\%} \text{ of } x + c$$

A recalculation of the $A_{1\,\mathrm{cm}}^{1\%}$ for $x + c$ at 470 requires isolation of the total leaf carotenoids (e.g., by TLC[33]) in diethyl ether ($A_{1\,\mathrm{cm}}^{1\%}$ for $x + c$ at λ_{max} = 2450) and measurement of the relative proportion of their absorption at 470 nm as compared to λ_{max} of $x + c$. The $A_{1\,\mathrm{cm}}^{1\%}$ value in diethyl ether (2450) is used as reference.

Acknowledgments

The establishment of the absorption coefficients was made possible by the kindness of Shimadzu Germany, Düsseldorf (Messrs. Steinfeld and Baumann), who for several weeks provided a new UV-260 UV-visible two-beam spectrometer; this is gratefully acknowledged. My particular thanks are due to Dr. Claus Buschmann for writing the programme for the HP 85 and to my son, Stefan Lichtenthaler, for programming the Apple IIe computer and doing most of the calculating and cross-checking. Finally I would like to thank Mr. E. Nagel for measuring the photoacoustic spectra of pigments as well as Mrs. U. Prenzel and U. Sieber for excellent assistance.

[35] α-Tocopherol and Plastoquinone Synthesis in Chloroplast Membranes

By JÜRGEN SOLL

Plant prenylquinones, α-tocopherol, plastoquinone-9, and phylloquinone, function in chloroplasts as membrane constituents, antioxidants, or electron carriers. For a long time biosynthetic studies were hampered by the fact that no chemical intermediates of the biosynthetic pathway or biochemically active organelles were available. Earlier studies[1,2] proposed pathways with a multitude of chemically possible intermediates. The detailed work[3] on the chemical synthesis of prenylquinones enabled others[4-6] to work out the most probable pathway (Fig. 1) in plastoquinone and tocopherol biosynthesis.

Chemical Synthesis of Prenylquinones

The small-scale synthesis of **IV** (Fig. 2) is described here[4] and can be applied to the synthesis of other prenylquinones (*I*, *II*, *III*, *V*, *VI*) without problems. In many cases methylquinones are not commercially available and have to be prepared in advance. The corresponding phenol (4 mmol) is dissolved in 10 ml methanol and oxidized by 10 mmol of Fremy's salt[7] $\{[(SO_3)_2NO]K_2\}$ in 120 ml water and 4 ml sodium acetate (1 *M*). The reaction is allowed to continue for 30 min and the quinone is then extensively extracted with diethyl ether. The organic solvent is evaporated and the quinone purified by column chromatography (silica gel 60, Merck, FRG) using $CHCl_3$ as developing solvent. Quinone (0.7 mmol) is dissolved in 1.5 ml benzene, 2.5 ml H_2O, and 250 mg $Na_2S_2O_4$ is added. The reduction is completed after 5 min, the quinol is washed with ice water, and dried in a desiccator. Freshly distilled BF_3–etherate (0.3 ml in 1 ml tetrahydrofuran) is added dropwise via a syringe to a solution of 1.1 mmol quinol, 200 mg Al_2O_3 (W-200 basic, Woelm-Pharma, FRG), and 1.1 mmol

[1] W. Janiszowska and J. F. Pennock, *Vitam. Horm. (N.Y.)* **34,** 77 (1976).

[2] D. R. Threlfall and G. R. Whistance, *in* "Aspects of Terpenoid Chemistry and Biochemistry" (T. W. Goodwin, ed.), p. 335. Academic Press, London, 1971.

[3] H. Mayer and O. Isler, this series, Vol. 18C, p. 241.

[4] J. Soll and G. Schultz, *Phytochemistry* **19,** 215 (1980).

[5] J. Soll, M. Kemmerling, and G. Schultz, *Arch. Biochem. Biophys.* **204,** 544 (1980).

[6] S. R. Morris and D. R. Threlfall, *Biochem. Soc. Trans.* **11,** 587 (1983).

[7] H. J. Teuber and W. Rau, *Chem. Ber.* **86,** 1036 (1953).

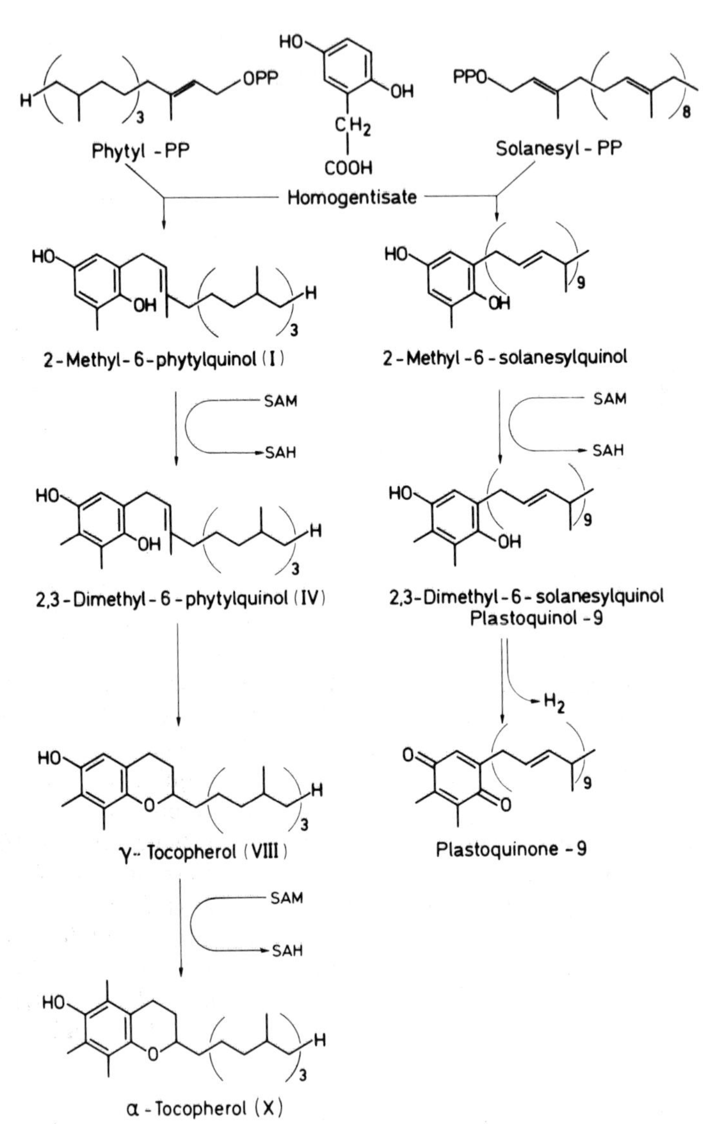

FIG. 1. Proposed pathway of α-tocopherol and plastoquinone-9 synthesis in spinach chloroplasts. From the available data these are the most likely intermediates to be involved in prenylquinone synthesis. A possible bypass in tocopherol synthesis might occur which leads from I via VII to VIII instead of I via IV to VIII (see text). SAM, *S*-Adenosylmethionine; SAH, *S*-adenosylhomocysteine.

Methyl-phytylquinols

Tocopherol

R_1	R_2	R_3		
CH_3	H	H	- 2-Methyl-6-	I
H	CH_3	H	- 2-Methyl-5-	II
H	H	CH_3	- 2-Methyl-3-	III
CH_3	CH_3	H	- 2,3-Dimethyl-5-	IV
CH_3	H	CH_3	- 2,5-Dimethyl-6-	V
CH_3	CH_3	CH_3	- Trimethyl-6-	VI

} phytylquinol

R_1	R_2	R_3		
CH_3	H	H	- δTocopherol	- VII
CH_3	CH_3	H	- γTocopherol	- VIII
CH_3	H	CH_3	- βTocopherol	- IX
CH_3	CH_3	CH_3	- αTocopherol	- X

FIG. 2. Nomenclature and identification of prenylquinols and tocopherols.

isophytol in 2 ml dry tetrahydrofuran. The mixture is stirred under N_2 in the dark for 35 hr. Residual BF_3 is hydrolyzed on ice, and the prenylated quinols extracted with diethyl ether, the ether solution dried, and the organic solvent evaporated. The resulting quinol (**IV**) is oxidized by 400 mg Ag_2O in dry diethyl ether. Prenylquinones are purified by column chromatography (Silica-gel 60, Merck) developed with petrol (bp 60–80°)–diethyl ether, 15 : 1 (system 1). Purity of the products is verified by thin-layer chromatography (precoated plates on glass, silica gel, G-1500 LS254, Schleicher and Schüll, FRG) in system 1. When the prenylquinones **I**, **II**, and **III** are to be synthesized, care has to be taken to separate the isomers properly. This can be achieved by repeated thin-layer chromatography as above. The succession of prenylquinones in this thin-layer chromatography system is shown in Fig. 3A. Quinones and

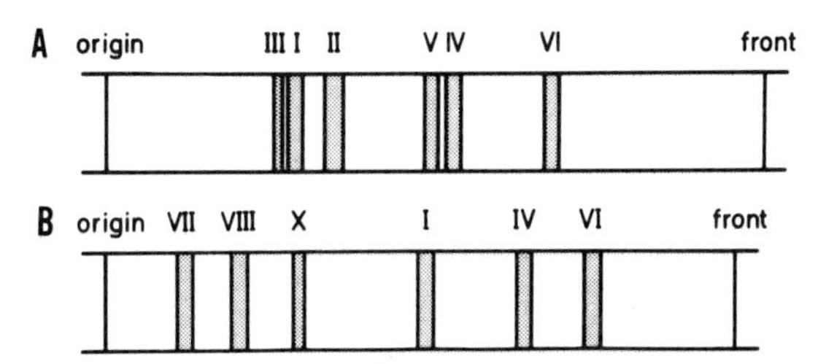

FIG. 3. Separation of prenylquinones and tocopherols by thin-layer chromatography. (A) The separation of mono-, di-, and trimethylphytylquinones on precoated thin-layer plates, silica gel G-1500, with diethyl ether : petrol (1 : 15, v : v) as developing solvent. The purification of prenylquinones and tocopherols using the same thin-layer plates but diethyl ether : petrol, (1 : 10, v : v) as solvent system. R_f values and separation are variable with silica gel plates obtained from different manufacturers. Precoated plates on glass give better resolution than those on plastic or aluminum foil (see also text).

quinols have a tendency to oxidize and polymerize during chromatography. We have obtained the best results using the systems described; other systems caused more oxidation, decomposition, and poor separation of quinones or prenylquinone isomers. All products should be stored and purified in the quinone form and not in the quinol form which is more susceptible to uncontrolled breakdown. Quinones and prenylquinones are detected on thin-layer plates with fluorescence indicator at 254 nm, while quinoles and prenylquinoles are visualized with $FeCl_3$/2,2′-dipyridyl 0.1% : 0.25% (w/w) in ethanol.

Chemicals Needed for the Enzymatic Assays

As can be seen from Fig. 1, homogentisate, polyprenyl diphosphate, and *S*-adenosylmethionine are necessary for the incubation. Unlabeled and labeled *S*-adenosylmethionine and homogentisate are commercially available, while phytyl diphosphate and solanesyl diphosphate have to be prepared.[8–10] Dry trichloroacetonitrile (15 mmol), 5 mmol ditriethylammonium phosphate,[10] and 30 ml dry acetonitrile are mixed in a round-bottom flask. Two millimoles of phytol in 15 ml acetonitrile is added dropwise over a 3-hr period. The mixture is stirred for another 12 hr, then 50 ml of acetone is added and concentrated ammonia is dropped into the solution until no further precipitation occurs. Precipitation occurs for 2 hr at 0°. The solid is repeatedly washed with 0.28 *M* ammonia in methanol to eliminate prenyl monophosphates. The resulting prenyl diphosphate is dried and used in the enzyme assay. Product analysis showed that this preparation is still heavily contaminated by inorganic phosphates which, however, do not interfere with the enzyme assays. If further purification is desired this can be achieved by recrystallization in $CHCl_3$–methanol.[8] [^{3}H]Homogentisate, labeled by tritium exchange service, has to be purified prior to use in the following system: silica gel precoated thin-layer plates on glass and toluene/methanol/acetic acid (80/20/4 v/v/v) as developing solvent.

Preparation of Chloroplasts and Chloroplast Components

Chloroplasts are isolated from spinach leaves by standard procedures[11] and further purified on silica sol gradients.[12] Chloroplast compo-

[8] C. N. Joo, C. E. Park, J. K. G. Kramer, and M. Kates, *Can. J. Biochem.* **51,** 1527 (1973).

[9] R. Widmaier, J. Howe, and P. Heinstein, *Arch. Biochem. Biophys.* **200,** 609 (1980).

[10] G. Popjak, J. W. Cornfarth, R. H. Cornfarth, R. Ryhage, and S. de Witt Goodman, *J. Biol. Chem.* **237,** 56 (1962).

[11] H. Nakatani and J. Barber, *Biochim. Biophys. Acta* **461,** 510 (1977).

nents, e.g., envelope and thylakoid membranes and soluble chloroplast protein, are prepared as described.[13]

Enzyme Assay for the Synthesis of Tocopherol and Its Intermediates in Chloroplasts

As outlined in Fig. 1, the synthesis of α-tocopherol comprises a number of reaction steps, catalyzed by the following enzymes: homogentisate decarboxylase-phytyltransferase; *S*-adenosylmethionine : methyl-6-phytylquinol methyltransferase; 2,3-dimethylphytylquinolcyclase; *S*-adenosylmethionine : γ-tocopherol methyltransferase (no EC numbers available). Of the precursors and cosubstrates used only homogentisate, polyprenyl diphosphate, and *S*-adenosylmethionine are water soluble. Prenylquinones, prenylquinols, and tocopherols are not water soluble and it is difficult to determine their real concentration in the test. They are either added in ethanol (no more than 1% ethanol final concentration in the enzyme assay) or in diethyl ether, which is evaporated to dryness prior to the assay. Introduction of the methyl groups into the aromatic moiety is only possible at the quinol stage and not in the quinone form.[4] The same is valid for the formation of the chromanol stage (**IV** → **VIII**). A photometrically adjusted amount of quinone (UV maxima, see Refs. 14 and 15; extinction coefficients, see Ref. 3) was dissolved in 1 ml methanol, reduced with a little solid $NaBH_4$ for 2 min, transferred to diethyl ether, and washed with H_2O. The diethyl ether is evaporated to dryness under N_2 in the reaction vials which were used later in the enzyme assay. Substrate concentrations described[4,14-16] are 50–100 μM prenyl diphosphate, 100 μM *S*-adenosylmethionine, 100–200 μM prenylquinol or tocopherol at pH 7.6–8.2 with $MgCl_2$ as cofactor (1–10 m*M*) and chloroplasts equivalent to 0.5–1 mg of chlorophyll. Other cofactors like cysteine, dithiothreitol (DTT), light, and Mn^{2+} do not seem to be necessary.[4,15,16] Increased solubilization of quinols and tocopherols can be achieved by detergents (Tween 80) which do not seem to inhibit the *S*-adenosylmethionine : γ-tocopherol methyltransferase.[17]

---

[12] G. Mourioux and R. Douce, *Plant Physiol.* **67,** 470 (1981).

[13] R. Douce and J. Joyard, *Adv. Bot. Res.* **7,** 1 (1979).

[14] J. Soll, G. Schultz, J. Joyard, R. Douce, and M. A. Block, *Arch. Biochem. Biophys.* **238,** 290 (1985).

[15] P. S. Marshall, S. R. Morris, and D. R. Threlfall, *Phytochemistry* **24,** 1705 (1985).

[16] B. Camara, F. Bardat, A. Seye, A. d'Harlingue, and R. Moneger, *Plant Physiol.* **70,** 1562 (1982).

[17] B. Camara and A. d'Harlingue, *Plant Cell Rep.* **4,** 31 (1985).

Identification and Purification of Labeled Products

The intermediates obtained in tocopherol synthesis are generally purified by repeated thin-layer chromatography[4,18,19] or HPLC.[15,20] The incubation mixture is extracted with $CHCl_3$: MeOH (1 : 2, v : v).[21] The chloroform phase contains prenylquinols, prenylquinones, tocopherols, other lipids, and pigments. Initial results have shown that prenylquinols are the products formed in this and the following reactions, if the products are analyzed under nonoxidizing conditions.[4,22] In general quinols are oxidized by air prior to thin-layer chromatography (for reasons, see above). A 25-μg aliquot of standard substances corresponding to the possible reaction products is added to the chloroform phase.

The first reaction in tocopherol and plastoquinone synthesis involves the prenylation of homogenisate with simultaneous decarboxylation (Fig. 1). The decarboxylation proceeds with stereochemical retention during the biosynthetic process.[23] While the chemical synthesis of monomethylprenylquinols using methylquinol, phytol, and BF_3 as described earlier yields a mixture of **I**, **II**, and **III**, the enzymatic prenylation of homogenisate yields only one product (**I**).[5,15] This is probably due to the directing influence of intermediates occurring during decarboxylation. Analysis of this reaction was done using spinach chloroplasts, [^{3}H]homogentisate, and phytyl diphosphate. The products were purified by thin-layer chromatography using two different systems in succession [first run, silica gel, petrol (bp 60–80°) : diethyl ether, 10 : 1 (system II); rechromatography, cellulose plates impregnated with 7% paraffin, acetone : H_2O, 85 : 15 (system III), or by HPLC (Lichrosorb Si 60, 5 μm, Merck, 0.06% dioxane in isooctane[15]]. Substances are recovered from thin-layer plates by elution of the zones in question twice with 1 ml of methanol. The methanol is evaporated under N_2 and the residual dissolved in acetone. Only **I** was found to be labeled, this very specific initial reaction excludes already many further intermediates.[5,15] This initial step of prenylquinone formation is followed by methylation of the aromatic moiety with *S*-adenosylmethionine as methyl group donor (see Fig. 4). Since the prenylation reaction yields only **I** the following methylation can occur only from two

[18] J. Soll and G. Schultz, *Biochem. Biophys. Res. Commun.* **91,** 715 (1979).

[19] H. K. Lichtenthaler, *in* "Lipids and Lipid Polymers in Higher Plants" (M. Tevini and H. K. Lichtenthaler, eds.), p. 231. Springer-Verlag, Berlin and New York, 1977.

[20] H. K. Lichtenthaler and U. Prenzel, *J. Chromatogr.* **135,** 493 (1977).

[21] E. G. Bligh and W. J. Dyer, *Can. J. Biochem. Physiol.* **37,** 911 (1959).

[22] K. G. Hutson and D. R. Threlfall, *Biochim. Biophys. Acta* **632,** 630 (1980).

[23] R. Krügel, K. H. Grumbach, H. K. Lichtenthaler, and J. Rètey, *Bioorg. Chem.* **13,** 187 (1985).

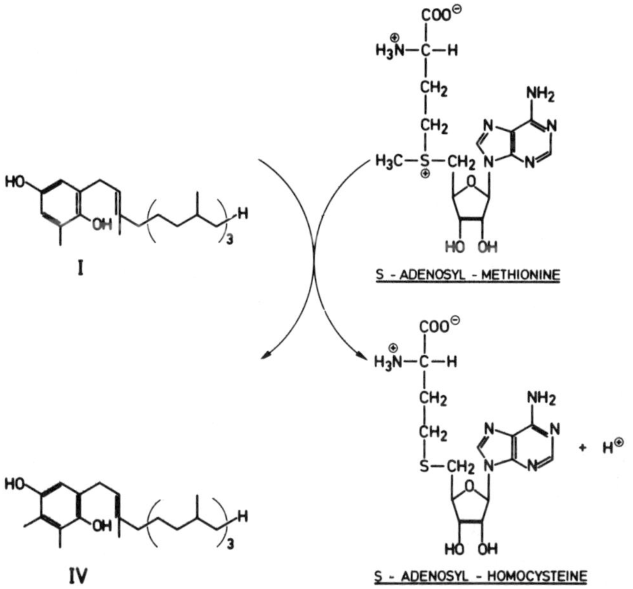

FIG. 4. Methylation of 2-methyl-6-phytylquinol by *S*-adenosylmethionine. Reaction products are 2,3-dimethyl-6-phytylquinol and *S*-adenosylhomocysteine.

different intermediates: (1) **I**, (2) after cyclization of **I** from **VII**. *In vitro* studies using isolated chloroplasts[4] demonstrated that methylation of **I** was about three times higher than from **VII**. This specificity is underlined also by the finding that **II** and **III** are methylated at 10 and 5% of the rate of **I**, respectively, with experimental conditions as described above.

The methylation product of **I** is **IV**, which is processed in a homologous sequence of events to form **X**. **X** formation takes place via **VIII**, which is methylated to form **X**. Products are purified on systems II and III (see Fig. 3). This sequence was supported from the data that **IV** is not methylated to **VI**.[4,15] In all reports known so far the cyclization of the prenylquinol to form the chromanol ring was not studied directly and it seems to be the slowest step in the α-tocopherol formation.

The reactions mentioned above described only one or two steps at a time of a series leading to α-tocopherol. If it is necessary to look at the whole sequence, all cosubstrates have to be included (homogentisate, phytyl diphosphate, *S*-adenosylmethionine). The incorporation rates vary from pmol to nmol/hr · mg chlorophyll for single reaction steps, which makes it obvious that only highly active chloroplast preparations can be used for these approaches. When these multiple step analyses are done the products formed are **I**, **IV**, **VII**, **VIII** and **X**,[4,15,24] confirming the results described earlier. It should be stressed again that it is difficult to separate the different possible isomers in tocopherol synthesis. Monomethylphytylquinols can only be separated by repeated thin-layer chromatography on precoated plates on glass or HPLC[4,15,19,20] ($\lambda_{max\ I}$ 254, $\lambda_{max\ II}$ 253, $\lambda_{max\ III}$ 249 nm). Compounds **IV** and **V** are separated by simple chromatography on precoated thin-layer plates on glass using system II. **VIII** and **IX** should be purified as nitroso derivatives[25] or by HPLC.[15,20] It is obvious from the literature and our own experience that successful separation of isomers depends strongly on the brand of chromatography plates used.

Plastoquinone Synthesis

Plastoquinone synthesis (Fig. 1) occurs essentially via reactions similar for tocopherol; homogentisate (20 μM) and solanesyl diphosphate (80 μM) are condensed to form the equivalent to **I**[5] (2-methyl-6-solanesylquinol), which is then methylated by *S*-adenosylmethionine (70 μM) to yield plastoquinol-9[5] (2,3-dimethylsolanesylquinol) (Fig. 1). The rates of synthesis are again in the picomolar range per hr · mg chlorophyll. Purification of the incubation products is done by two successive thin-layer chromatography systems (first system; system I; rechromatography, cellulose plates impregnated with 7% paraffin, acetone : H_2O, 90 : 10, system IV).

Localization of Prenylquinone Synthesis in Chloroplasts

Recently developed methods for the fractionation and purification of chloroplast components[13] enabled us[5,14,24,26] to localize all but one enzyme in tocopherol and plastoquinone synthesis at spinach chloroplast envelopes. Thylakoids or soluble chloroplast protein had no enzymatic activity. Recombination of membranes with soluble chloroplast extract did not increase prenylquinone synthesis. These observations are now extended to envelope membranes from pea chloroplasts (J. Soll, unpublished). All test and purification conditions were essentially as described for chloroplasts. The amount of membranes used was between 50 and 100 μg protein per assay. The available data do not demonstrate the enzyme responsible for the cyclization of **IV** to yield **VIII**. Though some enzymatic activity in plastoquinone synthesis is found associated with the thylakoid

[24] G. Schultz, J. Soll, E. Fiedler, and D. Schulze-Siebert, *Physiol. Plant.* **64,** 123 (1985).
[25] S. Marcinkiewicz and J. Green, *Analyst* **84,** 304 (1959).
[26] J. Soll, R. Douce, and G. Schultz, *FEBS Lett.* **112,** 243 (1980).

membrane it is probably due to contamination of this membrane fraction by envelopes.

The envelope forms a two-membrane barrier, which surrounds the chloroplast and is present at all stages of chloroplast development.[27] It is now possible to separate the two membranes into outer envelope and inner envelope membrane.[27,28] Applying these methods it is possible to align tocopherol and plastoquinone synthesis with the inner envelope membrane.[14] Again, all enzymes but the cyclization enzyme (**IV** → **VIII**) were demonstrated.

Purification of Enzymes Involved in Tocopherol Synthesis

A membrane fraction obtained from pepper (*Capsicum annuum*) chromoplasts was shown to catalyze α-tocopherol synthesis via the same intermediates as described for chloroplasts[16] (Fig. 1). The methods used were essentially as above. This membrane fraction was then used as a source for the enzyme purification. An acetone powder is obtained from the membranes at $-20°$ which is then solubilized in 0.1 *M* phosphate buffer (pH 7.0), 5 m*M* DTT, and Tween 80 (1 mg/ml), followed by $(NH_4)_2SO_4$ precipitation (20–60%). The protein is purified by column chromatography[29]: (1) blue Sepharose CL-6B, 50 m*M* KH_2PO_4, 1 m*M* DTT, 1 m*M* EDTA, pH 6.2; (2) blue Sepharose CL-6B, 50 m*M* Tris–HCl, 1 m*M* DTT, 1 m*M* EDTA, pH 8; (3) DEAE–Sephacel, 50 m*M* Tris-HCl, pH 7.6, 1 m*M* DTT eluted with a gradient of 0–0.4 *M* KCl. Analysis of the active fractions obtained, by SDS–polyacrylamide gel electrophoresis showed only one band at about 33,000 Da[29] (pH optimum, 8.2; K_m *S*-adenosylmethionine, 2.5 μM; K_m γ-tocopherol, 13.7 μM).[29]

Analysis of Envelope Membranes for Tocopherol and Plastoquinone

Determination should be done from fresh or deep-frozen material. Freeze-dried membranes contain less quinols and tocopherols than fresh membranes and more quinone and tocoquinone instead.[5,14,30,31] (see Table I). Membranes are extracted either by $CHCl_3$/MeOH (see earlier) or by

[27] R. Douce, M. A. Block, A. J. Dorne, and J. Joyard, *Subcell. Biochem.* **10,** 1 (1984).

[28] K. Cline, J. Andrews, J. Mersey, E. H. Newcomb, and K. Keegstra, *Proc. Natl. Acad. Sci. U.S.A.* **78,** 3595 (1981).

[29] A. d'Harlingue, F. Villat, and b. Camara, *C. R. Seances Acad. Sci., Ser. 3* **6,** 233 (1985).

[30] H. K. Lichtenthaler, U. Prenzel, R. Douce, and J. Joyard, *Biochim. Biophys. Acta* **641,** 99 (1981).

[31] G. Schultz, H. Bickel, B. Buchholz, and J. Soll, *in* "Chloroplast Development" (G. Akoyunoglou, ed.), p. 311. Balaban Int. Sci. Serv, Philadelphia, Pennsylvania, 1981.

TABLE I
PRENYL LIPID CONCENTRATION IN CHLOROPLAST MEMBRANES FROM SPINACH LEAVES[a]

Prenylquinone	Thylakoid	Envelope mixture	Inner envelope	Outer envelope	Retention time (min)
α-Tocopherol	1.1	2.8	6.7	9.8	5.6
α-Tocoquinone	0.24	0.2	—	—	3.8
Phylloquinone K_1	0.34	0.1	0.07	0.05	7.3
Plastoquinone-9	3.9[b]	1.2[b]	1.63	1.1	25.5
Plastoquinol-9	—		1.54	1.1	11.3
Total prenylquinones	5.5	4.3	10.0	12.1	—

[a] Adapted from Refs. 14, 30, and 31; values are expressed in μg/mg protein.
[b] Represents the sum of plastoquinone and plastoquinol.[30]

hexane/acetone (10 : 4, v : v).[30] The lipid extract is further analyzed by HPLC (RP8, 7-μm mesh, Merck) using methanol : water (95.7 : 4.3, v : v) as developing solvent (1.5 ml flow rate)[14,32] and two UV detectors set at 250 nm (to detect quinones) and 292 nm (to detect tocopherol and quinol), respectively.[14,33]

Acknowledgments

I would like to thank Prof. G. Schultz for his continuous support and encouragement during the progress of this work.

[32] H. K. Lichtenthaler, *in* "Handbook of Chromatography" (H. K. Mangold, ed.), p. 115. CRC Press, Boca Raton, Florida, 1984.
[33] D. R. Threlfall and T. W. Goodwin, *Biochem. J.* **103,** 573 (1967).

[36] Solubilization and Reconstitution of Carotenogenic Enzymes from Daffodil Chromoplast Membranes Using 3-[(3-Cholamidopropyl)dimethylammonio]-1-propane Sulfonate

By P. BEYER

Purified chromoplasts of daffodil flowers have been shown to be highly active in the *in vitro* synthesis of β-carotene with isopentenyl diphosphate (IPP) as a substrate.[1] Within these organelles, the sole site of carotenoid

[1] P. Beyer, K. Kreuz, and H. Kleinig, *Planta* **150,** 432 (1980).

accumulation is the concentrically stacked multilayer of lipid-rich membranes surrounding the plastid stroma (membranous-type chromoplast).[2] The topology of β-carotene synthesis takes place in close connection to these plastid membranes: 15-*cis*-phytoene, the first carotene in the carotenogenic sequence, is formed from IPP by the phytoene synthase complex, a peripheral membrane protein, which can be easily obtained in soluble form. The phytoene-metabolizing enzymes, on the other hand, are integral membrane proteins,[3] catalyzing one *cis-trans* isomerase, four dehydrogenase, and two cyclase reactions. These enzymatic reactions proceed with trace accumulation of *cis*-phytofluene as an intermediate and, as has been shown recently,[4] obey the criteria of "metabolic channeling." The enzymes participating thus can be expected to function in a very cooperative way in the daffodil chromoplast membranes.

A requirement for research on these membrane-bound carotenogenic reactions, which are poorly understood at present, is the solubilization of the individual enzymes in their native states, providing a prerequisite for their separation and their subsequent reconstitution into membrane structures (liposomes) in order to create a topology which permits enzyme interactions. Using the zwitterionic detergent 3-[(3-cholamidopropyl)dimethylammonio]-1-propane sulfonate (CHAPS),[5] whose well-solubilizing, nondenaturing and disaggregating properties are well known, such a procedure can be performed with the complete enzyme system in question while fully maintaining its biological activities. The experimental methodology is reported here.

Subfractionation of Purified Chromoplasts

Purified chromoplast pellets[2] are resuspended in incubation buffer (100 m*M* Tris–HCl, pH 7.2; 10 m*M* $MgCl_2$; 2 m*M* dithioerythritol) using a loose-fitting Dounce homogenizer and routinely homogenized by use of a French pressure cell at a pressure of 5000 psi. Homogenates can be stored at $-70°$ for at least 1 year without loss of biosynthetic activity. High-speed centrifugation (180,000 *g*, 4 hr) yields a weakly colored supernatant containing the phytoene synthase complex, whereas the phytoene-metabolizing and β-carotene-synthesizing enzyme system (*cis–trans* isomerase, dehydrogenase, and cyclase activities) remains bound to the membrane pellet.

[2] B. Liedvogel, P. Sitte, and H. Falk, *Cytobiologie* **12,** 155 (1976).
[3] K. Kreuz, P. Beyer, and H. Kleinig, *Planta* **154,** 66 (1982).
[4] P. Beyer, G. Weiss, and H. Kleinig, *Eur. J. Biochem.* **153,** 341 (1985).
[5] L. M. Hjelmeland, *Proc. Natl. Acad. Sci. U.S.A.* **77,** 6368 (1980).

Enzymatic Preparation of the Substrate [^{14}C]Phytoene

15-*cis*-[^{14}C]Phytoene, the substrate for *in vitro* experiments, is synthesized from [1-^{14}C]IPP (1.96 GBq mmol^{-1}, Amersham Buchler, Frankfurt, FRG) as a substrate using the supernatant of the 180,000 *g* centrifugation (see above). For incubation, 1-ml portions of the phytoene synthase-containing supernatant are supplemented with chromoplast membrane lipid (0.2 mg ml^{-1}) added as liposomes, 1 m*M* $MnCl_2$, 2 m*M* NADP, 20 μM FAD, 3 m*M* ATP, and incubated for 6 hr in the presence of 74 kBq [1-^{14}C]IPP. With 3 m*M* ATP present in the incubation assay, the formation of the alcoholic by-products (geranylgeraniol, prephytoenol), deriving from the prenyltransferase moiety of the phytoene synthase complex, is highly suppressed. Still contaminating amounts of by-products have to be removed by HPLC on a C_{18} reversed-phase column in the gradient system described earlier.[1] 15-*cis*-[^{14}C]Phytoene is detected by a column chromatography radiomonitor (Berthold, Wildbad, FRG), collected, and transferred to petroleum benzine by partition. Ideally up to 40% of the applied radioactive label is converted into 15-*cis*-[^{14}C]phytoene. Storage for several weeks is possible at $-20°$ without significant cis–trans isomerization, as can be judged by comparison with authentic cis–trans mixtures of phytoene on a HPLC aluminum oxide column.[4]

Alternatively, to avoid the formation of by-products, a purification step for the phytoene synthase complex is recommended. For this purpose we use a chromatofocusing procedure on a PBE gel bed (0.5 × 30 cm, Pharmacia, Freiburg, FRG) equilibrated with start buffer, 25 m*M* Tris–HCl, pH 7.4; 2.5 m*M* $MgCl_2$; 0.5 m*M* dithioerythritol; 20 vol% ethylene glycol. The sample (15 ml supernatant of the 180,000 *g* centrifugation) is equilibrated with the same buffer by dialysis and eluted from the column with the eluent polybuffer 74 (diluted 1 : 8, Pharmacia, Freiburg, FRG), pH 4, supplemented with additives as in the starting buffer. The reduction of the polarity in the starting buffer and in the eluent reduces the tendency for the formation of hydrophobic aggregates of the enzyme complex and provides better separation. It elutes at a pH of 5.3–5.1 from the column (Fig. 1) and is reequilibrated to the incubation buffer by dialysis. The enzyme preparation exhibits an enrichment of about 10-fold.

Preparation of 15-*cis*-[^{14}C]Phytoene-Containing Liposomes

The general strategy for the reconstitution experiment is to supply the lipophilic substrate 15-*cis*-[^{14}C]phytoene at the same site where the reinsertion (reconstitution) of the solubilized enzymes occurs, which is in the membrane structures. For this purpose liposomes are made from chromoplast membrane lipids, containing the applied label within their membranes.

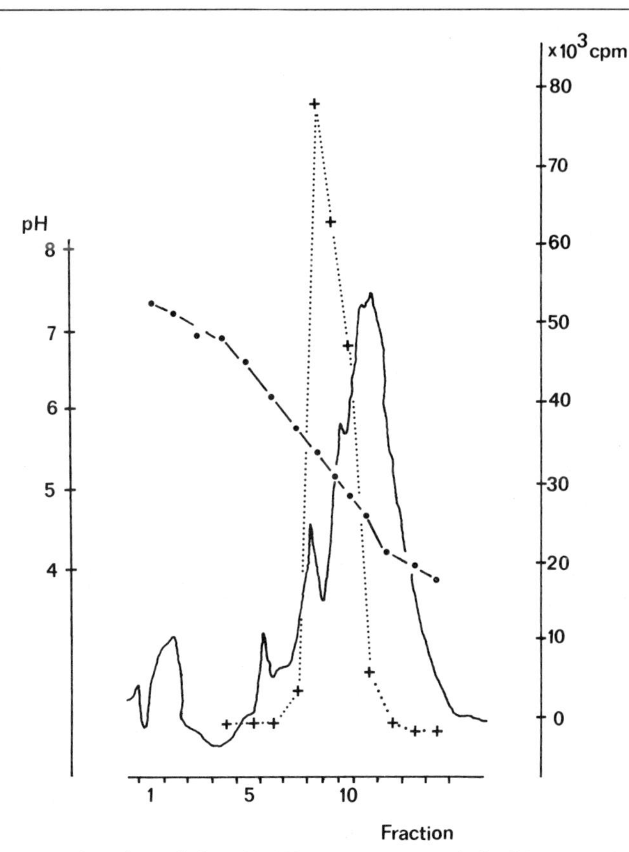

FIG. 1. Chromatofocusing of the 180,000 g supernatant. (—), $OD_{280\,nm}$; (●—●), pH gradient; (+···+), radioactivity in [^{14}C]phytoene after incubation of 0.5-ml aliquots (fraction size 5 ml) with 9.25 kBq [1-^{14}C]IPP.

Chromoplast membrane pellets are submitted to repeated chloroform/methanol (2/1) extractions and the combined extracts are washed with 1% cold aqueous NaCl solution. Residual amounts of protein are removed by subsequent ether precipitation. The extract is split into aliquots, each containing 0.46 μmol (0.26 mg) carotenoid, estimated spectrophotometrically using ε = 122,688 liters mol^{-1} cm^{-1} for lutein. This corresponds to 4.5 μmol (3.9 mg) glycerolipids, estimated via fatty acid determination according to Duncombe.[6] The aliquots are supplemented with 15-*cis*-[^{14}C]phytoene in petroleum benzine (350,000 cpm/aliquot), mixed, and dried under a stream of nitrogen. Two milliliters of incubation buffer is added per aliquot and prolonged sonications in the cold yield a turbid

[6] W. G. Duncombe, *Biochem. J.* **88,** 7 (1963).

suspension of still nonideal liposomes (ideality is reached in the reconstitution procedure, see below). The applied label is quantitatively retained within the liposome membranes as can be demonstrated by gel filtration. Each of the liposome aliquots is ready for a single reconstitution assay.

Membrane Solubilization

For membrane solubilization we routinely use a CHAPS concentration of 100 m*M* [more than 10 times the critical micelle concentration (CMC) in the given buffer system], in order to create maximal disaggregated protein species (ideally one copy per mixed micelle) by safely exceeding the CMC. The membrane material from 180,000 *g* centrifugation is resuspended with the aid of a Potter-Elvehjem glass homogenizer in incubation buffer to a final carotenoid concentration of 21 μmol (12 mg)/20 ml, corresponding to about 66 mg membrane protein and 182 mg glycerolipid. To this suspension solid CHAPS, synthesized according to Hjelmeland,[5] is added to a final concentration of 100 m*M*. This is placed in an ice bath and vigorously stirred for 45 min under nitrogen atmosphere. Nonsolubilized material is removed by centrifugation (120,000 *g*, 2 hr). The red, clear supernatant, designated the 100 m*M* solubilizate, is used for reconstitution experiments. Quantitative data for the solubilization under the conditions described here are given in Table I.

Reconstitution

A CHAPS concentration of 10 m*M* was chosen for reconstitution, as this detergent concentration allows for stability of mixed micelles and, to a high degree, also of liposomes. As a matter of fact, mixed micelles once formed are stable as long as the CMC is exceeded,[7] whereas liposomes can be expected to be solubilized to a very small extent only.

For lowering the detergent concentration in the 100 m*M* solubilizate, dialysis and gel filtration were found to be inadequate. Dialysis of the detergent is highly time consuming, whereas gel filtration on columns, preequilibrated with 10 m*M* CHAPS, produces relatively high volume rendering reconcentration of membrane constituents (mixed micelles) without reconcentration of the detergent (free detergent micelles) very difficult. As an alternative, we use a rapid ultrafiltration technique (Diaflo, Amicon, Witten, FRG) which has become possible with the availability of a hydrophilic ultrafiltration membrane (YM series, Amicon), exhibiting no adsorption phenomena with the substances in question (see Table I). From gel filtration techniques on columns, preequilibrated with 10 m*M* CHAPS, the apparent particle mass of mixed micelles is roughly 50–60

[7] L. M. Hjelmeland, *Dev. Biochem.* **2,** 29 (1978).

TABLE I
QUANTIFICATION OF PROTEINS, CAROTENOIDS, AND LIPIDS IN THE DIFFERENT FRACTIONS OF THE CHROMOPLAST MEMBRANE SOLUBILIZATION WITH 100 m*M* CHAPS

	Protein		Carotenoid		Lipid	
Fraction	(mg/fraction)	Solubilization (%)	(mg/fraction)	Solubilization (%)	(mg/fraction)	Solubilization (%)
Membrane pellet (180,000 *g*)	66.4		11.5		190.2	
100 m*M* solubilizate	33.0	49.7	4.4	38.3	120.8	63.5
20 m*M* solubilizate (Diaflo ultrafiltrated)	31.9		4.4		118.5	
Nonsolubilized material	29.2		5.5		67.9	

kDa, whereas the pure CHAPS micelle showed an equivalent volume in an apparent particle mass range of 5–10 kDa. A YM membrane with a particle mass cut-off of 30 kDa is highly effective in the complete separation of both types of micelles. Ultrafiltration of the 100 m*M* solubilizate demonstrated that the pure CHAPS micelle was present in great excess, as the colorless clear filtrate was 100 m*M* with respect to the detergent concentration. The red, protein- and lipid-containing mixed micelles were also quantitatively retained. The actual CHAPS concentration can be estimated rapidly and with sufficient accuracy by monitoring the refractive index of the solutions (Fig. 2).

For the reconstitution procedure the 100 m*M* solubilizate (20 ml) is diluted 1 : 5 with incubation buffer to a final CHAPS concentration of 20 mM. In order to avoid local detergent concentrations less than the CMC, the buffer is added dropwise under continuous stirring in ice. Reconcentration of the solubilized membrane constituents (mixed micelles) without reconcentration of the detergent (free detergent micelles) is performed by ultrafiltration to a volume of 26 ml. Thirteen individual samples are prepared (2 ml). They are diluted 1 : 1 with the 15-*cis*-[^{14}C]phytoene-containing liposome preparation (see above), thus resulting in a final CHAPS

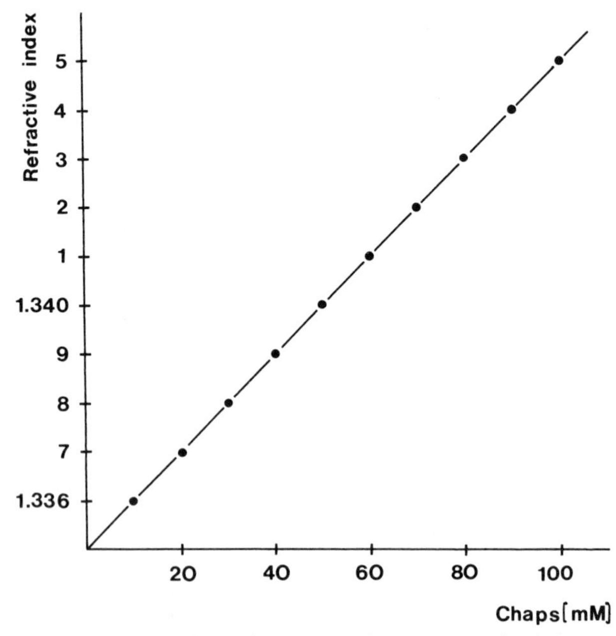

FIG. 2. Calibration curve: CHAPS concentration as determined from the refractive index.

concentration of 10 m*M*, as anticipated. Upon mixing, the liposome dispersion clears up, as ideal liposomes (small unilamellar vesicles, SUV)[8] with an average diameter of 27 nm[4] are spontanously formed. Incorporation of the solubilized membrane constituents into these liposomes is performed essentially according to Flügge and Heldt.[9] After mixing, the samples are immediately frozen in liquid nitrogen and thawed at 0° overnight. After a short sonication at low energy the samples are ready for incubation experiments.

The final 10 m*M* CHAPS concentration has to be adjusted precisely. Exceeding this concentration results in an almost total loss of the enzymatic activity in the reconstituted liposomes. In case of uncertainty, we recommend the dilution of the reconstituted preparation 1 : 2 or 1 : 3 with incubation buffer (the procedure does not change the relative enzyme–substrate concentrations), thus resulting in a final CHAPS concentration of 5 or 2.5 m*M* in the incubation assay. At final CHAPS concentrations lower than 1.25 m*M*, an almost total loss of the enzymatic activity is observed, which cannot be explained at present, so that a successful reconstitution can be monitored only in a final detergent concentration range of 1.25–10 m*M*.[4]

Incubation and Product Analysis

Incubations of the 4-ml samples are carried out at 26° in the dark, after supplementation with 1 m*M* $MnCl_2$, 2 m*M* NADP, 0.02 m*M* FAD. The reaction is stopped by addition of chloroform/methanol (2/1) and lipids are extracted.[10] For removal of excess glycerolipids the samples are dried and redissolved in acetone. The resulting pellet is removed by centrifugation. The concentrated acetone extracts are then subjected to HPLC analysis on a C_{18} reversed-phase column in the gradient system.[1] Radioactivity is monitored by a radiocolumn chromatography monitor and the individual radiolabeled fractions collected and quantified by color quenching using liquid scintillation counting. As a result, the reconstituted liposomes are very active in the formation of β-[^{14}C]carotene from 15-*cis*-[^{14}C]phytoene in contrast to the 100 m*M* solubilizate, which is biosynthetically inactive. It is obvious that the reconstitution of the solubilized enzymes into membrane structures reactivates the carotenogenic pathway. The typical time course of an incubation is given in Fig. 3. Nonreconstituted (protein-free) liposomes containing 15-*cis*-[^{14}C]phytoene, coincubated as a control, were free from enzymatic activities.

[8] *Proc. Conf. N.Y. Acad. Sci.*, p. 14 (1977).

[9] U. I. Flügge and H. W. Heldt, *Biochim. Biophys. Acta* **638,** 296 (1981).

[10] E. S. Bligh and W. J. Dyer, *Can. J. Biochem. Physiol.* **37,** 911 (1959).

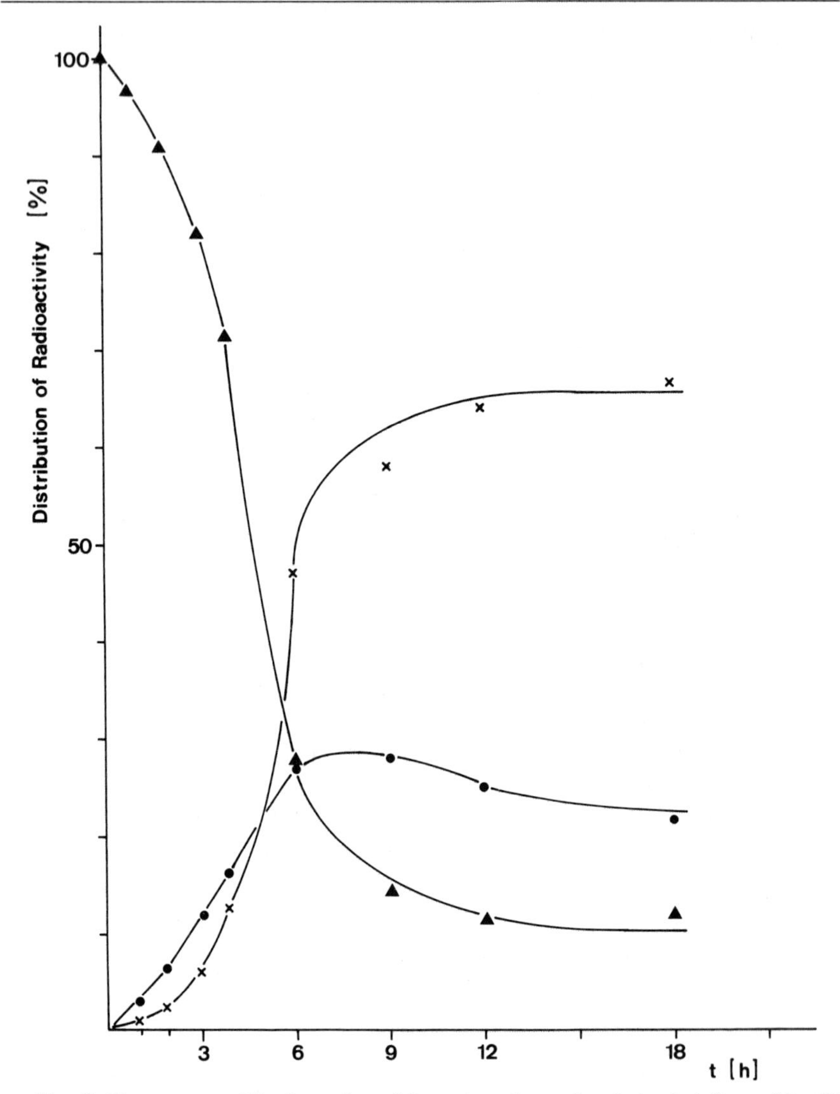

FIG. 3. Time course of the formation of β-carotene (×—×) and *cis*-phytofluene (●—●) at the expense of the radiolabeled substrate *cis*-phytoene (△—△) in the reconstituted liposomes.

Acknowledgments

The author is very indebted to Prof. Dr. H. Kleinig for his support and valuable discussions. The skillful participation of G. Weiss is gratefully acknowledged. These investigations were supported by Deutsche Forschungsgemeinschaft.

Section IV

Mitochondria

[37] Isolation of Plant Mitochondria: General Principles and Criteria of Integrity

By ROLAND DOUCE, JACQUES BOURGUIGNON, RENAUD BROUQUISSE, and MICHEL NEUBURGER

Considerable expertise is required in order to prepare mitochondria displaying the same biochemical and morphological characteristics as mitochondria *in vivo*. It is difficult if not impossible, at present, to prepare 100% intact and fully functional mitochondria; however, it is a simple matter to prepare a mitochondrial fraction (i.e., a subcellular fraction containing mitochondria). When plant tissue is homogenized in order to isolate mitochondria, the central vacuole is destroyed and numerous secondary products such as flavonoids, various colorless phenolic compounds, and terpenes are released into the medium with effects ranging from undesirable to devastating depending on the tissue and on the isolation technique. For example, once released, the various phenolic compounds and their oxidation products interact immediately with the membrane-bound proteins, giving rise either to an uncoupling of oxidative phosphorylation[1–3] or to an inhibition of the electron transport chain.[3,4] Likewise, during the course of the preparation of pine needle mitochondria, β-pinene (one of the main monoterpenes of the essential oil of pine needles) released from the resin canals adheres to the mitochondrial membranes and inhibits electron transport between complex I and complex IV.[5] Enzymes such as lipolytic hydrolases that liberate free fatty acids (linoleic and linolenic acids) are potentially the most troublesome contaminants in mitochondrial preparations from plants. These enzymes cause a rapid hydrolysis of mitochondrial phospholipids[6] leading to uncoupling of oxidative phosphorylation.[6] In addition, unsaturated fatty acids thus released are the best natural substrates for lipoxygenase (an enzyme that catalyzes the addition of O_2 to linoleic acid), producing harmful fatty acid hydroperoxides. In the presence of O_2 these hydroperoxides react strongly with all membrane proteins, including cytochromes, leading to a

[1] G. Stenlid, *Phytochemistry* **9,** 2251 (1970).

[2] P. Ravanel, M. Tissut, and R. Douce, *Phytochemistry* **20,** 2101 (1981).

[3] P. Ravanel, M. Tissut, and R. Douce, *Plant Physiol.* **69,** 375 (1982).

[4] P. Ravanel, M. Tissut, and R. Douce, *Plant Physiol.* **75,** 414 (1984).

[5] G. Pauly, R. Douce, and J. P. Carde, *Z. Pflanzenphysiol.* **104,** 199 (1981).

[6] R. Bligny and R. Douce, *Biochim. Biophys. Acta* **617,** 254 (1980).

complete destruction of the protein.[7,8] Furthermore, apart from problems associated with the presence of inhibiting substances, there is a need for strong shearing forces to disrupt the cell wall. These forces are also detrimental to the mitochondria.

Although many methods that are currently employed yield preparations containing an appreciable proportion of intact mitochondria,[9–11] they also result in substantial contamination by extramitochondrial components such as plastids (and/or plastid envelope) and peroxisomes and may lead, in consequence, to ambiguous and possibly misleading results. Finally, each plant organ presents its own difficulties for the preparation of mitochondria (there is no accepted standard method of isolation) and we must point out that all mitochondrial preparations from plant organs, such as roots or leaves, containing several cell types will contain a mixed population of mitochondria from the different cell types.[12]

Preparation of Washed Mitochondria from Plant Organs

If the investigations to be undertaken do not demand a particular species, difficulties can usually be minimized by careful choice of plant material (cauliflower buds, white and sweet potato tubers, mung bean hypocotyls, castor bean endosperm, maize roots, spinach and pea leaves, etc.).

Reagents

Extracting medium: 0.3 *M* mannitol; 4 m*M* L-cysteine; 1 m*M* EDTA; 0.1% (w/v) defatted bovine serum albumin (BSA); 0.6% (w/v) poly(vinylpyrrolidone) (PVP25, M_r ~25,000, Serva, Feinbiochemica, Heidelberg, FRG); and 30 m*M* 3-(*N*-morpholino)propanesulfonic acid (MOPS) buffer, pH 7.5 (MOPS buffer can be replaced by 20 m*M* pyrophosphate buffer). In the case of green leaf mitochondria L-cysteine should be replaced by 3 m*M* 2-mercaptoethanol

Washing medium: 0.3 *M* mannitol; 10 m*M* phosphate (or MOPS) buffer, pH 7.4 (this pH is chosen because it corresponds to the pH of the cytosolic compartment determined by ^{31}P nuclear magnetic resonance studies); 1 m*M* EDTA; and 0.1% (w/v) BSA

[7] F. Haurowitz, P. Schwerin, and M. M. Yenson, *J. Biol. Chem.* **140,** 353 (1941).

[8] J. Dupont, P. Rustin, and C. Lance, *Plant Physiol.* **69,** 1308 (1982).

[9] W. D. Bonner, Jr., this series, Vol. 10, p. 126.

[10] J. T. Wiskich, this series, Vol. 10, p. 122.

[11] G. Laties, this series, Vol. 31, p. 589.

[12] R. Douce, "Mitochondria in Higher Plants: Structure, Function, and Biogenesis." Academic Press, New York, 1985.

Procedure

Fresh material (0.5–1 kg) is cut into 3 liters of chilled extracting medium. Mitochondria require an osmoticum for the maintenance of their structure; sucrose, sorbitol, or mannitol are the most commonly used. (Under extreme hypotonic conditions water passes through the inner membrane into the matrix space, which enlarges. The inner membrane becomes extended with some material derived from the unfolding cristae and presses against the outer membrane, which having limited elasticity becomes ruptured.) Acidity, phenolic compounds, tannin formation, and oxidation products lead to the rapid inactivation of mitochondria. These difficulties are usually overcome by adding alkaline buffers, insoluble PVP,[13,14] and cysteine[9] to the extracting medium. It appears that PVP binds hydrogen ions as well as phenols. The presence of a mild reducing agent is required in order to minimize the concentration of oxidizing compounds and to prevent the oxidation of SH groups. The beneficial effect of BSA is due to its ability to bind free fatty acids[15] formed by the action of acyl hydrolase enzymes during extractions.[16,17] The well-known capacity of BSA to bind fatty acids by hydrophobic forces is consistent with its high content of hydrophobic amino acids.[18] BSA appears also to be an effective phenol and quinol scavenger.[19] Addition of EDTA to extraction media inhibits the activity of Ca^{2+}-dependent phospholipase D and membrane-bound lipolytic acyl hydrolases. In principle, a simple way to prevent oxidation of plant phenolic compounds is to bubble an inert gas heavier than air, such as argon, into the medium and to avoid the use of blenders which whip air into the homogenates. Other strategies to minimize the effects of vacuoles on plant mitochondria are to keep a low ratio of tissue to breakage medium[9] and to separate the organelles from soluble elements of the cellular brei as rapidly as possible.[20]

Two different techniques are commonly employed for breaking open tissue cells: (1) macerating for a short period, at a moderate speed, in a motor-driven blender [e.g., a Waring blender (Waring commercial blender), a Polytron (Kinematica GmbH, Lucerne, Switzerland), or a Moulinex (mixer-Moulinex, Alençon, France)] and (2) mortar and pestle grinding. Invariably, the more violent the homogenization, the lower the

[13] W. D. Loomis and J. Battaile, *Phytochemistry* **5,** 423 (1966).

[14] A. C. Hulme, J. D. Jones, and L. S. C. Wooltorton, *Phytochemistry* **3,** 173 (1964).

[15] R. F. Chen, *J. Biol. Chem.* **242,** 173 (1967).

[16] L. Dalgarno and L. M. Birt, *Biochem. J.* **87,** 586 (1963).

[17] R. Bligny and R. Douce, *Biochim. Biophys. Acta* **529,** 419 (1978).

[18] P. F. Spahr and J. T. Edsall, *J. Biol. Chem.* **239,** 850 (1964).

[19] W. D. Loomis, this series, Vol. 31, p. 528.

[20] J. M. Palmer and B. I. Kirk, *Biochem. J.* **140,** 79 (1974).

quality of the isolated mitochondria. The shear forces must be great enough to break the cells without stripping the outer membrane from the mitochondria. In other words, tissue disruption should be carried out very gently. A rapid and convenient method used routinely in our laboratory is to disrupt the tissue at low speed for 3–4 sec in a 1-gal Waring blender.[21,22] The blades should be routinely sharpened after several homogenizations so that cutting rather than shearing of the tissue occurs. The brei is rapidly squeezed through eight layers of muslin [muslin or cheesecloth, 60 cm large (Ruby, Voiron, France)] and 50-μm nylon netting [Nylon blutex, 50-μm aperture (Tripette et Renaud, Sailly-Saillisel, France)]. In order to increase the final yield of mitochondria the material retained by the muslin is pooled and mixed with 1 liter of extracting medium and blended with three 10-sec bursts of Polytron homogenizer (probe type, 35/2M) at half-full speed and filtered as above. The filtered suspension (from the blender and Polytron homogenizations) is then centrifuged (12 bottles, 500 ml) at 1500 g_{max} for 20 min (GS3 rotor, Sorvall). The pellet containing nuclei, chloroplasts (if present), intact cells, and other heavy cell fragments is discarded and the supernatant fluid is centrifuged for 20 min at 12,000 g_{max} (GS3 rotor, Sorvall). The supernatant is aspirated and each pellet is resuspended, using a spatula, by addition of 4 ml of washing medium. All the pellets are collected and lumps are disintegrated using a test tube and a loose-fitting pestle (Thomas tissue grinder) (three "up and down" strokes). The mitochondrial suspension thus obtained is centrifuged at 1500 *g* for 10 min (SS34 rotor, Sorvall) to pellet remaining heavy cell organelles such as intact plastids containing starch grains. The pellet is discarded and the supernatant is centrifuged for 20 min at 12,000 *g* (SS34, Sorvall). The pellet containing washed mitochondria is gently resuspended in a small volume of washing medium. The mitochondria are stored at approximately 50 mg protein/ml.

Mitochondria able to sustain rapid respiration rates may be isolated from a large variety of plant tissues provided that care is taken to avoid large-scale rupture of the two mitochondrial membranes. Inevitably a few ruptures occur during extraction and metabolically active preparations containing large numbers of intact mitochondria always contain a variable proportion of damaged or envelope-free mitochondria. The ruptured mitochondria exhibit a low rate of respiration inasmuch as cytochrome *c* (cyt *c*) is rapidly released into the medium when the outer membrane is damaged. In the past the fact that both types of mitochondria occur in the same preparation has not always received the attention it deserves and

[21] R. Douce, A. L. Moore, and M. Neuburger, *Plant Physiol.* **60,** 625 (1977).

[22] M. Neuburger, E. P. Journet, R. Bligny, J. P. Carde, and R. Douce, *Arch. Biochem. Biophys.* **217,** 312 (1982).

has occasionally led to erroneous conclusions concerning the permeability of the outer mitochondrial membrane to cyt *c*. The crucial point in the whole procedure is the grinding of the tissue. In order to obtain a maximum yield of intact mitochondria it is absolutely necessary to restrict the grinding procedure to a minimum. Longer blending improves the yield of recovered mitochondria but considerably increases the percentage of envelope-free mitochondria and of mitochondria that have resealed following rupture and the loss of matrix content. Under these conditions, it stands to reason that variable amounts of harmful vacuolar enzymes and nuclear material might have become confined within the matrix space.

Criteria for the Assessment of Mitochondrial Integrity

The criteria for the assessment of mitochondrial integrity, include (1) the respiratory rate in the presence of added ADP compared to the rate obtained following its expenditure,[9] (2) the ADP/O ratio,[9] (3) the latency of matrix enzymes such as fumarate hydratase {for these measurements, enzyme activities of samples of intact (in the presence of 0.3 *M* sucrose) and ruptured [by adding 0.025% (w/v) Triton X-100] mitochondria are measured. The activity in the ruptured preparation (T) minus that in the intact preparation (I) is called latent and is expressed as a percentage of the activity of the ruptured preparation ($100|T - I|/T$) to give latency}, and (4) the permeability of the outer membrane to cyt *c*.[23] In our experience the best criterion in critically assessing the proportion of organelles having an outer membrane in a mitochondrial population is to follow the oxidation of reduced (exogenous) cyt *c* by a mitochondrial preparation because an intact outer mitochondrial membrane provides a quasi-absolute barrier against penetration of cyt *c*.[23] The intactness of mitochondrial preparations can be rapidly evaluated by cyt *c* oxidation using an O_2 electrode.[22]

Reagents

Medium A: 0.3 *M* mannitol; 10 m*M* KCl; 5 m*M* $MgCl_2$; 0.1% (w/v) BSA; and 10 m*M* phosphate buffer, pH 7.2

Double-strength medium A: 0.6 *M* mannitol; 20 m*M* KCl; 10 m*M* $MgCl_2$: 0.2% (w/v) BSA; and 20 m*M* phosphate buffer, pH 7.2

Procedure

Mitochondria (0.5 mg protein) are added to isotonic medium (1 ml) containing medium A in a reaction vessel and to distilled water (0.5 ml) in

[23] R. Douce, E. L. Christensen, and W. D. Bonner, *Biochim. Biophys. Acta* **275,** 148 (1972).

another. The latter is then stirred for 10 sec before adding double-strength medium A (0.5 ml) to restore isotonic conditions following the osmotic shock. Under these conditions the outer membrane bursts and exogenous cyt *c* has access to the outer surface of the inner membrane where it is oxidized. The basic medium for cyt *c*-dependent O_2 uptake contained intact or swollen mitochondria—i.e., without outer membrane—(0.5 mg protein); 8 m*M* ascorbate; 1 ml medium A. The O_2 consumption is initiated with 30 μg cyt *c*. After 1 min, and in order to inhibit cytochrome *c* oxidase, 200 μM KCN is added. The cyanide-insensitive cyt *c*-dependent O_2 uptake which is very low corresponds to a nonenzymatic oxidation of reduced cyt *c* by molecular O_2. Consequently, the KCN-sensitive cyt *c*-dependent O_2 uptake is entirely attributable to cyt *c* oxidase functioning. Therefore it should be possible to estimate very rapidly the proportion of intact mitochondria in a given preparation from the relative rate of KCN-sensitive cyt *c*-dependent O_2 uptake by untreated and by osmotically shocked mitochondria in the same manner as for NADH- or succinate-dependent cyt *c* reduction.[23] The percentage of intact mitochondria present in the initial suspension is calculated from the formula

$$100 - (V_1/V_2)\ 100$$

where V_1 and V_2 represent the KCN-sensitive cyt *c*-dependent O_2 uptake rate in untreated and osmotically shocked mitochondria, respectively.

Even though the final preparation often contains more than 95% intact mitochondria, the material is not necessarily devoid of various extramitochondrial membranes and soluble enzymes (vacuolar enzymes, glycolytic enzymes, catalase, lipoxygenase, etc.). In addition, scrutinity by electron microscopy of the washed mitochondrial pellets obtained by differential centrifugation shows that the mitochondria are contaminated to varying extents by smooth vesicles of unknown origins and various cell organelles (plastids, microbodies, etc.). For example, mitochondria isolated from nongreen tissues such as cauliflower buds, etiolated shoots of *Faba* bean, etc., are heavily contaminated with plastids. The extent of plastid contamination can be easily evaluated either by the measurement of the latency of glucose-6-phosphate dehydrogenase activity[24] or by the absorption spectrum of a lipid extract (containing carotenoids) of the mitochondria. As a matter of fact plastids are capable of oxidizing hexose phosphate via the oxidative pentose phosphate pathway. Furthermore, in cells of photosynthetic tissues carotenoid pigments are exclusively localized in chloroplast membranes whereas in nonphotosynthetic tissues such as po-

[24] E. P. Journet and R. Douce, *Plant Physiol.* **79,** 458 (1985).

tato tubers these pigments are likely to be confined to the plastid envelope membranes[25] (the feature shared by the members of the plastid family is a pair of outer membranes known as the envelope).

For the measurement of glucose-6-phosphate dehydrogenase activity the standard assay solution contains 25 m*M* Tricine–NaOH (final pH 8.2); 0.3 *M* mannitol; 5 m*M* $MgCl_2$; 5 m*M* glucose 6-phosphate, and the reaction is initiated by adding 500 μM NADP. The final volume is 0.5 ml. Variations of absorbance at 25° is followed spectrophotometrically at 340 nm. For latency measurements glucose-6-phosphate dehydrogenase activity of samples of intact and ruptured (by adding 0.025% Triton X-100) organelles are measured.

Pigments are extracted from the mitochondrial fraction according to Jeffrey *et al.*[26] Pigments are dissolved in either ethanol or chloroform and absorption spectra are recorded on a spectrophotometer (Kontron, Uvikon 810).

In order to avoid the danger of long and fruitless controversy, the intact mitochondria obtained by differential centrifugation must be purified by isopycnic centrifugation in a sucrose[23] or silica sol[22,27] gradient or by partition in aqueous polymer two-phase systems.[28]

Purification of Potato Tuber Mitochondria by Isopycnic Centrifugation in Density Gradients of Percoll

The introduction of a nontoxic silica sol gradient material, Percoll [colloidal silica coated with poly(vinylpyrrolidone)], by Pharmacia has permitted the development of a rapid purification procedure utilizing isosmotic and low viscosity conditions.

Reagents

Medium B: 0.3 *M* sucrose; 10 m*M* phosphate buffer, pH 7.2; 1 m*M* EDTA; and 0.1% (w/v) BSA

Percoll medium: medium B containing 28% (v/v) Percoll

Procedure

Two centrifuge tubes are filled with 36 ml each of Percoll medium. The final concentration of Percoll and sucrose should be carefully determined according to the origin of mitochondria and to the nature of contamina-

[25] R. Douce and J. Joyard, *Adv. Bot. Res.* **7,** 1 (1979).

[26] S. W. Jeffrey, R. Douce, and A. A. Benson, *Proc. Natl. Acad. Sci. U.S.A.* **71,** 807 (1974).

[27] C. Jackson, J. E. Dench, D. O. Hall, and A. L. Moore, *Plant Physiol.* **64,** 150 (1979).

[28] P. Gardeström and I. Ericson, this volume [40].

tions present in the washed mitochondria which are extremely variable. Aliquots (3 ml sample, 30–40 mg protein) of the washed mitochondria are then layered on the Percoll medium. The tubes are placed in a precooled Sorvall SS34 angle-head rotor and centrifuged (at 3°) at 40,000 *g* for 30 min. In order to avoid the adverse effects of rapid acceleration on gradient stability an automatic rate controller is strongly advocated. At the conclusion of this step a continuous gradient of Percoll is obtained in each tube and intact mitochondria are recovered as a broad band located beneath a yellow layer at the top of the gradient. In addition, heavy organelles (peroxisomes) were recovered as a sharp dark band sitting above a clear colorless layer of Percoll at the bottom of the tube.[22] It is noteworthy that the substitution for sucrose by 0.3 *M* mannitol led to a marked shift, downward, of mitochondrial fraction along the gradient, and the mitochondria aggregated near the bottom of the tube just on top of the peroxisome fraction. In fact, we have observed by using density markers (density marker beds, Pharmacia) that the total density range covered by Percoll in sucrose is slightly different than that of Percoll in mannitol, sorbitol, or raffinose. The apparent density of Percoll-purified mitochondria is within the range 1.054 to 1.067. Consequently, the apparent average density of sucrose-purified mitochondria[23] is much higher than that of Percoll-purified mitochondria. The spectrum of the yellow layer at the top of the gradient shows a strong carotenoid absorption with the typical three-banded spectrum in the blue region. Since this fraction contains numerous small vesicles it is clear that unpurified potato mitochondria were appreciably contaminated by amyloplast envelope membranes. In fact, starch grains that exhibit a high sedimentation coefficient tear through the amyloplastid envelope at supraoptimal *g* forces and during blending. This leads to extensive contamination of the unpurified mitochondria by a variable proportion of amyloplast envelope membranes containing carotenoids.

The mitochondrial fraction is collected with a needle attached to a syringe. This fraction is well mixed with 15 times its volume of washing medium to dilute Percoll and mitochondria are sedimented at 12,000 *g* for 15 min. The brown-reddish pellet thus obtained is stored at approximately 50 mg protein/ml.

Potato tuber mitochondria separated from contaminating organelles and membrane vesicles on self-generated Percoll gradients showed markedly higher rates of α-ketoglutarate oxidation than did unpurified mitochondria (Table I). Assay of various markers (catalase, glucose-6-phosphate dehydrogenase, and carotenoids) showed that sedimentation of potato tuber mitochondria through a continuous Percoll gradient almost completely eliminates plastid and peroxisomal contamination (Table I). In

TABLE I
PURIFICATION OF POTATO TUBER MITOCHONDRIA[a]

Step	Washed mitochondria	Purified mitochondria
State-3 rate of α-ketoglutarate oxidation (nmol of O_2/min per milligram of protein)	160	220
Catalase (μmol of O_2/min per milligram of protein)	12	0.2
Carotenoids (μg/mg of protein)	0.1	0.008

[a] Washed mitochondria are prepared by differential centrifugation and subsequently purified on a Percoll gradient as described in the text. The mitochondrial band is removed from the gradient, diluted with suspending medium, and concentrated by centrifugation. α-Ketoglutarate oxidation is measured as O_2 consumed in the presence of 5 m*M* α-ketoglutarate, 200 μ*M* thiamin pyrophosphate, 200 μ*M* NAD^+ and 1 m*M* ADP, and catalase as O_2 consumed in the presence of 2.5 m*M* H_2O_2. If the purification procedure is repeated once (see text) mitochondrial catalase activity can be reduced to less than 0.02 μmol O_2/min per milligram of protein and carotenoids are almost undetectable.

addition the rate of KCN-sensitive cytochrome *c*-dependent O_2 uptake in intact mitochondria was only 2–5% of that recorded for osmotically shocked mitochondria, giving an apparent percentage of intactness of 95–98%. Respiratory control ratio and outer membrane integrity of the purified mitochondria do not change significantly during storage on ice for at least 20 hr nor do the rates of exogenous NADH oxidation.[29] However, during this time mitochondria lost their NAD^+ to the resuspending medium and O_2 uptake with NAD^+-linked substrates becomes increasingly dependent on added NAD^+.[29] This passive diffusion of NAD^+ out of the mitochondria which is strongly temperature dependent[30] occurs via a specific NAD^+ carrier because the NAD^+ analog NAP_4–NAD^+ almost completely prevents leakage of NAD^+ out of the mitochondria during their storage. Evidently the slow efflux of NAD^+ occurred via the same carrier which catalyzes NAD^+ uptake.[29]

Unfortunately even after centrifugation by isopycnic centrifugation in density gradients of Percoll mitochondria are still contaminated with plastids. The purification procedure therefore is repeated once—i.e., after

[29] M. Neuburger and R. Douce, *Biochem. J.* **216,** 443 (1983).
[30] M. Neuburger, D. A. Day, and R. Douce, *Plant Physiol.* **78,** 405 (1985).

pelleting mitochondria and layering the suspending pellet on top of a new Percoll gradient [medium B containing 27% (v/v) Percoll] in order to remove plastids containing small starch grains.[24] Again the final concentration of Percoll and osmoticum (sucrose, sorbitol, mannitol, raffinose, etc.), carefully determined, could be slightly modified according to the extent of plastid contamination in the crude mitochondrial pellet.

Isolation of Chlorophyll-Free Mitochondria from Pea Leaves

Since the report of a procedure to obtain fully functional and intact mitochondria from leaves of higher plants[21] several techniques for the purification of the mitochondria free from contaminating thylakoid membranes have been published. The methods used to date have included discontinuous Percoll gradient centrifugation of both mechanically prepared crude mitochondria[27] and those from broken protoplasts,[31] a combination of phase-partition and Percoll gradient centrifugation,[32] and linear sucrose gradient centrifugation.[33,34] We have developed a technique suitable for the large-scale preparation of intact, active, and chlorophyll-free mitochondria from pea leaves. The method employs a self-generating gradient of Percoll in combination with a linear gradient of PVP-25,[35] and a linear gradient of raffinose in a single step.

Reagents

Suspending medium: 0.3 *M* sucrose; 10 m*M* phosphate buffer, pH 7.2; 0.1% (w/v) BSA

Medium C: 0.3 *M* sucrose; 10 m*M* phosphate buffer, pH 7.2; 0.1% (w/v) BSA, and 28% (v/v) Percoll

Medium D: 0.3 *M* raffinose; 10% (w/v) poly(vinylpyrrolidone); 10 m*M* phosphate buffer, pH 7.2; 0.1% (w/v) BSA, and 28% (v/v) Percoll

Composite medium/gradient: using a small amount of medium C, fill the connecting passage between the two sides of a conical gradient maker (three-position gradient-forming devices, Kontron). Close the passageway, remove excess solution, and add 51 ml of medium C and 51 ml of medium D to the sides of the gradient, distal and proximal to the exit tube, respectively. Agitate medium D solution with a stirrer (Kontron) as actively as possible without splashing. Immediately open the passageway

[31] M. Nishimura, R. Douce, and T. Akazawa, *Plant Physiol.* **69,** 916 (1982).
[32] A. Bergman, P. Gardeström, and I. Ericson, *Plant Physiol.* **66,** 442 (1980).
[33] G. P. Arron, M. H. Spalding, and G. E. Edwards, *Plant Physiol.* **64,** 182 (1979).
[34] D. Nash and J. T. Wiskich, *Plant Physiol.* **71,** 627 (1983).
[35] D. A. Day, M. Neuburger, and R. Douce, *Aust. J. Plant Physiol.* **12,** 219 (1985).

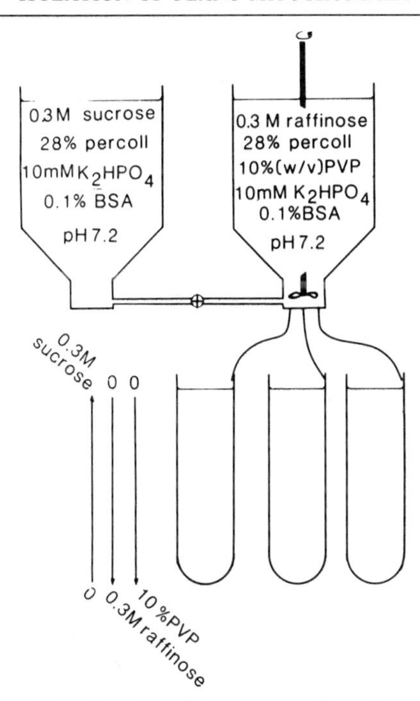

FIG. 1. A system to produce three identical linear gradients from 0 to 0.3 *M* raffinose, 0.3 to 0 *M* sucrose, and 0 to 10% (w/v) PVP. In addition this composite gradient contains 28% (w/v) Percoll.

between the two sides of the gradient maker and by using a peristaltic pump (minipuls 2, Gilson) achieve a flow rate of about 5 ml/min. Incoming solution should layer into three centrifuge tubes (Sorval SS34, angle head rotor) without turbulence on top of the rising level of poured solution (34 ml in each tube). The formation of linear concentration gradients (0–10% PVP25 gradient; 0.3–0 *M* sucrose gradient; 0–0.3 *M* raffinose gradient) requires instantaneous mixing of the contents of the "proximal" side of the gradient maker (Fig. 1).

Procedure

Pea (*Pisum sativum* L. var. "Douce Provence") plants are grown from seed in vermiculite for 15 days under a 12-hr photoperiod of warm white light from fluorescent tubes (20–40 $\mu E/m^2/sec$ of photosynthetically active radiation) at 22°. The plants can be grown at high density (20–30 plants/100 cm^2). The plants are watered every day with one-half strength Hoagland's nutrient solution. Approximately 300 g of fully expanded leaves is used. Washed mitochondria are prepared as above.

The mitochondrial pellet, heavily contaminated by thylakoids, is resuspended in approximately 6 ml of suspending medium. Two-milliliter aliquots are then layered over 34 ml of composite medium in a Sorvall SS-34 rotor tube and centrifuged for 45 min at 40,000 *g* (the automatic rate controller must be used). The mitochondria are found in a tight, brown band near the bottom of the tube while the thylakoids remain near the top of the tube. The mitochondrial fraction is carefully removed by siphoning from the bottom of the tube and pelleted, after diluting at least 5-fold with suspending medium, by centrifuging at 15,000 *g* for 20 min. The yield of mitochondria from 300 g of pea leaves is approximately 40 mg protein.

A critical step in this procedure is the second low-speed centrifugation during preparation of the washed mitochondria. This is necessary to remove intact chloroplasts and large thylakoid fragments which otherwise band with the mitochondria on the subsequent composite medium/gradient. It is also recommended that no more than 2 mg of chlorophyll be loaded on each 34-ml gradient to avoid some chlorophyll in the mitochondria fractions. Under these conditions, when the fractions containing mitochondria are collected and then concentrated by centrifugation, a trace of chlorophyll is in fact detected in the pellet but this is rarely more than 0.2% of the original chlorophyll present in the crude mitochondria. An average value of 0.1 μg chl/mg protein is found in the purified preparation (Table II); assuming a chlorophyll : protein ratio of 7 in the thylakoids,

TABLE II
PURIFICATION OF PEA LEAF MITOCHONDRIA[a]

Step	Washed mitochondria	Purified mitochondria
State-3 rate of glycine oxidation (nmol of O_2/min per milligram of protein)	60	200
Glycolate oxidase (nmol of O_2/min per milligram of protein)	90	30
Chlorophyll (μg mg^{-1} protein)	20	0.07

[a] Washed mitochondria are prepared by differential centrifugation and subsequently purified on a Percoll/PVP/raffinose gradient as described in the text. The mitochondrial band is removed from the gradient, diluted with suspending medium, and concentrated by centrifugation. Glycine oxidation is measured as O_2 consumed in the presence of 5 m*M* glycine and 1 m*M* ADP, and glycolate oxidase as O_2 consumed in the presence of 10 m*M* glycolate and 0.3 m*M* KCN. If the purification procedure is repeated once mitochondrial glycolate oxidase can be reduced considerably.

this represents only 0.06% of the total protein of the purified mitochondria.

A feature of pea leaf mitochondria purified on Percoll gradients is their very high rate of oxygen consumption with all substrates and their stability when stored on ice.[35] In addition, pea leaf mitochondria, in contrast with mitochondria from etiolated leaves, are capable of oxidizing glycine as substrate very rapidly (200–300 nmol O_2/min/mg protein) (Table II). Interestingly, the oxidation of glycine is strongly inhibited in the presence of 4 m*M* L-cysteine. In fact, this aminothiol, in which the two functional groups are in suitable proximity, reacts with pyridoxal-P, an important cofactor involved in serine hydroxymethyltransferase functioning. The product is a relatively stable complex containing a thiazolidine ring.[36]

[36] M. V. Buell and R. E. Hansen, *J. Am. Chem. Soc.* **82,** 6042 (1960).

[38] Purification of Plant Mitochondria on Silica Sol Gradients

By ANTHONY L. MOORE and MICHAEL O. PROUDLOVE

The close association between plant cell constituents means that, to isolate mitochondria which retain the properties found *in vivo*[1-3] but which are free of organellar or cytosolic contamination, there must be some degree of purification. By this process, extramitochondrial contaminants are removed, with minimal damage being done to mitochondrial membranes and maximum retention of matrix components. While differential centrifugation of tissue homogenates, both etiolated and green leaf,[4,5] will yield pellets highly enriched in mitochondria which may be adequate for many studies, there are other cases when further purification may be deemed essential.[2,6] In investigations into the intracellular distribution of enzyme activities[7] any cross-contamination of mitochondria, by

[1] R. Douce, "Mitochondria in Higher Plants: Structure, Function, and Biogenesis." Academic Press, New York, 1985.

[2] M. Neuburger, *Encycl. Plant Physiol., New Ser.* **18,** 7 (1985).

[3] A. L. Moore and P. R. Rich, *Encycl. Plant Physiol., New Ser.* **18,** 134 (1985).

[4] R. Douce, E. L. Christensen, and W. D. Bonner, *Biochim. Biophys. Acta* **275,** 148 (1972).

[5] R. Douce, A. L. Moore, and M. Neuburger, *Plant Physiol.* **60,** 625 (1977).

[6] A. L. Moore and M. O. Proudlove, *in* "Isolation of Membranes and Organelles from Plant Cells" (J. L. Hall and A. L. Moore, eds.), p. 153. Academic Press, New York, 1983.

[7] J. T. Wiskich and I. B. Dry, *Encycl. Plant Physiol., New Ser.* **18,** 281 (1985).

organelles such as microbodies and plastids, may lead to misleading conclusions about the functions of each. Similarly, studies on the constituents of the inner and outer mitochondrial membranes (for example, lipids,[8] proteins, and cytochromes)[9] or the matrix (for example, proteins, nucleic acids, and substrates/cofactors) require the mitochondrial preparation be as uncontaminated as possible, again to avoid misleading results and interpretations.

Principle

Purification of washed mitochondria isolated by differential centrifugation from etiolated mung bean (*Phaseolus aureus*) hypocotyl and green pea (*Pisum sativum*) leaf tissues[2,4–6,10] may be performed easily and rapidly using self-generated density gradients of the nontoxic, PVP-coated, silica sol Percoll (Pharmacia Fine Chemicals). Isotonicity of gradients is maintained by including 0.25 *M* sucrose throughout and Percoll gradients are self-generated in the centrifuge tubes (Pharmacia handbook). Such gradients have been successfully used to purify washed mitochondria from *Neurospora crassa*,[11] potato tubers,[12] avocado fruits,[13] and pea leaves.[14]

Equipment

Beckman J2-21 centrifuge, a JA-20 rotor, and four to six 50-ml centrifuge tubes

A commercial or homemade gradient former (2 × 25 ml chamber minimum) and a peristaltic pump (a Gilson minipuls HP2 may be used to form two gradients simultaneously)

A small, loose-fitting, 5-ml glass Potter-Elvehjem homogenizer

Washed Mung Bean Hypocotyl Mitochondria

Reagents and Media

Sucrose/buffer solution: 0.5 *M* sucrose; 0.2% (w/v) BSA; 20 m*M* MOPS, pH 7.2

[8] J. L. Harwood, *Encycl. Plant Physiol., New Ser.* **18,** 37 (1985).

[9] G. Ducet, *Encycl. Plant Physiol., New Ser.* **18,** 72 (1985).

[10] W. D. Bonner, this series, Vol. 10, p. 126.

[11] J. P. Schwitzguebel, I. M. Moller, and J. M. Palmer, *J. Gen. Microbiol.* **126,** 289 (1981).

[12] M. Neuburger, E. P. Journet, R. Bligny, J. P. Carde, and R. Douce, *Arch. Biochem. Biophys.* **217,** 312 (1982).

[13] F. Moreau and R. Romani, *Plant Physiol.* **70,** 1380 (1982).

[14] D. A. Day, M. Neuburger, and R. Douce, *Aust. J. Plant Physiol.* **12,** 119 (1985).

Mannitol/buffer solution: 0.6 *M* mannitol; 0.2% (w/v) BSA; 20 m*M* MOPS, pH 7.2

Percoll: as supplied by Pharmacia Fine Chemicals, Uppsala, Sweden

Gradient 1 solution: add 20 ml sucrose/buffer solution, 11.2 ml Percoll, and 8.8 ml distilled water to each centrifuge tube; mix by inversion

Gradient 2 solution: add 20 ml mannitol/buffer solution, 11.2 ml Percoll, and 8.8 ml distilled water to each 50-ml centrifuge tube; mix by inversion

Wash medium (WMa): 0.3 *M* mannitol; 0.1% (w/v) BSA; 1 m*M* EDTA; 10 m*M* MOPS, pH 7.2

Procedure. Washed mitochondria, 3 ml of 15–20 mg protein/ml wash medium, are layered onto each gradient (see Fig. 1); a MWS pellet from a 1.5-liter initial homogenization[6] should prove sufficient for three to four gradients. Centrifugation at 41,000 *g* for 25–30 min results in the self-generation of a Percoll gradient and three distinct bands within each tube. The uppermost band contains the majority of the lipoxygenase activity, the middle band is mainly mitochondria, and the lower band consists of peroxisomes (fractions 1–3, Neuburger *et al.*[12]). A Pasteur pipet is used to remove the mitochondrial fraction which is then diluted approximately 10-fold with wash medium and centrifuged at 11,000 *g* for 15 min. The pellet is then gently resuspended in a small volume of wash medium using two to three strokes of the glass homogenizer. If the mitochondria are considered to be of sufficient purity the volume of wash medium should be kept minimal (250–500 μl), final concentration of MPS to approximately 80 mg protein/ml. A further gradient purification, however, substituting mannitol for sucrose as the osmoticum, reduces both catalase and lipoxygenase activity significantly, by factors of 60 and 20, respectively.[12] For the second gradient the first MPS pellet is resuspended in 2–3 ml wash medium (using the glass homogenizer) and layered on top of the tube(s) containing gradient 2 solution. The gradient is generated by centrifuging at 41,000 *g* for 25–30 min and mitochondria form a tight band near the bottom of the centrifuge tube. Removal of this band, with a Pasteur pipet, and a 10-fold dilution with wash medium is followed by centrifugation at 11,000 *g* for 15 min to pellet the mitochondria. The pellet is gently resuspended in a small volume of wash medium (250–500 μl) using two to three strokes of the glass homogenizer. Final mitochondrial concentration should be approximately 60–80 mg protein/ml.

Washed Pea Leaf Mitochondria

Reagents and Media

PVP: soluble poly(vinylpyrrolidone) (Aldrich Chemical Co., Ltd., Gillingham, UK). Either PVP-40 or PVP-25 may be used; the former is

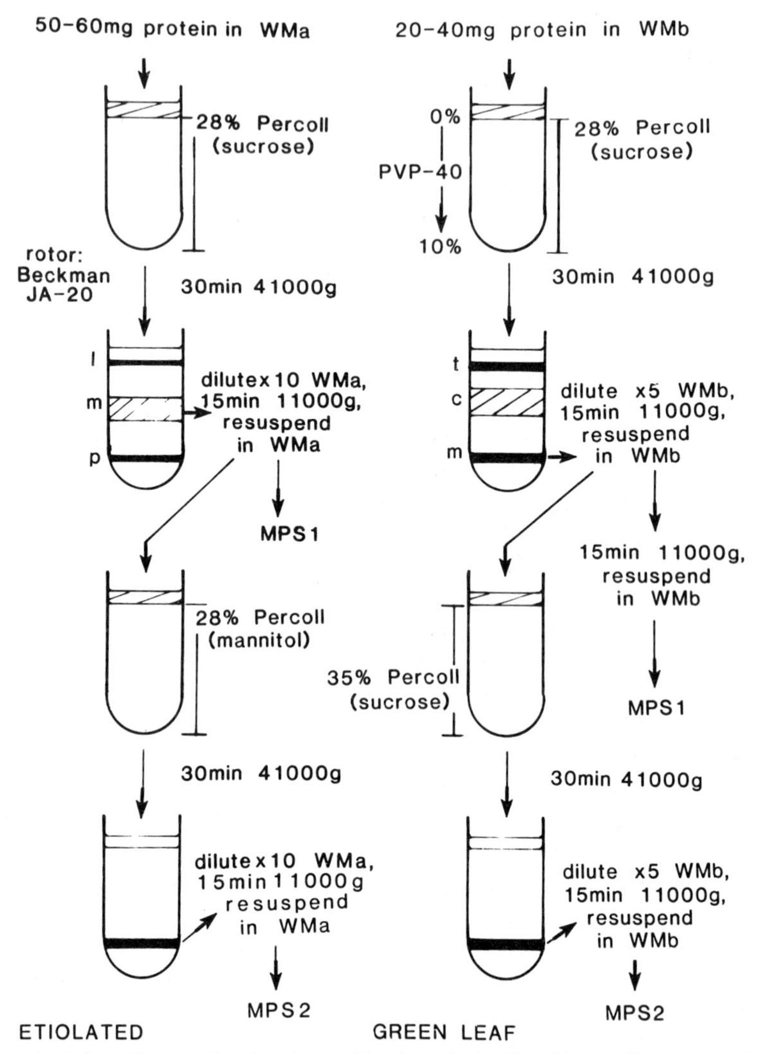

FIG. 1. A flow diagram showing the purification of mitochondria on silica sol gradients. Wash media for each type of tissue, WMa (etiolated) and WMb (pea leaf), are described in the text. Bands on the gradients enriched in lipoxygenase activity (l), thylakoids (t), mitochondria (m), chlorophyll (c), and peroxisomes (p) are as indicated. Purified mitochondria, MPS, are prepared from one or two gradient centrifugations.

slightly cheaper and both give similar results. Prepare a 50% (w/v) solution. This may take several hours to fully dissolve so it should be prepared the day before and stored at 4° overnight

Percoll: as supplied by Pharmacia Fine Chemicals, Uppsala, Sweden

Sucrose/buffer solution: 0.5 *M* sucrose; 0.2% (w/v) BSA; 20 m*M* KH_2PO_4, pH 7.5

Gradient medium 1: mix thoroughly 10 ml sucrose/buffer solution, 5.6 ml Percoll, 4 ml 50% (w/v) PVP-40, and 0.4 ml distilled water

Gradient medium 2: mix thoroughly 10 ml sucrose/buffer solution, 5.6 ml Percoll, and 4.4 ml distilled water

Wash medium (WMb): 0.3 *M* mannitol; 0.2% (w/v) BSA; 1 m*M* EDTA; 2 m*M* glycine; 10 m*M* MOPS, pH 7.5

Procedure. Equal volumes of gradient media 1 and 2 are pipetted into separate compartments of the gradient former; this will provide a solution which is homogeneous in Percoll [28% (v/v)] and 0 to 10% (w/v) in PVP, the gradient running from the top of the centrifuge tube to the base. Chamber A, which is the nearest to the centrifuge tube, contains gradient medium 1 (rapidly stirred) and is separated from chamber B, containing gradient medium 2, by a valve-controlled capillary tube. As the valve between chamber A and the centrifuge tube is opened so is that between chambers A and B, facilitating a gradual dilution of the solution in chamber A. A flow rate of 1–3 ml/min is controlled by a peristaltic pump and careful siphoning of the resultant solution down the side of the centrifuge tube leads to a linear PVP gradient.

Sufficient mitochondrial protein, for each 40 ml Percoll/PVP gradient, may be obtained by differential centrifugation[5,12,15,16] of an initial homogenate from 30–50 g of pea leaves in 400–500 ml isolation medium[6] (Sect. XXX); a fresh weight of 100–200 g of pea leaves is recommended for an adequate yield of purified mitochondria. For preparations, both washed and purified, 2 m*M* glycine should be included in the wash medium, to retain maximum rates of glycine decarboxylase activity (Sect. XXX) (I. B. Dry and J. T. Wiskich, personal communication). Washed pea leaf mitochondria (2–3 ml) are layered onto each gradient which is then centrifuged at 41,000 *g* for 30 min (see Fig. 1). Mitochondria are found in a band near the base of the centrifuge tube while thylakoids form a band near the top of the gradient. There may be some green coloration between these two bands, the actual amount depending on the quantity of chlorophyll initially loaded on to the gradient. The gradient is slowly aspirated away, down to the mitochondrial layer; this is removed by Pasteur pipet, diluted 5-fold with wash medium, and recentrifuged at 11,000 *g* for 15 min. The resulting pellet is often very loose (this may lead to loss of mitochondria when removing the supernatant), so it should be resuspended in 50–100 ml of wash medium and recentrifuged for 15 min at 11,000 *g*. Mitochon-

[15] I. B. Dry and J. T. Wiskich, *Aust. J. Plant Physiol.* **12,** 329 (1985).

[16] D. A. Day and J. T. Wiskich, *FEBS Lett.* **112,** 119 (1980).

dria are resuspended in a small volume of wash medium (250–500 μl) using two to three strokes of the glass homogenizer.

While these mitochondria are essentially free of chlorophyll contamination there may still be some peroxisomal activity present. To remove this, a further gradient purification may be performed. Instead of resuspending the mitochondria from the first gradient in a large volume they are, instead, resuspended in 2–3 ml of wash medium. Pipetting this onto gradient 2 and centrifuging at 41,000 *g* yields a tight, mitochondrial band near the base of the tube. After removing the gradient solution down to the mitochondrial band, by aspiration, this is decanted, using a Pasteur pipet, diluted 5-fold with wash medium, and the mitochondria pelleted by centrifuging at 11,000 *g* for 15 min. The pellet is gently resuspended in a small (250–500 μl) volume of wash medium using the glass homogenizer to give a final concentration of 60–80 mg protein/ml.

Properties

Mitochondria purified on Percoll gradients contain negligible levels of endogenous substrates, are devoid of galactolipids and carotenoids, and possess low levels of hydrolases such as lipolytic acyl hydrolases or lipoxygenase.[12] They show markedly higher rates of O_2 consumption in state 3 than do unpurified mitochondria and have increased ADP/O and respiratory control ratios. Membrane intactness studies, as determined by KCN-sensitive ascorbate–cytochrome *c*-dependent O_2 consumption or succinate : cytochrome *c*-oxidoreductase, reveal that mitochondria purified in such a manner possess a mean percentage of intactness of approximately 95%.[12] Thus it is evident that the physiological integrity of mitochondria is maintained during the purification process.

Acknowledgments

The financial support of the Science and Engineering Research Council and the Agricultural and Food Research Council is gratefully acknowledged.

[39] Isolation of Mitochondria from Leaves of C_3, C_4, and Crassulacean Acid Metabolism Plants

By GERALD E. EDWARDS and PER GARDESTRÖM

The isolation of mitochondria from leaves presents, in part, the same problems as isolation of mitochondria from other plant tissues. For a more complete discussion see Brouquisse *et al.*[1] The high mechanical resistance of the cell wall necessitates the use of powerful grinding methods. Prolonged grinding is harmful to mitochondria and other cell organelles, including chloroplasts. Because of this, the grinding time has to be kept to a minimum. This results in incomplete maceration of tissue and a low yield, which appears unavoidable if one wishes to obtain functionally intact mitochondria.

In leaf tissue additional difficulties are encountered. Inevitably, during grinding of the leaves many of the chloroplasts are broken, producing heterogeneous fragments of the thylakoid membrane. These thylakoid fragments are a major source of contamination of crude mitochondrial preparations which have been isolated from leaves by homogenization and differential centrifugation. Isolation of mitochondria from photosynthetic cells is most complicated in C_4 plants, where the parenchyma tissue is composed of two photosynthetic cell types. Some of the limitations of mechanical grinding of tissue can be overcome if the cells are separated and the cell wall removed by enzymatic digestion. The resulting protoplasts can be broken by very gentle methods which will allow organelles to be released in an intact state. This general approach to isolation of various cellular components is discussed by Nishimura[2] in a separate chapter.

The release of harmful substances from the vacuoles (phenolics, quinones, lipases) during tissue homogenization also makes it very difficult to get functional mitochondria from leaves of many plants. Even with species from which good mitochondrial preparations have been successfully obtained it is necessary to include protective agents in the homogenization medium. The addition of BSA (defatted) is particularly important; it may function to bind fatty acids which would otherwise attack membranes. Poly(vinylpyrrolidone) (PVP) has been used to reduce the deleterious effects of phenols. It may also be helpful to include a reducing agent

[1] R. Brouquisse *et al.*, this volume [37].
[2] N. Nishimura *et al.*, this volume [3].

like cysteine (or mercaptoethanol or ascorbate) to get functionally intact mitochondria from leaves of some species. As plant vacuoles often contain acidic compounds [particularly in Crassulacean acid metabolism (CAM) plants], it is advisable to include a relatively strong buffer to avoid acidification during extraction.

The following is a brief description of methods to isolate mitochondria from leaf tissue of three different photosynthetic groups of higher plants: C_3, C_4, and CAM plants. Recently, Gardeström and Edwards[3] reviewed known functions of mitochondria in leaves of these photosynthetic groups in relation to photosynthesis and respiration.

Isolation Procedures for C_3 Leaf Mitochondria

There has been recent interest in isolating mitochondria from the leaf mesophyll cells of C_3 plants, photosynthetic tissue in which atmospheric CO_2 is directly fixed through the reductive pentose phosphate pathway (with the 3-C phosphoglyceric acid as product). Through use of an appropriate grinding medium and a short grinding time, functionally intact mitochondria can be isolated from leaves of the C_3 species spinach (*Spinacea oleracea*) and pea (*Pisum sativum*) by differential centrifugation. It is reasonable to assume that mesophyll cells are the primary source of mitochondria isolated from C_3 leaf tissue by mechanical grinding. Epidermal tissue contains few mitochondria, and both epidermal and vascular tissue are resistant to breakage with mechanical grinding. Mitochondria isolated mechanically from leaves or from mesophyll protoplasts from C_3 leaves have a high capacity to oxidize glycine, a property specific to mitochondria of photosynthetic cells.[4,5] A suitable preparation procedure for spinach leaf mitochondria is described in Fig. 1. This method is based on procedures developed by Douce *et al.*,[6] Ebbighausen *et al.*,[7] and Gardeström *et al.*[8]

Grinding Medium

0.3 *M* sucrose, 5 m*M* glycine, 4 m*M* cysteine, 1 m*M* EDTA, 30 m*M* MOPS (morpholinopropanesulfonic acid), 0.2% BSA (defatted), 0.6% PVP (insoluble), pH 7.5

[3] P. Gardeström and G. E. Edwards, *Encycl. Plant Physiol., New Ser.* **18,** 314 (1985).
[4] P. Gardeström, A. Bergman, and I. Ericson, *Plant Physiol.* **65,** 389 (1980).
[5] M. Nishimura, R. Douce, and T. Akazawa, *Plant Physiol.* **69,** 916 (1982).
[6] R. Douce, A. L. Moore, and M. Neuberger, *Plant Physiol.* **60,** 625 (1977).
[7] H. Ebbinghausen, C. Jia, and H. W. Heldt, *Biochim. Biophys. Acta* **810,** 184 (1985).
[8] P. Gardeström, I. Ericson, and C. Larsson, *Plant Sci. Lett.* **13,** 23 (1978).

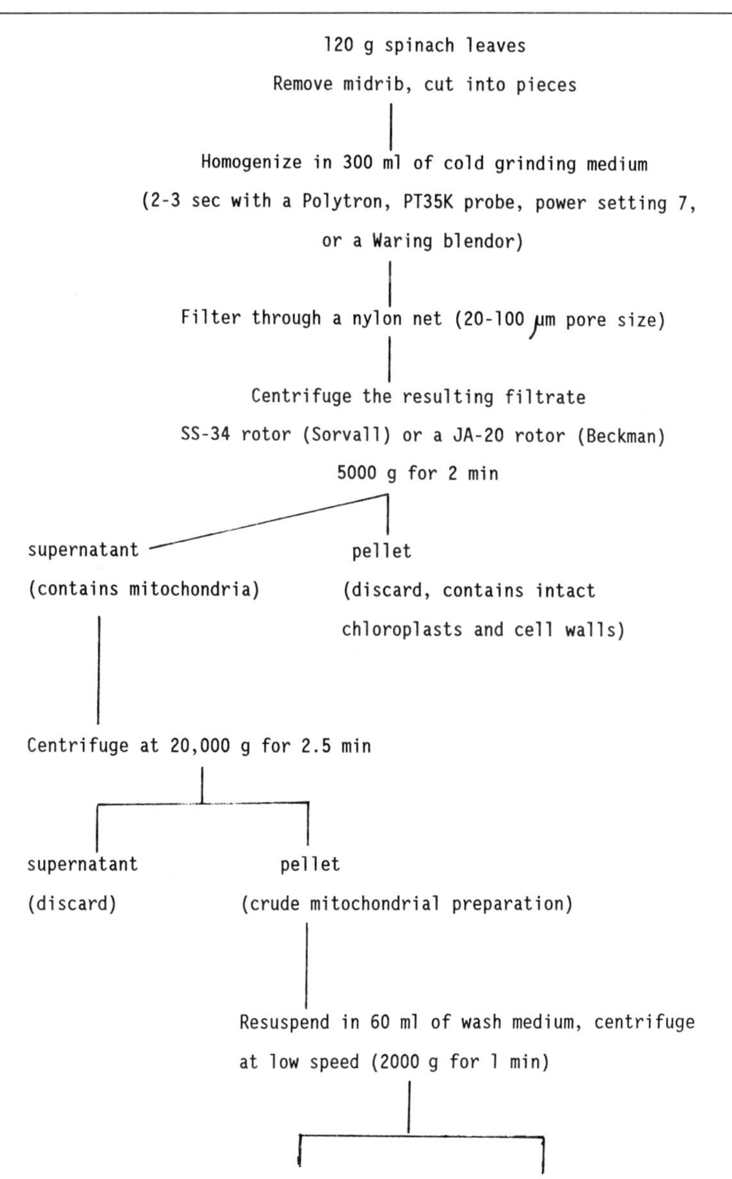

FIG. 1. Flow chart for isolation of mitochondria from leaves of the C_3 species spinach (*Spinacea oleracea*). Centrifugation times are given as time at full speed, not including acceleration and braking. Resuspension of mitochondrial pellets is accomplished using a Teflon/glass or glass homogenizer.

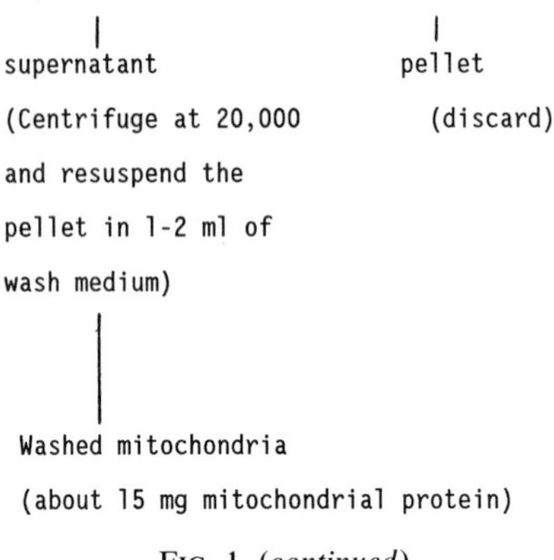

FIG. 1. (*continued*)

Wash Medium

0.3 *M* sucrose, 5 m*M* glycine, 1 m*M* EDTA, 0.1% BSA (defatted)

The volume that can be effectively homogenized per grinding depends on the type of homogenizer used. To get a good recovery of mitochondria from the starting material and to optimize the effectiveness of the grinding, it is important to predetermine the appropriate volume and tissue/volume ratio for a given homogenizer. In the procedure outlined in Fig. 1 short centrifugations at high speed are used; this will minimize the preparation time. If more starting material is used it is suitable to use a different rotor to accommodate larger volumes. In this case, the centrifugations have to be modified (typically 1000 *g* for 15 min for the first centrifugation and 10,000 *g* for 15 min for the second centrifugation). The crude mitochondrial pellet can either be further purified (see below), or, as indicated in Fig. 1, resuspended in the wash medium and taken through another differential centrifugation cycle.

The intactness of the mitochondria prepared by this method (Fig. 1) is better than 90%, as measured by the succinate–cytochrome-*c* oxidoreductase or ascorbate–cytochrome-*c* oxidoreductase assays.[1] The mitochondria will respire a variety of substrates at high rates with good respiratory control and ADP/O ratios. In the case of glycine oxidation, the inclusion of glycine (5 m*M*) in the isolation and storage media may give higher respiratory rates and better respiratory control.[7,9]

The method of isolation of mitochondria from protoplasts of C_3 plants (e.g., spinach, see Nishimura *et al.*[5]) offers no advantages over the me-

[9] G. H. Walker, D. J. Oliver, and G. Sarojini, *Plant Physiol.* **70,** 1465 (1982).

chanical method (e.g., with peas and spinach). Higher yields and equivalent respiratory control can be obtained with mitochondria isolated from these species using mechanical procedures, and less preparation time is required in comparison to procedures utilizing protoplasts. Isolation and purification of the cell type of interest (e.g., isolation of mesophyll protoplasts) prior to isolation of mitochondria does assure the purity of the source. This can be of particular value with some C_4 species.

Purification. In the crude mitochondrial fraction obtained by differential centrifugation (often referred to as "washed mitochondria"), typically about half of the protein is of nonmitochondrial (mainly thylakoid) origin. For some studies, like respiration, these washed mitochondria can be used. However, for many other types of studies (composition, enzyme localization, transport studies) the mitochondria must be purified in order to get valid results.

As already mentioned, a special problem in purification of leaf mitochondria is to get rid of cosedimenting chloroplast fragments which originate from chloroplasts broken during the homogenization. For further purification of the crude mitochondrial fraction obtained by differential centrifugation, a method of separation other than that based on sedimentation velocity must be used. Often, differences in density between the mitochondria and thylakoids are used as a basis for separation. For leaf mitochondria, silica sol density gradient centrifugation has been a very successful method.[10] Also, differences in surface properties of subcellular membranes can be used for purification through use of phase partitioning.[11] By combining silica sol gradient centrifugation and phase partition very pure mitochondria can be obtained from spinach.[12] In general, for further purification of the crude mitochondrial fraction obtained by differential centrifugation, procedures have to be slightly modified according to the particular plant material used to obtain maximum purity.

Isolation Procedures for C_4 Leaf Mitochondria

In C_4 plants, CO_2 assimilation occurs via two photosynthetic cell types, the mesophyll and the bundle sheath. Atmospheric CO_2 is initially fixed into C_4 dicarboxylic acids in the mesophyll cells; the C_4 acids then serve as donors of carbon to the reductive pentose phosphate pathway in bundle sheath cells. Therefore, to evaluate the functions of mitochondria in photosynthetic tissue in C_4 plants these cell types must be separated. C_4

[10] A. L. Moore and M. O. Proudlove, this volume [38].

[11] P. Gardeström and I. Ericson, this volume [40].

[12] A. Bergman, P. Gardeström, and I. Ericson, *Plant Physiol.* **66,** 442 (1980).

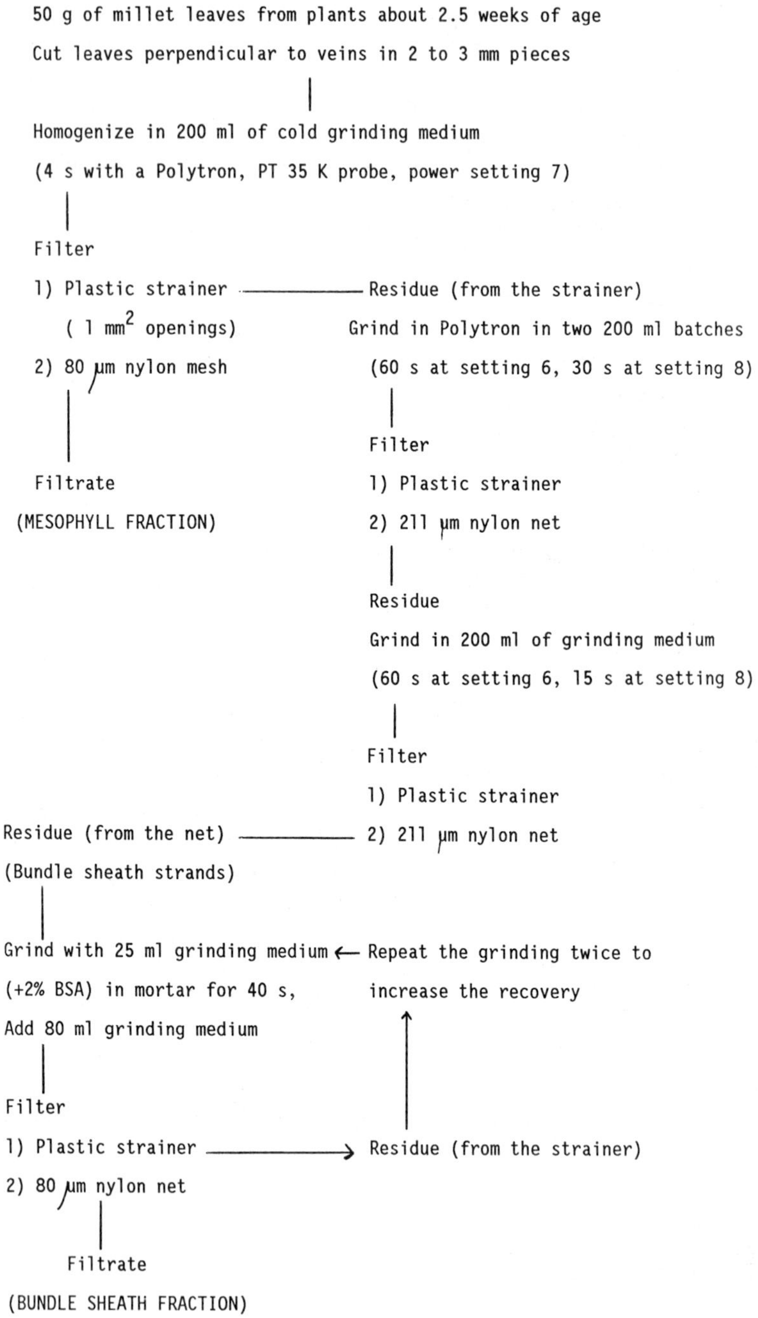

50 g of millet leaves from plants about 2.5 weeks of age

Cut leaves perpendicular to veins in 2 to 3 mm pieces

Homogenize in 200 ml of cold grinding medium

(4 s with a Polytron, PT 35 K probe, power setting 7)

Filter

1) Plastic strainer —— Residue (from the strainer)

(1 mm^2 openings)

2) 80 μm nylon mesh

Filtrate

(MESOPHYLL FRACTION)

Grind in Polytron in two 200 ml batches

(60 s at setting 6, 30 s at setting 8)

Filter

1) Plastic strainer

2) 211 μm nylon net

Residue

Grind in 200 ml of grinding medium

(60 s at setting 6, 15 s at setting 8)

Filter

1) Plastic strainer

Residue (from the net) —— 2) 211 μm nylon net

(Bundle sheath strands)

Grind with 25 ml grinding medium ← Repeat the grinding twice to increase the recovery

(+2% BSA) in mortar for 40 s,

Add 80 ml grinding medium

Filter

1) Plastic strainer → Residue (from the strainer)

2) 80 μm nylon net

Filtrate

(BUNDLE SHEATH FRACTION)

FIG. 2. Flow chart for the mechanical preparation of mesophyll and bundle sheath mitochondria from the C_4 species millet (*Panicum miliaceum*). (Also see the legend to Fig. 1.)

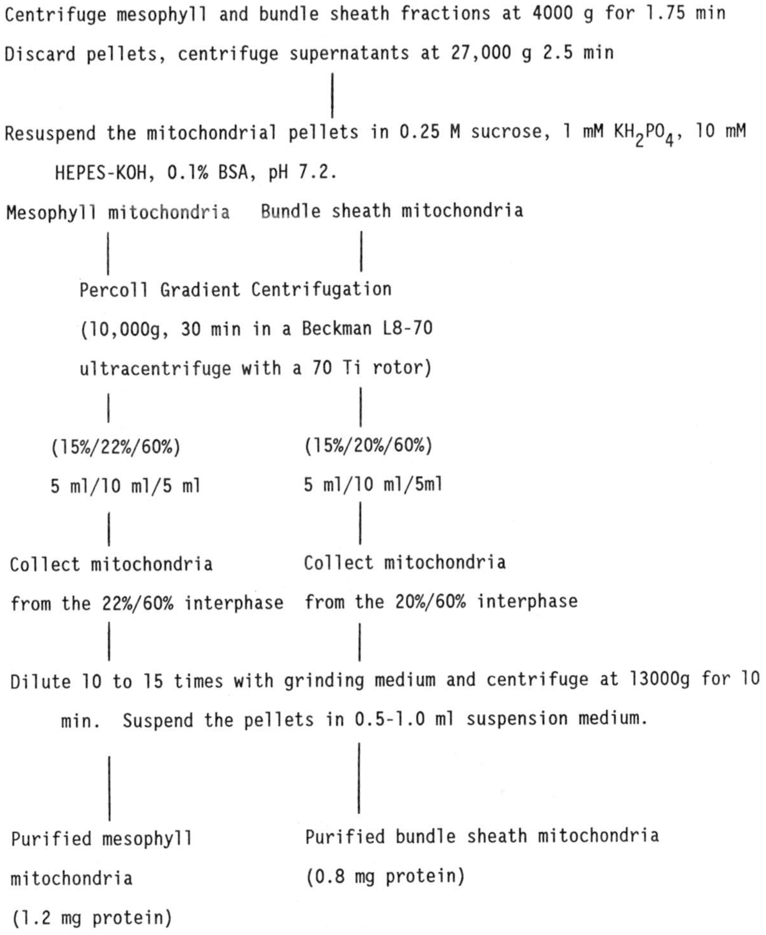

Centrifuge mesophyll and bundle sheath fractions at 4000 g for 1.75 min

Discard pellets, centrifuge supernatants at 27,000 g 2.5 min

Resuspend the mitochondrial pellets in 0.25 M sucrose, 1 mM KH_2PO_4, 10 mM HEPES-KOH, 0.1% BSA, pH 7.2.

Mesophyll mitochondria	Bundle sheath mitochondria
Percoll Gradient Centrifugation (10,000g, 30 min in a Beckman L8-70 ultracentrifuge with a 70 Ti rotor)	
(15%/22%/60%) 5 ml/10 ml/5 ml	(15%/20%/60%) 5 ml/10 ml/5ml
Collect mitochondria from the 22%/60% interphase	Collect mitochondria from the 20%/60% interphase

Dilute 10 to 15 times with grinding medium and centrifuge at 13000g for 10 min. Suspend the pellets in 0.5-1.0 ml suspension medium.

Purified mesophyll mitochondria (1.2 mg protein)	Purified bundle sheath mitochondria (0.8 mg protein)

FIG. 2. (*continued*)

plants are classified into three biochemical subgroups based on their different C_4 acid decarboxylation mechanisms: NAD-malic enzyme species, NADP-malic enzyme species, and phosphoenolpyruvate (PEP) carboxykinase species.

One means of isolating mitochondria from the two cell types is through a differential mechanical grinding of the leaf tissue. This method is dependent on the bundle sheath strands having more resistance to breakage than the mesophyll, so that in the initial grind the mesophyll cells are broken first. Bundle sheath strands are then isolated and further treated to obtain bundle sheath mitochondria. Choice of species and plant age are important factors, and modifications in the procedure for a given species may be required. It is necessary to assay marker enzymes for the two cell types (e.g., phosphoenolpyruvate carboxylase for mesophyll cells and ribulose-1,5-bisphosphate carboxylase for the bundle sheath cells) in order to assess the degree of cross-contamination. This is an adequate method if the cross-contamination is no greater than about 20% (i.e., 80% purity of each mitochondrial type). Figure 2 is a flow chart for mechanical isolation of mesophyll and bundle sheath mitochondria in millet (*Panicum miliaceum*) which was developed by Gardeström and Edwards[13] using the following media.

Grinding Medium

0.3 *M* sucrose, 25 m*M* HEPES–KOH (*N*-2-hydroxyethylpiperazine-*N'*-2-ethanesulfonic acid), 1 m*M* EDTA, 4 m*M* cysteine, 0.2% BSA (defatted), pH 7.8

Percoll Gradient

Percoll (for percentage see Fig. 2), 0.25 *M* sucrose, 1 m*M* K_2HPO_4, 10 m*M* HEPES–KOH, 0.1% BSA (defatted), pH 7.2

Suspension Medium

0.3 *M* sucrose, 10 m*M* KCl, 5 m*M* $MgCl_2$, 10 m*M* potassium phosphate, 0.1% BSA, pH 7.2

If, in some species, adequate separation is not achieved mechanically then use of a combined enzymatic–mechanical procedure can be employed to obtain pure bundle sheath strands. For example, a brief enzymatic digestion (e.g., 15–30 min) followed by mechanical grinding can be used to isolate pure bundle sheath strands,[14] from which mitochondria can be isolated as in Fig. 2. Alternatively, mesophyll protoplasts and bundle sheath strands can be enzymatically isolated and mitochondria then isolated from these preparations, as has been done by Ohnishi and Kanai[15] with maize and sorghum (NADP-malic enzyme-type species).

In some cases, both mesophyll and bundle sheath protoplasts have been isolated and used for preparation of mitochondria.[15] Since the two protoplast types can be isolated with high purity and little cross-contamination from some C_4 species,[15,16] the source of the mitochondria is assured. It has been used with good success with C_4 species belonging to the

[13] P. Gardeström and G. E. Edwards, *Plant Physiol.* **71,** 24 (1983).
[14] K. S. Chapman, J. A. Berry, and M. D. Hatch, *Arch. Biochem. Biophys.* **202,** 330 (1980).
[15] J. Ohnishi and R. Kanai, *Plant Cell Physiol.* **24,** 1411 (1983).
[16] G. E. Edwards, R. McC. Lilley, S. Craig, and M. D. Hatch, *Plant Physiol.* **63,** 821 (1979).

10 g of leaves of millet

ISOLATION OF PROTOPLASTS

Cut transversely into sections approximately 0.7 mm with a sharp razor

Incubate in 40 ml digestion medium for 2.5 to 3.5 h at 30 C

Discard enzyme medium, wash leaf segments several times

(20 to 25 ml per wash) with the sucrose medium

Filter through a plastic strainer (1-mm apertures)

and a 200 μm nylon mesh.

Pour filtrate in a Babcock flask to the neck

Overlay with 2 ml sucrose/sorbitol medium

+ 2 ml of sorbitol medium

Centrifuge at 300 x g for 5 to 10 min

Collect MESOPHYLL PROTOPLASTS from the upper interphase

(approximate yield = 500 μg chlorophyll)

Discard supernatant

Resuspend the pellet (crude BP preparation) in 16 ml of sucrose medium

containing 13% (w/v) dextran T-40

Add 4 ml of the crude BP preparation to each of 4 tubes (15 ml capacity),

overlay each with 2 ml of sucrose medium containing 8.7% dextran T-40, and 2

ml of sorbitol medium

FIG. 3. Flow chart for the preparation of mitochondria from mesophyll protoplasts (MP) and bundle sheath protoplasts (BP) of the C_4 species millet (*Panicum miliaceum*). (Also see the legend to Fig. 1.)

Centrifuge at 300 g for 10 min.

|

Collect BP from the upper interface with a Pasteur pipette, and dilute with 3 vol of sorbitol medium

|

Overlay the BP preparation on 10 ml of the sucrose/sorbitol medium

|

Centrifuge at 200g for 90 sec

supernatant	pellet
(discard)	(purified BUNDLE SHEATH PROTOPLASTS)
	(approximate yield = 100 ug chlorophyll)

ISOLATION OF MITOCHONDRIA

Resuspend protoplasts (about 200 ug chlorophyll/ml) in appropriate grinding media (see mechanical isolation for millet)

|

Mesophyll protoplasts

Break by several passages through a 20 um nylon net

|

Isolate mitochondria by differential centrifugation

(similar to Fig. 1, 3000g for 1 min, supernatant at 27,000g 6 min)

|

Approximate yield = 100 ug mitochondrial protein

assuming a recovery of 65% from MP and 0.3 mg mitochondrial protein per mg chlorophyll in MP[3,15]

Bundle sheath protoplasts

Extract with a Ten Broeck homogenizer

(three strokes)

|

Isolate mitochondria by differential centrifugation (see above)

|

Approximate yield = 40 ug mitochondrial protein

assuming a recovery from BP of 35% and 1.3 mg mitochondrial protein per mg chlorophyll in BP[3,15]

FIG. 3. (*continued*)

NAD-malic enzyme type (*Panicum miliaceum*) and PEP carboxykinase type (*Chloris gayana* and *Panicum maximum*) subgroups.[15] Figure 3 is a flow chart for isolation of mitochondria from millet using this method with the media shown below.

Digestion Medium

1% Cellulase "Onozuka" RS, 0.05% pectolyase Y-23, 0.5 *M* sorbitol, 1 m*M* $MgCl_2$, 1 m*M* $CaCl_2$, 0.1% BSA, 20 m*M* ascorbate, 10 m*M* MES–KOH [2-(*N*-morpholino)ethanesulfonic acid], pH 5.5

Sucrose Medium

0.5 *M* sucrose, 1 m*M* $MgCl_2$, 1 m*M* $CaCl_2$, 0.1% BSA, 5 m*M* HEPES–KOH, pH 7.0

Sorbitol Medium

0.5 *M* sorbitol, 1 m*M* $MgCl_2$, 1 m*M* $CaCl_2$, 0.1% BSA, 5 m*M* HEPES–KOH, pH 7.0

Sucrose/Sorbitol Medium

0.4 *M* sucrose, 0.1 *M* sorbitol, 1 m*M* $MgCl_2$, 1 m*M* $CaCl_2$, 0.1% BSA, 5 m*M* HEPES–KOH, pH 7.0

Isolation Procedures for CAM Leaf Mitochondria

In CAM plants CO_2 is fixed in the dark through PEP carboxylase and malic acid is formed and utilized as a carbon donor to the reductive pentose phosphate pathway in the subsequent light period. There are two subgroups of CAM plants based on their C_4 acid decarboxylases: malic enzyme type (having both NAD- and NADP-malic enzyme) and PEP carboxykinase type. In both types, the mitochondria may have a role in photosynthetic metabolism.[17] Functional mitochondria have been isolated from several malic enzyme type CAM species including *Sedum praealtum*[18] and *Kalanchoe daigremontiana*[19] but, to date, there are no reports of isolation from PEP carboxykinase type species. The approach to isolating mitochondria is similar to that used with C_3 species, with a few important modifications. Due to the high acidity of CAM tissue, a high concentration of buffer is required. As the buffer also serves as an osmoti-

[17] G. E. Edwards, J. G. Foster, and K. Winter, *in* "Crassulacean Acid Metabolism" (I. P. Ting and M. Gibbs, eds.), p. 92. Waverly Press, Baltimore, Maryland, 1982.

[18] G. P. Arron, M. H. Spalding, and G. E. Edwards, *Plant Physiol.* **64,** 182 (1979).

[19] D. A. Day, *Plant Physiol.* **65,** 675 (1980).

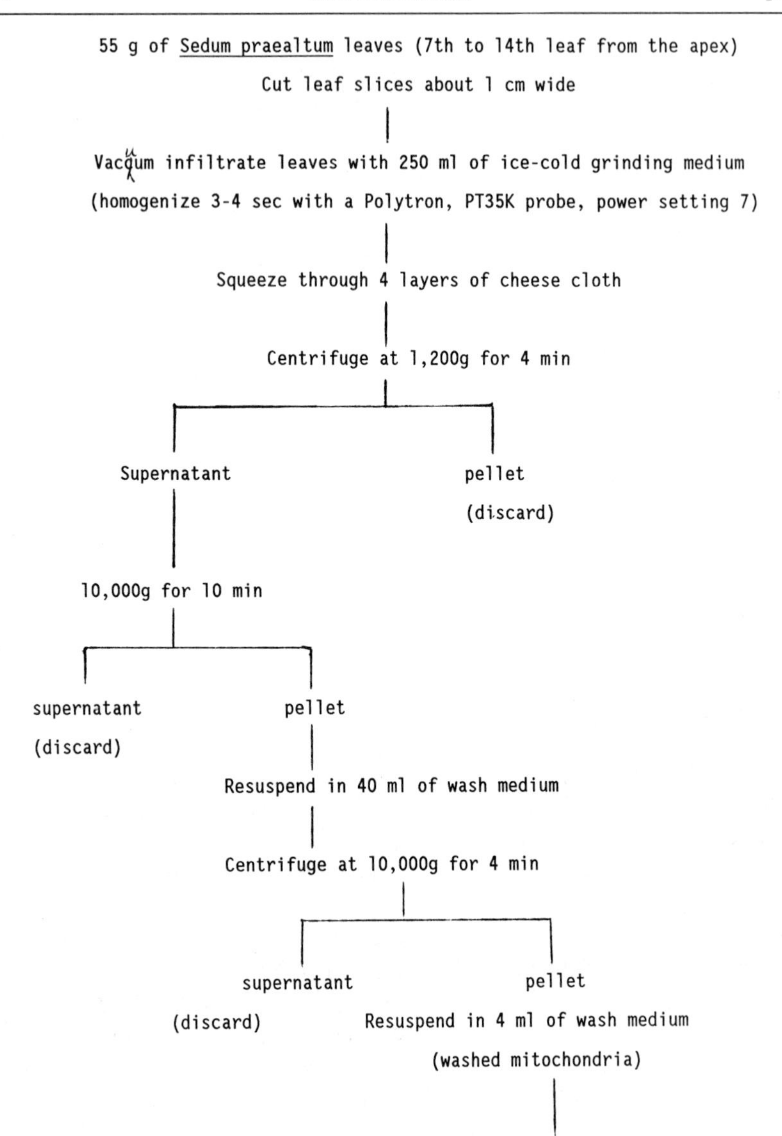

FIG. 4. Flow chart for the preparation of mitochondria from the CAM species *Sedum praealtum*. In this case most of the chlorophyll is removed by differential centrifugation. Following centrifugation on the linear sucrose gradient, the mitochondria appear in a band, while considerable proteinaceous material is found at the bottom of the centrifuge tube. (Also see the legend to Fig. 1.)

Overlay washed mitochondria on a 20-60% (w/v)
linear sucrose gradient

Centrifuge 25,000 rpm (82,000g) for 2 h
in an ultracentrifuge (e.g., Beckman SW 27.1 rotor)

Remove mitochondria from the gradient
with a large bore syringe
(density approximately 1.17 g cm^{-3})

Dilute to 0.4 M sucrose by drop wise addition of
10 mM TES buffer, pH 7.2, + 0.1 % BSA with slow stirring

Centrifuge at 10,000g for 10 min, resuspend
mitochondrial pellet in 1 ml of wash medium

Purified mitochondria
(about 4 mg protein)

FIG. 4. (*continued*)

cum, a corresponding adjustment in the level of polyol or sugar is needed. Also, addition of PVP is critical, and apparently protects the mitochondria through binding phenols; in some CAM tissue all of the protein and organelles aggregate and precipitate if PVP is not added. A procedure for isolation of mitochondria from a CAM plant is shown in Fig. 4 using the following media.

Grinding Medium

0.15 *M* mannitol, 4 m*M* cysteine, 5 m*M* EGTA, 2% PVP-40, 0.2% BSA (defatted), 0.25 *M* Tricine–KOH [*N*-tris(hydroxymethyl)methylglycine], pH 8.0

Wash Medium

0.4 *M* mannitol, 0.1% BSA (defatted), 25 m*M* TES [*N*-tris(hydroxymethyl)methyl-2-aminoethanesulfonic acid], pH 7.6

[40] Separation of Spinach Leaf Mitochondria according to Surface Properties: Partition in Aqueous Polymer Two-Phase Systems

By PER GARDESTRÖM and INGEMAR ERICSON

Introduction

When aqueous solutions of polymers, such as Dextran T-500 and polyethylene glycol 3350 (PEG 3350), are mixed at 4°, a two-phase system will form at polymer concentrations higher than about 5.6% (w/w) of each polymer. The phase system has a PEG-rich top phase and a Dextran-rich bottom phase. Both phases contain 80–90% water, thus rendering the phase system very suitable for separation of various biological materials. Membrane particles introduced into the phase system will, according to their surface properties, distribute between the bulk phases and the interface. Both hydrophilic and hydrophobic groups at the membrane surface are of importance in this respect. The distribution of a given particle depends on several factors such as type of polymers, polymer concentrations, salt composition, pH, and temperature. Since phase partition separates substances according to different parameters compared to centrifugation it will be a valuable complement to the latter technique. The method has been successfully used for separation and studies on macromolecules, cells, and subcellular fractions (e.g., plasma membrane vesicles,[1] inside-out thylakoid vesicles,[2] and inside-out submitochondrial particles[3]). For general discussions of applications and theoretical aspects of aqueous polymer two-phase systems see Albertsson[4] and recent reviews.[5,6]

In this chapter we describe how to design a phase system which in combination with differential centrifugation can be used to obtain purified, functionally intact spinach leaf mitochondria. Also, the application of some special techniques using substituted PEG is exemplified.

[1] C. Larsson, S. Widell, and P. Kjellbom, this volume [52].

[2] H.-E. Åkerlund and B. Andersson, this volume [26].

[3] I. M. Møller, A. C. Lidén, I. Ericson, and P. Gardeström, this volume [41].

[4] P.-Å. Albertsson, "Partition of Cell Particles and Macromolecules," 3rd ed. Almqvist & Wiksell, Stockholm, 1986.

[5] P.-Å. Albertsson, B. Andersson, C. Larsson, and H.-E. Åkerlund, *Methods Biochem. Anal.* **28,** 115 (1982).

[6] H. Walter, D. E. Brooks, and D. Fisher, eds., "Partition in Aqueous Two-Phase Systems." Academic Press, Orlando, Florida, 1986.

Materials

Chemicals

Dextran T-500 (M_r 500,000, Pharmacia, Uppsala, Sweden) is dissolved by gentle boiling to a 20% (w/w) stock solution. The Dextran powder is hygroscopic and will often contain 5–10% water. Thus, dried Dextran must be used to get a well-defined concentration. Alternatively the concentration of the Dextran solution can be measured by dry weight determination or by polarimetry.[5]

PEG 3350 (polyethylene glycol, M_r 3350, formerly designated PEG 4000, Union Carbide) is dissolved in water to give a 40% (w/w) stock solution. If the PEG has a yellowish color it should be recrystallized from ethanol before use: 300 g PEG is dissolved in 1.5 liters boiling absolute ethanol and about 10 g active charcoal is added. After filtration the clear solution is kept in the cold overnight. The precipitated PEG is collected by centrifugation. If necessary the procedure is repeated once before the PEG is heated at 80° for 1 hr to remove ethanol.

Substituted PEG was obtained from Aqueous Affinity (Arlöv, Sweden).

Preparation of Crude Mitochondria

A crude mitochondrial fraction is prepared from spinach leaves by differential centrifugation as described by Edwards and Gardeström,[7] and suspended in 0.3 *M* sucrose. Only sucrose should be used since the partition of membrane particles in the Dextran–PEG system is very sensitive to the salt composition. In this fraction about half of the protein is mitochondrial protein. The main contamination is thylakoid membrane fragments but peroxisomes are also present.

It is very important that the starch content of the leaves is low. Starch, which is associated with the thylakoid membranes, will drastically change the surface properties of these membranes and thus affect their partition in the phase systems.

How to Find a Suitable Phase System

Determination of the Critical Point

The minimum concentration of PEG and Dextran needed to obtain a two-phase system is called the critical point. As this depends on the

[7] G. E. Edwards and P. Gardeström, this volume [39].

TABLE I
PHASE SYSTEM FOR DETERMINATION OF THE CRITICAL POINT[a]

Amount (g)	Stock solution	Final concentration
2.40	20% (w/w) Dextran 500	6.0% (w/w)
1.20	40% (w/w) PEG 3350	6.0% (w/w)
2.00	1.2 mol/kg sucrose	0.3 mol/kg
0.40	0.1 mol/kg potassium phosphate, pH 7.8	5 mmol/kg
2.00	Distilled water	

[a] At 4° (total weight 8.0 g).

molecular weight of the polymers, it can be used to calibrate different batches of polymers.

A phase system is prepared in a test tube according to Table I. After bringing to the desired temperature, the solution is thoroughly mixed by inversions. A milky appearance of the mixture indicates the presence of two phases. Add 100 μl 0.3 *M* sucrose and mix again. This dilution and mixing is repeated until a clear solution is obtained during mixing. This indicates a composition below the critical point. For example, for the polymers used in Table I a total of 600 μl sucrose was added before the phase system was broken. Thus, the critical point at 4° was 8 g/8.6 g $\times$ 6% (w/w) = 5.6% (w/w). At room temperature the critical point was 6.1% (w/w) for the same polymers.

Effect of Polymer Concentration

Phase systems with different polymer concentrations were prepared according to Table II. Close to the critical point both mitochondria and thylakoid membranes partition almost totally to the top phase (Fig. 1a). At increasing polymer concentrations less material is found in the top phase.

This can be explained by the increased hydrophobicity of the top phase due to the increased concentration of PEG in that phase. A membrane with a hydrophilic surface will thus be pushed down to the more hydrophilic bottom phase.

Effect of Salt Concentration

Figure 1b shows the effect of increasing the KCl concentration at constant polymer concentrations. At low KCl concentration the material partitions mainly to the top phase. At KCl concentrations above

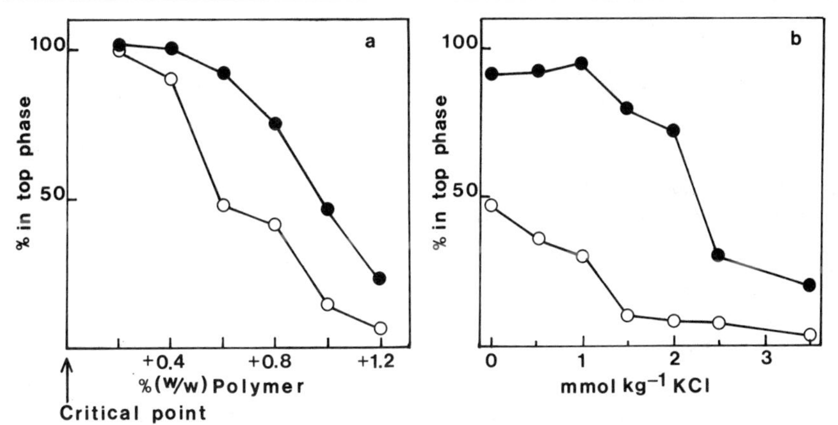

FIG 1. The effect of (a) polymer concentrations [equal concentration (w/w) of Dextran and PEG] and (b) salt composition at a polymer concentration of Dextran and PEG (0.6%, w/w), each above the critical point (Cp + 0.6%), on the partition of mitochondria (○—○, measured as cyt *c* oxidase activity), and thylakoid membrane fragments (●—●, measured as chlorophyll). Phase systems were prepared as described in Table II and 0.5 ml of a crude mitochondrial preparation from 120 g of leaves suspended in 8 ml 0.3 *M* sucrose was added to each phase system. The phase systems were mixed by 40–50 inversions and centrifuged at 600 *g* for 4 min in a swing-out rotor. After noting the phase volumes a sample from the top phase was diluted 1 : 1 with 0.3 *M* sucrose and used for the assays.

TABLE II
PHASE SYSTEMS WITH DIFFERENT POLYMER CONCENTRATIONS[a]

Stock solution[b]	Amount of stock solution (g)					
Dextran	1.16	1.20	1.24	1.28	1.32	1.36
PEG	0.58	0.60	0.62	0.64	0.66	0.68
Sucrose	0.875	0.875	0.875	0.875	0.875	0.875
Potassium phosphate	0.20	0.20	0.20	0.20	0.20	0.20
Water	0.685	0.625	0.565	0.505	0.445	0.385
Sample	0.5 ml	0.5 ml	0.5 ml	0.5 ml	0.5 ml	0.5 ml
Final concentration[c]						
Polymer	5.8	6.0	6.2	6.4	6.6	6.8
Cp + polymer	+0.2	+0.4	+0.6	+0.8	+1.0	+1.2

[a] Total weight, 4 g.

[b] Stock solutions used: 20% (w/w) Dextran 500, 40% (w/w) PEG 3350, 1.2 mol/kg sucrose, 0.1 mol/kg potassium phosphate (pH 7.8).

[c] Polymer, Final concentration each of Dextran and PEG (w/w); Cp + polymer, concentration (% w/w) of Dextran and PEG above the critical point (Cp). The final concentration of sucrose and potassium phosphate is 0.3 mol/kg and 5 mmol/kg, respectively.

3 mmol/kg very little mitochondria and thylakoid membranes are found in the top phase.

Phosphate buffer at pH 7.8 will create an interfacial potential with a negative bottom phase and a positive top phase since HPO_4^{2-} ions prefer the bottom phase. Negatively charged membranes will thus favor the top phase. Chloride has an opposite effect on the interfacial potential as compared to phosphate and will thus favor the partition of negatively charged membranes to the bottom phase.

Batch System

Leaf mitochondria are more sensitive than thylakoid membranes to an increase in the polymer concentrations (Fig. 1a). Mitochondria are also more strongly affected by increasing KCl concentrations (Fig. 1b). A system with polymer concentrations of 0.6% (w/w), each above the critical point, and with 1.5–2 mmol/kg KCl will thus be very suitable for batch purification of spinach leaf mitochondria.

Other types of leaf mitochondria or even mitochondria from spinach leaves grown under different conditions may require different conditions for purification by phase partition. For pea leaf mitochondria a phase system as described in Table II but with 5 mmol/kg KCl gave good results.[8]

The phase system for batch purification of spinach leaf mitochondria was obtained by varying, separately, the polymer concentrations and salt composition. A more general approach for the selection of a suitable phase system can be made, if several parameters are varied simultaneously, using experimental design.[9] By applying this technique a phase system [5.64% (w/w) Dextran 500, 6.96% (w/w) PEG 3350, 0.3 mol/kg sucrose, 9.79 mmol/kg potassium phosphate, pH 7.56, 1.48 mmol/kg KCl] was designed that gave similar purification as the system described in Table III.[10]

Purification of Spinach Leaf Mitochondria

Procedure

Prepare a 100-g phase system in a separation funnel according to Table III. Bring the system to the desired temperature and mix carefully. Let the system settle for several hours and separate the top from the bottom phase (discard the interface).

[8] D. Henriksson, unpublished (1981).

[9] L. Backman and V. Shanbhag, *Anal. Biochem.* **138,** 372 (1984).

[10] L. Backman, unpublished (1983).

TABLE III
COMPOSITION OF A 100-g PHASE SYSTEM FOR BATCH PURIFICATION OF LEAF MITOCHONDRIA

Amount (g)	Stock solution	Final concentration
31.0	20% (w/w) Dextran	6.2% (w/w) } (Cp + 0.6)
15.5	40% (w/w) PEG 3350	6.2% (w/w) } (Cp + 0.6)
25.0	1.2 mol/kg sucrose	0.3 mol/kg
5.0	0.1 mol/kg potassium phosphate, pH 7.8	5 mmol/kg
2.0	0.1 mol/kg KCl	2 mmol/kg
21.5	Water	

The crude mitochondrial pellets from 120 g spinach leaves are suspended in 10 ml top phase using a glass–Teflon homogenizer. This will result in a small dilution of the system as compared to the method used in Table II, but this is of no practical consequence if well-drained pellets are used. Eight milliliters of bottom phase is added and the system is carefully mixed. The bottom phase should be added as quickly as possible as prolonged storage of mitochondria in pure top phase will result in a reduction of ADP/O ratios obtained in the final preparation. Phase settling is accelerated by centrifugation in a swing-out rotor at 600 *g* for 3–4 min. The top phase containing mainly thylakoid membranes is removed, leaving the material on the interface. Fresh top phase is added and the extraction procedure is repeated twice. Additional extractions do not remove more of the chlorophyll from the bottom phase. The extracted bottom phase is diluted at least eight times with 0.3 *M* sucrose and any remaining intact chloroplasts are removed by a short centrifugation (5000 *g*, 2 min). Finally the mitochondria are collected from the supernatants by centrifugation at 11,000 *g* for 10 min. The mitochondrial pellet is suspended in a suitable medium. The total preparation time is about 1 hr, and the yield is about 6 mg mitochondrial proteins from 120 g leaves.

Properties of the Preparation

In the phase partition step more than 95% of the chlorophyll in the crude mitochondrial fraction is removed and mitochondria are enriched more than 10 times relative to thylakoid membranes.[11] The purification in the phase partition step is summarized in Table IV. On a protein basis the thylakoid membrane contamination is only a few percent. The intactness

[11] P. Gardeström, I. Ericson, and C. Larsson, *Plant Sci. Lett.* **13,** 23 (1978).

TABLE IV
PURIFICATION OF SPINACH LEAF MITOCHONDRIA BY DIFFERENTIAL CENTRIFUGATION (CRUDE) AND PHASE PARTITION (PURIFIED)

Mitochondrial fraction	Purification parameter	Enrichment[a]			
		Protein/Chl	Cyt ox[b]/Chl	Cyt ox/GO	Cyt ox/NADPH : cyt
Crude	Sedimentation velocity	15	20	7	4
Purified	Surface properties	100	270	6	15

[a] Recalculated data from A. Bergman, P. Gardeström, and I. Ericson, *Plant Physiol.* **66,** 442 (1980).

[b] Marker enzymes used were as follows: Mitochondria—cytochrome-*c* oxidase (Cyt ox); Peroxisomes—glycolate oxidase (GO); Microsomes—NADPH : cytochrome-*c* oxidoreductase (NADPH : cyt).

of the mitochondrial outer and inner membranes is about 90%, each based on latency of succinate : cytochrome-*c* oxidoreductase and NAD : isocitrate dehydrogenase, respectively. The mitochondria will actively respire with substrates such as NADH, NADPH, succinate, malate, and glycine. The highest respiratory control and ADP/O ratios are obtained with glycine and malate as substrates. The preparation described here is comparable to preparations obtained by differential centrifugation combined with Percoll gradient centrifugation with respect to purity and intactness of the mitochondria.[12,13] Inclusion of glycine during the preparation has been reported to improve respiratory parameters, especially with glycine.[12] A very pure, essentially chlorophyll-free preparation of spinach leaf mitochondria is obtained if the preparation obtained by centrifugation and phase partition is further purified by discontinuous Percoll density gradient centrifugation.[14]

Special Techniques

Affinity Partition

Ligands covalently coupled to PEG will affect the partition. The phase system used is such that in the absence of substituted PEG little material

[12] H. Ebbighausen, C. Jia, and H. W. Heldt, *Biochim. Biophys. Acta* **810,** 184 (1985).

[13] C. Jackson, J. E. Dench, D. O. Hall, and A. L. Moore, *Plant Physiol.* **64,** 150 (1979).

[14] A. Bergman, P. Gardeström, and I. Ericson, *Plant Physiol.* **66,** 442 (1980).

TABLE V
EFFECT OF SUBSTITUTED PEG ON THE DISTRIBUTION OF MITOCHONDRIA (NAD: ISOCITRATE DEHYDROGENASE—IDH), PEROXISOMES (CATALASE), AND THYLAKOID MEMBRANE FRAGMENTS (Chl)[a]

PEG 6000	Percentage in top phase[b]		
	Chl	IDH	Catalase
PEG	9	2	10
PEG—$(CH_2)_6$—COO^-	10	1	9
PEG—$(CH_2)_6$—NH_3^+	74	77	33
PEG—$(CH_2)_{10}$—NH_3^+	77	84	72

[a] P. Gardeström, A. Bergman, and I. Ericson, unpublished results (1978).

[b] Phase system: 6.1% (w/w) Dextran 500, 5.5% (w/w) PEG 3350, 0.6% (w/w) PEG 6000, 0.3 mol/kg sucrose, 5 mmol/kg potassium phosphate (pH 7.0), 3.5 mmol/kg KCl.

is present in the top phase. This can be achieved by high KCl concentration (Table V, Fig. 1b) or by high polymer concentrations (Fig. 1a). In Table V the negatively charged carboxyl-PEG does not increase the amount of membrane material in the top phase, whereas positively charged amino-PEG will attract material into the top phase. It is interesting to note that PEG—$(CH_2)_6$—NH_3^+ will attract the mitochondria more than peroxisomes into the top phase. This indicates some specificity of the ligand for mitochondria. PEG—$(CH_2)_6$—NH_3^+ has also been successfully used for separation of inside-out from right-side-out plant submitochondrial particles,[3] indicating specificity of the ligand for the outer surface of the membrane.

Cross-Partition

Two sets of phase systems, giving opposite interfacial potentials, are prepared. This can be achieved by using positively and negatively charged polymers (Fig. 2), or by using different salts. The membrane charge can then be studied by measurements of the pH dependence of the partition. At the isoelectric point of the membrane the distribution will be the same irrespective of the interfacial potential.[15] In Fig. 2 the partition

[15] I. Ericson, *Biochim. Biophys. Acta* **356,** 100 (1974).

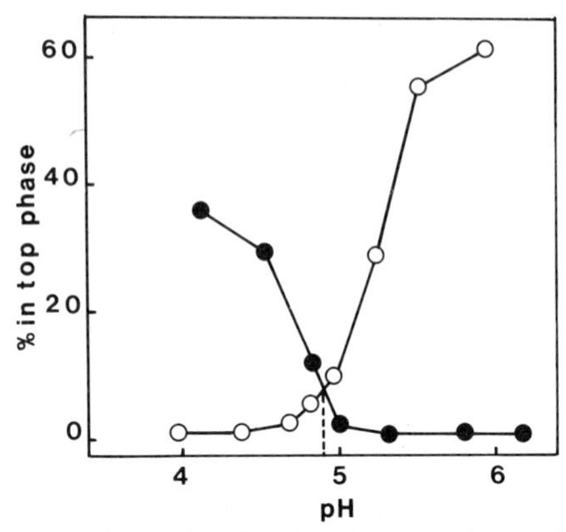

FIG. 2. Determination of the surface isoelectric point of spinach leaf mitochondria by cross-partition. Phase systems (2 g) were prepared with the following final concentration: 6.8% (w/w) Dextran 500, 5.1% (w/w) PEG 3350, 1.7% (w/w) trimethylamino-PEG 3350 (positively charged, ○—○) or 1.7% (w/w) sulfate-PEG 3350 (negatively charged, ●—●), 0.3 mol/kg sucrose, 2.5 mmol/kg potassium phosphate. The pH was adjusted by addition of citrate. Purified mitochondria were suspended in 0.3 *M* sucrose and added to phase systems. Phases were allowed to settle for 30 min. A sample was withdrawn from the top phase, diluted, and neutralized before the cytochrome-*c* oxidase activity was measured. The remaining phase system was diluted 1 : 1 with 0.3 *M* sucrose and the pH was measured at room temperature (A. Bergman, unpublished results 1982).

of spinach leaf mitochondria is plotted as a function of pH. The lines will cross (therefore the term cross-partition[5]) at pH 4.9, which thus is the surface isoelectric point for spinach leaf mitochondria.

Acknowledgments

We are grateful to Dr. A. Bergman for supplying the cross-partition data and to K. Smeds for technical assistance. Comments on the manuscript by Dr. C. Larsson and Dr. V. Shanbhag are gratefully acknowledged.

[41] Isolation of Submitochondrial Particles with Different Polarities

By IAN M. MØLLER, ANNIKA C. LIDÉN, INGEMAR ERICSON, and PER GARDESTRÖM

The isolation of membrane vesicles of well-defined polarities, namely, right-side-out (RO) and inside-out (IO), constitutes an important step in the study of any membrane system. For chloroplast thylakoid membranes, IO and RO vesicles have been purified from mixed populations of vesicles by phase partitioning and shown to be derived from stacked and unstacked regions, respectively. This has led to rapid advances in our understanding of the lateral and transverse distribution and function of protein complexes and lipids in the thylakoid membrane.[1] Likewise the purification of RO plasmalemma vesicles from a variety of plant organs and species by phase partitioning is proving to be very useful in the study of the properties of the plasmalemma.[2,3]

The inner membrane of mammalian mitochondria has been extensively studied and much is known about both faces of this membrane.[4] In spite of this, little is known about what forces govern the formation of cristae and the lateral distribution of proteins and lipids in relation to the cristae. Stacking of thylakoid membranes is electrostatically regulated[5] and we have shown that several properties of plant mitochondrial membranes are also electrostatically controlled.[6-8] One of these properties is probably cristae formation by analogy with thylakoid stacking (Fig. 1).

In the present chapter we describe (1) the purification of mitochondria from Jerusalem artichoke tubers; (2) how the polarity of inner membrane vesicles (submitochondrial particles, SMP) can be measured; (3) how SMP of different polarities can be generated by adjustment of the ionic conditions in the disruption medium; and (4) how RO- and IO-SMP can be partially purified from a mixed population of SMP by phase partitioning.

So far the methods described in this chapter have been applied only to mitochondria from potato and Jerusalem artichoke tubers, but we expect that similar results will be obtained with any type of mitochondria.

Reagents

Antimycin A (Sigma)
BAS: Bovine serum albumin (e.g., Sigma A-4503)
Cytochrome *c* (Boehringer-Mannheim)
Dextran 500 (Pharmacia)

[1] B. Andersson, C. Sundby, H.-E. Åkerlund, and P.-Å. Albertsson, *Physiol. Plant.* **65,** 322 (1985).
[2] C. Larsson, S. Widell, and P. Kjellbom, this volume [52].
[3] I. M. Møller and A. Bérczi, *FEBS Lett.* **193,** 180 (1985).
[4] A. N. Martonosi, ed., "The Enzymes of Biological Membranes," 2nd ed., Vol. 4. Plenum, New York, 1985.
[5] J. Barber, *Annu. Rev. Plant Physiol.* **33,** 261 (1982).
[6] I. M. Møller, C. J. Kay, and J. M. Palmer, *Biochem. J.* **223,** 761 (1984).
[7] K. Edman, I. Ericson, and I. M. Møller, *Biochim. J.* **232,** 471 (1985).
[8] C. J. Kay, I. Ericson, P. Gardeström, J. M. Palmer, and I. M. Møller, *FEBS Lett.* **193,** 169 (1985).

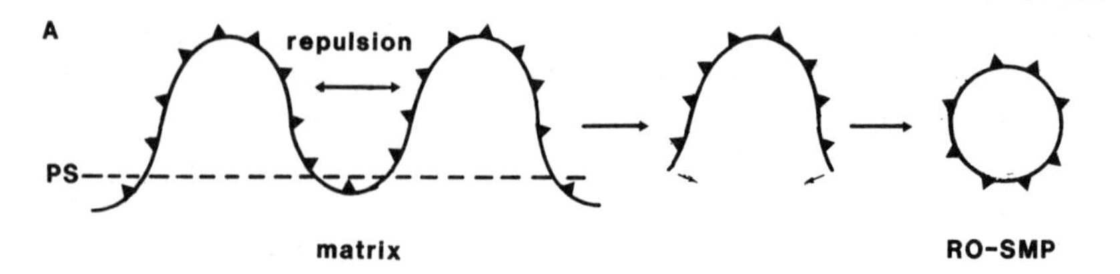

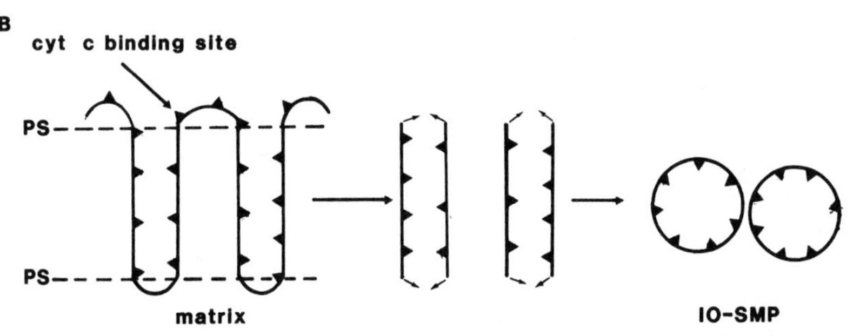

FIG. 1. Proposed mechanism for the formation of right-side-out submitochondrial particles (RO-SMP) under low-salt + EDTA conditions (A) and of IO-SMP under high-salt conditions (B). In (A) only one of the RO-SMP formed is shown. PS, plane of shear. The figure is reproduced from C. J. Kay, I. Ericson, P. Gardeström, J. M. Palmer, and I. M. Møller, *FEBS Lett.* **193,** 169 (1985).

HMDA–PEG 8000: Hexamethylenediamine–polyethylene glycol (Aqueous Affinity, Arlöv, Sweden)
MOPS: 3-(*N*-morpholino)propanesulfonic acid (Sigma or Boehringer-Mannheim)
NADH: (Boehringer-Mannheim)
PEG 3350: Polyethylene glycol 3350 (earlier designated PEG 4000) (Union Carbide)
Percoll (Pharmacia)
Oxaloacetate (Sigma)
Triton X-100 (Sigma)

Preparation of Mitochondria

Potato (*Solanum tuberosum* L.) and Jerusalem artichoke (*Helianthus tuberosus* L.) tubers can be purchased locally and are stored at 4° until used. Note, however, that a 23% Percoll gradient cannot be used for the purification of mitochondria from freshly harvested Jerusalem artichoke tubers (within ~4 weeks of harvest).

[9] W. D. Bonner, this series, Vol. 10, p. 126.

Mitochondria from potato tubers are isolated as described by Bonner[9] and purified by Percoll density centrifugation according to Neuburger *et al.*[10] The grayish-white mitochondrial band below the yellow amyloplast band is sedimented after dilution with the medium subsequently used in the disruption of the mitochondria (see below): 0.4 *M* sucrose, 5 m*M* K^+–MOPS, pH 7.0 (low-salt medium), plus 5 m*M* EDTA (low-salt + EDTA) *or* plus 50 m*M* $MgCl_2$ (high-salt medium). Lower concentrations of $MgCl_2$ (20 m*M*) will give similar results.

For the preparation of mitochondria from Jerusalem artichoke tubers, the tubers are peeled and cut into cubes, e.g., by use of a machine for making potato chips. These are covered with ice-cold homogenization medium [0.5 *M* sucrose, 10 m*M* K^+–MOPS, 5 m*M* EDTA, 4 m*M* cysteine, and 0.1% (w/v) BSA adjusted to pH 7.8] in a ratio of 1 g : 1.2–1.4 ml. Homogenization with a kitchen blender with a rotating knife, e.g., Waring blender or Krups 3 Mix 3000 for 15–20 sec gives a satisfactory yield of intact mitochondria. The slurry is strained through a 20-μm nylon net or eight layers of gauze and the mitochondria as well as heavier particles are rapidly removed from the homogenate by a 14,300 *g* spin for 15 min in an angle rotor. This is to minimize exposure to proteases, phenols, and other harmful substances released by the disruption of the tissue. The pellets are resuspended in a low-salt wash medium (0.4 *M* sucrose, 5 m*M* K^+–MOPS, pH 7.2, and 0.1% BSA) with a glass–Teflon homogenizer and diluted to 320 ml final volume for a 1-kg preparation. The heavy particles are pelleted either by a slow spin (1000 *g* for 10 min) or by rapid acceleration to 12,000 *g* followed by immediate deceleration. The loose pellets are discarded and the mitochondria are pelleted from the supernatant at 48,400 *g* for 2 min. The mitochondria are resuspended in wash medium and pelleted again (48,400 *g* for 2 min). When crude mitochondria are used for the preparation of SMP, the pellets are finally resuspended either in wash medium or in the medium used in the subsequent disruption of the mitochondria.

A typical preparation from 1 kg Jerusalem artichoke tubers will yield 100–150 mg mitochondrial protein. The mitochondria show 90–95% integrity of both outer and inner membranes as estimated by the latency of cytochrome oxidase and malate dehydrogenase, respectively (see below).

For the preparation of purified Jerusalem artichoke mitochondria,[11] the final 48,400 *g* pellets (see above) are resuspended in 0.3 *M* sucrose, 10

[10] M. Neuburger, E.-P. Journet, R. Bligny, J.-P. Carde, and R. Douce, *Arch. Biochem. Biophys.* **217,** 312 (1982).

[11] I. M. Møller and A. Lidén, *in* "Plant Mitochondria: Structural, Functional and Physiological Aspects" (A. L. Moore and R. B. Beechey, eds.). Plenum Press, New York, in press (1987).

mM potassium phosphate, 1 mM EDTA, and 0.1% (w/v) BSA (final volume ~10 ml for a 1 kg preparation), and 3 ml of this suspension is layered on top of 36 ml of the same medium containing 23% (v/v) Percoll. The gradient is generated by centrifuging at 40,000 *g* for 30 min in an angle rotor (Beckman J2-21 centrifuge, JA-20 rotor). The top ~25 ml of the gradient containing peroxisomes and broken mitochondria[11] is removed, e.g., by aspiration. The bottom 1 ml, which often contains significant amounts of catalase, is then removed with a Pasteur pipette and discarded. Finally, the bottom band (10–12 ml) is diluted to 80 ml in wash medium (3 × 10–12 ml and 240 ml for a 1 kg preparation) and the mitochondria are pelleted at 48,400 *g* for 3 min. The supernatant is decanted very carefully from the loose pellets, which sit on top of a Percoll cushion, and the mitochondria are resuspended in wash medium once more and pelleted at 48,400 *g* for 3 min. The purified mitochondria are either suspended in wash medium or in the medium used in the subsequent disruption of the mitochondria. The yield of purified mitochondria is 30–60 mg from 1 kg tubers, with an integrity of 95–98% for both membranes.[11]

The use of Jerusalem artichoke tubers has several advantages over potato tubers: (1) The polysaccharide in Jerusalem artichoke tubers, inulin, is water soluble. The removal of starch grains, which is a problem when working with potatoes, is thus avoided. (2) Amyloplast contamination is much less prominent so that crude mitochondria as isolated here can be used in many experiments. (3) For the same reason the yield of mitochondria is better with Jerusalem artichoke tubers.

Polarity Assay

Reduced cytochrome *c* interacts with cytochrome-*c* oxidase (EC 1.9.3.1) only on the cytoplasmic side of the inner mitochondrial membrane and it cannot cross an intact inner membrane. We have used these properties to assess the polarity of SMP. In the absence of detergent (intact membranes) only RO-SMP will give cytochrome oxidase activity (see Fig. 1) whereas cytochrome oxidase activity of all vesicles will be measured in the presence of 0.02% (w/v) Triton X-100 (solubilized membranes). Since too high concentrations of Triton inhibit cytochrome oxidase[12] one should check that the Triton X-100 concentration is optimal.

[12] I. M. Møller, A. Bergman, P. Gardeström, I. Ericson, and J. M. Palmer, *FEBS Lett.* **126,** 13 (1981).

Procedure

1. Cytochrome *c* is reduced by treatment with ascorbate. Excess ascorbate is removed on a Sephadex PD-10 column.[13]

2. Cytochrome *c* oxidase activity is measured as the decrease in A_{550} (21 mM^{-1} cm^{-1}) in a medium containing 0.3 M sucrose, 10 K^+–MOPS, pH 7.2, 5 mM $MgCl_2$, and 60 μM reduced cytochrome *c* plus or minus Triton X-100 (see below) and sample (10–30 μg protein/ml). The reaction is started by the addition of reduced cytochrome *c*.

3. Polarity is calculated as

$$\text{Percentage RO-SMP} = (\text{rate} - \text{Triton}) \times 100/(\text{rate} + \text{Triton})$$

The plus and minus Triton assays are best performed separately, at least for SMP, which are mainly RO because in this case the minus Triton rate is high.

The cytochrome *c* oxidase assay is very reproducible. It is most suitable for the assay of IO-SMP in which a large difference in the activities between plus and minus Triton is obtained.

The polarities calculated from polarity assays based on latency of enzyme activities are indirect since they are based on activity differences and not directly on differences in orientation of the membrane. The polarity values obtained should therefore not be overinterpreted. Membrane fragments adhering to the surface of the SMP can mask binding sites for cytochrome *c* on RO-SMP, making them behave as IO-SMP. The small difference in the activities measured plus and minus Triton for RO-SMP is another limitation in the polarity assay using cytochrome-*c* oxidase.

It would therefore be valuable if the cytochrome *c* oxidase assay could be complemented by an assay in which the access to an active site on the inner surface of the inner mitochondrial membrane is used. One such assay is the oxidation of succinate using ferricyanide as the electron acceptor (electron transport through the chain blocked with antimycin A). This polarity assay gave similar polarity as the cytochrome *c* oxidase assay on *Arum* SMP.[12] However, the succinate–ferricyanide assay has not worked satisfactorily on SMP from Jerusalem artichoke tubers, possibly due to permeability of the inner membrane to ferricyanide. Another problem is that succinate dehydrogenase is rather unstable and much activity is lost during the production of the SMP.[12]

In principle the above cytochrome *c* oxidase assay for the polarity of SMP is similar to the widely used assay for the integrity of mitochondria based on the oxidation of reduced cytochrome *c*.[10] However, in this in-

[13] I. M. Møller and J. M. Palmer, *Physiol. Plant.* **54,** 267 (1982).

tactness assay the outer membrane forms the cytochrome *c* permeability barrier which is disrupted by Triton.

Preparation of Submitochondrial Particles

All steps are performed at 0–4°.

RO-SMP

To obtain RO-SMP the mitochondria are suspended in (low-salt + EDTA) medium. Under these conditions the outer surface of the inner membrane is maximally negative[7] and cristae membranes will repulse each other (Fig. 1A). The mitochondria are treated in a French press at 200 MPa (a pressure of 17 MPa will also disrupt the mitochondria, but the yield appears to be smaller) at a flow rate of ca. 5 ml/min (the rate is not critical).

After disruption, samples are diluted ca. 10-fold with the medium used for disruption. Intact mitochondria and large membrane fragments are pelleted at 48,400 *g* for 10 min and discarded. The supernatant is centrifuged at 105,000 *g* for 60 min (however, see below) in an angle rotor. The final pellets are resuspended in 0.3 *M* sucrose, 5 m*M* K^+–MOPS, pH 7.2. The SMP are essentially RO (Table I), but some variability is observed between preparations (see below). The yield is usually 10–20% of the mitochondrial protein.

IO-SMP

To obtain IO-SMP the mitochondria are suspended in a high-salt medium where the surface potential is close to zero[7] and the cristae membranes can stack (Fig. 1B). The mitochondria are sonicated four times, 5 sec each time. For the sonication a Branson model B-30 Sonifier (setting 6) with a microprobe or an MSE sonicator (high power; amplitude 1) can be used without significant differences in the polarities or yields. Samples of 3–4 ml are used. The concentration of protein in the sample during disruption does not appear to affect the results appreciably. After sonication, the samples are diluted, centrifuged, and resuspended as described above. The SMP are >80% IO (Table I). The yield is normally 10–20% of the mitochondrial protein.

SMP of Intermediate Polarity

To obtain SMP of intermediate polarity, which are used in the separation of RO- and IO-SMP (see below), mitochondria in a low-salt medium

TABLE I
POLARITY OF SMP PRODUCED BY FRENCH PRESS TREATMENT OR BY SONICATION OF CRUDE JERUSALEM ARTICHOKE MITOCHONDRIA SUSPENDED IN MEDIA OF DIFFERENT IONIC COMPOSITION

Medium	Percentage RO-SMP	
	French press	Sonication
Low salt + EDTA	98 ± 1 (2)[a]	54 ± 1 (2)
Low salt	81 ± 8 (2)	42 ± 5 (12)
High salt (20 mM $MgCl_2$)	38 (1)	18 (1)

[a] Mean ± SD (number of independent preparations).

are sonicated, diluted, centrifuged, and resuspended as described. The polarity is 35–55% RO (Table I).

Size Distribution of SMP

In the description above the SMP are pelleted at 105,000 *g*; however, if the 105,000 *g* supernatant is spun at 300,000 *g* for 60 min differences in size distribution become apparent (Table II). French press treatment of mitochondria in (low-salt + EDTA) medium gives small vesicles which are predominantly RO. In contrast, sonication of mitochondria in high salt medium gives mainly large, inside-out vesicles. In general, RO-SMP are considerably smaller than IO-SMP.[14]

The variability in polarity for the (low salt + EDTA) SMP (98% RO in Table I and 68% RO in Table II) may be caused by small differences in size distribution between different preparations.

The SMP can be frozen in liquid nitrogen and will retain their polarity after thawing. Some loss of cytochrome *c* oxidase activity occurs (10–30%).

Assay for Matrix Inclusion

According to the model in Fig. 1 RO-SMP should enclose mainly matrix space whereas the IO-SMP will enclose intermembrane space. To test this one can measure the activity of a matrix enzyme, NAD^+-depen-

[14] P. J. Petit, K. A. Edman, P. Gardeström, and I. Ericson, *Biochim. Biophys. Acta* **890,** 377 (1987).

TABLE II
SEDIMENTATION AND POLARITY OF SMP FROM PURIFIED POTATO MITOCHONDRIA

Fraction	French press in low salt + EDTA		Sonication in high salt	
	Cyt. oxidase[a] (%)	RO (%)	Cyt. oxidase (%)	RO (%)
40,000 *g* pellet	5.6	39	56	18
105,000 *g* pellet	33	68	42	16
300,000 *g* pellet	22	84	2.3	58
300,000 *g* supernatant	39	82	<0.2	93

[a] Percentage of recovered cytochrome-*c* oxidase activity. Similar trends are observed with purified Jerusalem artichoke mitochondria.

dent malate dehydrogenase (EC 1.1.1.37). This is the most active Krebs cycle enzyme and it is easy to assay, as follows:

Sample (5–30 μg protein/ml) is added to a medium containing 0.3 *M* sucrose, 20 m*M* K^+–MOPS, pH 7.0, 1 m*M* oxaloacetate (adjusted to pH 6–7 with KOH; never exceed pH 8), 0.4 μM antimycin A, and 0.025% (w/v) Triton X-100. The concentration of Triton is not critical as with cytochrome *c* oxidase but should be above 0.015% to ensure rupture of all the membranes. The reaction is started by the addition of 0.2 m*M* NADH and followed at 340 nm (absorptivity 6.2 mM^{-1} cm^{-1}).

The specific activity of malate dehydrogenase in intact mitochondria lies in the range 3–15 μmol (mg protein)$^{-1}$ min^{-1} for both potatoes and Jerusalem artichokes. The activity is reduced by a factor 2–10 in the SMP depending on the disruption conditions. In general it is much higher in RO-SMP than in IO-SMP.[8,15]

The assay for malate dehydrogenase can also be used to determine the integrity of the inner mitochondrial membrane. In this case the assay is performed both in the presence and absence of Triton and the integrity calculated as

$$\text{Percentage intact} = \{[(\text{rate} + \text{Triton}) - (\text{rate} - \text{Triton})]/(\text{rate} + \text{Triton})\}\ 100 \quad (1)$$

[15] A. Lidén, M. Sommarin, and I. M. Møller, *in* "Plant Mitochondria: Structural, Functional and Physiological Aspects" (A. L. Moore and R. B. Beechey, eds.). Plenum Press, New York, in press (1987).

Separation of RO- and IO-SMP from a Population of Intermediate Polarity

Aqueous polymer two-phase systems have the ability to separate membrane vesicles with respect to differences in their membrane properties.[16] Here we use a two-phase system to discriminate between RO- and IO-SMP in a population of intermediate polarity. For a general treatment on the preparation and application of Dextran–PEG two-phase systems see Albertsson *et al.*[16]

The following procedure is used (all steps at 4°):

1. A two-phase system with a final composition of 6.6% (w/w) Dextran 500, 6.6% PEG 3350 (10% of which is HMDA–PEG 8000; this percentage depends strongly, however, on the batch of HMDA–PEG and a concentration curve should be performed with each batch), 0.3 mol sucrose/kg and 10 mmol potassium phosphate/kg, pH 8.0, is prepared according to the general guidelines in Albertsson *et al.*[16] The polymer concentration used is about 1% higher than the lowest polymer concentration that will form a two-phase system.[17] The final composition is calculated for a 40 g system. The system is made up to 36 g in a centrifuge tube and 4.0 ml of a suspension (ca. 2 mg protein/ml) of SMP of intermediate polarity (see above) is added.

An almost identical phase system has been used to purify IO-SMP from *Arum maculatum* spadix mitochondria.[12]

2. The two-phase system is thoroughly mixed by at least 30 inversions of the tube and centrifuged for 2 min at 600 *g* in a swing-out rotor to accelerate the phase separation.
3. The upper phase is carefully removed with a pipet, leaving the lower phase consisting of a thin layer of upper phase, the interface, and the bottom phase.
4. The upper phase and the lower phase are diluted separately with at least 2 vol of 0.3 *M* sucrose, 5 m*M* K^+–MOPS, pH 7.2, and centrifuged for 60 min at 105,000 *g*.
5. The pellets are suspended separately in the same medium as above and used for polarity assays or other experiments.

Starting with ca. 45% RO-SMP, the lower phase will contain 25% RO-SMP and 70% of the protein whereas the upper phase will contain 75% RO-SMP and 30% of the protein. Repeated extractions do not improve

[16] P.-Å. Albertsson, B. Andersson, C. Larsson, and H.-E. Åkerlund, *Methods Biochem. Anal.* **28**, 115 (1982).

[17] P. Gardeström and I. Ericson, this volume [40].

the polarity. Intact mitochondria are also preferentially distributed to the upper phase in phase systems containing HMDA-PEG.[17]

Perspectives

The transverse distribution of protein complexes is relatively well known for mammalian mitochondria but almost completely unexplored in plant mitochondria.[18] Furthermore, the inner membrane of plant mitochondria contains unique enzymes including an external NAD(P)H dehydrogenase(s), a rotenone-insensitive NADH dehydrogenase on the inner surface,[19] a cyanide-insensitive oxidase (the so-called alternative oxidase),[20] as well as several unique translocators.[18,21] The transverse localization of these proteins and the overall lateral distribution of proteins and lipids in the inner mitochondrial membrane should be greatly facilitated by the access to SMP of different polarities.

According to our model (Fig. 1), the IO-SMP originate from the stacked cristae regions, whereas the RO-SMP originate from the noncristae regions of the inner membrane. If this is correct, it provides a possibility to investigate the lateral distribution of protein complexes between different domains in the inner membrane. In fact the specific activity of cytochrome *c* oxidase and succinate dehydrogenase was 50% higher in IO-SMP than in RO-SMP both isolated from SMP of mixed polarity.[8] In addition, pure IO-SMP produced directly by sonication of Jerusalem artichoke mitochondria under high-cation conditions contain 4–5 times more cytochrome *c* oxidase than pure RO-SMP produced by French Press treatment under low-salt + EDTA conditions.[15] Thus, there are strong indications that the cristae regions of plant mitochondria are enriched in cytochrome c oxidase and succinate dehydrogenase and possibly other electron transport components. Proteins preferably located in the noncristae regions of the inner membrane would reasonably include transport proteins.

Finally the SMP of different polarity can be used for studying the interaction between matrix proteins and the matrix side of the inner membrane. The RO-SMP may "sample" the matrix from restricted parts of the matrix and IO-SMP may have specific matrix proteins adhering to their outer surface.[22]

[18] R. Douce, "Mitochondria in Higher Plants: Structure, Function, and Biogenesis." Academic Press, Orlando, Florida, 1985.

[19] I. M. Møller and W. Lin, *Annu. Rev. Plant Physiol.* **37,** 309 (1986).

[20] C. Lance, M. Chauveau, and P. Dizengremel, *Encyl. Plant Physiol., New Ser.* **18,** 202 (1985).

[21] H. Ebbighausen, Chen Jia, and H. W. Heldt, *Biochim. Biophys. Acta* **810,** 184 (1985).

[22] J. B. Robinson, Jr. and P. A. Srere, *J. Biol. Chem.* **260,** 10800 (1985).

There is still much to be learned about mitochondria in general and plant mitochondria in particular and SMP will provide an important tool in future studies.

Acknowledgments

We would like to thank Kristina Smeds for excellent technical assistance and K. Edman, C. J. Kay, A. Lundgren, P. Petit, and J. M. Palmer for valuable contributions to the project. This work was supported by grants from the Swedish Natural Sciences Research Council, and Carl Tesdorpfs Stiftelse.

[42] Isolation of the Outer Membrane of Plant Mitochondria

By CARMEN A. MANNELLA

The first successes in applying electron microscopy to sectioned biological tissue (~1950) brought many fundamental discoveries. Among these is the fact that the mitochondria are composed of two membranes, a convoluted *inner* membrane enveloped by a smooth *outer* membrane.[1] It was another 15 years before reliable procedures were developed for the fractionation of rat liver mitochondria into their component membranes.[2-4] Early attempts to subfractionate mitochondria were characterized by considerable confusion, due in large part to the use of uncontrolled lytic procedures. Successful separation and subsequent characterization of the two mitochondrial membranes were achieved only after conditions were defined to selectively disrupt the outer (and not the inner) membrane of the organelle. Two complementary kinds of criteria, ultrastructural and biochemical, were applied to assess the effects of different treatments on mitochondrial membrane integrity. Parsons and coworkers[5] showed that chemical agents which induce swelling of the matrix space of liver mitochondria (monitored by thin-section electron microscopy) generally cause breakage only of outer membranes; inner membranes become distended but appear to remain intact. Similarly, Schnaitman *et al.*[4] observed selective disruption of outer membranes

[1] F. S. Sjøstrand, *J. Ultrastruct. Res.* **59,** 292 (1977).
[2] D. F. Parsons, G. R. Williams, and B. Chance, *Ann. N.Y. Acad. Sci.* **137,** 643 (1966).
[3] M. Levy, R. Toury, and J. Andre, *Biochim. Biophys. Acta* **135,** 599 (1967).
[4] C. A. Schnaitman, V. G. Erwin, and J. W. Greenawalt, *J. Cell Biol.* **32,** 719 (1967).
[5] P. Wlodawer, D. F. Parsons, G. R. Williams, and L. Wojtczak, *Biochim. Biophys. Acta* **128,** 34 (1966).

when liver mitochondria were treated with low concentrations of the detergent digitonin. Enzymatic "integrity assays" were concurrently developed in several laboratories which correlate release or unmasking of particular enzyme activities with lysis of first outer, then inner membranes as mitochondria are exposed to successively greater digitonin concentrations[4,6] or osmotic shocks.[7,8] Extents of hypoosmotic swelling (in phosphate buffer) or digitonin treatment could thus be defined at which outer membrane lysis was optimal with minimal damage to inner membranes. Strategies to separate the two mitochondrial membranes following outer membrane rupture generally involved ultracentrifugtion of mitochondrial lysates on continuous or step sucrose gradients, taking advantage of the two membranes' very different buoyant densities (outer membranes, 1.08–1.14 g/cm^3; inner membranes *plus* matrix, 1.20–1.25 g/cm^3).

Shortly after consensus was reached on procedures for fractionation of liver mitochondrial membranes, similar approaches were applied to the separation of inner and outer membranes of plant mitochondria. Earliest successes were achieved with osmotic swelling techniques on mung bean and cauliflower mitochondria[8,9] and with digitonin treatment of potato mitochondria.[10] In this chapter, we will outline the steps involved in applying the former technique to the isolation of outer membranes from mitochondria, with emphasis on the specific problems encountered with plant mitochondria. For a comprehensive review of the composition and structure of plant outer mitochondrial membranes, the reader is referred to Mannella.[11]

Preliminary Considerations

Purity of Mitochondrial Fractions

Mitochondrial fractions isolated from plant tissue by differential centrifugation alone are generally contaminated with various nonmitochondrial membranes (microsomes, plastids, peroxisomes, etc.). Since the

[6] L. Wojtczak and H. Zaluska, *Biochim. Biophys. Acta* **193,** 64 (1969).

[7] R. Douce, E. L. Christensen, and W. D. Bonner, Jr., *Biochim. Biophys. Acta* **275,** 148 (1972).

[8] R. Douce, C. A. Mannella, and W. D. Bonner, Jr., *Biochim. Biophys. Acta* **292,** 105 (1973).

[9] F. Moreau and C. Lance, *Biochimie* **54,** 1335 (1972).

[10] D. Meunier, C. Pianeta, and P. Coulomb, *C.R. Hebd. Seances Acad. Sci.* **272,** 1376 (1971).

[11] C. A. Mannella, *in* "Higher Plant Cell Respiration" (R. Douce and D. A. Day, eds.), p. 106. Springer-Verlag, Berlin and New York, 1985.

outer mitochondrial membranes are isolated (following rupture) on the basis of buoyant density, it is essential that the low-density membrane contaminants in the mitochondrial fractions be removed first. This can be achieved by centrifugation of the mitochondrial fractions on sucrose step gradients immediately before outer membrane lysis.[7] Intact mitochondria are stopped by 1.45 *M* sucrose while separated outer membranes are collected on 0.9 *M* sucrose steps. Therefore a simple two-step (1.0/1.45 *M* sucrose) purification gradient may be sufficient to clean up the starting mitochondrial fractions. Recently, centrifugation of plant mitochondria on density gradients of Percoll (Pharmacia Fine Chemicals) has been shown to be at least as effective (if not more so) at separating plant mitochondria from contaminants.[12] In addition, Percoll techniques have several advantages over sucrose gradient centrifugation, such as shorter run times and maintenance of isotonicity in the solutions. It is unknown whether the silica sol might in any way interfere with subsequent isolation of outer membranes from plant mitochondria. The sol particles have a tendency to stick to the outer membranes of fungal mitochondria (C. A. Mannella, unpublished observations). However, electron micrographs of Percoll-purified plant mitochondria[12] show no similar adhering particles.

Choice of Mitochondrial Lysis Technique

If one is interested in obtaining inner mitochondrial membrane fractions which are free of outer membranes and which retain a degree of coupled respiratory activity (so-called "mitoplasts"), digitonin fractionation is the method of choice. (Inner membrane "ghosts" which result from hypoosmotic swelling of mitochondria can have considerable amounts of outer membranes still attached and are fully, irreversibly uncoupled.) Digitonin disrupts the outer membranes into small vesicles, typically 0.1–0.2 μm diameter,[4] which apparently detach from mitochondria more readily than the larger, osmotically ripped fragments. However, if one is interested in studying the composition (in particular, lipid content) or structure of the outer mitochondrial membrane, detergents should be avoided. (It is known that digitonin removes part of the cholesterol from the outer membranes of liver mitochondria.[13] In fact, the specificity of this detergent for sterols, which generally occur in mitochondria only in the outer membrane, is thought to be the basis for its selective disruption of this membrane at low concentrations.) Note also that, in the case of

[12] M. Neuberger, *in* "Higher Plant Cell Respiration" (R. Douce and D. A. Day, eds.), p. 7. Springer-Verlag, Berlin and New York, 1985.

[13] M. Levy, R. Toury, M.-T. Sauner, and J. Andre, *in* "Mitochondria: Structure and Function" (L. Ernster and Z. Drahota, eds.), p. 33. Academic Press, London, 1968.

plant mitochondria, the use of digitonin to isolate outer membrane fractions does not appear to offer any advantage in terms of the subsequent recovery of outer membranes on sucrose gradients (typical outer membrane yields based on marker enzyme recoveries: 30–40% for osmotically lysed mung bean[14] and potato[15] mitochondria, 43% for digitonin-lysed turnip mitochondria[16]).

Optimizing Hypoosmotic Swelling of Plant Mitochondria

In general, the outer membranes of plant (and fungal) mitochondria are more resistant to rupture by osmotic swelling than the liver membrane. Thus, swelling must be carried out at very low osmolarities and divalent cation chelators may need to be included in the swelling media. (Susceptibility of potato mitochondrial outer membranes to osmotic rupture is reduced in the presence of 30 μM $MgCl_2$ and increased by adding 170 μM EDTA to the medium.[14]) Also, preswelling of the matrix space (e.g., by warming the mitochondria to 37° for a few minutes) has been found to increase the extent of outer membrane lysis with fungal mitochondria, whose outer membranes are especially resistant to hypoosmotic rupture.[17] For plant materials grown under constant conditions (e.g., etiolated beans), optimal outer membrane lysis conditions are found to be very constant for different mitochondrial preparations. However, lysis conditions for mitochondria isolated from commercially obtained materials (e.g., potato tubers) can vary considerably for different batches and for the same batch during storage. Because of this variability, it may be necessary to monitor mitochondrial membrane lysis characteristics for such tissues prior to each outer membrane isolation (see below).

Techniques to Increase Outer Membrane Yields

It is uncertain whether the incomplete separation of outer and inner membranes from osmotically swollen mitochondria is due to the existence of contacts between the two membranes or simply to entanglement, i.e., wrapping of large outer membrane fragments about the inner membranes. In either event, Sottocasa *et al.*[18] showed that brief application of a shearing force (i.e., sonication) could increase the separation of inner and outer membranes following osmotic swelling of liver mitochondria. As in the

[14] C. A. Mannella and W. D. Bonner, Jr., *Biochim. Biophys. Acta* **413,** 213 (1975).
[15] C. A. Mannella, Ph.D. Thesis, University of Pennsylvania, Philadelphia (1974).
[16] D. A. Day and J. T. Wiskich, *Arch. Biochem. Biophys.* **171,** 117 (1975).
[17] C. A. Mannella, this series, Vol. 125, p. 595.
[18] G. L. Sottocasa, B. Kuylenstierna, L. Ernster, and A. Bergstrand, *J. Cell Biol.* **32,** 415 (1967).

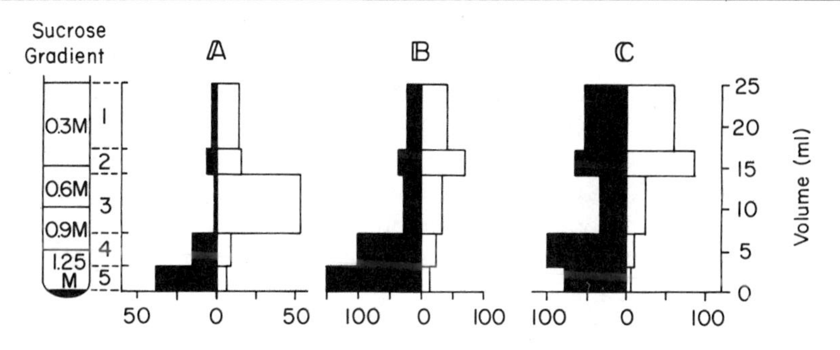

FIG. 1. Distribution of outer and inner membrane marker activites after centrifugation of disrupted mung bean mitochondria on sucrose step gradients (figure from Douce *et al.*[8]). The activities assayed (as described in text) are the antimycin A-insensitive (white blocks) and -sensitive (black blocks) NADH : cytochrome *c* oxidoreductases. (A) Hypoosmotically swollen mitochondria, partially contracted in 0.3 *M* sucrose. (B) Swollen–contracted mitochondria subsequently passed through a Yeda press. (C) Swollen–contracted mitochondria subsequently sonicated. (See Douce *et al.*[8] for details of the treatments.)

case of digitonin treatment, this procedure causes vesiculation of the outer membranes and, in addition, may result in loss of certain proteins from the membranes.[11,15] Figure 1 illustrates another drawback of such procedures in the case of mung bean mitochondria. While percentage recovery of outer membrane marker activities on sucrose gradients is increased following shearing of hypoosmotically treated mitochondria (by Yeda press or sonifier), levels of inner membrane markers are significantly increased in the same fractions. It may be possible to control shearing conditions for a particular kind of mitochondria such that outer membrane separation is enhanced while minimizing inner membrane disruption, but clearly care must be taken. In general, applying these techniques would appear to be justified only when outer membrane yields in their absence are very poor (i.e., 10% or less).

Establishing Outer Membrane Lysis Conditions

Principle

Mitochondria are incubated in media of progressively decreasing osmolarity and outer membrane lysis is followed as the unmasking of either succinate : cytochrome *c* oxidoreductase (EC 1.3.99.1, succinate dehydrogenase) or cytochrome *c* : oxygen oxidoreductase (EC 1.9.3.1, cytochrome-*c* oxidase). (Since the outer mitochondrial membrane is impermeable to cytochrome *c*, rates of reduction or oxidation of external cytochrome *c* by the inner membrane respiratory chain are a direct indicator of outer membrane breakage.[6,8,14,15]) Damage to the inner membrane is assessed by monitoring the release of a soluble matrix enzyme, malate dehydrogenase (EC 1.1.1.37), into the supernatant following centrifugation of the treated mitochondria. The optimal osmolarity is chosen as that which maximizes outer membrane rupture without significantly decreasing inner membrane integrity.

Materials

Gradient-purified mitochondria: suspension containing approximately 100 mg mitochondrial protein/ml of holding medium (0.3 *M* sucrose, 0.1% bovine serum albumin, 10 m*M* sodium phosphate buffer, pH 7.2)
Reaction medium: 0.3 *M* mannitol (or sucrose), 5 m*M* $MgCl_2$, 10 m*M* KCl, 10 m*M* sodium phosphate buffer (pH 7.2)
Swelling media: solutions of 100, 75, 50, 35, 20, 10, 5, and 0 m*M* sucrose containing (optional, see below) 0.25 m*M* EDTA and 0.25 m*M* EGTA (chelator solutions preadjusted to pH 7.2 with NaOH)
Succinate (sodium salt): 1 *M* in distilled water
Oxaloacetate (sodium salt): 0.1 *M* in distilled water
Cytochrome *c*: 1 m*M* in reaction medium
NADH: 0.1 *M* in 50 m*M* Na_2CO_3
Ascorbate (sodium salt): 1 *M* in distilled water
ATP (sodium salt): 0.1 *M* in 0.3 *M* mannitol, 10 m*M* Tris–HCl (pH 7.2)
KCN: 0.5 *M* in distilled water
Antimycin A: 0.5% (w/v) in ethanol
Uncoupler: 1 m*M* carbonyl cyanide *m*-chlorophenylhydrazone (CCCP) or carbonyl cyanide *p*-trifluoromethoxyphenylhydrazone (FCCP) in ethanol

Procedure

Mitochondrial aliquots (30 μl, approximately 3 mg protein) are added to centrifuge tubes containing 3 ml each of one of the swelling solutions (100, 75, . . . , 0 m*M* sucrose). After sitting 10–30 min on ice (with occasional gentle swirling), 1 ml is transferred from each suspension to a test tube containing 1 ml of reaction medium. The remaining suspensions are centrifuged (8000 *g* for 10 min) and the supernatants pipetted into clean tubes. Each uncentrifuged suspension is assayed for either succinate : cytochrome *c* oxidoreductase (spectrophotometrically) or cytochrome *c* oxidase (polarographically). The supernatants are assayed spectrophotometrically for malate dehydrogenase (as NADH : oxaloacetate oxidoreductase).

Succinate : Cytochrome c Oxidoreductase. Each suspension is allowed to come to room temperature and to each is added 50 μM cytochrome *c*, 1 m*M* KCN, 200 μM ATP, and 5 μM CCCP or FCCP. The suspension is transferred to a spectrophotometer cuvette and the reduction of cytochrome *c* is recorded at 550 nm (extinction coefficient = 21.0 mM^{-1} cm^{-1}) following addition of 10 m*M* succinate.

Cytochrome c Oxidase. Each suspension is transferred to an oxygen electrode cuvette and 5 μM CCCP or FCCP is added. After temperature equilibration to 25°, oxygen consumption is monitored following sequential additions of 8 m*M* ascorbate and 30 μM cytochrome *c*. The oxidation rate upon addition of 0.2 m*M* KCN to the reaction mixture prior to oxygen depletion is also recorded. This (small) nonenzymatic rate is subtracted from the previous steady state rate to yield the net enzymatic activity.[12]

Soluble Malate Dehydrogenase. One milliliter of each supernatant is mixed with 1 ml reaction medium (containing 1 m*M* KCN and 0.2 m*M* NADH) at room temperature in a UV spectrophotometer cuvette. The oxidation of NADH is recorded at 340 nm (extinction coefficient = 6.22 mM^{-1} cm^{-1}) following addition of 1 m*M* oxaloacetate.

The relative rates of cytochrome *c* reduction or oxidation and of NADH oxidation as functions of the final osmolarity of the swelling media are usually plotted together, as in Fig. 2 for mung bean mitochondria. [Note that the contribution of the mitochondrial suspension to the osmo-

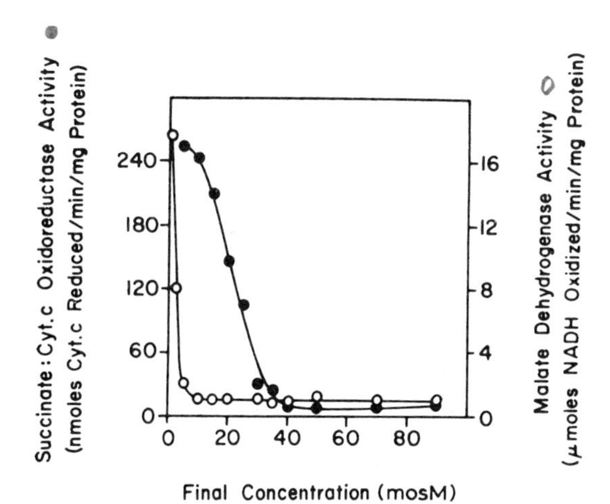

FIG. 2. Unmasking of succinate : cytochrome *c* oxidoreductase (closed circles) and release of malate dehydrogenase (open circles) from mung bean mitochondria suspended in media of decreasing osmolarity. (Figure from Douce *et al.*[8])

larity of the medium should be taken into account. For example, if 30 μl mitochondrial suspension were added to 3 ml of swelling medium, the suspension adds approximately 3 m*M* (= 0.3 *M* × 0.01) sucrose to the final concentration.] As Fig. 2 illustrates, there is a clear separation between the osmotic rupture of outer and inner membranes in the case of mung bean mitochondria. Succinate : cytochrome *c* activity is essentially fully unmasked at 10 m*M* sucrose, below which osmolarity malate dehydrogenase activity is released from the matrix in significant amounts. In cases in which outer membrane lysis is incomplete except at osmolarities which cause significant inner membrane damage, chelators may be added to the swelling media at levels indicated above. (Submillimolar concentrations of EDTA and EGTA increase the extent of outer but not inner membrane lysis at given hypoosmotic conditions.[15] Preswelling the mitochondria (e.g., by warming, see above) may also prove useful in the case of particularly lysis-resistant outer membranes.

Outer Membrane Isolation Procedure

Materials

Gradient-purified mitochondria: suspension containing approximately 100 mg mitochondrial protein/ml of holding solution (as above)

Lysis medium: Dilute solution of sucrose (concentration determined by the lysis characteristics of the mitochondria, see above) containing 0.25 m*M* EDTA and 0.25 m*M* EGTA (optional, see above), pH 7.2

Sucrose step solutions: 0.90 and 0.60 *M* sucrose solutions containing 10 m*M* sodium phosphate buffer, pH 7.2, and 0.25 m*M* EDTA + 0.25 *M* EGTA (the chelators are included in the gradient solutions if they are used in the lysis medium)

10 m*M* sodium phosphate buffer, pH 7.2

Procedure

The mitochondrial suspension is pipetted into 50–100 vol of lysis medium which is being stirred on ice. After 5–10 min (longer if so determined from lysis experiments) the suspension is layered atop 0.6–0.9 *M* sucrose step gradients and centrifuged in a swinging bucket rotor for 60 min at 40,000 *g*. The outer membranes appear as a colorless, light-scattering band at the 0.6–0.9 *M* sucrose interface, which can be collected with a long-stem Pasteur pipet. (For quantitative recovery, a gradient fractionator like the Isco model 185 can be used.) The mitochondrial inner membranes, along with remnants of outer membranes which did not separate from them, form a dark brown pellet at the bottom of the centrifuge tube.

To concentrate the outer membranes, the gradient fraction is diluted with 3 vol of 10 m*M* phosphate buffer (pH 7.2) and centrifuged for 90 min at 60,000 *g*. The resulting small, clear pellet should contain 3–7% of the total protein in the starting mitochondrial suspension.

Characterizing the Outer Mitochondrial Membrane Fractions

Marker Enzymes; Estimating Yield

An enzyme activity that appears to be common to all plant mitochondrial outer membranes is the rotenone- and antimycin A-insensitive NADH : cytochrome *c* oxidoreductase activity (EC 1.6.99.3) which was first described for the liver membrane.[18] In fact, the specific activity of this enzyme system is considerably higher in some plant outer mitochondrial membranes than in the corresponding animal membranes (350–700 vs 200 nmol cytochrome *c* reduced/min per milligram protein for plant and liver membranes, respectively[8,9,15,18]). In most cases, this activity can serve as a marker for estimating percentage recovery of mitochondrial outer membranes following the above isolation procedures, as in Fig. 1. (The activity is measured for the starting mitochondrial suspension and the final outer membrane fraction by a spectrophotometric technique like that described above for succinate : cytochrome *c* oxidoreductase. The reaction is initiated with 1 m*M* NADH instead of succinate and 0.5 μg/ml antimycin A is present.) In fact, antimycin A-insensitive NADH : cytochrome *c* oxidoreductase activity is not a true marker for the outer mitochondrial membrane since it is present on other plant (and animal) membranes, in particular microsomes. Experimentally it is found (by criteria described below) that contamination by microsomes of outer membrane fractions isolated from gradient-purified mitochondria is very low.[8,9,15] A more serious problem in the use of antimycin A-insensitive NADH : cytochrome *c* oxidoreductase as an outer membrane marker is associated with the NADH dehydrogenase located on the outer face of the inner membrane of plant mitochondria.[8] The reduction of cytochrome *c* associated with the inner membrane-bound NADH dehydrogenase is fully sensitive to antimycin A and rotenone. However, the enzyme detaches from the inner membrane upon hypoosmotic swelling of the mitochondria[8] and, in some tissues (e.g., potato and *Arum maculatum*), the released enzyme displays significant activity with cytochrome *c* as acceptor.[15,19] In the case of potato mitochondria, the total antimycin-insensitive NADH : cytochrome *c* oxidoreductase activity increases 3- to 4-fold upon osmotic

[19] N. D. Cook and R. Cammack, *Eur. J. Biochem.* **141,** 573 (1984).

rupture,[15] unlike mung bean mitochondria in which there is no change in activity.[8] One may still use the recovery of this activity as an indicator of outer membrane yield, provided comparisons are made between the total mitochondrial activity *before* hypoosmotic lysis and the activity in outer membrane fractions *after* pelleting, to minimize contributions from the soluble enzyme. [Note that the two enzyme activities that have proved to be very specific markers for liver mitochondrial outer membranes, monoamine oxidase and kynurenine hydroxylase (kynurenine monooxygenase), are both absent from the plant membrane.[9]]

Criteria for Evaluating Purity of the Fractions

Inner Membrane Contamination. The percentage recoveries in outer membrane fractions of respiratory chain activities, such as succinate : cytochrome *c* oxidoreductase, antimycin A-*sensitive* NADH : cytochrome *c* oxidoreductase (Fig. 1), or cytochrome *c* oxidase (assays described above), are good indicators of the extent to which the outer membrane fractions are contaminated by inner membranes.

Difference spectra obtained with scanning split-beam spectrophotometers[20] can also be used to determine levels of inner membrane contamination in outer mitochondrial membrane fractions. Outer membranes of plant and animal mitochondria contain a single *b*-type cytochrome that is the electron acceptor for the NADH dehydrogenase of these membranes and is fully reducible by NADH.[8,9,18] (The physiological terminal electron acceptor in this electron transfer chain is presently unknown for plants, although there is evidence that this system is involved in cholesterol side chain cleavage and aryl hydrocarbon hydroxylation in animal mitochondria.[21-23]) The outer membrane cytochrome *b* of plant mitochondria has a distinctive doublet β band (maxima at 551 and 557 nm) in NADH- or dithionite-reduced *minus* oxidized difference spectra obtained at liquid nitrogen temperature (Fig. 3). Detection in difference spectra of mitochondrial outer membrane fractions of *a*-type cytochromes (which have strong difference maxima near 600 nm) or of *b*- and *c*-type cytochromes reducible by substrates other than NADH would indicate the presence of inner membranes.

Microsomal Contamination. The specific activities of NADPH : cytochrome *c* oxidoreductase (EC 1.6.99.1, NADPH dehydrogenase) and glucose-6-phosphatase (EC 3.1.3.9) can be used to indicate levels of con-

[20] B. Chance, this series, Vol. 4, p. 273.
[21] A. Ito, S. Hayashi, and T. Yoshida, *Biochem. Biophys. Res. Commun.* **101,** 591 (1981).
[22] R. Natarajan and B. Harding, *Fed. Proc., Fed. Am. Soc. Exp. Biol.* **42,** 2065 (1983).
[23] T. Uemura and E. Chiesara, *Eur. J. Biochem.* **66,** 293 (1976).

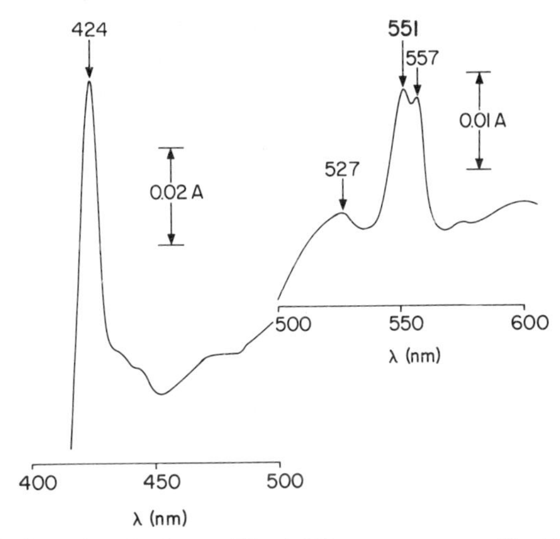

FIG. 3. Dithionite-reduced *minus* oxidized difference spectrum (liquid nitrogen temperature) of outer mitochondrial membranes from mung bean mitochondria. (Figure from Douce *et al.*[8])

tamination of the outer mitochondrial membrane fractions with nonmitochondrial membranes, e.g., microsomes. (The spectrophotometric assay for cytochrome *c* reduction by NADPH is exactly as described above for the NADH-dependent activity but the reaction is initiated with 1 m*M* NADPH. An assay for glucose-6-phosphatase has been described elsewhere in this series.[24]) These specific activities are compared to those of microsomes having the same buoyant density as the mitochondrial outer membranes, which can be prepared from the same tissue homogenates as used in mitochondrial isolation. (The first postmitochondrial supernatant is centrifuged at 20,000 *g* for 30 min; the pelleted membranes are centrifuged on a sucrose step gradient like that used in outer mitochondrial membrane isolation, and the membranes at the 0.6–0.9 *M* interface are collected.[14]) A rapid assessment of the presence of microsomes in outer membrane fractions may be provided by the ratio of antimycin A-insensitive reduction rates for cytochrome *c* with NADPH and NADH as electron donors. The ratio is 0.3–0.1 for plant microsomal membrane fractions and essentially zero in pure mitochondrial outer membrane fractions.[8,9]

Plant microsomes contain a NADH-reducible cytochrome *b* that is spectrally similar to the cytochrome *b* of outer mitochondrial membranes

[24] D. J. Morré, this series, Vol. 22, p. 130.

(but which, in the case of liver, has been shown to be immunologically distinct[25]). The microsomal membranes of some plant tissues contain additional cytochromes which can serve as microsomal markers in difference spectra of outer mitochondrial membrane fractions. For example, an ascorbate-reducible *b*-type cytochrome is found in mung bean and potato microsomes and cytochrome *P*-450 occurs in potato (but not mung bean) microsomes.[15]

Other Characteristics of Plant Mitochondrial Outer Membranes

The protein : phospholipid mass ratio in outer membranes isolated from plant mitochondria is approximately 2.[9,14,15] The prominent phospholipids are phosphatidylcholine, phosphatidylethanolamine, and (in the case of the cauliflower membrane) phosphatidylinositol.[26,27] Significant amounts of sterol have also been detected in both the mung bean and potato membranes.[14] A class of trypsin-resistant 30-kDa polypeptides accounts for about 50% of the total protein in the membranes (determined by SDS–polyacrylamide gel electrophoresis).[14] These polypeptides have been isolated by Nikaido[28] and co-workers and shown to have potent pore-forming activity in liposomes. Similar 30-kDa pore proteins are found in liver and fungal mitochondrial outer membranes and have been called VDAC (for voltage-dependent anion-selective channel) or mitochondrial porin.[29,30] In electron micrographs of negatively stained plant mitochondrial outer membranes, the membranes display disordered arrays of stain-accumulating subunits.[2] Structurally similar subunits can be induced to form *ordered* arrays in the outer membranes of *Neurospora* mitochondria.[17] These two-dimensional membrane crystals have been shown to be composed of VDAC, phospholipid, and sterol and are presently the focus of high-resolution electron microscopic investigations.[31,32]

[25] A. Ito, *Biochem. J.* **87,** 63 (1980).
[26] R. E. McCarty, R. Douce, and A. A. Benson, *Biochim. Biophys. Acta* **316,** 266 (1973).
[27] F. Moreau, J. DuPont, and C. Lance, *Biochim. Biophys. Acta* **345,** 294 (1974).
[28] H. Nikaido, this series, Vol. 97, p. 85.
[29] M. Colombini, *Nature (London)* **279,** 643 (1979).
[30] H. Freitag, W. Neupert, and R. Benz, *Eur. J. Biochem.* **123,** 639 (1982).
[31] C. A. Mannella and M. Colombini, *Biochim. Biophys. Acta* **774,** 206 (1984).
[32] C. A. Mannella, A. Ribeiro, and J. Frank, *Biophys. J.* **49,** 307 (1986).

[43] Characterization of Channels Isolated from Plant Mitochondria

By MARCO COLOMBINI

Characteristics of Channels

The term channel has been used by many investigators in a very broad and ill defined sense. In this chapter, a channel will be restricted to a structure in a membrane (usually a protein) that forms a continuous water-filled pathway through the membrane in at least one of its conformations. The channel allows the passage of water and a selected set of ions and/or molecules through the membrane. Therefore a channel simply facilitates the movement of substances through the membrane. Any net flux through the channel proceeds down a substance's electrochemical gradient (i.e., passive transport).

A particular channel can be characterized and distinguished from other channels based on its structural and functional characteristics. Structural characteristics include molecular weight, sedimentation coefficient, and detailed structural descriptions based on electron microscopy, X-ray diffraction, etc. Functional characteristics include single-channel conductance (often referred to as channel size), ion selectivity, nonelectrolyte selectivity (which yields an estimate of the effective size of the water-filled pathway), and gating properties. This chapter will deal with the functional aspects.

Procedures

Planar Phospholipid Membranes

There are two fundamental methods of generating planar phospholipid membranes: the painted films, first described by Mueller *et al.*,[1] and the monolayer method of Montal and Mueller.[2] The details of the techniques have evolved and continue to evolve with time and vary greatly from lab to lab. In both methods the two aqueous phases on either side of the membrane are easily accessible so that one can easily control and manipulate their composition and the transmembrane potential. An inert plastic partition separates the water phases. The membrane is generated in a

[1] P. Mueller, D. O. Rudin, H. T. Tien, and W. C. Wescott, *J. Phys. Chem.* **67,** 534 (1963).
[2] M. Montal and P. Mueller, *Proc. Natl. Acad. Sci. U.S.A.* **69,** 3561 (1972).

small hole in this partition so that any flow of matter from one aqueous phase to the other must occur through the membrane. An advantage of the monolayer method is that the resulting membrane contains much less organic solvent.

Figure 1 shows a chamber design being used by a number of labs for making membranes by the monolayer method. Two hemichambers milled out of Teflon are clamped together using two brass or stainless steel rods which penetrate both hemichambers, two plates on either end, and screws that screw into the rods and thus press the hemichambers together. Sandwiched between the hemichambers is the plastic partition which supports the membrane itself. A thin coat of Petrolatum (petroleum jelly) is used to seal the hemichambers together.

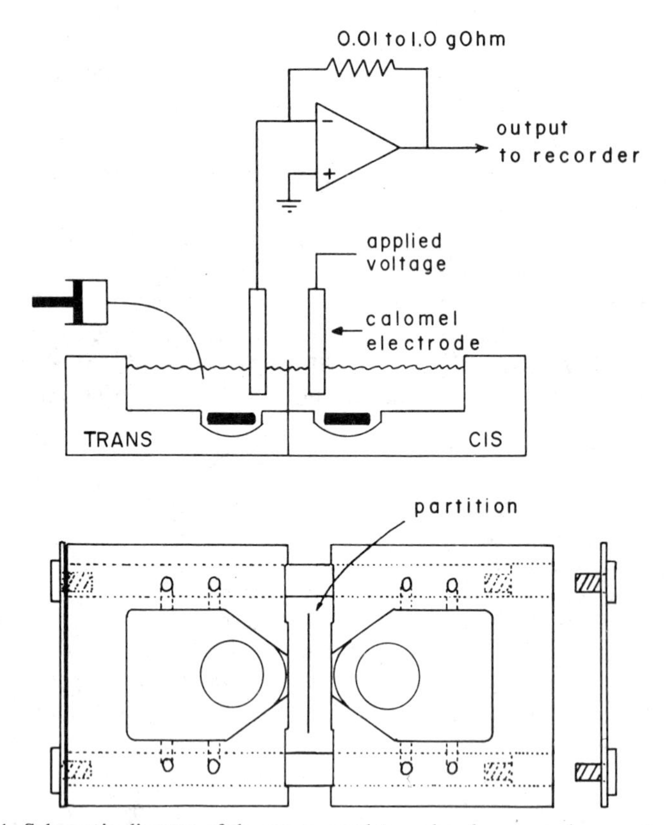

FIG. 1. Schematic diagram of the set-up used to make planar membranes. In the upper figure, the resistor shown is the feedback resistor and the triangle is the amplifier. The black bars on the bottom of the chamber are stir bars that are activated by a battery-powered DC motor that rotates a larger bar beneath the entire chamber. The lower figure is a top view of the chamber, essentially to scale, with overall dimensions of 2×3.7 in.

The plastic partitions are cut out of 10-μm-thick sheets of Saran (SaranWrap washed with petroleum ether and 2 : 1 chloroform/methanol works very well). The most straightforward way of making the small hole is to use a small press with accurately milled punch and die (punches as small as 100 μm are commercially available; Minitool Inc., Campbell, CA). However, when making holes in the very thin Saran sheets, the punch and die must have almost no clearance and must be perfectly aligned. Even so, better results are obtained when multiple sheets are punched at the same time. At times, tiny cracks form on the sides of the hole that make membrane formation difficult. An alternative method is to use a sharp hot needle (heated by being fastened to the head of a soldering iron) fixed to a jig that allows the position of the needle to be carefully controlled. The approach of the hot tip melts a hole in the Saran of variable size. The desired size is obtained by controlling the needle temperature and the length of contact. By this method, larger holes are more easily generated. A third method uses the high voltage of a Tesla coil to locally melt the Saran and generate the hole. With this method very small holes can be easily generated. The trick is to mark a spot on the Saran with a sharp felt pen and produce a tiny defect within this spot in the Saran with a sharp glass point. The spot with the defect is now placed between two wires pointed at each other (about a 2-mm gap). A dissecting microscope is used to observe the defect as it grows in size when very weak electric discharges are passed through it. The process continues until the desired hole size is reached (measured using a reticle in the eyepiece of the microscope). The procedure must be performed very slowly to avoid excessive wrinkling of the Saran around the hole.

Prior to membrane formation, the Saran partition is coated with a thin layer of Petrolatum by applying a 5% (w/v) solution of Petrolatum in petroleum ether (35–60°) using a cotton-tipped applicator. While this coating is drying an aqueous solution is injected into each compartment of the chamber to form a thin layer and 10 to 15 μl of a 1% (w/v) lipid solution in hexane is applied to the surface of each water phase (more can be applied later as indicated below). Care should be taken that the hexane from this layering process not come into contact with the coated Saran partition. Either the hemichambers are designed with a small Teflon step or U-shaped Teflon sheets are placed on either side of the partition so that the fluid level is not above this Teflon barrier. The volatile solvents vaporize in about 5 min. More aqueous solution is injected under the monolayers and, as the water level rises, the monolayers coat the partition with lipid. As the two monolayers (one from each side) rise past the hole in the partition, they form a bilayer, the membrane. Often, the stress of birth results in membrane breakage whereupon the water level on one side is lowered below the level of the hole and then raised once again until a complete membrane is formed.

Membrane formation is monitored by following the increase in capacitance of the system with an oscilloscope. As shown in Fig. 1, a simple voltage clamp is used to control the membrane potential and to measure the current flow across the membrane. The properties of the circuit are such that the current flowing across the membrane is equal to the current flowing across the feedback resistor. Hence the output from the amplifier is a voltage which, when divided by the value of the feedback resistor, yields the current through the membrane. During membrane formation a series of voltage steps (5- to 10-mV steps of 1-msec duration for a 0.1-gΩ resistor in the amplifier feedback loop) are applied and the current monitored (actually, what is being measured is the voltage output of the amplifier). Since the resistance of the newly formed membrane is usually very high, the resulting steady state current is extremely small but a large transient current occurs when the voltage is changed. This is the current needed to charge up the capacitive elements of the system (membrane, partition, etc.). As the membrane forms, the transient current increases dramatically while the steady state current does not visibly change. If the transient current increase is small, the hole may be occluded by an excessive coat of Petrolatum. When the steady state current increases dramatically, the membrane has usually expired.

Often in attempting to make membranes one observes that as the water level is rising through the hole in the partition, the level rapidly jumps to the top of the hole and simultaneously the current increases, indicating breakage of the membrane. If the level of water on one side is now lowered by removing solution from the subphase, one often observes that the current flow has not totally returned to normal despite the fact that the level appears to be below the hole. A jet of air from a micropipet is useful in this case to lower the adhering layer of water that is making contact with solution in the other compartment through the hole in the partition. This adherence is desirable, in moderation, because it indicates that the appropriate amount of lipid has been applied to the surfaces of the solutions. The absence of this behavior usually indicates insufficient lipid while an exaggerated amount of this indicates too much lipid.

Membrane formation and stability depend on thorough cleaning of the chamber. The following sequence works well: one wash (the chamber is filled with the solution followed by aspiration) with 2% sodium dodecyl sulfate, five washes with distilled water (soaking in water at this stage may help), five applications on each side of the partition of 2 : 1 chloroform/methanol (the solution is injected at the partition with a fire-polished Pasteur pipet and rapidly aspirated), one application of the chloroform/

methanol solution to the rest of the chamber, and gentle wiping of the chamber and partition with a disposable, low-lint tissue paper (e.g., Kimwipe).

Liposomes for Measurement of Channel-Mediated Nonelectrolyte Fluxes

The method is based on a technique developed by Bangham[3] to measure the ability of small molecules to penetrate lipid vesicles. It was modified[4] to determine the largest nonelectrolyte that could enter a liposome through channels in the surface membrane. This allows one to determine an effective radius for the channel's aqueous pore.

Liposomes are made which consist of 20 parts added lipid to 1 part membrane protein. A mixture consisting of 46 parts egg phosphatidylcholine (Sigma Chemical Co., St. Louis, MO) and 4 parts egg phosphatidic acid (Avanti Biochemicals Inc., Birmingham, AL) in chloroform is dried down under nitrogen. At the same time, mitochondrial membranes are centrifuged (at 4°) out of a dilute solution (1 m*M* KCl, 1 m*M* HEPES, Na^+ salt, pH 7.5) at 27,000 *g* for 10 min. (The mitochondrial membranes were isolated from hypotonically shocked mitochondria and stored at −20 or −70° in the above dilute solution supplemented with 15% DMSO. One milliliter of this stored material was diluted with 10 ml of the same medium but lacking DMSO prior to centrifugation.) The membrane pellet is suspended in the KCl–HEPES medium so as to achieve a protein concentration of about 0.5 mg/ml. This solution is then added to the dried lipid (the lipid must be free of solvent. This can be verified by exposing the residue to a vacuum, 20 mm Hg or below, and looking for signs of outgassing) in such proportions as to achieve the 20-to-1 ratio of added lipid to protein. The solution is flushed with nitrogen, and tightly capped with a Teflon-lined top. Vortex mixing is used to disperse the lipid followed by sonication in an ice bath (Laboratory Supplies Co., Hicksville, NY) until the suspension is somewhat clarified. The sonication should be performed in 10- to 15-sec bursts for a total of 2 to 3 min. This suspension is divided up into 0.2-ml aliquots, lyophilized, and stored desiccated at −20°. Each aliquot contains 2 mg added lipid and 0.1 mg protein.

To initiate an experiment, 0.4 ml of 8 m*M* KCl, 3 m*M* EDTA (Na^+ salt, pH 7.0) is added to one of the aliquots and liposomes are formed by gentle vortexing. The liposomes are then allowed to sit at room temperature for about 30 min. Fifty- to 100-μl aliquots of this suspension are added to 1 ml of the same KCl–EDTA medium and allowed to stabilize for 30–60

[3] A. D. Bangham, J. DeGier, and G. D. Greville, *Chem. Phys. Lipids* **1,** 225 (1967).

[4] M. Colombini, *J. Membr. Biol.* **53,** 79 (1980).

min prior to the initiation of the absorbance measurements at 400 nm. Stable liposomes show no change in absorbance with time. The addition of 80 μl of 100 mM permeable nonelectrolyte to the cuvette results in a biphasic curve of absorbance vs time. The absorbance increases transiently, indicating liposome shrinkage, followed by a slow absorbance decrease, indicating a slow reswelling resulting from nonelectrolyte penetration into the liposomes through the channels. Liposomes devoid of protein do not show the absorbance decrease (if this is not true then either the liposomes are unstable or the nonelectrolyte is too lipid soluble). The rate of absorbance decrease decreases with increasing nonelectrolyte molecular weight until no decrease is perceived, indicating that the molecular weight cut-off has been reached.

Cautionary Notes

This procedure does not work with all lipid mixtures. Some charge on the lipid is needed during the liposome formation process in order to generate multiwalled liposomes that have sufficient space between the walls to allow for good shrinkage. Some lipids do not form very stable liposomes, resulting in apparent shrinkage followed by reswelling even when an impermeant is added (perhaps the liposomes undergo a budding process). Some nonelectrolytes (such as α-cyclodextrin) cause the liposomes to aggregate, resulting in increased absorbance.

Measurement of Channel's Characteristics

Single Channel Conductance

Since the vast majority of channels allow at least some ions to traverse the membrane, the single-channel conductance (a measure of the ability of ions to traverse the channel) is used as a major way of distinguishing among channels. The large ion flux through channels allows an investigator to measure the ion flux through individual channels by inserting the channels into planar membranes.

These experiments are best performed in the presence of symmetrical solutions (i.e., identical solutions in each compartment), otherwise rectification may occur (the current and therefore the conductance of the channel would depend on the sign of the applied membrane potential). The value obtained for the single-channel conductance depends on the ionic composition of the medium. Channels usually insert individually into the membrane, resulting in stepwise current increases in the presence of an applied membrane potential. Using Ohm's law ($G = I/V$), these currents

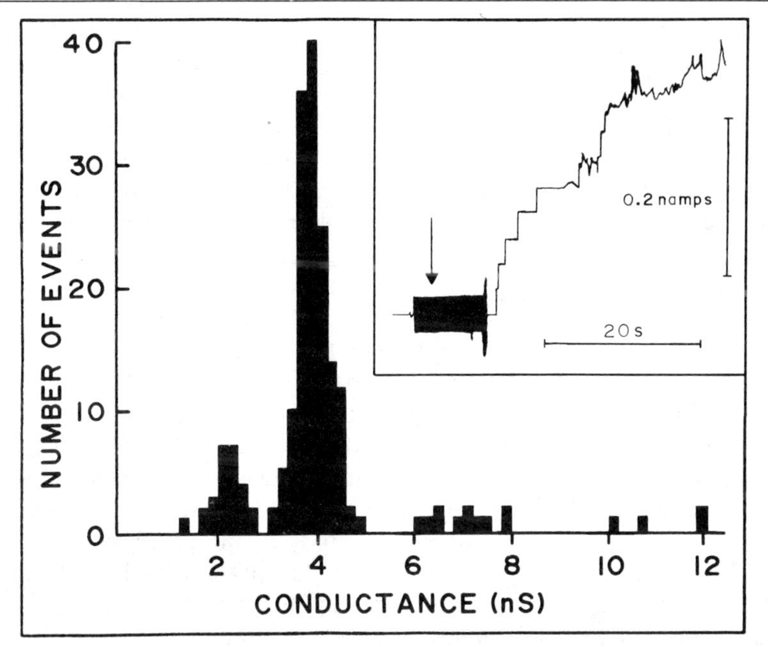

FIG. 2. Single-channel conductance increments associated with the insertion of corn channels into planar membranes. The water phases were 1.0 M KCl, 5 mM $CaCl_2$. The phospholipids used were from soybean purified according to Kagawa and Racker, *J. Biol. Chem.* **246,** 5477 (1971). The mitochondrial membranes were extracted with 1% (v/v) Triton X-100 and a 2-μl aliquot added, at the point indicated by the arrow (inset), to the cis side while stirring (noisy region). The current flow through the membrane was essentially zero prior to the addition. The applied voltage was −10 mV on the cis side so that upward deflections are negative currents. The conductance increments from a number of records like those shown in the inset were used to generate the histogram.

are converted into conductances. The conductance of a single channel (in one of its conformational states) is, to a good approximation, time independent. If only one type of channel is inserting, the current increments should be uniform. They are not identical to each other because some channels insert in different conformational states and because each channel probably has a slightly different three-dimensional structure, resulting in differences in conductance.

Figure 2 (inset) shows channels inserting into planar membranes as a result of the addition of detergent extracts of corn mitochondrial membranes to the water phase on one side of the membrane [usually 10 μl of a 1% (v/v) uncharged detergent extract to 4 ml of solution]. These channels were inferred[4] to be the plant equivalent of VDAC channels located in the outer mitochondrial membrane of animal cells because of their functional characteristics. The initial current increments were discrete increases characteristic of channels. Later the current increased in a ragged fashion due, perhaps, to the formation of other unspecified pathways for ion flow through the membrane. From the early increments (obtained from a number of membranes) a histogram of the conductance increments was generated (Fig. 2). The histogram reveals the presence of a minimum of three classes of events within the population. Although one might interpret the results as revealing the presence of three different channels, by analogy with VDAC isolated from fungal cells,[4,5] the results can be interpreted in terms of a single type of channel. The increments contained in the middle and major peak of the histogram are the channels inserting in the maximum conducting conformational (open) state. The smaller conductance increments are probably due to channels inserting in lower conducting (closed) states. The larger increments are probably due to the insertion of more than one channel at a time. The study of the gating properties of individual channels (see below) confirms this interpretation. Thus, the most probable single-channel conductance is given by the peak of the central subpopulation of conductance increments (i.e., 3.9 nS). This is very similar to the value reported for other sources of VDAC.

Channel Ion Selectivity

The single-channel conductance is a measure of the flow of charge through the channel per unit driving force. This charge is carried by ions in the solution. If one can determine which ions carry the charge and what fraction of the charge each type of ion carries, then one has a measure of the channel's ability to select among ions. In the simplest case, a single salt would be present in the medium (ignoring the contribution of H^+ and OH^- ions because of their relatively low concentration; their contribution can be tested by varying the salt concentration). The relative contribution of the cation and anion can be determined by inserting channels into the planar membrane in the presence of a salt gradient. With the membrane potential clamped at zero, flow of ions down their chemical gradient from the high concentration side to the low side proceeds if the channel is permeable to the ions. If both ions are equally permeable the resulting

[5] C. A. Mannella, M. Colombini, and J. Frank, *Proc. Natl. Acad. Sci. U.S.A.* **80,** 2243 (1983).

charge flow and current measured is zero. However, if the channel is more permeable to one ion than the other, the charge does not cancel out and a net current is measured. The voltage which brings this current to zero is the reversal potential. This potential inserted into the Goldman–Hodgkin–Katz equation (along with the ion activities) yields the permeability ratio for the ions present.

$$\frac{P_{Cl^-}}{P_{K^+}} = \frac{[KCl]_T - [KCl]_C \exp(\Delta VF/RT)}{[KCl]_T \exp(\Delta VF/RT) - [KCl]_C} \tag{1}$$

This method was applied to the corn channels resulting in a reversal potential of 10.2 mV for a salt gradient (1.0 *M* KCl vs 0.1 *M* KCl, 5 m*M* $CaCl_2$ on each side). This translates into a weak selectivity of Cl^- over K^+ of 1.7-fold. Normally one should check to see if the water gradient, which is formed in the process of generating the salt gradient, influences the observed reversal potential. The water gradient can be eliminated by using a nonelectrolyte and adding sufficient nonelectrolyte to balance the water activity (as measured with a vapor pressure osmometer). However, the introduction of the nonelectrolyte will change the activity of the salts. In addition, unless the reflection coefficients of the salt and the nonelectrolyte are the same, the correction for the water gradient is imperfect. Finally, the effect of solvent flux will be most evident in channels that are highly selective between cations and anions. Therefore, for VDAC, this correction was not performed.

Channel Selectivity among Nonelectrolytes

By using uncharged substances it is possible to distinguish between steric and electrostatic barriers to the flow of substances through a channel. The planar membrane, such a powerful tool to examine ion flow, is much less useful in the study of nonelectrolyte fluxes because its relatively small area results in very low fluxes. Lipid vesicles or liposomes have an enormous combined surface area yielding large total fluxes for the same number of channels per unit area. While radiolabeled nonelectrolytes are used to measure the ability of molecules to penetrate into vesicles, the technique described in Procedures does not require radiolabeled substrates.

When these experiments were performed on VDAC isolated from *Neurospora crassa,* the limiting size of the channel's aqueous pathway was given by the ability of inulin and polyethylene glycol 3400 to penetrate the liposomes while polyethylene glycol 6800 (now called PEG 8000) did not penetrate. An effective radius of 20 Å was estimated.

Gating Properties

Endocytic channel formers (made by the cell for insertion into its own membranes) are usually under control either by a chemical or voltage gating system. VDAC channels are voltage gated so that the channels exist in a high-conducting (open) state at low membrane potentials and low-conducting (closed) states at high membrane potentials. The voltage-gating properties can be easily studied once the channels are inserted into planar membranes. Using the circuit shown in Fig. 1, the membrane voltage can be varied continuously by applying a triangular voltage wave (produced by a function generator, Wavetek model 184) to the "applied voltage" point and monitoring the current flow (see the section, Planar Phospholipid Membranes). If the frequency of the triangular wave is sufficiently low so as to allow the channels to equilibrate between the various states, the voltage-dependent properties of the channel can be obtained in the form of a conductance/voltage plot.

Figure 3 shows how the conductance of corn VDAC channels varies with voltage. The channels close in a narrow voltage range and at both

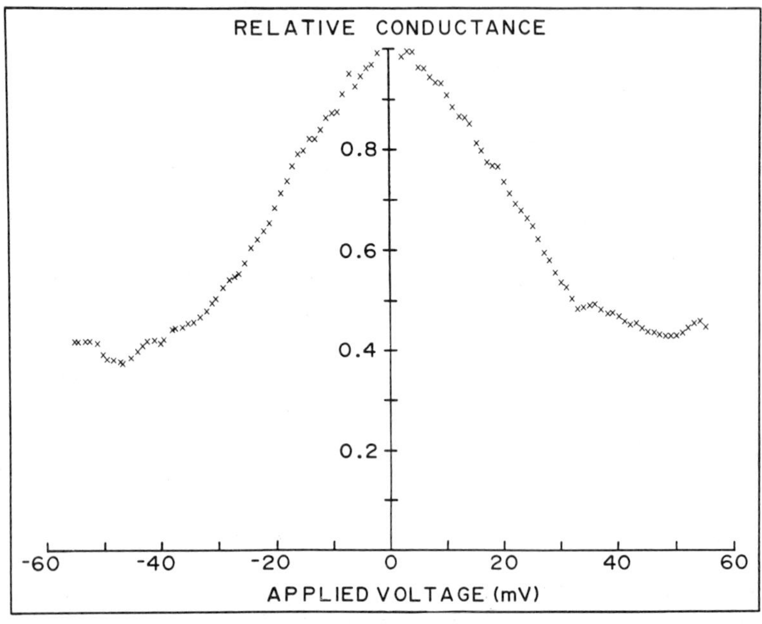

FIG. 3. Conductance–voltage relationship for corn VDAC channels. The voltage was changed linearly at 12 mV/min from high to low voltages. The figure was reprinted with permission from D. P. Smack and M. Colombini, *Plant Physiol.* **79,** 1094 (1985).

positive and negative potentials. The probability of corn VDAC being in the open state changes *e*-fold for 6.5 mV.

These are some of the basic functional properties that can be examined to characterize isolated channels. They are usually sufficient to describe the channel, see where it might fit into a classification scheme, and allow one to gain some insight into its biological function.

Acknowledgments

This work was supported by NIH Grant #GM28450 and ONR Grant #N00014-85-K-0651.

[44] Phosphoglycerides of Mitochondrial Membranes

By JOHN L. HARWOOD

Mitochondria with their outer membranes and convoluted inner membranes are one of the most complex of organelles in morphological terms. In spite of this, the lipid content of mitochondrial membranes is relatively simple. Phosphoglycerides predominate but there are smaller, though very important, quantities of other lipids such as quinones.

The structures of individual phosphoglycerides and their constituent fatty acids are discussed fully in relevant reviews[1,2] and will not be described further here.

Analysis of Mitochondrial Lipids

Much of the difference in mitochondrial lipid compositions which have been reported in the literature undoubtedly stems from the difficulty in purifying the organelles free from contamination. A careful assessment of their purity is essential before any meaningful analysis can be undertaken.[3] In nonphotosynthetic tissues, microbodies can be significant contaminants, with chloroplasts and/or their membranes in preparations from photosynthetic tissues. Some endoplasmic reticulum may be present, possibly due to continuities between mitochondria and these membranes

[1] J. L. Harwood, *in* "Biochemistry of Plants" (P. K. Stumpf and E. E. Conn, eds.), Vol. 4, p 1. Academic Press, New York, 1980.
[2] J. L. Harwood, *Encycl. Plant Physiol., New Ser.* **18,** 37 (1985).
[3] M. Neuburger and R. Douce, this volume [37].

in vivo.[4] Choice of tissue is also important and potato tubers, sycamore cells, bean shoots, and castor bean endosperm have all been used successfully. Suitable methods for the preparation of highly purified intact mitochondria are described elsewhere in this volume.[3]

The lipid analysis should be started immediately after isolation of the organelles—delay allows hydrolytic and oxidative processes to occur. The presence of large amounts of phosphatidic acid, nonesterified fatty acids, or unusual acyl short chain products is indicative of such degradation.

An aqueous suspension (1 ml) of the mitochondria is mixed with 3.75 ml chloroform–methanol (1 : 2, v/v). After 15 min, chloroform (1.25 ml) and 1.25 ml of 2 *M* KCl in 0.5 *M* potassium phosphate buffer (pH 7.4) are added and mixed. Once the two phases have separated the upper phase can be removed and the lower phase washed with synthetic upper phase. The presence of a high salt concentration in the upper phase prevents any loss of very polar phosphoglycerides during extraction.[5] Such phosphoglycerides can also be extracted efficiently with systems containing HCl.[6]

If degradative enzymes are a particular problem then the mitochondrial fraction can be heated with 2-propanol at 70° for 30 min before addition of chloroform (with 2-propanol substituting for methanol).[7]

The lipids in the chloroform phase are well separated by two-dimensional thin-layer chromatography using chloroform/methanol/water (65 : 25 : 4 by volume) and chloroform/acetone/methanol/acetic acid/water (10 : 4 : 2 : 2 : 1, by volume) in the first and second dimensions. Lipid classes are revealed with the nondestructive 0.2% (w/v) methanolic 8-anilino-4-naphthosulfonic acid spray and exposure to UV light. After drying the plates under N_2, the areas corresponding to individual lipids can be scraped for further analysis or quantitation.

Identification. Preliminary identification can be carried out by relative R_f values on TLC plates compared to standards and by the use of differential spray agents.[6,8] Fuller identification requires degradative analysis,[6,8] e.g., deacylation and separation of the deacylated moiety.[9]

[4] D. J. Morré, N. W. Lem, and A. S. Sandelius, *in* "Structure, Function and Metabolism of Plant Lipids" (P.-A. Siegenthaler and W. Eichenberger, eds.), p. 325. Elsevier, Amsterdam, 1984.

[5] J. Garbus, H. F. deLuca, M. E. Loomans, and F. M. Strong, *J. Biol. Chem.* **238,** 59 (1963).

[6] M. Kates, "Techniques of Lipidology." Elsevier, Amsterdam, 1972.

[7] K. L. Smith, G. W. Bryan, and J. L. Harwood, *J. Exp. Bot.* **36,** 663 (1985).

[8] W. W. Christie, "Lipid Analysis," 2nd ed. Pergamon, Oxford, 1982.

[9] R. M. C. Dawson, *Biochem. J.* **75,** 45 (1960).

TABLE I
COMPARISON OF THE FATTY ACID COMPOSITION OF SEPARATED MOLECULAR SPECIES OF PHOSPHATIDYLCHOLINE AND ITS DIACYLGLYCEROL MOIETIES[a]

Molecular species	Separation technique[b]	Percentage total fatty acids					
		16:0	18:0	18:1	18:2	18:3	Other
Monoenoic	I	30	18	49	1	—	2
	D	21	24	51	2	—	2
Dienoic	I	35	8	7	49	Trace	1
	D	33	11	5	50	Trace	1
Trienoic	I	4	3	42	49	Trace	2
	D	2	1	44	52	Trace	1
Tetraenoic	I	4	Trace	4	90	Trace	2
	D	1	1	1	94	3	Trace
Pentaenoic	I	2	2	3	52	41	Trace
	D	1	1	3	51	43	1

[a] From Harwood.[10]

[b] I, Intact molecular species separated by $AgNO_3$—TLC. D, separation of diacylglycerols produced by phospholipase C digestion.

Molecular Species Separation. Some intact phosphoglycerides can be separated into molecular species. Separations have been reported for phosphatidylethanolamine (PE)[10] and phosphatidylcholine (PC).[10,11] However, the lipids are usually converted to the corresponding diacylglycerol species by phospholipase C digestion. Great care must be taken (by fatty acid analysis) that the phospholipase digests all molecular species to equal extents. Some selection of the source of the phospholipase C is also necessary, the choice depending on which phosphoglyceride is to be hydrolyzed.[12] The diacylglycerols can be separated by argentation—silica gel TLC[10]—or by GLC[8] to give separations based on unsaturation or molecular weight. A comparison of the separation of intact phosphatidylcholine with that of its diacylglycerol moiety is shown in Table I. It is clear that cleaner fractions are obtained with the diacylglycerol separation but separation of the intact phosphoglycerides may be desired for certain radiolabeling studies (see Ref. 10). Although HPLC has not been applied to the separation of plant mitochondrial lipids, recent advances have in-

[10] J. L. Harwood, *Phytochemistry* **15,** 1459 (1976).

[11] C. Willemot and G. Verret, *Lipids* **8,** 588 (1975).

[12] D. Hanahan, *in* "The Enzymes" (P. D. Boyer, ed.), 3rd ed., Vol. 5, p. 82. Academic Press, New York, 1971.

cluded the successful separation of molecular species of PC and PE on μ Bondapak C_{18} columns.[13] Results obtained for the intact phospholipids were reasonably comparable to the separation of naphthylurethane derivatives of the diacylglycerols produced by phospholipase C digestion.[14] In the past a major problem with the application of HPLC to lipid separations has been the lack of a suitable detector system. Recent developments coupling a moving belt sampler to a flame ionization detector have overcome these problems to a large extent. Such systems have been applied to a number of different lipid classes including phosphatidylglycerol.[14a]

Quantitation. For many purposes phosphoglycerides can be conveniently quantitated by phosphorus analysis. We have found the method of Duck-Chong[15] to be particularly reliable and rather sensitive. The silica gel containing the lipid to be analyzed is heated with magnesium nitrate and then in 1 *M* HCl. The liberated inorganic phosphate is quantitated with a Malachite green–ammonium molybdate reagent in the presence of Triton X-100. The resultant color is read at 650 nm.

Alternatively, the phosphoglyceride can be measured by GLC quantitation of its fatty acids. This has the advantage that the fatty acyl composition of the lipid is then known. The silica gel containing the phosphoglyceride is heated in a sealed glass tube under N_2 with 2 ml 2.5% H_2SO_4–dry methanol for 2 hr at 70°. A known quantity of a fatty acid standard is included in the tube. Pentadecanoic acid is commonly used but the standard must be chosen so that it does not interfere with the sample's fatty acids on GLC. After methylation, the tubes are cooled and 5 ml of 5% NaCl added. The methyl esters are then quantitatively extracted with 3 × 3-ml aliquots of redistilled (60–80°) petroleum ether. The extracts should be separated by GLC as quickly as possible but, in any case, should be kept under N_2 and at −20° until analyzed. Antioxidants such as butylated hydroxytoluene cannot be added because they interfere with the GLC separation. Where the identity of the fatty acid components is unsure then they should be isolated and subjected to proper degradative analysis (see Ref. 1).

[13] C. Demandre, A. Trémolières, A.-M. Justin, and P. Mazliak, *Phytochemistry* **24,** 481 (1985).

[14] A.-M. Justin, C. Demandre, A. Trémolières, and P. Mazliak, *Biochim. Biophys. Acta* **836,** 1 (1985).

[14a] L. A. Smith, H. A. Norman, S. H. Cho, and G. A. Thompson, *J. Chromatogr.* **346,** 291 (1985).

[15] C. G. Duck-Chong, *Lipids* **14,** 492 (1979).

[16] T. M. Cheesbrough and T. S. Moore, *Plant Physiol.* **65,** 1076 (1980).

TABLE II
SOME EXAMPLES OF THE PHOSPHOGLYCERIDE COMPOSITIONS OF DIFFERENT PLANT MITOCHONDRIAL PREPARATIONS

	mol% total phosphoglycerides[a]						
Tissue	DPG	PA	PE	PE	PG	PI	PS
Broad bean root[b]	3	1	47	20	15	13	—
Castor bean endosperm[c]	5	—	50	33	2	10	—
Castor bean endosperm[d]	11	—	38	43	2	4	2
Cauliflower buds[e]	11	—	44	34	4	7	—
Lupin root[b]	9	—	33	24	13	13	9
Mung bean hypocotyls[f]	24	—	34	33	4	5	—
Mung bean hypocotyls[g]	14	—	36	46	1	3	—
Potato tuber[h]	8	5	43	30	3	7	3
Potato tuber[i]	17	Trace	35	37	1	10	—
Scorzonera tissue culture[j]	3	10	47	22	2	16	—
Sycamore tissue culture[g]	13	—	43	35	3	6	—
Sycamore tissue culture[k]	14	—	42	36	2	6	—

[a] DPG, Diphosphatidylglycerol; PA, phosphatidic acid; PC, phosphatidylcholine; PE, phosphatidylethanolamine; PG, phosphatidylglycerol; PI, phosphatidylinositol; PS, phosphatidylserine; [b] from Oursel *et al.*[18]; [c] from Ohmori and Yamada[19]; [d] from Donaldson and Beevers[20]; [e] from Moreau *et al.*[21]; [f] from McCarty *et al.*[22]; [g] from Bligny and Douce[17]; [h] from Meunier and Mazliak[23]; [i] from Journet *et al.*[24]; [j] from Douce[25]; [k] from Bligny *et al.*[26]

Phosphoglyceride Content of Plant Mitochondria

As mentioned above, the different methods which have applied to the separation of plant mitochondria have undoubtedly influenced the purity achieved and, hence, the final lipid analysis. Nevertheless, certain features are immediately obvious when one examines the different analyses (cf. Table II).[16–26] First, phosphatidylcholine and phosphatidylethanolamine are the main phosphoglyceride components. Second, phosphatidyl-

[17] R. Bligny and R. Douce, *Biochim. Biophys. Acta* **617,** 254 (1980).
[18] A. Oursel, A. Lamant, L. Salsac, and P. Mazliak, *Phytochemistry* **12,** 1865 (1973).
[19] M. Ohmori and M. Yamada, *Plant Cell Physiol.* **15,** 1129 (1974).
[20] R. P. Donaldson and H. Beevers, *Plant Physiol.* **59,** 259 (1977).
[21] F. Moreau, J. Dupont, and C. Lance, *Biochim. Biophys. Acta* **345,** 295 (1974).
[22] R. E. McCarty, R. Douce, and A. A. Benson, *Biochim. Biophys. Acta* **316,** 266 (1973).
[23] D. Meunier and P. Mazliak, *C.R. Hebd. Seances Acad. Sci.* **275,** 213 (1972).
[24] E.-P. Journet, J. L. Harwood, and R. Douce, unpublished results (1982).
[25] R. Douce, *C.R. Hebd. Seances Acad. Sci.* **259,** 2167 (1964).
[26] R. Bligny, F. Rebeille, and R. Douce, *J. Biol. Chem.* **260,** 9166 (1985).

TABLE III
COMPARISON OF THE PHOSPHOGLYCERIDE CONTENT OF THE INNER AND OUTER MEMBRANES OF CASTOR BEAN, MUNG BEAN, AND SYCAMORE MITOCHONDRIA

	mol% total phosphoglycerides[a]					
Tissue source	DPG	PC	PE	PG	PI	PS
Castor bean endosperm[b]						
Outer membrane	0	51	39	Trace	10	
Inner membrane	7	34	58	1	2	
Mung bean hypocotyl[c]						
Outer membrane	0	68	24	2	5	0
Inner membrane	17	29	50	1	2	0
Sycamore tissue culture[c]						
Outer membrane	0	54	30	5	11	0
Inner membrane	15	41	37	3	5	0

[a] For abbreviations see Table II.
[b] From Cheesbrough and Moore.[16]
[c] From Bligny and Douce.[17]

inositol is present in significant quantities. Third, the presence of diphosphatidylglycerol is characteristic of mitochondria (indeed this lipid can be used to assess mitochondrial contamination of other fractions).[16,17] Fourth, other phosphoglycerides are present in such small and/or erratic amounts that it is possible that their presence is due to contamination by other membranes or, in the case of phosphatidic acid, because of lipid degradation.[2]

When the inner and outer membranes are separated from mitochondrial preparations, their lipid compositions are found to be very different (Table III). The outer membrane is relatively enriched in phosphatidylcholine and phosphatidylinositol while the inner membrane is relatively enriched in phosphatidylethanolamine and is the exclusive site of diphosphatidylglycerol.[17]

The presence of other acyl lipid components in plant mitochondria has not been convincingly demonstrated (see Harwood[2]). However, of other lipid types the electron transport component ubiquinone is particularly important. Recently, the Q_{10} isoprenalog has been shown to be the main component in many plant types, but the Q_9 homolog is major in some species. The Q_8 and Q_{11} homologs are sometimes minor constituents.[27]

[27] S. Schindler and H. K. Lichtenthaler, *in* "Structure, Function and Metabolism of Plant Lipids" (P.-A. Siegenthaler and W. Eichenberger, eds.), p. 273. Elsevier, Amsterdam, 1984.

Fatty Acyl Distribution of Mitochondrial Lipids

Individual acyl lipids from plant tissues usually contain characteristic fatty acid compositions (see Harwood[1]). Phosphatidylinositol in mitochondria is usually enriched in saturated fatty acids (Table IV) as it is in other membranes.[1] Phosphatidylglycerol, where present, has also been found to contain high amounts of palmitate while diphosphatidylglycerol is the richest in polyunsaturated fatty acids. The two major phosphoglycerides, phosphatidylcholine and phosphatidylethanolamine, have very similar fatty acid compositions—again as noted in other cell fractions.[10,14] The major fatty acids of these two lipids are linoleate, α-linolenate, and palmitate. Oleate and stearate are significant but minor components (Table IV).

There are some differences in the fatty acyl content of a given phosphoglyceride present in the inner or the outer membrane. Generally, higher saturation is found in the outer membrane.[17,21] For example, in Table V it can be seen that mung bean phosphatidylcholine contains twice as high a percentage of saturated fatty acids in the outer compared to the inner mitochondrial membrane. The increase in palmitate was at the expense of the polyenoic fatty acids, linoleate, and α-linolenate.

TABLE IV
FATTY ACYL COMPOSITION OF DIFFERENT MITOCHONDRIAL PHOSPHOGLYCERIDES

		Fatty acid content (% total)				
Tissue	Lipid[a]	16:0	18:0	18:1	18:2	18:3
Mung bean hypocotyl[b]	DPG	3	1	7	39	50
	PC	10	2	10	38	40
	PE	20	2	4	41	33
	PG	63	4	6	11	15
	PI	32	3	9	30	26
Potato tuber[c]	DPG	5	1	1	65	37
	PC	12	3	Trace	64	21
	PE	9	2	Trace	70	18
	PI	35	4	Trace	37	19
Sycamore tissue culture[d]	DPG	7	1	5	62	25
	PC	24	2	3	47	23
	PE	28	2	3	52	14
	PG	61	4	5	24	6
	PI	51	2	3	37	8

[a] For abbreviations see Table II.
[b] From Bligny and Douce.[17]
[c] From Journet *et al.*[24]
[d] From Douce.[26]

TABLE V
COMPARISON OF THE FATTY ACID COMPOSITION OF PHOSPHATIDYLCHOLINE AND PHOSPHATIDYLETHANOLAMINE IN THE INNER AND OUTER MITOCHONDRIAL MEMBRANE[a]

Lipid and source	Membrane	Fatty acids (% total) 16:0	18:0	18:1	18:2	18:3
Phosphatidylcholine						
Mung bean hypocotyl	Inner	9	2	10	41	40
	Outer	17	4	10	32	37
Sycamore tissue culture	Inner	24	2	4	46	24
	Outer	29	2	5	45	19
Phosphatidylethanolamine						
Mung bean hypocotyl	Inner	18	2	4	43	33
	Outer	29	3	6	33	29
Sycamore tissue culture	Inner	28	2	3	51	16
	Outer	31	2	4	50	12

[a] From Bligny and Douce.[17]

Transmembrane Distribution of Phosphoglycerides

It is increasing clear that natural membranes contain an asymmetric distribution of their lipid components. This property has been demonstrated for castor bean mitochondria. There are a number of possible errors in the assessment of transmembrane distributions. These may be due to substrate specificity with enzyme probes and to inaccessibility of certain phosphoglycerides to various probes, as well as to transbilayer movement of lipids or leaky membrane preparations. In order to reduce the possible errors due to these factors, it is very important to use more than one method for the measurement of transbilayer asymmetry and to include appropriate controls.[28] In the castor bean work, Cheesbrough and Moore[16] were careful to use both phospholipase A_2 and chemical probes, as well as quantitating the inaccessible phosphoglycerides.

Phospholipase A_2 (*Naja naja*) was used to study the distribution of a range of lipids which had been previously labeled from [^{14}C]acetate. (It was shown that the distribution of radioactively labeled phosphoglycerides was the same as that for the same lipids measured by phosphate assay.) The results were corroborated for phosphatidylethanolamine by using the nonpermeable reagent TNBS (2,4,6-trinitrobenzenesulfonic acid) to label the outer leaflet and the permeable compound DNFB (dini-

[28] J. A. F. op den Kamp, *in* "Membrane Structure" (J. B. Finean and R. H. Michell, eds.), p. 83. Elsevier, Amsterdam, 1981.

TABLE VI
SIDED DISTRIBUTION OF PHOSPHOGLYCERIDES IN MITOCHONDRIAL MEMBRANES FROM CASTOR BEAN ENDOSPERM[16]

Lipid[a]	Membrane	Distribution (mol%)		
		Outer leaflet	Inner leaflet	Inaccessible
PC	Outer membrane	25	30	45
	Inner membrane	21	20	59
PE	Outer membrane	44	32	24
	Inner membrane	70	18	12
PI + PS	Outer membrane	50	25	25
	Inner membrane	0	100	0

[a] For abbreviations see Table II.

tro-3-fluorobenzene) to check results with membrane vesicles ruptured by detergent.[16]

The results (Table VI) showed that the outer membrane had little asymmetry except for phosphatidylserine plus phosphatidylinositol. However, even then the large amount of inaccessible lipid made firm conclusions difficult. In contrast, phosphatidylethanolamine was concentrated in the outer leaflet of the inner membrane while phosphatidylserine plus phosphatidylinositol were exclusively in the inner leaflet. The small amounts of accessible phosphatidylcholine make conclusions about its transmembrane distribution premature.

Phosphoglyceride Metabolism in Plant Mitochondria

Phosphoglyceride metabolism in plant mitochondria has been reviewed.[2] Synthetic enzyme activities which have been positively detected in plant mitochondria are listed in Table VII. It will be seen that synthesis of phosphatidate by acylation of glycerol 3-phosphate, the methylation pathway for phosphatidylcholine, and phosphatidate cytidylyltransferase and phosphatidylglycerol synthesis are significant activities. It might also be anticipated that diphosphatidylglycerol synthesis will be found to be localized in mitochondria—so far there is only a preliminary report of its presence.[29]

There have been a few reports of lipid degradative enzymes in mitochondria (see Harwood[2]). The presence of a Ca^{2+}-stimulated nonspecific

[29] T. S. Moore, *Annu. Rev. Plant Physiol.* **33,** 235 (1982).

TABLE VII
PHOSPHOGLYCERIDE SYNTHETIC ENZYMES DETECTED IN PLANT MITOCHONDRIA[a]

Enzyme	Approximate percentage of cellular activity in mitochondria
sn-Glycerol-3-phosphate acyltransferase	13
Choline phosphotransferase (CDP-choline : 1,2-diacylglycerol cholinephosphotransferase)	2
Phosphatidylethanolamine methyltransferase (*S*-adenosyl-L-methionine : phosphatidylethanolamine *N*-methyltransferase)	10
Ethanolaminephosphotransferase (CDP-ethanolamine : 1,2-diacylglycerolethanolaminephosphotransferase)	2
Phosphatidate cytidylyltransferase (CTP : phosphatidate cytidylyltransferase)	25
Glycerol-3-phosphate : CDP-diacylglycerol phosphatidyltransferase + phosphatidylglycerophosphatase (phosphatidylglycerolphosphate phosphohydrolase)	50
Phosphatidylglycerol : CDP-diacylglycerol phosphatidyltransferase	Present

[a] For more details, see Moore.[29]

acyl hydrolase in potato mitochondria[30] is important since failure to prevent its activity will lead to misleading lipid analyses. In contrast to the presence of this enzyme and of various lipases in mitochondria,[2] lipoxygenase[31] and the enzymes of β-oxidation appear to be absent from higher plant mitochondria.[32] In contrast, a report has described the synthesis of fatty acids by a modified reversed β-oxidation in mitochondria from *Euglena gracilis*.[33]

In addition, as mentioned in the section on Analysis it is very important to ensure that lipolytic acyl hydrolase (see Burns *et al.*[34]) and phospholipase D enzymes, which are also likely to be present in most plant tissues, are inactivated. If these enzymes are active during mitochondrial isolation, or contaminate the final isolated fraction, then enzyme damage and the appearance of large amounts of normally minor lipids will result.

[30] R. Bligny and R. Douce, *Biochim. Biophys. Acta* **529,** 419 (1978).

[31] M. Neuberger, E.-P. Journet, R. Bligny, J.-P. Carde, and R. Douce, *Arch. Biochem. Biophys.* **217,** 312 (1982).

[32] B. Gerhardt, *in* "Structure, Function and Metabolism of Plant Lipids" (P.-A. Siegenthaler and W. Eichenberger, eds.), p. 189. Elsevier, Amsterdam, 1984.

[33] H. Inui, K. Miyatake, Y. Nakano, and S. Kitaoka, *Eur. J. Biochem.* **142,** 121 (1984).

[34] D. D. Burns, T. Galliard, and J. L. Harwood, *Phytochemistry* **19,** 2281 (1980).

Breakdown of mitochondrial lipids due to the action of acyl hydrolases or phospholipases has also been implicated in certain *in vivo* situations such as cold shock[35] or in bean roots infected with *Rhizoctonia solani*.[36]

Changes in Mitochondrial Lipids Caused by Growth Temperature

Mitochondrial membrane lipids will undoubtably be affected by the same environmental parameters which alter the composition of other plant membranes (see Harwood[37]). However, to date only the effects of temperature have been followed in any detail. In general, as the plant's growth temperature is lowered, fatty acid unsaturation increases and there is an increase in the lipid/protein ratio of membranes.[37] The rise in lipid content may be due to an increase in the activity of synthetic enzymes. Changes in lipid class proportions have also been noted.[38] The reason for the increase in unsaturation has attracted considerable work. Possible causes include the induction of specific enzymes and the modification of desaturase activity by changes in the enzyme's environment or in supplies of one or more substrates (see Harwood[2]). Experiments with sycamore cells showed that fatty acid unsaturation was a function of oxygen tension rather than temperature[39] and recent results demonstrated that large fatty acid changes could be produced by manipulating oxygen tension whereas Arrhenius discontinuities of mitochondrial respiration were unaltered.[26] These data cast doubt on an increase in fatty acid unsaturation as a device for ensuring membrane function at lowered temperatures.

Such an increase in fatty acyl unsaturation has been suggested to be a device for preventing freezing injury. However, correlations between unsaturation and chilling resistance have not been very convincing.[37] Demonstrations that particular molecular species of phosphatidylglycerol can be correlated with susceptibility to chilling damage in chloroplast membranes[40] are important in that they show the importance of examining the fine details of membrane composition. Thus, it may be necessary to analyze changes in molecular species and positional distribution of fatty acids in mitochondrial lipids in order to reveal a possible connection between these compounds and chilling resistance/sensitivity.

[35] V. K. Vojnikov, G. B. Luzova, and A. M. Korzun, *Planta* **158,** 194 (1983).

[36] G. V. Sednina, V. A. Aksenova, and G. A. Khartina, *Fizio. Rast.* **28,** 1030 (1981).

[37] J. L. Harwood, *in* "Structure, Function and Metabolism of Plant Lipids" (P.-A. Siegenthaler and W. Eichenberger, eds.), p. 543. Elsevier, Amsterdam, 1984.

[38] S. Kimura, M. Kanno, Y. Yamada, K. Takahashi, H. Murashige, and T. Okamoto, *Agric. Biol. Chem.* **46,** 2895 (1982).

[39] F. Rebeille, R. Bligny, and R. Douce, *Biochim. Biophys. Acta* **620,** 1 (1980).

[40] N. Murata and J. Yamaya, *Plant Physiol.* **74,** 1016 (1984).

[45] Ubiquinone Biosynthesis in Plant Mitochondria

By FRIEDHELM LÜTKE-BRINKHAUS and HANS KLEINIG

The biosynthesis of ubiquinones, starting from isopentenyl diphosphate and 4-hydroxybenzoate as natural precursors, requires several enzymatic steps such as prenyl transfer, 4-hydroxybenzoate : polyprenyl transfer, hydroxylation, decarboxylation, O-methylation, and methylation. In animal and yeast cells these reactions occur within the mitochondria.[1,2] It has been shown recently that plant mitochondria also perform several reactions in the biosynthesis of the ubiquinones.[3] The biosynthesis of plant ubiquinones and the compartmentation of this pathway within the plant cell can be studied by incorporation of [1-^{14}C]isopentenyl diphosphate, 4-hydroxy[*carboxy*-^{14}C]benzoic acid, and *S*-[*methyl*-^{14}C]-adenosylmethionine into ubiquinone intermediates by isolated mitochondria from potato tubers (*Solanum tuberosum*).

Preparation of Mitochondria and Mitochondrial Subfractions from Potato Tubers

Potato tubers (1 kg) are cut and homogenized in a meat grinder in 2 liters extraction medium containing 0.3 *M* sorbitol, 20 m*M* KH_2PO_4, pH 7.5, 0.5 m*M* EDTA, 20 m*M* HEPES, 4 m*M* cysteine, 0.1% bovine serum albumin, 1% poly(vinylpyrrolidone) 40. The homogenate is filtered through three layers of nylon cloth (40-μm mesh) and the filtrate centrifuged for 5 min at 3000 *g*. The supernatant is then recentrifuged again for 15 min at 13,000 *g*. The crude mitochondria are purified by centrifugation in a preformed linear 18–36% Percoll gradient (total volume 28 ml, 0.3 *M* sucrose, 10 m*M* KH_2PO_4, pH 7.2, 1 m*M* EDTA, 0.1% bovine serum albumin). After resuspending the crude mitochondria in extraction medium, 2-ml aliquots are layered on top of each of the three gradients. After centrifugation for 30 min at 40,000 *g* the mitochondrial fraction (visible as a gray band in the middle of the gradient) is collected, diluted with 15 times their volume of washing medium (0.3 *M* sorbitol, 10 m*M* KH_2PO_4, pH 7.2, 1% bovine serum albumin), and sedimented for 20 min at 14,000 *g*.

[1] K. Momose and H. Rudney, *J. Biol. Chem.* **247,** 3930 (1972).
[2] J. Casey and D. R. Threlfall, *Biochim. Biophys. Acta* **530,** 487 (1978).
[3] F. Lütke-Brinkhaus, B. Liedvogel, and H. Kleinig, *Eur. J. Biochem.* **141,** 537 (1984).

For preparation of mitochondrial membrane and matrix subfractions the isolated mitochondria are resuspended in 1.5 ml of incubation buffer containing 100 m*M* Tris/HCl, pH 7.6, 10 m*M* $MgCl_2$, 2 m*M* dithioerythritol and disrupted by sonification, 2 × 3 sec. After centrifugation for 1 hr at 120,000 *g* the supernatant containing the soluble matrix enzymes is removed and the membranes resuspended in 1 ml incubation buffer.

Incubation Conditions

Using the precursors [1-^{14}C]isopentenyl diphosphate, 4-hydroxy[*carboxy*-^{14}C]benzoic acid, and *S*-[*methyl*-^{14}C]adenosylmethionine for ubiquinone biosynthesis in isolated mitochondria most of the biosynthetic intermediates can be covered. In parallel incubations the precursors are applied in combinations using one labeled and one or two unlabeled substrates, respectively, in equimolar concentrations. Thus, from the ratios of incorporated radioactivities information may be obtained about the chemical composition of the various products observed.

To localize the corresponding enzymatic activities within the mitochondria a mitochondrial supernatant subfraction containing the soluble enzymes and a membrane subfraction are incubated separately and after recombination in the presence of the substrates.

The incubation mixture contains 100 m*M* Tris/HCl, pH 7.6, 10 m*M* $MgCl_2$, 2 m*M* dithioerythritol. All incubations are performed in the presence of 1 m*M* $MnCl_2$ and 10 m*M* NaF in a total volume of 0.5 ml at 25° for 3 hr. The reactions are started by the addition of substrates, 18.8 μM in either case: [1-^{14}C]isopentenyl diphosphate (18.5 kBq), 4-hydroxy[carboxy-^{14}C]benzoic acid (18.5 kBq), *S*-[*methyl*-^{14}C]adenosylmethionine (20.2 kBq), plus unlabeled substrates, respectively, in the same concentrations. The reactions are stopped by addition of chloroform/methanol (2 : 1) and the lipids are extracted.[4] In order to avoid permeation problems for the substrates isolated mitochondria are disrupted using a French pressure cell (27.6 MPa) before incubation.

Identification of Products

Radioactive incubation products are separated by thin-layer silica gel plates using the solvent system petroleum benzene/ether/acetone (4 : 1 : 1) and subsequently by HPLC.[5] Polyprenyl chain lengths are determined using an HPLC method.[6] 4-Hydroxy-3-polyprenylbenzoic acids and hy-

[4] E. S. Bligh and W. J. Dyer, *Can. J. Biochem. Physiol.* **37,** 911 (1959).

[5] P. Beyer, K. Kreuz, and H. Kleinig, *Planta* **150,** 435 (1980).

[6] P. Beyer and H. Kleinig, this series, Vol. 110, p. 299.

TABLE I
INCORPORATION OF ISOPENTENYL DIPHOSPHATE, 4-HYDROXYBENZOIC ACID, AND *S*-ADENOSYLMETHIONINE IN POLYPRENOLS AND UBIQUINONE INTERMEDIATES BY ISOLATED DISRUPTED POTATO TUBER MITOCHONDRIA[a]

Precursor	Total incorporation	Distribution of radioactivity (total = 100%)									
		Polyprenols	2-Decaprenylphenols	4-Hydroxy-3-decaprenylbenzoic acid	Hydroxydecaprenylphenol	Dihydroxydecaprenylphenol	Methoxy-4-hydroxy-5-decaprenylbenzoic acid	Methoxy-2-decaprenylphenol	Methoxymethyl-2-decaprenylphenol	Methoxyhydroxydecaprenylphenol	Methoxydihydroxy decaprenylphenol
[1-^{14}C]IPP + 4-HB	33.4	16.0	12.3	71.7	—	—	—	—	—	—	—
4-[^{14}C]HB + IPP	2.6	—	—	100.0							
[1-^{14}C]IPP + 4-HB/SAM	41.7	13.5	5.1	15.0	5.9	3.1	11.7	2.2	1.9	2.7	1.5
4-[^{14}C]HB + IPP/SAM	1.8	—	—	59.0	—	—	41.0				
[^{14}C]SAM + IPP/4-HB	1.2	—	—	—	—	—	53.3	9.3	16.6	3.5	7.3

[a] The chain lengths of free polyprenols are mainly C_{15} and C_{20}; that of the ubiquinone intermediates preferentially C_{50}.

TABLE II

INCORPORATION OF [1-^{14}C]ISOPENTENYL DIPHOSPHATE, 4-HYDROXYBENZOIC ACID, AND *S*-ADENOSYLMETHIONINE IN POLYPRENYL DIPHOSPHATES, POLYPRENOLS, AND UBIQUINONE INTERMEDIATES BY MITOCHONDRIAL SUBFRACTIONS AND RECOMBINED SUBFRACTIONS[a]

		Distribution of Radioactivity (total = 100%)								
		Polyprenyl diphosphates			Polyprenols					
Subfractions	Total incor-poration	C_{15}	C_{20}	C_{50}	C_{15}	C_{20}	C_{50}	Prenyl-phenols	Hydroxy-prenyl-benzoic acid	Methylated products
Supernatant	70.0	48.0	33.2	3.5	8.0	5.0	1.7	—	—	—
Membranes	1.4									
Supernatant + membranes	22.5	15.8	9.0	—	34.8	16.3	—	8.0	8.2	8.0

[a] The formation of ubiquinone intermediates is reduced in the recombined subfractions, when compared with the original system (see Table I, line 3).

droxypolyprenylphenols with different chain lengths show the same retention times and 2-polyprenylphenols delayed retention times compared with the corresponding polyprenols, when using this separation system. Polyprenyl diphosphates are extracted by EDTA/butanol,[7] hydrolyzed using alkaline phosphatase,[8] and analyzed as polyprenols.

For the identification of the ubiquinone intermediates three criteria are used as follows: (1) cochromatography with standards, (2) incorporation of radioactivity from the different substrates and radioactivity ratios, and (3) chemical modification followed by rechromatography (acetylation using acetic anhydride, silylation using chlorotrimethylsilan, reduction using $NaBH_4$).

For instance, for methoxy-4-hydroxy-5-decaprenylbenzoic acid

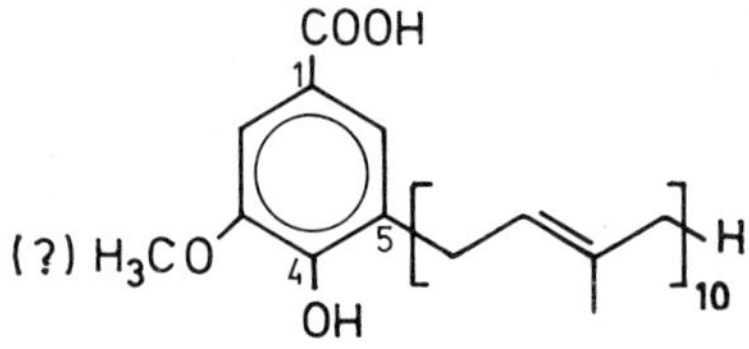

a radioactivity ratio of 10 : 1 : 1 is found when [1-^{14}C]isopentenyl diphosphate, 4-hydroxy[*carboxy*-^{14}C]benzoic acid, and *S*-[*methyl*-^{14}C]adenosylmethionine are used as substrates. This means that this compound still has the carboxy group and contains a C_{50} polyprenyl side chain and a methyl group incorporated from *S*-[*methyl*-^{14}C]adenosylmethionine. The free hydroxyl group present is demonstrated by acetylation or silylation which causes a shift in polarity on HPLC and thin-layer chromatography.

The ubiquinone intermediates found when potato tuber mitochondria are incubated under the conditions outlined above are shown in Table I. In Table II the incorporation pattern of [1-^{14}C]isopentenyl diphosphate plus unlabeled 4-hydroxybenzoic acid and *S*-adenosylmethionine in ubiquinone intermediates by mitochondrial subfractions are given. True quinoid compounds are detected only in trace amounts.

[7] B. E. Allen and D. V. Banthorpe, *Phytochemistry* **20,** 35 (1981).
[8] G. R. Daleo and R. Pont Lezica, *FEBS Lett.* **74,** 247 (1977).

[46] Purification of Complexes II and IV from Plant Mitochondria

By MASAYOSHI MAESHIMA, TSUKAHO HATTORI, and TADASHI ASAHI

The enzyme complexes (complexes I–IV) of the electron transport system in higher plant mitochondria have not yet been isolated in pure form except for complex IV (cytochrome-*c* oxidase). Complex III (cytochrome bc_1 complex) has been isolated in a pure form from sweet potato root mitochondria but the purified preparation shows no ubiquinol–cytochrome-*c* reductase activity.[1]

Efforts to isolate higher plant complex II have not been successful. Instead, succinate dehydrogenase, a component of complex II, has been purified in a considerably high degree of purity from sweet potato root mitochondria.[2] As is well known, succinate dehydrogenase catalyzes the reduction of dyes such as phenazine methosulfate (PMS) plus 2,6-dichloroindophenol (DCIP) with succinate [Eq. (1)], that is, a part of the

$$\text{Succinate} + \text{dye} \rightarrow \text{fumarate} + (\text{dye})H_2 \quad (1)$$

$$\text{Succinate} + \text{ubiquinone} \rightarrow \text{fumarate} + \text{ubiquinol} \quad (2)$$

reaction catalyzed by complex II [Eq. (2)]. From our experience it appears that the higher plant complex II seems to be easily dissociated into two portions, succinate dehydrogenase and the portion associated with the mitochondrial inner membrane. Higher plant succinate dehydrogenase appears to be solubilized in the form of complex II from the submitochondrial particles with detergents[2–4] but becomes a form free from the membrane-associated portion of complex II during purification.[2,3] This contrasts with the case of mammalian complex II, in which the dissociation into succinate dehydrogenase and the membrane-associated portion occurs only by treatment of the complex with chaotropic reagents.[5]

Complex IV, which catalyzes the following reaction [Eq. (3)], has been purified from sweet potato root and pea shoot mitochondria.[6,7]

$$2\ \text{Ferrocytochrome}\ c + 2\ H^+ + \tfrac{1}{2}O_2 \rightarrow 2\ \text{ferricytochrome}\ c + H_2O \quad (3)$$

[1] T. Nakajima, M. Maeshima, and T. Asahi, *Agric. Biol. Chem.* **48,** 3019 (1984).
[2] T. Hattori and T. Asahi, *Plant Cell Physiol.* **23,** 515 (1982).
[3] N. Nakayama and T. Asahi, *Plant Cell Physiol.* **22,** 79 (1981).
[4] J. J. Burke, J. N. Siedow, and D. E. Moreland, *Plant Physiol.* **70,** 1577 (1982).
[5] Y. Hatefei, this series, Vol. 53 [6].
[6] M. Maeshima and T. Asahi, *Arch. Biochem. Biophys.* **187,** 423 (1978).
[7] M. Matsuoka, M. Maeshima, and T. Asahi, *J. Biochem.* (*Tokyo*) **90,** 649 (1981).

The purified sweet potato complex has a high specific activity, whereas the pea complex is inactivated during purification. The sweet potato complex, but not the pea one, can be solubilized with a high concentration of deoxycholate from submitochondrial particles which have been washed with a low concentration of the detergent. Pea complex IV can be solubilized from the deoxycholate-washed submitochondrial particles only with high concentrations of Triton X-100; in such a case, the complex is very difficult to isolate in an active form.

Here we describe the procedures for purifying succinate dehydrogenase and complex IV from sweet potato root mitochondria as well as properties of the purified enzymes.

Succinate Dehydrogenase Portion of Complex II

Assay Method

Principle. Succinate dehydrogenase activity is assayed by spectrophotometrically following the reduction of DCIP coupled to succinate oxidation with PMS as a mediator, as described previously.[8]

Reagents

1. 0.3 *M* potassium phosphate buffer, pH 7.8, containing 0.6% bovine serum albumin and 30 m*M* KCl. On the day of use, KCN is dissolved in the buffer at a final concentration of 0.58 mg/ml
2. 9 m*M* PMS. The solution should be stored at −20° in the dark or prepared on the day of use
3. 0.75 m*M* DCIP
4. 0.6 *M* sodium succinate

Procedure. In a 0.3-ml spectrophotometric cuvette with a 10-mm light path are placed 50 μl of potassium phosphate buffer, pH 7.8, containing bovine serum albumin and KCN, 20 μl of succinate, and 110 μl of an enzyme preparation plus water. When a mitochondrial or submitochondrial suspension is assayed for the activity, the cuvette is kept at 25° for 30 min to activate the enzyme; usually severalfold activation is observed for such a suspension. Next, 20 μl of DCIP is added to the cuvette and the reaction is started by adding 100 μl of PMS. The decrease in absorbance at 600 nm is followed at 25° as a function of time. The rate of succinate oxidation is calculated using an extinction coefficient of $\varepsilon_{\text{red-ox}} = 21\ \text{m}M^{-1}\ \text{cm}^{-1}$ for DCIP at 600 nm.

All solutions used should be kept at 25° before being added to the

[8] J. L. Hedrick and A. J. Smith, *Arch. Biochem. Biophys.* **126,** 155 (1968).

cuvette. Special caution should be directed to nonenzymatic reduction of the dyes if the enzyme preparations contain reductants such as dithiothreitol, 2-mercaptoethanol, and ascorbate. For such enzyme preparations, it should be conditioned that the succinate-dependent reduction of DCIP is much faster than the reduction of DCIP in the absence of succinate and the rate of the nonenzymatic reduction should be determined by a control assay.

The activity of complex II (succinate : ubiquinone oxidoreductase) is assayed as follows (see also Hatefei and Stiggall[9]); to a 0.3-ml spectrophotometric cuvette with a 10-mm light path are added 50 μl of 0.3 *M* potassium phosphate buffer, pH 7.4, 20 μl of 0.6 *M* sodium succinate, 30 μl of 0.1% (w/v) Triton X-100, 20 μl of 1.5 m*M* EDTA, 110 μl of enzyme solution plus water, and 5 μl of 0.75 m*M* DCIP. The reaction is started by adding 5 μl of 2.4 m*M* ubiquinone-2 (ethanol solution), and the decrease in absorbance at 600 nm is followed as a function of time.

Purification Procedures

All steps described below are carried out at 0–5°. Mitochondria are isolated from the parenchymatous tissue of sweet potato (*Ipomoea batatas,* Kokei No. 14) roots, which have been stored at 13–15° after harvest, as described previously.[6] Briefly, a homogenate of the tissue is fractionated by differential centrifugation and the crude mitochondrial fraction obtained is centrifuged on a discontinuous sucrose density gradient (40 and 60%, w/v). The fraction in the interface between 40 and 60% sucrose solutions is diluted with 3 vol of 5% sucrose and centrifuged at 10,000 *g* for 30 min to obtain a mitochondrial pellet. Commercially available deoxycholic acid is recrystallized from its solution in 70% ethanol after the solution is decolorized with active charcoal.

Step 1: Preparation of Submitochondrial Particles. The mitochondrial pellet from 1.6 kg of the tissue is suspended in 15 ml of 50 m*M* Tris–HCl buffer, pH 7.8 (buffer A). The suspension is subjected twice to sonic oscillation at 20 kHz for 30 sec and centrifuged at 105,000 *g* for 1 hr. The pellet is suspended in buffer A at a protein concentration of 3 mg/ml. The suspension can be stored at −70° for up to a week.

Step 2: Solubilization of Succinate Dehydrogenase. To 20 ml of the submitochondrial particle suspension is added 0.4 ml of 5% (w/v) sodium deoxycholate prepared by neutralizing a recrystallized deoxycholic acid solution with NaOH. Then the suspension is kept standing in an ice bath for 30 min with occasional agitation and centrifuged at 105,000 *g* for 1 hr. The supernatant is passed through a Sephadex G-25 column (2.5 × 30 cm)

[9] Y. Hatefei and D. Stiggall, this series, Vol. 53 [5].

preequilibrated with buffer A to remove the detergent, and the void volume fraction (25 ml) is used for the next purification step.

Step 3: Ammonium Sulfate Fractionation. To the enzyme solution from step 2 is added 12.5 ml of saturated ammonium sulfate solution (saturated at 4° and adjusted to pH 8.3 with NH_4OH) to make the solution 33% saturated in ammonium sulfate with continuous stirring. The mixture is kept in an ice bath for 20 min and then centrifuged at 10,000 *g* for 20 min. The supernatant is adjusted to 50% saturation by adding another 12.5 ml of saturated ammonium sulfate solution with stirring. The mixture is kept cold and centrifuged as described above. The precipitate is dissolved in 2 ml of 10 m*M* potassium phosphate buffer, pH 7.8 (buffer B), containing 5 m*M* dithiothreitol and the solution is dialyzed against buffer B for 4 hr.

Step 4: DEAE–Cellulose Column Chromatography. The enzyme solution from step 3 is diluted to 10 ml with buffer B and applied to a DEAE–cellulose (DE-52, Whatman) column (1 × 4 cm) preequilibrated with buffer B. The column is washed with 15 ml of buffer B containing 80 m*M* NaCl, then the enzyme is eluted from the column with 20 m*M* potassium phosphate buffer, pH 6.2, containing 20 m*M* sodium succinate. The flow rate is kept constant at about 10 ml/hr. Elution of the enzyme begins when the pH of the eluate decreases to 6.8.

Note. The enzyme is so labile that the purification should be performed as quickly as possible. If the preparation from step 2 or 3 as well as the final preparation must be stored, it should be kept at −80° under strict anaerobic conditions.

The final preparation still contains significant amounts of contaminating proteins, but can be used for analysis of the subunit composition and isolation of the subunits. The large subunit has been isolated from the preparation by successive polyacrylamide gel electrophoresis; first, of the preparation on a gel under nondenaturating conditions and then of the enzyme band detected by active staining on a gel containing sodium dodecyl sulfate (SDS).[8]

Properties

Succinate dehydrogenase is solubilized probably in the form of complex II from higher plant submitochondrial particles with detergents, although the final preparation from sweet potato root tissue shows no succinate-DMB (2,3-dimethoxy-5-methyl-6-pentyl-1,4-benzoquinone) reductase activity. Preliminary experiments in our laboratory have shown that a solubilized fraction from castor bean endosperm submitochondrial particles with Triton X-100 shows an activity of succinate : ubiquinone oxidoreductase. Burke *et al.*[4] have solubilized the enzyme from mung bean shoot mitochondria with Triton X-100 and fractionated it with ammonium sulfate. Their preparation showed activity for reduction of DCIP with succinate in the absence of PMS, which suggested that the enzyme was in the form of complex II. However, it did not show other enzymatic properties characteristic of complex II from other sources; that is, the DCIP-reducing activity was not inhibited by 2-thenolyltrifluoroacetone, and the preparation exerted no activity for reduction of DMB with succinate.

There are two forms of succinate dehydrogenase in sweet potato root mitochondria.[2] The two forms are separated from each other by polyacrylamide gel electrophoresis (Fig. 1) or hydroxyapatite column chromatography. They differ from each other only in the charge of protein and not in the K_m value for succinate (0.29 mM), the optimal pH (7.8–8.0), the molecular weight [110,000 for the solubilized fraction determined by Furguson's plot analysis[8] of polyacrylamide gel electrophoresis under nondenaturating conditions followed by staining for PMS reductase activ-

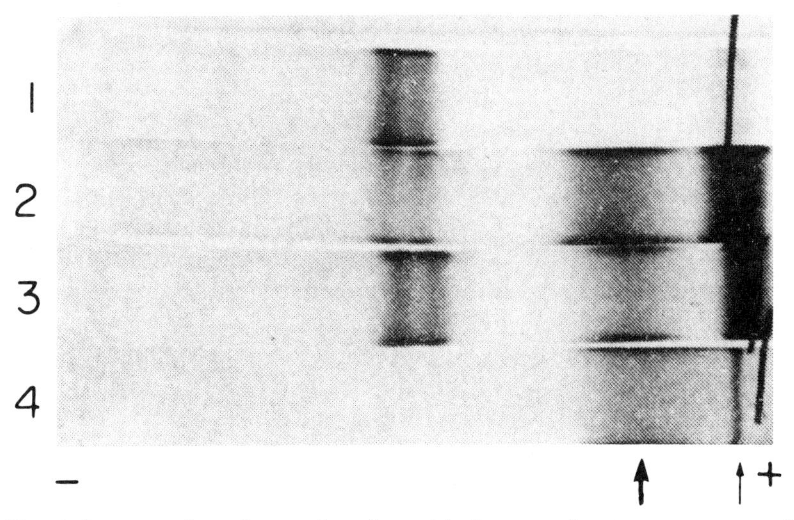

FIG. 1. Presence of two forms of succinate dehydrogenase in sweet potato root mitochondria. The enzyme was solubilized from the submitochondrial particles with 0.8 M $NaClO_4$ (track 1), 0.2% Triton X-100 (track 2), and 0.1% deoxycholate (track 3). The solubilized fractions were electrophoresed on a 7% polyacrylamide slab gel, and the gel was stained for succinate–PMS reductase activity by being incubated in 114 mM potassium phosphate buffer, pH 7.8, containing 43 mM succinate, 0.37 mM PMS, and 0.28 mM nitroblue tetrazolium. Track 4: the same as track 3 except that succinate was omitted from the staining solution. The thick and then arrows indicate the band produced nonenzymatically by dithiothreitol present in the samples and the position of tracking dye, respectively.

ity (see legend to Fig. 1)], or the subunit composition. It remains obscure how they differ with respect to function and whether their existence is unique for sweet potato root mitochondria.

Sweet potato succinate dehydrogenase is composed of two subunits, large and small subunits,[2] like the enzyme from other sources.[5] The large subunit with a molecular weight of 65,000 has been proved to be a flavoprotein. The small subunit has a molecular weight of 26,000 and probably corresponds to the iron–sulfur protein subunit of the mammalian enzyme. The preparation of Burke *et al.*[4] contained two additional major polypeptides when analyzed by SDS–polyacrylamide gel electrophoresis. The polypeptides may be subunits of the membrane-associated portion of complex II and thus both or either one of them may be a *b*-type cytochrome.

Complex IV (Cytochrome-c Oxidase)

Assay Method

Principle. The rate of oxidation of reduced cytochrome *c* is measured spectrophotometrically as described for the complex from other sources.[10,11]

Reagents

1. Reduced cytochrome *c*. A few grains of solid sodium dithionite is dissolved in an aqueous cytochrome *c* (horse heart, Boehringer-Mannheim) solution. Then the solution is passed through a Dowex 1-X8 (OH form, Dow Chemical) column washed with water to remove the reducing reagent. The final concentration of reduced cytochrome *c* is adjusted to 1 mg/ml
2. Soybean lecithin micelle suspension. A soybean lecithin (Wako Pure Chemical) suspension (10 mg/ml) in 10 m*M* Tris–acetate buffer, pH 7.5, containing 0.2% deoxycholate and 1 m*M* EDTA is subjected twice to sonic oscillation at 20 kHz for 30 sec. The suspension is then centrifuged at 10,000 *g* for 30 min and the supernatant is used as the micelle suspension
3. 30 m*M* potassium phosphate buffer, pH 7.0, containing 0.1% (w/v) Triton X-100

Procedure. A suitably diluted enzyme solution is mixed with an equal volume of soybean lecithin micelle suspension. When mitochondrial and

[10] B. Erred, M. D. Kamen, and Y. Hatefi, this series, Vol. 53 [8].

[11] M. S. Rubin and A. Tzagoloff, this series, Vol. 53 [12].

submitochondrial preparations are assayed for enzyme activity, they may not be mixed with the micelle suspension. To a 0.3-ml cuvette with a 10-mm light path are added 50 μl of the enzyme–soybean lecithin micelle mixture and 150 μl of potassium phosphate buffer, pH 7.0, containing Triton X-100. The reaction is started by adding 100 μl reduced cytochrome *c* to the cuvette and the decrease in absorbance at 550 nm is followed at 25° as a function of time.

All solutions should be kept at 25° before being added to the cuvette. Since the oxidation of reduced cytochrome *c* follows first-order kinetics for the concentration of reduced cytochrome *c*, the activity of complex IV should be calculated as described previously[10] when being determined exactly. However, when various preparations are being compared with respect to the activity, the initial rates of the oxidation of reduced cytochrome *c* may be referred to provided that the initial concentrations of reduced cytochrome *c* in all cuvettes are the same.

Purification Procedures

Unless otherwise specified, all manipulations described below (see Table I) are performed at 0–4°. Sweet potato root mitochondrial pellet and recrystallized deoxycholic acid are prepared as described for the purification of succinate dehydrogenase.

Step 1: Preparation of Submitochondrial Particles. The mitochondrial pellet is suspended in 0.1 *M* KCl, and the suspension is subjected twice to sonic oscillation at 20 kHz for 30 sec and centrifuged at 80,000 *g*

TABLE I
PURIFICATION OF SWEET POTATO COMPLEX IV (CYTOCHROME *c* OXIDASE)[a]

Fraction	Total protein	Total units	Recovery (%)	Specific activity (U/mg of protein)	Purification (*n*-fold)
Submitochondrial particles[b]	144	466	100	3.24	1
Deoxycholate extract	11.5	292	62.7	25.4	7.8
First DEAE–cellulose eluate	3.70	337	72.3	91.1	28.1
Second DEAE–cellulose eluate	2.14	289	62.0	135	41.7
$(NH_4)_2SO_4$ fraction	1.46	214	45.9	147	45.4

[a] One unit of the enzyme activity is defined as the amount of enzyme which oxidizes 1 μmol of reduced cytochrome *c* per minute at the initial rate under the assay conditions described in the text.

[b] The particles were prepared from 3.2 kg of sweet potato root tissue and stored at approximately −80° until use.

for 1 hr. The pellet is suspended in 50 m*M* Tris–acetate buffer, pH 7.5 (buffer C), containing 0.25 *M* sucrose at a final protein concentration of 3 mg/ml. Thirty to 70 ml of the suspension is used for the following procedures.

Step 2: Solubilization of Complex IV. To the submitochondrial particle suspension are added solid KCl and 5% (w/v) potassium deoxycholate (a recrystallized deoxycholic acid solution is neutralized with KOH) at final concentrations of 1 *M* and 0.4 mg/mg of protein, respectively. The suspension is then centrifuged at 80,000 *g* for 1 hr and the brownish pellet is suspended in buffer C to make the volume up to half of that of the original submitochondrial particle suspension. Solid KCl and potassium deoxycholate (5%) are again added to the suspension at final concentrations of 0.1 *M* and 0.76%, respectively. The suspension is kept standing for 30 min with continuous stirring and then centrifuged at 80,000 *g* for 1 hr. The supernatant is mixed with an equal volume of buffer C containing 0.1% (w/v) Triton X-100.

Step 3: First DEAE–Cellulose Column Chromatography. The solution from step 2 is applied to a DEAE–cellulose (DE-52, Whatman) column (2.2 × 6 cm) preequilibrated with buffer C containing 0.1% Triton X-100. The column is washed with 30 ml of buffer C containing 0.1% Triton and 0.1 *M* KCl, then the complex is eluted from the column with 20 ml of buffer C containing 0.1% Triton X-100 and 0.3 *M* KCl.

Step 4: Second DEAE–Cellulose Column Chromatography. Yellow-green fractions from step 3 are diluted with 4 vol of buffer C containing 0.1% Triton X-100 and then applied to a DEAE–cellulose column (1.5 × 4 cm) preequilibrated with buffer C containing 0.1% Triton X-100. The column is washed with 20 ml of buffer C containing 0.1% Triton X-100 and 0.3 *M* KCl. Deep green fractions eluted with the latter buffer are collected.

Step 5: Ammonium Sulfate Fractionation. To the deep green fractions from step 5 are added saturated ammonium sulfate solution (saturated at 4° and adjusted to pH 7.0 with NH_4OH) and 20% sodium cholate at final concentrations of 33% saturation and 2%, respectively, with continuous stirring. The mixture is kept standing for 30 min and then centrifuged at 80,000 *g* for 10 min. The supernatant is mixed with saturated ammonium sulfate solution and 10% Triton X-100 to final concentrations of 43% saturation and 0.1%, respectively, with stirring. The mixture is kept standing for 30 min and centrifuged in a swing rotor (Hitachi RPS-50 rotor) at 80,000 *g* for 10 min. The insoluble materials floating at the top of the tube and the pellet are both dissolved in buffer C containing 0.1% Triton X-100.

Some final preparations from step 5 contain small amounts of contami-

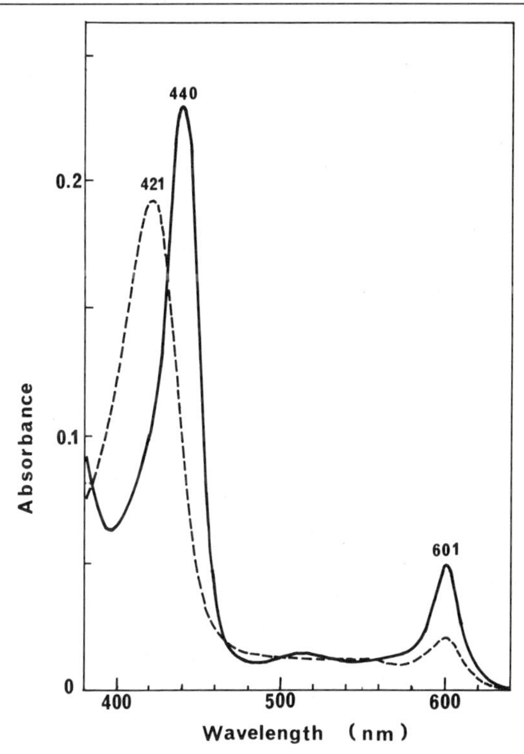

FIG. 2. Absorption spectra of sweet potato complex IV (cytochrome-*c* oxidase). The enzyme solution (0.288 mg/ml) after sucrose density gradient centrifugation of the enzyme preparation from step 5 was used for taking the spectra of the air-oxidized (---) and sodium dithionite-reduced forms (—) of the enzyme.

nating proteins. When a completely pure preparation of the complex must be obtained (for instance, to prepare antibody against the complex), the final preparation should be subjected to sucrose density gradient centrifugation as described below. The complex, however, loses its enzymatic activity during the centrifugation. The solution (0.7 ml) from step 5 is layered on a linear sucrose density gradient (4.5 ml; 5–20%, w/v, sucrose in buffer C containing 0.1% Triton X-100). The gradient is centrifuged in a swing rotor (Hitachi RPS-50 rotor) at 180,000 *g* for 9 hr, and the green fractions are collected.

The subunits of the complex can be isolated from the preparation from step 4 by the following procedures.[12] About 1 ml (2 mg of protein) of the preparation is incubated in a 2% SDS solution containing 2% 2-mercap-

[12] M. Maeshima and T. Asahi, unpublished.

toethanol and 2 *M* urea at 60° for 30 min. It is then electrophoresed on a 10% polyacrylamide slab gel (15 × 17 cm, 1 mm thick) containing 0.1% SDS and 6 *M* urea. After electrophoresis, the gel is stained with Coomassie brilliant blue R, then destained with a 7% acetic acid solution containing 10% methanol and 10% 2-propanol. The stained bands are cut out from the gel, minced with a spatula, and incubated in 70% formic acid at 4° for 2 days with shaking. The gel pieces are removed by filtration through glass wool in a pipet tip, and the filtrate is lyophilized. The dried materials are dissolved in 0.2 ml of 0.1% SDS.

FIG. 3. SDS–urea polyacrylamide gel electrophoresis of sweet potato complex IV (cytochrome-*c* oxidase) subunits. The enzyme solution after sucrose density gradient centrifugation of the enzyme preparation from step 5 was incubated in 2% SDS, 2% 2-mercaptoethanol, and 2 *M* urea at 60° for 20 min. Electrophoresis was run with a 10% polyacrylamide (acrylamide : bisacrylamide, 10 : 1) gel containing 0.1% SDS and 8 *M* urea.

Properties

The purified preparation contains 10–13 nmol of heme *a*/mg of protein and no heme *b* and *c*. As shown in Fig. 2, the dithionite-reduced form of the complex has maxima at 440 nm (the Soret band) and 601 nm (α-band). The specific activity of the final preparation ranges from 120 to 150 μmol heme *c* oxidized/min per milligram of protein at the initial rate under the assay conditions described above. The final preparation contains 1.0 μg of phosphorus (corresponding to 25 g of phospholipids) per milligram of protein but requires exogenously added phospholipids for its full activity. Experiments with a partially purified preparation of pea complex IV have suggested that the longer and more unsaturated the acyl groups of phospholipids are, the more effective they are in the activation.[13]

The activity of the purified enzyme decreases as the concentration of potassium phosphate in the assay medium is raised over 15 m*M*. KCl and Triton X-100 inhibit the activity, giving 50% inhibition at 50 m*M* and 0.3%, respectively. The activity can be retained when the preparation is kept standing at −70°, but it decreases rapidly at 4° (about 70% of the activity is lost overnight).

The subunit composition of sweet potato and pea complex IV has been analyzed by electrophoresis on polyacrylamide gels containing SDS and urea. Both complexes contain at least five nonidentical subunits (Fig. 3). The molecular weights of the subunits for the sweet potato complex are as follows: subunit I = 39,000, subunit II = 33,500, subunit III = 26,000, subunit IV = 20,000, and subunit V = 5700. Recent study in our laboratory has shown that subunit V is separated into three polypeptides by two-dimensional polyacrylamide gel electrophoresis, and thus the higher plant complex IV seems to be composed of at least seven subunits.

[13] M. Matsuoka and T. Asahi, *Agric. Biol. Chem.* **46,** 1405 (1982).

Section V

Peroxisomes and Glyoxysomes

[47] Isolation of Glyoxysomes and Purification of Glyoxysomal Membranes

By EUGENE L. VIGIL, TUNG K. FANG, and ROBERT P. DONALDSON

The term microbody, formerly employed to describe a single membrane structure present in electron micrographs of animal and plant cells as discussed elsewhere,[1] has given way to a more universal and functional nomenclature of peroxisomes and glyoxysomes,[2,3] which encompasses similar organelles in animal and plant cells. Glyoxysomes, as specialized peroxisomes, are subcellular organelles in germinating oil seeds which are involved in the conversion of stored triglyceride to sucrose.[4] Glyoxysomes not only contain the glyoxylate cycle enzymes but also H_2O_2-generating oxidases and catalase, enzymes characteristic of peroxisomes. Peroxisomes are membrane-enclosed organelles found in almost all eukaryotic cells where they participate in a variety of processes, such as photorespiration in leaves, long chain fatty acid oxidation in mammalian tissues, and methanol metabolism in yeast.[5,6] Purified peroxisomes or glyoxysomes are essential for metabolic studies to determine the functions of these organelles in eukaryotic cells.

Furthermore, the membrane of this type of organelle is of particular interest because of its capacity to accept electrons (reducing equivalents) released within the organelle during oxidative metabolism[7–9] and as a site for transport of metabolites or import of newly synthesized proteins.[10] With purified membranes it is now possible to address fundamental problems of electron shuttling during oxidative metabolism and determine lipid and protein composition. Information from the latter will undoubtedly be important in resolving questions of glyoxysome and peroxisome

[1] E. L. Vigil, *Subcell. Biochem.* **2,** 237 (1973).
[2] A. H. C. Huang, R. N. Trelease, and T. S. Moore, Jr., "Plant Peroxisomes." Academic Press, New York, 1983.
[3] P. Bock, R. Kramer, and M. Pavelka, "Peroxisomes and Related Particles in Animal Tissues." Springer-Verlag, Berlin and New York, 1980.
[4] T. G. Cooper and H. Beevers, *J. Biol. Chem.* **244,** 3514 (1969).
[5] N. E. Tolbert, *Annu. Rev. Biochem.* **50,** 133 (1981).
[6] H. Kindl and P. B. Lazarow, *Ann. N.Y. Acad. Sci.* **386,** 1 (1982).
[7] D. B. Hicks and R. P. Donaldson, *Arch. Biochem. Biophys.* **215,** 280 (1982).
[8] M. Huttinger, M. Pavelka, H. Goldenberg, and R. Kramar, *Histochemistry* **71,** 259 (1981).
[9] T. K. Fang, R. P. Donaldson, and E. L. Vigil, *Planta* (in press).
[10] C. Gietl and B. Hock, *Planta* **162,** 261 (1984).

biogenesis, in view of the unresolved controversy over the origin of membrane and matrix components.[11]

Early studies in Beevers' laboratory[4,12] describe and discuss the innovations that led to a general procedure used by many investigators for isolating glyoxysomes from storage tissue of germinating oil seeds. Procedures for isolation of glyoxysomes from plant cells often require separation of glyoxysomes from mitochondria and plastids. Furthermore, isolation of glyoxysomes may be complicated by adherence of endoplasmic reticulum or lipid bodies to the glyoxysomes.[13] Glyoxysomes are osmotically active and require consideration of osmotic environment during homogenization and isolation.

We have extended and modified the earlier procedures for castor bean endosperm from Beevers' laboratory and present here a method for isolating a pure fraction of glyoxysomes for direct analysis or further separation into matrix and membrane fractions. Glyoxysomes are obtained from endosperm homogenates by centrifugation on sucrose density gradients in a vertical rotor. The membranes are then separated from matrix proteins by osmotic shock and washing with sodium carbonate.[11] The glyoxysomal and membrane preparations are evaluated by enzyme measurements and electron microscopy.

Methodology

General Considerations

Isolation of a highly enriched population of glyoxysomes involves three critical stages: (1) homogenization in a slightly hypertonic buffered medium, (2) separation of organelles from a low-speed supernatant on sucrose density gradient, and (3) isolation of the glyoxysomal band relatively free of contaminating mitochondria by careful selection of the glyoxysomal fraction from the gradient. Purity can be determined by comparing the enzymatic and morphological composition of the glyoxysomal fraction to the endoplasmic reticulum (ER) and mitochondrial fractions. The glyoxysomal preparation can also be checked for intactness by enzyme assay in an isotonic medium with or without a disrupting detergent. Once glyoxysomes have been isolated and assessed for biochemical and morphological homogeneity, they can be used for separation of the limiting membrane from matrix.

[11] Y. Fujiki, S. Fowler, H. Shio, A. L. Hubbard, and P. B. Lazarow, *J. Cell Biol.* **93,** 93 (1982).

[12] R. W. Breidenbach and H. Beevers, *Biochem. Biophys. Res. Commun.* **27,** 462 (1967).

[13] E. L. Vigil, *in* "Isolation of Membranes and Organelles from Plant Cells" (J. L. Hall and A.L. Moore, eds.), p. 211. Academic Press, London, 1983.

Plant Material

Castor bean (*Ricinus communis* L.) seeds (about 400) are soaked for 10 min in diluted bleach (0.75% sodium hypochlorite, final concentration) and then rinsed in tap water. To each of two plastic pans, approximately 20 × 30 × 10 cm (washed with the dilute hypochlorite), the following are added in sequence: 4 liters of vermiculite, half the seeds evenly distributed over the vermiculite, another 2 liters of vermiculite spread over the seeds, and 2.5 liters of tap water to moisten the contents of each pan. The pans are loosely covered and placed in a dark germination chamber at 30° for 4.5 days. At this time seedlings of uniform size and development are selected. The endosperm should be firm and the extending hypocotyl root between 7 to 9 cm in length. The endosperm tissue is cleared of adhering yellow cotyledons with a small blunt spatula. The pooled pieces of endosperm are rinsed in tap water and collected on a wet paper towel on ice. A total of 80 g of endosperm from about 100 seeds is generally used for isolating glyoxysomes and, if desired, ER and mitochondria.

Glyoxysome Isolation Procedure

An equal amount of chilled endosperm (40 g) is placed in each of two 10-cm specimen dishes (heavy glass) on ice with 40 ml of grinding medium (0.4 *M* sucrose, 0.15 *M* Tricine, pH 7.5, 10 m*M* KCl, 1 m*M* $MgCl_2$, 1 m*M* sodium EDTA, pH 7.5) in each. The endosperm tissue is homogenized by chopping with four razor blades (inserted in rigid plastic tubing in pairs held about 5 cm apart) for 10 min to yield pieces of tissue less than 0.5 mm. The chopping is best performed by two persons so as not to delay centrifugation of the homogenate. The homogenate is immediately strained through four layers of cheese cloth (or sheer polyester curtain material with ca. 0.1-mm openings) held in a funnel, collected in four 50-ml centrifuge tubes (on ice), and centrifuged at 2600 rpm (1200 *g*) for 10 min in a refrigerated centrifuge. The supernatant is decanted into a flask on ice. The floating layer of fat is held back with a pipet or cotton swab. A few milliliters of supernatant are left behind so as not to disturb the pellet.

Nine milliliters of the 1200 *g* supernatant is layered carefully on the top of a combination step and linear sucrose gradient. There is usually a sufficient amount of supernatant for eight gradients. The gradients, which may be made 12 hr or more in advance and kept refrigerated, are prepared in 34-ml polyallomer tubes. Each linear gradient is formed using 10 ml of 1 *M* and 10 ml of 2 *M* sucrose, each containing 1 m*M* EDTA, pH 7.5, and mixed by using a two-chamber mixer or a pump system, e.g., the ISCO model 570 gradient maker. After the linear gradient is formed, the following are carefully layered on top: 2 ml 0.63 *M* sucrose, 2 ml 0.53 *M* sucrose, and, last, 9 ml of the 1200 *g* supernatant. The tubes are balanced in pairs, placed in a vertical rotor (Du Pont/Sorvall TV 850), and centrifuged. Centrifugation involves slow acceleration to 2000 rpm in about 3.5 min, normal acceleration to 25,000 rpm, 25 min at 25,000 rpm (50,000 *g*), normal deceleration to 2000 rpm, and slow deceleration to stop (coasting without braking). (The tubes may also be centrifuged in a swinging bucket rotor for 2 hr or in a fixed angle rotor.) The serious problem of gradient reorientation needs to be considered regardless of the rotor used.

After centrifugation the bands should be clearly visible in the tubes. Near the bottom of the tube is the glyoxysomal band, above that the slightly darker mitochondrial band, in the middle of the tube is the diffuse ER band, and at the interface between the top of the gradient and the supernatant is a fourth band. A milky fat layer is at the top. A streak on the inside (toward the axis of the rotor) of the tube indicates incomplete removal of the fat from the supernatant after the initial low-speed centrifugation. A streak on the outside of the tube represents contamination by the low-speed pellet.

The gradients are fractionated in a cold box or cold room using an ISCO model 180 equipped with a UA5 UV absorbance monitor. This instrument punctures the bottom of the tube and allows the gradient to be displaced upward through the UV monitor by 2.2 *M* sucrose. The cytosolic fraction is the first 9 ml followed sequentially by ER, mitochondria, and glyoxysome peaks. The tubes are advanced manually so that the ER, mitochondrial, and glyoxysomal bands are collected separately in tubes. The collection of glyoxysomes is begun only after the plotter on the UV monitor indicates that about half the amplitude of the glyoxysomal peak has been reached. Thus, the first part of the glyoxysomal band is not collected to minimize mitochondrial contamination. The densities of the fractions can be estimated with a refractometer (ER, 1.11 to 1.12 g/cm^3 or 0.9 *M* sucrose; mitochondria, 1.18 g/cm^3, 1.4 *M*; glyoxysomes, 1.23 g/cm^3, 1.8 *M*). The enriched glyoxysomal fraction from each gradient is about 2 ml. The pooled volume of 16 ml from eight gradients yields about 10 mg of glyoxysomal protein from 80 g of endosperm.

Once the glyoxysomal fractions from several gradients have been collected and pooled, they can be used for direct enzyme analysis (Table I), morphological analysis with electron microscopy (Figs. 1 and 2), or subjected to osmotic shock in aqueous solutions varying in ion concentration for isolation of glyoxysomal sacs (Fig. 3) or ghosts (Fig. 4). We distinguish sacs from ghosts based on whether or not some matrix remains associated with the membrane. In our experience, treatment of purified glyoxysomes with water or 0.2 *M* KCl yields glyoxysomal sacs (Fig. 3); whereas treatment with 0.1 *M* sodium carbonate[11] yields clean glyoxysomal ghosts (Fig. 4) (Table II).

TABLE I
ORGANELLES ISOLATED FROM CASTOR BEAN ENDOSPERM[a]

Enzyme	U/mg protein		
	Glyoxysomes	Mitochondria	ER
Isocitrate lyase[b]	2008	45	33
Alkaline lipase[c]	688	79	195
Cytochrome-*c* oxidase[d]	52	3253	111
Choline phosphotransferase (phosphorylcholine-glyceride transferase)[e]	39	43	721

[a] Fractions were obtained by sucrose-density gradient centrifugation in a vertical rotor. Data from T. K. Fang, R. P. Donaldson, and E. L. Vigil, *Planta* (in press).
[b] In nmol/min.
[c] In dF/min. dF, Change in fluorescence.
[d] In nmol/min.
[e] In pmol/min.

TABLE II
ENZYME ACTIVITIES IN MEMBRANE PELLETS

Enzyme	Water	KCl		Na_2CO_3
		0.2 *M*	1.0 *M*	
Catalase	12	36	58	0.04
Malate dehydrogenase	25	31	23	1.60
Malate synthase	96	29	26	0.07
Isocitrate dehydrogenase	25	30	19	0.00
Isocitrate lyase	31	37	43	0.00
Palmitoyl-CoA oxidation	84	46	22	0.00
β-hydroxyacyl-CoA dehydrogenase	56	30	16	0.60
Protein	54	48	44	5.00

[a] Glyoxysomal fractions were diluted with water, KCl, or carbonate and centrifuged as described. The enzyme activities are expressed as percentage in the pellet relative to the total in the original fraction. Data from T. K. Fang, R. P. Donaldson, and E. L. Vigil, unpublished results.

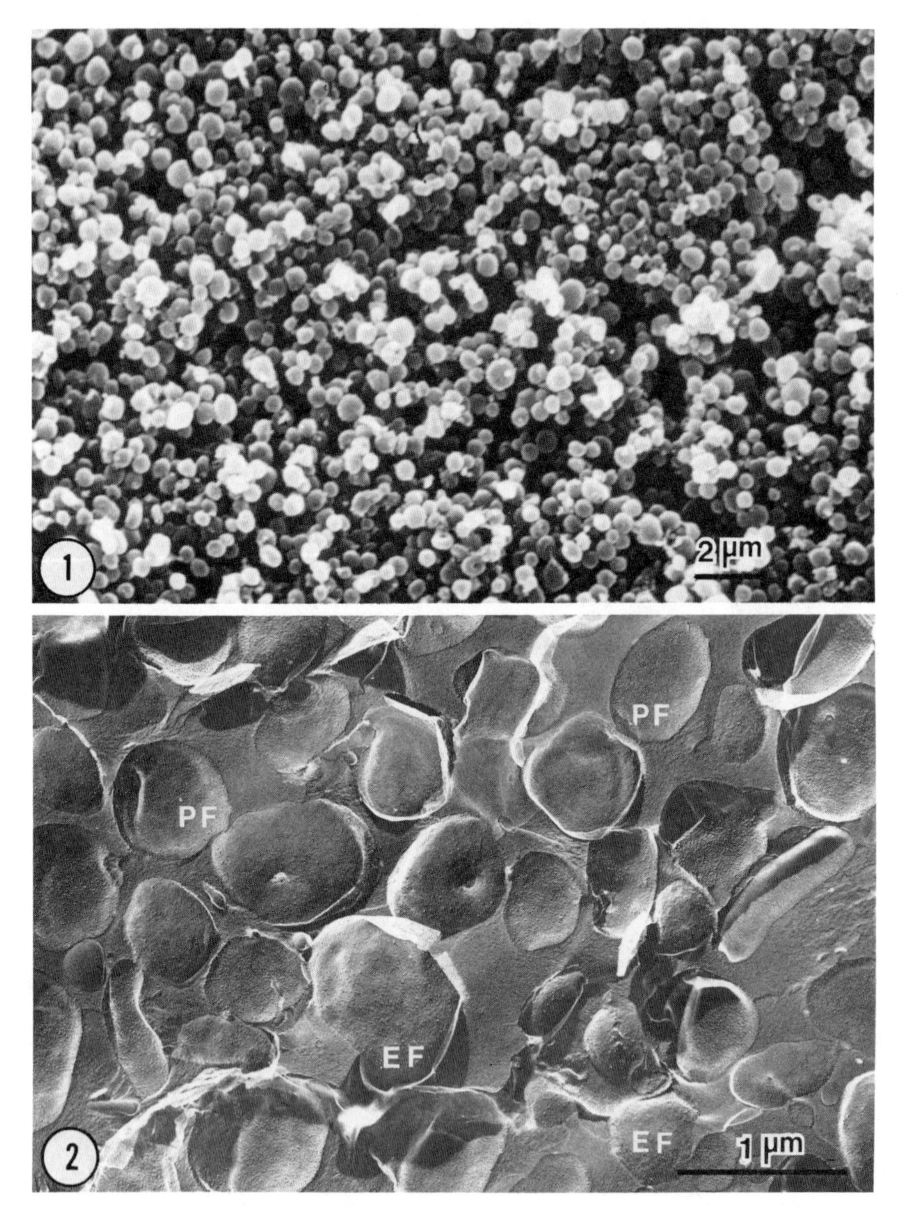

FIG. 1. Isolated glyoxysomes as viewed with SEM. The preparation is very homogeneous with little evidence of contamination by other organelles. Magnification: ×5000.

FIG. 2. Freeze-fracture replica of glyoxysomes. Preparation is similar to Fig. 1. EF, Ectoplasmic fracture face; PF, protoplasmic fracture face. Magnification: ×23,400.

Membrane Isolation Procedures

Membrane preparation begins by diluting the combined glyoxysomal fractions collected from eight gradients (about 16 ml) with 2 vol (e.g., 32 ml) of 0.63 *M* sucrose, 1 m*M* EDTA, pH 7.5. This is mixed well, held on ice for 30 min, then centrifuged at 120,000 *g* for 30 min (40,000 rpm, Du Pont/Sorvall T865 rotor). A sample of the supernatant is reserved for enzyme assays. The pellets are combined and homogenized in a 5-ml Potter-Elvehjem homogenizer with 1 ml of 0.1 *M* sodium carbonate, pH 11.5. An additional 8 ml of carbonate is added to the resuspension and this is centrifuged at 120,000 *g* for 30 min. The transluscent brown pellet is the glyoxysomal membrane ghosts. Alternatively, the combined glyoxysomal fraction may be diluted with 2 vol of 0.3 *M* KCl or water and centrifuged once, as above, to yield glyoxysomal sacs.

Purity of Isolated Glyoxysomes and Membranes

The degree of contamination of the glyoxysomal fraction by mitochondria and ER is determined by cytochrome-*c* oxidase[14] and cholinephosphotransferase (phosphorylcholine-glyceride transferase),[15] respectively. The activities for these two enzymes (per milligram protein) are compared to those in mitochondrial and ER fractions to estimate the percentage purity.

Enzyme Activities

In our experience the pooled sample from eight gradients is sufficient for an assessment of the enzymatic and morphological purity for intact glyoxysomes and isolated membranes. The enzymes which are assayed include glyoxysomal isocitrate lyase,[4] NADP-isocitrate dehydrogenase,[4] β-oxidation[4] as modified,[9] catalase,[16] malate dehydrogenase,[17] malate synthase,[18] β-hydroxyacyl-CoA dehydrogenase,[19] NADH–cytochrome-*c* reductase,[7] NADPH–cytochrome-*c* reductase,[7] NADH–ferricyanide reductase,[7] ER cholinephosphotransferase (phosphorylcholine-glyceride transferase),[15] and mitochondrial cytochrome-*c* oxidase.[14] Protein is determined by one of two methods.[20,21]

[14] B. Miflin and H. Beevers, *Plant Physiol.* **53,** 870 (1974).

[15] J. M. Lord, T. Kagawa, and H. Beevers, *Proc. Natl. Acad. Sci. U.S.A.* **69,** 2429 (1972).

[16] B. P. Gerhardt and H. Beevers, *J. Cell Biol.* **44,** 94 (1970).

[17] A. H. C. Huang and H. Beevers, *J. Cell Biol.* **58,** 379 (1973).

[18] J. A. Miernyk, R. N. Trelease, and J. S. Choinski, Jr., *Plant Physiol.* **63,** 1068 (1979).

[19] P. Overath, E. M. Raufuss, W. Stoffel, and W. Ecker, *Biochem. Biophys. Res. Commun.* **29,** 28 (1967).

[20] O. H. Lowry, N. J. Rosebrough, A. L. Farr, and R. J. Randall, *J. Biol. Chem.* **193,** 265 (1951).

[21] M. Bradford, *Anal. Biochem.* **72,** 248 (1976).

The measurements of enzyme activities and protein in the sucrose gradient fractions depend on careful sampling technique. The fraction must be well mixed. The more viscous sucrose fractions, mitochondria and glyoxysomes, cannot be accurately sampled using air displacement-type pipettors. A pipet which employs a positive displacement plunger, such as a Microman (Gilson) or a Wiretrol (Drummond), is best.

Alkaline lipase is an enzyme which indicates the presence of castor bean endosperm glyoxysomal membranes. It is not a specific marker enzyme because some activity is associated with the ER.[22] Also, it is not found in glyoxysomes obtained from other oil seeds.[23] Alkaline lipase measurement is based on the fluorometric procedure.[22] The 3-ml reaction mixture consists of 83 m*M* glycine–NaOH buffer, pH 9, 5 m*M* dithiothreitol, 0.083 m*M* *N*-methylindoxyl myristate, and about 0.05 mg glyoxysomal protein. *N*-methylindoxyl myristate is obtained from Research Organics, Inc., Cleveland, Ohio, and is dissolved in ethylene glycol monomethyl ether (22.3 mg/2.5 ml, 0.01 ml of this per assay). The reaction is followed in a spectrofluorimeter (such as a Perkin-Elmer 204) set at 430 nm excitation and 500 nm emission to monitor increase in fluorescence. Activity is expressed as change in fluorescence (arbitrary units) per minute per milligram protein in each fraction.

Choline phosphotransferase (phosphorylcholine-glyceride transferase) is the marker enzyme for the ER. The measurement involves the incorporation of [*methyl*-^{14}C]-CDP-choline (New England Nuclear, ca. 40 Ci/mol) into endogenous diglyceride.[15] The 0.25 ml reaction mixture contains 40 m*M* Tris–HCl, pH 7, 20 m*M* $MgCl_2$, 50 nCi [*methyl*-^{14}C]-CDP-choline, and 0.01 to 0.05 mg organellar protein. Each sample is incubated for exactly 30 min at 30° in a 2-ml screw-cap microcentrifuge tube (Sarstedt, West Germany). The reaction is stopped by addition of 1 ml cold ethanol. The tubes are microcentrifuged (ca. 12,000 rpm) for 5 min and the ethanol is transferred to a 10-ml glass screw-cap culture tube. Another 1 ml of ethanol is added to each microcentrifuge tube which is again microcentrifuged before transfer of the ethanol wash to the respective 10-ml glass tube. To each 2.25 ml ethanol–sample extract is added 1.5 ml chloroform and about 2.5 ml 1 *M* KCl (use a wash bottle). This is mixed and centrifuged at about 1000 rpm for 1 min to separate the phases. The upper aqueous phase is removed using a suction flask with care to leave the chloroform layer. The chloroform extract is washed three more times, once again with 2.5 ml of 1 *M* KCl and twice with 2.5 ml water. The chloroform extract is transferred to a scintillation vial and allowed to dry

[22] S. Muto and H. Beevers, *Plant Physiol.* **54,** 23 (1974).

[23] A. H. C. Huang, *Plant Physiol.* **55,** 870 (1975).

overnight in a fume hood. Scintillation fluid is added and the nanomole of CDP-choline incorporated per minute per milligram organellar protein can be calculated from the Ci/mol value provided by the manufacturer.

Cytochrome oxidase activity (for mitochondrial contamination) is determined in a modified reaction mixture[14] containing 50 m*M* potassium phosphate, pH 7.5, and 0.5 mg/ml reduced cytochrome *c*. (The cytochrome *c* is reduced by adding a few crystals of sodium hydrosulfite such that the absorbance ratio, $A_{550\text{ nm}}/A_{565\text{ nm}}$, is greater than 10.) The oxidation of cytochrome *c* in 1 ml of the mixture containing about 0.0075 mg mitochondrial protein or 0.05 mg glyoxysomal or ER protein is followed at 550 nm and the velocity of the reaction calculated from the extinction coefficient, 21.1 cm/m*M*.

Morphologic Analysis

Examination of isolated glyoxysomes requires care in handling and processing through osmication. Details are presented elsewhere[9,13] for both intact glyoxysomes and glyoxysomal ghosts. The main consideration for intact glyoxysomes is to maintain the osmotic environment above 1.4 *M* sucrose, following initial fixation at 2 *M* sucrose. Although the osmotic environment is not as critical for membrane preparations since the organelles have already been subjected to osmotic shock, we, nevertheless, add sucrose (0.5 *M*) to the glutaraldehyde solution for initial fixation. The advantage of this approach is to minimize alteration in the structure of the organelles during fixation and mimic the osmotic environment used during the treatment with KCl or carbonate.

Once samples are fixed in glutaraldehyde they can be osmicated directly in 2% osmium tetroxide with addition of an equivalent concentration of sucrose used for rinsing samples after glutaraldehyde fixation, that is 1.4 *M* for intact glyoxysomes and 0.5 *M* for isolated membrane sacs or ghosts. Following postosmication, the samples are rinsed briefly in water and dehydrated in ethanol with embedding in epoxy resin.[24] We have found that using the ferricyanide method of fixation[25] enhances contrast of organelle structure[9] (Figs. 3 and 4). Cytochemical localization of the marker enzyme catalase[26] is made within the different glyoxysomal preparations, taking into account sucrose concentration and preincubation in diaminobenzidine.[27] Examination of different portions of the fixed pellet with or without cytochemical staining for catalase provides qualitative

[24] A. R. Spurr, *J. Ultrastruct. Res.* **26,** 31 (1969).
[25] P. K. Hepler, *Eur. J. Cell Biol.* **26,** 102 (1981).
[26] S. Angermuller and H. D. Fahimi, *Histochemistry* **71,** 34 (1981).
[27] C. Bieglmayer, J. Graf, and H. Ruis, *Eur. J. Biochem.* **37,** 553 (1973).

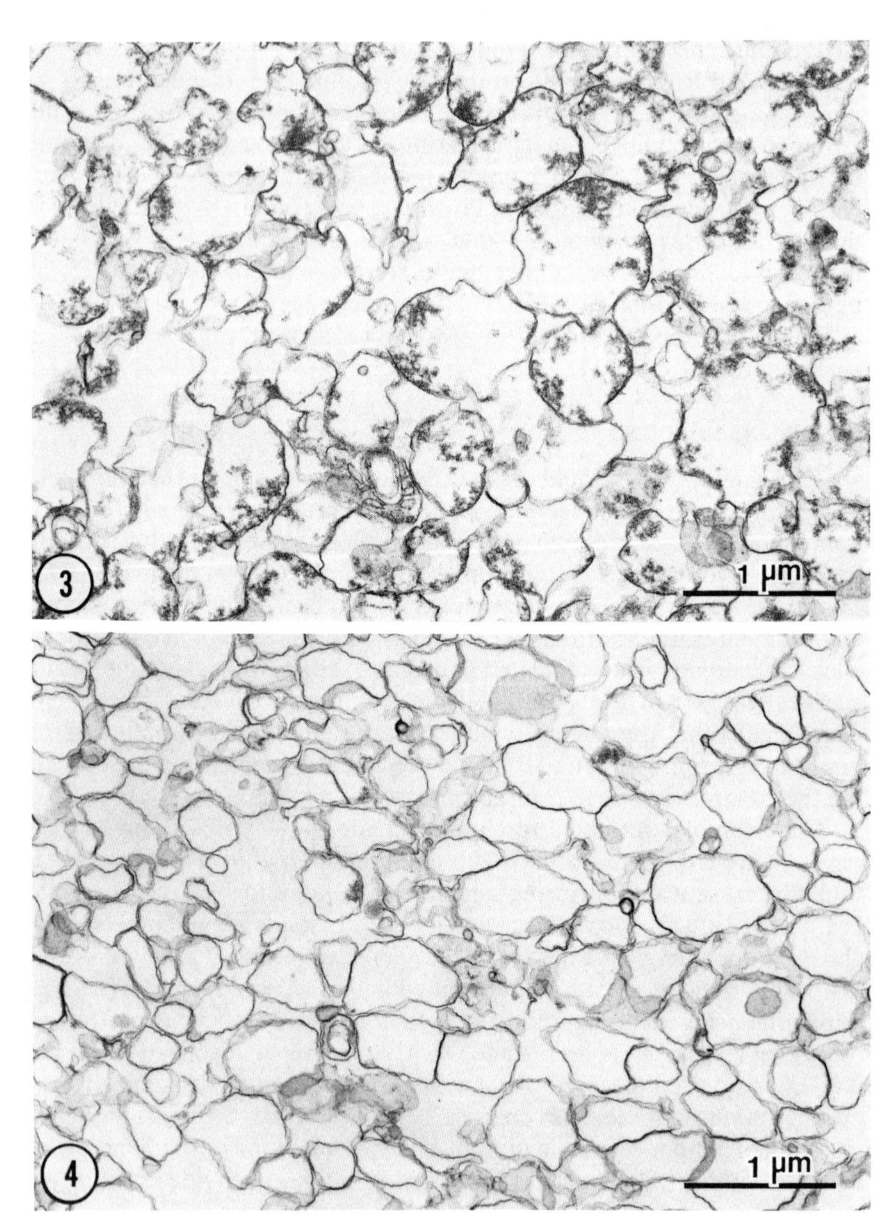

FIGS. 3 and 4. Portions of pellets following treatment with 0.2 *M* KCl (Fig. 3) or 0.1 *M* carbonate (Fig. 4). Matrix still evident inside glyoxysomal sacs (Fig. 3) is totally absent from glyoxysomal ghosts (Fig. 4). Magnification: ×21,000.

information on the purity. Quantitative stereological methods can be used[28,29] to obtain specific information on the amount of contamination by mitochondria, ER, and other organelles present in the enriched pellet of glyoxysomes.

Samples to be examined with scanning electron microscopy (SEM) are dehydrated in ethanol after glutaraldehyde fixation and cryofractured under liquid nitrogen. The fractured tissue is returned to 100% ethanol, critical point dried, and sputter-coated with gold–palladium before viewing in a SEM (Fig. 1). The procedure for preparing freeze-fracture replicas following glutaraldehyde fixation is that of Steere[30] (Fig. 2). Given the high concentration of sucrose in the glyoxysomal fraction, it is possible to prepare freeze-fracture replicas of unfixed samples to compare with fixed material.

Criteria for Purity and Intactness

Evaluation of isolated glyoxysomes follows criteria for other organelles.[31] Essentially, the isolated organelles should have similar morphology to that observed *in situ* and the pellet be free of contamination by other organelles or organisms, i.e., bacteria. Biochemical analyses provide accurate information on enzyme complement as well as amount of contamination by other organelles which may not be readily evident through morphological examinations. For example, the amount of ER contamination of glyoxysomal preparations, as determined by the specific activity of cholinephosphotransferase (phosphorylcholine-glyceride transferase), falls in the range of 5% (Table I) to 8%.[32] The specific activity of cytochrome-*c* oxidase, as a measure of mitochondrial contamination in glyoxysome fractions, is around 2% (Table I).

Studies concerning metabolite transport or import of newly synthesized protein require starting with intact and highly purified preparations of glyoxysomes. Organelles isolated using sucrose gradients often are intact (Table III), but the high sucrose interferes with measurements. Percoll allows isolation of intact glyoxysomes in 0.25 *M* sucrose,[33–36] how-

[28] M. A. Williams, "Quantitative Methods in Biology." North-Holland Publ., Amsterdam, 1977.

[29] E. R. Weibel, "Stereological Methods," Vol. 1. Academic Press, New York, 1979.

[30] R. L. Steere, *in* "Current Trends in Morphological Techniques" (J. E. Johnson, Jr., ed.), Vol. 2, p. 131. CRC Press, Boca Raton, Florida, 1981.

[31] P. H. Quail, *Annu. Rev. Plant Physiol.* **30,** 425 (1979).

[32] R. P. Donaldson and H. Beevers, *Protoplasma* **97,** (1978).

[33] C. Sauter, E. Kollmannsberger, and B. Hock, *Fachz. Lab.* **26,** 550 (1982).

[34] Z. Liang, C. Yu, and A. H. C. Huang, *Plant Physiol.* **70,** 1210 (1982).

[35] M. R. Schmitt and G. E. Edwards, *Plant Physiol.* **70,** 1213 (1982).

[36] J.-P. Schwitzguebel and P.-A. Siegenthaler, *Plant Physiol.* **75,** 670 (1984).

TABLE III
LATENCY OF ENZYME ACTIVITIES IN GLYOXYSOMES FROM A SUCROSE GRADIENT[a]

Enzyme	Intact (nmol/min/ml)	Broken (nmol/min/ml)	Latency (%)
Malate synthase	0.48	26.60	98.2
Hydroxyacyl-CoA dehydrogenase	0.23	1.52	84.9
Malate dehydrogenase	0.08	0.28	71.4

[a] The activities were measured in the presence of 1.9 *M* sucrose. The glyoxysomes were broken by including 0.2% Triton X-100 in the sucrose-containing reaction mixture. From R. P. Donaldson, R. E. Tully, O. A. Young, and H. Beevers, *Plant Physiol.* **67,** 21 (1981).

ever difficulties in reproducibility have been encountered with some tissues (Anderson and Butt, personal communication).

The methods described here yield glyoxysomes from castor bean endosperm relatively free of contaminating organelles and membranes (ghosts) relatively free of matrix (Table II). We recommend that centrifugation of sucrose gradients in a vertical rotor be used as a rapid procedure for isolating peroxisomes and glyoxysomes from other tissues. Furthermore, we suggest using carbonate washing[11] as a method to obtain glyoxysomal/peroxisomal ghosts free of matrix protein.

Acknowledgments

We thank N. Chaney (SEM), E. Erbe (freeze fracture), L. C. Frazier (TEM and sample preparations), G. Kaminski (TEM and sample preparations), and C. Pooley (photography). This work was funded by NSF PCM 8216051.

[48] Peroxisomes and Fatty Acid Degradation

By BERNT GERHARDT

Peroxisomes have been isolated from a wide range of plant tissues and are considered to be present in virtually all cells of higher plants. Plant peroxisomes have been subcategorized according to their formerly known metabolic function as glyoxysomes, leaf peroxisomes, and unspecialized peroxisomes, i.e., peroxisomes of unknown function. Fatty acid β-oxidation is now known to be a general, basic metabolic function of the higher plant peroxisome.[1]

Peroxisomal β-oxidation was first discovered in glyoxysomes,[2] which are organelles found only in fat-storing tissues of germinating seeds. Since glyoxysomes have already been treated in this series,[3,3a] the subject of this chapter will be procedures for studying fatty acid degradation by nonglyoxysomal peroxisomes. The peroxisome population of nonfatty tissues comprises approximately 1% of the organelle protein of those tissues.[4]

With the exception of the isolation on Percoll gradients, all methods described in this chapter have been successfully applied to peroxisomes from mung bean hypocotyls. Peroxisomes from spinach and pea leaves, maize seedling roots, and potato tubers have also been studied using these methods.[1,5]

Isolation of Peroxisomes

The isolation of glyoxysomes from castor bean endosperm[3,3a] and of leaf peroxisomes[6,7] has been described in this series. Therefore, isolation procedures outlined here will be restricted to the isolation of peroxisomes from mung bean hypocotyls on sucrose gradients and to the isolation of leaf and root peroxisomes on Percoll gradients.

All procedures are carried out at 0–4°. Small pieces of plant material are gently ground with ice-cold homogenization medium in a chilled mortar. Only mortars with a glazed surface are suitable. The relative contents of sucrose (w/w) or Percoll (v/v) in the solutions used for the construction of density gradients are adjusted refractometrically. After centrifugation the density gradients are separated into 1-ml fractions and the location of the peroxisomes (and other organelles) is determined by assaying the fractions for marker enzyme activities. The peroxisomes are collected as a whole fraction when the location of the peroxisomes in the gradient is known and a peroxisomal band is visible.

[1] B. Gerhardt, *Physiol. Veg.* **24,** 397 (1986).
[2] T. G. Cooper and H. Beevers, *J. Biol. Chem.* **244,** 3514 (1969).
[3] H. Beevers and R. W. Breidenbach, this series, Vol. 31, p. 565.
[3a] A. H. C. Huang, this series, Vol. 72, p. 783.
[4] B. Gerhardt, "Microbodies/Peroxisomen pflanzlicher Zellen." Springer-Verlag, Wien and New York, 1978.
[5] B. Gerhardt, *Planta* **159,** 238 (1983).
[6] N. E. Tolbert, this series, Vol. 23, p. 665.
[7] N. E. Tolbert, this series, Vol. 31, p. 734.

Isolation of Peroxisomes from Mung Bean Hypocotyls

Mung bean seedlings are grown in the dark for 2.5–3 days. Hypocotyls (20 g) are homogenized in 20 ml of homogenization medium containing 170 m*M* Tricine–NaOH, pH 7.5, 1 *M* sucrose, 10 m*M* KCl, 1 m*M* $MgCl_2$, 1 m*M* EDTA, 10 m*M* mercaptoethanol, and bovine serum albumin (9 mg/ml). The homogenate is squeezed through four layers of cheesecloth and centrifuged at 600 *g* for 10 min. From the 600 *g* supernatant, the peroxisomes are concentrated by pelleting a crude organelle fraction at 12,000 *g* (15 min). This fraction is gently resuspended in 2 ml of homogenization medium using a glass rod wrapped with cotton wool on its tip. The resuspended crude organelle preparation is layered onto a previously prepared sucrose gradient. The gradient is constructed from precooled sucrose solutions which are prepared by dissolving the sucrose in 1 m*M* EDTA, pH 7.5. A continuous gradient is formed from 20 ml of 30% and 15 ml of 60% sucrose, or alternatively a discontinuous gradient is constructed using, sequentially, 3 ml of 60%, 5 ml of 57%, 7 ml of 53%, 7 ml of 46%, 6 ml of 43%, and 6 ml of 35% sucrose. The discontinuous gradient has a loading capacity higher than that of the continuous gradient. The crude organelle fraction obtained from up to 50 g of hypocotyls can be applied to the discontinuous gradient. The continuous or discontinuous gradients are centrifuged at 83,000 *g* (average) for 3 hr. The peroxisomes are recovered in the discontinuous gradient at the interface between the 46 and 53% sucrose layer.

Peroxisomes from mung bean hypocotyls can also be isolated using a modification[8] of the procedure described by Schuh and Gerhardt.[9] The crude organelle fraction is prepared as described above and layered onto a discontinuous gradient composed of, in sequence, 5 ml of 60%, 5 ml of 53%, 5 ml of 43%, 7 ml of 38%, 7 ml of 35%, and 3 ml of 30% sucrose. The gradient is centrifuged for 7.5 min at an acceleration of 2700 revolutions $\times$ min^{-2} using a Beckman L2 65B ultracentrifuge and a SW 27 rotor (final centrifugal force 61,000 *g*, average). The plastids are very effectively separated from the other organelles. The organelle band (approximately 4 ml) stretching from the interface between the 30 and 35% sucrose layer into the upper part of the 35% sucrose layer is transferred onto a second discontinuous sucrose gradient. This gradient consists of 3 ml of 60%, 7 ml of 55%, 10 ml of 48%, 7 ml of 43%, and 5 ml of 35% sucrose. It is centrifuged at 83,000 *g* (average) for 3 hr. The peroxisomes are recovered at the interface between the 48 and 55% sucrose layer. Peroxisomal fractions isolated by this two-gradient method are uncontaminated by plastids and show, if at all, extremely low contamination by mitochondria. The

[8] H. Gerbling and B. Gerhardt, *Planta,* in press (1987).

[9] B. Schuh and B. Gerhardt, *Z. Pflanzenphysiol.* **114,** 477 (1984).

yield of peroxisomes however, amounts to only 1–3% based on catalase activity.

Isolation of Nonglyoxysomal Peroxisomes on Percoll Gradients

Before use, the Percoll stock solution (100 ml) is dialyzed against charcoal (5 g of charcoal suspended in 1250 ml of H_2O).[10] The stock solution becomes diluted due to water uptake during dialysis. It is adjusted to 80% Percoll with H_2O. An isosmotic Percoll solution is then prepared by adding 10.3 g of sucrose, 179 mg of Tricine, and 37 mg of EDTA to 100 ml of the dialyzed 80% Percoll solution. The isosmotic 80% Percoll solution is adjusted to pH 7.5. When required it is diluted with a solution consisting of 0.3 *M* sucrose, 1 m*M* EDTA, and 10 m*M* Tricine–NaOH, pH 7.5.

Leaf Peroxisomes. Peroxisomes are isolated from leaf tissue following a procedure developed by Betsche and Müller.[10] Five to 10 g of young leaves are ground with 2.5 vol of homogenization medium consisting of 0.1 *M* Tricine–NaOH, pH 7.5, 0.25 *M* mannitol, 1 m*M* EDTA, 15 m*M* mercaptoethanol, and bovine serum albumin (1 mg/ml). The homogenate is squeezed through four layers of cheesecloth and centrifuged at 2400 *g* for 5 min. From the 2400 *g* supernatant, the peroxisomes are concentrated by pelleting a crude organelle fraction at 12,000 *g* (15 min). This fraction is resuspended in 2 ml of homogenization medium and layered onto a preformed continuous Percoll gradient. The gradient is constructed from 16 ml of isosmotic 15% and 16 ml of isosmotic 50% Percoll solution over a cushion (5 ml) of isosmotic 80% Percoll. The gradient is placed in a Sorvall SS 34 rotor and centrifuged in a Sorvall RC-5C centrifuge using the automatic rate control to avoid the formation of disturbances within the gradient during acceleration and deceleration. The speed is set at 20,000 rpm (48,000 *g*), the run time at 5–8 min. The exact centrifugation time within the given limits depends upon the plant species from which the peroxisomes are isolated.

Root Peroxisomes. Peroxisomes isolated from roots on sucrose or Percoll gradients are commonly heavily contaminated by mitochondria. The separation is improved[11] if the density of the mitochondria is artificially increased by intramitochondrial deposition of an insoluble tetrazolium salt produced by the activity of succinic dehydrogenase.[12] Roots (15 g) from 3-day-old maize seedlings are ground in 30 ml of medium A supplemented with 4 m*M* cysteine. Medium A contains 0.1 *M* Tricine–

[10] T. Betsche and G. Müller, personal communication (1984).

[11] B. Gerhardt, *in* "Structure, Function and Metabolism of Plant Lipids" (P.-A. Siegenthaler and W. Eichenberger, eds.), p. 189. Elsevier, Amsterdam, 1984.

[12] G. Davis and F. Bloom, *Anal. Biochem.* **51,** 429 (1973).

NaOH, pH 7.5, 0.3 *M* mannitol, 1 m*M* EDTA, and bovine serum albumin (1 mg/ml). The homogenate is squeezed through four layers of cheesecloth and centrifuged at 600 *g* for 15 min. From the 600 *g* supernatant a crude organelle fraction is pelleted at 12,000 *g* (15 min). This fraction is resuspended in 2 ml of medium A and pelleted again (12,000 *g*, 15 min). It is then resuspended and incubated in 4 ml of medium A containing additionally 80 m*M* succinate and 10 m*M* iodonitrotetrazolium. After 30 min incubation at 30°, 4 ml of medium A is added to the incubated mixture and the organelles are pelleted at 12,000 *g* (15 min). They are washed once with 4 ml of medium A, finally resuspended in 2 ml of medium A, and layered onto a preformed continuous Percoll gradient. The gradient is constructed from 15 ml of isosmotic 60% Percoll solution and 15 ml of 0.3 *M* sucrose dissolved in a 1 m*M* EDTA solution containing also 10 m*M* Tricine–NaOH, pH 7.5. The gradient is centrifuged in a fixed angle rotor (Sorvall SS 34 rotor) at 40,000 *g* for 20 min.

Contamination

The peroxisomal fractions isolated on sucrose gradients are normally contaminated by mitochondria and especially by plastids, those isolated on Percoll gradients by mitochondria. With respect to the topic of this chapter, mitochondrial contamination of the peroxisomal fractions has to be considered in particular although mitochondrial β-oxidation activity has not yet unequivocally been demonstrated for higher plant cells.[1] A contribution of contaminating mitochondria to the β-oxidation enzyme activities measured in a peroxisomal fraction is checked by the following method.[5,13] The activity of a given β-oxidation enzyme as well as that of a peroxisomal marker enzyme (catalase, glycolate oxidase) is assayed in the peroxisomal and mitochondrial fraction from one and the same gradient. For both fractions, the ratio of the β-oxidation enzyme activity to the peroxisomal marker enzyme activity is then calculated. If this ratio determined for a sufficient number of gradients does not significantly differ (according to a statistical analysis) between the peroxisomal and mitochondrial fraction, the β-oxidation enzyme activity in the peroxisomal (and mitochondrial) fraction is considered to be associated only with the peroxisomes.

Assay Methods

Thiol reagents such as dithioerythritol (DTE), dithiothreitol (DTT), mercaptoethanol, and cysteine react with acyl-CoA's in a nonenzymatic

[13] B. Gerhardt, *FEBS Lett.* **126,** 71 (1981).

reaction.[14] Therefore, care has to be taken if thiol reagents are included in assay mixtures for measuring β-oxidation pathway activities. On principle, thiol reagents are not included in the reaction mixtures described here.

Enzyme Assays

$$R{-}CH_2{-}CH_2{-}COOH + ATP + CoASH \rightleftharpoons R{-}CH_2{-}CH_2{-}COSCoA + AMP + PP_i \quad (1)$$

$$R{-}CH_2{-}CH_2{-}COSCoA + O_2 \rightarrow R{-}CH{=}CH{-}COSCoA + H_2O_2 \quad (2)$$

$$R{-}CH{=}CH{-}COSCoA + H_2O \rightleftharpoons R{-}CHOH{-}CH_2{-}COSCoA \quad (3)$$

$$R{-}CHOH{-}CH_2{-}COSCoA + NAD^+ \rightleftharpoons R{-}CO{-}CH_2{-}COSCoA + NADH + H^+ \quad (4)$$

$$R{-}CO{-}CH_2{-}COSCoA + CoASH \rightleftharpoons R{-}COSCoA + CH_3{-}COSCoA \quad (5)$$

All five enzymes of fatty acid activation (reaction 1) and β-oxidation (reactions 2–5) can be assayed spectrophotometrically using a single-beam recording spectrophotometer with zero offset capability, a full-scale absorbance of 0.1 or 0.2, and a recorder speed of 0.5–2.0 cm/min. The reaction mixtures of 1 ml total volume in self-masking microcuvettes (1-cm light path) contain 10–100 μl of the peroxisomal fraction (approximately 5–50 μg of protein). The reactions are started by adding the substrate to the reaction mixtures, unless otherwise stated. For measurements of unspecific absorbance changes, reactions without substrate are followed in parallel cuvettes. Reaction rates are calculated for the test reaction after subtracting the absorbance change of the blank from that of the test. Rates of NAD^+ reduction and NADH oxidation, respectively, are calculated using the extinction coefficient 6220 M^{-1} cm^{-1} for NADH at 340 nm.

Acyl-CoA Synthetase (Reaction 1). The enzyme is routinely assayed by a coupled assay method.[8] The acyl-CoA formed by the synthetase is oxidized by added acyl-CoA oxidase (reaction 2) and the H_2O_2 formation in the oxidase reaction is determined by a coupled peroxidatic reaction.[15] The reaction mixture contains 175 mM Tris–HCl, pH 8.5, 10 mM ATP, 7.5 mM $MgCl_2$, 0.6 mM CoASH, 50 μM FAD, 13 mM *p*-hydroxybenzoic acid, 1 mM 4-aminoantipyrine, 1 mM NaN_3, 0.1 U acyl-CoA oxidase (from *Candida* species; Sigma), 5.3 U of horseradish peroxidase (Sigma, type II), and 0.1 mM palmitic acid. The palmitic acid is given to the reaction mixture dissolved in 10 μl acetone. The reaction is started by the addition of CoASH. The absorbance increase at 500 nm is monitored. An

[14] G. B. Stokes and P. K. Stumpf, *Arch. Biochem. Biophys.* **162,** 638 (1974).

[15] C. C. Allain, L. S. G. Poon, W. Richmond, and C. Fu, *Clin. Chem.* (*Winston-Salem, N.C.*) **20,** 470 (1974).

absorbance increase of 0.51 corresponds to the consumption of 0.1 μmol H_2O_2/ml in the coupled peroxidatic reaction.[16]

In certain cases (e.g., determination of substrate specificity) acyl-CoA synthetase cannot be assayed by coupling reaction 1 and 2 and following the formation of H_2O_2. Acyl-CoA synthetase may then by assayed by a spectrophotometric assay which either determines the AMP formation by a coupled reaction sequence[17] or couples an assay for pyrophosphate[18] to reaction 1. ATP hydrolase and pyrophosphatase activity, respectively, can substantially interfere with these assays.

Acyl-CoA synthetase is also assayed by a direct method which determines the rate of [^{14}C]palmitoyl-CoA formation from [^{14}C]palmitic acid.[19] Substrate and product are separated by thin-layer chromatography.[8] The reaction mixture contains, in a total volume of 0.5 ml, 175 m*M* Tris–HCl, pH 8.5, 10 m*M* ATP, 7.5 m*M* $MgCl_2$, 0.02% Triton X-100 (to inhibit acyl-CoA oxidase[5]), 10–30 μg protein contained in ≤200 μl of the peroxisomal fraction from a sucrose gradient, and 0.175 m*M* [1-^{14}C]palmitic acid (0.7 mCi/mmol). The added palmitic acid is dissolved in 10 μl acetone. The reaction is started by adding CoASH to a final concentration of 1 m*M*. After a reaction time of 10 min an aliquot (10–50 μl) of the reaction mixture is immediately spotted on a thin-layer cellulose plate (20 × 20 cm; 0.1 mm coating thickness) and chromatographed in a butanol : glacial acetic acid : water system (5 : 2 : 3, by volume) at room temperature. Developing the chromatogram takes 6–7 hr. After drying the chromatogram at room temperature, the chromatographic process is repeated once. Radioactive spots on the chromatogram are detected with a thin-layer chromatogram scanner. For locating palmitoyl-CoA, an aliquot of a reaction mixture is chromatographed which contains unlabeled instead of labeled palmitic acid and is mixed with labeled palmitoyl-CoA after incubation. For locating palmitic acid, an aliquot of the test reaction mixture is spotted on the thin-layer plate prior to incubation. The R_f values are approximately 0.60 and 0.95 for palmitoyl-CoA and palmitic acid, respectively. The formed palmitoyl-CoA is quantified either by integration of the palmitoyl-CoA peak of the scanner record or by counting the palmitoyl-CoA in a liquid scintillation counter.

Before transferring the palmitoyl-CoA spot from the thin-layer plate into scintillation vials the cellulose is covered with a celluloid layer.[20] A celluloid solution is prepared by dissolving 66 g of cellulose acetate and

[16] D. J. Hryb and J. F. Hogg, *Biochem. Biophys. Res. Commun.* **87,** 1200 (1979).
[17] T. Tanaka, K. Hosaka, and S. Numa, this series, Vol. 71, p. 334.
[18] J. Edwards, T. ap Rees, P. M. Wilson, and S. Morrell, *Planta* **162,** 188 (1984).
[19] J. Bar-Tana, G. Rose, and B. Shapiro, this series, Vol. 35, p. 117.
[20] P. Leusing, Doctoral Thesis, University of Münster (1985).

20 g of camphor in a solvent prepared from 750 ml acetone, 250 ml *n*-propanol, and 28 ml diethylene glycol. Five to 7 ml of the celluloid solution is spread over the cellulose layer of the thin-layer plate. After drying, the area (1 × 1 cm) of the palmitoyl-CoA spot is cut out. The obtained piece of celluloid to which the cellulose firmly adheres is dissolved in 0.5 ml dimethyl sulfoxide and then 4.5 ml scintillation fluid is added.

The radioactive assay and the three coupled spectrophotometric assays for measuring acyl-CoA synthetase activity give essentially identical results if the assays are performed with palmitic acid as substrate.[8]

Acyl-CoA Oxidase (Reaction 2). The enzyme is assayed by following the H_2O_2 formation in a coupled assay which determines the H_2O_2 in a peroxidatic reaction.[15] The assay mixture consists of 175 m*M* Tris–HCl, pH 8.5, 50 μ*M* FAD, 13 m*M* *p*-hydroxybenzoic acid, 1 m*M* 4-aminoantipyrine, 1 m*M* NaN_3 (to inhibit the peroxisomal catalase), 5.3 U of horseradish peroxidase (Sigma, type II), and 50 m*M* palmitoyl-CoA. The dye formation is monitored at 500 nm. An absorbance increase of 0.51 corresponds to the consumption of 0.1 μmol H_2O_2/ml in the coupled peroxidatic reaction.[16]

Triton X-100 at concentrations ≥0.01% inhibits the acyl-CoA oxidase[5] and is therefore not included in the reaction mixture. When acyl-CoA oxidase activities are measured by the spectrophotometric assay and a polarographic assay which determines directly the oxygen uptake by the acyl-CoA oxidase, essentially identical results are obtained,[5] demonstrating the reliability of the coupled spectrophotometric assay.

Enoyl-CoA Hydratase (Crotonase; Reaction 3). The enzyme is assayed by a coupled assay method monitoring the absorbance increase at 340 nm due to the NAD^+ reduction in the coupled 3-hydroxyacyl-CoA dehydrogenase reaction (reaction 4).[21] The reaction mixture consists of 175 m*M* Tris–HCl, pH 9.5, 2 m*M* EDTA, 2 m*M* KCN, 0.3 m*M* NAD^+, 3 U of 3-hydroxyacyl-CoA dehydrogenase (from procine heart; Sigma type III),[22] 0.05% Triton X-100, and 0.1 m*M* crotonyl-CoA.

3-Hydroxyacyl-CoA Dehydrogenase (Reaction 4). The enzyme is assayed in the backward direction, i.e., in the direction of reduction of 3-oxoacyl-CoA to 3-hydroxyacyl-CoA.[23] The reaction is followed by the absorbance decrease at 340 nm due to NADH oxidation. The reaction mixture contains 50 m*M* potassium phosphate, pH 6.8, 2 m*M* EDTA, 0.25

[21] J. R. Stern and A. del Campillo, *J. Am. Chem. Soc.* **75,** 2277 (1953).

[22] A 3-hydroxyacyl-CoA dehydrogenase preparation has to be used which does not contain the bifunctional protein exhibiting 3-hydroxyacyl-CoA dehydrogenase and enoyl-CoA hydratase activities.

[23] F. Lynen, L. Wessely, O. Wieland, and W. Rueff, *Angew. Chem.* **64,** 687 (1952).

mM NADH, 2 mM KCN, bovine serum albumin (1 mg/ml), 0.05% Triton X-100, and 0.1 M acetoacetyl-CoA.

3-Oxoacyl-CoA Thiolase (Reaction 5). The enzyme is assayed by trapping the formed acetyl-CoA with citrate synthase and oxaloacetate which is generated from malate in the NAD^+-dependent malate dehydrogenase reaction.[2] Activity is therefore measured by following NAD^+ reduction at 340 nm. The reaction mixture consists of 100 mM potassium phosphate, pH 7.5, 0.25 mM $MnCl_2$, 2 mM CoASH, 0.3 mM NAD^+, 3 mM L-malate, 12 U of malate dehydrogenase, 2.2 U of citrate synthase, 2 mM KCN, 0.05% Triton X-100, and 40 μM acetoacetyl-CoA.

Assay of Peroxisomal Overall β-Oxidation

In contrast to the known mammalian mitochondrial β-oxidation system, the peroxisomal β-oxidation system is insensitive to KCN. Therefore, peroxisomal β-oxidation is assayed by measuring acyl-CoA- or fatty acid-dependent, KCN-insensitive acetyl-CoA production or NADH formation. The NADH formation results from reaction 4.

Determination of NADH Formation. The low yield and β-oxidation activity of nonglyoxysomal peroxisomes normally requires an amplification of the formed NADH by a cycling method.[24]

Step 1: The reaction mixture (total volume, 1 ml) for the β-oxidation is prepared in small test tubes. It contains 175 mM Tris–HCl, pH 8.5, 0.15 mM NAD^+, 0.15 mM CoASH (to allow thiolytic cleavage of the formed 3-oxoacyl-CoA which otherwise may inhibit reaction 4),[25] 2 mM KCN, 20–80 μg protein contained in ≤ 300 μl of the peroxisomal fraction from a sucrose gradient, and 25 μM palmitoyl-CoA (or 0.1 mM palmitic acid, 10 mM ATP, 7.5 mM $MgCl_2$, and a CoASH concentration raised to 0.6 mM). The reaction is run at 25° for 15 min.

Step 2: The β-oxidation reaction is terminated by adding 1 ml of 0.2 N NaOH to the reaction mixture which is then boiled for 4 min to destroy the remaining NAD^+. The mixture is cooled to room temperature and centrifuged at 4500 g for 10 min. An aliquot (0.5 ml) of the supernatant is neutralized by adding an equal volume of 0.2 N HCl. During neutralization the sample is vigorously shaken to avoid destruction of NADH by local acidification.

Step 3: An aliquot (0.1–0.3 ml) of the neutralized sample is used in the cycling assay for NADH. The cycling assay is performed as described in this series.[26] The cycling reaction proceeds for 10 min. It is recommended

[24] I. Papke and B. Gerhardt, unpublished results (1984).

[25] T. Osumi, T. Hashimoto, and N. Ui, *J. Biochem.* (*Tokyo*) **87,** 1735 (1980).

[26] H. Matsumura and S. Miyachi, this series, Vol. 69, p. 465.

to use black test tubes and an alcohol dehydrogenase purified from bound NAD^+/NADH for the cycling reaction.

A blank which does not contain palmitoyl-CoA (or palmitic acid) in the β-oxidation mixture (step 1) is also carried through steps 1 to 3. If no standard curve has been prepared,[26] rates of NADH formation are calculated using an internal standard prepared as follows: 50 pmol NADH dissolved in 0.1 ml H_2O is included in a cycling reaction mixture (step 3) prepared with an aliquot of the neutralized blank. This standard is carried through step 3 in parallel with the sample and the blank.

The rate of NADH formation is increased when a system for trapping the formed acetyl-CoA is included in the β-oxidation mixture (step 1). Such a system consists of 0.5 m*M* aspartate, 0.5 m*M* 2-oxoglutarate, 0.1 m*M* pyridoxalphosphate, 0.02 U of L-aspartate : 2-oxoglutarate aminotransferase, and 0.2 U of citrate synthase.

Determination of Acetyl-CoA Formation. The reaction mixture (total volume, 0.6 ml) for the β-oxidation contains 175 m*M* Tris–HCl, pH 8.5, 0.15 m*M* NAD^+, 2 m*M* KCN, 0.1 m*M* [U-^{14}C]palmitic acid (2 mCi/mmol), 10 m*M* ATP, 7.5 m*M* $MgCl_2$, 1 m*M* CoASH, and 20–50 μg protein contained in ≤200 μl of the peroxisomal fraction from a sucrose gradient. The added palmitic acid is dissolved in 10 μl acetone. After a reaction time of 15 min at 25° an aliquot (10–50 μl) of the reaction mixture is immediately spotted on a thin-layer cellulose plate. The chromatogram is developed and acetyl-CoA and palmitic acid are detected correspondingly to the procedures described in the radioactive acyl-CoA synthetase assay. The R_f values for palmitic acid and acetyl-CoA are approximately 0.95 and 0.45, respectively. The formed acetyl-CoA is quantified correspondingly to the palmitoyl-CoA in the acyl-CoA synthetase assay.

Protein Determination

For specific activity measurements, protein in fractions from sucrose gradients is determined by a modification[5] of the Lowry method.[27] Protein in fractions from Percoll gradients is determined by following the procedure of Vincent and Nadeau.[28]

Acknowledgment

Studies in the author's laboratory were supported by the Deutsche Forschungsgemeinschaft.

[27] O. H. Lowry, N. J. Rosebrough, A. L. Farr, and R. J. Randall, *J. Biol. Chem.* **193,** 265 (1951).

[28] R. Vincent and D. Nadeau, *Anal. Biochem.* **135,** 355 (1983).

[49] Proteins and Phospholipids of Glyoxysomal Membranes from Castor Bean

By HARRY BEEVERS and ELMA GONZÁLEZ

For an accurate analysis of protein and phospholipid components of membranes of an organelle the first requirement is that the organelle fraction should be uncontaminated with other cell constituents. In the particular case of glyoxysomes from castor bean endosperm this requirement is close to being realized, since this plant material yields a remarkably good separation of major organelles in a single linear sucrose density gradient. On such a gradient the glyoxysomes are found as a clear separate protein band at equilibrium density 1.25 g/ml and they are virtually uncontaminated by other organelles or membranes, as revealed by measurements of marker enzymes and the uniformity of structures in electron micrographs. The glyoxysomes are seen in such pictures as spherical organelles surrounded by a single unit membrane. There are no membranes within the amorphous matrix.[1]

Analyses for protein and lipid membrane constituents have been performed on peak fractions of glyoxysomes identified by the coincident distributon of constituent enzymes (alkaline lipase, catalase, enzymes of β-oxidation, and particularly malate synthase and isocitrate lyase, the distinctive enzymes of the glyoxylate cycle).[2]

The second requirement for an accounting of membrane constituents is that the matrix components are completely removed. In the early work on distribution of enzymes within glyoxysomes it was shown that disruption by osmotic shock (by diluting the preparation from 54 to 18% sucrose) removed ~60% of the protein.[3,4] Most of the content of several glyoxysomal enzymes was solubilized by this procedure, but the membranes, recovered as ghosts, retained 50% or more of several enzymes and 90% of malate synthase and citrate synthase. These enzymes were recovered with the membranes (marked by phosphatidyl[^{14}C]choline prelabeled by a prior exposure of the tissue to [^{14}C]choline) when the ghosts were recentrifuged on a sucrose gradient; the equilibrium density was 1.21 g/ml.[3] However, when the membranes were subsequently exposed to 0.2 *M* KCl the bulk of each of the above enzyme activities was solubi-

[1] R. W. Breidenbach, A. Kahn, and H. Beevers, *Plant Physiol.* **43,** 705 (1968).
[2] H. Beevers and R. W. Breidenbach, this series, Vol. 31A, p. 565.
[3] A. H. C. Huang and H. Beevers, *J. Cell Biol.* **58,** 379 (1973).
[4] C. Beigelmayer, J. Graf, and H. Ruis, *Eur. J. Biochem.* **39,** 379 (1973).

lized, whereas almost all of the phosphatidylcholine was recovered in the pelleted fractions.[3] Thus it was clear that none of the enzymes studied was an integral membrane protein although a fraction of the total content of several enzymes remains even after osmotic breakage and salt treatment.

The pellet, representing the crude membrane fraction, contains ca. 65% protein by weight and this has been used for several analytical procedures. It is emphasized that although the phospholipid analyses are not affected by whatever nonmembrane proteins that remain in the crude membrane pellet, these have to be reckoned with in the protein accounting. We have shown, for example, that as much as 30% of the glyoxysomal catalase remains with the crude glyoxysomal membrane fraction, but that this is removed by 0.06% deoxycholate which does not solubilize integral membrane proteins.[5] The catalase so removed behaves in other respects like the catalase extracted with other matrix proteins in the preparation of the crude glyoxysomal membrane. In the meantime an additional glyoxysomal enzyme was discovered and this enzyme, the alkaline lipase, in contrast to others present on the crude membrane fraction, is retained on the membranes during osmotic treatment, in extraction with 0.1 *M* KCl and 0.06% deoxycholate,[6] and in extraction with 0.1 *M* Na_2CO_3 (E. Gonzalez, unpublished). Thus this enzyme, subunit M_r 62,000, is an integral protein of the glyoxysomal membrane and it has been purified.[7]

Lipid Analysis of the Crude Membrane Fraction

Table I shows the lipid components extracted by the conventional chloroform : methanol (2 : 1, v/v) solvent. Phospholipids in the extract were separated by TLC on silica gel 60 plates (E. Merck) (chloroform : methanol : acetic acid : water, 65 : 50 : 5 : 3, v/v) and neutral lipids in benzene : diethyl ether : ethanol : acetic acid (50 : 40 : 2 : 0.2, v/v).[8] In addition to the phospholipids, which accouned for 84 mol% of the total fatty acids in the membrane, there was a notable amount of free fatty acid, much smaller amounts of di- and triacylglycerols, as well as stigmasterol and β-sitosterol. The major phospholipids were phosphatidylcholine, phosphatidylethanolamine, and phosphatidylinositol, with much smaller amounts of phosphatidylglycerol and phosphatidic acid and phosphatidylserine. An analysis of the bilayer distribution of the major phospho-

[5] M. Maeshima and H. Beevers, unpublished (1984).

[6] S. Muto and H. Beevers, *Plant Physiol.* **54,** 23 (1974).

[7] M. Maeshima and H. Beevers, *Plant Physiol.* **79,** 489 (1985).

[8] R. P. Donaldson and H. Beevers, *Plant Physiol.* **59,** 259 (1977).

TABLE I
LIPID COMPOSITION OF CRUDE GLYOXYSOMAL MEMBRANE[a]

Constituent	nmol/mg membrane protein
Phosphatidylcholine	342.6
Phosphatidylethanolamine	193.8
Phosphatidylinositol	56.9
Phosphatidylglycerol[b]	15.7
Phosphatidylserine	9.0
Cardiolipin	3.3
Free fatty acid	97.9
Diacylglycerol	6.1
Triacylglycerol	5.3
Stigmasterol	4.1
β-Sitosterol	2.4

[a] R. P. Donaldson and H. Beevers, *Plant Physiol.* **59,** 259 (1977).
[b] Includes phosphatidic acid.

lipids based on susceptibility to phospholipase A_2 has been made.[9] Each of the phospholipids has a distinctive set of fatty acid substituents (Table II) but in all three, linoleic acid is the major substituent, followed by palmitic and oleic acids. Linolenic acid is a minor component.

Polypeptides in Highly Purified Glyoxysomal Membranes

KCl solutions have been used commonly to remove peripheral proteins from membranes, including those obtained after osmotic breakage of glyoxysomes.[3] However, careful analyses have shown that this is not completely effective.[10] The carbonate procedure (0.1 *M* Na_2CO_3, pH 11.5) developed by Fujiki *et al.* to purify peroxisomal membranes from rat liver[11] has been applied to castor bean organelles and shown to be superior to extraction by KCl. Membrane pellets so obtained are washed in water before solubilization in sample buffer for SDS–PAGE or in 0.1% trifluoroacetic acid and 50% acetonitrile for reversed-phase HPLC.[12] An-

[9] T. Cheeseborough and T. S. Moore, *Plant Physiol.* **65,** 1076 (1980).
[10] R. P. Donaldson and H. Beevers, *Protoplasma* **97,** 317 (1978).
[11] Y. Fujiki, A. L. Hubbard, B. Fowler, and P. B. Lazarow, *J. Cell Biol.* **93,** 97 (1982).
[12] E. Gonzalez, R. P. Donaldson, and V. L. Alvarez, unpublished.

TABLE II
FATTY ACID SUBSTITUENTS OF MAJOR PHOSPHOLIPIDS OF CRUDE GLYOXYSOME MEMBRANE[a]

	mol% total fatty acid				
Lipid	16:0	18:0	18:1	18:2	18:3
Phosphatidylcholine	22	9	17	44	5
Phosphatidylethanolamine	26	8	9	52	4
Phosphatidylinositol	35	10	8	38	4

[a] R. P. Donaldson and H. Beevers, *Plant Physiol.* **59,** 259 (1977).

other approach which has been used to separate integral polypeptides depends on their differential solubility in liquid detergent. This phase separation in Triton X-114, however, is a lengthy procedure requiring careful control of the temperature of the sample solutions.[13] It has been used previously to separate the integral polypeptides of organelles from castor bean.[14]

Figure 1 shows the separation by SDS–PAGE of polypeptides remaining after extraction of crude glyoxysomal membranes with (A) 1.0 *M* KCl, (B) 0.1 *M* Na_2CO_3, and (C) Triton X-114. The patterns are generally similar and although the resolution of SDS–PAGE is inadequate to allow an accounting of all of the membrane polypeptides, 22 bands can be identified. These constituents are listed by molecular weight in Table III, which also includes separate entries for the 62K and two 58K polypeptides which are not completely resolved in Fig. 1. It is clear from these results that the carbonate procedure was as effective as extraction in Triton X-114 and less complicated. Another advantage of the carbonate procedure is that it does not completely disrupt the membranes; membrane sheets are still observed in electron micrographs.[11,12]

Carbonate-extracted glyoxysomal membranes have been examined in a two-dimensional separation system in which proteins are separated on the basis of their hydrophobicity by reversed-phase HPLC and the peaks further resolved by molecular weight on SDS–PAGE.[12] For chromatography on HPLC, 300 to 500 μg of the protein was injected into a Brownlee RP 300 C_{18} column (Rainin Instrument Co., Emeryville, CA) with column guard or a Vydac C-4 column (Western Analytical Co., Temecula, CA). Elution from the reversed-phase column was effected by a linear acetoni-

[13] C. Bordier, *J. Biol. Chem.* **256,** 1604 (1981).
[14] M. J. Conder and J. M. Lord, *Plant Physiol.* **72,** 547 (1983).

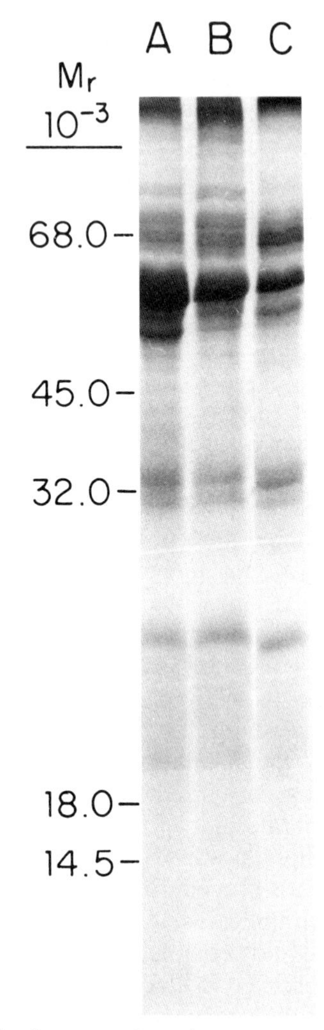

FIG. 1. Polypeptides of the glyoxysomal membrane. For this comparison isolated glyoxysomes were osmotically ruptured and the membranes preliminarily washed by 0.5 *M* KCl. Subsequently, three membrane aliquots were washed by 1.0 *M* KCl (A), 0.1 *M* Na_2CO (B), or extracted in Triton X-114 (C). Samples of the remaining protein (50 μg) were loaded on SDS–polyacrylamide gels and run under reducing conditions. Polypeptide bands were stained with Coomassie blue R-250.

trile gradient in 0.1% trifluoroacetic acid in HPLC grade water. Rate of flow was generally 1 ml/min and the eluting proteins were monitored at 205 nm by a variable wavelength detector.[12] Freeze-dried protein fractions were solubilized in sample buffer and loaded onto acrylamide gels in

TABLE III
POLYPEPTIDE COMPOSITION OF THE GLYOXYSOMAL MEMBRANE[a]

Size (from Fig. 1) ($M_r \times 10^{-3}$)	Ref. to glycoconjugate
90.0	*b, c*
84.0	*a*
80.0	*a*
77.5	
74.5	
73.0	
70.0	*a, b*
67.5[d]	*c*
65.5[d]	*a*
64.8	
62.0[d]	
60.5	*c*
58.0 (2×)	
56.5[d]	*b*
53.7	*a*
50.7	
47.0	*a, b*
44.5	
43.8	*b*
36.5[d]	
34.5	
24.8[d]	
18.0	*a*

[a] E. Gonzalez, R. P. Donaldson, and V. L. Alvarez, unpublished.
[b] U. Bergner and W. Tanner, *FEBS Lett.* *131,* 68 (1981).
[c] J. M. Lord and L. M. Roberts, *Ann. N.Y. Acad. Sci.* **386,** 362 (1982).
[d] Abundant components.

the usual way.[15] Polypeptides can be described by their elution time from the HPLC column (or percentage acetonitrile) and by their molecular weight (from SDS–PAGE).

HPLC analysis has shown that a few of the polypeptides which persisted after extraction in 1.0 *M* KCl (47K, 59K, and 65K) were removed from the membranes by 0.1 *M* Na_2CO_3. The two-dimensional separation

[15] U. K. Laemmli, *Nature (London)* **227,** 680 (1970).

has allowed the resolution of some complex bands seen in Fig. 1. For example, the 58K components are resolved because one is much more hydrophobic than the other. The five bands in the 54K–62K cluster (Fig. 1) are resolved by SDS–PAGE only after prior separation on HPLC. Thus the 62K alkaline lipase[7] can be resolved.

Approximately 12 of the glyoxysomal membrane polypeptides have been identified as glycoproteins.[12,16,17] Table III indicates their probable identity. Although alkaline lipase was initially thought not to be glycosylated,[7] recent work suggests that it is conjugated to an oligosaccharide of the high mannose type.[18]

[16] U. Bergner and W. Tanner, *FEBS Lett.* **131,** 68 (1981).
[17] J. M. Lord and L. M. Roberts, *Ann. N.Y. Acad Sci.* **386,** 362 (1982).
[18] E. Gonzalez, M. D. Brush, and M. Moeshima, *in* "Peroxisomes in Biology and Medicine" (H. D. Fahimi and H. Sies, eds.), pp. 194–198. Springer-Verlag, Heidelberg, 1987.

Section VI

Nuclei, Endoplasmic Reticulum, and Plasma Membrane

[50] Isolation of Nuclei from Soybean Suspension Cultures

By Paul Keim

Nuclei have been isolated from plants in order to purify nuclear DNA,[1,2] to study RNA transcription *in vitro*,[3-9] and to study DNA synthesis *in vitro*.[10] A variety of tissue types have been used as sources of nuclei including, embryos,[3] whole seedlings,[8] and tissue culture cells.[5,10] Different protocols have been used, but all involve breaking the cell wall and cell membrane, and then separating the nuclei from cytoplasmic components.

Breaking the cell wall has been accomplished by mechanical means (blenders, mortar and pestle, etc.) or by forming protoplasts with enzymatic digestion. After breaking the cell wall nonionic detergents, such as Triton X-100, have been used to solubilize the plasma membrane.[3,8,10] Triton has been reported to disrupt the nuclear membrane,[8] but does not appear to inhibit the nuclear processes being studied.[4,7,8,10]

Nuclei have been purified from the crude lysate by filtration and differential centrifugation. Filtering removes unbroken cells and larger debris while differential centrifugation helps separate small debris and soluble cytoplasmic contaminants. These techniques have been adequate for DNA isolation and *in vitro* DNA synthesis,[10] but *in vitro* transcription studies were plagued with ribonuclease contamination. Further purification on Percoll density gradients removes this nuclease.[3]

Generally, divalent cations such as Ca^{2+} and Mg^{2+} are added to the lysis and purification buffers in order to stabilize the nuclei. When these ions are present many nucleases are active. One innovation has been to stabilize the nuclei with polyamines such as spermine and spermidine.[5] This allows the addition of chelators to sequester metal ions and thus inhibit nucleases during the isolation procedure.

[1] H. Stern, this series, Vol. 12B, p. 100.
[2] D. E. Cress, P. J. Jackson, A. Kadouri, Y. E. Chu, and K. G. Lark, *Planta* **143,** 241 (1978).
[3] D. S. Luthe and R. S. Quatrano, *Plant Physiol.* **65,** 305 (1980).
[4] D. S. Luthe and R. S. Quatrano, *Plant Physiol.* **65,** 309 (1980).
[5] L. Willmitzer and K. G. Wagner, *Exp. Cell Res.* **135,** 69 (1981).
[6] T. F. Gallagher and R. J. Ellis, *EMBO J.* **1,** 1493 (1982).
[7] J. Silverthorne and E. M. Tobin, *Proc. Natl. Acad. Sci. U.S.A.* **81,** 1112 (1984).
[8] E. Mosinger and E. Schafer, *Planta* **161,** 444 (1984).
[9] E. Mosinger, A. Batschauer, E. Schafer, and K. Apel, *Eur. J. Biochem.* **147,** 137 (1985).
[10] R. Roman, M. Caboche, and K. G. Lark, *Plant Physiol.* **66,** 726 (1980).

METHODS IN ENZYMOLOGY, VOL. 148

The nuclear purity and biochemical integrity has been judged by a variety of methods. One of the easiest is microscopic examination which detects unlysed cells and larger cellular debris. While nuclei may appear pure by microscopic examination, more sensitive enzyme assays can detect cytoplasmic contamination. Ribonucleases can be detected by assays and are removed from the nuclear preparation after Percoll density gradient centrifugation.[3] (Absolutely RNase-free nuclei may not be necessary as ribonuclease inhibitors are available that can be used during *in vitro* transcription.)[8]

Some cytoplasmic DNAs can be separated from nuclear DNA on cesium chloride gradients. This makes it is possible to detect and quantitate the amount of contamination by these DNAs (Fig. 1). However, this technique has limited sensitivity and DNA hybridization[11] can detect cytoplasmic DNA where cesium chloride profiles indicated none.[12]

The most important criterion of nuclear integrity is the product of the *in vitro* reactions. It has been found that nuclei isolated with Percoll gradients synthesize full-length transcripts.[4] In addition, this transcription is selective in accordance with the developmental status of the cells contributing the nuclei. For example, when cells or tissue are treated with specific light regimes to alter their gene expression, the transcription in isolated nuclei reflects the *in vivo* gene expression.[6-9] Likewise, DNA is synthesized *in vitro* as small Okazaki fragments which are processed into mature DNA.[10] This is the same mechanism as DNA synthesis *in vivo*.[13] When cells are treated with UV light they switch from replicative to repair DNA synthesis.[13] Nuclei isolated from UV-treated cells show reduced DNA synthesis activity, consistent with a switch to repair synthesis.[10] The fact that isolated nuclei are accurately reflecting their *in vivo* synthetic status has justified the use of *in vitro* studies to examine the biochemical basis for cellular phenomena.

Presented here are two protocols for isolating nuclei from cell suspension cultures. The first uses protoplasting enzymes to remove the cell wall, followed by lysis of the protoplasts with Triton X-100.[10] The yield of nuclei is high (40%) and free of most cytoplasmic DNAs (Fig. 1). Because of these attributes most DNA replication studies have been done with this protocol. An alternative procedure, which uses a milk homogenizer to break the cell wall, is necessary when cell cultures fail to form protoplasts. The yield of nuclei is low (ca. 1%) and cytoplasmic DNA contamination is higher (Fig. 1). This protocol has been used mainly for isolating DNA from nonprotoplasting cell lines.

[11] E. Southern, *J. Mol. Biol.* **98,** 503 (1975).

[12] P. Keim, personal observations (1984).

[13] A. Kornburg, "DNA Replication." Freeman, San Francisco, California, 1980.

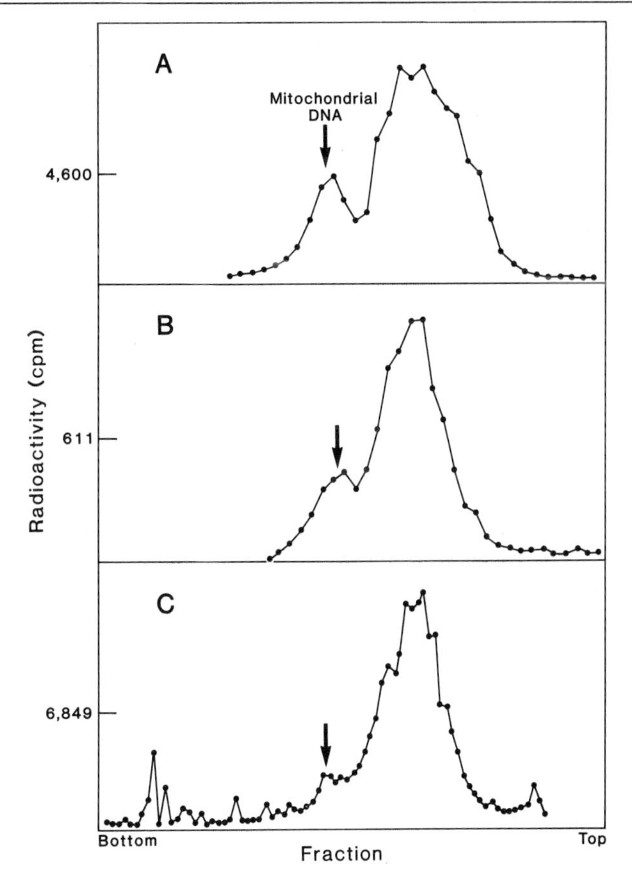

FIG. 1. Hoechst dye cesium chloride isopycnic equlibrium centrifugation of soybean DNA. Soybean cells were radioactively labeled with tritiated thymidine 24 hr prior to lysis. DNA was isolated by three different techniques and banded on cesium chloride gradients in the presence of Hoechst dye (see text for details). (A) The total cellular DNA was isolated by lyzing protoplasts directly onto the gradients. (B) DNA was isolated by protocol II. (C) DNA was isolated by protocol I. The location of mitochondrial DNA was determined by molecular hybridization.

Protocol I. Isolating Nuclei from Protoplasts

1. Cells growing in suspension with a 24-hr doubling time are diluted to a density of 1.5×10^5/ml in a volume of 400 ml.

2. Two days later the cells are removed from the media by centrifugation (5 min at 500 *g*) and resuspended in 400 ml of protoplasting media containing 1% cellulase Onozuka R10 (Kinki Yakult, Nishinomiya, Japan) and 0.02% Rhozyme HP150 (Rohm and Haas, Philadelphia, PA) at 28°.

This medium uses KCl as an osmoticum to maintain the forming protoplasts.

Protoplasting Medium (PM)

200 m*M* KCl
2 m*M* NH_4NO_3
3 m*M* $CaCl_2$
1 m*M* KH_2PO_3
50 m*M* Sucrose

3. The cells are incubated at 28° with shaking. The cells should be completely protoplasted in 1 to 3 hr.

4. The protoplasts are filtered through a 60-μm nylon mesh to remove any unprotoplasted cell clumps and concentrated by centrifugation (500 *g* for 5 min).

5. The pellets are resuspended in 50 ml of 33% Percoll in PM. The 50 ml are divided into two tubes and overlayed with 5 ml of PM. Centrifuge at 2000 *g* for 10 min. Healthy protoplasts band at the 0/20% interface. This band is removed with a pipet.

6. The protoplasts are diluted with PM to 100 ml and pelleted at 2000 *g* for 10 min. The pellet is resuspended in 100 ml of PM and sedimented as before. The final protoplast pellet is resuspended in 24 ml of ice cold LB. All subsequent steps are on ice.

Lysis Buffer (LB)

25 m*M* Tris, pH 8.0
3 m*M* $CaCl_2$
60 m*M* KCl
200 m*M* Sorbitol

7. The protoplast suspension is brought to 0.1% Triton X-100 and shaken vigorously to break the cell membrane. Lysis should be monitored by phase-contrast microscopy. If the lysis is incomplete more Triton X-100 can be added (ca. 0.1%) followed again by vigorous mixing.

8. The lysate is diluted wth 2.5-fold excess LB. Unbroken cells and large debris are then cleared from the lysate by low-speed centrifugation (100 *g* for 15 sec). The nuclei are concentrated from this supernatant by centrifugation at 2000 *g* for 10 min. The nuclear pellets are resuspended in 50 ml of LB and resedimented at 2000 *g* for 5 min. This is the final nuclear pellet and is resuspended in 1 ml of LB. If these nuclei are to be used for DNA preparation, resuspend the nuclei in TES and proceed with step 8 of protocol II (see below).

9. The *in vitro* DNA synthesis reaction is initiated by mixing the nu-

clear suspension with an equal volume of a reaction mixture. Both nuclei and reaction mixture are briefly warmed to 28° and then combined to initiate the reaction. The total reaction volume should be 2 ml.

Reaction Mixture

50 m*M* Tris–HCl, pH 8.0
24 m*M* $MgCl_2$
14 m*M* EGTA
400 m*M* Sucrose
10 m*M* Dithiothreitol
2 m*M* ATP
300 μM each dGTP, dCTP, and dATP
6 μM [α-^{32}P]TTP, specific activity 8 Ci/mmol

Comments on Procedure I

Cells. The culture must be in midlog growth for a good yield of protoplasts and to obtain a high level of *in vitro* DNA synthesis. A 400-ml culture at a density of 4.0×10^5 cells/ml yields 6.4×10^7 nuclei (40% yield).

Labeled Deoxynucleotide Triphosphates. For many experiments [α-^{32}P]TTP is desirable. Only fresh label can be used since older preparations (greater than a week) accumulate inhibitors of DNA synthesis. Tritiated TTP doesn't accumulate these inhibitors and is stable for several months. In a standard reaction (3 μM [α-^{32}P]TTP at a specific activity of 8 Ci/mmol) 2.4×10^6 cpm are incorporated into DNA. *In vitro* nascent DNA can be used as a hybridization probe. To do this, the specific activity of the TTP must be increased about 100-fold.

Nuclease Contamination. In step 6 the protoplasts are lysed with Triton X-100. If lysis is not complete, the remaining protoplasts may copurify with the nuclei and contribute nucleases to the reaction mix. In such cases the *in vitro* product and template DNA maybe degraded to a small size (Fig. 2).

Protocol II. Isolation of Nuclei Using a Milk Homogenizer

1. A culture in mid to late log phase is sedimented at 500 *g* for 5 min. The supernatant is discarded.
2. The cells are resuspended in the original volume of ice-cold LB. All subsequent steps are at 4° or on ice.
3. The cells are forced though a precooled homogenizer (Sargent-Welch Scientfic Company, Cat. No. S-61616).

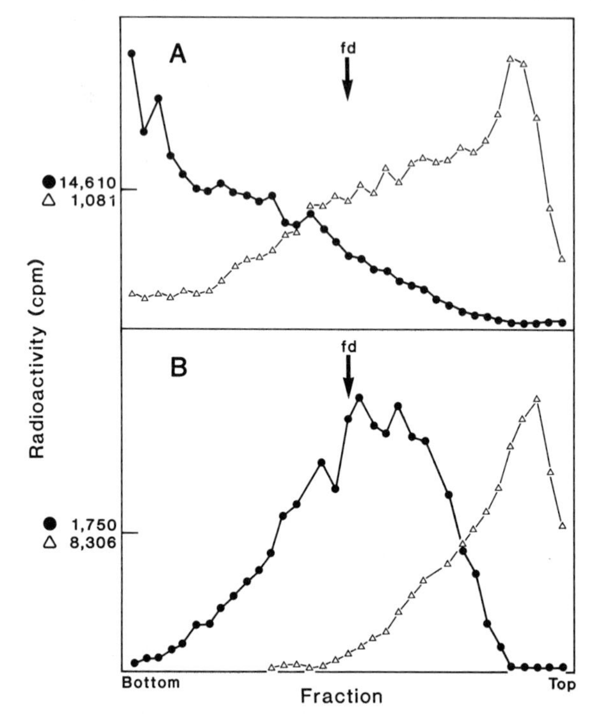

FIG. 2. Sedimenation velocity on alkaline sucrose gradients. DNA from two *in vitro* reactions was analyzed by sedimentation velocity.[2] Degradation can be observed in both the bulk DNA (●) and the *in vitro* synthesized DNA (△) when the nuclei are contaminated with nucleases (B). The fd phage DNA marker is 16 S.

4. The lysate is filtered though a 60-μm mesh. Unbroken cell clumps will catch on the screen.

5. The nuclei are sedimented at 2000 g for 10 min.

6. The pellet is resuspended in one-quarter of the original volume of LB and sedimented at 2000 g for 10 min. This wash is repeated once more.

7. If these nuclei are to be used for *in vitro* DNA synthesis, they are resuspended in 1 ml of LB. Subsequent steps are identical to protocol 1 starting at step 8.

8. If the nuclei are to be used as a DNA source, resuspend the nuclei in 1 ml TES.

TES

50 m*M* Tris, pH 8.0
20 m*M* EDTA
150 m*M* NaCl

9. Sodium dodecyl sulfate (SDS) is added to a concentration of 1.0% (w/v) and heated to 65° for 5 min. The nuclei are lysed by this treatment.

10. Predigested Pronase is Added to a concentration of 1.5 mg/ml and digested at 37° for 1 hr.

11. DNA is further purified by CsCl equilibrium centrifugation. Add 8.7 g of cesium chloride to 6.8 g of sample and centrifuge at 40,000 rpm for 40 hr in a Beckman Ti-50 rotor.

Comments on Procedure II

Low Yield. The breakage of cells by this technique is very inefficient. It is possible to increase the breakage by treating cells with protoplasting enzymes as in steps 1, 2, and 3 of protocol I. Even if particular cell lines won't form protoplasts, this treatment weakens their cell wall and increases the breakage by the homogenizer.

Analysis of in Vitro DNA Synthesis

Reaction Kinetics. In order to monitor DNA synthesis, periodically remove 50-μl aliquots from the reaction and mix with 0.5 ml of stop buffer (0.5 *N* NaOH, 0.5 mg/ml salmon sperm DNA, 10 mg/ml SDS, 20 m*M* sodium pyrophosphate). Add 110 μl 100% trichloroacetic acid (w/v) per sample to precipitate the DNA. Filter the precipitate onto nitrocellulose disks which have been soaked in saturated sodium pyrophosphate for greater than 24 hr. Dry the disks and determine the radioactive incorporation by liquid scintillation counting.

Hoechst Dye, Cesium Chloride Isopycnic Equilibrium Centrifugation. Different classes of DNAs (e.g., ribosomal RNA genes, 5 S RNA genes, and mitochondrial DNA) which have unique GC contents can be separated from the main band nuclear DNA using Hoechst dye 33258 on cesium chloride gradients[14] (Fig. 1). Add 0.4 μg of Hoechst dye/μg of DNA and then bring the sample volume to 7.2 ml with TES. Mix the sample with 8.7 g of cesium chloride. Centrifuge in a Beckman Ti50 rotor (or equivalent) for 40 hr at 40,000 rpm.

Acknowledgment

This work was supported by Grant AI 10056 to K.G.L. from the National Institute of Allergy and Infectious Disease.

[14] A. Bendich, personal communication (1983).

[51] Isolation of the Plasma Membrane: Membrane Markers and General Principles

By DONALD P. BRISKIN, ROBERT T. LEONARD, and THOMAS K. HODGES

The plasma membrane defines the outer surface of the protoplast of plant cells; serving as a barrier between the external environment and the numerous processes taking place in the cytoplasm. In this capacity, the plasma membrane is involved in the transport of solutes between the cell exterior and the cytoplasm.[1] This transport of solutes is essential for maintaining the conditions required for cell metabolism and growth. In addition, this membrane may serve as the receptor for both hormonal and environmental signals[2] and may also be involved in certain host–pathogen interactions. As would be expected from its close proximity to the cell wall, the plasma membrane is intimately involved in the synthesis of cellulose and the assembly of cell wall microfibrils.[2] In order to study these processes at the most basic level, it is often desirable to have an *in vitro* system of plasma membrane vesicles isolated from plant cells.

Depending on the type of tissue from which plasma membrane vesicles are to be isolated and the nature of the experiments to be performed with the membrane preparations, the techniques for the isolation of plasma membrane fractions may vary. This chapter contains three techniques for the isolation of plasma membrane-enriched fractions from plant tissue. The first technique was developed for the preparation of plasma membrane fractions from young, nongreen tissues, and the second technique was developed for use with storage tissues where it is particularly useful for the production of plasma membrane fractions in bulk quantities. The third procedure utilizes aqueous polymer partitioning and has broad application, but is particularly well suited for situations where it is desirable to obtain vesicles with a uniform sidedness or for use with leaf tissues where contamination by chloroplast membranes can be a problem. In addition to methods for the isolation of plasma membrane fractions, protocols are also described for the membrane markers most frequently used during plant membrane isolations. Although data from specific tissue sources will be presented, the three procedures should be considered as

[1] R. T. Leonard, *in* "Advances in Plant Nutrition" (P. B. Tinker and A. Lauchli, eds.), p. 209. Praeger, New York, 1984.

[2] R. T. Leonard and T. K. Hodges, *in* "The Biochemistry of Plants" (P. K. Stumpf and E. E. Conn, eds.), Vol. 1, p. 163. Academic Press, New York, 1980.

general approaches which can be modified or optimized for use with other cell or tissue types.

Membrane Markers

In order to identify the plasma membrane and assess contamination by other membrane components during a membrane isolation procedure, specific markers are required. Several reviews are available describing the concept and use of membrane markers in plant cell fractionation.[3-5] Ideally, these markers should represent either a unique enzyme activity or intrinsic property associated with a membrane component. However, in practice this has proved more difficult because most markers have a primary location in one organelle and a secondary location elsewhere.[4] For marker enzyme activities, this can often be resolved through the use of selective inhibitors or by modifying the enzyme reaction conditions. The following marker assays are frequently used during plant plasma membrane isolations.

Plasma Membrane

Potassium-Stimulated ATPase.[6] Assays are generally carried out in a 1-ml reaction volume containing 3 mM ATP, 3 mM $MgSO_4$, 30 mM Tris–MES buffer (MES titrated with Tris to pH 6.5), 50 mM KCl (when present), and 15 to 35 μg of membrane protein. The ATP substrate is present as the Tris salt (pH 6.5) converted from the sodium salt by treatment with Dowex 50-W exchange resin (H^+ form).[6] The assay is generally performed at 38° under conditions that lead to minimal hydrolysis of ATP with product formation that is linear with time and protein concentration. Potassium stimulation represents the difference in activity between assays performed in the absence and presence of 50 mM KCl. As shown in Fig. 1, this component of activity correlates closely with the presence of plasma membrane vesicles as determined by electron microscopy (see below).

With all phosphohydrolase-based marker enzymes, it is important to assess the presence of nonspecific phosphatases which can hydrolyze ATP and interfere with the marker assay. This can be evaluated by exam-

[3] D. J. Moore, C. D. Cline, R. Coleman, W. H. Evans, H. Glaumann, D. R. Headon, E. Reid, G. Siebert, and C. C. Windell, *Eur. J. Cell Biol.* **20,** 195 (1979).

[4] G. Nagahashi, *in* "Modern Methods of Plant Analysis" (H. F. Linskens and J. F. Jackson, eds.). Springer-Verlag, Berlin and New York (in press).

[5] P. H. Quail, *Annu. Rev. Plant Physiol.* **30,** 425 (1979).

[6] T. K. Hodges and R. T. Leonard, this series, Vol. 32, p. 392.

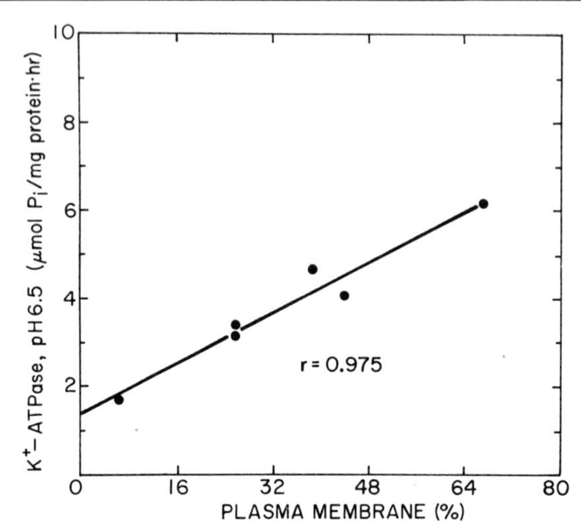

FIG. 1. Correlation between pH 6.5, K^+-stimulated ATPase activity and plasma membrane content judged by PACP staining technique. The data are from various fractions during the isolation of a corn root plasma membrane fraction. (From Leonard and Vander-Woulde.[20])

ining phosphate release from *p*-nitrophenol phosphate which is substituted for the nucleoside phosphate in the assay. In addition, it has been shown that many phosphatases present in plant membrane preparations can be inhibited by molybdate.[7,8] Therefore, this agent could be included in an assay for ATPase (0.1–1 m*M* final concentration) to reduce the activity of contaminating phosphatases or the molybdate-sensitive component could be used as an additional indicator for phosphatase activity.

The released inorganic phosphate can be determined using the method of Ames.[9] Following sufficient incubation time, each assay tube is removed from the water bath and 2.6 ml of a combined stopping and color reagent is added. The reagent is produced by mixing six parts 0.42% (w/v) ammonium molybdate in 1 *N* H_2SO_4 with one part 10% (w/v) ascorbic acid. After incubation at room temperature for 20 min, the optical density is determined at 700 nm.

Vanadate-Sensitive ATPase.[7] Assays for this component of ATPase are carried out as above with 50 m*M* KCl but in the presence and absence of 50 μ*M* Na_3VO_4. Vanadate-sensitive ATPase, which is strongly corre-

[7] S. R. Gallagher and R. T. Leonard, *Plant Physiol.* **70,** 1335 (1982).

[8] R. A. Leigh and R. R. Walter, *Planta* **150,** 222 (1980).

[9] B. N. Ames, this series, Vol. 8, p. 113.

lated with the plasma membrane,[7] represents the difference in activity observed in the presence and absence of the inhibitor. It should be noted that high concentrations of vanadate interfere with some assays for inorganic phosphate,[7] but not with procedures using ascorbic acid as reductant.[9]

Glucan Synthase II.[10–12] The assay[11] is carried out in a 0.1-ml assay volume containing 50 m*M* Tris–HCl (pH 8), 2 m*M* Na_4EDTA, 0.5 m*M* dithiothreitol (DTT), 20 m*M* cellobiose, and 0.15 μCi of UDP-D-[U-^{14}C]glucose plus 0.5 m*M* unlabeled UDP-glucose and 50 μl of membrane suspension. The reaction is carried out for 25 min at 25° and then terminated by boiling for 5 min. After adding a small amount of powdered cellulose, the contents of each tube are transferred to a filter (Whatman GF/B) and then washed three times with 2 ml each of boiling water. Then the filters are washed with 2.5 ml of methanol : chloroform (2 : 1), 2.5 ml of absolute methanol, and finally with 3 ml of water. The filters are dried and radioactivity is determined by liquid scintillation spectroscopy.

Cellulase.[11,13,14] Cellulase is assayed by measuring the decrease in viscosity of a solution of carboxymethyl-cellulose (types 7HF and 7HP, Hercules Powder Co., Wilmington, DE). The reaction is carried out in the presence of two parts carboxymethyl-cellulose [1.33% (w/v) in 20 m*M* sodium phosphate buffer, pH 6.1] and one part membrane suspension. Following a 20-hr incubation, viscosity is measured by drainage time through a calibrated 0.1-ml pipet. Using the conversion factors of Almin *et al.*[15] these values are expressed in terms of relative cellulase activity. It should be noted that the unequivocal association of this marker with the plasma membrane is still uncertain.

Naphthylphthalamic Acid (NPA) Binding.[16–18] The procedure of Lembi *et al.*[16] is described. Membrane aliquots (0.4 ml, 0.3–3 mg protein) are mixed with 0.025 μg [^{3}H]NPA (about 5000 cpm) in 4.6 ml of distilled water (final NPA concentration is 6 n*M*). A control set of samples containing additional unlabeled NPA (10 μ*M* final concentration) to correct for

[10] P. M. Ray, *Plant Physiol.* **59,** 594 (1977).

[11] W. S. Pierce and D. L. Hendrix, *Planta* **146,** 161 (1979).

[12] W. J. VanderWoude, C. A. Lembi, D. J. Morre, J. I. Kindinger, and L. Ordin, *Plant Physiol.* **54,** 333 (1974).

[13] L. N. Lewis and J. E. Varner, *Plant Physiol.* **46,** 194 (1970).

[14] D. E. Koehler, R. T. Leonard, W. J. VanderWoude, A. E. Linkins, and L. N. Lewis, *Plant Physiol.* **58,** 324 (1976).

[15] K. E. Almin, K. E. Eriksson, and C. Jansson, *Biochim. Biophys. Acta* **139,** 248 (1967).

[16] C. A. Lembi, D. J. Morre, K. St. Thomson, and R. Hertel, *Planta* **99,** 37 (1971).

[17] K. Trillmich and W. Michalke, *Planta* **144,** 119 (1979).

[18] J. C. Roland, *in* "Electron Microscopy and Cytochemistry of Plant Cells" (J. L. Hall, ed.), p. 1. Elsevier/North-Holland, Amsterdam, 1978.

nonspecific binding is also carried through the assay. Both sets of tubes are centrifuged at 80,000 *g* for 15 min and the difference in radioactivity associated with the pellet between the low NPA (specific binding) and high NPA (nonspecific binding) treated samples is determined.

Phosphotungstic Acid–Chromic Acid (PACP) Staining.[18–21] The procedure of Nagahashi *et al.*[21] is described. Membrane fractions are diluted to 17 ml with a solution of 2.5% glutaraldehyde in 100 m*M* potassium phosphate (pH 7.2) in a cellulose nitrate thin-walled centrifuge tube. The samples are then centrifuged at 80,000 *g* (25,000 rpm) for 30 min in a Beckman SW 28.1 rotor. The pelleted samples (in glutaraldehyde) are then held at ice temperature for an additional hour. The bottom of the centrifuge tube containing the pellet is then cut off, and the pellet is post-fixed with 1% buffered OsO_4 for 1.5 hr at 0 to 2°. The pellets are dehydrated through a graded acetone series and infiltrated by a dropwise addition of Spurr's epoxy resin over 24 hr.

The PACP stain is applied to sections of 900 Å thickness (silver–gold). The sections are treated with periodic acid by flotation on a 1% solution for 30 min in a moist chamber followed by a 5-min rinse in distilled water. The sections are stained by flotation on a solution of 1% phosphotungstic acid in 10% chromic acid for 7 to 10 min in a moist chamber. The specific staining of plasma membrane vesicles in thin sections is used in morphometric determinations to calculate the plasma membrane content on a percentage basis.

Tonoplast

Nitrate-Sensitive ATPase.[22,23] The hydrolysis of ATP is measured in the presence of 3 m*M* ATP (Tris salt), 3 m*M* $MgSO_4$, 30 m*M* Tris–MES, pH 8, and 50 m*M* KCl or KNO_3. The release of inorganic phosphate is determined as described above and nitrate-sensitive ATPase represents the difference in activity observed between assays with KCl or KNO_3. It should be noted that the unique localization of this marker on the tonoplast is uncertain.

Endoplasmic Reticulum

NADH or NADPH–cytochrome-c Reductase, Cyanide Insensitive.[6] The assay is carried out at 25° in a 3-ml reaction volume containing 0.1 ml

[19] T. K. Hodges, R. T. Leonard, C. E. Bracker, and T. W. Keenen, *Proc. Natl. Acad. Sci. U.S.A.* **69,** 3307 (1972).

[20] R. T. Leonard and W. J. VanderWoulde, *Plant Physiol.* **57,** 105 (1976).

[21] G. Nagahashi, R. T. Leonard, and W. W. Thomson, *Plant Physiol.* **61,** 993 (1978).

[22] K. A. Churchill and H. Sze, *Plant Physiol.* **76,** 490 (1984).

[23] R. J. Poole, D. P. Briskin, A. Kratky, and R. M. Johnstone, *Plant Physiol.* **74,** 549 (1984).

of membrane suspension, 0.1 ml of 50 mM NaCN, 0.2 ml of 0.45 mM cytochrome *c*, and 2.5 ml of 50 mM sodium phosphate buffer, pH 7.5. The reaction is started by the addition of 0.1 ml of either 3 mM NADH or NADPH. The reduction of cytochrome *c* is measured by the change in optical density at 550 nm. The rate of cytochrome *c* reduction is calculated using an extinction coefficient of 18.5 mM^{-1} cm^{-1} at 550 nm. The activity observed with NADPH is generally lower than that observed with NADH and greater specificity of this marker for the endoplasmic reticulum can be achieved by determining the component of activity that is insensitive to 1 μM antimycin A.[5]

Mitochondria (Inner Membrane)

Cytochrome-c Oxidase.[6] The enzyme assay is performed as for NAD(P)H–cytochrome-*c* reductase except that the oxidation of reduced cytochrome *c* is observed. Cytochrome *c* is reduced by the addition of a molar excess of sodium dithionite. Excess dithionite is oxidized by aerating the solution for a few minutes. The reaction solution contains 0.1 ml of membrane suspension, 0.1 ml of 0.3% digitonin or Triton X-100, and 2.7 ml of 50 mM sodium phosphate, pH 7.5. The reaction is started by the addition of 0.1 ml of reduced cytochrome *c* and the change in optical density is observed at 550 nm. The rate of cytochrome *c* oxidation is calculated as above.

Azide-Sensitive ATPase.[7] The assay is carried out at pH 8.5 as for plasma membrane ATPase with 50 mM KCl in the presence and absence of 1 mM NaN_3. The difference in activity as a result of the inhibitor represents the sensitive component of the activity.

Golgi Membrane Vesicles

Latent IDPase.[24] Latent IDPase assays are carried out as for ATPases in a 1-ml reaction volume containing 3 mM IDP (sodium salt), 3 mM $MgSO_4$, 30 mM Tris–MES, pH 7.5, 50 mM KCl, and 15–35 μg of membrane protein. The activity is determined on freshly isolated membranes and after 6 days of storage at 2–4°.[5] Latent IDPase represents the difference in activity observed upon cold storage.

Triton-Stimulated UDPase.[25] The assay is carried out as for ATPases in a 1-ml reaction volume containing 3 mM UDP (Tris salt, pH 6.5), 3 mM $MnSO_4$, 30 mM Tris–MES, pH 6.5, 15 to 35 μg of membrane protein, and 0.03% (v/v) Triton X-100 (when present). In order to prevent interference of Triton X-100 with the Ames phosphate determination, the color reagent

[24] P. M. Ray, T. L. Shininger, and M. M. Ray, *Proc. Natl. Acad. Sci. U.S.A.* **60,** 605 (1969).
[25] G. Nagahashi and A. P. Kane, *Protoplasma* **112,** 167 (1982).

should also contain 1.5% (w/v) SDS (phosphate free). Triton-stimulated UDPase represents the difference in activity in the presence and absence of the detergent.

Glucan Synthase I.[12] The assay is carried out as described for glucan synthase II except that 1.5 μM UDP-glucose and 10 mM $MgCl_2$ are present in the reaction solution.

Isolation Procedures

Technique 1—Differential and Density Gradient Centrifugation

This technique[6] was developed for young, nongreen tissue such as primary roots or dark-grown stem segments, where the isolation of large amounts (>20 mg protein) of plasma membrane is not required.

Homogenization. Gentle homogenization of plant roots, coleoptiles, or young stems can be carried out using a chilled mortar and pestle. The homogenization medium contains 0.25 M sucrose, 3 mM EDTA, 25 mM Tris–MES, pH 7.8, and 1 mM dithiothreitol (DTT) or dithioerythritol (DTE). The tissue (in medium) can be cut into small pieces with chilled scissors or a razor blade and homogenized (without abrasives) for 1 to 2 min using about 4 ml of homogenizing medium per gram of tissue. The homogenate is then filtered through four layers of cheesecloth or one layer of Miracloth. It is desirable to check the pH of the filtrate and adjust to 7.8, if necessary, with concentrated Tris buffer.

Differential Centrifugation. The filtered homogenate is centrifuged at 13,000 g for 15 min to pellet large organelles such as mitochondria. A microsomal membrane pellet is obtained by centrifuging the supernatant at 80,000 g for 30 min, and the pelleted membranes are suspended in a small volume of 0.25 M sucrose, 1–5 mM Tris–MES, pH 7.2, and 1 mM DTT (suspension buffer).

Density Gradient Centrifugation. Microsomal membrane components are separated by centrifugation on linear or discontinuous sucrose density gradients. In the development of an isolation procedure, preliminary linear or discontinuous gradients encompassing a wide range of density [e.g., 15 to 45% (w/w) sucrose] should be used to determine the buoyant density of the various cellular components. The gradient medium is typically buffered with 1–5 mM Tris–MES, pH 7.2, and contains 1 mM DTT or DTE. Linear gradients should be centrifuged to density equilibrium at 115,000 g in a swinging bucket rotor. Long centrifugation times (8–15 hr) ensure that the membranes have reached their isopycnic densities; however, this can pose a problem if desired enzyme activities are labile. The gradients can be fractionated from the top or bottom and the sucrose gradient profile can be determined using a refractometer.

A discontinuous gradient which is useful for preliminary isolations consists of 6.4 ml each of 38, 34, 30, 25, and 20% (w/w) sucrose layered over 4 ml of 45% (w/w) sucrose in a 38-ml centrifuge tube.[6] Two milliliters of microsomal membranes in suspension buffer is layered on the gradient which is centrifuged at 115,000 *g* for 2 hr. Membranes banding at each interface are removed using a Pasteur pipet and analyzed for membrane markers.

Based upon the results obtained with a preliminary linear or discontinuous gradient, a preparative density gradient can be designed to isolate a plasma membrane-enriched fraction. For both oat[19] and corn[20] roots, a simplified preparative gradient consisting of 6 to 8 ml of 34% (w/w) sucrose layered over 28 to 30 ml of 45% (w/w) sucrose has been used with a Beckman SW 27 (or 28) rotor. The gradients were centrifuged for 2 hr at 80,000 *g* and the plasma membrane preparation represented membranes banding at the 34%/45% interface.

Typical Results and Comments. The distribution of plasma membranes (marked by K^+-stimulated ATPase) and mitochondrial membranes (marked by cytochrome-*c* oxidase) following differential and density gradient (34%/45% interface) centrifugation of a corn root homogenate is shown in Table I. In addition, a mitochondrial fraction was obtained by centrifuging a 13,000 *g* pellet on a similar gradient used to isolate plasma membranes. It is apparent from the data in Table I that large amounts of plasma membrane vesicles are lost during low-speed centrifugation. However, this loss is necessary to reduce contamination by mitochondria in the plasma membrane fraction. It is possible to maximize the yield of plasma membrane without significant increase in mitochondrial contamination by adjustments in the speed and time of the low-speed centrifugation.[26]

The final plasma membrane fraction normally shows about a 3-fold enrichment in pH 6.5, K^+-stimulated ATPase with a substantial reduction in the total activity of mitochondrial markers. It appears that major and minor contaminants in the plasma membrane fraction are Golgi membranes and endoplasmic reticulum (Fig. 2). The distribution of tonoplast is uncertain, but the density of tonoplast vesicles is generally lower than plasma membrane vesicles so that it should not be a significant contaminant in this fraction.

Technique 2—Modifications for Plant Storage Tissue

In order to isolate large quantities of plasma membranes from plant cells, storage tissue can be employed. Examples of storage tissues which

[26] G. Nagahashi and K. Hiraike, *Plant Physiol.* **69,** 546 (1982).

TABLE I
K^+-STIMULATED ATPase, CYTOCHROME-*c* OXIDASE, TOTAL PROTEIN PERCENTAGE PLASMA MEMBRANE IN VARIOUS MEMBRANE FRACTIONS FROM CORN ROOTS[a]

Fraction	K^+-Stimulated ATPase[b]				Cytochrome *c* oxidase			
	pH 6.5		pH 9					
	Specific activity (μmol P_i/mg hr)	Total activity (%)[c]	Specific activity (μmol P_i/mg hr)	Total activity (%)[c]	Specific activity (μmol/mg min)	Total activity (%)[c]	Total protein (%)[d]	Plasma membrane (%)[c]
1,000 g	1.53	6	2.46	6.8	0.24	6.6	12.3	
1,000–8,000 g	3.34	25.4	8.57	45.7	1.36	70.7	23.6	26
8,000–13,000 g	4.04	19.8	5	17.2	0.46	15.2	15.1	44
13,000–40,000 g	3.13	40	2.45	22	0.08	5.9	35.8	26
40,000–80,000 g	1.71	8.9	2.30	8.4	0.05	1.6	13.2	6
Mitochondria	4.68	22.9	10.87	37.3	2.05	61.8	13.7	39
Plasma membranes	6.16	17.3	3.41	6.7	0.15	2.6	7.9	67

[a] Adapted from Leonard and VanderWoulde.[20]

[b] Derived from data in Table I by subtracting activity in the presence of Mg^{2+} plus K^+ and that in presence of Mg^{2+} alone.

[c] Expressed as percentage of total K^+-stimulated ATPase activity or total milligrams protein recovered in the fractions obtained by differential centrifugation up to 80,000 *g*.

[d] Surface area of PACP-stained membrane divided by the total membrane surface area × 100.

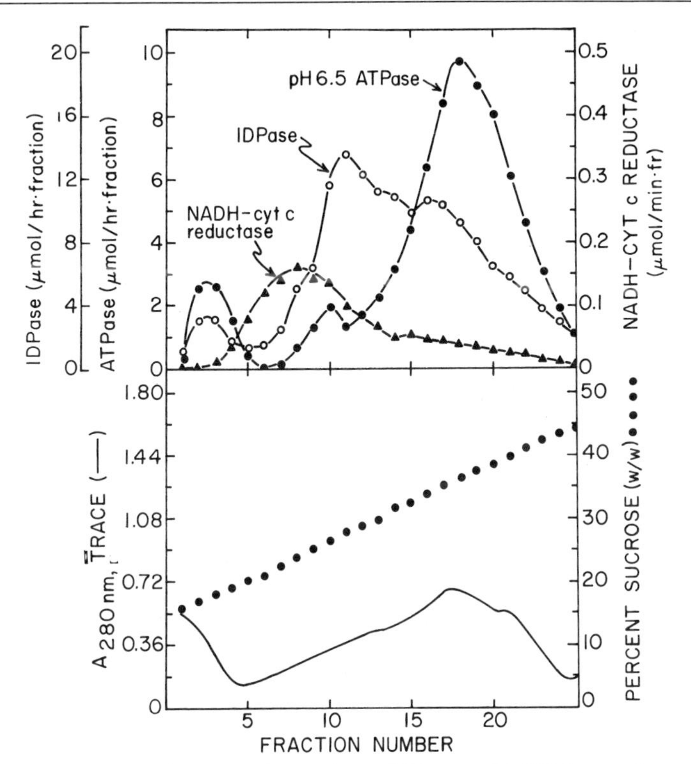

FIG. 2. Distribution of pH 6.5 ATPase, latent IDPase, and NADH–cytochrome-*c* reductase when a microsomal pellet (13,000 to 80,000 *g*) from corn roots was centrifuged for 15 hr on a linear sucrose gradient. (From Leonard and VanderWoulde.[20])

have been used in plasma membrane isolations include red beet[27] and sugarbeet storage root.[28] The method is similar to technique 1 except that a mechanized homogenization is used to facilitate the processing of large quantities of tissue and the homogenizing medium is modified to include additional protectants. In addition, high-capacity centrifuge rotors are used to allow large-scale preparations. With this technique plasma membrane fractions can be isolated from kilogram quantities of starting tissue and the yield of the plasma membrane fraction can exceed 35 mg of protein.

Homogenization. Beet storage tissue is cut into sections (approximately 3 cm^3) and kept at 0–2°. The tissue is homogenized using an Oster vegetable juice extractor in a medium containing 0.25 *M* sucrose, 3 m*M*

[27] D. P. Briskin and R. J. Poole, *Plant Physiol.* **71,** 350 (1983).

[28] D. P. Briskin and W. R. Thornley, *Phytochemistry* **24,** 2797 (1985).

EDTA, 0.5% (w/v) poly(vinylpyrolidone) (40,000), 15 m*M* 2-mercapto-ethanol, 70 m*M* Tris–HCl, pH 8, and 5 m*M* DTE. One milliliter of medium is used per gram of tissue and all operations are carried out at 2–4°. The homogenate is filtered through four layers of cheesecloth and transferred to chilled centrifuge tubes or bottles.

Differential Centrifugation. A large-scale low-speed centrifugation can be carried out using a Sorvall GSA rotor which can accommodate six 250-ml bottles. The homogenate is centrifuged at 13,000 *g* (9000 rpm) for 15 min (actual time at 13,000 *g*). The supernatant is decanted and the microsomal membranes are pelleted by centrifugation at 80,000 *g* for 30 min. This can be accomplished using a Beckman type 35 or Ti 45 rotor (32,000 rpm) which holds six 90-ml bottles. The microsomal pellet is suspended in a small volume of suspension medium.

Treatment with KI. In both red beet[27] and sugarbeet,[28] the microsomal membranes contain substantial levels of nonspecific phosphatase activity. Since many of the frequently used membrane markers rely on phosphohydrolase activity, this interfering activity must be removed. This has been accomplished by washing the microsomal membranes from beet storage tissue with 0.2 *M* KI, which removes a substantial amount of phosphatase activity (shown by hydrolysis of *p*-nitrophenol phosphate) relative to ATPase activity (Fig. 3). This was carried out by the slow addition of an equal volume of 0.5 *M* KI in suspension buffer to the membrane suspension. Following incubation on ice for 20 min, the membranes are recovered by centrifugation at 80,000 *g* for 30 min. The KI-washed microsomal pellet is then suspended in a small volume of suspension medium and layered onto a sucrose gradient.

Density Gradient Centrifugation. Further resolution of the microsomal membranes can be accomplished by using density gradient centrifugation as described in technique 1 except with particular care not to overload the gradient. For red beet, a simplified preparative sucrose gradient consisting of 4 ml of 25% (w/w) sucrose layered over 8 ml of 35% (w/w) sucrose has been used to obtain a plasma membrane fraction.[27] The gradient solutions were buffered with 1 m*M* Tris–MES, pH 7.2, and contained 1 m*M* DTE. One milliliter of membrane suspension was applied to the gradients which were centrifuged at 115,000 *g* for 3 hr in a Beckman SW 41 Ti rotor. The plasma membrane fraction represented those membranes which banded at the 25%/35% sucrose interface. For sugarbeet a similar procedure was used except that the gradient solutions consisted of 30% (w/w) and 38% (w/w) sucrose.[28]

Typical Results and Comments. The distribution of phosphohydrolase activities is shown during the isolation of a plasma membrane fraction from sugarbeet taproot (Table II). An examination of inhibitor-sensitive components indicates that the majority of the ATP hydrolytic activity

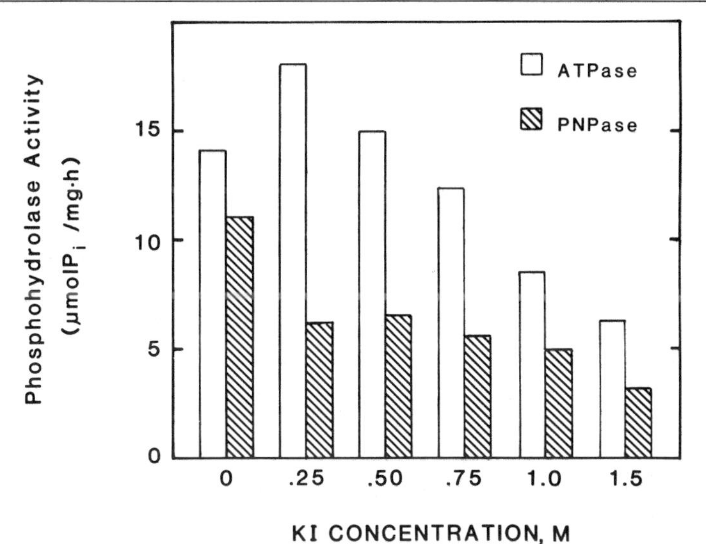

FIG. 3. Effect of KI on the levels of ATP and *p*-nitrophenol phosphate hydrolysis by a microsomal membrane fraction from red beet. The membranes were treated with the indicated concentration of KI for 20 min and then centrifuged at 80,000 *g* for 30 min to collect the membranes. (From Briskin and Poole.[27])

TABLE II
DISTRIBUTION OF INHIBITOR-SENSITIVE ATPase AND PROTEIN DURING THE ISOLATION OF A PLASMA MEMBRANE FRACTION FROM SUGARBEET[a]

	Fraction			
Activity	Homogenate	Microsomal	KI extracted	Plasma membrane
ATPase[b] (μmol Pi/hr mg)				
Control	39.7 (100)	27.8 (100)	32.8 (100)	47.4 (100)
ΔNa_3VO_4	25.2 (64.5)	19.8 (71.2)	24.3 (74.1)	43.3 (91.4)
ΔKNO_3	4.1 (10.3)	4.7 (16.9)	0 (0)	0.7 (1.5)
ΔNaN_3	4.9 (12.3)	2.7 (9.7)	0.2 (0.6)	0 (0)
ΔNa_2MoO_4	22.9 (57.7)	14.9 (53.6)	5.2 (15.9)	3.2 (6.8)
Protein (mg/fraction)	354.0	36.3	29.5	12.8

[a] Adapted from Briskin and Thornley.[28]

[b] Control ATPase assay contained 3 m*M* ATP, 3 m*M* $MgSO_4$, 50 m*M* KCl, 30 m*M* Tris–MES, pH 7.0, and 20 to 40 μg protein. Inhibitor-sensitive components of ATPase activity represent the difference between the control activity and the activity in the presence of 50 μM Na_3VO_4 (ΔNa_3VO_4) or 50 mM KNO_3 (ΔKNO_3) or 10 m*M* NaN_3 (ΔNaN_3) or 500 μM Na_2MoO_4 (ΔNa_2MoO_4). Values in parentheses represent the percentage of the control activity.

present in the final plasma membrane fraction is vanadate sensitive (plasma membrane ATPase) with only a minor contribution by nitrate-sensitive (tonoplast) activity and no measurable contribution by azide-sensitive (mitochondria) activity. The effect that treatment with KI had upon reducing phoshatase activity (molybdate-sensitive activity) is also apparent from these data. Treatment with 150 m*M* KCl was also effective in removing phosphatase activities during the isolation of plasma membrane fractions from barley roots.[29] In comparison with membranes isolated from corn roots, the plasma membrane bands at a lower density on linear sucrose gradients and is well separated from both mitochondrial and endoplasmic reticulum membranes (Fig. 4).[29a] It should be noted that the treatment with KI may alter the equilibrium sucrose densities of membrane vesicles compared to untreated membranes. As in the previous example, plasma membrane preparations produced in this manner contain Golgi membranes (Triton-stimulated UDPase) as a major contaminant.

Technique 3—Aqueous Polymer Two-Phase Isolation

A major problem associated with both of the previously presented techniques is that it is often difficult to resolve chloroplast thylakoid membranes from plasma membrane vesicles using the combination of differential and density gradient centrifugation.[4] Therefore, optimal results using these techniques have been limited to nongreen tissue. Recently, this problem has been solved by replacing sucrose gradients as a means to resolve microsomal membranes with an aqueous polymer phase system.[30] This system separates membranes based on surface charge rather than by intrinsic density. An additional benefit of this method is that it may allow the production of a plasma membrane preparation with a defined sidedness.[31]

Homogenization. The tissue is homogenized as described for technique 1; however, additional protectants may need to be added to the media to prevent oxidation and browning when using green tissue. This occurs because chloroplasts are a major source of polyphenol oxidase activity.[32] These protectants include mercaptoethanol, bovine serum albumin, or poly(vinylpyrolidone). Typical concentrations of these protectants are 15 m*M*, 0.1% (w/v) and 0.5% (w/v), respectively. However, the

[29] F. M. Dupont and W. J. Hurkman, *Plant Physiol.* **77,** 857 (1985).

[29a] D. P. Briskin, W. R. Thornley, and R. J. Poole, *Arch. Biochem. Biophys.* **236,** 228 (1985).

[30] C. Larsson, *in* "Isolation of Membranes and Organelles from Plant Cells" (J. L. Hall and A. L. Moore, eds.), p. 278. Academic Press, London, 1983.

[31] T. K. Hodges and D. Mills, this series, Vol. 118, p. 41.

[32] K. C. Vaughn and S. O. Duke, *Physiol. Plant.* **60,** 106 (1984).

exact concentrations of these protectants to be used with a given tissue source may need to be determined empirically.

Differential Centrifugation. A microsomal fraction is obtained as described for techniques 1 and 2.

Aqueous Polymer Resolution of Microsomal Membranes. The phase separation technique involves mixing microsomal membranes with a mix-

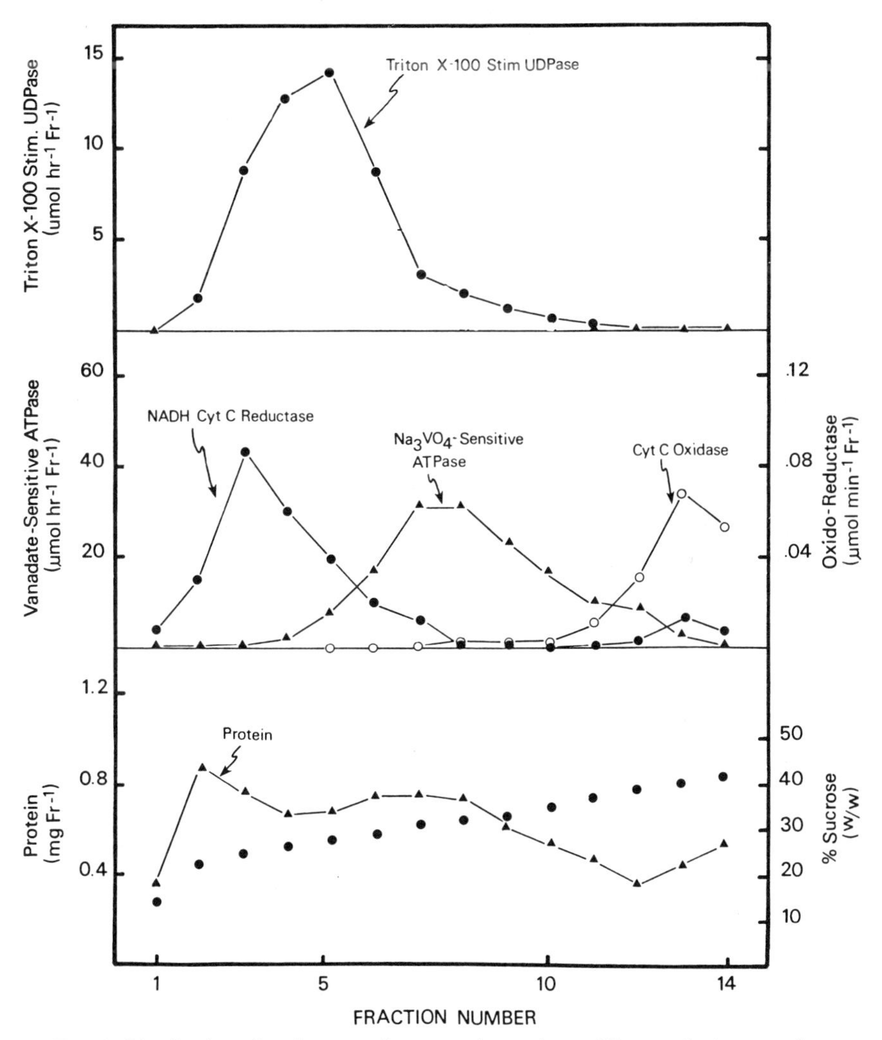

FIG. 4. Distribution of various membrane markers when a KI-treated microsomal membrane fraction from sugarbeet was centrifuged on a linear sucrose gradient for 3 hr. Fr, Fraction. (From Briskin *et al.*[29a])

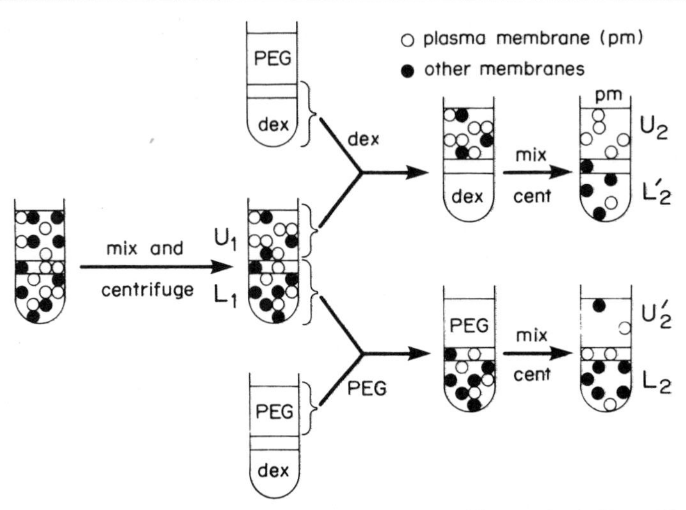

FIG. 5. Schematic diagram of the aqueous polymer partition method. (From Hodges and Mills.[31])

ture of Dextran T-500 and polyethylene glycol 3350 (PEG) containing NaCl or KCl. Since surface charge may vary for microsomal membranes from different plant species it is necessary to test a range of polymer and salt solutions for each tissue source. The preparation and evaluation of the phase system was discussed in a previous chapter in this series.[31] The membrane phase system is inverted and returned upright 20 times and then centrifuged at 1000 *g* for 5 min in a swinging bucket rotor. Following phase separation, the upper and lower phases can be repartitioned for further purification (Fig. 5). This can be repeated additional times for even further purification.[33] Finally, the membranes from a given phase can be recovered by dilution with homogenizing medium and centrifugation at 100,000 *g* for 60 min.

Typical Results and Comments. Data adapted from Kjelbom and Larsson[33] are presented for the isolation of plasma membrane vesicles from barley and spinach leaves which are relatively free of contamination by chloroplast membranes (Table III). The membranes from a microsomal pellet are mixed with a polymer system consisting of 6.2% (w/w) Dextran T-500, 6.2% (w/w) polyethylene glycol 3350, 0.33 *M* sucrose, 3 m*M* KCl, 5 m*M* potassium phosphate, pH 7.8, and phase separation was facilitated by centrifugation at 1500 *g*. The upper phase was then repartitioned twice in this system (U_3) in order to produce a chlorophyll-free preparation. The

[33] P. Kjelbom and C. Larsson, *Physiol. Plant.* **62,** 501 (1984).

TABLE III

TOTAL PROTEIN, CHLOROPHYLL, AND MARKER ENZYME ACTIVITIES IN THE PLASMA MEMBRANE ($U_3 + U_{3'}$) AND THE INTRACELLULAR MEMBRANE (L_1) FRACTIONS OBTAINED BY PHASE PARTITION OF MICROSOMAL FRACTIONS (MF) FROM LEAVES OF BARLEY AND SPINACH[a]

			Chlorophyll		Glucan synthase II		Cytochrome-*c* oxidase		NADPH–cytochrome-*c* reductase	
Species	Fraction	Protein (mg)	(mg)	(%)	(nmol min^{-1})	(%)	(nmol min^{-1})	(%)	(nmol min^{-1})	(%)
Barley	MF	58	6.7	100	12.5	100	7400	100	470	100
	$U_3 + U_{3'}$	1.3	ND[b]		4.7	38	42	0.57	6.5	1.4
	L_1	28	3.1	46	5.4	43	5100	68.0	280	59.6
Spinach	MF	49	4.0	100	63	100	19000	100	610	100
	$U_3 + U_{3'}$	4.0	ND		37	59	92	0.5	29	4.7
	L_1	22	2.8	70	11	17	14200	75	360	59.0

[a] Adapted from Kjelbom and Larsson.[33]

[b] ND, Not detectable.

final membrane fraction was enriched in plasma membranes (glucan synthase II) relative to mitochondrial (cytochrome-*c* oxidase) and endoplasmic reticulum membranes (NADPH–cytochrome-*c* reductase). For spinach, the final plasma membrane fraction had a greater total content of plasma membrane vesicles than barley; however, this occurred with a greater amount of contamination by endoplasmic reticulum. The amount of chlorophyll in both of these plasma membrane fractions produced by partitioning was undetectable. The distribution of tonoplast, however, was not determined.

Storage of Plasma Membrane Fractions

Due to the length of time required to isolate plasma membrane fractions, it is often desirable to store the preparations for a period of time following isolation. This is especially true when membrane fractions can be produced in bulk quantities and then used at a later time. At least for studies on the plasma membrane ATPase, it has been possible to store membranes by freezing in liquid nitrogen and then either storing the membranes in liquid nitrogen or in an ultralow freezer (−80°). This storage method allows full retention of ATPase activity up to several months.

[52] Preparation of High-Purity Plasma Membranes

By Christer Larsson, Susanne Widell, and Per Kjellbom

Introduction

Partitioning in aqueous Dextran–polyethylene glycol two-phase systems as a method for plasma membrane purification was introduced in 1981[1,2] and has turned out to be quite superior to the traditional method, sucrose gradient centrifugation, in use since 1972.[3] The reason for the

[1] S. Widell and C. Larsson, *Physiol. Plant.* **51,** 368 (1981).

[2] T. Lundborg, S. Widell, and C. Larsson, *Physiol. Plant.* **52,** 89 (1981).

[3] T. K. Hodges, R. T. Leonard, C. E. Bracker, and T. W. Kennan, *Proc. Natl. Acad. Sci. U.S.A.* **69,** 3307 (1972).

superiority of the former method is that it separates membrane vesicles according to surface properties[4] and not according to size or density as is achieved with centrifugation methods. While the density of the plasma membrane (1.14–1.17 g cm^{-3}) on a sucrose gradient is very close to that of mitochondrial membranes (1.18–1.20), Golgi membranes (1.12–1.15), rough endoplasmic reticulum (1.15–1.17), and thylakoids (1.16–1.18), its surface properties, or at least the properties of its outer surface, seem to be very different from those of all intracellular membrane surfaces. Thus, with optimal composition of the two-phase system, about 90% of the plasma membranes are obtained in the upper phase, while about 90% of the various intracellular membranes are recovered at the interface or in the lower phase.

By repartitioning the initial upper phase with fresh lower phase virtually all contaminating membranes are removed and even chlorophyll-free preparations may be obtained from leaves of both mono- and dicotyledons.[5] The purity of these preparations can be estimated to be about 95%,[6,7] which should be compared to the 50–75% achieved with sucrose gradient centrifugation in comparative studies[8,9] (see below).

An additional advantage is that phase partitioning produces mainly sealed membrane vesicles which are right side out,[10] whereas sucrose gradient centrifugation has been reported to produce leaky vesicles with uncertain orientation.[8,9,11] Sealed vesicles with a defined sidedness are an invaluable tool in, for example, studies on transport and hormone binding.

Phase partition is easily scaled up and the method is rapid; plasma membranes may be obtained within 2 to 4 hr from homogenization. The method has now been used with a wide variety of species and organs, and seems to work successfully regardless of the plant material used (for a review, see Ref. 6). The method is presented below, using the preparation of plasma membranes from light-grown oat (*Avena sativa* L.) leaves as an example.

[4] P.-Å. Albertsson, B. Andersson, C. Larsson, and H.-E. Åkerlund, *Methods Biochem. Anal.* **28,** 115 (1982).

[5] P. Kjellbom and C. Larsson, *Physiol. Plant.* **62,** 501 (1984).

[6] C. Larsson, *in* "Modern Methods in Plant Analysis, New Series" (H. F. Linskens and J. F. Jackson, eds.), Vol. 1, p. 85. Springer-Verlag, Berlin and New York, 1985.

[7] A. S. Sandelius, C. Penel, G. Auderset, A. Brightman, M. Millard, and D. J. Morré, *Plant Physiol.* **81,** 177 (1986).

[8] T. K. Hodges and D. Mills, this series, Vol. 118, p. 41.

[9] A. Bérczi and I. M. Møller, *Physiol. Plant.* **68,** 59 (1986).

[10] C. Larsson, P. Kjellbom, S. Widell, and T. Lundborg, *FEBS Lett.* **171,** 271 (1984).

[11] S. K. Randall and A. W. Ruesink, *Plant Physiol.* **73,** 385 (1983).

Procedure

Chemicals

Dextran T-500 is purchased from Pharmacia Fine Chemicals, Uppsala, Sweden, and polyethylene glycol 3350 (earlier designated 4000) from Union Carbide, New York.

Stock solutions of 20% (w/w) Dextran and 40% (w/w) polyethylene glycol are prepared and stored in the cold or frozen. A problem is that Dextran is hygroscopic (containing 5–10% water) and the exact concentration can therefore be determined only with a polarimeter (specific rotation $+199°$ ml g^{-1} dm^{-1}, measured at 589 nm).[4,6] If a polarimeter is not available, the Dextran should be stored dry and relatively large stock solutions prepared to ensure reproducible experiments.

Finding the Optimal Phase Composition

By varying the polymer concentrations and the salt composition of the phase system, optimal conditions for separation are obtained. This is most conveniently done by preparing a series of 3.6-g phase systems (10% of the phase mixture in Table I) in test tubes varying either of the parameters. The effects of increasing polymer and chloride concentrations on the partitioning of a number of membrane markers are shown in Figs. 1 and 2. The plasma membrane markers are distributed almost exclusively to the upper phase even at polymer and chloride concentrations where the markers for intracellular membranes are found mainly in the lower phase. From these experiments a polymer concentration of 6.2% (w/w) Dextran T-500 and 6.2% (w/w) polyethylene glycol 3350 and a KCl concentration of 3 m*M* were chosen for the preparation described below. Note that the effects of increasing polymer and chloride concentrations are synergistic. Thus, 3 m*M* chloride has a stronger effect at 6.2% of each polymer than at 5.8% (results not shown).

Batch Preparation

A bulk phase system (50–300 g), to provide the fresh upper and lower phases used in the batch procedure (Fig. 3), is prepared in a separating funnel by weighing and pipetting from stock solutions (Table I). After temperature equilibration in the cold room at 4°, the phase system is well shaken and allowed to settle overnight. The upper phase and the lower phase are collected and stored separately in the cold or frozen. A 27-g

TABLE I
PHASE MIXTURE AND PHASE SYSTEM USED IN PLASMA MEMBRANE PURIFICATION

Step	Phase mixture	Phase system
20% (w/w) Dextran T-500	11.16 g	93.00 g
40% (w/w) Polyethylene glycol 3350	5.58 g	46.50 g
Sucrose (solid)	3.05 g	33.89 g
0.2 *M* Potassium phosphate, pH 7.8	0.675 ml	7.50 ml
2 *M* KCl	0.041 ml	0.45 ml
Add H_2O to a final weight of:	27.00 g	300.00 g

phase mixture to be used for the first separation step is prepared similarly (Table I) in a centrifugation tube.

Oat leaves (110 g) are homogenized (using an ordinary kitchen homogenizer) 3 × 20 sec in 275 ml of 0.5 *M* sucrose, 50 m*M* HEPES–KOH, pH 7.5, 5 m*M* ascorbic acid, 1 m*M* DTT, and 0.6% (w/v) poly(vinylpyrroli-

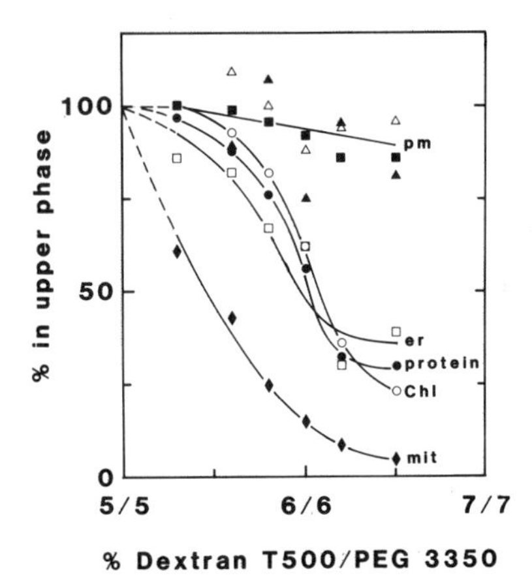

FIG. 1. Effects of increasing polymer concentrations on the partition of different membranes in a microsomal fraction from oat leaves. Markers used were as follows: Cytochrome-*c* oxidase (◆) for mitochondria (mit), chlorophyll (○) for chloroplast thylakoids (Chl), antimycin A-resistant NADPH–cytochrome-*c* reductase (□) for endoplasmic reticulum (er) (but see text), and for the plasma membrane (pm) total K^+-stimulated, Mg^{2+}-dependent ATPase (■), total vanadate-inhibited K^+,Mg^{2+}-ATPase (▲), and latent K^+/Mg^{2+}-ATPase (△). The partition of all membranes has been extrapolated back to the critical point {the minimum polymer concentrations to obtain a two-phase system which was 5.0% of each polymer at 4° for the polymer batches used here (compare Gardeström and Ericsson, this volume [40] where all material is in the upper phase)}. Apart from polymers, the phase systems contained 0.33 *M* sucrose and 5 m*M* potassium phosphate, pH 7.8. Temperature, 4°.

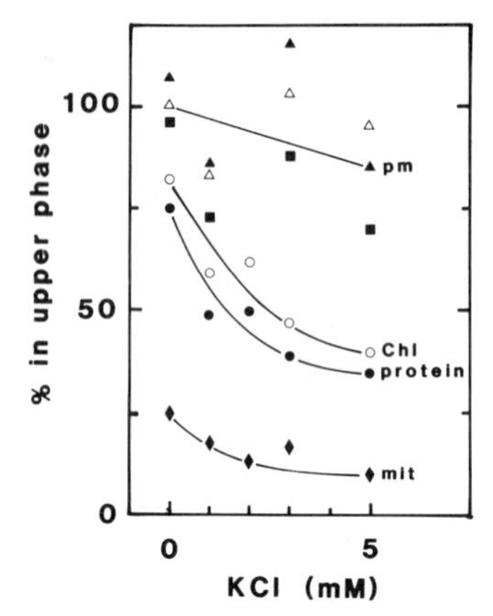

FIG. 2. Effects of increasing Cl^- concentrations on the partition of different membranes in a microsomal fraction from oat leaves. Markers as in Fig. 1, except that the latency of the vanadate inhibition of the K^+,Mg^{2+}-ATPase (△) was used instead of the latency of the whole activity. The phase systems contained 5.8% Dextran, 5.8% polyethylene glycol, 0.33 *M* sucrose, 5 m*M* potassium phosphate, and 0–5 m*M* KCl. Temperature, 4°.

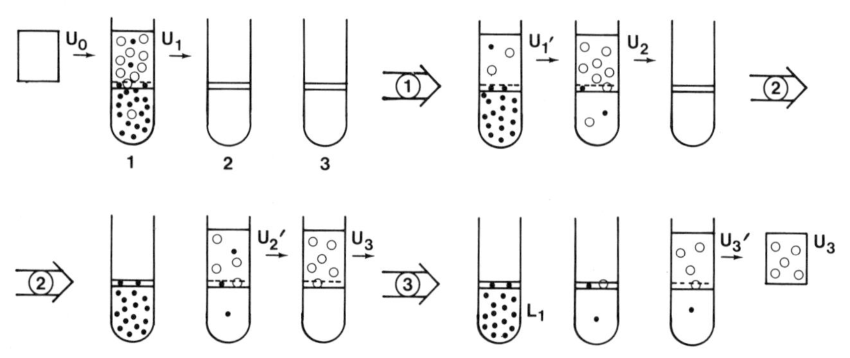

FIG. 3. Separation of plasma membranes (○) and intracellular membranes (●) by a batch procedure of three steps. The microsomal fraction suspended in the medium used in the phase system is added to a preweighed phase mixture to give a phase system of the desired final composition. The phase system is well mixed (by, e.g., 20 to 30 inversions of the tube) and phase settling is facilitated by centrifugation in a swinging bucket centrifuge for a few minutes. Then, 90% (indicated by the dashed line) of the upper phase (U_1) is removed with a Pasteur pipet, taking care not to disturb the interface, and repartitioned twice with fresh lower phase (tubes 2 and 3) to increase the purity, giving U_3. To increase the yield of plasma membranes, the lower phase may be reextracted, in sequence, with fresh upper phase (U_0), giving $U_{3'}$. Fresh upper and lower phases are obtained from a bulk phase system of identical composition prepared separately. In addition to fresh lower phase, tubes 2 and 3 also contain a small volume of upper phase corresponding to the 10% left in tube 1. This is to keep a constant upper phase volume throughout the procedure. Note that there is a true equilibrium between the upper and lower phase and the volumes of fresh phases are therefore not critical and can thus be added by eye.

done). The homogenate is filtered (240-μm nylon cloth) and most of the chloroplasts and mitochondria are pelleted at 10,000 *g* for 10 min. A microsomal pellet is obtained from the supernatant by centrifugation at 50,000 *g* for 30 min. This pellet is suspended to a total volume of 10 ml in 0.33 *M* sucrose, 3 m*M* KCl, 5 m*M* potassium phosphate, pH 7.8, and 9.0 g (30–40 mg protein) of the suspension is added to the 27.0-g phase mixture (Table I) to give a 36.0-g phase system with a final composition of 6.2% (w/w) Dextran T 500, 6.2% (w/w) polyethylene glycol 3350, 0.33 *M* sucrose, 3 m*M* KCl, 5 m*M* potassium phosphate, pH 7.8. The phase system is thoroughly mixed by 20–30 inversions of the tube and phase settling is facilitated by centrifugation in a swinging bucket centrifuge at 1500 *g* for 5 min. (A clearly visible interface should be obtained. If not, spin for another 5 min without remixing). Further processing is carried out according to the scheme in Fig. 3. The final phases (U_3 and $U_{3'}$) are diluted at least 2-fold with a suitable medium and the plasma membranes are collected by centrifugation at 100,000 *g* for 30 min. If the lower phase material (intracellular membranes) is to be saved it should be diluted at least 10-fold before pelleting. Note that the composition of the phases, and thereby the partitioning, is temperature dependent. Therefore, keep a constant (e.g., 4°) temperature throughout.

Properties of the Fractions

Using the batch procedure two fractions containing purified plasma membrane (U_3 and $U_{3'}$) and one fraction containing intracellular membranes depleted of plasma membrane (L_1) are obtained. In the experiment shown in Table II 5.8% of the protein and about 50% of the plasma membrane markers were recovered in U_3 whereas $U_{3'}$ contained 3.5% of the protein and about 25% of the plasma membrane markers. This signified a 10-fold enrichment of plasma membrane markers in U_3 and a 7-fold enrichment in $U_{3'}$ (Table III). Thus, $U_{3'}$ was not as pure as U_3 and assuming 95% purity for U_3 (see below) gives a purity of about 75% for $U_{3'}$. Note that antimycin A-resistant NADPH–cytochrome-*c* reductase is far from an ideal marker for the endoplasmic reticulum. Even if most of the activity was recovered in the lower phase (L_1, Table II) a substantial amount was recovered in U_3 and the activity was in fact enriched in the plasma membrane (Table III). This is because not only the endoplasmic reticulum but also the plasma membrane contains cytochrome *P*-450 and hence also NADPH–cytochrome-*P*-450 reductase[12] (assayed as NADPH–cytochrome-*c* reductase).

[12] P. Kjellbom, C. Larsson, P. Askerlund, C. Schelin, and S. Widell, *Photochem. Photobiol.* **42,** 779 (1985).

TABLE II

TOTAL PROTEIN, CHLOROPHYLL, AND MARKER ENZYME ACTIVITIES IN MICROSOMAL (MF), PLASMA MEMBRANE (U_3 AND $U_{3'}$), AND INTRACELLULAR MEMBRANE (L_1) FRACTIONS

	Fraction							
		U_3		$U_{3'}$		L_1		
Component	MF, amount[a]	Amount	(%)	Amount	(%)	Amount	(%)	Recovery (%)
Protein	31.3	1.8	5.8	1.1	3.5	19.6	63	72
K^+,Mg^{2+}-ATPase[b]	1.82	0.97	53	0.50	27	0.20	11	91
K^+-ATPase[c]	0.51	0.28	55	0.12	24	0.02	4	83
K^+,Mg^{2+}-$ATPase_{van}$[d]	0.72	0.49	68	0.17	24	0.10	14	106
GS II	0.088	0.048	54	0.020	23	0.012	14	91
Cytochrome-*c* oxidase	14.7	0.067	0.5	0.051	0.3	8.6	59	60
NADPH–cytochrome-*c* reductase	0.44	0.13	30	0.04	9	0.18	42	81
Chlorophyll	4.81	0.001	0.02	0.002	0.04	3.59	75	75

[a] Protein and chlorophyll are given in milligrams, all enzyme activities are given in micromoles $minute^{-1}$.

[b] Total K^+-stimulated, Mg^{2+}-dependent ATPase (assayed in the presence of 0.05% Triton X-100).

[c] The K^+-stimulation of the K^+,Mg^{2+}-ATPase.

[d] The vanadate inhibition of the K^+,Mg^{2+}-ATPase.

TABLE III

SPECIFIC ACTIVITIES AND ENRICHMENT OF MARKER ENZYME ACTIVITIES IN MICROSOMAL (MF), PLASMA MEMBRANE (U_3 AND $U_{3'}$), AND INTRACELLULAR MEMBRANE (L_1) FRACTIONS

Enzyme	MF, activity[a]	U_3 Activity	U_3 Enrichment[b]	$U_{3'}$ Activity	$U_{3'}$ Enrichment	L_1 Activity	L_1 Enrichment
K^+,Mg^{2+}-ATPase	0.058	0.54	9.3	0.45	7.8	0.010	0.17
K^+-ATPase	0.016	0.16	10.0	0.11	6.9	0.001	0.06
K^+,Mg^{2+}-ATPase$_{van}$	0.023	0.27	11.7	0.15	6.5	0.005	0.22
GS II	0.0028	0.027	9.6	0.018	6.4	0.0006	0.21
Cytochrome-*c* oxidase	0.47	0.037	0.08	0.046	0.10	0.44	0.94
NADPH–cytochrome-*c* reductase	0.014	0.072	5.1	0.036	2.6	0.009	0.64

[a] Specific activities are micromoles (mg protein)$^{-1}$ minute^{-1}.

[b] Enrichment is relative to the specific activities of the microsomal fraction.

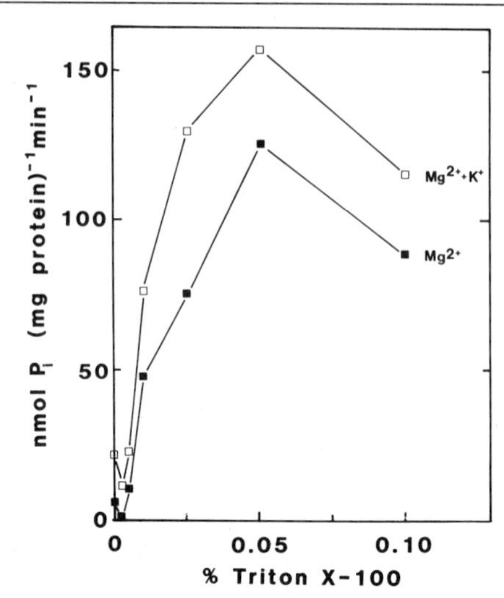

FIG. 4. The K^+-stimulated, Mg^{2+}-dependent ATPase activity of isolated plasma membrane vesicles (cauliflower inflorescences, redrawn from Larsson *et al.*[10]) shows a high latency, which stresses the importance of measuring the ATPase in the presence of detergent to obtain the full activity. For each species and batch of detergent, the latency should be titrated as shown above, since too high a detergent concentration may be inhibitory.

Marker Enzymes

ATPase activities were always measured on pelleted fractions and essentially according to Gallagher and Leonard,[13] using 1 m*M* azide and 0.1 m*M* molybdate to suppress mitochondrial ATPase and unspecific phosphatases, respectively. A rather low pH (6.0) was chosen to further enhance plasma membrane–ATPase activity relative to other ATPases. The ATPase activities were measured in isosmotic (0.33 *M* sucrose) medium with and without 0.05% Triton X-100, since a large part of the plasma membrane ATPase activity is latent[8–10] due to the right-side-out orientation of the plasma membrane vesicles (Fig. 4). Total ATPase activity is represented by the activity obtained in the presence of 0.05% detergent, while latent activity is given by the difference between the activities assayed in the presence and absence of Triton X-100. Vanadate inhibition was achieved with 0.1 m*M* orthovanadate.

Glucan synthase II (GS II) was assayed according to Ray[14] at pH 7.0. Note, however, that the GS II assay has recently been greatly improved by including digitonin, spermine and Ca^{2+} in the assay medium,[15] and this latter procedure is therefore recommended.

[13] S. R. Gallagher and R. T. Leonard, *Plant Physiol.* **70,** 1335 (1982).

[14] P. M. Ray, *in* "Plant Organelles" (E. Reid, ed.), p. 135. Ellis Horwood, Chichester, 1979.

[15] H. Kauss and W. Jeblick, *Plant Sci.* **48,** 63 (1987).

Cytochrome-*c* oxidase and antimycin A-resistant NADPH-cytochrome-*c* reductase were measured according to Hodges and Leonard,[16] in the presence of 0.01% Triton X-100.

Protein was determined according to Bearden[17] using 0.01% Triton X-100 to solubilize membrane proteins.

Comparison of Phase Partition with Other Methods

Specific staining with phosphotungstic acid or silicotungstic acid[18] seems to be the only universal marker for the plant plasma membrane and the only one which permits a real estimation of the purity of the preparations. Based on this staining[5,19–21] we have estimated the purity to be higher than 90% and often close to 100% plasma membrane,[6] and similar purities were recently reported by Buckhout and Hrubec[22] (93%) for plasma membrane from maize roots, and by Sandelius *et al.*[7] (>90%) for plasma membrane preparations obtained both by phase partitioning and free flow electrophoresis. Assuming 95% purity of the preparations produced by phase partition, a purity of 50% is obtained for oat root plasma membrane prepared by sucrose gradient centrifugation (calculated from the data on specific K^+-ATPase activity reported by Hodges and Mills,[8] Table III, in a comparative study on the two methods). Similarly, a purity of 95 and 75%, respectively, for the two methods can be calculated for wheat root plasma membrane.[9] Thus, it is obvious from comparative studies, that phase partition produces plasma membrane preparations with a much higher purity than sucrose gradient centrifugation.

Practical Advice

To begin your study of phase partitioning, choose some green material like spinach, oat, or maize leaves. This will enable you to follow the partitioning of the intracellular membranes by eye. Do a full-scale experiment with the green material using the conditions given above for oat. After three steps of the batch procedure, dilute the phases and pellet the membranes. Assay for appropriate markers. If the U_3 fraction is green, make a series of small-phase systems with increasing polymer concentra-

[16] T. K. Hodges and R. T. Leonard, this series, Vol. 32, p. 392.

[17] J. C. Bearden, Jr., *Biochim. Biophys. Acta* **533,** 525 (1978).

[18] J. C. Roland, *in* "Electron Microscopy and Cytochemistry of Plant Cells" (J. L. Hall, ed.), p. 1. Elsevier, Amsterdam, 1978.

[19] S. Widell, T. Lundborg, and C. Larsson, *Plant Physiol.* **70,** 1429 (1982).

[20] S. Widell and C. Larsson, *Physiol. Plant.* **57,** 196 (1983).

[21] J. Hellergren, S. Widell, T. Lundborg, and A. Kylin, *Physiol. Plant.* **58,** 7 (1983).

[22] T. J. Buckhout and T. C. Hrubec, *Protoplasma* **135,** 144 (1986).

tions above those used. Do not change the ionic composition. Choose the lowest polymer concentration where chlorophyll is well partitioned into the lower phase (compare Fig. 1). If, on the other hand, the U_3 fraction is whitish but the yield is very low, make a corresponding experiment with polymer concentrations below those used. When the procedure is working well the yield in U_3 of protein and plasma membrane markers should be about twice that in $U_{3'}$ (compare Table II). If the purity is still not satisfactory, the chloride concentration may be increased further (Fig. 2), or further partition steps may be added. For instance, to obtain chlorophyll-free plasma membrane from barley leaves five steps were needed (giving U_5 and $U_{5'}$), whereas three steps were enough for spinach leaves[5] using phase systems of identical composition. Note that before pooling U_3 and $U_{3'}$ it should be checked that the latter has a purity which is satisfactory (cf. Table III). If the yield is important but the purity of $U_{3'}$ is not sufficient, $U_{3'}$ alone may be repartitioned an extra time before pooling the plasma membrane fractions. The phase composition worked out with green material will most probably also suit your material of interest.

Acknowledgments

We wish to thank Mrs. Adine Karlsson and Mrs. Ann-Christine Holmström for skillful technical assistance. We are also indebted to Dr. Marianne Sommarin, Department of Plant Physiology, Lund, Sweden, and Dr. Thomas J. Buckhout, Beltsville Agricultural Research Center, Maryland, for valuable advice concerning the ATPase assays. This work was supported by the Swedish Natural Science Research Council, the Swedish Council for Forestry and Agricultural Research, and the Carl Tesdorpf Foundation.

[53] Possible Approaches to Surface Labeling of the Plasma Membrane

By J. L. HALL

Several reviews have concluded that there is no unequivocal enzymatic or morphological marker for the plasma membrane of higher plants.[1-3] More recently interest has centered on cation-sensitive and

[1] P. H. Quail, *Annu. Rev. Plant Physiol.* **30,** 425 (1979).

[2] J. L. Hall, *in* "Isolation of Membranes and Organelles from Plant Cells" (J. L. Hall and A. L. Moore, eds.), p. 55. Academic Press, London, 1983.

[3] D. S. Perlin and R. M. Spanswick, *Plant Physiol.* **65,** 1053 (1980).

vanadate-inhibited ATPase activity.[4] A similar activity also appears to be associated with secretory vesicles[5] while a vanadate-sensitive ATPase has been reported for chloroplast envelopes.[6] An alternative approach to the search for intrinsic markers is to attempt to tag the plasma membrane of intact cells with nonpenetrating labels. These may be monitored during the subsequent cell fractionation procedures by making use of their radioactive, fluorescent, or electron-opaque properties. There are several requirements for such labels.[7,8] They should (1) not penetrate the cell membrane or do so only very slowly, (2) react irreversibly (preferably covalently) with a membrane constituent under physiological conditions, (3) not seriously alter membrane structure or activity, (4) be detectable in small amounts. In practice, great care is needed to ensure that only the plasma membrane is labeled since the protein exposed externally at the plant cell surface constitutes no more than 2–3% of the total cell protein.[1] Thus any labeling of internal membranes from damaged or leaky cells can confuse the apparent selective labeling of the surface membrane. This is particularly a problem when excised tissue pieces are used,[9–11] so that many workers have turned to isolated protoplasts in an attempt to overcome these difficulties. The use of protoplasts should allow rapid and even access of the label to the surface of all cells in a preparation, although the potential leaky nature of protoplasts[3] must be carefully assessed in all systems.

A wide range of techniques have been used to label specifically the surface of animal cells and some, such as enzymatic iodination, have been explored in considerable detail.[8] However, there have been relatively few attempts to label the higher plant plasma membrane, and there is no widely applicable method available at the present time that has been shown to selectively tag the surface membrane of a variety of different cells. The optimum conditions and possibility of internal labeling must be carefully assessed for each method and cell system, and there is considerable potential for the development of new techniques. Thus this chapter cannot attempt to give specific details of methodologies that have wide application. Instead it will review the range of methods that have been

[4] H. Sze, *Annu. Rev. Plant Physiol.* **36,** 175 (1985).

[5] L. L. W. Binari and R. H. Racusen, *Plant Physiol.* **71,** 594 (1983).

[6] D. R. McCarty, K. Keegstra, and B. R. Selman, *Plant Physiol.* **76,** 584 (1984).

[7] D. F. H. Wallach and P. S. Lin, *Biochim. Biophys. Acta* **300,** 211 (1973).

[8] A. L. Hubbard and Z. A. Cohn, *in* "Biochemical Analysis of Membranes" (A. H. Maddy, ed.), p. 427. Chapman & Hall, London, 1976.

[9] J. L. Hall and R. M. Roberts, *Ann. Bot. (London)* [N.S.] **39,** 983 (1975).

[10] P. H. Quail and A. Browning, *Plant Physiol.* **59,** 759 (1977).

[11] E. A.-H. Baydoun and D. H. Northcote, *J. Cell Sci.* **45,** 147 (1980).

employed with plant cells and tissues, and discuss their potential advantages and pitfalls as a guide to further studies on this subject.

Enzymatic Iodination

There are a number of enzyme-mediated labeling systems[8] which utilize the lack of penetration of cells by large exogenously applied enzymes to label specific groups in the plasma membrane. The lactoperoxidase-mediated iodination method was the first reported enzyme-mediated system,[12] is the most widely used with animal cells,[8] and is the surface labeling technique that has received the most attention from plant biologists. The method relies on the iodination of reactive aromatic rings (e.g., tyrosine residues) at the cell surface by electrophilic forms of iodine (I^+) generated in the presence of $Na^{125}I$, H_2O_2, and lactoperoxidase. Early reports using maize coleoptile segments reported a preferential labeling of the plasma membrane,[13,14] although experiments with excised roots suggested that considerable labeling of intracellular proteins occurred.[9] This could be due to penetration by the lactoperoxidase or be a result of the considerable intracellular peroxidase activity found in plant cells.[15,16]

Another major source of artifact that can arise with the iodination of tissue segments comes from the labeling of extracellular material that subsequently behaves as a membrane fraction.[10] When attempts were made to label excised *Cucurbita* hypocotyl segments, autoradiography indicated that the ^{125}I was almost exclusively associated with an amorphous film (perhaps originating from the phloem) overlying the cut ends and along the outer surface of the cuticle.[10] This material may associate with membrane fractions after homogenization. Similar problems can presumably arise with other surface labeling techniques and emphasizes the need for careful checks (e.g., autoradiographic) to verify that selective surface labeling does occur.

Some of these problems may be alleviated by the use of isolated protoplasts together with immobilized enzymatic or chemical iodination procedures. Enzymobeads (Bio-Rad Laboratories) and iodination beads (New England Nuclear) contain lactoperoxidase and glucose oxidase covalently attached to inert supports and are considered to allow controlled, gentle labeling. $Na^{125}I$, glucose (to generate H_2O_2 by glucose oxidase), and the protein to be iodinated are added. This approach has been used to label

[12] D. R. Phillips and M. Morrison, *Biochemistry* **10,** 1766 (1971).

[13] T. Hendricks, *Plant Sci. Lett.* **7,** 347 (1976).

[14] R. Yu, J. Carter, and T. Osawa, *J. Exp. Bot.* **27,** 294 (1976).

[15] J. L. Hall and R. Sexton, *Planta* **108,** 103 (1972).

[16] D. W. Galbraith and D. H. Northcote, *J. Cell Sci.* **24,** 295 (1977).

the surface proteins of protoplasts isolated from ryegrass (*Lolium multiflorum*) endosperm[17] and *Zea mays* coleoptiles,[18] although in both cases there was some evidence that intracellular proteins were also labeled, either from broken cells or by penetration of the protoplasts by $Na^{125}I$ and reaction with intracellular peroxidase.

Iodination can also be achieved by chemical catalysis. Iodogen (1,3,4,6-tetrachloro-3α,6α-diphenylglycoluril) (Pierce Chemical Co) is described as a gentle, effective solid-phase iodinating reagent for soluble and membrane proteins.[19] It is virtually insoluble in water and so can be used to coat the surface of the reaction vessel by solubilization in and then removal of a solvent. The iodination is terminated by simply decanting the reaction solution from the solid iodogen. The iodogen procedure has been applied to ryegrass protoplasts,[17] and in this and some animal systems[20] appears to be more effective than procedures based on lactoperoxidase. However, the work on ryegrass protoplasts and unpublished results in this laboratory using barley leaf protoplasts[21] indicate that intracellular proteins are labeled, presumably by penetration of $Na^{125}I$.

An alternative chemical method employes chloramine-T,[20] but this is a powerful oxidant and can have harsh effects. A more recent addition to the range is Iodo-beads (Pierce Chemical Co.) that employs *N*-chlorobenzenesulfonamide (sodium salt) derivatized uniform nonporous polystyrene beads, but there are no reports of an application to plant cells.

These immobilized systems are clearly worthy of further investigation, although optimal conditions for labeling, washing, protoplast lysis, and fractionation will have to be carefully analyzed for each system and careful checks for intracellular iodination included.

Diazotized Sulfanilic Acid

This reagent reacts with histidine and tyrosine residues on membrane proteins in a mild reaction that occurs at neutral pH and is considered to have little effect on membrane function. It was initially developed for animal systems,[22] but has been applied to both intact plant tissues and to isolated protoplasts. A preferential labeling of the plasma membrane was indicated when used with intact maize roots, although the results were not unequivocal and the cell wall was thought to present an obstacle to specific labeling.[11] With isolated protoplasts, both diazotized [^{35}S]sulfanilic

[17] A. Schibeci, G. B. Fincher, B. A. Stone, and A. B. Wardrop, *Biochem. J.* **205,** 511 (1982).
[18] B. Brummer and R. W. Parrish, *Planta* **157,** 446 (1983).
[19] P. J. Fraker and J. C. Speck, *Biochem. Biophys. Res Commun.* **80,** 849 (1978).
[20] M. A. K. Markwell and C. F. Fox, *Biochemistry* **17,** 4807 (1978).
[21] R. T. Jackson, Ph.D. Thesis, University of Southampton (1986).
[22] H. C. Berg and D. Hirsh, *Anal. Biochem.* **66,** 629 (1975).

acid[16] and diazotized [^{125}I]iodosulfanilic acid[3] have been employed and both have provided evidence for a preferential labeling of the plasma membrane. However, it should be pointed out that although labeling was coincident with ATPase activity, plasma membrane-specific characteristics of this enzyme (e.g., cation and vanadate sensitivity) were not investigated and no independent verification of specific surface labeling (e.g., autoradiography) was supplied.

Pyridoxal Phosphate-Tritiated Borohydride

This procedure was first applied to virus particles,[23] and later to animal tissue culture cells.[24] Pyridoxal phosphate permeates some membranes very slowly and generates Schiff bases with protein amino groups, under physiological conditions, which are then tritiated by reduction with tritiated borohydride. Work with animal cells clearly indicates the importance of establishing the optimal surface labeling conditions for each new cell or tissue system investigated since some membranes are penetrated more easily than others.[8]

However, the limited experience with intact plant tissues[1,9] indicates a greater permeability of plant membranes to phosphate esters and so a lack of selective surface labeling. There is clearly scope for a more rigorous study of optimal conditions for a wider range of tissues and also of the potential use of isolated protoplasts.

Concanavalin A

Concanavalin A (Con A) is a sugar-specific lectin that binds to and agglutinates plant protoplasts.[25,26] It is specific for glucosyl and mannosyl residues and binds to glycoproteins and glycolipids on the plasma membrane. There is evidence for Con A-binding sites on the endoplasmic reticulum,[27,28] although they appear to be absent from chloroplasts, mitochondria, and microbodies.[29]

Labeled Con A has been used in two ways. It has been applied to intact protoplasts as a potential plasma membrane marker in subsequent cell fractionation or to isolated membrane fractions to study such aspects

[23] D. B. Rifkin, R. W. Compans, and E. Reich, *J. Biol. Chem.* **247,** 6432 (1972).

[24] R. C. Hunt and J. C. Brown, *Biochemistry* **13,** 22 (1974).

[25] F. A. Williamson, L. C. Fowke, F. C. Constabel, and O. L. Gamborg, *Protoplasma* **89,** 305 (1976).

[26] J. Burgess and P. J. Linstead, *Planta* **130,** 73 (1976).

[27] R. L. Berkowitz and R. L. Travis, *Plant Physiol.* **68,** 1014 (1981).

[28] M. A. Hartmann, P. Benveniste, and J. C. Roland, *Plant Sci. Lett.* **30,** 239 (1983).

[29] S. E. Frederick, B. Nies, and P. J. Gruber, *Planta* **152,** 145 (1981).

as the sidedness of membrane vesicles. When protoplasts from carrot cell suspensions were labeled with [^{14}C]acetyl-Con A, a single labeled membrane fraction was isolated after fractionation on both continuous and discontinuous Renografin gradients, and was tentatively identified as plasma membrane.[30] Since Con A is sugar specific and binding is not covalent, it was necessary to use a gradient medium such as Renografin. This appears to be the only thorough investigation of this potentially useful technique. However, ferritin-labeled Con A has been used in conjunction with electron microscopy to study the sidedness of isolated vesicles. Membrane vesicles in putative plasma membrane fractions from both soybean[31] and maize coleoptiles[28] showed extensive labeling on the outer surface. The labeling pattern with maize coleoptile vesicles was not altered by detergent treatment, suggesting that the vesicles were right side out. This was also suggested by the binding of radiolabeled Con A to membrane fractions from cultured carrot cells.[32] Thus these results suggest that labeled Con A may be a useful method for identifying the plasma membrane when applied either to protoplasts or to isolated membrane fractions. However, binding sites may exist on other membranes and the possibility of dissociation and rebinding must always be considered.

Lanthanum

When isolated protoplasts from tobacco leaves were incubated with $LaCl_3$ or $La(NO_3)_3$, electron-opaque lanthanum deposits appear to bind irreversibly to the plasma membrane with no detectable penetration and internal staining.[33,34] Lanthanum ions probably compete for calcium-binding sites and are known to penetrate the plasma membrane only very slowly or not at all.[35] Fractionation of prelabeled tobacco protoplasts produced lanthanum-stained membrane fragments, often as large sheets, and no detectable staining of identifiable cell organelles.[34] This suggests that redistribution of lanthanum during cell fractionation does not occur. However, lanthanum is not easily assayed quantitatively. Neutron activation analysis has been employed,[36] analysis of electron micrographs may be used, while fluorescence techniques are currently under investigation in this laboratory.

[30] W. F. Boss and A. W. Ruesink, *Plant Physiol.* **64,** 1005 (1979).
[31] R. L. Travis and R. L. Berkovitz, *Plant Physiol.* **65,** 871 (1980).
[32] S. K. Randall and A. W. Ruesink, *Plant Physiol.* **73,** 385 (1983).
[33] A. R. D. Taylor and J. L. Hall, *Protoplasma* **96,** 113 (1978).
[34] A. R. D. Taylor and J. L. Hall, *Plant Sci. Lett.* **14,** 139 (1979).
[35] R. T. Leonard, G. Nagahashi, and W. W. Thomson, *Plant Physiol.* **55,** 542 (1975).
[36] A. R. D. Taylor and J. L. Hall, *in* "Advances in Protoplast Research," p. 463. Hung. Acad. Sci., Budapest, 1980.

Positively Charged Beads

This method relies on the binding of the negatively charged surfaces of intact cells to positively charged beads. Surfaces still exposed are then neutralized, the cells ruptured, and the cellular debris washed away; the beads covered with surface membrane are then collected by simple centrifugation. This approach was first applied to erythrocytes using glass beads coated with a layer of poly(lysine).[37,38] A modification of this method was developed for use with both yeast protoplasts and *Dictyostelium*,[39,40] and employs positively charged silica microbeads (10–50 nm in diameter) to coat the cells. High yields of plasma membrane (based on vanadate-sensitive ATPase as a marker) were obtained from yeast protoplasts and these appeared to be essentially free of other organelles.[39]

This method has now been applied to higher plants using protoplasts isolated from the apical regions of etiolated *Pisum sativum* stems.[41] Briefly, stable protoplasts are isolated and concentrated by standard procedures. The protoplasts are mixed with the microbeads and then separated from unbound beads by low-speed centrifugation. Exposed surfaces are neutralized by the addition of dextran sulfate. The protoplasts are then washed, lysed, and the pellet, containing the bead/plasma membrane sheets, separated by further low-speed centrifugation. This technique yields a fraction containing long sheets of membranes with beads coating only one surface. The fraction was enriched in vanadate-sensitive ATPase activity, while marker activities for other cellular membranes were very low or not detectable. This approach clearly merits wider investigation. In this laboratory it has been employed with protoplasts isolated from *Cucumis* hypocotyls. Similar membrane sheets are obtained, although considerable difficulty has been experienced in isolating such a fraction free of high cytochrome oxidase activity. It is possible that some internal membranes are trapped in the large sheets during lysis and centrifugation.

UDPglucose

Radioactive UDPglucose, a substrate for callose-synthesizing activity which is presumed to be located at the plasma membrane, has been em-

[37] B. S. Jacobson, J. Cronin, and D. Branton, *Biochim. Biophys. Acta* **506,** 81 (1978).

[38] D. I. Kalish, C. M. Cohen, B. S. Jacobson, and D. Branton, *Biochim. Biophys. Acta* **506,** 97 (1978).

[39] R. Schmidt, R. Ackermann, Z. Kratky, B. Wasserman, and B. Jacobson, *Biochim. Biophys. Acta* **732,** 421 (1983).

[40] L. K. Chaney and B. S. Jacobson, *J. Biol. Chem.* **258,** 1062 (1983).

[41] D. R. Polonenko and G. A. Maclachlan, *J. Exp. Bot.* **35,** 1342 (1984).

ployed as a surface label with pea stem segments.[42] The exogenously applied glucose was assumed to be incorporated into callose polymers that remain attached to the plasma membrane during subsequent homogenization and fractionation. Certainly the distribution of K^+-ATPase activity and other markers suggested that the plasma membrane was selectively tagged. However, subsequent work with a similar tissue[43] did not find such a clear coincidence with plasma membrane markers and there was some evidence of a mitochondrial origin.

Miscellaneous Approaches to Surface Labeling

There are a variety of other approaches to the selective labeling of the plasma membrane which have been only briefly explored, or have yet to receive attention.

For example, radioactive fusicoccin could be employed as this toxin is considered to act at the plasma membrane. There is some evidence that fusicoccin binds selectively to the plasma membrane when added to isolated membrane fractions from maize coleoptiles, although the peak obtained after gradient centrifugation was quite broad and coincided with that for cytochrome oxidase.[44] Another toxin, helminthosporaside, is thought to bind to a protein at the external surface of sugar cane cells.[45] Fractionation studies, however, were not unequivocal since a detailed study of marker distributions was not undertaken.

Two other approaches seem well worth consideration, one involving affinity labeling and the other the use of monoclonal antibodies to plant plasma membrane antigens. For example, binding sites for auxin are thought to occur on the plasma membrane. This raises the possibility of using an auxin analog with a chemical reactive group which will become irreversibly attached at the binding site. Recent studies in this laboratory using the diazonium salt of the tritiated auxin analog chloramben[46] to label barley leaf protoplasts have yielded promising results.[21]

The monoclonal antibody approach is potentially very exciting.[47] It involves the generation of specific antibodies to plasma membrane antigens from initially impure membrane fractions. Screening of a range of antibodies requires surface binding tests using isolated protoplasts. The plasma membrane antibodies can then be tested against membrane fractions separated by standard gradient and phase partition procedures.

[42] R. L. Anderson and P. M. Ray, *Plant Physiol.* **61,** 723 (1978).
[43] W. S. Pierce and D. L. Hendrix, *Planta* **146,** 161 (1979).
[44] U. Dohrmann, R. Hertel, and H. Kowalik, *Planta* **140,** 97 (1978).
[45] G. A. Strobel and W. M. Hess, *Proc. Natl. Acad. Sci. U.S.A.* **71,** 1413 (1974).
[46] M. A. Venis, *Planta* **134,** 145 (1977).
[47] P. M. Norman, V. P. M. Wingate, M. S. Fitter, and C. J. Lamb, *Planta* **167,** 452 (1986).

[54] Isolation of Endoplasmic Reticulum: General Principles, Enzymatic Markers, and Endoplasmic Reticulum-Bound Polysomes

By J. MICHAEL LORD

Plant cells contain a well-defined endoplasmic reticulum (ER) which, like its mammalian cell counterpart, has several characteristic functions including protein synthesis, glycosylation, and lipid synthesis. These functions account, in part, for the observation that the cytoplasmic surface of the ER is predominantly studded with ribosomes (rough ER). Certain areas of the ER membrane are free of attached ribosomes (smooth ER) and are therefore not directly involved in protein synthesis.[1]

General Principles

The ER is a complex, convoluted network of interconnecting membrane channels and sheets, and is inevitably damaged during even the gentlest and briefest cellular homogenization. As a result the ER is conveniently fragmented into a large number of small, closed vesicles called microsomes which can be readily isolated by differential or density gradient centrifugation. Differential centrifugation of postmitochondrial supernatants at high speed has been successfully used to isolate mammalian microsomes,[2] but is generally less effective when applied to plant cell homogenates. Plant "microsomal" pellets prepared in this way are often heavily contaminated with other membranous material derived from Golgi, tonoplast, thylakoid, or plasma membrane. Sucrose density gradient centrifugation has become the preferred method for isolating plant microsomes. Smooth microsomes exhibit a relatively narrow equilibrium density range in such gradients. In contrast, rough-surfaced microsomes have a wider range of equilibrium densities. Because these membrane vesicles have heavier ribosomes attached, these densities are greater than for smooth microsomes. Isolation of rough microsomes by sucrose density gradient centrifugation is often facilitated by deliberately stripping off the attached ribosomes by using homogenization buffers and gradients

[1] M. J. Chrispeels, *in* "The Biochemistry of Plants" (P. K. Stumpf and E. E. Conn, eds.), Vol. 1, p. 389. Academic Press, New York, 1980.

[2] C. M. Redman and D. D. Sabatini, *Proc. Natl. Acad. Sci. U.S.A.* **56,** 608 (1966).

containing EDTA.[3] These ribosome-denuded vesicles then revert to the mean buoyant density of smooth microsomes and a relatively homogeneous microsomal population has been generated. The isolated microsomes retain most of the functional properties of the ER from which they are derived.

Isolation of ER Membranes

Methods and assays which have been employed to isolate and characterize ER membranes from germinating castor bean endosperm tissue are described here. With appropriate modifications such as using the most suitable homogenization method for the tissue under investigation, a similar approach has been successfully utilized to isolate ER membranes from a range of plant tissues.

Ten castor bean endosperm halves, removed from seedlings after 3 days' germination at 30° in the dark, are homogenized by chopping for 5–10 min with a single razor blade in 3 ml of grinding medium contained in a Petri dish on ice. The grinding medium consists of 150 m*M* Tricine, pH 7.5, 10 m*M* KCl, 1 m*M* EDTA, pH 7.5, 1 m*M* $MgCl_2$, and 12% (w/w) sucrose. The crude homogenate is filtered through four layers of cheesecloth and the residue is extracted with a further 2 ml of grinding medium. The volume of the filtered homogenate is adjusted to 5 ml with grinding medium, and cell debris is removed by centrifuging at 270 *g* and 2° for 10 min. The supernatant is layered directly onto a sucrose density gradient. The gradient consists of 25 ml of sucrose solution increasing linearly in concentration from 30–60% (w/w) sucrose, topped with a 5 ml layer of 20% (w/w) sucrose, formed over a 2-ml cushion of 60% sucrose. Gradients are formed in 38.5-ml centrifuge tubes. All sucrose solutions contain 1 m*M* EDTA, pH 7.5. Gradients are centrifuged for 3 hr at 25,000 rpm (83,000 g_{av}) and 2° in an SW27 rotor on a Beckman ultracentrifuge. After centrifugation, 1.0-ml fractions are collected using an ISCO model 185 density gradient fractionator. The ER membrane band is situated at the interface between the 20 and 30% sucrose layers. This gradient distribution can be confirmed using a variety of biochemical and enzymatic markers, several of which are described below. Some of these markers occur to a certain extent in other cellular membranes; their presence in a particular fraction does not necessarily identify that fraction as ER membranes. These markers are considerably enriched in the ER membrane, however, and a major peak of activity reliably identifies microsomes, particularly when the distribution coincides with that of an enzyme uniquely present

[3] J. M. Lord, T. Kagawa, T. S. Moore, and H. Beevers, *J. Cell Biol.* **57**, 659 (1973).

in ER membranes such as the glycolipid glycosyltransferases with a synthetic role in protein N-glycosylation.

Electron Transport Enzymes and Cytochromes

NADH- and NADPH-cytochrome-*c* reductases are found in the ER membrane. Mitochondria also contain NADH-cytochrome-*c* reductase but this can be specifically inhibited, and hence differentiated from the ER enzyme, by including antimycin A in the assay. ER membranes also contain cytochrome b_5 and cytochrome *P*-450 whose presence can be demonstrated by difference spectra. The assay for NADH-cytochrome-*c* reductase contains, in a final volume of 1.0 ml, 20 m*M* potassium phosphate, pH 7.2, 0.2 m*M* NADH, 0.02 m*M* cytochrome *c*, and 10 m*M* potassium cyanide. The reaction is started by adding ER membrane suspension and the reduction of cytochrome *c* is followed spectrophotometrically as an absorbance increase at 550 nm.

When appropriate, antimycin A is added in a small volume (5 or 10 μl) of ethanol to give a final concentration of 1 μM.

Difference spectra are produced by adding the microsomal suspension to each of two 1-ml cuvettes which are placed in a dual beam recording spectrophotometer. A wavelength scan is made to ensure a level baseline. Cytochrome b_5 is demonstrated by the difference spectrum between the untreated reference sample and the second sample which has been reduced by the addition of a few crystals of dithionite, 0.2 m*M* NADH, or 0.2 m*M* NADPH. The presence of cytochrome *P*-450 is demonstrated by recording spectral changes resulting from carbon monoxide binding to the dithionite-reduced cytochrome. Carbon monoxide is bubbled through the sample cuvette for 30 sec and the spectrum is recorded against a dithionite-reduced reference sample.

Phospholipid Biosynthesis

The active sites of several phospholipid synthesizing enzymes have been localized to the cytoplasmic surface of the ER.[4] From these and other data it has been inferred that newly synthesized phospholipids are initially inserted into the cytoplasmic surface monolayer of the ER membrane. Assembly of the phospholipid bilayer requires the translocation of some of these phospholipids to the monolayer facing the ER lumen, a step thought to be achieved by a phospholipid transporter protein.[5]

[4] R. M. Bell, L. M. Ballas, and R. A. Coleman, *J. Lipid Res.* **22,** 391 (1981).
[5] W. R. Bishop and R. M. Bell, *Cell (Cambridge, Mass.)* **42,** 51 (1985).

The ER is not the exclusive site of synthesis of all the phospholipids found in the various intracellular membrane fractions of enkaryotic cells.[6] For example, chloroplasts actively synthesize phosphatidylglycerol[7] and mitochondria synthesize both phosphatidylglycerol and cardiolipin.[8]

Other major membrane phospholipids such as phosphatidylcholine and phosphatidylethanolamine are synthesized by enzymes present in the ER membrane. Whether such enzymes are exclusively located in the ER membrane is controversial, and a small proportion of the total cellular activity has been ascribed to Golgi-derived fractions in certain plant tissues.[9] It is clear, however, that the ER is the major cellular location of phospholipid biosynthetic enzymes. Cholinephosphotransferase, the enzyme which catalyzes phosphatidylcholine synthesis from CDPcholine and a 1,2-diglyceride, is therefore a useful ER membrane marker.[3]

Cholinephosphotransferase can be assayed in a reaction mixture which contains, in a final volume of 0.5 ml, 100 m*M* Tris–HCl, pH 7.0, 10 m*M* $MgCl_2$, 0.1 μCi CDP[*methyl*-^{14}C]choline (specific activity 30 Ci/mol; Amersham, UK), and enzyme. After incubation at 30° for periods up to 1 hr, the reaction is stopped by adding 2 ml of ethanol. The precipated protein is removed by centrifugation and the pellet is extracted with a further 2 ml of ethanol. After centrifugation, the ethanol phases are combined and mixed with 3 ml of chloroform. The organic phase is washed twice with 5-ml portions of 2 *M* KCl and twice with 5-ml portions of water to ensure complete removal of unreacted CDP[^{14}C]choline. The residual chloroform phase, which contains the phosphatidyl[^{14}C]choline, is transferred to a scintillation vial, evaporated to dryness, and assayed for ^{14}C-labeled phospholipid content after adding an appropriate scintillant.

It is not usually necessary to add exogenous 1,2-diglyceride to the assay mixture. Cholinephosphotransferase activity is inhibited by most detergents, which cannot therefore be used to solubilize endogenous 1,2-diglyceride.

While the conventional assay described above can be used to locate ER membranes after cellular fractionation, for example among collected gradient fractions, an alternative approach is to incubate the crude homogenate with CDP[^{14}C]choline prior to fractionation.[10] This approach takes advantage of the fact that phosphatidyl[^{14}C]choline is not only syn-

[6] M. D. Marshall and M. Kates, *Biochim. Biophys. Acta* **260,** 558 (1972).

[7] T. S. Moore, *Plant Physiol.* **54,** 164 (1974).

[8] M. J. Montague and P. M. Ray, *Plant Physiol.* **59,** 225 (1977).

[9] J. M. Lord, T. Kagawa, and H. Beevers, *Proc. Natl. Acad. Sci. U.S.A.* **69,** 2429 (1972).

[10] J. M. Lord, *Plant Physiol.* **57,** 218 (1976).

thesized by an ER enzyme but is concomitantly inserted into the cytoplasmic monolayer of the ER membrane.[5]

Adjust the Mg^{2+} content of the crude homogenate to 10 m*M* and add 0.25 to 0.5 μCi of CDP[^{14}C]choline. Incubate at 30° for 30 min before fractionating by sucrose density gradient centrifugation. The phosphatidyl[^{14}C]choline content of each collected gradient fraction can be determined by chloroform extraction exactly as described for the choline phosphotransferase assay. Problems may be encountered with this approach if the original crude homogenates contain high levels of soluble phospholipid exchange protein activity.[11] In this case there may be extensive redistribution of newly synthesized phosphatidyl[^{14}C]choline from the ER membrane to the membranes of other organelles prior to gradient fractionation. This can be prevented by preparing a total particulate fraction by differential centrifugation which is then carefully resuspended in fresh homogenization buffer before incubation with CDP[^{14}C]choline.

Monosaccharide Lipid Synthesis

Protein N-glycosylation is one of the major and exclusive biosynthetic functions of the ER.[12] The initial glycosylation step involves the *en bloc* transfer of a mannose-rich oligosaccharide (—$GlcNAc_2Man_9Glc_3$) from a lipid (dolichol pyrophosphate) carrier to the side-chain amino group of appropriate asparagine residues. This transfer is catalyzed by a membrane-bound enzyme whose active site is exposed on the luminal surface of the ER membrane.[13] Protein glycosylation occurs cotranslationally shortly after the target asparagine residue in the nascent polypeptide chain appears on the luminal side of the ER membrane.[14] This oligosaccharide transfer step is difficult to assay *in vitro*, but several of the enzymatic reactions involved in assembling the oligosaccharide lipid are both exclusive to the ER membrane and are readily assayed. Individual sugars are converted to nucleotide sugars in the cytoplasm and are then transferred to dolichol monophosphate, which is tightly bound to the ER membrane. Individual lipid-linked sugars are used to assemble the lipid-linked oligosaccharide. Enzymes which transfer the sugar moieties from either GDPmannose or UDP-*N*-acetylglucosamine to form dolichol monophos-

[11] K. W. A. Wirtz, *Biochim. Biophys. Acta* **334,** 95 (1974).

[12] S. Kornfeld, *in* "The Glycoconjugates" (M. I. Horowitz, ed.), Vol. 3, Part A, p. 3. Academic Press, New York, 1982.

[13] J. A. Hanover and W. J. Lennarz, *J. Biol. Chem.* **225,** 3600 (1980).

[14] C. G. Glabe, J. A. Hanover, and W. J. Lennarz, *J. Biol. Chem.* **255,** 9236 (1980).

phate mannose or dolichol pyrophosphate *N*-acetylglycosamine can be assayed as follows:

One milliliter of microsomal suspension is adjusted to 10 m*M* $MgCl_2$ and 2 m*M* 2-mercaptoethanol and the reaction is started by adding 0.025–0.25 μCi of GDP[^{14}C]mannose or UDP-*N*-[^{14}C]-acetylglucosamine (Amersham, UK; specific activity 200 Ci/mol) and is allowed to proceed at 30° for 30 min. The reaction is stopped by adding 4 ml of chloroform/methanol (2 : 1) followed by 1 ml of water. Mix well, separate the phases, and remove the lower organic phase which contains the dolichol phosphate ^{14}C-labeled sugar. Remove nonlipid material from the organic phase by washing with an equal volume of chloroform/methanol/water (3 : 48 : 47). Transfer the washed organic phase to a scintillation vial, evaporate to dryness, and determine [^{14}C]monosaccharide lipid content after adding an appropriate scintillant.

Endogenous dolichol monophosphate in the ER membrane is often rate limiting and incorporation into monosaccharide lipid may be significantly increased by adding exogenous dolichol monophosphate (up to 20 μg) and adjusting the reaction mixture to 0.1% (v/v) Triton X-100.[15]

Purity of the Isolated ER Membranes

The isolation method described here entails trapping ribosome-denuded microsomes at a sucrose interface during density gradient centrifugation. This ensures that virtually all the microsomes are recovered as a tight band and that marker enzyme activity profiles give a pleasingly sharp peak. What is not known, of course, is the extent to which this band is contaminated with other membranous components. It is therefore essential that ER-containing fractions are assayed for the presence of markers characterizing such components. Golgi membranes are among the likely contaminants. Glycoprotein glycosyltransferases such as galactosyl- and fucosyltransferases, which transfer the sugar moieties from UDPgalactose or GDPfucose to glycoprotein, are useful Golgi markers.[16]

One useful approach to further identify ER membranes on sucrose gradients is the Mg^{2+}-induced density shift.[3] The smooth microsomes obtained on the EDTA-containing sucrose gradients described above are predominantly derived from rough ER which has been stripped of attached ribosomes. Confirmation that these vesicles are of rough ER origin

[15] M. J. Conder and J. M. Lord, *Plant Physiol.* **72,** 547 (1983).

[16] J. R. Munro, S. Nariamham, S. Wetmore, J. R. Riordan, and H. Schachter, *Arch. Biochem. Biophys.* **169,** 269 (1975).

can be obtained by including excess Mg^{2+} in homogenization buffer and sucrose solutions in order to maintain the ribosome–membrane association. In this case ER membranes show a characteristic Mg^{2+}-induced shift to higher sucrose densities after gradient centrifugation, giving a broad peak of marker enzyme activities. A corresponding peak of ribosomal RNA should also be seen. The concentration of Mg^{2+} used to generate this density shift is critical. It must be high enough to prevent ribosome dissociation from the ER membrane without giving nonspecific membrane aggregation. Whenever a Mg^{2+}-induced density shift is observed for ER membrane markers, it is essential to simultaneously demonstrate that other organelle fractions still peak at their characteristic equilibrium densities. In the case of castor bean endosperm homogenates, ribosome–membrane association is maintained by including 3 m*M* $MgCl_2$ in all sucrose solutions used for gradient construction.

The typical distribution of ER enzymatic markers and RNA when castor bean endosperm homogenates are fractionated in the absence or presence of Mg^{2+} is shown in Fig. 1.

ER-Bound Polysomes

Messenger RNA molecules which are being actively translated in a cell are found in polyribosomes or polysomes, structures formed by several ribosomes lying closely spaced on a single mRNA molecule. The 80 S ribosomes not involved in translation at any particular time exist individually in the cytoplasm as monosomes. Polysomes exist in two different forms in eukaryotic cells depending on the particular type of protein they are actively engaged in synthesizing. In general, polysomes synthesizing proteins destined to occur and function in the cell cytoplasm themselves occur as free cytoplasmic structures. In contrast, polysomes translating mRNAs encoding proteins which must be inserted into or transferred across a membrane are themselves associated with the ER membrane.[17] These attached polysomes define the regions of rough ER.

The purpose of isolating ER-bound polysomes is often to identify the proteins encoded by the engaged mRNA molecules. This goal is frequently complicated by the susceptibility of the polysomes to degradation during isolation. The mRNA strand is the most vulnerable component for degradation by endogenous ribonuclease activity. Care must be taken during tissue homogenization and polysome isolation to minimize or eliminate the threat posed by ribonuclease. In addition exogenous ribonuclease must not be introduced into the system via reagents or glassware.

[17] G. Blobel, *Proc. Natl. Acad. Sci. U.S.A.* **77,** 1476 (1980).

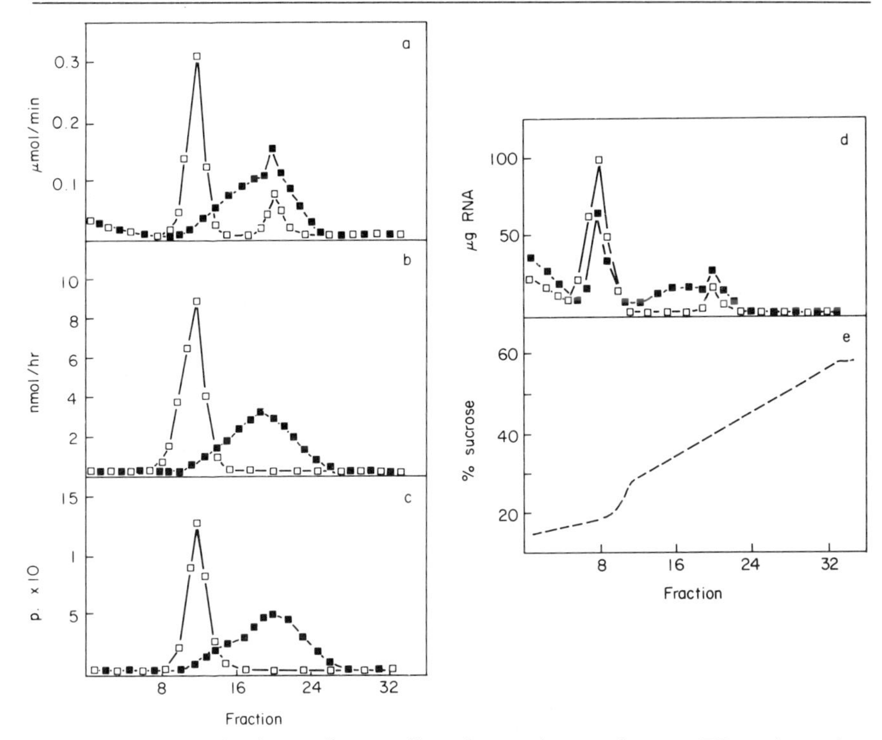

FIG. 1. Sucrose density gradient profiles of castor bean endosperm ER markers after organelle separation in gradients prepared without added $MgCl_2$ (open squares) or in gradients containing 3 mM $MgCl_2$ (closed squares). (a) NADH–cytochrome-*c* reductase, (b) cholinephosphotransferase, (c) mannosyltransferase, (d) RNA, (e) sucrose concentration. In both types of gradient, mitochondria and glyoxysomes peak in fractions 20 and 28, respectively.

Several compounds have been added to extraction buffers in order to inhibit ribonuclease activity. Such additions have included diethyl pyrocarbonate and heparin, which inhibit ribonuclease activity; ethylene glycol bis(2-aminoethyl ether)tetraacetic acid (EGTA), which chelates Ca^{2+} and thus inhibits Ca^{2+}-activated ribonuclease; exogenous ribonuclease-free RNA to dilute out the endogenous substrate; Mg^{2+} and buffers of high ionic strength and high pH, all of which inhibit ribonuclease.[18] Naturally occurring inhibitors such as human placental ribonuclease inhibitor are particularly effective.[19] All solutions which come into contact with poly-

[18] E. Davis and B. A. Larkins, *in* "The Biochemistry of Plants" (P. K. Stumpf and E. E. Conn, eds.), Vol. 1, p. 413. Academic Press, New York, 1980.

[19] P. Blackburn, G. Wilson, and S. Moore, *J. Biol. Chem.* **252,** 5904 (1977).

somes should have been sterilized and all glassware, etc., should have been washed in boiling 1% diethyl pyrocarbonate.

Sucrose often contains significant ribonuclease activity and it is advisable to use ribonuclease-free sucrose in extraction buffers and for the construction of sucrose gradients.

Polysomes are isolated from homogenates by centrifugation. If necessary, the attached mRNA molecules can be characterized by completing previously initiated polypeptides after adding the polysomes to an appropriate readout system, or by translating the mRNA extracted from the polysomes in an initiating cell-free system.

Polysome Isolation and Characterization

Freeze the tissue in liquid nitrogen and crush to a frozen powder using a sterile mortar and pestle. Transfer the powder to a second, ice-chilled mortar and continue the grinding for a few minutes after adding 5 vol of extraction buffer containing 200 m*M* Tris–HCl, pH 8–6, 60 m*M* KCl, 30 m*M* $MgCl_2$, and 25 m*M* EGTA. Filter the homogenate through two layers of Miracloth (prewetted with 0.05% diethyl pyrocarbonate and autoclaved) and centrifuge for 10 min at 1000 *g*. Overlay the 1000 *g* supernatant on a sucrose bilayer consisting of equivalent volumes of 50% (w/v) and 25% (w/v) sucrose solution in extraction buffer contained in a centrifuge tube. Centrifuge for 2 hr at 26,000 rpm in a Beckman SW28 rotor and collect the rough microsome fraction from the 25%/50% sucrose interface. Dilute the membrane fraction with an equivalent volume of extraction buffer containing 2% (v/v) Triton X-100 to release the bound polysomes from the ER membrane. Overlay the detergent-treated microsome fraction over 50% (w/v) sucrose in extraction buffer and centrifuge for at least 2 hr at 100,000–130,000 *g* in a swinging bucket rotor on an appropriate ultracentrifuge to sediment the released polysomes.

Polysome profiles can be obtained by gently resuspending the pellet in 2 ml of 40 m*M* Tris–HCl, pH 8–6, 20 m*M* KCl, and 10 m*M* $MgCl_2$ using a Pasteur pipet or a small glass tissue homogenizer. Layer the polysome suspension over a linear sucrose gradient consisting of 10 ml of sucrose solution increasing linearly in concentration from 15% (w/v) to 50% (w/v) sucrose in resuspension buffer. Centrifuge for 1–2 hr at 39,000 rpm in an SW 41-Ti rotor and then analyze the gradient using an ISCO model 640 gradient fractionator to continuously measure the absorbance at 254 nm.

[55] Phosphoglyceride Synthesis in Endoplasmic Reticulum

By THOMAS S. MOORE, JR.

The endoplasmic reticulum of plants is thought to play a major role in membrane phosphoglyceride synthesis,[1,2] similar to the situation in animals,[3] although it is known that the Golgi apparatus, mitochondria, and plastids[1,2] also play a role. The production of phospholipids by an *in vitro* microsomal fraction was first demonstrated by Sastry and Kates in 1966,[4] and such fractions, probably consisting of endoplasmic reticulum (ER) plus various other membranes, have been reported to synthesize a variety of phospholipids.[1,2]

The purity of the organelle fractions in many studies is uncertain, and so the relative roles of the ER, Golgi apparatus, and other membranes cannot be clearly argued for most plant systems. The castor bean endosperm system has the advantage of a reduced contribution from the Golgi apparatus as well as clean separation of the other organelles present on sucrose density gradients, and the clear compartmentation of most of the phospholipid synthesis into the endoplasmic reticulum of that tissue supports a major role for that organelle in synthesis of many organelle membranes.[5] For these reasons, the enzyme assays described below are based primarily on work with castor bean endosperm (see Fig. 1).

EC 2.7.8.2 Cholinephosphotransferase (CDPcholine : 1,2-Diacylglycerol Cholinephosphotransferase)

CDPcholine + *sn*-1,2-diacylglycerol → phosphatidylcholine + CMP

This activity usually is the most active of the phospholipid synthesizing enzymes,[6] which may be a reflection of its product being the most

[1] T. S. Moore, *Annu. Rev. Plant Physiol.* **33,** 235 (1982).
[2] T. S. Moore, *in* "Structure, Function and Metabolism of Plant Lipids" (P.-A. Siegenthaler and W. Eichenberger, eds.), p. 83. Elsevier Science, Amsterdam, New York, 1984.
[3] R. M. Bell and R. A. Coleman, *Annu. Rev. Biochem.* **49,** 459 (1980).
[4] P. S. Sastry and M. Kates, *Can. J. Biochem.* **44,** 459 (1966).
[5] T. S. Moore, this series, Vol. 71, p. 596.
[6] T. S. Moore and G. D. Troyer, *in* "Biosynthesis and Function of Plant Lipids" (W. W. Thomson, J. B. Mudd, and M. Gibbs, eds.), p. 16. Am. Soc. Plant Physiol., Rockville, Maryland, 1983.

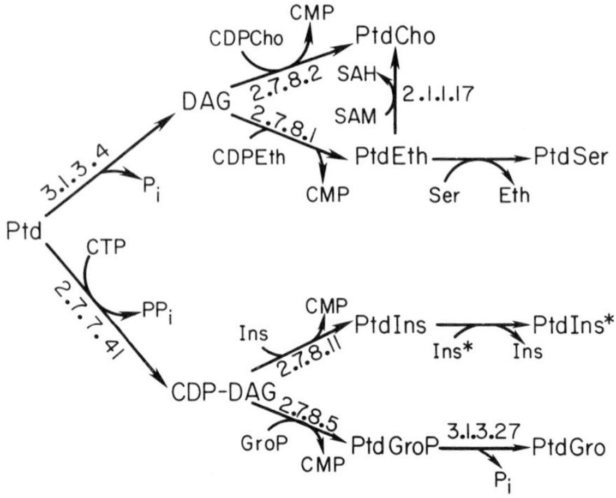

FIG. 1. Outline of the pathways of phosphoglyceride synthesis in plant endoplasmic reticulum. The enzymes are indicated by their nomenclature numbers, which correlate with the text. The phosphatidate phosphatase (EC 3.1.3.4) has not been adequately described in plant endoplasmic reticulum, and so is not discussed. The two exchange enzymes (producing PtdIns and PtdSer) have not been assigned a number. Abbreviations: CDP-DAG, CDP-diacylglycerol; DAG, diacylglycerol; Ptd, phosphatidate; PtdCho, phosphatidylcholine; PtdEth, phosphatidylethanolamine; PtdGro, phosphatidylglycerol; PtdGroP, phosphatidylglycerol-P; PtdIns, phosphatidylinositol; PtdSer, phosphatidylinositol.

abundant phospholipid of many membranes.[7,8] The activity was first characterized from plants in 1971.[9]

Assay Method

Principle. The chloroform-soluble product radiolabeled from the water-soluble CDPcholine is extracted from the reaction mixture following the assay and its radioactivity determined.[10]

Reagents

Tris–HCl, 50 m*M*, pH 7.5
Dithiothreitol, 5 m*M*
$MgCl_2$, 100 m*M*
Diacylglycerol, from egg phosphatidylcholine, 2 m*M*

[7] C. Hitchcock and B. W. Nichols, "Plant Lipid Biochemistry." Academic Press, New York, 1971.

[8] M. Kates, *Adv. Lipid Res.* **8,** 225 (1970).

[9] K. A. Devor and J. B. Mudd, *J. Lipid Res.* **12,** 403.

[10] T. S. Moore, *Plant Physiol.* **57,** 382 (1976).

CDP[1,2-^{14}C]choline, 1 mM, 5 mCi/mmol
Chloroform–methanol–H_2O (1 : 2 : 0.3, by volume)
Chloroform
KCl, 1.0 M

Assay Procedure. Equal volumes of buffer, dithiothreitol, $MgCl_2$, and diacylglycerol are mixed and 0.4-ml aliquots placed into test tubes. Fifty-microliter aliquots of enzyme are added and the reaction started by the addition of 50 μl of the CDPcholine. The assay is run for 30 min with shaking in a water bath at 30° and is stopped by the addition of 3.3 ml of the chloroform–methanol–water mixture followed by mixing with a vortex mixer.

Extraction Procedure. One milliliter of chloroform and 3 ml of 2 M KCl are added to each tube, followed by vigorous mixing. The tubes are centrifuged at about 2500 rpm in a tabletop centrifuge. Following this the upper aqueous layer is removed and discarded. The lower chloroform layer is washed twice more with 1- to 3-ml aliquots of 1 M KCl, aspirating and discarding the aqueous layer each time.

Radioactivity Measurement. The chloroform extract is placed into scintillation vials and dried completely. Radioactivity is measured by standard techniques.

Properties

Substrate Requirements. The apparent K_m of the castor bean enzyme for CDPcholine is about 10 μM,[10] and similar results have been obtained with other tissues.[9] Estimates have not been made for diacylglycerol, since it occurs in the membrane; some stimulation of activity may be achieved by including this substrate in the assay mixture, however.[10]

pH and Metal Ion Requirements. The optimal pH generally is from 7.5[10] to 8.0,[9,10] and a divalent cation is required. Mg^{2+} is preferred over Mn^{2+} by the castor bean enzyme,[10] but the two give equal activities with spinach.[9] Ca^{2+} inhibits the activity.[11,12]

Inhibitors. Dithiothreitol promotes and sulfhydryl reagents inhibit the reaction to varying degrees, depending on the tissue source.[9,10]

EC 2.7.8.1 Ethanolaminephosphotransferase (CDPethanolamine : 1,2-Diacylglycerol Ethanolaminephosphotransferase)

CDPethanolamine + *sn*-1,2-diacylglycerol → phosphatidylethanolamine + CMP

[11] J. M. Lord, *Plant Physiol.* **57,** 218 (1976).
[12] A. Oursel, A. Trémolières, and P. Mazliak, *Physiol. Veg.* **15,** 377 (1977).

This enzyme catalyzes a reaction similar to the cholinephosphotransferase and with a similar high level of activity.[13–15] It has not, however, been as thoroughly examined and demonstrated not to be catalyzed by the same enzyme that synthesizes phosphatidylcholine.[16]

Assay Method

Principle. The reaction is conducted in the presence of the radioactive, water-soluble substrate, followed by extraction of the chloroform soluble product and measurement of its radioactivity.[15]

Reagents

MES, 50 m*M*, pH 6.5
$MgCl_2$, 15 m*M*
Dithiothreitol, 5 m*M*
CDP[1,2-^{14}C]ethanolamine, 75 μM, 3.0 mCi/mmol
Chloroform–methanol–H_2O (1 : 2 : 0.3, by volume)
Chloroform
KCl, 1.0 *M*

Assay Procedure. Equal quantities (0.1 ml each) of buffer, $MgCl_2$, and dithiothreitol are combined with 0.1 ml of H_2O in a test tube. The enzyme is added, the mixture preincubated at 37°, and the reaction started by the addition of 0.05 ml CDPethanolamine. The final volume is 0.5 ml. The tubes are incubated with shaking at 37° for 30 min and the reaction stopped by the addition of 3.3 ml of chloroform–methanol–H_2O.

Extraction and Radioactivity Measurement. The extraction and radioactivity measurements were as described for the cholinephosphotransferase assay above.

Properties

Substrate Requirements. The apparent K_m of the enzyme from castor bean endosperm for CDPethanolamine is 6.0 μM, while that for spinach is 20 μM.[14,15] Maximum velocities are achieved at 50 to 100 μM CDPethanolamine.[14,15] CDPcholine is a strong competitive inhibitor of ethanolamine incorporation.[14,15] Diacylglycerol requirements have not been defined due to its natural occurrence in the membranes.

Metal and pH Requirements. Mg^{2+} is most effective for the castor

[13] M. O. Marshall and M. Kates, *Can. J. Biochem.* **52,** 469 (1974).
[14] B. A. Macher and J. B. Mudd, *Plant Physiol.* **53,** 171 (1974).
[15] S. A. Sparace, L. K. Wagner, and T. S. Moore, *Plant Physiol.* **67,** 922 (1981).
[16] J. M. Lord, *Biochem. J.* **151,** 451 (1975).

bean enzyme, being effective at 3 m*M*,[15] but for spinach Mn^{2+} has been found to be best, giving an optimal reaction between 0.6 and 2 m*M*.[14] The optimum pH ranges from 6.5 for castor bean to 7.0–8.0 for spinach.[14,15]

EC 2.1.1.17 Phosphatidylethanolamine Methyltransferase (*S*-Adenosyl-L-Methionine : Phosphatidylethanolamine *N*-Methyltransferase)

Phosphatidylethanolamine + 3 *S*-adenosyl-L-methionine →
phosphatidylcholine + *S*-adenosylhomocysteine

In castor bean endoplasmic reticulum this reaction occurs with about 10% of the activity of the cholinephosphotransferase.[10] It also has been described from spinach leaf[13] and potato tuber.[17] It is not certain how many enzymes are involved in this three-step methylation.

Assay Method

Principle. *S*-Adenosyl-L-[*methyl*-^{14}C]methionine is fed as the aqueous substrate and the radioactive phosphatidylcholine product is extracted into chloroform, followed by measuring its radioactivity.[10]

Reagents

Tris–HCl, 250 m*M*, pH 9.0
Phosphatidylethanolamine, 1.0 m*M*
S-Adenosyl-L-[*methyl*-^{14}C]methionine, 1.0 m*M*, 1 mCi/mmol
HCl, concentrated
Chloroform–methanol–H_2O (1 : 2 : 0.3, by volume)
Chloroform
KCl, 1.0 *M*

Assay Procedure. Aliquots of buffer, phosphatidylethanolamine, and *S*-adenosylmethionine (0.1 ml each) are added to assay tubes along with sufficient water to give a final volume (with enzyme) of 0.5 ml. The tubes are preincubated at 37° for 5 min, after which the reaction is started by the addition of enzyme and the tubes incubated with shaking in a water bath at 37° for 30 min. The reaction is stopped by the addition of 50 μl of HCl, followed by 3.3 ml of the chloroform–methanol–H_2O mixture, and then incubated at room temperature for 1 hr.

Extraction and Radioactivity Measurement. The radioactive product is extracted into chloroform and the radioactivity measured as described for the cholinephosphotransferase assay.

[17] W. Tang and P. A. Castelfranco, *Plant Physiol.* **43,** 1232 (1968).

Properties

Substrate Requirements. The castor bean endoplasmic reticulum reaction(s) demonstrates an apparent K_m for *S*-adenosylmethionine of 31 μM; phosphatidylethanolamine stimulates only about 33%[10] and does not stimulate the spinach activity.[14] The single- and double-methylated lipid intermediates stimulate the reaction in both tissues.[10,14]

pH Requirement. The optimal pH for the enzyme is around 8.0 to 9.0 in both castor bean and spinach.[10,14]

Products. The product of the reaction under these conditions is primarily phosphatidylcholine, but the mono- and dimethylated intermediates also may occur.[10,14]

Phosphatidylethanolamine : L-Serine Phosphatidyltransferase

$$\text{Phosphatidylethanolamine} + \text{L-serine} \rightarrow \text{phosphatidylserine} + \text{ethanolamine}$$

Phosphatidylserine generally occurs only in small concentrations in plant tissues, but may play an important role.[18] The reaction described here is an exchange-type reaction (exchanging serine for ethanolamine), but another reaction utilizing CDPdiacylglycerol and L-serine also has been described in plants.[13]

Assay Procedure

Principle. The Ca^{2+}-stimulated exchange reaction is measured by the incorporation of the water-soluble L-serine into the chloroform-soluble lipid product.[19]

Reagents

HEPES, 200 m*M*, pH 7.8
$CaCl_2$, 10 m*M*
L-[3-^{14}C]Serine, 2.0 m*M*, 2.5 mCi/mmol
Chloroform–methanol–H_2O (1 : 2 : 0.3, by volume)
Chloroform
KCl, 1.0 *M*

Assay Procedure. Aliquots (0.1 ml each) of the buffer, $CaCl_2$, and L-serine, plus an appropriate amount of water to give a final volume (including enzyme) of 0.5 ml, are added to assay tubes. The tubes are preincu-

[18] N. Murata, N. Sato, and N. Takahashi, *in* "Structure, Function and Metabolism of Plant Lipids" (P.-A. Siegenthaler and W. Eichenberger, eds.), p. 153. Elsevier, Amsterdam, 1984.

[19] T. S. Moore, *Plant Physiol.* **56,** 177 (1975).

bated at 30° for 5 min and the reaction started by adding enzyme. Further incubation at 30°, with shaking in a water bath, is performed for 60 min. The reaction is stopped by the addition of 3.3 ml of chloroform–methanol–H_2O followed by incubation on ice for 30 min.

Extraction and Radioactivity Determination. The extraction and measurement of radioactivity in the product were as described for the cholinephosphotransferase assay.

Properties

Substrate Requirements. The K_m for serine is about 20 μM.[19] Ethanolamine was found to compete with L-serine but D-serine and choline did not.[19] Phosphatidylethanolamine stimulated the reaction, but it may be nonspecific since CDPdiacylglycerol did also.[19]

Metal and pH Requirements. Calcium is required, with the optimum concentration being 2 mM.[19] The best pH at that Ca^{2+} concentration is 7.8, with a shift to a lower pH at higher Ca^{2+} concentrations.[19]

Products. Both phosphatidylserine and phosphatidylethanolamine occurred as products of the reaction, suggesting the presence of a decarboxylase enzyme in the endoplasmic reticulum.[19]

EC 2.7.7.41 Phosphatidate Cytidylyltransferase (CTP : Phosphatidate Cytidylyltransferase)

$$\text{CTP} + \text{phosphatidate} \rightarrow \text{CDPdiacylglycerol} + \text{PP}_i$$

This enzyme produces a lipid which serves as a precursor to the synthesis of both phosphatidylglycerol and phosphatidylinositol in the endoplasmic reticulum. CDPdiacylglycerol occurs at only very low concentrations in membranes.[20]

Assay Method

Principle. The enzyme utilizes the water-soluble substrate, CTP, labeled in the cytidine ring and produces the chloroform-soluble CDPdiacylglycerol which is extracted and its radioactivity measured.[20]

Reagents

MES, 50 mM, pH 6.5
$MnCl_2$, 37.5 mM
Phosphatidate, 12.5 mM, derived from egg lipids
[^{3}H]CTP, 0.6 mM, 32.2 mCi/mmol

[20] K. F. Kleppinger-Sparace and T. S. Moore, *Plant Physiol.* **77**, 12 (1985).

Chloroform–methanol–H_2O (1 : 2 : 0.3, by volume)
Chloroform
KCl, 1.0 *M*

Assay Procedure. One-tenth-milliliter fractions of buffer, $MnCl_2$, phosphatidate, and CTP are added to an assay tube, along with sufficient water to make a final volume (after enzyme addition) of 0.5 ml. The reaction is initiated by the addition of enzyme. This mixture is incubated at 37° for 30 min in a shaker waterbath. The reaction is stopped by the addition of 3.3 ml of the chloroform–methanol–water solution.

Extraction and Radioactivity Measurement. These are carried out as described for the cholinephosphotransferase assay above.

Properties

Substrate Requirements. The apparent K_m of the castor bean enzyme for CTP is about 17 μM,[20] which is lower than those reported for other plant microsomal activities.[21] The best stimulation is apparent with mixed phosphatidates, but endogenous lipid substrate prevents an estimation of the precise requirements.[20]

pH and Metal Ion Requirements. Manganese is strongly preferred over Mg^{2+} and was optimal at 7.5 m*M*. The best pH is 6.5.[20]

Detergent Effects. Both deoxycholate and Triton X-100 strongly inhibit the enzyme at concentrations as low as 0.01% (w/w).[20]

EC 2.7.8.5 CDPdiacylglycerol–Glycerol-3-phosphate 3-Phosphatidyltransferase (CDPdiacylglycerol : *sn*-Glycerol-3-phosphate 3-Phosphatidyltransferase)
EC 3.1.3.27 Phosphatidylglycerophosphatase (Phosphatidylglycerophosphate Phosphohydrolase)

sn-1-Glycerol 3-phosphate + CDPdiacylglycerol →
3-*sn*-phosphatidyl-1′-*sn*-glycerol 3′-phosphate + CMP

3-*sn*-Phosphatidyl-1′-*sn*-glycerol 3′-phosphate → 3-*sn*-phosphatidyl-1′-*sn*-glycerol + P_i

These two enzymes have not been separated from plant tissues, but the reaction is strongly pulled to the production of phosphatidyglycerol and so assays measure the production of that final product.[22] The reac-

[21] J. Bahl, T. Guillot-Salomon, and R. Douce, *Physiol. Veg.* **8,** 55 (1970).
[22] T. S. Moore, *Plant Physiol.* **54,** 164 (1974).

tions were first demonstrated in spinach[4] and have since been studied further in spinach[23] as well as in cauliflower[24] and castor bean.[22]

Assay Method

Principle. Radioactive *sn*-glycerol 3-phosphate is incorporated into a chloroform-soluble product, which is extracted and the radioactivity measured.[22]

Reagents

Tris–HCl, 2.0 *M*, pH 7.3
$MnCl_2$, 40 m*M*
Triton X-100, 0.6% (w/w)
CDPdiacylglycerol, 0.8 m*M*
sn-[U-^{14}C]Glycerophosphate, 4.0 m*M*, 1.9 mCi/mmol
Chloroform
Methanol
KCl, 1.0 *M*

Assay Procedure. Twenty-five microliters each of buffer, $MnCl_2$, detergent, and the two substrates are added to assay tubes along with sufficient water to give a final volume of 0.2 ml following further additions. The tubes are preincubated for 5 min at 30° in a water bath. Enzyme is added to start the reaction and the tubes are incubated, with shaking, for an additional 1.5 hr. The reaction is stopped by adding 2.0 ml of methanol followed by incubating on ice for 1 hr.

Extraction Procedure. Two milliliters of chloroform is added to each assay tube and mixed. The emulsion is broken by centrifugation at 2500 rpm in a tabletop centrifuge. The upper aqueous layer is removed and discarded, and the lower chloroform layer washed twice with 5.0-ml aliquots of the KCl solution, followed by twice with distilled water. Each time the tubes are mixed, centrifuged, and the upper layer removed.

Radioactivity Measurement. This is as described in the section on cholinephosphotransferase.

Properties

Substrate Requirements. Calculated K_m values for glycerol-P range from 50 to 250 μM,[22,23] while that for CDPdiacylglycerol is about 2–3 μM.[22,23]

[23] M. O. Marshall and M. Kates, *Biochim. Biophys. Acta* **260,** 558 (1972).
[24] R. Douce and J. Dupont, *C.R. Hebd. Seances Acad. Sci., Ser. D* **268,** 1657 (1969).

Metal and pH Requirements. Mn^{2+} stimulates the reaction best at 5 m*M*, and Mg^{2+} is only partially effective.[22] The pH optimum is 7.3.[22,23]

Detergent Effects. Triton X-100 stimulates at concentrations up to 0.075% (w/w).[22]

Products. The major product is phosphatidyglycerol, but the phosphorylated intermediate is found in low quantities in some cases.[22,23]

EC 2.7.8.11 CDPdiacylglycerol–Inositol 3-Phosphatidyltransferase (CDPdiacyglycerol : *myo*-Inositol 3-Phosphatidyltransferase)

CDP-*sn*-1,2-diacylglycerol + *myo*-inositol → phosphatidylinositol + CMP

This enzyme activity has now been described from several plant species[1] and solubilized from soybean.[25] It was first described from cauliflower by Sumida and Mudd.[26]

Assay Method

Principle. Radioactive *myo*-inositol is incorporated into phosphatidylinositol, which can be extracted into chloroform and its radioactivity determined.[17]

Reagents

Tris–HCl, 250 m*M*, pH 8.5
$MnCl_2$, 7.5 m*M*
CDPdipalmitoylglycerol, 5.0 m*M*
myo-[2-^{3}H]Inositol, 6.0 m*M*, 1.79 mCi/mmol
Chloroform–methanol–H_2O (1 : 2 : 0.3, by volume)
Chloroform
KCl, 1.0 *M*

Assay Procedure. A final volume of 0.5 ml is achieved by the addition of 0.1 ml each of buffer, $MnCl_2$, CDPdipalmitoylglycerol, and radioactive *myo*-inositol, along with sufficient water to achieve the volume after enzyme addition. The contents of the tubes are mixed, the tubes preincubated 5 min in a water bath at 37°, and the reaction started by addition of enzyme. The tubes are then incubated for 30 min at 37°, following which the reaction is terminated by addition of chloroform–methanol–H_2O. The radioactive product is extracted and the radioactivity measured as described for the cholinephosphotransferase assay.

[25] M. L. Robinson and G. M. Carman, *Plant Physiol.* **69,** 146 (1982).
[26] S. Sumida and J. B. Mudd, *Plant Physiol.* **45,** 712 (1970).

Properties

Substrate Requirements. Reported K_m values for *myo*-inositol range from 0.045 to 0.3 m*M*,[26,27] and for CDPdiacylglycerol from 0.027 to 1.35 m*M*.[26,27] Excess levels of CDPdiacylglycerol can inhibit the reaction.[27]

Metal and pH Requirements. A divalent cation is absolutely required, and Mn^{2+} is strongly preferred.[26,27]

Detergent Effects. Stimulation of activity can be achieved in some cases by the addition of low concentrations of detergents,[27] but inhibition also may result.[27]

Phosphatidylinositol : *myo*-Inositol Phosphatidyltransferase

Phosphatidylinositol + *myo*-inositol* → phosphatidylinositol* + *myo*-inositol

This unusual activity has been reported only twice with plant tissues.[28,29] The role of the enzyme is unknown.

Assay Method

Principle. The incorporation of radioactive *myo*-inositol into phosphatidylinositol, under conditions which minimize the activity of CDP-diacylglycerol : *myo*-inositol phosphatidyltransferase, is followed.[28]

Reagents

HEPES, 250 m*M*, pH 8.0
$MnCl_2$, 125 m*M*
myo-[2-^{3}H]Inositol, 2.5 m*M*, 1.79 mCi/mmol
Chloroform–methanol–H_2O (1 : 2 : 0.3, by volume)
Chloroform
KCl, 1.0 *M*

Assay Procedure. Buffer, $MnCl_2$, and radioactive inositol are added to each assay tube (0.1 ml of each), along with adequate water to give a final volume with enzyme of 0.5 ml. The tubes are mixed, preincubated at 37° for 5 min, and the reaction started by adding enzyme. The tubes are shaken in a water bath at 37° for 1 hr, followed by the addition of 3.3 ml of chloroform–methanol–H_2O to stop the reaction.

Extraction and Radioactivity Measurement. The radioactive product is extracted into chloroform and the radioactivity measured as described for the cholinephosphotransferase.

[27] J. C. Sexton and T. S. Moore, *Plant Physiol.* **62,** 978 (1978).
[28] J. C. Sexton and T. S. Moore, *Plant Physiol.* **68,** 18 (1981).
[29] D. J. Morré, B. Gripshover, A. Monroe, and J. T. Moore, *J. Biol. Chem.* **259,** 15364 (1984).

Properties

Substrate and Phospholipid Effects. The apparent K_m for *myo*-inositol is 26 μM.[28] Phosphatidylinositol from yeast and castor bean endosperm stimulates the reaction, but that from soybean does not[28]; however some other phospholipids can also stimulate the response.[28]

Metal and pH Requirements. A divalent cation is absolutely required, and Mn^{2+} is strongly preferred.[28] The pH optimum is 8.0.[28]

Detergent Effects. Triton X-100 stimulates the reaction at concentrations up to 0.025% (w/w).[28]

Nucleotide Effects. CMP, CDP, and CTP all stimulate the reaction 15-fold at 40 μM, while CDPcholine, CDPethanolamine, and CDPdiacylglycerol all stimulate to a lesser extent.[28]

Acknowledgments

Much of the work described here was supported by grants from the National Science Foundation, with preparation of this manuscript being partially supported by NSF Grant PCM-8402001.

Section VII

General Physical and Biochemical Methods

[56] Electron Microscopy of Plant Cell Membranes

By JEAN-PIERRE CARDE

Every plant cell is highly compartmentalized. Compartment boundaries are represented by one or two biological membranes which control the flow of water, ions, and metabolites between bordering compartments. In addition to the usual organelles encountered in all eukaryotic cells, plant cells contain vacuoles and plastids. The vacuolar compartment, more often very developed in differentiated cells, is bounded by the tonoplast. Plastids are surrounded by a double envelope and contain an inner membrane system whose development and organization reflect the functional specialization of these organelles in cell metabolism.

The lipid/protein pattern (ratio, composition) is specific for each plant cell membrane. Interspecific and intraspecific variations occur also. In addition, cell membranes are structurally and functionally asymmetric. However, in spite of these molecular differences, the fluid protein lipid bilayer of plant cell membranes displays the same basic organization when observed in transmission electron microscopy (TEM) after convenient processing of plant material even if this appearance is not absolutely representative of its actual structure.[1] Therefore, the dark–light–dark or railroad construction seems to be the univocal representation of plant cell membranes in TEM as well as animal cell membranes.[2] The optimal attainment of this structural configuration may be considered as one of the purposes of electron microscopy of plant membranes. In this respect, TEM study of cell membranes is just a usual application for plant cell description. But, owing to the physicochemical nature of membranes, it is clear that obtaining reliable and constant information on their organization requires distinct, adequate material processing. A number of methodological improvements for TEM have been achieved over the years, but most difficulties arising in membrane studies have been overcome only recently. This chapter is an attempt to summarize a number of technical points which have to be taken into account in routine studies of plant membranes by transmission electron microscopy. Other excellent and informative techniques—ultracryotomy, freeze-etching, low-temperature methods,[3] surface or enzymatic markers—will not be considered here

[1] R. B. Luftig, E. Wehrli and P. N. McMillan, *Life Sci.* **21,** 285 (1977).
[2] J. D. Robertson, *Biochem. Soc. Symp.* **16,** 3 (1959).
[3] D. E. Fernandez and L. A. Staehelin, *Planta* **165,** 455 (1985).

because they represent an extensive field which is beyond the scope of this chapter.

Preparation of Plant Material

The aim of preparative techniques is to realize the best conditions for preservation and visualization of the membranes. At the same time a fine description of their structural organization and interrelations and, in some instances, a cytochemical characterization of some structural components are also necessary.

Routinely, chemical fixation of plant material, dehydration, and embedding in resin are needed. Ultrathin sections are then submitted to adequate staining or specific cytochemical tests. Processing of the material is essentially the same for intact tissues or subcellular fractions.

Standard Fixation Procedure

Adequate fixation of plant material is obviously the most critical step in membrane studies. Potassium permanganate[4] was favored by plant cytologists for early studies using electron microscopy, and indeed, in well-fixed material, permanganate fixation results in a sharp delineation of cell membranes.[5,6] This fixation is, however, difficult to perform under controlled conditions and has drastic effects on the soluble proteins and nucleic acids of the cell. Short fixation times with potassium permanganate (10 min) have been used recently by Hurkmann *et al.*[7] to stain specifically the etioplast membranes from soybean hypocotyls.

Sequential fixation with glutaraldehyde[8] and osmium tetroxide (G/Os) is now a widely used combination for all kinds of material with very large variations in concentration of the fixative, optimal pH, choice of the buffer vehicle, presence of additional osmoticums (salts, sucrose), and length and temperature of the fixation stages. Some of these conditions have been revised recently.[9-11]

The standard fixation procedure given below is usually the best compromise: (1) Perform the first fixation by glutaraldehyde, 2.5% (v/v) in

[4] J. H. Luft, *J. Biophys. Biochem. Cytol.* **2,** 799 (1956).

[5] H. H. Mollenhauer and D. J. Morré, *Cytobiologie* **13,** 297 (1976).

[6] D. G. Robinson, *Eur. J. Cell Biol.* **23,** 22 (1980).

[7] W. J. Hurkmann, D. J. Morré, C. E. Bracker, and H. H. Mollenhauer, *Plant Physiol.* **64,** 398 (1979).

[8] D. D. Sabatini, K. Bensch, and R. J. Barrnett, *J. Cell Biol.* **17,** 19 (1963).

[9] G. R. Bullock, *J. Microsc.* (*Oxford*) **133,** 1 (1984).

[10] J. Coetzee and C. F. Van Der Merwe, *J. Microsc.* (*Oxford*) **135,** 147 (1984).

[11] J. Coetzee and C. F. Van Der Merwe, *J. Microsc.* (*Oxford*) **137,** 129 (1985a).

phosphate Na/Na buffer, 100 mM, pH 7.2, 4 hr at 4°; the osmolarity of this fixative (205 mOsm)[12] may be adjusted with sucrose to the osmolarity of the tissue. (2) Rinse with buffer 30 min at 4° (2 × 15 min). (3) Postfix with osmium tetroxide, 1% (w/v) in phosphate buffer, 100 mM, pH 7.2, 4 hr at 4°. (4) Rinse with buffer overnight at 4°.

For optimal prefixation of the specimen, the usual requirements for plant materials have to be fulfilled, i.e., immersion without delay of very thin slices of material under partial pressure to ensure good penetration of the fixative. Reagent grade glutaraldehyde (with an absorbing peak at 235 nm) is as good as analytical grade (EM-distilled) glutaraldehyde for structural studies. The pH, the buffer, and the concentration of the glutaraldehyde suggested above cause least extraction of substances.[10] Ice-cold fixation seems preferable to fixation at room temperature.[13] The degree of cross-linking by glutaraldehyde increases considerably with rather long fixation (4 hr).[14] Owing to the reducing ability of glutaraldehyde, it is necessary to completely remove excess aldehyde before treatment with osmium tetroxide.

Overnight postfixation by osmium tetroxide has no advantage but it has been empirically shown many times that a long buffer rinse after osmium tetroxide before starting dehydration may significantly improve membrane structure preservation. At this stage pitfalls in membrane fixation are usually due to overfixation by glutaraldehyde (temperature, length) which results in some blurring of the membrane structure or to underfixation by osmium tetroxide which gives rise to gross artifacts in membrane organization. Other undesirable effects of prolonged primary fixations with glutaraldehyde have been described.[15] The extraction of structural components may be favored by the presence of high concentration of sucrose (400 mg/ml^{-1}) in the glutaraldehyde solution.[16] Simultaneous fixation by mixtures of glutaraldehyde and osmium has been advocated[17] but in every case membrane preservation is improved by additional fixative treatments (see below).

Glutaraldehyde reacts only with the protein moieties of the membranes[14] and has no effect on the phospholipid components, except for phospholipids with amino groups.[18] Glutaraldehyde fixation is therefore

[12] J. Coetzee and C. F. Van Der Merwe, *J. Microsc.* (*Oxford*) **138,** 99 (1985b).

[13] B. Mersey and M. E. McCully, *J. Microsc.* (*Oxford*) **114,** 49 (1978).

[14] D. Hopwood, *in* "Fixation in Histochemistry" (P. J. Stoward, ed.), p. 47. Chapman & Hall, London, 1973.

[15] M. E. Lawrence and J. V. Possingham, *Biol. Cell* **52,** 77 (1984).

[16] D. J. Goodchild and S. Craig, *Aust. J. Plant Physiol.* **9,** 689 (1982).

[17] W. W. Franke, S. Krein, and R. M. Brown, *Histochemie* **19,** 162 (1969).

[18] R. C. Roozemond, *J. Histochem. Cytochem.* **17,** 482 (1969).

ineffective in adequate preservation of membranes when used alone before dehydration and embedding in hydrophobic resins. Consequently, after glutaraldehyde fixation and embedding in either acrylic or epoxy resins, cytosolic or organelles proteins, as well as nucleic acids, are satisfactorily preserved but cell membranes are extracted and represented by clear, empty spaces in the cytoplasm surrounding the organelles. Osmium tetroxide interacts with the unsaturated lipids of cell membranes but it has been suggested[19] that, in addition, osmium has a possible protein-unsaturated lipid cross-linking role in systems where these components are intimately mixed. Even after embedding in hydrophobic resins, osmium fixation alone results however in a poor definition of the substructural organization of plant cell membranes.

Results. In addition to a fine overall preservation of cell structure, adequate fixation of thin tissue slices by glutaraldehyde and osmium tetroxide used sequentially results in most instances in the formation of a dark–light–dark construction of plant cell membranes. The thickness of the trilaminar leaflet ranges between 4 and 10 nm. Except in a few cases,[20] the plasmalemma appears more often as a "thick" membrane (8 to 10 nm),[21] as do the Golgi vesicles.[22] Tonoplast structure is often well preserved too. Less satisfactory results are obtained with "thinner" membranes (4 to 6 nm), e.g., nuclear envelope, endoplasmic reticulum, dictyosome stacks, and mitochondrial membranes.[23] Finally, the G/Os sequence does not result in adequate fixation of plastid membranes (thylakoids and envelope membranes) and of peroxisomal or glyoxysomal membranes.

Therefore, the reaction of each cell membrane to this fixation procedure, in terms of obtaining a sharp construction, is not uniform and, on the whole, this standard procedure is not adequate for a correct fixation of most thin membranes and does not prevent a substantial leaching of structural components in the course of further processing. These difficulties have to be overcome by complementary fixative treatments.

This situation is of particular importance for morphometric studies on membrane fractions where only plasma membrane vesicles (identified also by PACP staining) are well delineated among various badly defined membrane vesicles of unknown origin.

[19] A. J. Nielson and W. P. Griffith, *J. Histochem. Cytochem.* **26,** 138 (1978).

[20] M. J. Powell, C. E. Bracker, and D. J. Morré, *Protoplasma* **111,** 87 (1982).

[21] J. C. Roland, *Int. Rev. Cytol.* **36,** 45 (1973).

[22] F. M. Twohig, M. S. Thesis, Purdue University, West Lafayette, Indiana (1974).

[23] D. J. Morré, J. C. Roland, and C. A. Lembi, *Proc. Indiana Acad. Sci.* **79,** 96 (1970).

Additional Treatments

Two routine postfixative treatments bring significant improvements in membrane fixation after the G/Os sequence. These include uranyl acetate and tannic acid.

Uranyl Acetate. Uranyl salts were first proposed to improve the fixation of bacteria membranes.[24] Similar effects had been observed on *Plasmodium* sporozoites and on various eukaryotic cells.[25] Uranyl acetate is now used as a routine postfixative treatment for membrane structure studies.[20] Osmicated blocks are thoroughly washed with distilled water and soaked overnight in 0.5% uranyl acetate in water (pH 3.9). The use of uranyl nitrate or of Tris–maleate-buffered solutions has no practical advantage.

Although often regarded as an *en bloc* staining method giving enhanced contrast of membranes,[26] the posttreatment of osmicated blocks with uranyl salts is truly an additive fixation which reduces the loss of membrane material during dehydration. The extraction of phospholipids in ethanol baths is markedly reduced.[27] Better retention of phospholipid material within the membranes improves the definition of thin membranes on ultrathin sections. On the basis of membrane thickness, morphometric methods can be used to determine the origin of membrane vesicles after homogenization of the plant material and subcellular fractionation on sucrose density gradient.[20]

However, it is very likely that the leaching of membrane proteins during the course of dehydration and embedding is not reduced, which explains the relative disorganization of very thin and fragile membranes (endoplasmic reticulum). This suggestion is supported by the overall aspect of the ground cytoplasm, mitochondria matrix, or plastid stroma which indicates that some extraction of soluble proteins occurs in uranyl-treated blocks as compared with untreated blocks (Fig. 20).

Tannic Acid. Tannic acid (TA) as an additional fixative for electron microscopy has been introduced by Futaesaku *et al.*[28] and recognized to be of special value for EM studies of membranes.[29] Various experimental

[24] A. Ryter and E. Kellenberger, *Z. Naturforsch., B: Anorg. Chem., Org. Chem., Biochem., Biophys., Biol.* **13B,** 597 (1958).

[25] J. A. Terzakis, *J. Ultrastruct. Res.* **22,** 168 (1968).

[26] M. G. Farquhar and G. E. Palade, *J. Cell Biol.* **26,** 263 (1965).

[27] M. T. Silva, J. M. Santos Mota, J. V. C. Melo, and F. Carvalho Guerra, *Biochim. Biophys. Acta* **233,** 513 (1971).

[28] Y. Futaesaku, V. Mizuhira, and H. Nakamura, *Int. Cong. Histochem. Cytochem.*, p. 155 (1972).

[29] N. Simionescu and M. Simionescu, *J. Cell Biol.* **70,** 608 (1976).

procedures have been proposed using TA alone, either before or after glutaraldehyde prefixation,[30] in conjunction with glutaraldehyde or as a posttreatment after a secondary fixation by osmium tetroxide (or methylamine tungstate[31]). In plant material, a combination of glutaraldehyde and tannic acid has been used successfully for improved fixation of thylakoid membranes[32] and plasmodesmata membranes,[33] thus permitting the observation of membrane subunits which are assumed to be representative of protein membrane conformation *in situ*.

A number of experimental procedures using tannic acid in tissue fixation have been carefully described by Simionescu and Simionescu,[29] who demonstrated that the best results are obtained with tissues sequentially fixed by glutaraldehyde, osmium tetroxide, and tannic acid, in conjunction with lead citrate staining of ultrathin sections. From the experience accumulated in our laboratory with various plant materials, the provisional conclusion may be drawn that the original procedure of Simionescu and Simionescu (G/Os/TA) is very well suited to routine studies on plant membranes.

After a sequential fixation with glutaraldehyde and osmium tetroxide, the osmicated specimens are thoroughly washed with Na/Na phosphate buffer or cacodylate buffer, pH 7.2, and soaked 30 min at 20° in 1% tannic acid in the same buffer. The tissue blocks are then rinsed with water and dehydrated as usual. Membrane or organelle pellets have to be resuspended in the TA solution at this step. Ultrathin sections are stained with uranyl and lead.

Interestingly, membrane thickness is increased after TA treatment, especially the "thin" endomembranes, indicating that the leaching of structural components which occurs during dehydration and embedding is greatly reduced by TA. In addition, this result confirms a differential behavior of the plasma membrane after G/Os fixation and supports the concept that pictures of endomembranes after G/Os are not representative of their native organization.

In addition to Mallinckrodt tannic acid, which contains low-molecular-weight galloylglucoses (LMGG), TA's from various sources have been used. Most of them gave comparable results to LMGG in terms of penetration and contrast. Inferior results are obtained with other phenolic molecules (gallic acid).[29]

[30] S. Fujikawa, *J. Ultrastruct. Res.* **84,** 289 (1983).

[31] M. M. Chestnut and A. L. Cohen, *J. Microsc.* (*Oxford*) **133,** 323 (1984).

[32] P. Olesen, *Biochem. Physiol. Pflanz.* **172,** 319 (1978).

[33] P. Olesen, *Planta* **144,** 349 (1979).

Under these conditions, all plant cell membranes are well preserved and highly contrasted after lead citrate staining, permitting routine use of very thin sections, thereby potentially increasing resolution (see below). The trilaminar appearance of membranes is clearly obtained in intact cells or with isolated membranes such as nuclear envelope, endoplasmic reticulum, and mitochondrial membranes, which are hardly definite after the standard G/Os procedure (Figs. 1–5, 19). Moreover, a more striking and somewhat unexpected result concerns plastid membranes. The thylakoid membranes are sharply delineated (Fig. 2). In addition, using the G/Os/TA technique, it was possible to demonstrate very clearly the dark–light construction of the plastid envelope membranes[34] (Fig. 6).

In intact cells, the G/Os/TA sequence overcomes most of the failures of G/Os fixation alone and affords a new tool for investigating the intermembrane relationships which could exist within the plant cell between different organelles and membrane compartments, a question which is still controversial. It is possible, for example, to demonstrate the existence of tiny tubules connecting the inner face of wrapping endoplasmic reticulum and the plastid outer envelope membrane in terpene-producing cells of *Foeniculum* (Fig. 7), the membrane connections between mitochondria and plastids, and the structural relationships between Golgi apparatus and endoplasmic reticulum in a number of plant cells (Fig. 8).

The quality of purified organelles fractions is better ascertained with G/Os/TA, which results in a spectacular improvement in the preservation of structural components of isolated organelles and particularly of the limiting membranes of mitochondria and peroxisomes,[35] chloroplasts,[36] and leucoplasts[37] (Figs. 9–12).

After TA treatment of isolated membrane vesicles, membrane structure is recovered at its best and the vesicles may be considered as truly representative of the fraction at the end of the preparative procedures. Consequently, every disturbance in membrane configuration is assumed either to be present in the initial membrane compartment or to occur during tissue homogenization, sedimentation on density gradients, or phase partition. The adequate maintenance of the membrane thickness both in intact cells and isolated vesicles is helpful to determine the origin of membrane fragments and to evaluate their relative enrichment by way

[34] J. P. Carde, J. Joyard, and R. Douce, *Biol. Cell* **44,** 315 (1982).

[35] M. Neuburger, E. P. Journet, R. Bligny, J. P. Carde, and R. Douce, *Arch. Biochem. Biophys.* **217,** 312 (1982).

[36] A. J. Dorne, J. Joyard, M. A. Block, and R. Douce, *J. Cell Biol.* **100,** 1690 (1985).

[37] M. Gleizes, G. Pauly, J. P. Carde, A. Marpeau, and C. Bernard-Dagan, *Planta* **159,** 373 (1983).

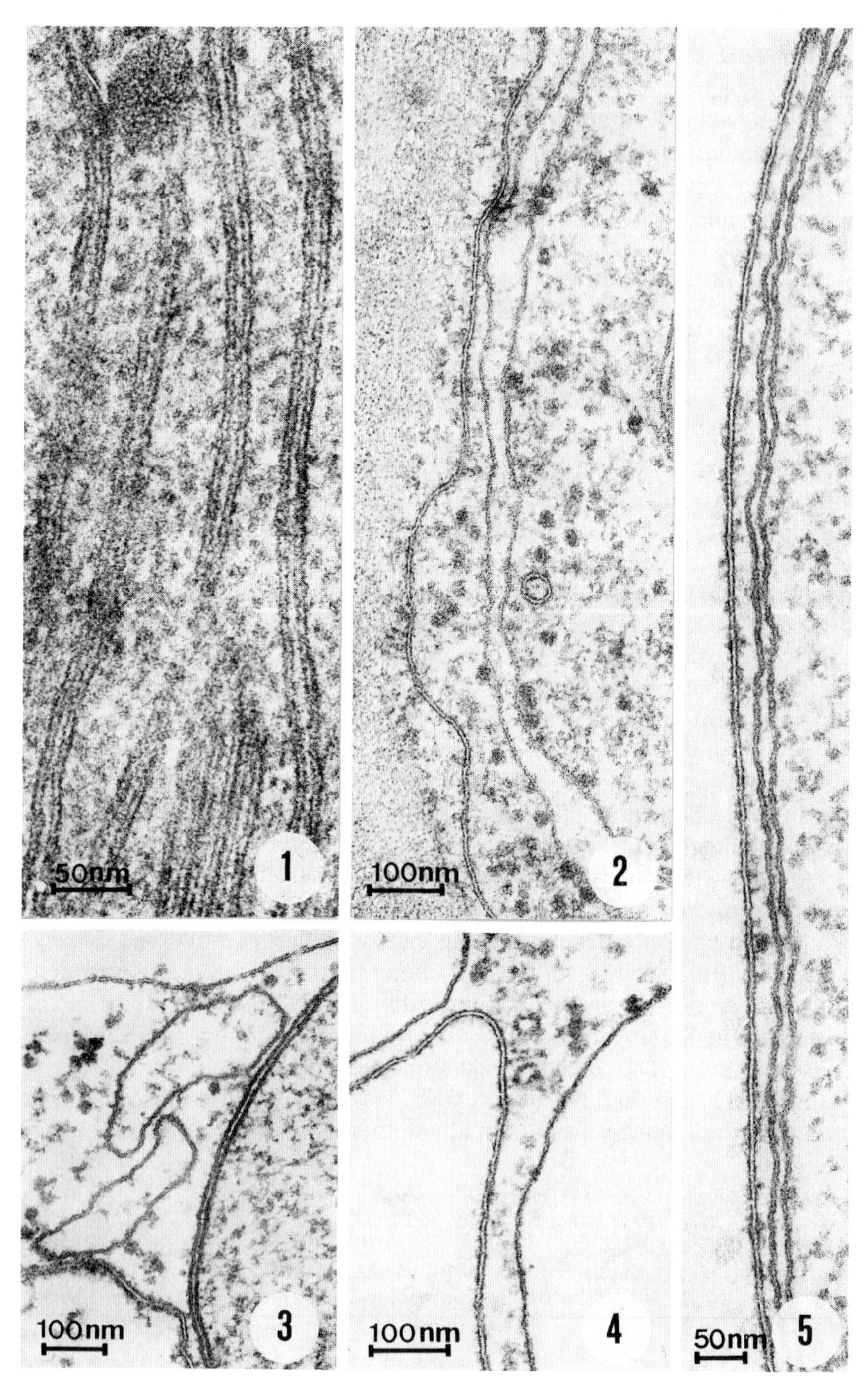

FIGS. 1 to 5. See Legends on p. 609.

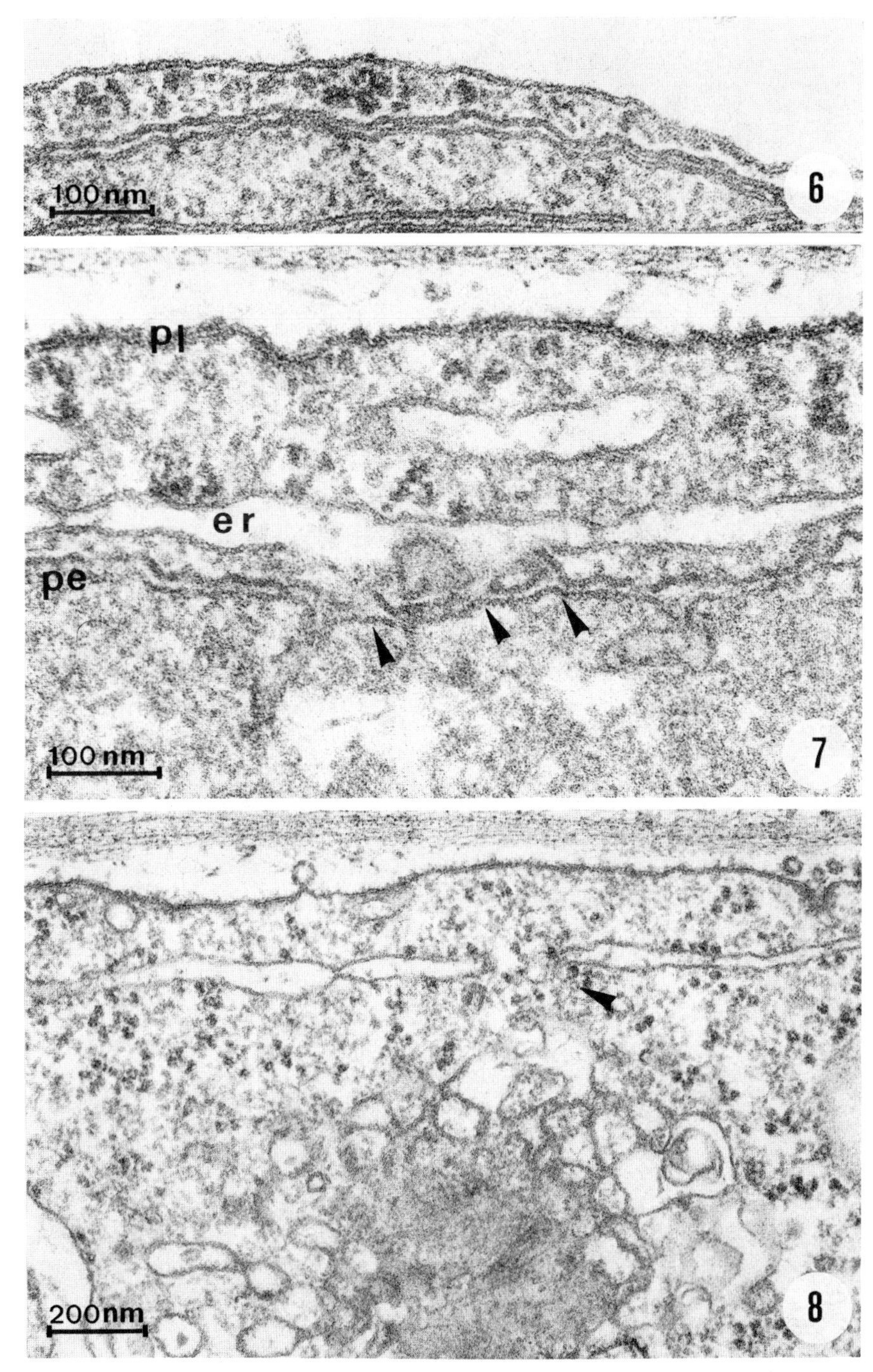

FIGS. 6 to 8. See legends on p. 609.

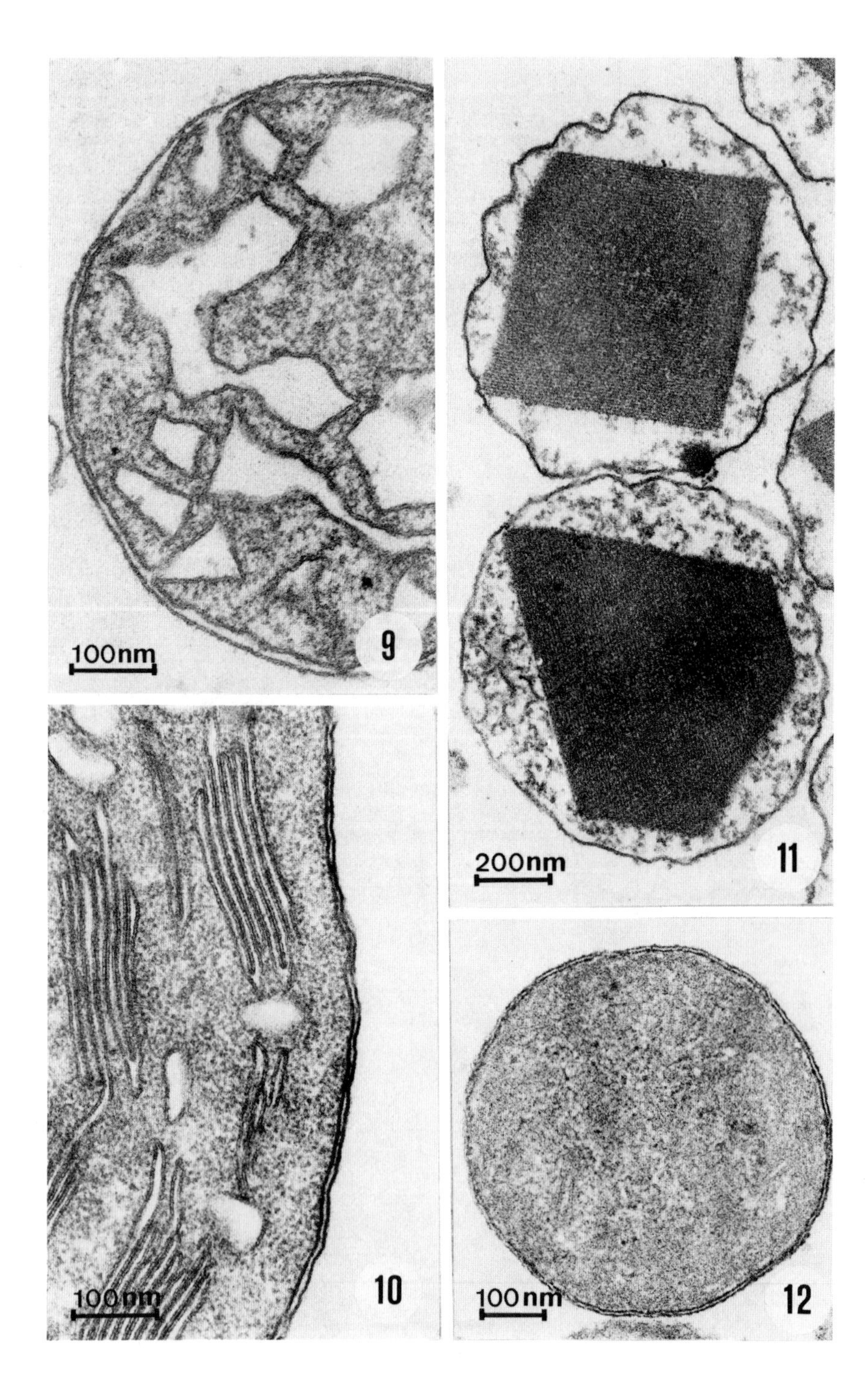
100nm
9
100nm
10
200nm
11
100nm
12

of morphometric studies of membrane fractions purified from microsomal pellets[38] (Figs. 13–15).

The use of TA after G/Os results at the same time in contrast enhancement and improved preservation and resolution of plant cell membranes. Both effects of TA have not been clearly correlated in a number of previous papers. However, there are a number of indications that, in addition to its mordanting action, TA acts as a true fixative with effects complementary to glutaraldehyde and OsO_4. The contrast-enhancing effect of TA is related to a better retention of membrane structural components which are otherwise extracted during the dehydration and embedding steps. TA interacts with phosphatidylcholine, but not with phosphatidylethanolamine, to form a complex[39] and the enhancement of membrane contrast by TA in the presence of OsO_4 is interpreted as being at least in part due to the multivalent capacity of binding to reactive sites of choline, as well as with OsO_4. Interestingly, PC is the major phospholipid of plastid outer membrane, whereas PE is present in trace amounts.[40]

[38] P. Baudhuin, this series, Vol. 32B, p. 3.

[39] M. Kalina and D. C. Pease, *J. Cell Biol.* **74,** 742 (1977).

[40] R. Douce and J. Joyard, *in* "The Biochemistry of Plants" (P. K. Stumpf, ed.), Vol. 4, p. 321. Academic Press, New York, 1980.

FIG. 1. Chloroplast thylakoid membranes of a spinach leaf fixed with G/Os/TA. ×204,000.

FIG. 2. Membrane system in a resin cell of pine seedling, fixed with G/Os, then with TA (30 min). ×95,000.

FIG. 3. Membrane systems in a resin cell of pine seedling, fixed with a mixture of glutaraldehyde and osmium, then treated with tannic acid; membrane preservation is very good. ×83,000.

FIG. 4. Plasmalemma and tonoplast in a mesophyll cell of *Pinus pinaster* hypocotyl fixed with G/Os, then with TA (10 min). ×110,000.

FIG. 5. Plasmalemma and chloroplast envelope in a mesophyll cell of *Pinus pinaster* hypocotyl fixed with G/Os, then with TA (30 min). ×131,000.

FIG. 6. Plastid envelope and tonoplast in a spinach leaf cell fixed with G/Os/TA. ×140,000. From Carde *et al.*[34] with permission.

FIG. 7. Structural connections (arrowheads) between leucoplast envelope (pe) and endoplasmic reticulum (er) in a resin duct of *Foeniculum dulce* fixed with G/Os/TA. pl, Plasma membrane. ×150,000.

FIG. 8. Structural connections (arrowhead) between a dictyosome stack and endoplasmic reticulum in a resin duct cell of *Foeniculum dulce* fixed with G/Os/TA. ×66,000.

FIGS. 9 to 12. Isolated plant cell organelles after G/Os/TA fixation.

FIG. 9. Soybean mitochondria. ×120,000.

FIG. 10. Spinach chloroplasts. ×102,000.

FIG. 11. Soybean peroxisomes. ×55,000.

FIG. 12. Calamondin leucoplast vesicles. ×90,000.

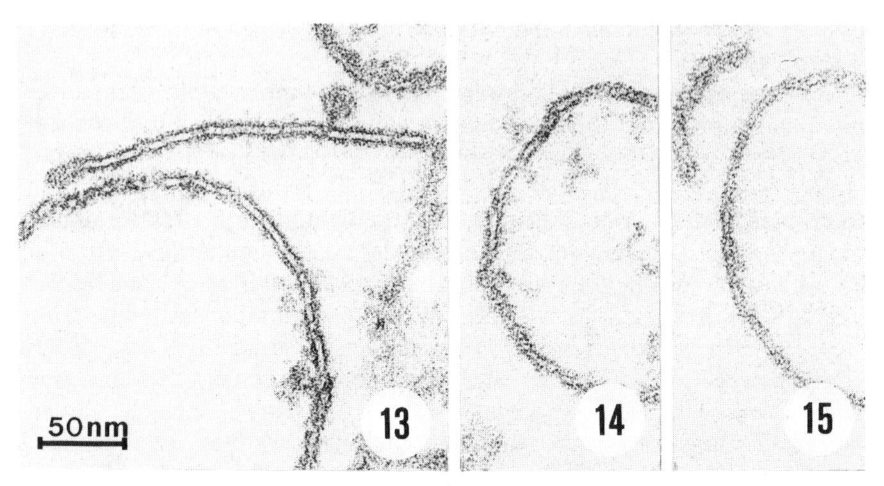

FIGS. 13 to 15. Membrane vesicles from calamondin fruit peel after G/Os/TA fixation. The thickness of the membrane leaflet decreases from Fig. 13 to Fig. 15. ×235,000.

Interactions between TA and cell proteins have also been demonstrated. The size of microtubules is increased after TA treatment.[41,42] The visualization of the erythrocyte membrane skeleton is improved by tannic acid, which reacts with membrane components to bring a marked increase in size, as shown by freeze-etching studies.[30] It is suggested that the size increase by tannic acid is restricted to the membrane proteins which are located outside the membranes and are then exposed directly to the fixative. It is assumed also that TA reacts with membrane glycoproteins.[43]

The complementary fixative action on membrane proteins could also explain the marked improvement in membrane preservation when TA is used after OsO_4, without previous aldehydic fixation or after simultaneous fixation with G/Os mixture (Fig. 3).[44] As a matter of fact, it has been demonstrated[19] that osmium tetroxide reacts rapidly with phenols containing *o*-dihydroxy groups to give very stable chelate complexes.

In addition, TA acts as a mordant between osmicated structures and lead,[29] although in some instances contrast enhancement is obtained without staining. Triple fixations (G/Os/TA) or tetrafixations (G/TA/Os/TA or G/Os/TA/Os) have been suggested[45] in order to improve the overall

[41] L. G. Tilney, J. Bryan, D. J. Bush, K. Fujiwara, M. S. Mooseker, D. B. Murphy, and D. H. Snyder, *J. Cell Biol.* **59,** 267 (1973).

[42] R. W. Seagull and I. B. Heath, *Eur. J. Cell Biol.* **20,** 184 (1979).

[43] B. Van Deurs, *J. Ultrastruct. Res.* **50,** 185 (1975).

[44] J. P. Carde and J. Launay, *Biol. Cell.* **38,** 13a (1980).

[45] P. Park, S. Yamamoto, K. Kohmoto, and H. Otani, *Can. J. Bot.* **60,** 1796 (1982).

contrast of stained or unstained ultrathin sections of plant material embedded in Spurr resin.

The use of the G/Os/TA fixative sequence is therefore particularly suitable to studies on plant membranes. However, several drawbacks are inherent to this method. TA penetrates only two or three layers of cells, and complete impregnation by the postfixative has to be excluded for tissue slices, but is obtained on membrane or organelle fractions resuspended in TA after OsO_4 fixation. Longer incubation times or higher temperatures do not improve the penetration. The fixative effect is therefore different from one cell to the other and generally either complete or absent. Nonpenetrated cells display the usual appearance of G/Os-treated cells. TA penetrates only the outer layers of tissue that suffer "microwoundings" during the preparation of tissue slices.[46] Overfixation effects are observed in some cases and the membrane structure cannot be resolved. As indicated above, cytosolic or organelles proteins are more coarsely aggregated than after G/Os, due to the additive fixation effect and, in addition, ribosomes are difficult to see. Finally, a number of cytochemical tests (e.g., PACP staining) are hindered by tannic acid postfixation.

Taking into account these restrictions, which at the present time preclude the use of this technique as an universal method of plant cell fixation, a sequential treatment of plant material by glutaraldehyde, osmium tetroxide, and tannic acid should be used every time that membrane studies are the main goal of an experiment, owing to the spectacular improvement in plant membrane preservation and staining. Surprisingly, however, this method has seldom been routinely used on plant material, in spite of its advantage, even when using light microscopy.[47]

Other Fixation Procedures

Osmium/Ferricyanide and Osmium/Ferrocyanide Mixtures. Following the work of Karnovsky,[48] mixtures of OsO_4 with potassium ferricyanide or potassium ferrocyanide have been used by several workers as postfixatives after glutaraldehyde treatment in order to enhance selectively the contrast of some animal or plant cell membranes and to improve the resolution of their substructure. A slight variation of these methods was proposed by Hepler[49] to obtain an overall contrast of the nuclear envelope–endoplasmic reticulum membranes.

[46] D. C. Cottell and A. C. B. Hooper, *J. Microsc.* (*Oxford*) **139,** 331 (1985).
[47] M. Dauwalder, W. G. Whaley, and R. C. Starr, *J. Ultrastruct. Res.* **70,** 318 (1980).
[48] M. J. Karnovsky, *Proc. 11th Annu. Meet. Am. Soc. Cell Biol.*, p. 146 (1971).
[49] P. K. Hepler, *J. Cell Biol.* **86,** 490 (1980).

A specific contrast of plastid membranes (thylakoids and envelope) is obtained[34] using an OsO_4/potassium ferricyanide mixture according to Mollenhauer and Droleskey.[50] After a prefixation with 2.5% glutaraldehyde buffered with 100 m*M* Na/Na phosphates, 4 hr at 4°, the plant material is rinsed with the buffer containing 1.65% potassium ferricyanide, 2 hr at 4°, then postfixed by a mixture in equal amounts of aqueous 2% OsO_4 and aqueous 3.3% potassium ferricyanide, 2 hr at 4°, and rinsed overnight. The contrast of plastid membranes is selective and enhanced by lead citrate staining. Cytoplasmic or chloroplastic ribosomes are faintly visible after counterstaining (Fig. 22).

The osmium/ferricyanide method gives quite different results with the conditions used by Hepler.[49,51] Plant tissues are fixed in 2% glutaraldehyde in 50 m*M* cacodylate buffer with 5 m*M* $CaCl_2$ for 1 or 2 hr at room temperature. After three rinses in buffer plus $CaCl_2$ for 30 min, they are postfixed in 2% OsO_4 plus 0.8% potassium ferricyanide in cacodylate buffer and $CaCl_2$ for 1 or 2 hr at room temperature. In these conditions (lower ferricyanide concentration, presence of Ca^{2+}), the nuclear envelope and endoplasmic reticulum (NE–ER) are selectively stained, giving clear pictures of the overall distribution of these membrane systems within a dividing[49] or resting[51] plant cell. However, the contrast improvement depends upon an accumulation of electron-dense material at the inner membrane leaflet and in the cisternal space but the proper contrast of the bounding membranes is not clearly improved, nor is their resolution. In this respect, Hepler's method is comparable to osmium impregnation methods (see below), but with a good overall preservation of cell structure.

According to Hepler, the presence of divalent cations with cacodylate buffer is necessary to obtain the selective staining of NE–ER and this reaction would be a marker of calcium-sequestering membranes. However, comparable results have been obtained in pollen grains and pollen tubes without addition of $CaCl_2$ to the fixative and buffer.[52]

Further improvement in contrast and resolution of the plastid membranes is obtained in plant material fixed by the OsO_4–ferrocyanide method of Karnovsky[48] with the potamium-osmium-cyanide complex (POCC) modification of Hoshino *et al.*[53] After glutaraldehyde prefixation and phosphate buffer rinses, the plant material is treated with a mixture of equal amounts of aqueous 2% OsO_4 (w/v) and aqueous 3% potassium

[50] H. H. Mollenhauer and R. E. Droleskey, *J. Ultrastruct. Res.* **72,** 385 (1980).

[51] P. K. Hepler, *Eur. J. Cell Biol.* **26,** 102 (1981).

[52] M. Cresti and C. J. Keijzer, *J. Submicrosc. Cytol.* **17,** 615 (1985).

[53] Y. Hoshino, W. A. Shannon, Jr., and A. M. Seligman, *Acta Histochem. Cytochem.* **9,** 125 (1976).

ferrocyanide (w/v), 2 hr at 4°, then treated with aqueous 2% OsO_4, 2 hr at 4°, and rinsed overnight. Plastid membranes are selectively contrasted especially after lead citrate staining. Cytoplasmic or organelles proteins and ribosomes are no longer visible. Interestingly, the trilaminar structure of the plastid envelope membranes is clearly demonstrated by this post-treatment, which seems to indicate that the contrast enhancement is also related to a better preservation of membrane structural components. An additional treatment of the specimen with TA after POCC abolishes the specific staining.

The osmium tetroxide–potassium ferrocyanide method (OsFeCN) buffered with cacodylate in the presence of $CaCl_2$ has been applied to a number of different plant cells.[54] The OsFeCN method enhanced the staining of most membrane types of the cell, but usually only the membranes—not the lumen—of the ER were stained. The authors concluded that this method was not reliable for specific contrast of ER and NE, nor as an indicator for Ca^{2+}-sequestering compartments, but was interesting as a general method of section contrast, with caution in interpreting the results.

The chemical bases of membrane staining by osmium tetroxide–ferri(o)cyanide mixtures have been established.[55] Osmium complexes are formed. The greater reactivity and concentration of the Os intermediates in the system leads to more Os deposition than OsO_4 alone. These osmium complexes are chelated by appropriately placed donor atoms in the macromolecular tissue matrix and chelation is responsible for the immobilization of Os and the observed staining pattern in electron micrographs. Osmium and iron compounds must be present simultaneously to give enhanced contrast. Tannic acid as a chelating agent could be used *before* treatment with the OsFeCN mixtures.

Impregnation Methods. Impregnation of plant tissue blocks by osmium tetroxide or osmium tetroxide–iodide systems results in an overall contrast of several plant membranes.[56,57] An adequate impregnation is obtained after prolonged treatments in the osmium solution and/or high temperature. The zinc iodide–osmium tetroxide (ZIO) method of Harris[58] usually gives reproducible results.

After a prefixation in 2.5% glutaraldehyde buffered with 100 m*M* Na/Na phosphates, the plant material is thoroughly rinsed and soaked 1–24 hr in the ZIO mixture at 25 or 60°. The ZIO mixture is prepared by mixing

[54] E. Schnepf, K. Hausmann, and W. Herth, *Histochemie* **76,** 261 (1982).

[55] D. L. White, J. F. Mazurkiewicz, and R. J. Barrnett, *J. Histochem. Cytochem.* **27,** 1084 (1979).

[56] F. Marty, *C.R. Hebd. Seances Acad. Sci., Ser. D* **227,** 1317 (1973).

[57] N. Poux, *C.R. Hebd. Seances Acad. Sci., Ser. D* **276,** 2163 (1973).

[58] N. Harris, *Planta* **141,** 121 (1978).

equal volumes of aqueous 2% osmium tetroxide and a freshly prepared solution of ZnI. The zinc iodide is obtained by mixing 3 g zinc powder in 20 ml water and adding 1 g resublimed iodine. After 5 min of stirring the zinc is filtered off and the solution used immediately. The material is then washed and dehydrated as usual. Gold sections are observed at 80 kV. Thick sections (0.5–2 μm) are observed with a high-voltage electron microscope (HVEM).[59,60] The spatial organization of impregnated cells is better evaluated on stereo pairs.

In higher plant cells, ZIO impregnation results in a dark staining of nuclear envelope, endoplasmic reticulum, Golgi apparatus, plastid lamellae, and mitochondria. In the same conditions, the plasma membrane, the tonoplast, and the plastid envelope are not contrasted. This technique is therefore of special value to study the organization of the endomembrane compartment: tubular or cisternal forms of the ER,[61,62] fenestrated reticulum wrapping the plastids,[63] connections between nuclear envelope and endoplasmic reticulum and Golgi apparatus, dictyosome differentiation,[64] structure of the Golgi-endoplasmic reticulum-lysosomal system (GERL),[65] and nuclear pore distribution.[58] The staining of mitochondria is inconsistent. Reliable results are obtained only in the outer part of the specimens. Optimal conditions have to be determined for each material.

The use of ZIO has been extended[66–68] to other OsO_4–iodide systems involving various divalent cations (Cd, Li, Ca, Rb, Cs, K) in order to obtain a differential staining of some membrane compartments according to the experimental procedure. Membrane staining is not the same in specimens previously fixed with glutaraldehyde. In addition, the staining pattern is also controlled by the cation–iodide combination used. Staining variations are observed on the tonoplast, plastid lamellae, dictyosomes, endoplasmic reticulum, and mitochondria cristae.

It has been shown by X-ray microanalysis that the dense deposit after ZIO impregnation contains Zn and Os, but not iodine.[69] However, the presence of iodine in the mixture is necessary to obtain a staining effect.[70]

[59] F. Marty, *J. Histochem. Cytochem.* **28,** 1129 (1980).
[60] C. R. Hawes, *Micron* **12,** 227 (1981).
[61] N. Harris, *Planta* **146,** 63 (1979).
[62] N. Harris and M. J. Chrispeels, *Planta* **148,** 293 (1980).
[63] J. Charon, Ph.D. Thesis, University of Paris (1983).
[64] M. Dauwalder and W. G. Whaley, *J. Ultrastruct. Res.* **45,** 279 (1973).
[65] F. Marty, *Proc. Natl. Acad. Sci. U.S.A.* **75** (2), 852 (1978).
[66] F. Carrapico and M. S. Pais, *C.R. Hebd. Seances Acad. Sci.* **292,** 131 (1981).
[67] F. Carrapico, F. Madalena-Costa, and M. S. Pais, *J. Microsc. (Oxford)* **134,** 193 (1984).
[68] F. Butor and F. Marty, *Biol. Cell* **52,** 87 (1984).
[69] J. Gilloteaux and J. Naud, *Histochemistry* **63,** 227 (1979).
[70] M. Maillet, *Z. Mikrosk.-Anat. Forsch.* **70,** 397 (1963).

The chemistry of the reaction mechanism is dependent of the local physicochemical conditions and reveals the physiological functions governing the microenvironment of the reactive structures. In some instances, the ZIO technique might be useful in revealing some Ca^{2+} storage sites in the cells.[69]

At this time, impregnation techniques have not been used on membrane subcellular fractions. They could represent, however, an interesting way to separate membrane vesicles at the EM level on the basis of their staining pattern, insofar as the staining specificity remains after tissue homogenization. Preliminary results using the technique of Harris on membrane fractions from *Citrus* fruit (L. Belingheri, unpublished results) indicate that "thick" membrane vesicles interpreted as plasma membrane are not stained, whereas at the same time a number of "thin" membrane vesicles contain dense deposits attached to the membrane (Fig. 18). Only the trilaminar structure of the thick vesicles is well delineated. These results seem to indicate that the staining pattern is conserved after tissue homogenization. Therefore, this technique may be helpful to ascertain the origin of membrane fragments by using different experimental procedures (prefixation, nature of the cation).

Dehydration, Embedding, and Sectioning of the Plant Material

Before sectioning, the plant material is treated according to the above fixative procedures. It is then dehydrated and embedded in Epon, a hydrophobic resin.

Dehydration and Embedding

The fixed plant specimens are rinsed with distilled water and dehydrated by 30, 40, and 50% ethanol, 5 min each, then by 60, 70, 80, and 90% ethanol for 10 min each. The dehydration is completed by 100% ethanol (3 × 10 min), 100% ethanol and propylene oxide (1 : 1) for 15 min, propylene oxide (2 × 15 min). Tissue blocks are then progressively infiltrated by a mixture of propylene oxide and resin (2 : 1) in capped vials. When the objects sink to the bottom of the vial, the caps are removed and the solvent eliminated by placing the vials under a sucking hood (3 hr). The specimens are then placed in fresh resin overnight at room temperature and embedded in gelatin capsules or plastic molds filled with fresh resin. The resin is polymerized 36 hr or more at 58–60°.

This embedding procedure is easy to use and gives reproducible results. However, a number of technical points have to be taken in account. Ethanolic dehydration is perfectly suited to Epon embedding. The

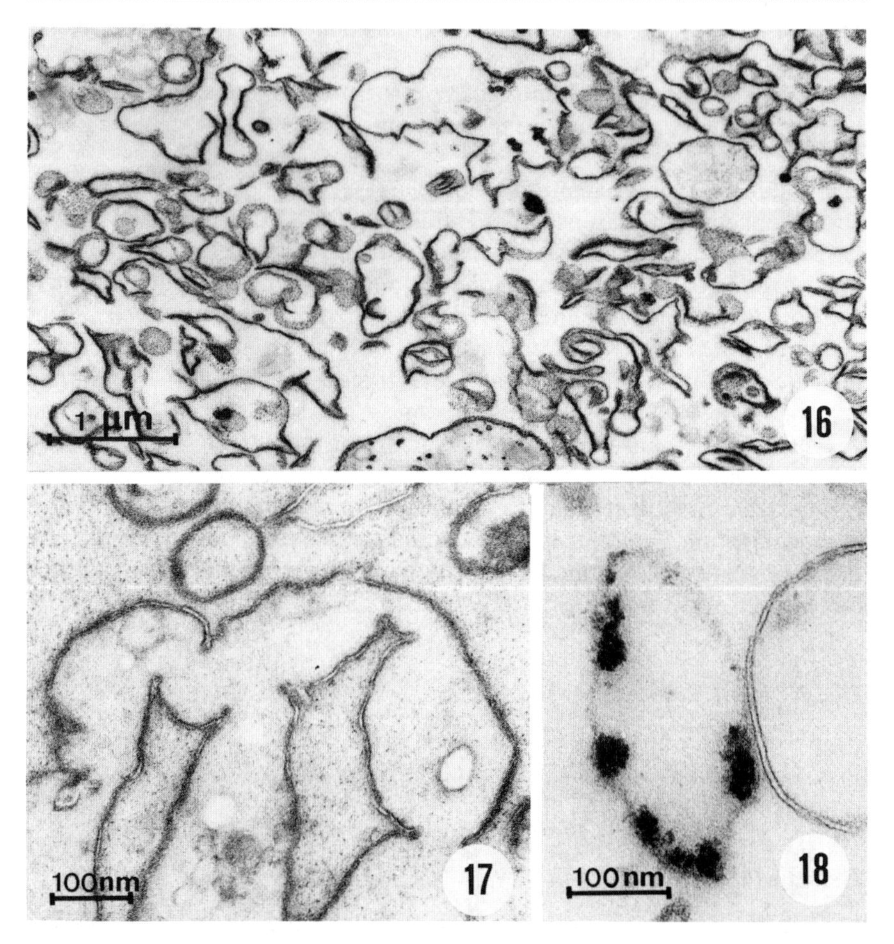

FIG. 16. Plasmalemma-enriched fraction from *Saccharomyces cerevisiae* after G/Os fixation and PATAG staining. ×17,500. From Blanchardie *et al.*[74] with permission.

FIG. 17. Plasmalemma vesicle from *Saccharomyces cerevisiae* after G/Os fixation and PACP staining. ×110,000. From Blanchardie *et al.*[74] with permission.

FIG. 18. Membrane vesicles from calamondin fruit peel after ZnI/OsO_4 impregnation: a dense deposit is attached only to the thin membrane vesicles. ×137,000.

dehydration scheme presented above may be shortened. Insofar as very small pieces are used for optimal fixation, no cell shrinkage results from the use of fast dehydration procedures. Shortening the stages in strong ethanol in order to minimize the extraction of the lipid compounds of the membranes does not improve membrane preservation.

At the same time epoxy resins provide good preservation of cell structure, good cutting qualities, and high electron contrast. Glycidyl ether (formerly Epon 812) is usually assumed as the best compromise because of its rather low viscosity and therefore good penetration rate, excellent cutting qualities, and resistance to the electron beam. Spurr's resin is advocated by many workers but often gives erratic results. Stock solutions of diglycidyl ether and hardeners are prepared by mixing 77 g of diglycidyl ether and 100 g dodecenylsuccinic anhydride (DDSA) (solution A) and 124 g of diglycidyl ether and 110 g of methyl nadic anhydride (MNA) (solution B). These solutions have a long life when kept at 4°. The final resin is prepared by mixing, very thoroughly, equal weights of solutions A and B plus 1.5% (w/w) tridimethylaminomethylphenol (DMP 30). The final resin plus accelerator may be kept several months at −20°. Some batches of dodecenylsuccinic anhydride are inadequate for satisfactory embedding, leading to soft "chattering" blocks with lower staining qualities, and have to be discarded. Use only DDSA specially distilled for EM.

Other commercial epoxy or acrylic resins do not bring significant improvements to the routine procedure described above.

Sectioning

Clear pictures of membrane ultrastructure and relationships require the use of high magnifications of the EM (up to ×100,000) with good resolution and contrast. Adequate resolution cannot be obtained from the usual silver or silver-gray sections (50–70 nm thick) designed for overall surveys of cell organization at low to medium magnifications of the electron microscope.

Sectioning must fit the aim of the work, which is to obtain ultimate resolution on ultrathin sections of biological material embedded in epoxy resins. Trimming of very small pyramids (300 × 300 μm) is helpful for thin sectioning. Only sections with interference colors ranging from light gray to "black" are actually useful. The thickness of these sections is lower than 40 nm. Perfectly adequate results are obtained with glass knives, but diamond knives give easier and more reproducible results. Scratches on the knife edge are quite undesirable. The very thin sections are collected for optimal support and resolution on naked 600 or 1000 mesh grids, with a square or hexagonal support mesh (Gilder grids, London). Sections are of course very fragile and can break when collected, owing to the surface tension, thereby giving rise to many holes. This is of no consequence because it is not necessary to observe the entire specimen. The grids are thoroughly cleaned with acetone before use.

When very thin sections are used, contrast is a limiting factor. Section contrast is markedly improved by the use of G/Os/TA fixation procedure and lead staining.

For high-resolution studies, sections have to be observed as soon as possible after cutting—in any case within 24 hr—because a gradual blurring of the details occurs after keeping the sections.[71]

Staining the Sections

Routine Staining

A sequential staining with uranyl acetate and lead is perfectly convenient for high magnification studies of plant cell membranes. Routinely, sections are stained first with saturated ethanolic uranyl acetate (14%, w/v), 30 min at 20°, diluted in half with distilled water just before use, rinsed twice with 50% ethanol, and dried on filter paper. After dipping in water,[71] the grids are immersed in alkaline lead citrate[72] without dilution, 10 min at 20°, rinsed twice with distilled water, and dried. Lead citrate staining is performed in a closed vial containing sodium hydroxide to trap atmospheric carbon dioxide. For both stains, the grids are submersed in single drops of the stains on a sheet of Parafilm.

Some difficulties may arise during this double-staining procedure. Inadequate uranyl acetate staining results in a patchy cloudiness of the sections hardly visible at low magnifications of the EM. These very fine precipitates (likely insoluble alkaline uranyl salts) may occur during dilution of the stock solution with water or during natural aging of the ethanolic solution. It is necessary to use only deionized water and to discard the stock uranyl solutions once this staining problem appears. The quality of lead citrate staining is controlled first by the pH of the solution (12 ± 0.1).[72] Large precipitates indicate contamination of the stock solution by lead carbonate but do not prevent an adequate staining of the sections.

This staining procedure is very effective with sections of G/Os, G/Os/TA, as well as ferricyanide/Os or ferrocyanide/Os fixed material.

Contrasted sections have to be observed as soon as possible after staining, since the binding of heavy metal ions to macromolecules is unstable. This is the case for membranes where the particulate grains of stains migrate outside the dark leaflets, leading to granulous unadequate pictures.

Cytochemical Stainings

A limited number of cytochemical procedures may be performed on ultrathin sections of G/Os fixed specimens, in order to detect certain chemical groups present in the membranes.

[71] H. H. Mollenhauer, *Stain Technol.* **49,** 305 (1974).

[72] E. S. Reynolds, *J. Cell Biol.* **17,** 208 (1963).

PATAG Staining. A positive staining with the periodic acid–thiocarbohydrazide–silver proteinate (PATAG) method of Thiery[73] reveals the free glycol groups and depends on the nature of the constituent sugars and their linkage to proteins and lipids.

The sections collected on gold grids are first oxidized for 20–30 min on 1% (w/v) periodic acid at 22°, then thoroughly rinsed with water. The grids are then floated on 0.2% (w/v) thiocarbohydrazide in 20% acetic acid, 4 hr (or more) at 22°, rinsed with 10% acetic acid (2 × 10 min) and water (1 hr), and stained with 1% aqueous silver proteinate, 30 min at 22°, in the dark. The grids are then rinsed with water and dried. The reactive structures are covered with small electron-dense silver grains.

In addition to the cell wall, which contains very reactive glucosidic polymers (except callose), this method is useful to contrast several membrane systems, either *in situ* or in isolated vesicles. A positive reaction is obtained with the plasma membrane of higher plant cells[21] and yeasts.[74] In the latter case this technique, in conjunction with PACP and morphological markers, has been used to check a plasma membrane-enriched fraction, owing to the high glucomannan content of the plasma membrane (PM). Golgi vesicles containing cell wall precursors are also heavily contrasted.

Plastid membranes contain large amounts of galactolipids which are strongly PATAG reactive. A slight modification of the above method allows the selective contrast of these membranes. Bleaching and oxidation of ultrathin sections by 1% aqueous periodic acid are performed at low temperature (30 min at 4°). In these conditions of mild oxidation, the thylakoids, prolamellar bodies, and envelope membranes are well contrasted whereas plasma membrane and Golgi vesicles are only faintly contrasted (Fig. 21).[34]

PACP Staining. Phosphotungstic acid in chromic acid solution (PTA/CrO_3 or PACP) is assumed to be a specific stain of plasma membrane on sections of plant tissue or of isolated membrane fractions.[75] Sections of glutaraldehyde–osmium-fixed material are collected on gold grids and bleached by 1% (w/v) periodic acid for 30 min; after a thorough washing in distilled water, the grids are transferred to 1% (v/v) phosphotungstic acid in 10% (v/v) aqueous chromic acid for 10 min before further thorough washing in three to four changes of distilled water. All steps are carried out at room temperature. Ultrathin sections are observed at 60 kV. Staining results in a contrast enhancement of the PM, with sharp delineament

[73] J. P. Thiery, *J. Microsc. (Paris)* **6,** 987 (1967)

[74] P. Blanchardie, J. P. Carde, and C. Cassagne, *Biol. Cell.* **30,** 127 (1977).

[75] J. C. Roland, C. A. Lembi, and D. J. Morré, *Stain Technol.* **47,** 195 (1972).

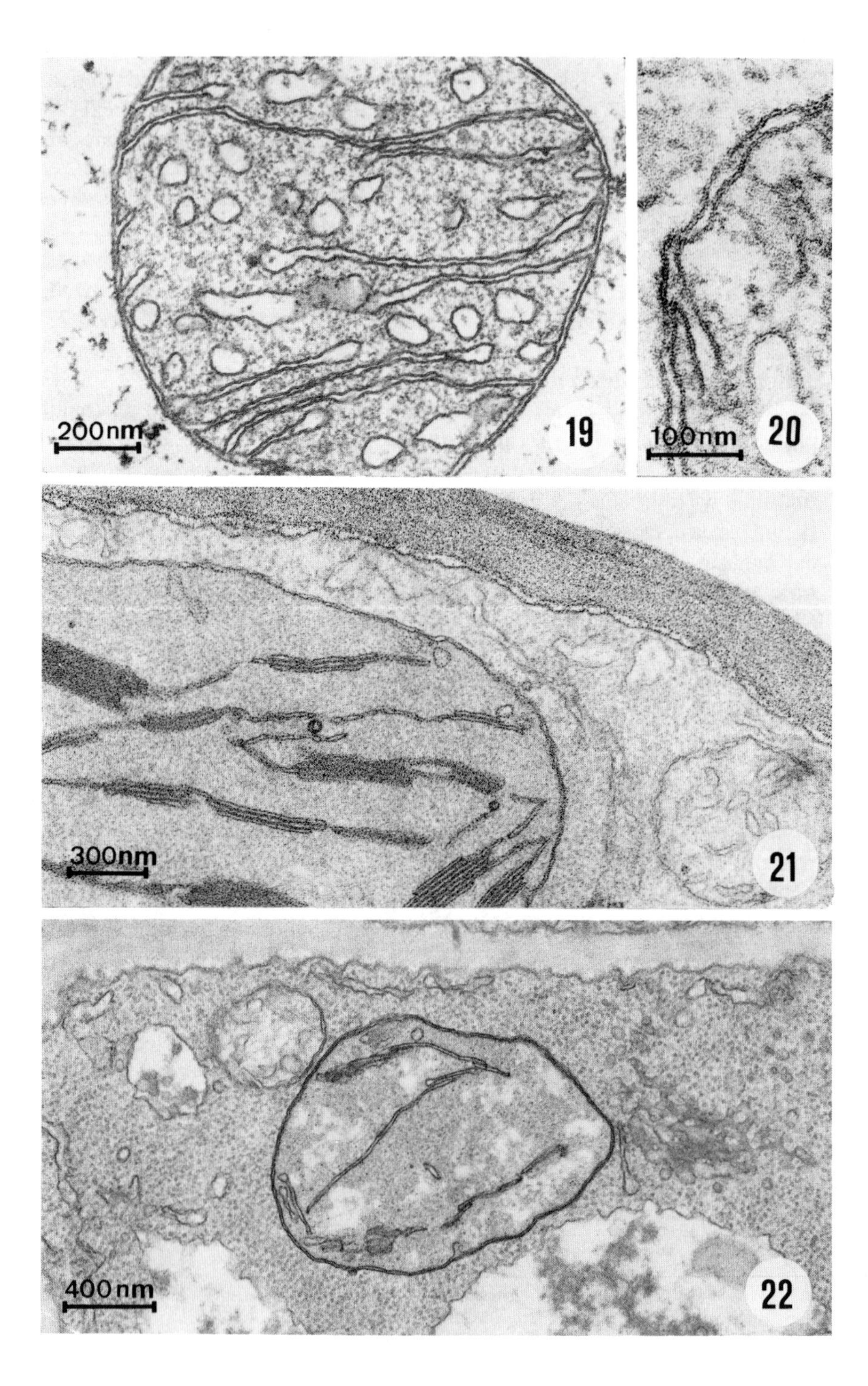
200nm
19
100nm
20
300nm
21
400nm
22

of the trilaminar structure (Fig. 17). Both leaflets of the membrane may be symmetrically contrasted. Other cell membranes are hardly visible. Sodium silicotungstate may be used instead of PTA.[73] Although conclusive results have been obtained by many workers on various materials[76–79] for unequivocally identifying and quantifying the plasma membrane, the reliability of this technique on tissue sections or membrane fractions has been seriously questioned.[80] As a matter of fact, prolamellar body membranes, lipid droplets, and ribosomes are sometimes strongly stained and these non-plasma membrane components also have the capacity to appear PTA/CrO_3 positive in subcellular fractions.[81]

However, the reliability of PACP seems to depend to a large extent upon the operating conditions which have been carefully reviewed.[82,83] The intensity and specificity of the staining are controlled by length of oxidation (30 to 50 min), staining with PACP (5 to 12 min), and washing (50 min). The optimal parameters are somewhat variable according to the plant material and different for intact cells and membrane fractions. PACP is highly specific for the plasma membrane in a variety of cell types from both mature and meristematic regions of the soybean root which precludes a putative relationship between the specificity or intensity of staining and a change in PM composition or ultrastructure.[84]

In addition, PACP staining and concanavalin A–ferritin binding are complementary with respect to topological specificity.[83]

[76] C. A. Lembi, D. J. Morre´, K. S. Thomson, and R. Hertel, *Planta* **99,** 37 (1971).

[77] W. J. Van Der Woude, C. A. Lembi, D. J. Morré, J. I. Kindinger, and L. Ordin, *Plant Physiol.* **54,** 330 (1974).

[78] C. Cassagne, R. Lessire, and J. P. Carde, *Plant Sci. Lett.* **7,** 127 (1976).

[79] M. A. Hartmann-Bouillon, P. Benveniste, and J. C. Roland, *Biol. Cell.* **35,** 183 (1979).

[80] J. L. Hall and T. J. Flowers, *J. Exp. Bot.* **27,** 658 (1976).

[81] P. H. Quail and J. E. Hughes, *Planta* **133,** 169 (1977).

[82] R. L. Berkowitz and R. L. Travis, *Plant Physiol.* **63,** 1191 (1979).

[83] R. L. Travis and R. L. Berkowitz, *Plant Physiol.* **65,** 871 (1980).

[84] O. W. Galbraith and D. H. Northcote, *J. Cell Sci.* **24,** 195 (1977).

FIG. 19. Mitochondrion of *Portulaca oleracea* leaf after G/Os/TA fixation: the envelope and inner cristae are sharply delineated. ×60,000.

FIG. 20. Mitochondrion of *Allium porrum* leaf fixed with G/Os, then soaked in uranyl acetate: the membrane structure is not as well preserved as in Fig. 19. ×140,000.

FIG. 21. Mesophyll cell of *Spinacia* leaf after G/Os fixation and PATAG staining in mild oxidative conditions: the plastid membranes are well contrasted by silver deposits but the endoplasmic reticulum or mitochondria are not. ×39,000.

FIG. 22. Spinach leaf cell after G/OsO_4/ferricyanide fixation and lead staining: the plastid membranes are distinctly contrasted. ×31,500.

From various experiments in our laboratory, and others, it may be concluded that PACP staining is a valuable marker for PM, insofar as no more convenient procedure has been devised for a given plant material. However, the staining intensity is sometimes very low. This poor reactivity cannot be interpreted at this time as due to the composition of the PM or to changes occurring during preparation of subcellular fractions.

Operating the Electron Microscope and Photographic Procedures

When working on ultrathin sections of resin-embedded biological material, with a recent TEM, the limiting factor in obtaining ultimate resolution is by no means the electron microscope itself but depends upon previous processing of the specimens. At the present time, the point-to-point resolution of most TEM (often less than 0.2 nm) is far better than necessary for optimal resolution of plant cell membranes.

However, it must be remembered that high-resolution observations on membranes require an optimization of the electron microscope adjustments, i.e., correct alignment of the lens system, choice and cleanliness of condenser and objective apertures, stability of the electron beam, presence of an effective anticontamination device, correction of lens astigmatism (condensers, objective), and optimal photographic records. The astigmatism of the objective lens is corrected at high magnifications of the EM ($\times$100,000).

Insofar as very thin sections are used, an accelerating voltage of 60 kV usually represents the best compromise between optimal resolution and contrast. It is necessary to avoid excessive exposure to the electron beam of the sections, which are very fragile. A "wobbler" is an aid for focusing even at high magnifications up to 50,000. Optimal contrast without significant loss of resolution is obtained after a slight defocusing from the focus, judged by eye or according to a calibrated curve. Focal series are helpful in some instances.

The choice of the photographic emulsion is governed by the necessity to obtain at the same time highly contrasted pictures without loss of fine details and reasonably short exposure times (1–2 sec). This result is obtained either with commerical emulsions specially devised for EM (Kodak SO-163), or with line emulsions adequately processed.[85]

[85] J. Bohonek, *Zeiss Inf., Mem.* N°. 2 (1983).

[57] Two-Dimensional Electrophoresis in the Analysis and Preparation of Cell Organelle Polypeptides

By R. REMY and F. AMBARD-BRETTEVILLE

Two-dimensional electrophoresis implies a fractionation of complex protein mixtures according to different parameters in each dimension. The first dimension may correspond to an electrophoresis under nondenaturating or mild denaturating conditions, which separates proteins on the basis of charge to molecular weight ratio, but also to an isoelectric focusing (IEF) in which proteins are fractionated according to their charge.

In the second dimension, separation of proteins is generally performed according to their molecular weight under strong denaturating conditions by sodium dodecyl sulfate gel electrophoresis (SDS–PAGE). This chapter will deal only with the description of the two-dimensional method combining IEF in the first dimension and SDS–PAGE in the second dimension.

Since O'Farrell[1] first described two-dimensional gel electrophoresis as a high-resolution technique for the separation of *Escherichia coli* proteins, this technique has been used in a wide field of research which currently includes molecular biology, genetic as well as clinical investigations. However, when applied to polypeptides of plant organelles and especially to thylakoid membranes, this technique often appears not resolutive enough because some membrane proteins with high hydrophobicity run as streaks; this is probably why only a few groups have used the two-dimensional electrophoresis for chloroplast protein studies.[2–5] In particular, as it has been reported by Roscoe and Ellis,[2] the major protein of thylakoid membranes, which is associated with pigments and lipids and corresponds to the light-harvesting antenna (LHCP) for photosystem II, ran as a continuous streak in the IEF dimension. In the laboratory, we achieved the two-dimensional separation of the different constitutive polypeptides of LHCP and succeeded in distinguishing between its phos-

[1] P. H. O'Farrell, *J. Biol. Chem.* **250,** 4007 (1975).

[2] T. J. Roscoe and R. J. Ellis, in "Methods in Chloroplast Molecular Biology." Elsevier/North-Holland Biomedical Press, 1015 Amsterdam, 1982.

[3] I. Novak-Hofer and P. A. Siegenthaler, *Biochim. Biophys. Acta* **468,** 461 (1977).

[4] A. Boschetti, E. Sauton-Heiniger, J. C. Schaffner, and W. Eichenberger, *Physiol. Plant* **44,** 134 (1978).

[5] C. W. Gilbert and D. E. Buetow, *Plant Physiol.* **67,** 623 (1981).

phorylated and nonphosphorylated polypeptides.[6,7] Protein streaking problems were greatly reduced. This was done as suggested by Perdew *et al.*[8] for polypeptide analysis of trout liver microsomes, by replacing Triton X-100 or Nonidet P-40 in the solubilization buffer and in the electrofocusing gel by the zwitterionic detergent 3-[(3-cholamidopropyl)dimethylammonio]-1-propane sulfonate (CHAPS) first synthesized by Hjelmeland.[9]

This chapter provides complete directions for the solubilization of thylakoid membranes and the preparation of the first and second dimension gels. Although this procedure has been developed specifically for thylakoid membranes, application to other chloroplast proteins and mitochondrial proteins gave us good results.

Outline of Procedures

Chemicals

Acrylamide (twice crystallized), *N,N'*-methylenebisacrylamide, *N, N, N', N'*-tetramethylenediamine, 3-[(3-cholamidopropyl)dimethylammonio]-1-propane sulfonate (CHAPS), sodium dodecyl sulfate (SDS) are from Serva

Ammonium persulfate, glycine, sodium hydroxide, tris(hydroxymethyl) aminomethane (Tris), urea are from Merck

2-Mercaptoethanol, Bromphenol blue, and Coomassie brilliant blue R-250 are from Sigma

Acetic acid, 2-propanol, 1-butanol, and methanol are from Labosi

Ampholytes, pH 5–8, are from LKB

Sample Preparation

In order to remove the pigments and part of the lipids, the first step before the solubilization of thylakoid membrane involves an acetone treatment. To a thylakoid pellet containing about 0.7 to 1 mg of protein, 10 ml of 80% (v/v) ice-cold acetone containing 35 m*M* mercaptoethanol is added. After a 1-hr incubation on ice, the precipitated proteins are centrifuged at 10,000 *g* for 10 min. To obtain complete depigmentation, one or

[6] R. Rémy and F. Ambard-Bretteville, *Physiol. Veg.* **23,** 381 (1985).
[7] R. Rémy and F. Ambard-Bretteville, *FEBS Lett.* **188,** 43 (1985).
[8] G. H. Perdew, H. W. Schaup, and D. P. Selivonchick, *Anal. Biochem.* **135,** 453 (1983).
[9] L. M. Hjelmeland, *Proc. Natl. Acad. Sci. U.S.A.* **77,** 6368 (1980).

two additional washes with the same acetone mixture may be necessary. The final pellet is dried rapidly under a nitrogen stream.

When analyzing intact organelles, nucleic acids may be present in the sample. In this case, the protein pellet can be treated for 30 min at room temperature by 500 μl of nuclease solution (10 m*M* Tris, pH 7.4, 5 m*M* $MgCl_2$, pancreatic RNase and DNase at 50 μg/ml). Proteins are then pelleted again in 80% acetone. However, it must be borne in mind that these enzymes will later appear as contaminants.

Sample Solubilization

Two solubilization procedures are described below and designated as A and B. Procedure A is a modification of that first described by Rubin and Leonardi,[10] where CHAPS is employed instead of Nonidet P-40. The depigmented pellet is successively treated with (1) 75 μl of a solution containing 0.05 g of SDS, 0.5 ml of mercaptoethanol, and 2.85 g of urea to make 5 ml with glass-distilled water (pH 7.4). Homogenize with a glass rod until clarification is achieved. Heating must be avoided to minimize artifactual spots due to carbamylation of proteins. (2) Add 25 μl of a solution containing 1 ml of 10% CHAPS and 0.1 ml of ampholytes (pH 5–8). Homogenize. (3) Add 100 μl of a solution containing 1 ml of 10% CHAPS (w/v), 0.1 ml of ampholytes (pH 5–8), and 2.85 g of urea to make 5 ml. All these 3 solutions are stored at 4° for no longer than 2 weeks. Since urea crystallizes at low temperature, solutions are clarified before use by thawing at no more than 35°. In order to avoid numerous freezing and thawing cycles of solutions containing urea, it is best to dispense these solutions in small volumes of 1 ml.

Procedure B allows to separate by IEF a sample which has been previously solubilized for SDS–PAGE in the Laemmli system.[11] Although this procedure is slightly less resolutive than procedure A (occasional presence of streaks in two-dimensional separations), it has the advantage that samples can be boiled, which sometimes increases their solubilization and moreover reduces the risk of their alteration due to proteases or other enzymes. It involves two steps:

1. Solubilization according to Laemmli,[11] i.e., for a pellet as above containing about 1 mg of proteins, add 50 μl of a solution containing 0.0625 *M* Tris–HCl buffer (pH 6.8), 2% SDS, 10% glycerol, 0.5% mercaptoethanol, and 0.001% Bromphenol blue. Homogenize the sample for 2 min in boiling water.

[10] R. W. Rubin and C. L. Leonardi, this series, Vol 96, p. 184.

[11] U. K. Laemmli, *Nature (London)* **227,** 680 (1970).

2. After cooling the sample to room temperature, add 150 μl of a solution containing 2.85 g urea, 0.25 ml ampholytes (pH 5–8), and 200 mg CHAPS adjusted to 5 ml with distilled water. It is important to note that the ratio of the final percentage of CHAPS over that of SDS must be maintained at a very high level in order to avoid excessive streaking on the gel. In procedures A and B, this ratio is 6. A similar ratio of around 8 was first recommended by Ames and Nikaido[12] for solubilization mixtures composed of Nonidet P-40 and SDS.

First Dimension Isoelectric Focusing—Gel Preparation and Running Conditions

IEF separations are carried out in gel rods rather than slabs because large volumes of sample (up to 60 μl) are easier to load and no special equipment is required. Any commercially available or workshop-made tube gel apparatus can be used.

IEF gels (cylindrical, 10 cm long and 3 mm i.d.) are cast essentially according to the method of O'Farrell[1] with the following modifications: the acrylamide/bisacrylamide ratio is increased (25 instead of 17.5) and Nonidet P-40 is replaced by CHAPS. For a final volume of 10 ml, the IEF mixture is made of 5.5 g of urea, 2 ml of glass-distilled water, 2 ml of CHAPS (10%), 1.5 ml of a 26% acrylamide solution (25 g acrylamide + 1 g bisacrylamide), and 400 μl of ampholytes (pH 5–8). The ampholyte composition can be altered according to the desired result.

After complete dissolution of the urea at 30–35° under gentle stirring, polymerization is initiated by 20 μl of a 10% ammonium persulfate solution and 15 μl of TEMED. We do not recommend the previous degassing of the small volume of urea solution to avoid crystallization problems. Tubes are filled using a very thin Pasteur pipet so that the gels have the same length. The gel mixture is then carefully overlayed with about 20 μl of distilled water and allowed to polymerize at 20–25° for at least 2 hr. Gels are run 4 to 5 hr after polymerization. Prerunning was found unnecessary since it did not improve the resolution. It is better to centrifuge the samples before loading the gels in order to remove possible traces of unsolubilized materials. Up to 200 μg of proteins can be loaded onto each gel with a Gilson Pipetman or Hamilton syringe. Samples are then overlayed with 20 μl of a solution containing 3 *M* urea and 1% ampholytes. The IEF gels are run at 400 V for 16 hr and then at 500 V for 1 hr with degassed 0.02 *M* NaOH in the upper cathode chamber and 0.01 *M* H_3PO_4 in the lower anode chamber.

[12] G. Ames and K. Nikaido, *Biochemistry* **15,** 616 (1976).

After the run, the IEF gels are extruded from the tubes by using the pressure of a 20-ml syringe filled with water connected to the electrophoresis tube via a short piece of Tygon tubing. At this point, the IEF gel can be (1) stained, (2) cut into small pieces in order to control the pH gradient, (3) equilibrated and frozen at −80° wrapped into aluminum foil until further use, or (4) equilibrated and run immediately on a second dimension gel. This last alternative is now described. The equilibration of the IEF gels is performed for 20 to 30 min at room temperature in O'Farrell's buffer "O" without mercaptoethanol.[1]

Second Dimension SDS Electrophoresis

Apparatus. A workshop-built slab gel electrophoresis apparatus made of Plexiglas allows us to run two slab gels in parallel. As shown in Fig. 1,

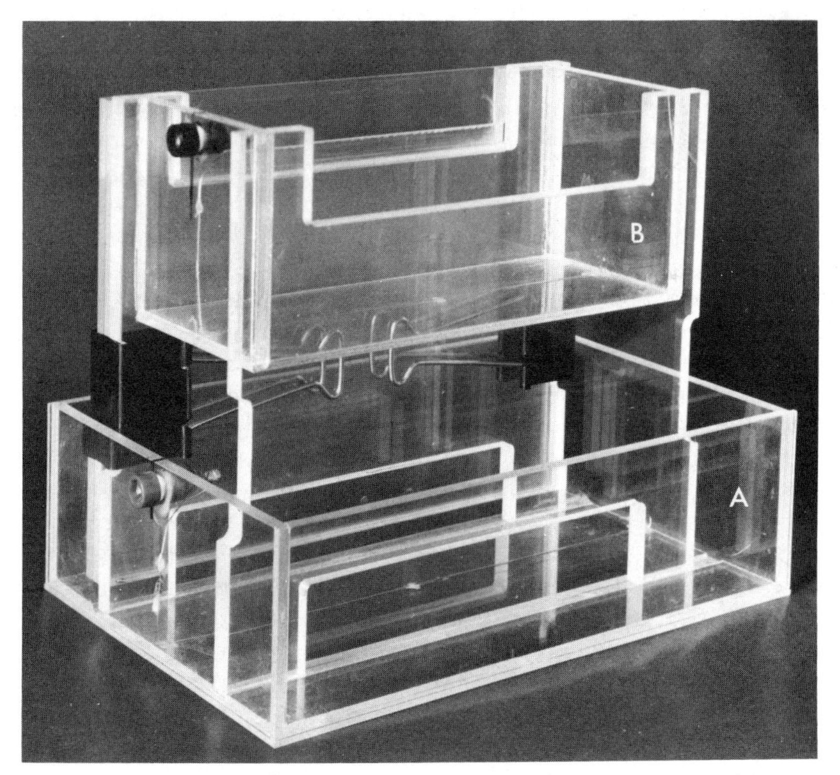

FIG. 1. Photograph of the apparatus used for the second dimension. Only one set of glass plates is assembled on the back of the apparatus. A, Lower anodic chamber; B, upper cathodic chamber.

this apparatus is composed of two distinct parts: a lower anode chamber (A) and an upper part consisting of the cathodic chamber (B) to which two gel plates are clamped on either side. Glass plates used in making the slab gels are separated by 1-mm-thick rubber spacers and made as described by O'Farrell[1] except that they are wider (29 × 16 cm instead of 16 × 16 cm). This improvement, which allows the running of two IEF gels on the same slab gel, solves the problems due to slight variations of migration in the second dimension from one gel to another.

Acrylamide slab gels are made according to the discontinuous system described by Laemmli[11] using a 4.6% (w/v) acrylamide concentration in the stacking gel and 12% in the separating gel. Choosing a uniform acrylamide concentration rather than a gradient for the separating gel provides more reproducible results between different runs.

A typical protocol (shown here for a 12% acrylamide, 0.32% bisacrylamide gel) is as follows: combine 20 ml of a mixture of acrylamide–bisacrylamide [30–0.8%, (w/v) in water], 6.25 ml of 3 *M* Tris–HCl buffer, pH 8.8, 0.5 ml of 10% SDS, 0.375 ml of 10% ammonium persulfate, distilled water to 50 ml. The mixture is degassed and 25 μl TEMED added. Once the gel solution is rapidly loaded, 1-butanol is overlayed to give a smooth surface during polymerization. After polymerization, 1-butanol is removed and the gel surface rinsed with distilled water.

For the stacking gel (2.5 cm high), a typical protocol is as follows: combine 3 ml of acrylamide–bisacrylamide mixture (30–0.8%, w/v), 5 ml of 0.5 *M* Tris–HCl buffer, pH 6.8, 0.2 ml of 10% SDS, 0.2 ml of 10% ammonium persulfate, distilled water to 20 ml. After degassing 20 μl TEMED is added. The stacking gel mixture is poured between the glass plates to a level 2 mm above the base of the notch and overlayed with 1-butanol. This is done so that the level of the polymerized stacking gel reaches the surface of the notch or protudes slightly because some unpolymerized gel mixture is extruded from the gel during polymerization, thus lowering the initial level.

The gel is overlayed with 1-butanol, as before, to produce a flat surface. After polymerization, the gel is rinsed with distilled water, covered with 0.065 *M* Tris–HCl buffer, pH 6.8, and allowed to stand overnight. Thus IEF gels and SDS slab gels are prepared the same day. IEF gels are run overnight and the second dimension performed the following day. The focusing gels, which have been previously equilibrated, are rinsed with distilled water and rapidly drained. Using latex gloves, the IEF gels are then deposited one by one along the top of the stacking gel while a hot agarose solution [1% (w/v) agarose in 0.0625 *M* Tris–HCl, pH 6.8, at 80–100°] is simultaneously poured with a Pasteur pipet. Care should be taken not to trap air bubbles under the IEF gels.

The second dimension can be run after a few minutes, the time necessary to fill the electrode chambers with running buffer (3.05 g Tris, 14.42 g glycine, and 1.0 g SDS/liter) and to expel bubbles remaining at the bottom of the slab gels. Several drops of 1% (w/v) Bromphenol blue are added to the upper reservoir as tracking dye. Gels are run at constant voltage (100 V) for about 4–5 hr until the dye front reaches the bottom of the gel.

Staining of the Gel

At present, the two most common staining procedures are those using Coomassie brilliant blue R-250 and silver nitrate. The sensitivity of Coomassie blue is known to detect about 50 ng protein/mm^2, while silver stain is reported to be 20–200 times more sensitive. We will now describe how the two procedures can be used successively on the same gel. The gels are first immersed in a staining solution composed of 0.25% (w/v) Coomassie brilliant blue R-250 in a mixture of methanol–acetic acid–distilled water (5/1/5, v/v). After 30 min of gentle shaking, the staining solution is removed. The gels are then rinsed with distilled water and immersed in a destaining solution composed of methanol–acetic acid–water (25/8/67, v/v/v). Methanol can be replaced by 2-propanol, which is less toxic. The destaining time can be considerably shortened by using small pieces of inert sponge foam rubber, which regenerate the destaining solution.

Gels thought to be slightly underloaded can be subsequently silver-stained. They are immersed in a mixture of 50% methanol, 10% acetic acid at 50° until full destained. They are then washed in 5% methanol and 7% acetic acid before the 10% glutaraldehyde fixation and all subsequent silver staining operations described by Oakley *et al.*[13] are performed. Note that after silver staining, however, further electrophoretical protein elution or transfer to nitrocellulose sheets is impossible.

Examples of Typical Separations

Chloroplast Proteins

Using the procedure described here, Fig. 2 shows a typical protein map obtained with pea thylakoids. The pH gradient ran from approximately 5 to 7, a range which was found to reveal most polypeptides from the sample. More than 70 spots are resolved by Coomassie blue staining and if silver staining had been performed about 150 spots could have been

[13] B. L. Oakley, D. R. Kirsch, and N. R. Morris, *Anal. Biochem.* **105,** 361 (1980).

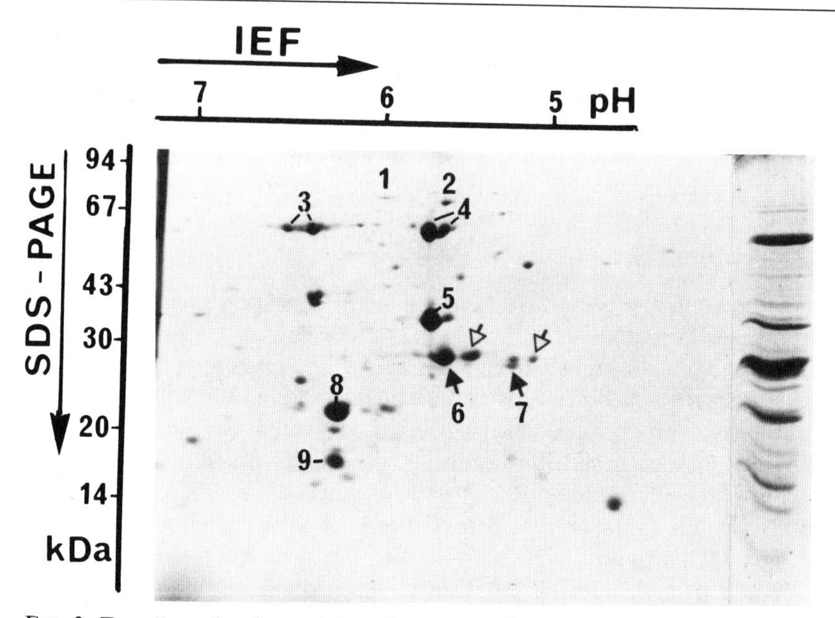

FIG. 2. Two-dimensional separation of thylakoid proteins from pea (var. "Kelvedon Wonder"). On the right is a conventional one-dimensional SDS separation of the same sample. Proteins were stained with Coomassie blue. Numbers represent an attempt to identify some polypeptides: 1 and 2, apoproteins of *P*-700-CP_1 complex; 3 and 4, respectively, α and β subunits of CF_1 ATPase; 5, a group of 33-kDa polypeptides, including cytochrome *f* and a polypeptide of the O_2-evolving complex; 6 and 7, apoproteins of LHCP (filled arrows indicate nonphosphorylated forms; open arrows, phosphorylated ones); 8 and 9, respectively, two polypeptides of 24 and 18 kDa of the oxygen-evolving complex.

counted.[6] Polypeptides of LHCP are clearly resolved for the first time, allowing distinction between their phosphorylated and nonphosphorylated forms. Further details on this point have been reported earlier.[7]

Mitochondrial Proteins

Figure 3 shows a two-dimensional protein map of mitochondria isolated from the inflorescence of cauliflower and purified on a discontinuous sucrose gradient as described by Douce *et al.*[14] The Coomassie blue-stained gel reveals more than 200 spots. As for thylakoid membranes, most polypeptides appear between pH 5 and 7. Up to now, unfortunately, none of these spots has been identified with certainty. As shown at the

[14] R. Douce, E. L. Christensen, and W. D. Bonner, Jr., *Biochim. Biophys. Acta* **275**, 148 (1972).

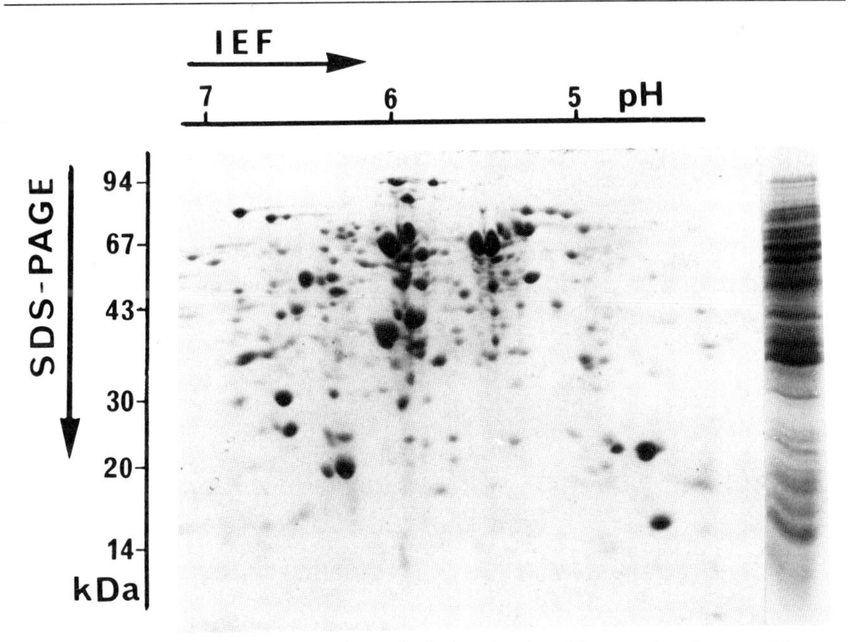

FIG. 3. Two-dimensional separation of whole mitochondria proteins from cauliflower. Coomassie blue staining.

right-hand side of Fig. 3, a large number of proteins cannot be properly resolved by conventional one-dimensional electrophoresis.

Concluding Remarks

The method presented here should be a valuable instrument for genetic and physiological analysis of organelle polypeptides. However, it must be borne in mind that some proteins may escape detection for the following reasons: (1) loss during the acetone precipitation step through failure to precipitate or irreversible precipitation; (2) loss during isoelectric focusing, particularly for basic proteins which in the IEF/SDS system are not well separated. This problem may be overcome either by extending the pH gradient with more basic ampholytes or by using a nonequilibrium pH gradient electrophoresis in the first dimension as described by O'Farrell *et al.*[15]; (3) insufficient quantity, which is the main problem with

[15] P. Z. O'Farrell, H. M. Goodman, and P. H. O'Farrell, *Cell* (*Cambridge, Mass.*) **12,** 1133 (1977).

plant organelle polypeptides where a few proteins make up a large percentage of the total. This problem is encountered in chloroplasts with LHCP for thylakoid proteins and with ribulose-bisphosphate carboxylase for stromal proteins. If minor proteins must be analyzed, it is necessary to overload the gels, thus causing a loss of the technique's resolving power. In spite of these difficulties, the two-dimensional technique should provide substantial improvements in the knowledge of organelles' molecular biology. Progress has already been made and obviously its wide use in many laboratories should allow the fulfillment of all its potential in this field of investigation.

[58] Plant Membrane Sterols: Isolation, Identification, and Biosynthesis

By MARIE-ANDRÉE HARTMANN and PIERRE BENVENISTE

Whereas only a few sterols are found in animals, an impressive array of sterols is encountered in the plant kingdom.[1,2] Cholesterol (**1**),[3] well known as the major sterol of animal cells, has a C_8 side chain and a Δ^5-

[1] L. J. Goad and T. W. Goodwin, *in* "Progress in Phytochemistry" (L. Reinhold and Y. Liwschitz, eds), p. 113. Wiley, New York, 1972.

[2] W. R. Nes, *Adv. Lipid Res.* **15**, 233 (1977).

[3] Nomenclature: (**1**) cholesterol: cholest-5α-en-3β-ol; (**2**) sitosterol: (24*R*)-24-ethylcholest-5α-en-3β-ol; (**3**) stigmasterol: (24*S*)-24-ethylcholesta-5α,*E*-22-dien-3β-ol; (**4**) campesterol: (24*R*)-24-methylcholest-5α-en-3β-ol; (**5**) 24-methylene cholesterol: ergosta-5,24(28)-dien-3β-ol; (**6**) isofucosterol: stigmasta-5,*Z*-24(28)-dien-3β-ol; (**7**) 24-methylene lophenol: 4α-methyl-5α-ergosta-7,24(28)-dien-3β-ol; (**8**) cycloeucalenol: 4α,14α-dimethyl-9β,19-cyclo-5α-ergost-24(28)-en-3β-ol; (**9**) obtusifoliol: 4α,14α-dimethyl-5α-ergosta-8,24(28)-dien-3β-ol; (**10**) 24-ethylidene lophenol: 4α-methyl-5α-stigmasta-7,*Z*-24(28)-dien-3β-ol; (**11**) 24-methylene cycloartanol: 4,4,14α-trimethyl-9β,19-cyclo-5α-ergost-24(28)-en-3β-ol; (**12**) cycloartenol: 4,4,14α-trimethyl-9β,19-cyclo-5α-cholest-24-en-3β-ol; (**13**) α-amyrin: urs-12-en-3β-ol; (**14**) β-amyrin: olean-12-en-3β-ol; (**15**) α-spinasterol: (24*S*)-24-ethyl-5α-cholesta-7,22-dien-3β-ol; (**16**) stigmast-7-enol: (24*R*)-24-ethyl-5α-cholest-7-en-3β-ol; (**17**) cholest-7-en-3β-ol; (**18**) (24*R*)-24-methyl-5α-cholesta-7,22-dien-3β-ol; (**19**) (24*R*)-24-methyl-5α-cholest-7-en-3β-ol; (**20**) 24-methylpollinastanol: 14α-methyl-9β,19-cyclo-5α-ergostan-3β-ol; (**21**) 24-methylene pollinastanol: 14α-methyl-9β,19-cyclo-5α-ergost-24(28)-en-3β-ol; (**22**) 24-dihydrocycloeucalenol: 4α,14α-dimethyl-9β,19-cyclo-5α-ergostan-3β-ol; (**23**) 31-norcyclobranol: 4α,14α-dimethyl-9β,19-cyclo-5α-ergost-24(25)-en-3β-ol; (**24**) cyclofontumienol: 4α,14α-dimethyl-9β,19-cyclo-5α-stigmast-*Z*-24(28)-en-3β-ol; fenpropemorph: 4-[3-(4-*tert*-butylphenyl)-2-methylpropyl]-2,6-dimethylmorpholine.

double bond. The sterols produced by most plant taxa also have a Δ^5-bond, but there is usually and additional C_1 or C_2 unit at C-24 of the side chain that is regarded as the distinguishing feature of phytosterols. Sitosterol (**2**),[3] stigmasterol (**3**),[3] and campesterol (**4**),[3] are generally cited as the typical plant sterols, and they are certainly the most frequently reported sterols in tracheophytes.[1,2] It is now recognized that cholesterol is widespread in higher plants, but only as a minor component.[1,2] The introduction of an alkyl group at C-24 of the side chain renders this carbon chiral and thus two epimers are possible. It has been shown by proton magnetic resonance (^{1}H NMR) spectroscopy that the 24-ethyl sterols which represent about 80% of total sterols in vascular plants are the 24α-epimers, sitosterol and stigmasterol, whereas the 24-methyl sterols occur as mixtures of both 24α (campesterol)- and 24β (22,23-dihydrobrassicasterol)-epimers.[2] In relation to the sterol ring system, most higher plants contain Δ^5-sterols as major compounds and have small amounts of Δ^7-sterols which are precursors.[2,4] The predominance of Δ^7-sterols as terminal products appears to be restricted to a few plant families (Chenopodiaceae, Cucurbitaceae, Theaceae).[2] The various 4,4-dimethyl and 4α-methyl sterols are usually minor compounds in plant sterol mixtures, as expected of possible precursors.

In addition to free sterols (FS),[5] higher plants contain sterol derivatives in which the 3-hydroxyl group of the sterol is either esterified by a long chain fatty acid to give steryl esters (SE) or β-linked to the 1-position of a monosaccharide, usually glucose, to form either steryl glycosides (SG) or acylated steryl glycosides (ASG) in which the 6-position of the hexose is esterified by an acyl group.[6] The existence of water-soluble complexes of sterols, which cannot be extracted by organic solvents, has been also reported.[7,8] The nature of these complexes, which represent 0.1 to 1% of total sterols,[8] is still hypothetical.

Free sterols are usually the major sterolic compounds, but the predominance of steryl glycosides in some members of Solanaceae[9] and that

[4] L. J. Goad, *in* "Lipids and Lipid Polymers in Higher Plants" (M. Tevini and H. K. Lichtenthaler, eds.), p. 146. Springer-Verlag, Berlin and New York, 1977.

[5] Abbreviations: FS, free sterols; SE, steryl esters; SG, steryl glycosides; ASG, acylated steryl glycosides; ER, endoplasmic reticulum; PM, plasma membrane; GA, Golgi apparatus; Mit, mitochondria; COI, cycloeucalenol-obtusifoliol isomerase; AdoMet-CMT, *S*-adenosyl-L-methionine cycloartenol-C_{24}-methyltransferase; UDPGlc-SGT, UDP-glucose-sterol-β-D-glucosyltransferase; SG-AT, SG-6′-*O*-acyltransferase.

[6] C. Grunwald, *Annu. Rev. Plant Physiol.* **26,** 209 (1975).

[7] R. J. Pryce, *Phytochemistry* **10,** 1303 (1971).

[8] M. Rohmer, G. Ourisson, and R. Brandt, *Eur. J. Biochem.* **31,** 172 (1972).

[9] R. Dupéron, M. Thiersault, and P. Dupéron, *C.R. Séances Acad. Sci. Ser. III* **296,** 239 (1983).

of SE in tobacco leaves[10] and seedlings of *Pinus palustris*[11] has been observed.

The broad outline of the sterol biosynthetic pathway in plants is now well understood although some points remain to be elucidated. For valuable background information about free sterol biosynthesis, the reader is referred to the excellent reviews available.[2,4,12,13] The distribution and biosynthesis of sterol derivatives (SE, SG, and ASG) have also been reviewed several times.[6,14,15]

Our concern in this chapter is to examine some less known cellular aspects of sterol biochemistry. First, an attempt will be made to summarize the present state of knowledge about the intracellular distribution of free sterols and their derivatives in higher plants. Whereas the presence of these compounds in crude membrane fractions, especially in microsomes, has been known for a long time, little information about sterol composition of well-characterized membrane fractions from higher plant tissues is available. Very few comparative studies have been done. Moreover it should be pointed out that the word "microsomes" has often been misinterpreted as meaning the endoplasmic reticulum (ER). Microsomes sedimenting at 100,000 *g* are composed of a mixture of vesicles originating from various membrane types, among them ER, plastid envelopes, membranes derived from Golgi apparatus (GA), tonoplast, and plasma membrane (PM). Here we present a detailed analysis of sterols and sterol derivatives of well-characterized (by a rich array of both morphological and biochemical markers) membrane fractions from etiolated maize coleoptiles, with a special focus on a PM-rich fraction. Then data about the different cellular sites of plant sterol biosynthesis in the same material will be given. Finally, we will show that it is possible to modify the sterol profile of higher plant membranes using sterol biosynthesis inhibitors. Results that we have obtained with maize roots treated by fenpropimorph will be given. Such a technique should provide an interesting and challenging way for a comprehensive understanding of the role of sterols and their derivatives in higher plant cells.

[10] C. Grunwald, *Plant Physiol.* **45,** 663 (1970).

[11] C. Van Vu and R. H. Biggs, *Physiol. Plant.* **42,** 344 (1978).

[12] T. W. Goodwin, *Annu. Rev. Plant Physiol.* **30,** 369 (1979).

[13] P. Benveniste, *Annu. Rev. Plant Physiol.* **37,** 275 (1986).

[14] J. B. Mudd, *in* "Biochemistry of Plants" (P. K. Stumpf, ed.), Vol. 4, p. 509. Academic Press, New York, 1980.

[15] M. Axelos and C. Péaud-Lenoël, *in* "Plant Carbohydrates" (F. A. Loewus and W. Tanner, eds.), p. 613. Springer-Verlag, Berlin and New York, 1982.

Isolation and Characterization of Higher Plant Membranes

General Comments

A review of various methods available for isolating and identifying specific plant subcellular components is beyond the scope of this chapter. However, as the reliability of results is closely dependent on the degree of purity of isolated membranes, much attention has to be paid to the accurate characterization of the different membrane types and to quantitation of cross-contamination. Let us highlight briefly some problems encountered in plant fractionation studies. First, the choice of plant material is of importance. Thus, the difficulty of preparing membrane fractions free of contamination from thylakoids of broken chloroplasts has resulted in the use of etiolated or storage tissues for most studies. The second criterion of importance is the reliability of the chosen markers, either morphological or biochemical, for identifying and quantitating the various subcellular components in the isolated fractions. It should be kept in mind that absolute markers (i.e., completely restricted to and uniformly distributed throughout the whole population of a given component) may be the exception rather the rule. Whereas chloroplast and mitochondrial membranes have been studied extensively, much less work has been done to characterize other plant membranes such as ER, GA, tonoplast, and PM (the so-called endomembrane system). During mechanical disruption of plant tissues, all these membranes are fragmented, giving rise to a heterogeneous population of morphologically indistinguishable smooth-surfaced vesicles that are difficult to separate accurately. An excellent review of the various markers currently in use and their reliability has been published by Quail[16] and should be profitably consulted. In general, the use of several markers for identifying a specific subcellular component is strongly recommended. A last point to be taken into consideration is the too-often neglected problem of the sidedness and integrity of membrane vesicles in isolated fractions. Membranes are organized asymmetrically and thus vesiculation of membranes can proceed in two ways, resulting in closed vesicles that expose either the original external side (right-side-out vesicles) or the internal one (inside-out vesicles). Most of the substrates for membrane-bound enzymes do not permeate membranes and thus could not reach the active site of enzymes located on the inner face of vesicles. These cautions should be taken into account to avoid a misinterpretation of negative results.

[16] P. H. Quail, *Annu. Rev. Plant Physiol.* **30,** 425 (1979).

Isolation and Identification of Membrane Fractions from Zea mays

The procedure that we have developed for the separation and identification of the different membrane fractions from maize coleoptiles has been described in detail.[17–19] In short, maize seeds (*Zea mays*, cv. LG 11) were grown in the dark at 25°. Coleoptiles were excised after 6 days. After mechanical disruption of tissues in an appropriate buffered medium [0.5 *M* mannitol, 2 m*M* EDTA, 10 m*M* 2-mercaptoethanol, 0.5% (w/v) bovine serum albumin, 0.1 *M* Tris–HCl, pH 8.0], membrane fractions were isolated by differential and isopycnic sucrose density gradient centrifugation. Both step[17] and continuous[18,19] sucrose gradients were used. Specific subcellular components were identified by assays for the following markers: ER, NADH- and NADPH–cytochrome-*c* reductases, insensitive to antimycin A, and cinnamic-4-hydroxylase; GA, β-glucan synthase of high affinity for UDPglucose (GS I); PM, β-glucan synthase of low affinity for UDPglucose (GS II), K^+-stimulated ATPase (pH 6.5), *N*-naphthylphthalamic acid binding, and phosphotungstic acid and periodic acid–thiocarbohydrazide–silver proteinate staining in electron microscopy. Succinate cytochrome-*c* reductase was used as a marker for inner mitochondrial membranes, and UDPgalactose diacylglycerol galactosyltransferase and carotenoids as markers for plastid envelopes.

Three main membrane fractions were used in most of our studies: a light fraction (*d*: 1.10 g/ml) rich in ER, but also containing plastid envelopes, a heavy one (*d*: 1.17 g/ml) highly enriched in PM and a fraction of purified mitochondria (*d*: 1.18 g/ml). In a few cases, a fraction of intermediate density (*d*: 1.15 g/ml) enriched in Golgi membranes was employed.

Sterol Analysis

Extraction of Free Sterols and Sterol Derivatives

If both free sterols and sterol conjugates of membrane fractions are to be analyzed, then the total lipid must be extracted three times with 6 vol of dichloromethane–methanol (2 : 1, v/v). The combined solvent extracts are dried over anhydrous sodium sulfate and evaporated to dryness. The addition of dichloromethane to the organic phase just before filtering through sodium sulfate is recommended to better eliminate water traces. The total lipid extract is then applied to Merck (HF 254) silica gel thin-layer plates. TLC is performed using dichloromethane–methanol–water

[17] M. A. Hartmann, G. Normand, and P. Benveniste, *Plant Sci. Lett.* **5,** 287 (1975).
[18] M. A. Hartmann and P. Benveniste, *Phytochemistry* **17,** 1037 (1978).
[19] M. A. Hartmann, P. Benveniste, and J. C. Roland, *Biol. Cell.* **35,** 183 (1979).

(85 : 15 : 0.5, v/v) as developing solvent. The four groups of sterol compounds are separated into SG (R_f 0.35), ASG (R_f 0.55), FS as the mixture of 4,4-dimethyl, 4α-methyl, and 4-demethyl sterols (R_f 0.7), and SE together with hydrocarbons and squalene (R_f 0.90). Bands are located by light spraying with a 0.1% solution of berberin hydrochloride in ethanol and observed under a UV lamp (340 nm). They are then scraped off and compounds eluted from silica with dichloromethane–methanol (4 : 1, v/v) for SG and ASG and with dichloromethane only for FS and SE.

If only FS are to be analyzed, membrane fractions can be extracted three times with 3 vol of *n*-hexane or petroleum ether. The organic phases are dried over sodium sulfate and evaporated to dryness. In this case, lipid extracts contain FS and SE.

Hydrolysis of Steryl Glycosides and Steryl Esters

To obtain sterols from SG and ASG, the corresponding fractions can be treated by heating for 4 hr with 5 ml of 1% H_2SO_4 in ethanol under nitrogen atmosphere. After addition of 1 vol water, the reaction medium is neutralized using 1 *N* NaOH. Sterols are then extracted three times with 3 vol of hexane.

To obtain sterols from SE, the corresponding fraction can be hydrolyzed by heating for 1 hr with 5 ml of 6% (w/v) KOH in methanol and pyrogallol. After addition of 1 vol water, the nonsaponifiable matter is extracted three times with 3 vol of hexane. Sterols (4,4-dimethyl, 4α-methyl, and 4-demethyl sterols) are then separated by TLC as described below.

Sterol Analytical Procedure

A wide range of excellent analytical techniques have been developed in recent years, but this section will describe only methods that we have employed. References to literature will be given when needed.

Free sterols and sterols liberated from sterol conjugates are subjected to TLC with dichloromethane as the developing solvent (two runs). Such a chromatography separates sterols according to the degree of methyl group substitution at C-4. In order of increasing polarity, they are separated into SE together with squalene (R_f 0.95), 4,4-dimethyl sterols (R_f 0.4), 4α-methyl sterols (R_f 0.35), and 4-demethyl sterols (R_f 0.25). Bands are scraped off and compounds eluted with dichloromethane. The three classes of compounds are usually acetylated at room temperature for 16 hr with acetic anhydride-pyridine (2 : 1, v/v). Reagents in excess are evaporated to dryness and the crude acetates purified by TLC using dichloromethane (one run). Each of the three classes of acetates can be analyzed by gas liquid chromatography (GC) and the total amount of sterols in each class quantified (see below). Further separation can be achieved by using analytical TLC on 10% $AgNO_3$-silica gel plates. By use of such a procedure, sterols differing by the degree of unsaturation and position of double bonds in either the ring system or the side chain can be separated. A migration is performed for 15 hr with cyclohexane–toluene as the developing solvent in the proportions 7 : 3 (v/v) to separate the 4,4-dimethyl and the 4-demethyl steryl acetates and 3 : 2 (v/v) to separate the 4α-methyl steryl acetates. In order of decreasing polarity, the 4-demethyl steryl acetates are separated into 24-methylene cholesteryl (**5**)[3] acetate, isofucosteryl (**6**) acetate, and the mixture of sitosteryl (**2**), stigmasteryl (**3**), and campesteryl (**4**) acetates; the 4α-methyl steryl acetates are separated into 24-methylene lophenyl (**7**) acetate, the mixture of cycloeucalenyl (**8**) and obtusifolyl (**9**) acetates, and 24-ethylidene lophenyl (**10**) acetate; the 4,4-dimethyl steryl acetates are separated into 24-methylene cycloartanyl (**11**) acetate, cycloartenyl (**12**) acetate and the mixture of α(**13**)- and β(**14**)-amyrin acetates. A list of typical R_f values for some 4,4-dimethyl and 4-demethyl steryl acetates is given in Goad.[20]

The methods that we currently use for identification of sterols are mainly GC, GC–mass spectrometry (MS), and ^{1}H NMR spectroscopy. In former experiments, GC was carried out at 270° on glass columns (1.5 m × 3 mm) packed with either 1% OV-17 or 1% SE-30; the carrier gas was N_2 (30 ml/min). All sterol analyses are now performed using a Carlo Erba chromatograph (model 4160) equipped with a flame ionization detector and a glass capillary column (WCOT, 25 m × 0.25 mm) coated with OV-1. The temperature program used includes a fast rise from 60 to 230° (30°/min), then a slow rise from 230 to 280° (2°/min). An internal standard of cholesterol is usually used. The total amount of sterols in each class is quantified using a computing integrator. The relative retention times (RRT) of acetates of the various sterols are given in Schmitt *et al.*[21,22] For additional data on contribution of some molecular features to GC retention times, the reader is referred to Nes.[23] GC–MS is carried out at 70 eV and separation of sterols is performed on a glass capillary column (OV-1). Some of the mass spectra are detailed in Schmitt and Benveniste[21] and Nes.[23]

Other techniques are becoming increasingly important as means for isolating and purifying sterols from plants such as high-pressure liquid

[20] L. J. Goad, this series, Vol. 111, p. 311.

[21] P. Schmitt and P. Benveniste, *Phytochemistry* **18,** 445 (1979).

[22] P. Schmitt, P. Benveniste, and P. Leroux, *Phytochemistry* **20,** 2153 (1981).

[23] W. R. Nes, this series, Vol. 111, p. 3.

TABLE I
STEROL COMPOSITION OF ETIOLATED MAIZE COLEOPTILES

Sterol	(μg/g)[a]	(%)	Relative percentage of sterols					
			2	**3**	**4**[b]	**1**	**6**	Other sterols
FS[c]	1300	(79)	24	52	21	1	1.5	0.5
SE	70	(4)	41	28	20	4	2	5
SG	80	(5)	36	40	17	5	1	1
ASG	200	(12)	38	41	16	3	1	1

[a] Micrograms of sterols per gram of dry weight.

[b] This fraction contains four different sterols with similar relative retention times: (1) a mixture of campesterol (24α-epimer) and 22,23-dihydrobrassicasterol (24β-epimer) in equal amounts as shown by F. Sheid, M. Rohmer, and P. Benveniste [*Phytochemistry* **21,** 1959 (1982)] and (2) a mixture of 5α-ergosta-5,23-dien-3β-ol and 5α-ergosta-7,23-dien-3β-ol, which represent about 11% of the total 4-demethyl fraction according to F. Scheid and P. Benveniste [*Phytochemistry* **18,** 1207 (1979)].

[c] FS and sterols from SE were analyzed as acetates, and sterols from SG and ASG as alcohols by GC (1% OV-17).

chromatography (HPLC) and ^{1}H NMR spectroscopy. For the details of the use of silicic and reversed-phase HPLC columns, two recent reviews can be profitably consulted.[23,24] For a list of ^{1}H NMR spectra of various sterols, see Nes.[23]

Intracellular Distribution of Free Sterols and Sterol Derivatives

Before starting a sterol analysis of membrane fractions, the complete sterol composition of whole tissue under study needs to be determined. Such an analysis for etiolated maize coleoptiles is given in Table I.

Distribution of Free Sterols

All the membrane fractions from maize coleoptiles contain free sterols (Table II). Sitosterol (**2**), stigmasterol (**3**), and campesterol (**4**) are the major compounds as in whole tissues (Table I). The same sterols are present in all the fractions. No sterol is specifically associated with a given type of membrane. Thus a great homogeneity in relative sterol composition of all the membrane fractions is exhibited. A similar behavior

[24] R. J. Rodriguez and L. W. Parks, this series, Vol. 111, p. 37.

TABLE II
RELATIVE STEROL COMPOSITION OF MEMBRANE FRACTIONS FROM MAIZE COLEOPTILES

	ER[a]			PM			Mit		
Sterol	FS	SG	ASG	FS	SG	ASG	FS	SG	ASG
2[b]	25[c]	38	38	24	37	37	27	61	40
3	46	43	44	48	41	42	53	24	40
4	25	16	17	25	20	19	18	11	15
Other sterols	3	3	1	3	2	2	2	4	3

[a] Fractions are designated according to their major subcellular component.
[b] **2**, Sitosterol; **3**, stigmasterol; **4**, campesterol.
[c] Relative percentages.

related to phospholipid composition of plant membranes has been reported.[25]

By contrast, when amounts of FS are expressed as micrograms per milligram of protein, profound differences between fractions are observed (Table III). The PM-rich fraction appears to contain the highest concentration of FS and mitochondria the lowest. Differences are even greater when sterol to phospholipid molar ratios are considered. This ratio is five times as high in PM as in ER or mitochondria (Table III). These results are in full agreement with previous results[17] and those of other authors.[26,27] They clearly demonstrate that like PM of animal cells[28] and fungi,[29] the plant PM is also characterized by a high content of free sterols and a high sterol to phospholipid molar ratio.

Thus all the membrane fractions from maize coleoptiles (ER, PM, and Mit) contain FS. In the case of mitochondria, these compounds might be associated mainly with the external membrane.[30,31] The presence of FS in the nuclear envelope[32] and in glyoxysomes[33] has also been reported. The

[25] P. Mazliak, *in* "Lipids and Lipid Polymers in Higher Plants" (M. Tevini and H. K. Lichtenthaler, eds.), p. 48. Springer-Verlag, Berlin and New York, 1977.
[26] T. K. Hodges, R. T. Leonard, C. E. Bracker, and T. W. Keenan, *Proc. Natl. Acad. Sci. U.S.A.* **69,** 3307 (1972).
[27] J. W. Hardin, J. H. Cherry, D. J. Morré, and C. A. Lembi, *Proc. Natl. Acad. Sci. U.S.A.* **69,** 3146 (1972).
[28] A. Colbeau, J. Nachbaur, and P. M. Vignais, *Biochim. Biophys. Acta* **249,** 462 (1971).
[29] G. A. Scarborough, *Methods Cell Biol.* **20,** 117 (1978).
[30] C. A. Manella and W. D. Bonner, *Biochim. Biophys. Acta* **413,** 213 (1975).
[31] J. Méance, P. Dupéron, and R. Dupéron, *Physiol. Veg.* **14,** 745 (1976).
[32] E. I. Philipp, W. W. Franke, T. W. Keenan, J. Stadler, and E. D. Jarasch, *J. Cell Biol.* **68,** 11 (1976).
[33] R. P. Donaldson and H. Beevers, *Plant Physiol.* **59,** 259 (1977).

TABLE III
LIPID COMPOSITION OF MEMBRANE FRACTIONS FROM MAIZE COLEOPTILES

Lipid	ER	PM	Mit
FS[a]	80 ± 10	200 ± 20	21 ± 8
SG[a]	1.0	7.0	2.0
ASG[a]	11.0	18.0	5.0
Phospholipids[b]	820 ± 10	315 ± 30	260 ± 35
Sterols/PL, molar ratio	0.18	1.15	0.15

[a] Micrograms sterols per milligram of protein.
[b] Micrograms per milligram of protein.

problem of the presence of FS in chloroplasts remained to be investigated.

In several articles, the occurrence of sterols in more or less purified chloroplast preparations had been mentioned.[14,34] In a general way, chloroplasts appeared to be poor in sterols compared to microsomal and even mitochondrial fractions. One could wonder whether the present sterols were normal components of chloroplasts or resulted from contaminations of chloroplasts by other subcellular membranes richer in sterols.

Several chloroplast subfractions (envelopes, thylakoids, and stroma) were isolated from highly purified chloroplasts of spinach leaves by Joyard according to Joyard and Douce.[35] Microsomes were also prepared from the same material. All these fractions were assayed for their FS content. Sterols were identified by GC and GC–MS. Given the very low amounts of both purified envelopes and sterols to be analyzed, blank controls were carried out in order to appreciate the reliability of results.

Table IV shows that chloroplast subfractions contain very low amounts of FS compared to microsomes. Because of the high quantity of plastid membranes, especially thylakoids, compared to other subcellular membranes in photosynthetic cells, microsomes were probably strongly contaminated by thylakoid fragments. Thus, the sterol concentration of microsomes, expressed in micrograms per milligram of protein, was certainly very underestimated. The envelopes contain low but significant amounts of FS (7 μg/mg protein). In thylakoids, these compounds are barely detectable (0.5 μg/mg protein) and in stroma they are totally absent. If now the relative sterol composition is considered, the presence of the same sterols in all the fractions is observed. These sterols are similar

[34] R. P. Poincelot, *Arch. Biochem. Biophys.* **159,** 134 (1973).
[35] J. Joyard and R. Douce, *Physiol. Veg.* **14,** 31 (1976).

TABLE IV[a]
STEROL COMPOSITION[b] OF CHLOROPLAST SUBFRACTIONS AND MICROSOMES FROM SPINACH LEAVES

Fractions	Sterols (μg/mg protein)	Relative percentage of sterols[c]							Sterols : galactolipids, molar ratio
		1	**15**	**16**	**17**	**18**	**19**	**X**	
Envelope	7 ± 2	3	31	43	12	2	6	3	0.021
Thylakoids	0.5 ± 0.2	10	60	18	4	1	5	2	0.0032
Stroma	1	nd	nd	nd	nd	nd	nd	nd	—
Microsomes	14 ± 3	1	70	22	3	1	2	1	—

[a] These data have been previously reported by R. Douce, M. A. Block, A. J. Dorne, and J. Joyard [*in* "Subcellular Biochemistry" (D. B. Roodyn, ed.), Vol. 10, p. 33. Plenum, New York, 1985].
[b] Sterols were analyzed with the collaboration of Bouvier-Navé.
[c] **1**, Cholesterol; **15**, α-spinasterol; **16**, stigmast-7-enol; **17**, Δ^7-cholesterol; **18**, (24*R*)-24-methyl-5α-cholesta-7,22-dien-3β-ol; **19**, (24*R*)-24-methyl-5α-cholest-7-en-3β-ol; **X**, unidentified sterols; nd, not determined.

to those of whole leaves. Major sterols are α-spinasterol (**15**)[3] and stigmast-7-enol (**16**). An interesting observation is the different relative sterol composition of the envelope compared to microsomes and thylakoids. The major sterol in envelopes is stigmast-7-enol whereas it is α-spinasterol in microsomes and thylakoids. These results clearly indicate that FS are normal components of the chloroplast envelope. A similar conclusion has been given elsewhere,[34,36] but without sterol analytical determination. Sterols have also been found in chromoplasts.[37]

Distribution of Sterol Derivatives

Because sterol conjugates are usually present in much lower amounts than FS, it is relatively difficult to quantify them accurately. Moreover, sterol moieties have first to be released from sterol conjugates by acid or basic hydrolysis. For steryl glycoside analysis, problems related to the yield and the reproducibility of acid hydrolysis are encountered and a decomposition of some labile SG and ASG during this process can be observed.[38] Such problems could be overcome by using HPLC, which would be applicable to analysis of not only FS but also to SG and ASG.[38]

As shown in Table III, SG and ASG were found in all the membrane fractions from maize coleoptiles, but in much lower amounts than FS. ASG are the major form of steryl glycosides. For both SG and ASG, the

[36] C. H. Moeller and J. B. Mudd, *Plant Physiol.* **70,** 1554 (1982).
[37] B. Liedvogel and H. Kleinig, *Planta* **133,** 249 (1977).
[38] J. Kesselmeier, W. Eichenberger, and B. Urban, *Dev. Plant Biol.* **9,** 233 (1984).

highest concentrations, expressed as micrograms sterols per milligram of protein, are associated with the PM-rich fraction and the lowest concentrations with mitochondria. For these organelles, steryl glycosides would be mainly located in the internal membrane.[31] The presence of SG and ASG in the chloroplast envelope has been suggested,[35] but further work needs to be done to quantify them accurately.

GC analysis of sterols released from SG and ASG indicates that these compounds contain the same 4-demethyl sterols (i.e., sitosterol, stigmasterol, and campesterol) than those of free forms, but in different proportions. Whereas stigmasterol is the major sterol of FS, a significant enrichment of both SG and ASG in sitosterol is observed in all the fractions (Table II). The relative sterol composition of SG and ASG of mitochondria is slightly different from that of other fractions.

The presence of steryl glycosides in various fractions from other plants has also been reported.[15,39] Thus it can be concluded that SG and ASG, as well as FS, are really specific membrane components in higher plant cells.

Very little work has been done with SE. Their distribution in membrane fractions from maize coleoptiles has not been determined. They represent about 4% of total sterols. SE have been found in different membrane fractions from various plants,[39–41] but the problem of their occurrence as genuine membrane components of higher plant cells needs further consideration.

Cellular Sites of Sterol Biosynthesis

Free Sterol Biosynthesis

As already outlined (see the introduction), the free sterol biosynthetic pathway in higher plants is now well known. Let us recall that downstream to farnesyl pyrophosphate all the steps leading to final 4-demethyl sterols are catalyzed by membrane-bound enzymes. This pathway is characterized by specific features such as cyclization of 2,3-oxidosqualene to give cycloartenol (**12**) or β-amyrin (**14**), opening of the 9β,19-cyclopropane ring of cycloeucalenol (**8**), and double alkylation at C-24 of the side chain, which could be considered as phylogenetic markers. The enzymatic reactions involved in these steps have been studied extensively in our laboratory.[13]

[39] P. Dupéron, M. Charifi-Khoub, and R. Dupéron, *C.R. Seances Acad. Sci. Ser. III* **300,** 367 (1985).

[40] W. Janiszowska and Z. Kasprzyk, *Phytochemistry* **16,** 473 (1977).

[41] M. A. Hartmann, M. Ferne, C. Gigot, R. Brandt, and P. Benveniste, *Physiol. Veg.* **11,** 209 (1973).

In this section, we report the intracellular distribution of some of these enzyme activities. The three membrane fractions from maize coleoptiles enriched, respectively, in ER, PM, and mitochondria, were tested for their ability to synthesize 24-methylene cycloartanol (**11**) from cycloartenol (**12**) and *S*-adenosyl-L-methionine (AdoMet), and obtusifoliol (**9**) from cycloeucalenol (**8**). These two reactions are catalyzed by the AdoMet-cycloartenol-C24-methyltransferase (AdoMet-CMT) and the cycloeucalenol-obtusifoliol isomerase (COI), respectively. They were measured according to Hartmann and Benveniste.[18]

Figure 1 shows that both AdoMet-CMT and COI are located exclusively in the light fraction (*d*: 1.10 g/ml) and correlate with NADH-cytochrome-*c* reductase and cinnamate hydroxylase activities (see Fig. 1 of Ref. 18), providing evidence that free sterol biosynthesis occurs in ER membranes. When membranes are separated using gradient media containing 4 m*M* $MgCl_2$ to prevent stripping of ribosomes from rough ER, the AdoMet-CMT activity is shifted to a higher density exactly as are the

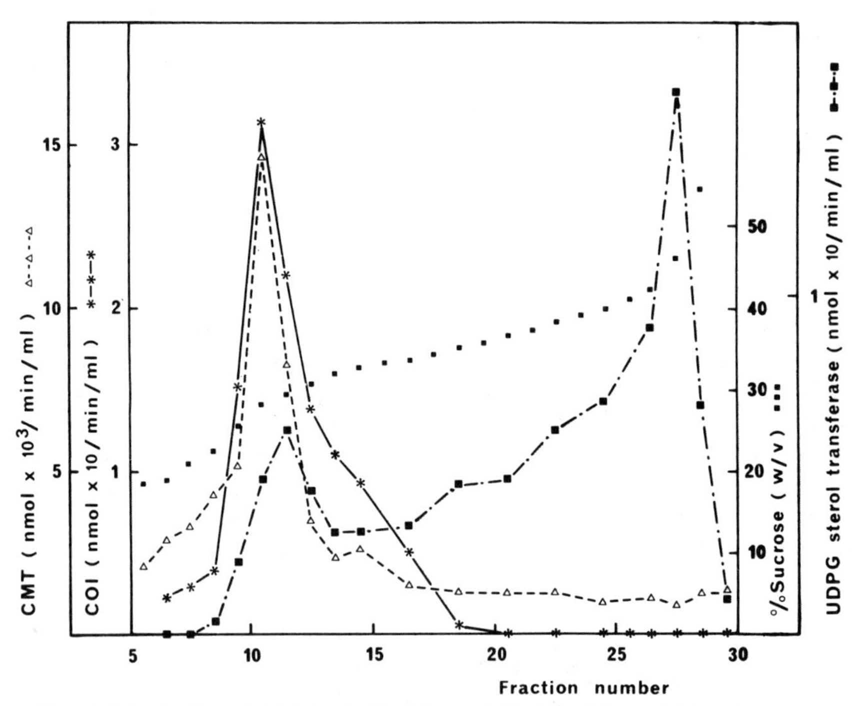

FIG. 1. Distribution of AdoMet-CMT, COI, and UDPGlc-SGT activities after sucrose density gradient centrifugation of microsomes from maize coleoptiles. Data are from Hartmann and Benveniste.[18]

activities cited above (see Table 3 of Ref. 19). By contrast, UDPgalactose diacylglycerol galactosyltransferase, an enzyme bound to the plastid envelope, is not shifted by Mg^{2+} ions, indicating that AdoMet-CMT is not associated with this type of membrane. No AdoMet-CMT or COI activity is detected in PM (Fig. 1) or in mitochondria.[42] Thus both enzyme activities are exclusively located in ER in agreement with previous experiments done with discontinuous sucrose gradients for membrane separation.[42] These latter had also shown that the distribution of the 2,3-oxidosqualene-cycloartenol cyclase correlated exactly with those of both AdoMet-CMT and COI (M. A. Hartmann and P. Fouteneau, unpublished results). The presence of the farnesyl pyrophosphate-squalene synthase in smooth and rough ER from bean leaves had also been previously reported.[41]

All these results clearly indicate that free sterol biosynthesis occurs in ER in higher plant cells like in animal cells.[43] It should be pointed out that the phospholipid biosynthesis takes place also in ER.[44] Therefore ER may play an important role in membrane biogenesis.

Biosynthesis of Steryl Glucosides

The UDPglucose-sterol-β-D-glucosyltransferase (UDPGlc-SGT) involved in the formation of SG has been characterized in several higher plant tissues and shown to be a membrane-bound enzyme.[14,15,45–49] It catalyzes the incorporation of glucose from UDPGlc into free sterols to give SG. These compounds can then serve as acceptors of endogenous acyl groups to form ASG, a reaction driven by the SG-6′-*O*-acyltransferase (SG-AT). Both reactions are tightly coupled, at least in cell-free systems.

Membrane fractions from maize coleoptiles were assayed for their ability to synthesize *in vitro* SG and ASG from UDP[U-^{14}C]glucose. The protocol used was that of Hartmann and Benveniste.[18] After extraction and TLC of organic extracts under conditions described above, radioactivity was shown to be associated only with SG and ASG. As ASG are formed by enzymatic acylation of SG, UDPGlc-SGT activity is given by the measure of the ^{14}C radioactivity incorporated into both SG and ASG.

[42] M. A. Hartmann, P. Fonteneau, and P. Benveniste, *Plant Sci. Lett.* **8,** 45 (1977).

[43] C. J. Chesterton, *J. Biol. Chem.* **243,** 1147 (1968).

[44] J. M. Lord, T. Kagawa, T. S. Moore, and H. Beevers, *J. Cell Biol.* **57,** 659 (1973).

[45] W. Eichenberger, *in* "Lipids and Lipid Polymers in Higher Plants" (M. Tevini and H. K. Lichtenthaler, eds.), p. 169. Springer-Verlag, Berlin and New York, 1977.

[46] C. Grunwald, *Lipids* **13,** 697 (1978).

[47] Z. A. Wojciechowski, *Biochem. Soc. Trans.* **11,** 565 (1983).

[48] E. Quantin, M. A. Hartmann, F. Schuber, and P. Benveniste, *Plant Sci. Lett.* **17,** 193 (1980).

[49] P. Bouvier-Navé, P. Ullmann, D. Rimmele, and P. Benveniste, *Plant Sci. Lett.* **36,** 19 (1984).

The most active synthesis of steryl glycosides was found to be associated with the membrane fraction of high density (*d*: 1.17 g/ml) (see Fig. 1), characterized by a specific positive reaction toward phosphotungstic acid and periodic acid–thiocarbohydrazide–silver proteinate in electron microscopy and a high content in free sterols (Table II and Ref. 18). Moreover, the distribution of UDPGlc-SGT activity was shown to correlate with that of *N*-naphthylphthalamic acid binding sites and that of GS II and K^+-stimulated ATPase (pH 6.5) activities but not with that of GS I, a presumed marker for GA.[18] Mitochondria were found not to be active.[19]

Thus these results, in agreement with previous data,[42] clearly indicate that glycosylation of sterols occurs mainly in PM, a subcellular site different from that (ER) where free sterol biosynthesis takes place (see above). Until now, a participation of GA to sterol glycosylation cannot be excluded.[50,51] A synthesis of SG and ASG in amyloplast membranes has also been reported.[52] Anyway, PM appears to be the major site of sterol glycosylation. Such a localization was later confirmed by ourselves in experiments with cotyledons of *Phaseolus aureus*[53] and maize roots,[54] and by others.[55–58] Thus the use of UDPGlc-SGT as a marker for PM should be recommended.[57,58]

The topology of the UDPGlc-SGT in the transverse plane of PM was investigated. When subjected to homogenization, PM is extensively disrupted and gives a mixture of fragments susceptible to have different orientations. In the case of maize coleoptiles, we have shown that all the PM vesicles have the same right-side-out orientation.[59,60] Studies using nonpermeant, group-specific reagents connected to various treatments altering membrane integrity clearly indicate that the enzymatic sites of UDPGlc-SGT for UDPGlc are exposed on the cytoplasmic surface of PM.[48,61,62]

[50] M. Lercher and Z. A. Wojciechowski, *Plant Sci. Lett.* **7,** 337 (1976).

[51] D. J. Bowles, L. Lehle, and H. Kauss, *Planta* **134,** 177 (1977).

[52] D. S. Catz, J. S. Tandecarz, and C. E. Cardini, *Plant Sci. Lett.* **38,** 179 (1985).

[53] M. A. Hartmann, B. Walter, and E. Lazar, *Plant Cell Rep.* **1,** 56 (1981).

[54] M. A. Hartmann, A. Grandmougin, and P. Benveniste, *C.R. Seances Acad. Sci. Ser. III* **301,** 601 (1985).

[55] E. A. H. Baydoun and D. H. Northcote, *J. Cell Sci.* **45,** 147 (1980).

[56] C. Chadwick and D. H. Northcote, *Biochem. J.* **186,** 411 (1980).

[57] T. J. Bruckhout, L. Heyder-Caspers, and A. Sievers, *Planta* **156,** 108 (1982).

[58] A. Chanson, E. McNaughton, and L. Taiz, *Plant Physiol.* **76,** 498 (1984).

[59] M. A. Hartmann, A. Ehrhardt, and P. Benveniste, *Plant Sci. Lett.* **30,** 227 (1983).

[60] M. A. Hartmann, P. Benveniste, and J. C. Roland, *Plant Sci. Lett.* **30,** 239 (1983).

[61] E. Quantin-Martenot, P. Benveniste, M. A. Hartmann, and P. Bouvier-Navé, *Plant Sci. Lett.* **29,** 305 (1983).

[62] M. M'Voula-Tsieri, P. Benveniste, E. Quantin-Martenot, and M. A. Hartmann, *in* "Biogenesis and Function of Plant Lipids" (P. Mazliak, P. Benveniste, C. Costes, and R. Douce, eds.), p. 431. Elsevier, Amsterdam, 1980.

The mechanism of the glucosylation reaction and the determination of its kinetic parameters is currently under study in our laboratory. First data suggest that a sequential mechanism applies to SG synthesis.[63]

In the presence of an endogenous acyl donor, SG can serve as a substrate for the SG-AT to give ASG. Membrane fractions from maize were tested for this enzyme activity. They were incubated in the presence of [^{14}C]SG without addition of acyl donors. Results showed that SG-AT was also associated mainly with PM but compared to UDPGlc-SGT, a more important part of the activity was recovered in the ER-rich fraction. Mitochondria were not very active. However, the origin of acyl groups transferred to SG was not known and experiments were done with fractions varying both in their acyl pattern and their efficiency as acyl donors. Further work with delipidated preparations is therefore needed to confirm such a distribution.

As shown above, SG and ASG are genuine components of almost all membranes from higher plant cells. As they are mainly synthesized in PM, they need to be translocated toward other membranes. The way this translocation occurs still remains unknown.

Biosynthesis of Steryl Esters

Very few data related to SE biosynthesis are available. The acyl donors are diacylglycerols.[14] Preliminary data from *in vivo* labeling studies suggest that SE are synthesized in ER membranes (M. A. Hartmann, unpublished results).

Modification of Sterol Composition of Higher Plant Membranes

Whereas much has been done on identification and biosynthesis of sterols, very little is still known about their functions and metabolic turnover. It is now well established that cholesterol plays an important structural role in the lipid core of biological membranes. The high concentrations of cholesterol in some membranes, especially PM, also have an influence on membrane functions such as passive transport, carrier-mediated transport, and activity of membrane-bound enzymes.[64] Thus cholesterol would regulate membrane fluidity and it is known that such a role is played by the bulk of sterols present in membranes.[65] In addition to this structural role, a very small fraction of cholesterol would be involved in a "metabolic" function[65] which is not yet precisely defined. The structural

[63] P. Ullmann, D. Rimmele, P. Benveniste, and P. Bouvier-Navé, *Plant Sci. Lett.* **36,** 29 (1984).

[64] R. A. Demel and B. De Kruyff, *Biochim. Biophys. Acta* **457,** 109 (1976).

[65] K. E. Bloch, *CRC Crit. Rev. Biochem.* **14,** 47 (1983).

and metabolic roles of 24-methyl and 24-ethyl sterols in higher plant membranes are still more poorly understood.

To investigate these matters, we have adopted a strategy consisting of qualitative and quantitative modification of the sterol composition of membranes from plant cells. For this purpose, we have treated plant cells with inhibitors of enzymes involved in plant sterol biosynthesis. Depending on the targets of these inhibitors in the biosynthetic pathway, two effects were expected: (1) a decrease in the sterol content of membranes, and (2) a change in their sterol profile by replacing the final sterols with biosynthetic intermediates. But that could be true only if the target enzyme is downstream to cycloartenol to allow an accumulation of intermediates structurally similar to the final Δ^5-sterols.[66] Results obtained by applying such a strategy to suspension cultures of plant cells and whole plants have been recently reported.[66–69] Thus when maize caryopses are grown with fenpropimorph, a drug known to interfere with ergosterol biosynthesis in fungi,[70,71] a dramatic increase of 9β,19-cyclopropyl sterols and an almost complete disappearance of the Δ^5-sterols present in the control plant are observed. The main target of this fungicide in the plant is the COI.[72] We report here the intracellular distribution of these 9β,19-cyclopropyl sterols in maize roots treated with fenpropimorph for 7 days.

Experimental. Maize (*Zea mays,* cv. LG 11) caryopses were grown in moist vermiculite in the dark at 25°. The vermiculite was daily soaked with 0.51 water with or without (R,S)-fenpropimorph (20 mg/liter). Roots were excised after 7 days and membrane fractions isolated as described above. Membranes were characterized by assays for the following markers: ER, NADH-cytochrome-*c* reductase, insensitive to antimycin A, AdoMet-CMT and COI; PM, K^+-stimulated ATPase (pH 6.5) and UDPGlc-SGT. Sterols were extracted and analyzed as already reported (see above).

[66] A. Rahier, P. Navé, P. Schmitt, and P. Benveniste, *in* "Physiologische Schlüsselprozesse in Pflanze und Insect" (P. Böger, ed.), p. 125. Universitätsverlag Konstanz GmbH, 1985.

[67] P. Schmitt, A. Rahier, and P. Benveniste, *Physiol. Veg.* **20,** 559 (1982).

[68] M. Bladocha and P. Benveniste, *Plant Physiol.* **71,** 756 (1983).

[69] P. Benveniste, M. Bladocha, M. F. Costet, and A. Ehrhardt, *in* "Journal of the Annual Proceedings of the Phytochemical Society of Europe" (A. M. Boudet, G. Allibert, G. Marigo, and P. J. Lea, eds.), Vol. 24, p. 283. Oxford Univ. Press, London and New York, 1984.

[70] T. Kato, *in* "Pesticide Biochemistry: Human Welfare and the Environment" (J. Miyamoto and P. C. Kearney, eds.), Vol. 3, p. 33. Pergamon, Oxford, 1983.

[71] A. Kerkenaar, *in* "Pesticide Biochemistry: Human Welfare and the Environment" (J. Miyamoto and P. C. Kearney, eds.), Vol. 3, p. 123. Pergamon, Oxford, 1983.

[72] A. Rahier, P. Schmitt, B. Huss, P. Benveniste, and E. H. Pommer, *Pestic. Biochem. Physiol.* **25,** 112 (1986).

Results. The qualitative and quantitative sterol analysis of membrane fractions from maize roots treated and not treated with fenpropimorph is shown in Table V. In membrane fractions from control plants, major sterols are the Δ^5-sterols present in most higher plants, i.e., sitosterol (**2**), stigmasterol (**3**), and campesterol (**4**), which represent 95 and 98% of total sterols in ER- and PM-rich fractions, respectively. The highest concentration of sterols is found in PM, in agreement with what we have reported above. In both membrane fractions from treated plants, Δ^5-sterols are present only in low amounts and are replaced with 9β,19-cyclopropyl sterols, among them cycloeucalenol (**8**), the substrate of COI, 24-methylpollinastanol (**20**), and 24-dihydrocycloeucalenol (**22**). An interesting observation is that in treated roots, the highest concentration of sterols is found in the ER-rich fraction and not in PM, which contains only 25 μg sterols/mg of protein.

Thus treatment of maize roots with fenpropimorph induces a complete modification of both qualitative and quantitative intracellular distri-

TABLE V
STEROL COMPOSITION OF MEMBRANE FRACTIONS FROM MAIZE ROOTS TREATED AND NONTREATED BY FENPROPIMORPH[a]

	ER		PM	
Sterol	Control (%)	Treated (%)	Control (%)	Treated (%)
Sitosterol (**2**)	15	Traces[b]	14	1
Stigmasterol (**3**)	54	4	56	11
Campesterol (**4**)	23	1	27	3
Isofucosterol (**6**)	2	0	1	0
24-Methyl pollinastanol (**20**)	0	31	0	38
24-Methylene pollinastanol (**21**)	0	6	0	5
Obtusifoliol (**9**)	Traces	0	Traces	0
Cycloeucalenol (**8**)	3	31	1.5	23
24-Dihydrocycloeucalenol (**22**)	1.5	17	Traces	13
31-Norcyclobranol (**23**)	0	2	0	0
Cyclofontumienol (**24**)	0	4	0	4
Cycloartenol (**12**)	Traces	2	Traces	1
24-Methylene cycloartanol (**11**)	Traces	2	Traces	1
Δ^5-Sterols	94	5	98	15
9β,19-Cyclopropyl sterols	5	95	2	85
μg sterols/mg protein	30 ± 3	50 ± 5	60 ± 6	25 ± 2.5

[a] Data given as percentage of total sterols. Data are taken from M. A. Hartmann, A. Grandmougin, and P. Benveniste [*C.R. Seances Acad. Sci. Ser. III* **301,** 601 (1985)].
[b] Traces: less than 1%.

bution of sterols. An almost complete disappearance of the normally present Δ^5-sterols and an accumulation of 9β,19-cyclopropyl sterols are observed. These results clearly indicate that the cyclopropyl sterols are really integrated in two well-characterized membrane fractions, enriched, respectively, in ER and PM, and they can replace Δ^5-sterols despite important structural differences. These 9β,19-cyclopropyl sterols might therefore fill at least partially some functions normally devoted to the bulk of Δ^5-sterols in nontreated plants, especially those involved in the control of microviscosity homeostasis, which is known to be of importance for an optimal operation of membrane-bound enzymes. As shown above, PM is particularly concerned, and different physiological consequences are expected. Effects of changes in the sterol profile on polar lipid composition, on membrane fluidity, and on the activity of PM-bound enzymes (K^+-dependent ATPase and β-glucan synthase) are currently under study.

[59] Separation of Molecular Species of Plant Glycolipids and Phospholipids by High-Performance Liquid Chromatography

By J. KESSELMEIER and E. HEINZ

Introduction

Glycolipids and phosphoglycerolipids together with sterols, steryl glycosides, and acylsteryl glycosides contribute to the physical characteristics of membranes or the membrane areas in which they are formed. Rapid separation and quantitation of these compounds and their molecular species is advantageous. HPLC has become an essential tool in the study of many natural products. Its limited applications in the study of plant lipids may be due to problems of quantitation rather than to difficulties in separation. Various techniques of quantitation, such as flame ionization,[1] attachment of chromophores,[2] and refractometry,[3,4] have been reported for plant lipids. Alternatively, fractions have been analyzed by

[1] O. S. Privett, K. A. Dougherty, W. L. Erdahl, and A. Stolyhwo, *J. Am. Oil Chem. Soc.* **50,** 516 (1973).
[2] G. I. Nonaka and Y. Kishimoto, *Biochim. Biophys. Acta* **572,** 423 (1979).
[3] W. S. M. G. van Kessel, M. Tiemann, and R. A. Demel, *Lipids* **16,** 58 (1981).
[4] R. Yamauchi, M. Kojima, M. Isogai, K. Kato, and Y. Ueno, *Agric. Biol. Chem.* **46,** 2847 (1982).

phosphate[5] or fatty acid analysis subsequent to their separation by HPLC.[6] In the case of steryl glycosides and acylsteryl glycosides hydrolysis to the corresponding free sterols was necessary.[7]

We separated the molecular species of several of the above-mentioned lipids by HPLC without derivatization, whereas lipids having negatively charged residues had to be reacted with diazomethane for satisfactory resolution. The concomitant quantitation at 200 nm depends on the presence of double bonds in the different molecular species. Therefore, completely saturated species can be detected only after conversion of phospho- and glycolipids to the corresponding *sn*-1,2-diacylglycerols containing a chromophoric acyl substituent at the *sn*-3 position.

Materials and Methods

Isolation of Lipid Classes

Leaves or isolated membrane fractions were extracted with boiling 2-propanol/chloroform/methanol (1/2/1). After washing with 0.45% (w/v) NaCl the lipids were fractionated by preparative TLC on silica gel plates. The choice of the solvent system depends on the lipid components to be separated. Monogalactosyldiacylglycerol (MGDG), steryl glycosides (SG), and acylsteryl glycosides (ASG) were isolated in $CHCl_3$/methanol (90/10) and digalactosyldiacylglycerol (DGDG) using $CHCl_3$/methanol/H_2O (70/30/4). In this solvent phosphatidylglycerol (PG) and sulfoquinovosyl diacylglycerol (SQD) are not well separated and are scraped off together. Rechromatography in $CHCl_3$/methanol/acetic acid/H_2O (85/15/10/3.5) separates the faster running PG from SQD. Lipids were detected under UV after spraying with anilinonaphthalene sulfonate [ANS, 0.2%, (w/v) in methanol]. Separated lipids were eluted with and dissolved in methanol. $CHCl_3$ should be avoided for this purpose, since its subsequent evaporation leads to peaks interfering with the 200-nm detection. MGDG, DGDG, and SG, eluted with distilled methanol and dissolved in methanol, could be analyzed directly without chemical alteration. ASG was converted to SG by addition of 20 μl of a 20 m*M* solution of sodium methoxide in methanol to 100 μl of ASG solution. The mixture was incubated at 35° for 15 min. SG obtained from ASG was purified by TLC as described above.

[5] E. N. Ashworth, J. B. St. John, M. N. Christiansen, and G. W. Patterson, *J. Agric. Food Chem.* **29,** 879 (1981).

[6] D. V. Lynch, R. E. Gundersen, and G. A. Thompson, Jr., *Plant Physiol.* **72,** 903 (1983).

[7] J. Kesselmeier, W. Eichenberger, and B. Urban, *Plant Cell Physiol.* **26,** 463 (1985).

Formation of Methyl Esters from PG and SQD

SQD and PG had to be converted to the methyl ester derivatives by reaction of the protonated acid groups with diazomethane.[8] For this purpose the SQD and PG zones scraped off from the TLC plates were resuspended in $CHCl_3$/methanol (2 : 1, 4 ml) and shaken with dilute HCl (0.1 *N*, 1 ml). After a short centrifugation the lower phase was withdrawn and mixed with diazomethane solution (in CH_2Cl_2) until the yellow color persisted. The resulting methyl ester derivatives of both lipids (SQD-Me and PG-Me) were purified by TLC in $CHCl_3$/methanol (9/1). Recoveries of SQD-Me and PG-Me from SQD and PG were found between 60 and 90%.[9] The lipids were eluted with and dissolved in methanol and injected onto the column.

Formation of p-Anisoyldiacylglycerols from Glycerolipids

For detection of saturated species MGDG, DGDG, SQD, and PG were converted to their corresponding 1,2-diacyl-3-*p*-anisoyl-*sn*-glycerols: PG was hydrolyzed by phospholipase C,[10] whereas the diacylglycerol moieties of MGDG, DGDG, and SQD were released by a recently described chemical method[11]: The glycosyldiacylglycerol (0.1–10 mg) recovered from the preparative TLC run was dissolved in methanol (1 ml) and oxidized by addition of periodic acid (45 mg HJO_4 in 1 ml of methanol). After 90 min at room temperature in the dark the oxidation mixture was diluted with $CHCl_3$ (4 ml) and washed with NaCl solution [0.45% (w/v) 1.5 ml]. After a short centrifugation and removal of the upper phase, the lower organic solution was washed with aqueous ethylene glycol [5% (w/v) 1.5 ml] to destroy residual periodate and washed again with NaCl solution. The final lower phase was removed, mixed with a few drops of 2-propanol, and concentrated to dryness in a stream of nitrogen. The residue was dissolved in 0.5 ml of 1,1-dimethylhydrazine solution [1% (w/v) in 2-propanol/$CHCl_3$/acetic acid/water (3.5/3/1.5/1) which is prepared by mixing 3 ml of this solvent mixture with 37.3 μl of dimethylhydrazine]. This solution is kept in a stoppered centrifuge tube in the dark for 4 hr in the case of DGDG and for 20–24 hr in the case of MGDG and SQD. Then diethyl ether (5 ml) is added and the mixture washed twice with aqueous KH_2PO_4 (50 m*M*, 2 ml) and dried by addition of anhydrous Na_2SO_4. The ether is removed in a stream of nitrogen and the released diacylglycerols purified by TLC in diethyl ether/petroleum ether (3/1). Pure 1,2-diacyl-

[8] A. P. Tulloch, E. Heinz, and W. Fischer, *Hoppe-Seyler's Z. Physiol. Chem.* **354,** 879 (1973).

[9] J. Kesselmeier and E. Heinz, *Anal. Biochem.* **144,** 319 (1985).

[10] R. F. A. Zwaal and B. Roelofsen, this series, Vol. 32, Part B, p. 154.

[11] F. J. Heinze, M. Linscheid, and E. Heinz, *Anal. Biochem.* **139,** 126 (1984).

glycerols are recovered from the corresponding zone (R_f 0.5; resuspension of the scrapings in diethyl ether followed by short centrifugation) in yields of 30–50% (based on recovery of labeled acyl groups).

The diethyl ether is removed by a stream of nitrogen and the residual diacylglycerols are dissolved in pyridine (0.3 ml). After addition of *p*-anisoyl chloride (20 μl, 80% solution in toluene) the mixture is kept overnight at room temperature with the exclusion of moisture. Then petroleum ether/diethyl ether (1/1, 5 ml) is added and the mixture washed three times with aqueous 50 m*M* KH_2PO_4 to remove most of the pyridine. After drying with anhydrous Na_2SO_4 and solvent removal the residue is chromatographed in petroleum ether/diethyl ether/acetic acid (50/40/0.5). The 1,2-diacyl-3-*p*-anisoyl-*sn*-glycerols (R_f of about 0.7) are recovered by elution in diethyl ether. After removal of this solvent they are redissolved in methanol.

High-Performance Liquid Chromatography

Lipid samples, dissolved in methanol, were injected directly onto the column. Samples in $CHCl_3$/methanol have to be diluted with methanol to $CHCl_3$ proportions below 10% prior to injection. Alternatively, samples were taken to dryness in a stream of nitrogen and redissolved in methanol to give concentrations in the range of 0.5–2.5 nmol per species in a volume of 20 μl. Samples were injected via a Rheodyne rotary valve equipped with a 20-μl loop. Separations were accomplished by gradient elution using a 125 × 4.6 mm reversed-phase column (Spherisorb C_6, 5 μm, prepacked by Bischoff, FRG). The binary linear gradient was run within 30 min at a flow rate of 1 ml/min from 50 to 100% (for MGDG, DGDG, SQD-Me, PG-Me, and SG) or from 70 to 100% (for *p*-anisoyldiacylglycerols) of acetonitrile (HPLC-grade, Baker) in water. Eluted species were detected at 200 nm with the exception of *p*-anisoyldiacylglycerols, which were detected at 250 nm.

Separation

The lipids MDGD, DGDG, SQD, PG, SG, and SG from ASG were first isolated by conventional TLC, but even for this step HPLC can be used.[12–14] Their subsequent resolution into molecular species was

[12] D. Marion, G. Gandemer, and R. Douillard, *in* "Structure, Function and Metabolism of Plant Lipids" (P. A. Siegenthaler and W. Eichenberger, eds.), p. 139. Elsevier, Amsterdam, 1984.

[13] E. Deleens, N. Schwebel-Dugue, and A. Trémolières, *FEBS Lett.* **178,** 55 (1984).

[14] C. Demandre, A. Trémolières, A. M. Justin, and P. Mazliak, *Phytochemistry* **24,** 481 (1985).

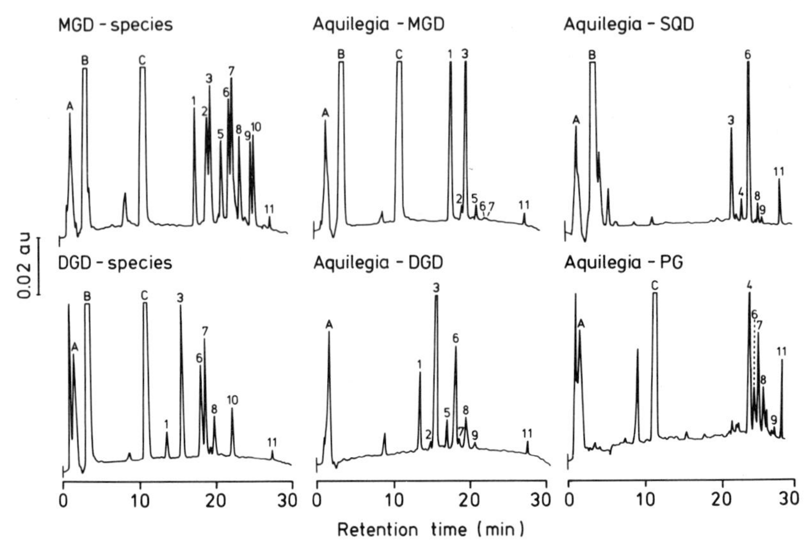

FIG. 1. Separation of major thylakoid lipids from *Aquilegia alpina* into molecular species by reversed-phase HPLC as detected at 200 nm. All molecular species marked with numbers have been identified by fatty acid analysis and are characterized by first naming the *sn*-1 bound acyl group. References from MGDG and DGDG are shown at the left. Peak labels represent the following compounds: A, methanol, 2-propanol; B, chloroform; C, antioxidant; 1, 18 : 3/16 : 3; 2, 18 : 3/16 : 2; 3, 18 : 3/18 : 3; 4, 18 : 3/3-*trans*-16 : 1; 5, 18 : 3/18 : 2; 6, 18 : 3/16 : 0; 7, 18 : 2/18 : 2, but in PG 7, 18 : 2/3-*trans*-16 : 1; 8, 18 : 2/16 : 0 (plus 18 : 1/16 : 1 in reference compounds); 9, 18 : 1/16 : 0; 10, 18 : 1/18 : 1; 11, contaminations from TLC plates. Reproduced from Kesselmeier and Heinz.[9]

achieved by reversed-phase HPLC with gradient elution on a short C_6 column. The same linear gradient of acetonitrile in water was suitable for all lipids, although the acid groups of SQD and PG had to be methylated. The molecular species were identified by cochromatography of reference compounds, by fatty acid analysis, and by GLC and GLC–MS.[7,9] Figure 1 shows the separation of reference compounds as well as of natural mixtures of the four glycerolipids. The resolution of steryl glycosides (or SG from ASG) is shown in Fig. 2. The elution sequence of intact SG species follows exactly that of free sterols, thus showing that the glycosidic residue has no influence on the HPLC separation.[7] For more information about the influence of various methyl groups, double bonds, and branching on the affinity of free sterols to column material see DiBussolo and Nes.[15] The *p*-anisoyl derivatives[16] of diacylglycerols released from glyco-

[15] J. M. DiBussolo and W. R. Nes, *J. Chromatogr. Sci.* **20,** 193 (1982).

[16] M. Batley, N. H. Packer, and J. W. Redmond, *Biochim. Biophys. Acta* **710,** 400 (1982).

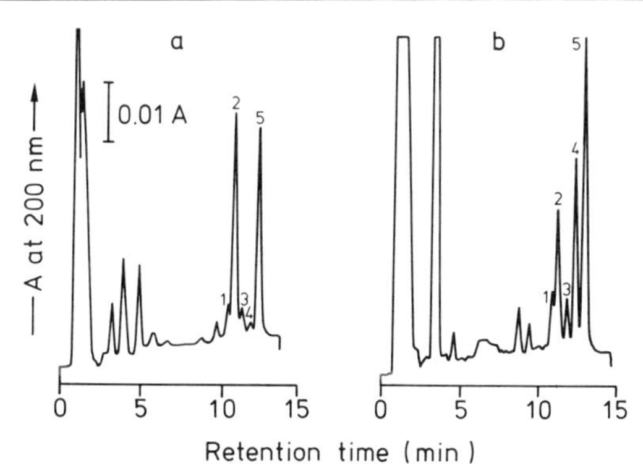

FIG. 2. Separation of intact sterol glycosides from oat (a) seeds and (b) leaves into molecular species by reversed-phase HPLC. All molecular species marked with numbers have been identified by GLC–MS. 1, Cholesterol glycoside plus Δ^7-avenasterol glycoside; 2, Δ^5-avenasterol glycoside; 3, campesterol glycoside; 4, stigmasterol glycoside and Δ^7-stigmastenol glycoside; 5, sitosterol glycoside. Reproduced from Kesselmeier *et al.*[7]

and phospholipids were separated on the same column with a similar gradient. Representative chromatograms including reference species are shown in Fig. 3. Due to the attached chromophore fully saturated species such as 18:0/18:0 and 16:0/16:0 are detected, whereas they escaped detection at 200 nm when chromatographed as underivatized lipids. This may be of importance, because the 16:0/16:0 species of PG may have particular significance with respect to cold hardiness.[17]

In our system (but see discussion below) the elution sequence of glycerolipid molecular species is influenced by the sum of carbon atoms in both acyl groups and by the total number of double bonds. When the relative retention times of isolated reference species from different lipids are plotted as a function of carbon atoms and double bond numbers in acyl pairs, straight lines are obtained for all lipids (Fig. 4). For the corresponding sets of 18/16 and 18/18 combinations with different numbers of cis-double bonds pairs of parallel lines are formed; 18/18 combinations were clearly separated from 18/16 combinations. The only exception we noticed are the 18:3/16:0 and 18:2/18:2 combinations of anisoyldiacylglycerols. Apart from this exception, it is possible to identify an unknown species from its relative retention time which intersects with only one of the two lines. However, it should be pointed out that this identification yields only the sum of carbon atoms and double bonds from both acyl groups. This has been demonstrated with 18:1/16:1 and 18:2/16:0 species (Fig. 1). When this behavior is extended to other species as suggested

[17] N. Murata, *Plant Cell Physiol.* **24,** 81 (1983).

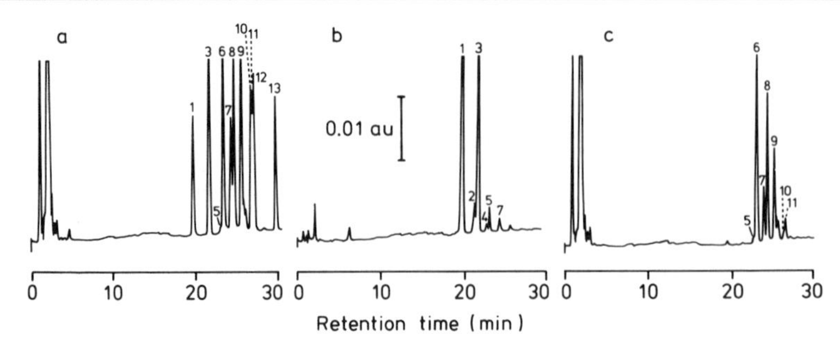

FIG. 3. Separation of diacylglycerols released by chemical or enzymatic methods from intact glyco- and phospholipids and converted to *p*-anisoyl derivatives. (a) Reference species; (b) diacylglycerols from MGDG of *Aquilegia*; (c) diacylglycerols from PG of *Aquilegia*. Identification of peaks: 1, 18 : 3/16 : 3; 2, 18 : 3/16 : 2; 3, 18 : 3/18 : 3; 4, 18 : 3/16 : 1; 5, 18 : 3/18 : 2; 6, 18 : 3/3-*trans*-16 : 1; 7, 18 : 3/16 : 0 plus 18 : 2/18 : 2; 8, 18 : 2/3-*trans*-16 : 1; 9, 18 : 2/16 : 0 (plus 18 : 1/16 : 1 in references); 10, 16 : 0/16 : 0; 11, 18 : 1/16 : 0; 12, 18 : 1/18 : 1; 13, 18 : 0/18 : 0. Reproduced from Kesselmeier and Heinz.[9]

by the linearity of the plots in Fig. 4, only four molecular species of the most common ones are separated as single species (18 : 3/16 : 3, 18 : 3/18 : 3, 18 : 1/16 : 0, 18 : 1/18 : 1), whereas the others coelute in seven groups containing two or three different species. In the case of PG-Me and SQD-Me (Fig. 4b) only one common line for 18/16 combinations was constructed, since the relative retention times were nearly identical; 18/18 combinations were not available and it should be noted that during MS and GLC analysis 18/18 combinations were detected only rarely in PG and SQD from several plants.[17–19] In the case of PG from leaves this may be due to the fact that most of this lipid is localized and synthesized in chloroplasts and therefore consists exclusively of 18/16 combinations.[20] During HPLC of anisoyldiacylglycerols a separation of the 18 : 3/3-*trans*-16 : 1 from the 18 : 3/16 : 1 and 18 : 3/16 : 0 species was achieved (Fig. 3), showing an influence of both unsaturation and bond geometry in agreement with reversed-phase HPLC of fatty acid isomers.[21–23] A similar elution behavior is observed for the 18 : 2/3-*trans*-16 : 1. Therefore, a tentative line for 3-*trans*-16 : 1-containing species is included in Fig. 4c.

[18] H. P. Siebertz, E. Heinz, M. Linscheid, J. Joyard, and R. Douce, *Eur. J. Biochem.* **101,** 429 (1979).
[19] M. Nishihara, K. Yokota, and M. Kito, *Biochim. Biophys. Acta* **617,** 12 (1980).
[20] S. A. Sparace and J. B. Mudd, *Plant Physiol.* **70,** 1260 (1982).
[21] M. Özcimder and W. E. Hammers, *J. Chromatogr.* **187,** 307 (1980).
[22] L. Svensson, L. Sisfontes, G. Nyborg, and R. Blomstrand, *Lipids* **17,** 50 (1982).
[23] R. Wood and T. Lee, *J. Chromatogr.* **254,** 237 (1983).

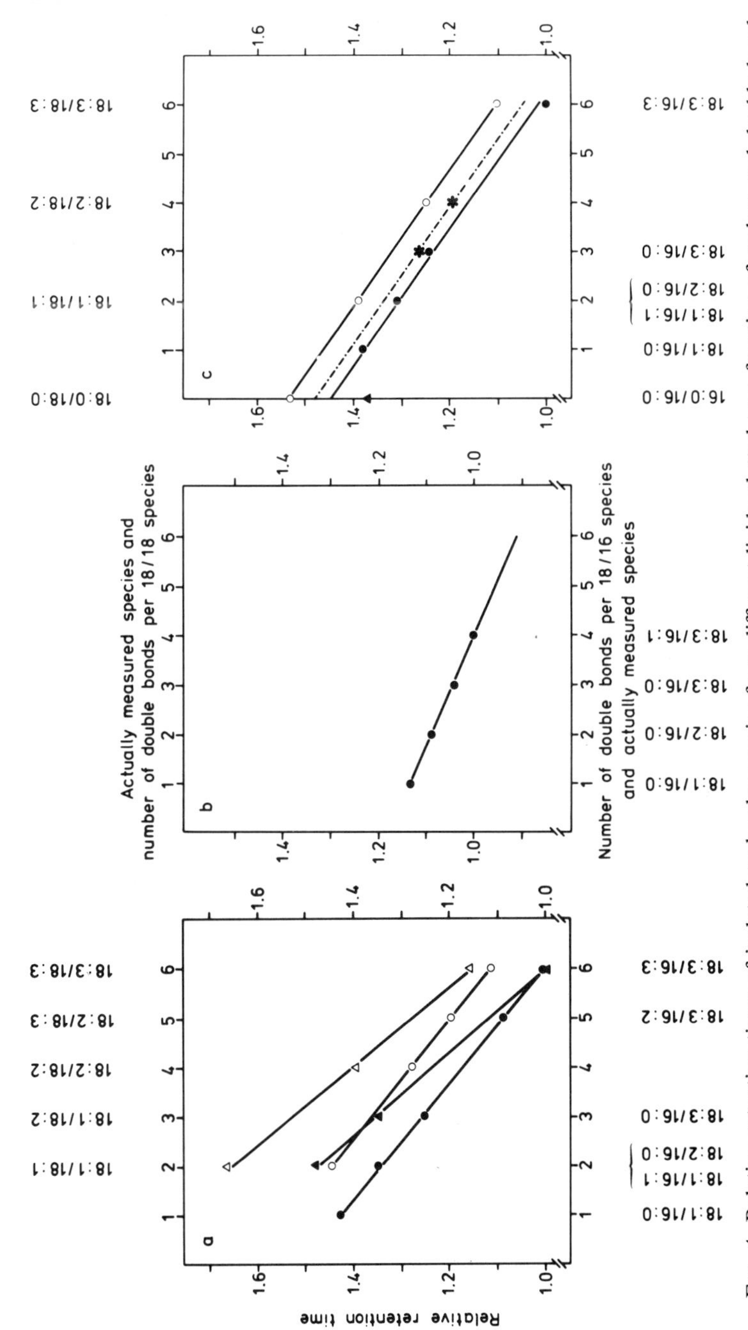

FIG. 4. Relative retention times of isolated molecular species from different lipids plotted as a function of carbon and double bond numbers in acyl pairs. In (a) and (c) the absolute retention times of the 18 : 3/16 : 3 species were set as 1.00, whereas in (b) it was the 18 : 3/16 : 1 combination. The numbers at the upper abscissa refer to double bonds in 18/18 combinations, whereas double bond numbers of 18/16 species are indicated at the lower abscissa. (a) ●, 18/16, and ○, 18/18 combinations of MGDG; ▲, 18/16, and △, 18/18 combinations of DGDG; (b) 18/16 species of PG-Me and SQD-Me (18/18 combinations were not available); (c) ●, 18/16, and ○, 18/18 combinations of *p*-anisoyldiacylglcerols. ▲, 16 : 0/16 : 0; *, 18 : 3/3-*trans*-16 : 1 and 18 : 2/3-*trans*-16 : 1. Reproduced from Kesselmeier and Heinz.[9]

Molecular species of glyco- and phospholipids can also be separated isocratically and without derivatization on a μBondapak C_{18} column.[14] With methanol/water/acetonitrile (90.5/7/2.5) containing 20 m*M* choline chloride the various species are eluted in the same order as described above, but 18:2/18:2 and 16:0/18:3 combinations are not separated, which is possible with the gradient system (see Fig. 1). On the other hand, the resolution of these two species as chromophore-carrying diacylglycerol derivatives was not possible, neither as *p*-anisoyl derivatives in our gradient system nor as naphthylurethanes in isocratic separations with a solvent mixture similar to that used for intact lipids.[24] Similarly, the chromatography of PG results in separations which in both systems are less clean than those achieved for other lipids.

A solution of many of these difficulties may be the conversion of glycerolipids to dinitrobenzoyl derivatives of their constituent diacylglycerols. As recently shown[25] isocratic chromatography of these derivatives separates for example 18:1/16:1 from 16:0/18:2, 18:1/18:1 from 18:0/18:2, and even 18:0/20:4 from 20:4/18:0.

Quantitation

Elution profiles of intact lipids were recorded at 200 nm, which restricts the detection to species with at least one double bond per molecule. For detection of fully saturated species the lipids were converted to their *p*-anisoyldiacylglycerols, allowing a detection of the chromophore at 250 nm. Since even highly unsaturated acyl groups do not absorb at this wavelength, elution profiles of *p*-anisoyldiacylglycerols measured at 250 nm reflect molar proportions. In contrast to the derivatized lipids, the underivatized forms require calibration for each species (Fig. 5). The response factors for individual acyl pairings may be obtained by two methods. Isolated single species may be subjected to HPLC analysis after appropriate quantitative calibration by colorimetry or GLC of fatty acid methyl esters. For this purpose each species is required in pure form and sufficient quantity, and both absolute and relative response factors are obtained. But since many applications are only concerned with the proportions of species within a given lipid, relative response factors for underivatized lipids are of equal importance. They may be obtained by an alternative and more widely usable method for which individual species are not required in isolated form. First, these lipids are converted to *p*-

[24] A. M. Justin, C. Demandre, A. Trémolières, and P. Mazliak, *Biochim. Biophys. Acta* **836,** 1 (1985).

[25] M. Kito, H. Takamura, H. Narita, and R. Urade, *J. Biochem.* (*Tokyo*) **98,** 327 (1985).

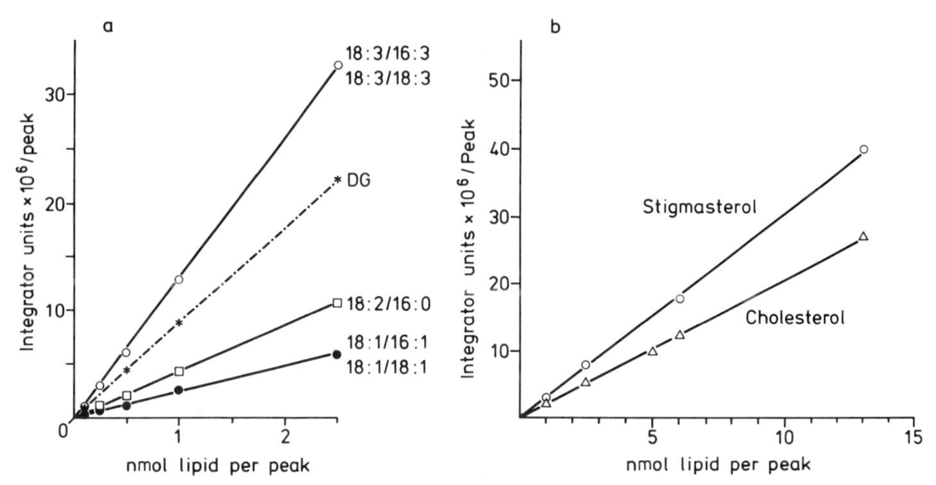

FIG. 5. Detector sensitivity in relation to the unsaturation of molecular species. (a) Detector response at 200 nm for colorimetrically calibrated species of MGDG (all combinations indicated) and of DGDG (all except 18:3/16:3 and 18:2/16:0) as well as for crystalline *p*-anisoyl derivatives of 1,2-dipalmitoyl- and 1,2-distearoyl-*sn*-glycerols (DG, detected at 250 nm). (b) Detector response for cholesterol (one double bond) and stigmasterol (two double bonds). Reproduced from Kesselmeier and Heinz[9] and Kesselmeier *et al.*[7]

anisoyldiacylglycerols, and their elution profile is recorded at the two wavelengths of 200 and 250 nm. The absorbance at 250 nm is exclusively caused by anisoyl residues and yields molar proportions, whereas both anisoyl and unsaturated acyl groups contribute to the absorbance at 200 nm. By recording the elution of pure *p*-anisoyl methyl ester at these two wavelengths it was found that the absorbance of the anisoyl group is 1.29 times higher at 200 than at 250 nm. This ratio was used to subtract the contribution of the anisoyl residue from the 200-nm absorbance of the various *p*-anisoyldiacylglycerol peaks. By correlating these corrected 200 nm values with those measured for the same sample at 250 nm according to the formula $(E_{200\text{ nm}} - E_{250\text{ nm}} \times 1.29)/E_{250\text{ nm}}$ and by setting this ratio for the 18:1/18:1 combination as 1.00, relative response factors for individual acyl pairs are obtained. These factors in turn are used for analysis of molecular species of underivatized lipids at 200 nm. The different peak areas or intensities of an elution profile of an intact lipid are divided by the appropriate factors and the corrected intensities converted to molar proportions.

Table I compares relative response factors obtained by the two different methods: calibration of several isolated galactolipid molecular species by anthrone colorimetry versus calculation from the absorbances of ani-

TABLE I
RELATIVE RESPONSE FACTORS FOR DIFFERENT MOLECULAR SPECIES OF INTACT LIPIDS[a]

Molecular species	Response factors based on	
	200 nm/250 nm ratios	Colorimetric calibration
18:1/18:1	1.00	1.00
18:1/16:0	0.49	—
18:2/16:0	2.05	1.83
18:3/16:0	3.45	—
18:1/16:1	1.13	1.00
18:2/18:2	4.12	—
18:3/18:3	6.49	5.65
18:3/16:3	6.48	5.65

[a] Absorbances recorded at 200 nm are converted to molar proportions by dividing with the appropriate factors. For comparison factors obtained by direct colorimetric calibration using isolated species of MGDG and DGDG are also shown. Details are given in the text. Reproduced from Kesselmeier and Heinz.[9]

soyldiacylglycerols at the two wavelengths. As expected, the factors increase with increasing number of double bonds and factors of acyl pairs can be calculated by combining the factors of the two constituent acyl groups. The suitability of these factors to quantitate intact lipid species at 200 nm was confirmed by analyzing different leaf lipids by two methods: recording of underivatized species at 200 nm as compared to analysis of *p*-anisoyldiacylglycerols prepared from the same sample and measured at 250 nm.[9] The two methods show good agreement with regard to the major molecular species in each lipid, whereas larger differences are observed for several minor components. This variability may be due to impurities, which elute with or instead of lipid peaks and result in proportionally large errors for minor constituents.

Calibration of steryl glycosides was achieved by using free cholesterol and stigmasterol as standards (Fig. 5), eluted with the same gradient system as used for the underivatized acyl lipids.[7] The peak area per nanomole cholesterol was identical to that found for steroidal saponins with one double bond.[26] This shows that glycosyl residues do not affect the

[26] J. Kesselmeier and D. Strack, *Z. Naturforsch., C: Biosci.* **36C,** 1072 (1981).

detection at 200 nm and supports the general suitability of response factors in intact lipids irrespective of the substituent at C-3. Therefore, ratios of nanomoles per peak area as measured for free sterols can also be used for the corresponding species of steryl glycosides.

Acknowledgments

The financial support by the Deutsche Forschungsgemeinschaft is gratefully acknowledged. We thank Miss A. Lührs, Miss K. Wrage, Miss U. Laudenbach, and Dr. R. Haas for experimental help.

[60] Movement of Phospholipids between Membranes: Purification of a Phospholipid Transfer Protein from Spinach Leaf

By JEAN-CLAUDE KADER, CHANTAL VERGNOLLE, and PAUL MAZLIAK

The cytosolic fractions from various animal and plant tissues contain proteins which facilitate the *in vitro* movement of phospholipids between intracellular membranes (mitochondria or microsomes) or between sonicated lipid vesicles.[1-5] These proteins, formerly called phospholipid exchange proteins,[1] facilitate an exchange of intact phospholipid molecules between membranes. Depending on membrane properties, however, this exchange may be not a strict one-for-one process. Under certain conditions, a net transfer of phospholipids from one membrane to another is observed, explaining the new generic name of phospholipid transfer proteins.[2] These proteins could participate in membrane biogenesis by facilitating the *in vivo* distribution of newly synthesized phospholipids to the various membranes of the cell. After the detection in plant tissues (potato tuber and cauliflower florets)[6] of proteins having properties of phospholipid transfer, the first partial purification of a plant phospholipid transfer protein has been achieved from potato tuber.[7] Active, partially purified

[1] K. W. A. Wirtz, *Biochim. Biophys. Acta* **344,** 95 (1974).
[2] K. W. A. Wirtz, *in* "Lipid-Protein interactions" (P. C. Jost and O. H. Griffiths, eds.), p. 151. Wiley, New York, 1982.
[3] D. B. Zilversmit and M. E. Hughes, *Methods Membr. Biol.* **7,** 211 (1976).
[4] H. H. Kamp and K. W. A. Wirtz, this series, Vol. 32, p. 140.
[5] D. B. Zilversmit and L. W. Johnson, this series, Vol. 34, p. 262.
[6] J. C. Kader, *Cell Surf. Rev.* **3,** 127 (1977).
[7] J. C. Kader, *Biochim. Biophys. Acta* **380,** 31 (1975).

proteins have also been obtained from seeds.[8] The purification to homogeneity of phospholipid transfer proteins from higher plants has been achieved starting from maize seed,[9,10] castor bean seed,[11] and spinach leaf.[12] The protein from spinach leaf, which was the first to be purified from a photosynthetic tissue, is also able to facilitate transfer of phospholipids toward chloroplasts. For the purification of phospholipid transfer protein from leaf, it was essential to overcome the difficulties due to the presence of lipases and proteases in crude cytosolic extracts.[12] The purification procedures as well as the assays for measuring the activity of the spinach protein are described here. Similar methods have been used for other plant tissues.[10,11]

Assay Method

Principle of Assay. Transfer of [^{3}H]phosphatidylcholine from liposome vesicles to mitochondria is determined with and without added phospholipid transfer protein. After incubation, the mitochondria were collected by centrifugation and the transfer of phosphatidylcholine is determined by measuring the radioactivity recovered in the mitochondria. The values are corrected for eventual cross-contamination of mitochondria by intact liposomes by introducing into liposomes a trace of cholesteryl [1-^{14}C]oleate, which is not transferred by the protein.

Preparation of Liposomes. Liposomes are prepared by mixing in a flask the following lipids dissolved in chloroform: 2.6 μmol of egg yolk phosphatidylcholine (purchased from Sigma), 4 nmol of cholesteryl [1-^{14}C]oleate (1.25 GBq mmol^{-1}, 2.7 $\times$ 10^4 dpm; purchased from Amersham), and 100 nmol of [^{3}H]phosphatidylcholine (500 MBq mmol^{-1}, 2.8 $\times$ 10^4 dpm). The latter phospholipid was obtained by incubating potato tuber slices with methyl[^{3}H]choline chloride (555 GBq mmol^{-1}; purchased from Amersham) as described earlier.[9] The solvent was evaporated under a stream of nitrogen while rotating the flask in order to obtain a thin film of lipids. Five milliliters of buffer A (0.25 *M* sucrose, 1 m*M* EDTA, 10 m*M* Tris–HCl, pH 7.2) and three glass beads were added; the mixture was vigorously shaken to disperse the lipids. The dispersion was then irradiated ultrasonically in an MSE sonifier at a 150 W output for 30 min at 15–

[8] M. Yamada, T. Tanaka, J. C. Kader, and P. Mazliak, *Plant Cell Physiol.* **19,** 173 (1978).

[9] D. Douady, M. Grosbois, F. Guerbette, and J. C. Kader, *Biochim. Biophys. Acta* **710,** 143 (1982).

[10] D. Douady, F. Guerbette, M. Grosbois, and J. C. Kader, *Physiol. Veg.* **23,** 373 (1985).

[11] M. Yamada, I. Nishida, S. Watanabe, and J. C. Kader, *in* "Structure, Function and Metabolism of Plant Lipids" (P.-A. Siegenthaler and W. Eichenberger, eds.), p. 291. Elsevier, Amsterdam, 1984.

[12] J. C. Kader, M. Julienne, and C. Vergnolle, *Eur. J. Biochem.* **139,** 411 (1984).

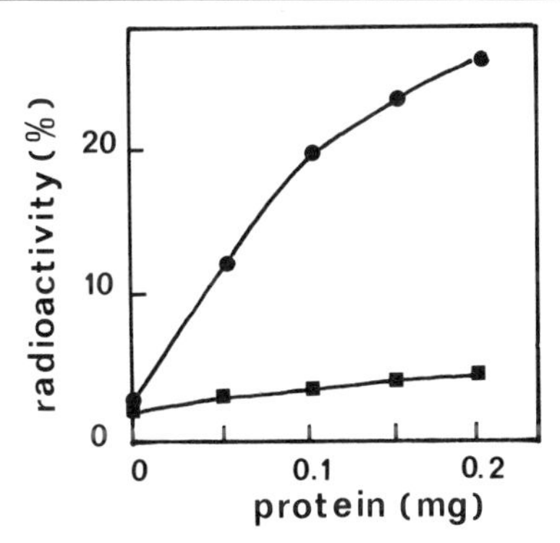

FIG. 1. Transfer of phosphatidylcholine from liposomes to mitochondria in the presence of spinach protein. Liposomes made from [^{3}H]phosphatidylcholine and cholesteryl [1-^{14}C]oleate were incubated with mitochondria (2 mg of protein) and increasing amounts of spinach protein (step 4) for 30 min at 30°. The ^{3}H label (●—●) and the ^{14}C label (■—■) recovered in mitochondria are expressed as a percentage of the radioactivity of the initial liposomes.

20° under nitrogen. The sonicated suspension can be used within 1 day when kept at 4°.

Isolation of Mitochondria. Mitochondria were isolated from potato tuber by sucrose gradient centrifugation as described earlier.[8] Potato mitochondria were suspended in buffer A at a final concentration of 10 mg of protein ml^{-1} (determined by the Lowry method[13]). Mitochondria were 92% intact as judged by succinate–cytochrome-*c* oxidoreductase activity.[14] Mitochondria may be stored at −20° up to 2 weeks before use.

Transfer Assay. Liposomes (0.5 ml) were incubated with potato mitochondria (2 mg of protein) with or without added protein factors in a total volume of 3 ml of buffer A. The incubations were carried out in polyethylene scintillation vials placed in a shaking bath. After 30 min of incubation at 30°, 5 ml of cold buffer A was added; the mixture was then centrifuged 10 min at 12,000 *g*; the pellets, containing the mitochondria, were suspended in 0.3 ml of 1% Triton X-100. After addition of 10 ml of toluene scintillator mixture containing 33% of Triton X-100, the radioactivity of mitochondria as well as that of supernatants—which contained the liposomes—was measured.

Calculation of Transfer Activity. When mitochondria were incubated with increasing amounts of phospholipid transfer proteins, the ^{3}H label recovered in the organelles increased, whereas the ^{14}C label remained constant as shown in Fig. 1. This experiment indicates that a transfer of

[13] O. H. Lowry, N. J. Rosebrough, A. L. Farr, and R. Randall, *J. Biol. Chem.* **193,** 265 (1951).

[14] R. Douce, E. L. Christensen, and W. D. Bonner, *Biochim. Biophys. Acta* **275,** 148 (1972).

[^{3}H]phosphatidylcholine occurred between liposomes and mitochondria. The transfer activity can be calculated as follows: The radioactivity of phosphatidylcholine transferred (P) is equal to ($M - mL/l$), where M and m are, respectively, the ^{3}H and ^{14}C labels of mitochondria after incubation for 30 min and L and l are the ^{3}H and ^{14}C labels of liposomes before incubation. The transfer of phosphatidylcholine may be expressed as percentage ($100P/L$) or as nanomoles of phosphatidylcholine transferred in 30 min (P/S, where S is the specific radioactivity of liposomal lipids before incubation, expressed in dpm nmol^{-1}). The specific activity of protein factions is expressed as nanomoles of phosphatidylcholine transferred in 30 min/mg of protein. One unit of activity is defined as the amount of protein able to transfer 1 nmol of phosphatidylcholine/min.

Purification Procedure

The various steps in the purification procedure are presented in Table I. All steps were performed at 4°.

Step 1: Ammonium Sulfate Precipitation. Spinach leaves (5 kg), obtained from farms near Paris, deribbed and washed several times with distilled water, were ground in a Waring blender (2 min at high speed) in 3 liters of buffer B (0.4 *M* sucrose, 1 m*M* EDTA, 1 m*M* cysteine chloride, 10 m*M* 2-mercaptoethanol, 1 m*M* phenylmethylsulfonyl fluoride, 100 m*M* Tris–HCl, pH 7.5). The homogenate was centrifuged for 20 min at 9000 *g* in a Sorvall centrifuge. The pH of the supernatant was adjusted to 5.1 in order to eliminate the residual membranes. After centrifugation at 9000 *g* for 20 min and pH adjustment of the supernatant to pH 7.5, solid ammonium sulfate was added (52 g/100 ml); the mixture was stirred 3 hr and centrifuged at 9000 *g* for 30 min. The protein pellets were suspended in a minimal volume of buffer C (3 m*M* sodium azide, 8 m*M* 2-mercaptoethanol, 10% glycerol 10 m*M* sodium phosphate, pH 7.2). The suspension (about 150 ml) was dialyzed 3 hr against 10 liters of buffer C.

TABLE I
PURIFICATION OF PHOSPHOLIPID TRANSFER PROTEIN FROM SPINACH LEAF

Step	Protein (mg)	Specific activity[a]	Activity (units)	Recovery (%)	Purification factor
Ammonium sulfate	2400	0.04	96		—
Sephadex G-75	760	0.46	349	100	11
DEAE–Trisacryl	550	0.60	330	94	15
CM–Sepharose (CM_2)	27	3.50	94	27	87

[a] Nanomoles of phosphatidylcholine transferred per 30 min.

Step 2: Gel Filtration. The dialyzed protein (2.4 g) was loaded in two halves on two Sephadex G-75 columns (Pharmacia) (5 × 70 cm) equilibrated with buffer C. Fifteen-milliliter fractions were collected at a flow rate of 90 ml hr^{-1}. The phospholipid transfer activity was found in low-molecular-mass fractions eluted with 1.5 to 2.6 vol of the column. At this step, an 11-fold purification was obtained (Table I).

Step 3: Anion Exchange Chromatography. The active fractions were chromatographed on a DEAE–Trisacryl (LKB) (2.5 × 30 cm) equilibrated with buffer C. The unbound fractions, containing the major part of the transfer activity (94%), were kept.

Step 4: Cation-Exchange Chromatography. The unbound fractions were applied to a Carboxymethyl-Sepharose CL-6B (CM-Sepharose) (Pharmacia)(2.5 × 20 cm) column equilibrated with buffer C. The column was eluted with a linear gradient of phosphate made with 1 liter of buffer C and 1 liter of 3 m*M* sodium azide, 8 m*M* 2-mercaptoethanol, 10% glycerol, 0.25 *M* potassium phosphate, pH 7.2. Fifteen-milliliter fractions were collected with a flow rate of 60 ml hr^{-1}. The fractions were dialyzed against buffer C before transfer assays. A major peak of activity (CM_2) was eluted at about 0.2 *M* potassium phosphate (Fig. 2); two other peaks, although less active (CM_1 and CM_0), were observed. The active fractions of the CM_2 peak were pooled and dialyzed 3 hr against 20 vol of buffer C. This final step resulted in an 87-fold purified protein.

Properties

Purity and Molecular Mass. The final protein preparation exhibits a unique band after electrophoresis on SDS–polyacrylamide gel according to Chua.[15] The apparent molecular mass is 9000. A similar value was obtained with phospholipid transfer proteins isolated from other plants.[10,11,16] In contrast, higher values have been obtained for animal (from 11.2 kDa to 32 kDa),[1-3,17] yeast[18] (35 kDa), or *Rhodopseudomonas sphaeroides*[19] (27 kDa) phospholipid transfer proteins.

Isoelectric Point. The spinach transfer protein is basic, with an isoelectric point near 9.0 determined by chromatofocusing. Maize and castor bean transfer proteins are also basic.[10,11,16]

Stability. The spinach transfer protein, when kept at −20°, loses only

[15] N. H. Chua, this series, Vol. 69, p. 434.

[16] J. C. Kader, *Chem. Phys. Lipids* **38,** 51 (1985).

[17] J. C. Kader, D. Douady, and P. Mazliak, *in* "Phospholipids" (J. N. Hawthorne and G. B. Ansell, eds.), p. 279. Elsevier, Amsterdam, 1982.

[18] G. Daum and F. Paltauf, *Biochim. Biophys. Acta* **794,** 385 (1984).

[19] S. P. Tai and S. Kaplan, *J. Biol. Chem.* **259,** 12178 (1984).

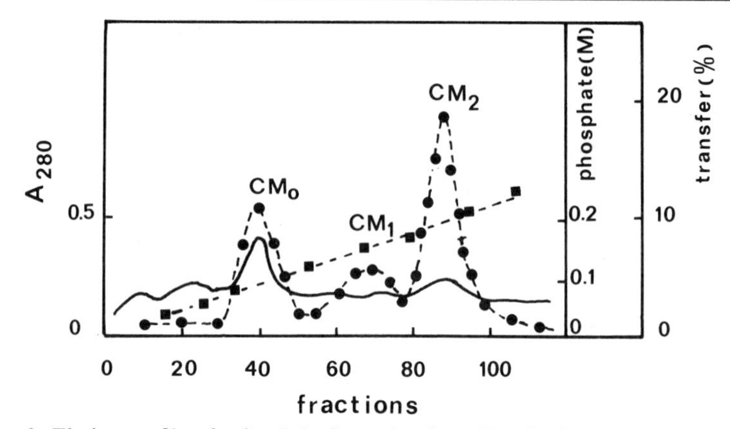

FIG. 2. Elution profile of spinach leaf proteins from CM–Sepharose chromatography. Unbound proteins (0.5 g) from step 3 of the purification procedure were loaded onto a CM–Sepharose column eluted at a flow rate of 60 ml hr^{-1} with a gradient of phosphate (1 liter of buffer C and 1 liter of 3 mM sodium azide, 8 mM 2-mercaptoethanol, 10% glycerol, 0.25 M potassium phosphate, pH 7.2). The absorbance (—) of the 15-ml fractions was measured at 280 nm. The shape of the gradient (■--■) was established from conductivity measurements. The transfer activity (●--●) is expressed as a percentage of liposomal phosphatidylcholine transferred in the presence of 2 ml of each fraction. Fractions 84 to 94 were pooled at the CM_2 purified protein.

25% of its activity per month. The protein is more stable when stored at −20° in 50% glycerol.

Specificity for Phospholipids. In addition to phosphatidylcholine, the spinach transfer protein is also able to transfer phosphatidylinositol, phosphatidylglycerol, and, to a lesser extent, phosphatidylethanolamine. A similar nonspecific characteristic has also been found for other plant[17] or animal[2] transfer proteins.

Specificity for Membranes. The spinach protein is active with various mixtures of natural and artificial membranes: liposome–mitochondria, liposome–liposome, liposome–microsome. The recent observations that the spinach protein is active with intact chloroplasts[12] and with chloroplast envelope membranes[20] open new perspectives for the study of the physiological roles of these proteins in photosynthetic cells.

[20] M. Miquel, M. A. Block, J. Joyard, A. J. Dorne, J. P. Dubacq, J. C. Kader, and R. Douce, *in* "Structure, Function and Metabolism of Plant Lipids" (P. A. Siegenthaler and W. Eichenberger, eds.), p. 295. Elsevier, Amsterdam, 1984.

[61] Pesticides and Lipid Synthesis in Plant Membranes

By C. ANDING

The study of membranous lipids in plants still remains a complex research area. Whereas a fair amount of knowledge is currently available on biosynthesis and the distribution through the membrane of terpene lipids, especially of sterols, carotenoids, and their respective precursors, the actual part they play often remains speculative. The subject of polar lipids would appear to be more difficult to study. These products are classified according to their structure and their distribution within the membranes has been elucidated. A large number of assumptions still exist as to their biosynthesis and role. The question of waxes remains virtually an enigma.

These studies could be furthered via the use of inhibitors likely to interfere with the metabolism of lipids. Such inhibitors exist—the pesticides. Some of them, described in this chapter (see Fig. 1), are of potential interest in clarifying biosynthetic processes as well as the role of membranous lipids.

Pesticides and the improvement of analytical methods have opened the way for in-depth studies whose global approach is now becoming specific. However, despite these steps forward, only a few studies on an enzymatic level are available to date. They are based on specific analyses and on the knowledge of enzymatic mechanisms. The sterol biosynthesis pathway is a good example. It will not be detailed here and the relevant chapter by Hartmann and Benveniste [58] in this volume should be consulted.

Overall Methods

Various methods make it possible to take a global view of the action of pesticides on lipidic constituents.

Lipid Synthesis

Incorporation of Radioactive Precursors. Dinoben inhibits the incorporation of radioactivity in the neutral lipids, the phospholipids, and glycolipids of soybean seeds germinated for 24 hr at 25° in a solution of 0.5 ml water containing 0.25 μCi [1-^{14}C]acetate.[1]

[1] R. K. Muslih and D. L. Linscott, *Plant Physiol.* **60,** 730 (1977).

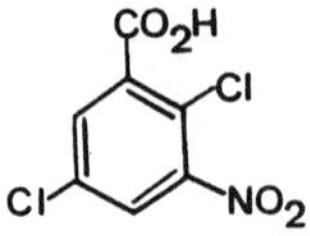

Dinoben

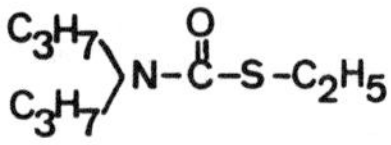

EPTC

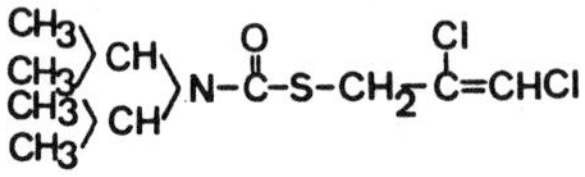

Diallate

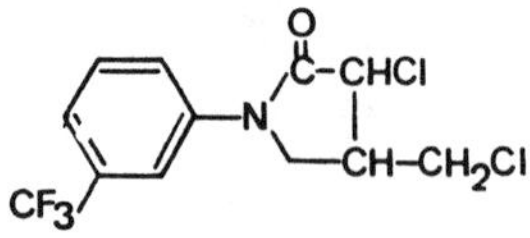

R-40244

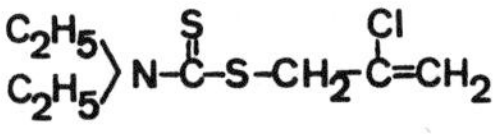

CDEC

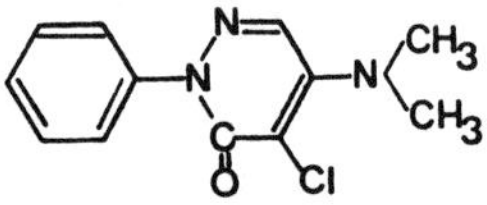

SAN-9785 (BASF 13-338)

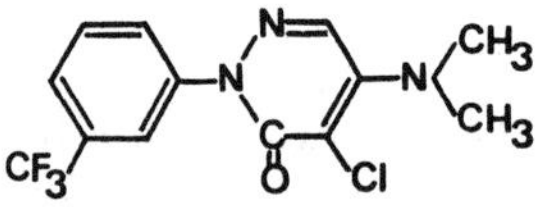

SAN-6706

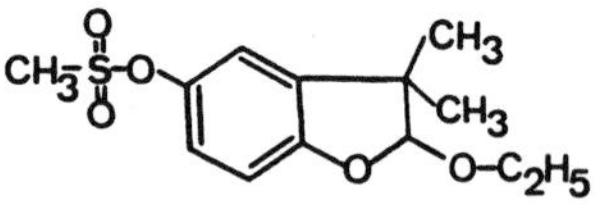

Ethofumesate

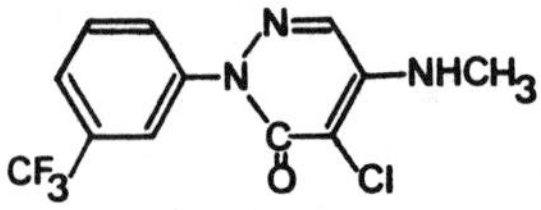

Norflurazon

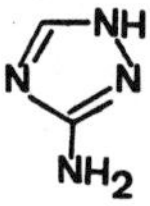

Amitrole

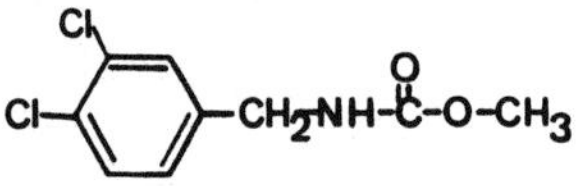

Dichlormate

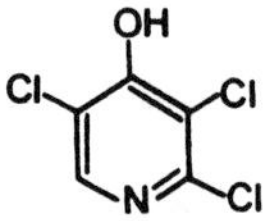

Pyriclor

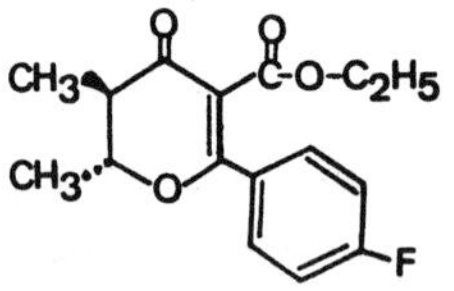

LS-76 1243

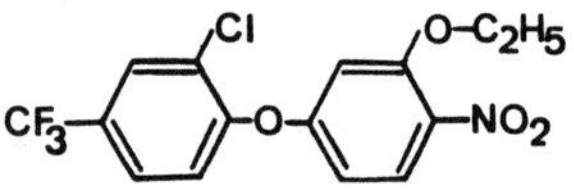

Oxyfluorfen

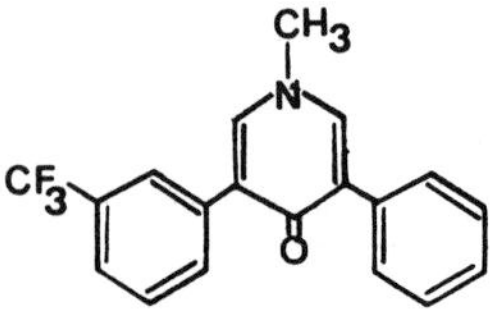

Fluridone

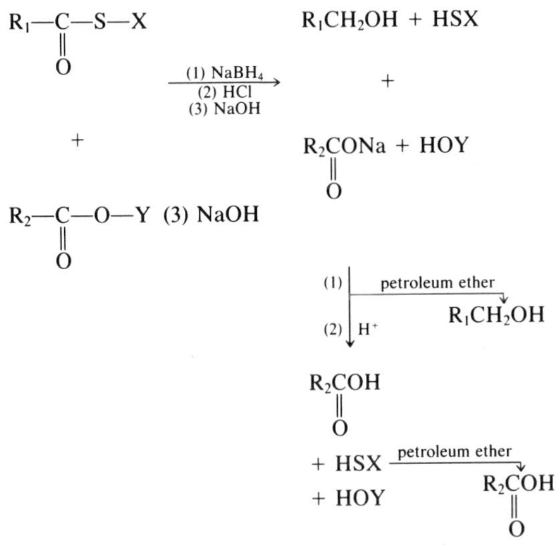

$$R_1-\underset{\underset{O}{\|}}{C}-S-X \quad + \quad R_2-\underset{\underset{O}{\|}}{C}-O-Y \xrightarrow[\substack{(2)\ HCl \\ (3)\ NaOH}]{(1)\ NaBH_4} R_1CH_2OH + HSX \quad + \quad R_2\underset{\underset{O}{\|}}{C}ONa + HOY$$

(3) NaOH

$$\xrightarrow[(2)\ H^+]{(1)\ \text{petroleum ether}} R_1CH_2OH$$

$$R_2\underset{\underset{O}{\|}}{C}OH + HSX + HOY \xrightarrow{\text{petroleum ether}} R_2\underset{\underset{O}{\|}}{C}OH$$

The incorporation of radioactivity is equally inhibited in subcellular fractions, whether mitochondrial (nonpurified), microsomal, and soluble, obtained from germinated soybean seeds. The mitochondrial fraction is a pellet obtained at 21,600 *g* from the crude homogenate: it can therefore include plastids (proplasts or etioplasts).

Similar results are obtained with [1-^{14}C]malonate. However, dinoben is not seen inhibiting the incorporation of [1-^{14}C]acetyl-CoA and [1-^{14}C]malonyl-CoA in these experiments.

EPTC (80 μM) and diallate (5 μM) reduce the incorporation of [2-^{14}C]acetate and [2-^{14}C]malonate in the fatty acids of (nonpurified) spin-

FIG. 1. Some pesticides interfering with lipid synthesis in plant membranes: Dinoben, 3-nitro-2,5-dichlorobenzoic acid; EPTC, 3-ethyldipropyl thiocarbamate; diallate, *S*-(2,3-dichloroallyl)diisopropyl thiocarbamate; R-40244, 1-(*m*-trifluoromethylphenyl)-3-chloro-4-chloromethyl-2-pyrrolidinone; CDEC, β-chloroallyldiethyl dithiocarbamate; SAN-9785, 4-chloro-5-(dimethylamino)-2-phenyl-3(2*H*)pyridazinone; SAN-6706, 4-chloro-5-(dimethylamino)-2-(α,α,α-trifluoro-*m*-tolyl)-3(2*H*)-pyridazinone; ethofumesate, 2-ethoxy-2,3-dihydro-3,3-dimethyl-5-benzofuranylmethane sulfonate; norflurazon, 4-chloro-5-(methylamino)-2-(α,α,α-trifluoro *m*-tolyl)-3(2*H*)-pyridazinone; amitrole, 3-amino-*S*-triazole; dichlormate, 3,4-dichlorobenzylmethyl carbamate; pyriclor, 2,3,5-trichloro-4-pyridinol; LS 76-1243, ethyl-*trans*-5,6-dimethyl-2(4-fluorophenyl)-5,6-dihydropyran-4-one 3-carboxylate; oxyfluorfen, 2-chloro-4-trifluoromethylphenyl-3′-ethoxy-4′-nitrophenyl ether; fluridone, 1-methyl-3-phenyl-5-[3-(α,α,α-trifluoro-*m*-tolyl)](1*H*)-pyridinone.

ach chloroplasts.[2] In this experiment, total lipids are reduced by $NaBH_4$. This method makes it possible to distinguish between thioester and oxygen ester lipids, since the reduction of the carbonyl group of the acyl thioesters provides the fatty alcohols. These alcohols are extracted following hydrolysis while keeping the acids released by the oxygen esters as salts.

This method can only apply when Y is not a fatty alkyl residue itself. It is not applicable to a lipid extract of whole plants, namely due to the presence of cutin.

The distinction between thioester and oxygen ester lipids, the use of various concentrations of [^{14}C]acetate and [^{14}C]malonate (and the former can act as the latter's precursor), as well as the use of an inhibitor should prove to be a powerful technique in the overall study of the biosynthesis of fatty acids and that of polar lipids.

Global Analysis of Fatty Acids in Polar Lipids. EPTC induces a variation in the incorporation of fatty acids in the global fraction of polar lipids in soybean leaves.[3] The test is carried out using soybean plantlets kept for 19 days in a nutrient medium (0.5 Hoagland and Arnon) containing 0 or 2.6 μM of the pesticide. Lipids are extracted from the three trifoliate leaves ($CHCl_3$/methanol, 2 : 1, v/v). The extract is washed (0.2 vol water) and lipids are separated by thin-layer chromatography (silica gel G acetone/benzene/H_2O, 91 : 30 : 8, v/v/v). This developing solvent, however, only allows for a median separation of the various lipids and more particularly of phospholipids. Following chromatography, fatty acids are analyzed by GLC (10% DGS, temperature ranging from 70 to 220°, 4°/min) once saponification (0.5 N NaOH, 65°, 5 min), methylation (14% BF_3/methanol, 65°, 5 min), and extraction (petroleum ether + 10 ml H_2O) from the derivatization medium have been achieved. The considerable drop in the linolenic acid content (both as μg/g FW and as percentage total fatty acids) would seem to imply that EPTC interferes with the biosynthesis of polyunsaturated fatty acids. This is partly confirmed when looking at the fatty acid content of other lipid classes.

A similar methodology also revealed that the action of R-40244 caused the total amount of membranous lipids and the linolenic acid content to decrease in the stem of *Lolium multiflorum* (7-day treatment).[4]

Analysis of Waxes. Diallate causes a considerable reduction in the amount of cuticular waxes.[5] The experiment was carried out on pea plants which had undergone a 7-day vapor effect treatment with the product

[2] R. E. Wilkinson and A. E. Smith, *Weed Sci.* **23,** 100 (1975).

[3] R. E. Wilkinson, A. E. Smith, and B. Michel, *Plant Physiol.* **60,** 86 (1977).

[4] J. B. St. John, *Pestic. Biochem. Physiol.* **23,** 13 (1985).

[5] G. E. Still, D. G. Davis, and G. L. Zander, *Plant Physiol.* **46,** 307 (1970).

(diallate concentration: unknown). Waxes are extracted by immersing the leaves in $CHCl_3$ (for 10 sec). The extract is concentrated and the residue amount is expressed as mg/g DW of leaf.

Thiocarbamate compounds (Diallate, EPTC, CDEC) also inhibit the incorporation of [1-^{14}C]acetate in the cuticular lipids of pea leaf fragments during a 90-min treatment.[6] The principle of these experiments is the rapid extraction of surface lipids by a $CHCl_3$/methanol mixture (2 : 1, v/v) followed by counting the radioactivity of the extract. Separation of these lipids by thin-layer chromatography (silica gel G in benzene) shows a decrease in the radioactivity of alkanes and secondary alcohols.

Degradation of Lipids

Diphenyl ethers and phenoxybenzoic acids induce a peroxidative degradation of polyunsaturated fatty acids. This may result in some membranous destruction: this work and the peroxidation mechanism are summarized in the literature.[7,8] This pesticidal action on the degradation of fatty acids has always been studied on a overall scale since its contribution to the process remains unknown. The best testing method is the quantification by GLC of the hydrocarbons formed during the peroxidative process. Another method consists of the cellular efflux assessment of a radioactive probe (e.g., $^{14}CH_3–NH_2$), enabling this phenomenon to be correlated with membranous degradation.

Specific Methods

These methods employ finer and/or more suitable analytical techniques so as to specify the action of a pesticide on a given site or a biosynthetic pathway.

Analysis of Phospholipids

Pyridazinones induce a specific variation in the composition of phospholipidic fatty acids. The kinetic investigation of a system, based on the distribution of $^{14}CO_2$ radioactivity in fatty acids gives an accurate result in the case of SAN-9785 (BASF 13-338) and SAN-6706. After 48 hr, the total radioactivity shows to be greater in the Δ^3-hexadecenoic acid than in the palmitic acid contained in the phosphatidylglycerol when compared to the

[6] P. E. Kolattukudy and L. Brown, *Plant Physiol.* **53,** 903 (1974).

[7] P. Böger, *Z. Naturforsch., C: Biosci.* **39C,** 468 (1984).

[8] G. L. Orr and F. D. Hess, *in* "Biochemical Responses Induced by Herbicides" (D. E. Moreland, J. B. St. John, and F. D. Hess, eds.), p. 131. Am. Chem. Soc., Washington, D.C., 1982.

control, the pattern used being leaf disks of *Vicia faba* (the testing device is described in the following paragraph).[9]

It would be interesting to look into the plastid aspects of these results.

Using the same plant material, a completely different result (higher palmitic acid to Δ^3-hexadecenoic acid ratio) was observed following analytical tests irrespective of kinetic assessment.[10] Analytical tests with SAN-6706 were nevertheless conducted on treated plants for 3 weeks. This period seems too long to have any significant connection with a direct mode of action. The advantage of a short treatment period and the use of kinetic tests aimed at identifying an inhibitor's site of action must be emphasized here.

Analysis of Galactolipids

SAN-9785 (BASF 13-338) drastically impairs the composition of galactosyldiacylglycerol fatty acids.

This action was studied on *Vicia faba* leaf disks according to the kinetic metabolism of fatty acids.[9] In order to prevent SAN-9785 from interfering with CO_2 fixation and the intermediary stages of fatty acids synthesis, disks of 3-week-old leaves were placed first without pesticides in a $^{14}CO_2$ (1 mCi) atmosphere for 10 min. This was followed by a 15-min chase (normal atmospheric conditions). The disks were then placed in various Petri dishes containing (or not) the pesticide (1 m*M*). After 2 hr of treatment, the disks were left to float on water (control) or a 0.15 *M* sucrose (pesticide-treated) solution (sucrose prevents degeneration of leaf-treated tissues and does not alter the metabolism of fatty acids).

At different time intervals (not exceeding 48 hr) the lipids were extracted from the disks ($CHCl_3$/methanol, 2 : 1, v/v). Nonlipid contaminants were removed.[11] The lipids were separated by thin-layer chromatography on silica gel impregnated with ammonium sulfate (acetone/benzene/H_2O, 91/30/8, v/v/v). The purified lipids were transesterified using a 0.5 *M* solution of MeONa, then extracted with hexane from the reaction mixture. The fatty acid esters were analyzed by a combination of GLC/fraction collector allowing for their quantification and the determination of their radioactivity. GLC analysis was done at 190° (glass column, 10% Apolar 10C on Chromosorb Q or 15% Reoplex 400 on Chromosorb W).

It appears from this study that SAN-9785 (BASF 13-338) causes radio-

[9] N. W. Lem and J. P. Williams, *Plant Physiol.* **68,** 944 (1981).

[10] M-U. Khan, N. W. Lem, K. R. Chandorkar, and J. P. Williams, *Plant Physiol.* **64,** 300 (1979).

[11] J. P. Williams and P. A. Merrilees, *Lipids* **5,** 367 (1970).

activity to accumulate in the linoleic acid and to inhibit the incorporation of radioactivity into the linolenic acid. The same results apply to mono- and digalactosyldiacylglycerol.

Similar conclusions were drawn following GLC analysis of galactolipids isolated from stems of wheat plantlets treated for 4 days.[12]

SAN-9785 (BASF 13-338) is therefore involved in the desaturation process of linoleic acid. But it is still not proved if the enzymatic reaction or the enzyme level is involved.

Analysis of Waxes

Ethofumesate inhibits the formation of cuticular waxes and especially that of *n*-nonocosane and of *n*-nonocosan-15-one. This is achieved by incorporating ethofumesate in a soil (0.84 kg/ha) where cabbage seeds had been allowed to germinate.[13] At the six-leaf stage, waxes from leaves 4 and 5 were isolated (washing with $CHCl_3$). The extract was concentrated and analyzed by GLC at 320° (3% Dexil 300). In this experiment, the inhibitory action of EPTC on the *n*-nonocosan-15-one deposit was inferior to that of ethofumesate.

Analysis of Carotenes

Various classes of pesticides inhibit the biosynthesis of β-carotene. This also results in xanthophyll inhibition.

Several studies lead to this conclusion. The methods used are usually quite simple since they involve spectrophotometer analysis of the organic extract containing the carotenes. It is equally possible to separate the different carotenes before analysis by thin-layer chromatography [silica gel G, toluene/petroleum ether (bp 100 to 140°), 15 : 85, v/v].

HPLC should become a powerful method of analysis in the future. This technique is illustrated by the case of SAN-6706 acting on the carotenoids of *Chlorella fusca*.[14]

The carotenoid's sensitivity to light and to oxidation is a major factor to take into account. Unfortunately, technical precautions used to get around this difficulty are seldom mentioned in the literature.

Most of the results were obtained using fluridone and norflurazon; these pesticides cause phytoene and phytofluene to accumulate in wheat plantlets treated for 7 days.[15] Applied to the same plant material and to

[12] J. B. St. John, *Plant Physiol.* **57,** 38 (1976).

[13] J. R. C. Leavitt, D. N. Duncan, D. Penner, and W. F. Meggitt, *Plant Physiol.* **61,** 1034 (1978).

[14] M. M. Tan Tawy and L. H. Grimme, *Pestic. Biochem. Physiol.* **18,** 304 (1982).

[15] P. G. Bartels and C. W. Watson, *Weed Sci.* **26,** 198 (1978).

treated radish plants (leaves and chloroplasts), SAN-6706 produces the same effects.[16,17] Other pesticide inhibitors of β-carotene biosynthesis are amitrole, dichlormate, and pyriclor.

LS 76-1243 also interferes with the biosynthesis of β-carotene in wheat plantlets (personal results).

The biosynthetic pathway of carotenoids have recently been investigated through *in vitro* systems. Some of them are intended to specify the action of the pesticides inhibiting the biosynthesis of β-carotene. This is the case of the *Phycomyces blakesleanus* or the *Aphonacaspa* system.[18,19] The only isolated plant pattern providing a description of pesticide action is the chromoplast of the narcissus flower.[20] In this instance, SAN-6706 (0.05 m*M*) is observed to inhibit the incorporation of [1-^{14}C]IPP in β-carotene.

All these studies tend to conclude that one desaturation stage in the biosynthetic pathway of β-carotene has been inhibited.

Degradation of Sulfoquinosovyldiglyceride

Oxyfluorfen, a diphenyl ether, degrades sulfoquinosovyldiglyceride in the alga *Scenedesmus acutus*.[21] This degradation is unlikely to be specific but rather connected to the peroxidation of fatty acids mentioned above (see Degradation of Lipids). The ^{35}S-labeled sulfolipid is extracted with acetone, then purified by thin-layer chromatography (silica gel G $CHCl_3$/methanol/H_2O, 65/25/4, v/v/v). The radioactivity corresponding to the sulfolipid fraction is then assessed.

Concluding Remarks

These examples reveal that certain pesticides interfere with plant cell lipids. This metabolic interference (biosynthesis or degradation) should account further for their functional role. New methods were used to demonstrate these facts. It would appear from these results that pesticides, in addition to other benefits, continue to be of basic scientific interest.

[16] A. Ben-Aziz and E. Koren, *Plant Physiol.* **54,** 916 (1974).
[17] K. H. Grumbach and G. Britton, *Phytochemistry* **22,** 1937 (1983).
[18] G. Sandmann, P. M. Bramley, and P. Böger, *Pestic. Biochem. Physiol.* **14,** 185 (1980).
[19] I. E. Clarke, G. Sandmann, P. M. Bramley, and P. Böger, *FEBS Lett.* **140,** 203 (1982).
[20] P. Beyer, K. Krentz, and H. Kleinig, *Planta* **150,** 435 (1980).
[21] G. Sandmann and P. Böger, *Plant Sci. Lett.* **24,** 347 (1982).

[62] Rapid Filtration Technique for Metabolite Fluxes across Cell Organelles

By YVES DUPONT and MARIE-JOSÉ MOUTIN

Introduction

The rapid filtration technique which is described in this chapter may apply directly to the study of a large number of biochemical systems. Our concern here is to describe fast measurements of ions or metabolite fluxes through biomembranes.

The most popular technique to measure ion fluxes in isolated membrane fractions has been, so far, the manual filtration technique. In this procedure the vesicular fraction, in which transport or efflux is to be assayed, is immobilized in an adequately chosen filter. After evacuation of excess buffer, the reaction medium containing the radioactive substrate to be transported or the washing solution (in case of efflux measurement) is allowed to pass through the filter, the time of reaction being equal to the duration of the filtration. This procedure presents several important advantages: first, the volume of the reactor (the filter) is low, about 30 μl, and the amount of the biological sample can be as high as several milligrams, giving a concentration in the reaction volume of more than 10 mg/ml. Thus even a very low concentration factor can be detected with this technique. The second advantage is that the incubation medium around the vesicular preparation is completely washed out during the first fractions of a second of filtration and replaced by the filtration medium facilitating the control of the transport reaction.

The major inconvenience of the filtration technique is that it is limited to slow transport processes since the shortest feasible time is around a few seconds (depending of the skill of the manipulator). Only in one instance was this inconvenience overcome, by slowing down the transport reaction at a low temperature and by performing the filtration experiment in the presence of cryoprotective solvents.[1]

In some selected cases true initial transport rates have been obtained in the millisecond range by stopped flow techniques or with the aid of specific transport inhibitors and the combined use of rapid mixing and filtration techniques.

The continuous or stopped flow techniques consist of rapid mixing of

[1] Y. Dupont, *Biochim. Biophys. Acta* **688,** 75 (1982).

the vesicular preparation with the substrate to be transported or with the washing solution when an efflux is to be measured, mixing being completed within a few milliseconds. Progress of the transport reaction may be followed by optical methods (fluorescence or absorbance changes) or by performing a second mixing with a transport inhibitor or a denaturing agent after a preset time. These methods are therefore restricted to the cases where the transported substrate is optically active or where the reaction under study can be stopped by an inhibitor before examination of the reaction mixture with a slower method. This method has been applied with success to the calcium transport in sarcoplasmic reticulum vesicle,[2,3] the ADP/ATP carrier of mitochondria,[4,5] and to the study of the acetylcholine receptor in subsynaptic vesicles from the electric organ of eel or *Torpedo*.[6,7] This technique, however, implies a very good knowledge of the inhibitor mechanism and especially of its rate of inhibition. Furthermore, only irreversibly transported ions can be evaluated. All information on transiently and reversibly bound ions, which may be very important in the initial phase of the transport process and crucial for the understanding of the carrier mechanism, is lost.

The rapid filtration technique described in this chapter is a great improvement over the above-mentioned fast reaction techniques. The obvious advantage is that it allows temporal resolution of noncovalent binding and of transport processes without implying the use of a specific carrier inhibitor. It is, however, restricted to the cases where the enzyme or membrane fragments can be immobilized or adsorbed within a filter.

Description of the Technique

The apparatus used is shown schematically on Fig. 1. Before filtration the enzyme suspension or the membrane fragments are equilibrated in the preincubation medium. For each single point one aliquot of the enzyme suspension is deposited on a porous substrate. This sample layering is made with the aid of a special filter holder permitting the deposited sample to cover a reproducible surface of the filter (about 200 mm^2). Excess

[2] S. Verjovski-Almeida, M. Kurzmack, and G. Inesi, *Biochemistry* **17,** 5006 (1978).

[3] M. Sumida, T. Wang, F. Mandel, J. P. Froehlich, and A. Schwartz, *J. Biol. Chem.* **253,** 8772 (1978).

[4] H. Nohl and M. Klingenberg, *Biochim. Biophys. Acta* **503,** 155 (1978).

[5] C. Duyckaert, C. M. Sluse-Goffart, J. P. Fux, F. E. Sluse, and C. Liebecq, *Eur. J. Biochem.* **106,** 1 (1980).

[6] J. P. Hess, D. J. Cash, and H. Aoshima, *Annu. Rev. Biophys. Bioeng.* **12,** 443 (1983).

[7] T. Heidmann, J. Bernhardt, E. Neumann, and J. P. Changeux, *Biochemistry* **22,** 5452 (1983).

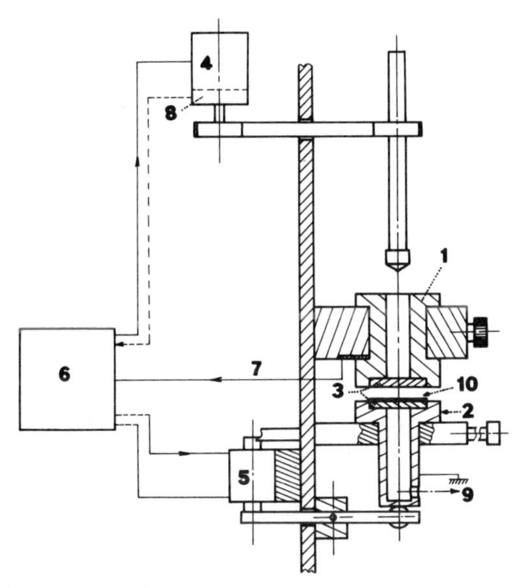

FIG. 1. Schematic drawing of the rapid filtration apparatus. (1) Injection device; (2) filter holder; (3) porous stainless steel disks; (4) stepping motor; (5) solenoid; (6) electronic controller; (7) detection of the contact between the filter holder and the injection device; (8) rotation sensor; (9) vacuum; (10) filter.

preincubation buffer is evacuated from the bottom of the filter holder with a vacuum pump. This lets the enzyme equilibrate with the filter hydration water, which amounts to about 30 μl for 0.45-μm Millipore filters. In the next phase the filter and its holder are positioned a few millimeters under an injection device containing the buffer used in the filtration sequence.

The sequence which follows is automatically driven by the electronic part of the apparatus. The filter is applied strongly against the porous end of the injection device during a preset time and the filtration buffer is forced to pass through the filter during this contact. This phase ends with a physical separation between the filter and the injection device. This movement of the filter is necessary to prevent undesirable contact between the enzyme and the filtration medium except during the filtration experiment. Last, the filter is removed and its reacted content evaluated with appropriate techniques (scintillation counting, spectrophotometry, etc). Duration of filtration may be selected from 10 msec to seconds and buffer flow rate from 10 μl/sec to 5 ml/sec. The shortest measurable point is dependent on the charge of the filter and on the nature of the preincubation buffer, it may vary from 10 to 50 msec. At the highest rate of filtration this corresponds to 2 to 10 times the filter hydration volume.

Description of the Instrument

The Filter Holder. The filter holder is made in stainless steel with a porous stainless steel disk in its central part, permitting evacuation of preincubation and filtration buffers. The filter holder is fitted on a moveable deck. The forward position allows the operator to deposit the sample on the filter. In its backward position, the filter holder is fitted a few millimeters below the injection device; a lever is used to move it upward and to apply strongly the filter against the lower end of the injection device. The lever is attached to a solenoid which may be activated manually or by the electronic controller during an automatic filtration sequence.

The Injection Device. The injection device has the general shape of a stainless steel syringe. It is designed to inject the filtration buffer through the filter over a reproducible surface; to do so, its lower part is flat with a porous stainless steel disk in its center. The piston of the syringe is finely driven by a stepping motor through a bolt and a screw. Filling of the injection device is made from the top of the syringe body. In the filling sequence the piston is in its uppermost position, it is constructed so as to eliminate air bubbles during its downward movement.

The Controller. This part of the apparatus contains a microprocessor and the power supplies which serve to drive the stepping motor and actuate the filter holder solenoid over a predetermined automatic sequence.

Experimental Conditions

Choice of Filters

The choice of filter type and its pore size is very critical and the overall performance is very much dependent on it.

This choice mainly depends on the nature of the enzymes, on the membrane preparations, or the organelles to be investigated. Some will necessitate pore size as small as 0.2 μm while others will work well with 1.2- or 2-μm filters. This choice is very critical since use of low-porosity filters limits the flow rate and thus the time resolution of the filtration experiment. Moreover, low-porosity filters will be saturated by lower amounts of enzyme which results in a decrease of the signal-to-noise ratio. Thus, taking the highest possible porosity of the filter will result, first, in shortening the dead time of the filtration and, second, will allow measurement of transport at higher concentrations of substrate.

Adjustment of the Flow Rate

Figure 2 shows the effect of varying the flow rate and the filtration time on the background. The decrease of the number of counts for the shortest times is due to incomplete washing of the filter. It is clear that washing at least 5 to 10 times is necessary for the filter to be in equilibrium with the filtration buffer and for the preincubation buffer to be completely eliminated. Thus, faster flow rates are necessary at the shortest times. The tabulation below is an indication of the flow rates which should be used.

Filtration duration (msec)	Recommended flow rate (ml/sec)	Filtered volume (μl)
5–20	10 to 5	50–100
20–50	4	100–200
50–200	3	150–600
200–500	2	400–1 ml
500–1000	1.5	750–1.5 ml
>1 sec	1	>1 ml

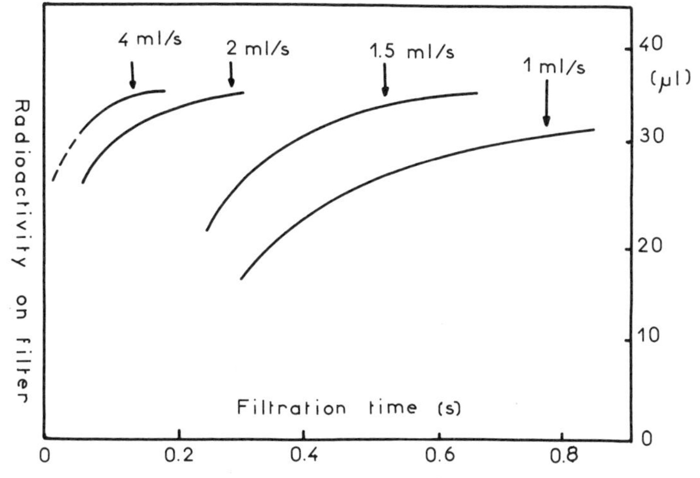

FIG. 2. Measurement of the radioactivity retained in the filter in the absence of enzyme as a function of time and flow rate. The scale on the right ordinate gives the apparent volume of the filter (in microliters) calculated from the retained radioactivity. These points were obtained with ^{45}Ca and 0.45-μm Millipore filters.

Blank Evaluation

The rapid filtration technique prevents any buffer washing after the filtration. Evaluation of bound or transported substrate implies subtraction of the background from the measured value. This background can be evaluated in blank measurements performed under the same conditions as the sample.

Three methods can be used, as follow:

1. With less than 0.5 mg enzyme, an adequate blank can be obtained with an unloaded filter and the value subtracted from the sample. The reproducibility of the rapid filtration experiment allows very accurate blanks to be obtained.
2. For higher loads, water is excluded from the filter by the presence of enzyme especially for microsomes or vesicles. The method in (1) may lead to wrong estimation of the blank. It is recommended to use, as blank, a filter loaded with the same amount of enzyme as in the sample, but containing inhibited or inactivated enzyme.
3. Finally, the best method is to use double labeling and to add to the filtration medium a nontransported radioactive label like tritiated sucrose. The amount of tritium retained in the filter after filtration will give a direct measurement of the filter apparent volume. The measured volume will give directly the contribution of the background on the transported substrate. The greatest advantage of this last method is that the blank and transport measurements are performed on the same filter.

Some Selected Examples

The experiments which will be described in this section illustrate some applications of the rapid filtration technique.

ADP/ATP Transport in Potato Mitochondria

The nucleotide carrier of mitochondria performs a very specific exchange across the inner mitochondrial membrane. Various authors have tried to use rapid techniques for the evaluation of the initial rate of transport in intact or reconstituted mitochondria. Most used were inhibitor stop and filtration,[4,5] the best resolution obtained being around 0.5 sec with a filtration apparatus developed by Klingenberg.[8] Most of the inhibitor-stop experiments were performed by using carboxyatractyloside (CATR) as inhibitor and conclusions were limited by the restricted knowl-

[8] M. Klingenberg, *Eur. J. Biochem.* **76,** 553 (1977).

edge of CATR action.[9] Rapid filtration experiments have been attempted to elucidate the mechanism of action of the nucleotide carrier.

In the experiments reported in Dupont,[10] a complete resolution of the initial phase of ADP binding and transport by the ADP/ATP carrier was obtained down to the millisecond range without using an inhibitor. A very pronounced multiphasic process was observed with a fast initial binding, lasting 100 msec, followed by a lag phase of about 500 msec and resumption of the uptake at a rate slower than the initial phase.

P_i–P_i Exchange in Intact Mitochondria

The kinetics of phosphate–phosphate exchange mediated by the P_i carrier of the mitochondrial membrane were investigated by Ligeti *et al.*[11] using the rapid filtration technique described here. Rat liver mitochondria were used and the P_i–P_i exchange was measured down to about 100 msec. This limitation in time resolution was due to very unfavorable experimental conditions: high radioactivity, important background, and high concentration of protein necessitated by the high K_m of the exchange process. In spite of these limitations the rapid filtration technique was applied and the initial rate of P_i exchange was obtained for the first time at physiological temperatures.

Calcium-Induced Calcium Release in Sarcoplasmic Reticulum

Calcium release from the sarcoplasmic reticulum membrane of skeletal muscle triggers the contraction of the muscle fiber. Elucidation of the mechanism of release is a challenging research subject. Much effort has been given to study of this mechanism *in vitro*, using purified sarcoplasmic reticulum vesicles. Before the rapid filtration technique was available, the kinetics of calcium release were studied mainly by using calcium indicators and rapid mixing techniques. In addition to possible artifacts arising from the dye, the major technical difficulty was the necessity to rapidly change the medium around the vesicles from uptake or loading conditions to efflux conditions. If the vesicles were actively loaded the energy donor (ATP) was always present in the efflux conditions and able to affect the rate of release; if the vesicles were passively loaded, the high Ca^{2+} concentration in the loading medium (usually in the low millimolar range or higher) had to be rapidly reduced to the desired concentration (the low micromolar range) by addition of EGTA at the beginning of the

[9] M. Neuburger, E. P. Journet, R. Bligny, J. P. Carde, and R. Douce, *Arch. Biochem. Biophys.* **217,** 313 (1982).

[10] Y. Dupont, *Anal. Biochem.* **142,** 504 (1984).

[11] E. Ligeti, G. Brandolin, Y. Dupont, and P. V. Vignais, *Biochemistry* **24,** 4423 (1985).

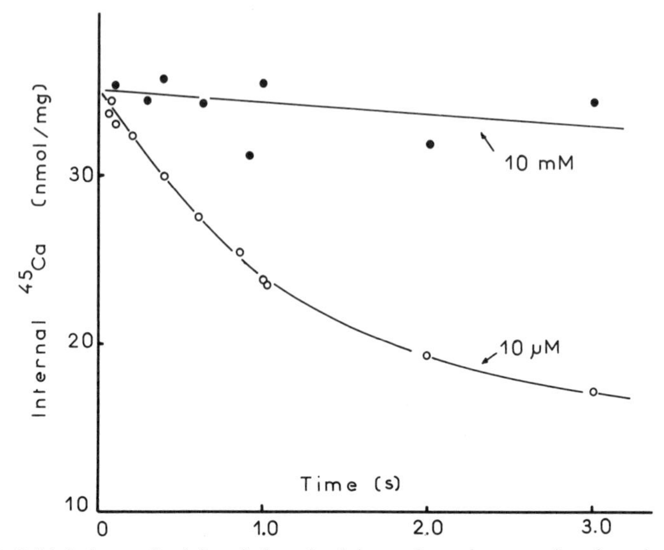

FIG. 3. Initial phase of calcium-induced calcium release in sarcoplasmic reticulum vesicles. Concentrated vesicles (10 mg/ml) were preincubated for 4 hr in 20 m*M* HEPES–K, pH 6.8, 100 m*M* KCl, and 10 m*M* [^{45}Ca]Ca^{2+}. For each point an aliquot of the suspension was diluted 20-fold for a period of 30 sec in the same medium but with cold Ca^{2+}. One milliliter of the dilution was layered on the filter (0.65 μm Millipore) before rapid filtration. Filtration buffer contained 20 m*M* Hepes–K, pH 6.8, 100 m*M* KCl, and 10 m*M* Ca^{2+} (●) or Ca–EGTA buffer, giving a free concentration of Ca^{2+} = 10 μ*M* (○).

release reaction, a procedure which usually produced important pH artifacts. The rapid filtration technique does not suffer from these inconveniences. Figure 3 shows a measurement of the calcium-induced calcium release using heavy sarcoplasmic reticulum vesicles. Vesicles were passively loaded in 10 m*M* Ca^{2+}. Release was initiated by washing the vesicles with a buffer containing 10 μ*M* Ca^{2+}.

Conclusions

The use of the rapid filtration technique has been described and illustrated in this chapter. It has been found that this method can be used to measure transport processes down to very short times, 10 to 20 msec in the best cases. This is more than 10 times better than the faster filtration technique reported in Klingenberg,[8] and more than a 100-fold improvement over the generally used manual filtration procedure. The rapid filtration technique is clearly very well adapted to the study of ion or metabolite transport.

The rapid filtration process described in this chapter is protected by a patent owned by the Centre d'Etude Nucleaire de Grenoble. A commercial version is manufactured by Bio-Logic Co. (Zirst, rue des Preles, Meylan, France).

Acknowledgment

This work was supported by a CNRS-ATP grant: Conversion de l'energie dans les membranes biologiques.

[63] Application of Nuclear Magnetic Resonance Methods to Plant Tissues

By R. G. RATCLIFFE

Introduction

Nuclear magnetic resonance (NMR) spectroscopy is a versatile technique with an ever-increasing range of applications in physics, chemistry, biology, and medicine. For the plant scientist, NMR offers not only an analytical technique for determining the structures of compounds extracted from plant cells, for example the molecular structures of secondary metabolites and cell wall polysaccharides, but also a number of methods for studying intact plant tissues. The latter approach encompasses several different techniques including low-resolution ^{1}H NMR, high-resolution multinuclear NMR, solid-state NMR, and magnetic resonance imaging. With the exception of the solid state approach, which is effectively restricted to freeze-dried samples by the need for magic angle spinning at rotor speeds of the order of kilohertz, these techniques aim to obtain information on functioning tissues, i.e., by virtue of the noninvasive nature of the technique to obtain information *in vivo*. In the context of plant cell membranes, these methods can be used in principle to define the physical nature and composition of membrane-bound compartments, to observe metabolic events directly within these regions, and to monitor metabolic processes that are mediated by the membranes. The early development of this work has been comprehensively described[1] and selected topics have been reviewed.[2-4] The following introductory paragraphs

[1] B. C. Loughman and R. G. Ratcliffe, *in* "Advances in Plant Nutrition" (P. B. Tinker and A. Lauchli, eds.), Vol. 1, p. 241. Praeger, New York, 1984.
[2] J. K. M. Roberts, *Annu. Rev. Plant Physiol.* **35,** 375 (1984).
[3] B. C. Loughman, *Proc. 7th Sch. Biophys. Membr. Transp.* **1,** 249 (1985).
[4] F. Martin, *Physiol. Veg.* **23,** 463 (1985).

summarize the current level of interest in the application of the various NMR techniques to plant tissues.

The low-resolution ^1H NMR method, which is used to study the water fractions in plant tissues and which has been used extensively, for example, in studies of frost hardiness and membrane permeability, is exciting less interest than previously. Recent applications of the method include further work on the relation between the tissue water content and the spin-lattice relaxation of the water resonance in plant tissues[5] and a study of water relaxation in maize roots in the presence of paramagnetic ions.[6] The declining interest in this technique can be partly attributed to the difficulty of making unambiguous assignments in multicompartment systems.[7]

In the presence of linear magnetic field gradients, low-resolution ^1H NMR can be used to produce three-dimensional ^1H images of heterogeneous objects. The contrast in the image of a living system usually arises from the spatial dependence of the water content and the variation of the water relaxation rates between different tissues.[8] At the present time, the clinical applications of the technique dominate research in magnetic resonance imaging, but since ^1H images can also be obtained from plant tissues and since imaging systems for small objects, which would be suitable for studying the germination of seeds or water exchange in plant roots, are commercially available it is likely that physiological investigations of plant tissues will be undertaken in the near future. Preliminary observations of H_2O–D_2O exchange in maize roots were reported at a recent meeting.[9]

Solid state NMR techniques are also becoming more widely available and cross-polarization, magic angle spinning ^{13}C NMR spectra can be recorded routinely in many laboratories. For example, the characteristic ^{13}C resonances of carbohydrate, lignin, and protein can be used to investigate the composition of freeze-dried plant tissues.[10,11] Qualitative information is readily obtained, but a full quantitative interpretation of the data is often complicated by the way in which polarization transfer is used to generate intensity in the spectrum and this is a significant limitation of the

[5] S. Kaku, M. Iwaya-Inoue, and L. V. Gusta, *Plant Cell Physiol.* **25,** 875 (1984).

[6] G. Bačić and S. Ratković, *Biophys. J.* **45,** 767 (1984).

[7] P. S. Belton and R. G. Ratcliffe, *Prog. Nucl. Magn. Reson. Spectrosc.* **17,** 241 (1985).

[8] P. G. Morris, "Nuclear Magnetic Resonance Imaging in Medicine and Biology." Oxford Univ. Press (Clarendon), London and New York, 1986.

[9] G. Bačić, unpublished results reported at the 218th conference of the Society for Experimental Biology, Norwich, England (1985).

[10] D. S. Himmelsbach, F. E. Barton, III, and W. R. Windham, *J. Agric. Food Chem.* **31,** 401 (1983).

[11] G. E. Maciel, J. F. Haw, D. H. Smith, B. C. Gabrielsen, and G. R. Hatfield, *J. Agric. Food Chem.* **33,** 185 (1985).

technique. The other solid state NMR method of interest to plant physiologists is the double cross-polarization technique for monitoring ^{13}C- and ^{15}N-labeled carbon–nitrogen bonds. This technique has been used extensively in studying the nitrogen metabolism of soybean cotyledons, most recently in an investigation of the degradation of allantoin,[12] but the wider application of the technique has been restricted by the difficulty of implementing double cross-polarization on commercial instruments.

High-resolution NMR methods can be used to detect freely mobile, low-molecular-weight tissue metabolites *in vivo*[13] and the application of these methods to plant tissues is being actively pursued in an increasing number of laboratories. As in the earlier development of these methods for biochemical investigations in microorganisms and animal tissues, it is ^{31}P NMR which is most readily exploited and detailed studies of the phosphorus metabolism in a variety of tissues, including seedling root tips and plant cell suspension cultures, are underway.[1,2] The method can be extended to a number of other magnetic nuclei, including ^{1}H, (^{13}C), ^{14}N, ^{23}N, and ^{39}K at natural abundance and ^{13}C and ^{15}N in labeled systems, although the potential applications of the different nuclei vary because of the different magnetic properties of the isotopes.

The high-resolution multinuclear NMR methods and, to a lesser extent, the low-resolution ^{1}H NMR method are the most suitable NMR techniques for investigating plant tissues at the cellular and subcellular levels. This chapter is concerned with the practical details of implementing these methods and with the principles of compartmental analysis in plant tissues by NMR. Familiarity with the basic principles of NMR and with the interpretation of the measurable spectroscopic quantities is assumed throughout: background material on the NMR technique,[14] its biochemical applications,[15] and its use in the study of living systems[13,16] is readily available elsewhere.

Practical Considerations

Choosing a Tissue

NMR spectra are observed by placing the sample in a strong, homogeneous magnetic field and applying pulses of radiofrequency power to

[12] G. T. Coker, III and J. Schaefer, *Plant Physiol.* **77,** 129 (1985).

[13] D. G. Gadian, "Nuclear Magnetic Resonance and its Applications to Living Systems." Oxford Univ. Press (Clarendon), London and New York, 1982.

[14] M. L. Martin, J. J. Delpuech, and G. J. Martin, "Practical NMR Spectroscopy." Heyden, London, 1980.

[15] G. R. Moore, R. G. Ratcliffe, and R. J. P. Williams, *Essays Biochem.* **19,** 142 (1983).

[16] J. K. M. Roberts and O. Jardetzky, *Biochim. Biophys. Acta* **639,** 53 (1981).

induce transitions between the nuclear magnetic energy levels. Low magnetic field strengths (<2.5 T), which are suitable for low-resolution ^{1}H studies, are generated by electromagnets while higher field strengths, which are essential for multinuclear high-resolution studies, are obtained from superconducting magnets. The receiver coil in the probehead of a conventional spectrometer is typically 2–3 cm long and is designed to accommodate a sample tube of a particular diameter, usually between 5 and 25 mm. It is clear that the dimensions of the probehead and the geometry of the magnet place an immediate restriction on the size and nature of the sample. Plant cell suspensions and excised tissues are the most convenient samples for high-resolution studies, the typical requirement for a ^{31}P spectrum being a fresh weight of ~0.8 g in a 10-mm-diameter tube. Intact plants such as seedlings are much more difficult to handle in the confined space of a conventional high-resolution probehead[17] and purpose-built probeheads are required for investigations of this kind. In contrast the greater flexibility of the usual probehead configuration in an electromagnet, together with the high sensitivity of ^{1}H NMR and the high water content of most tissues, makes it possible to record low-resolution ^{1}H spectra of different parts of intact seedlings with little difficulty.[6,18]

Inhomogeneities in the magnetic field experienced by the sample can cause unacceptable line broadening in high-resolution spectra. For a tissue sample, the difference between the magnetic susceptibilities of the intracellular solutions and the surrounding air can be a potent source of line broadening and so it is essential to suspend the plant tissue in an aqueous medium. Since the aim is to use NMR to monitor functioning tissues, it is important to ensure that the tissue does not deteriorate in the NMR tube and for actively metabolizing tissues with a high respiration rate, for example the meristematic tissue in root tip samples, it is important to ensure an adequate supply of oxygen. Sections of excised tissue can be aerated by circulating an aerated buffer solution through the NMR tube.[19] The same arrangement can be used for plant cell suspensions,[20] although in some cases it may be preferable to bubble oxygen into the sample directly.

Other factors to consider when choosing a tissue sample, and interpreting the spectra, include the cell water content, the likely concentra-

[17] M. J. Kime, B. C. Loughman, R. G. Ratcliffe, and R. J. P. Williams, *J. Exp. Bot.* **33,** 656 (1982).

[18] S. Ratković, G. Bačić, C. Radenović, and Z. Vučinić, *Stud. Biophys.* **91,** 9 (1982).

[19] R. B. Lee and R. G. Ratcliffe, *J. Exp. Bot.* **34,** 1213 (1983).

[20] W. A. Hughes, S. M. Bociek, J. N. Barrett, and R. G. Ratcliffe, *Biosci. Rep.* **3,** 1141 (1983).

tion of detectable metabolites, the relative volumes of the cytoplasmic and vacuolar fractions, the presence of intercellular air spaces, the line broadening effects of free paramagnetic ions, and the cellular heterogeneity of the sample. Tissues that present particular difficulties include germinating seeds, where the low water content and/or intrinsic variations in the magnetic susceptibility lead to very broad ^{31}P resonances of limited utility,[17,21,22] and leaf tissue, where intercellular air spaces and free paramagnetic ions cause serious line broadening problems.[23,24] Vacuum infiltration of the air spaces improves ^{31}P leaf spectra in some cases[24] but the technique apparently lacks generality.[25]

The large size of the vacuolar compartment in mature plant cells hinders the detection of cytoplasmic metabolites (Fig. 1) and to monitor cytoplasmic events by NMR it is necessary to use samples containing a high cytoplasmic fraction. In an NMR study of the intracellular distribution of inorganic phosphate in pea root tips,[26] it was estimated that the cytoplasm occupied ~19% of the tissue volume in 5-mm root tips in comparison with ~5% in the fully vacuolated mature tissue. It is for this reason that much of the high-resolution NMR work on root tissue is carried out with root tips.[1,2] However, tissues of this kind are inherently heterogeneous at the cellular level and in principle, at least, a synchronously dividing suspension of cultured plant cells would seem to offer a better experimental system. NMR studies have been reported on asynchronous cell cultures from at least 10 different species, including work on *Acer pseudoplatanus*,[27] *Nicotiana tabacum*,[28] *Rosa damascena*,[29] and *Catharanthus roseus*,[30] but there appears to have been no attempt so far to use synchronously dividing cells. Unfortunately many cultured plant cells are fully vacuolated and the cytoplasmic contributions to the ^{31}P spectra of most plant cell suspensions compare unfavorably with the corresponding root tip spectra. In contrast very good ^{31}P spectra have been reported for *Elaeis guineensis* cultures with a very low average degree of vacuolation.[20]

[21] D. C. McCain, *Biochim. Biophys. Acta* **763,** 231 (1983).

[22] M. Delfini, R. Angelini, F. Bruno, F. Conti, M. E. Di Cocco, A. M. Giuliani, and F. Manes, *Cell. Mol. Biol.* **31,** 385 (1985).

[23] C. Foyer, D. Walker, C. Spencer, and B. Mann, *Biochem. J.* **202,** 429 (1982).

[24] J. C. Waterton, I. G. Bridges, and M. P. Irving, *Biochim. Biophys. Acta* **763,** 315 (1983).

[25] C. Foyer and C. Spencer, *Annu. Rep. Res. Inst. Photosynth., Sheffield Univ.*, p. 23 (1984).

[26] R. B. Lee and R. G. Ratcliffe, *J. Exp. Bot.* **34,** 1222 (1983).

[27] J. B. Martin, R. Bligny, F. Rébeillé, R. Douce, J. J. Leguay, Y. Mathieu, and J. Guern, *Plant Physiol.* **70,** 1156 (1982).

[28] V. Wray, O. Schiel, J. Berlin, and L. Witte, *Arch. Biochem. Biophys.* **236,** 731 (1985).

[29] T. M. Murphy, G. B. Matson, and S. L. Morrison, *Plant Physiol.* **73,** 20 (1983).

[30] P. Brodelius and H. J. Vogel, *J. Biol. Chem.* **260,** 3556 (1985).

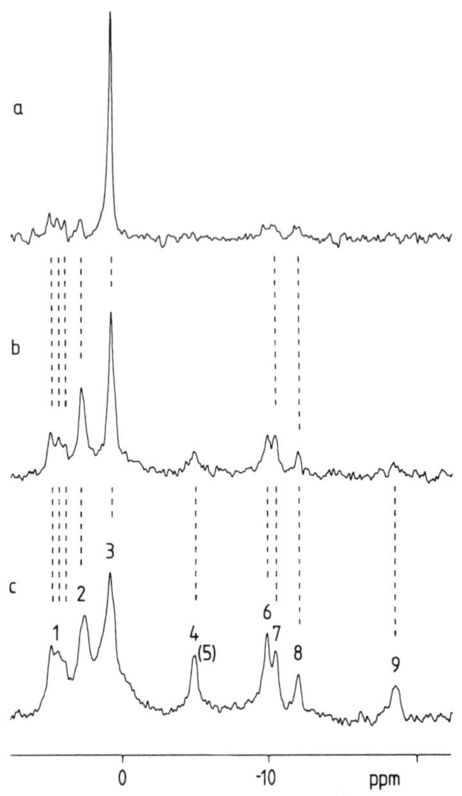

FIG. 1. 121.49 MHz, ^{31}P spectra of maize (*Zea mays*) root tissues: (a) 2.5-cm root segments, measured from the root tip and excluding the first 5 mm, accumulated in 1 hr with a 90° pulse and a 20-sec recycle time; (b) 5-mm root tips accumulated under the same conditions as in (a); and (c) 5-mm root tips accumulated in 30 min with a 90° pulse and a 0.8-sec recycle time. The assignment of the labeled resonances is as follows: 1, a composite resonance with glucose 6-phosphate on the downfield edge; 2, cytoplasmic P_i; 3, vacuolar P_i; 4, 6, and 9, the γ-, α-, and β-phosphates, respectively, of magnesium-bound nucleoside triphosphate, principally ATP; 7 and 8, UDPglucose, with NAD contributing to peak 7. The nucleoside diphosphate resonances are not detectable here, but would appear at peaks 5 and 6. A comparison of the fully relaxed spectra in (a) and (b) shows that the vacuolar P_i resonance dominates the spectrum of the fully vacuolated root segments, while a comparison of (b) and (c) shows the marked effect on the relative intensities of the resonances caused by changing the recycle time.

Accumulating Spectra

NMR is an insensitive analytical technique and with the usual signal averaging it may take several minutes or longer, depending on the magnetic nucleus observed, to detect 1 μmol of material. This lack of sensitivity is unimportant in low-resolution ^{1}H studies, because of the abundance of the water fractions, but it is crucially important in high-resolution studies of intracellular metabolites. A cursory glance through the recent plant NMR literature (particularly ^{31}P studies) will reveal numerous examples of poor quality spectra in which the signal-to-noise ratio is barely adequate for qualitative, let alone quantitative, deductions. When NMR is used to characterize a tissue in a steady state, increasing the number of scans improves the signal to noise and the insensitivity of the technique may be relatively unimportant, but if the aim is to detect intracellular changes spectroscopically, then the sensitivity problem becomes acute since it is the rate of acquisition of signal to noise which determines the maximum rate of change that can be detected.

Instrumental factors that affect the signal-to-noise ratio include the magnetic field strength and the design of the probehead and the crucial importance of these factors in determining the quality of the spectra from living systems has been emphasized by Gadian.[13] The sensitivity of the NMR experiment increases with increasing field strength and for most purposes the minimum acceptable field strength for multinuclear high-resolution studies of plant tissue is 7 T, corresponding to a ^{1}H frequency of 300 MHz. Spectrometers operating at this frequency or slightly higher are widely available in chemistry and biochemistry laboratories, but it should be noted that the probeheads on multipurpose high-resolution spectrometers are not necessarily the most suitable for recording tissue spectra. In particular, the commonly available variable frequency probeheads with Helmholtz coils are significantly less sensitive, perhaps by a factor of 2, than selective frequency probeheads with solenoidal radiofrequency coils.[13] Despite this advantage, solenoid coils have been used only occasionally in high-resolution plant NMR studies[30,31] and most spectra are recorded with the conventional, but suboptimal, Helmholtz configuration. The sensitivity can also be improved by increasing the sample volume, provided sufficient tissue is available, although the gain in signal to noise in going from a 10-mm-diameter sample tube to a 20-mm-diameter tube may be partly offset by a deterioration in the line shape. The sample size variable is not readily exploited, since each tube with a different diameter requires its own radiofrequency coil, and so most high-resolution plant tissue NMR work is done using a conventional 10-mm-diameter tube.

The inherent sensitivity of the observed nucleus is another important factor in determining the spectral signal-to-noise ratio. At natural abundance on a scale relative to ^{1}H, the sensitivities of the nuclei that are

[31] P. S. Belton, R. B. Lee, and R. G. Ratcliffe, *J. Exp. Bot.* **36,** 190 (1985).

commonly observed in biological systems fall in the order: ^{1}H, 1.0; ^{23}Na, 9.25×10^{-2}; ^{31}P, 6.63×10^{-2}; ^{14}N, 1.01×10^{-3}; ^{39}K, 4.73×10^{-4}; ^{13}C, 1.76×10^{-4}; ^{15}N, 3.85×10^{-6}. As a result of this variation, and taking into account the different relaxation properties of the nuclei, the concentration thresholds for convenient detection are approximately: ≤ 1 m*M*, ^{1}H; 1 m*M*, ^{31}P; 5 m*M*, ^{23}Na; 10 m*M*, ^{14}N; 20 m*M*, ^{39}K; $\geqslant 20$ m*M*, ^{13}C and ^{15}N. These concentrations, which are averaged over the total volume of the sample tube within the detecting coil of the probehead, correspond to a sample fresh weight of 1 g and a total accumulation time for the spectra of 15–20 min. They provide only a rough guide to the usefulness of the individual nuclei since there are other factors that may affect the acquisition of the spectra and their subsequent interpretation. For example, labeling increases the usefulness of ^{13}C and ^{15}N NMR: natural abundance studies of ^{15}N are impracticable and natural abundance studies of ^{13}C are restricted to storage pools containing high concentrations of carbohydrates or fatty acids.[4] The most commonly used nucleus in high-resolution studies of plant tissue is ^{31}P, reflecting the good magnetic properties of the nucleus, the relative ease with which the spectra can be interpreted, and the key metabolic importance of the detectable metabolites.

The principles governing the acquisition of high-resolution Fourier transform NMR spectra from solutions[14] are also applicable to plant tissues and it is only necessary to make a few comments here in relation to the operation of the spectrometer. Although there are many pulse sequences that can be used to generate spectra from solutions,[32] plant tissue NMR studies employ relatively few. The basic pulse NMR experiment, i.e., a single excitation pulse followed by the collection of the free induction decay, with options for decoupling and magnetization transfer, simple sequences for measuring longitudinal and transverse relaxation rates, and the occasional two-dimensional experiment, suffice for most purposes. Maximizing the signal to noise within the constraints of the pulse sequence requires a knowledge of the approximate longitudinal relaxation times for the observed resonances and a calibration of the transmitter power. The radiofrequency characteristics of the probehead depend on the nature of the sample and so the probehead should be tuned and matched with the tissue sample *in situ* before calibrating the transmitter. The pulse length corresponding to a 90° pulse in a tissue sample can be measured conveniently using the method proposed by Haupt.[33] In the ^{31}P spectra of plant tissues, where there is a spread of an order of magnitude in the longitudinal relaxation times of the observed resonances, the acquisition conditions can have a marked effect on the relative intensities of the

[32] C. J. Turner, *Prog. Nucl. Magn. Reson. Spectrosc.* **16,** 311 (1984).

[33] E. Haupt, *J. Magn. Reson.* **49,** 358 (1982).

resonances and while this can be turned to advantage,[17] due allowance must be made for the saturating effects of short recycle times in quantitative analyses. It is preferable to use fully relaxed spectra in quantitative experiments but this may lead to inconveniently long accumulation times in the presence of a slowly relaxing resonance, such as that from vacuolar inorganic phosphate.

Brief mention should be made of a number of other practical details, including (1) field frequency locking—this is usually unnecessary on a modern high-field spectrometer since the small instabilities in the magnetic field are masked by the width of the tissue resonances; (2) shimming—the magnetic field homogeneity is worse in a stationary, heterogeneous tissue sample than in a spinning solution, but it can be optimized by monitoring the free induction decay of the sample water resonance, which is usually readily detectable even on a probehead that cannot be tuned to the ^{1}H frequency; (3) decoupling—the usual techniques are applicable, e.g., broadband ^{1}H decoupling for ^{13}C and ^{15}N spectra and selective frequency gated irradiation for water suppression in ^{1}H spectra; (4) referencing—in the absence of a suitable chemical shift reference within the tissue or in the circulating medium surrounding the tissue, it may be necessary to supply a reference in a coaxial capillary, e.g., methylene diphosphonic acid and its tetraethyl ester are used for this purpose in ^{31}P spectra.[17]

Interpreting the Data

The resonances in NMR spectra have characteristic energies (chemical shifts), line shapes, and relaxation times. These properties are influenced by the local electron density and by the interactions with neighboring atoms experienced by the detected nucleus and they can be interpreted in terms of the chemical environment and the motional state of the nucleus. In addition, the resonances have characteristic intensities which can be related to the quantity of the nucleus in the sample. It follows that NMR spectra can be used to obtain detailed information about molecular structures and their dynamics and this is the reason for the extensive use of NMR techniques in chemistry and biochemistry.

The resonances in the high-resolution spectra of plant tissues (Figs. 1 and 2) arise from the freely mobile ions and molecules that are present in the tissue at concentrations above the detection limits discussed previously. With the possible exception of the ^{1}H spectra, where the high sensitivity of the technique might lead in principle to a contribution from an abundant macromolecule in the same way that hemoglobin dominates the simplest ^{1}H spectra of erythrocytes,[34] the observed resonances are

[34] F. F. Brown, I. D. Campbell, and P. W. Kuchel, *FEBS Lett.* **82,** 12 (1977).

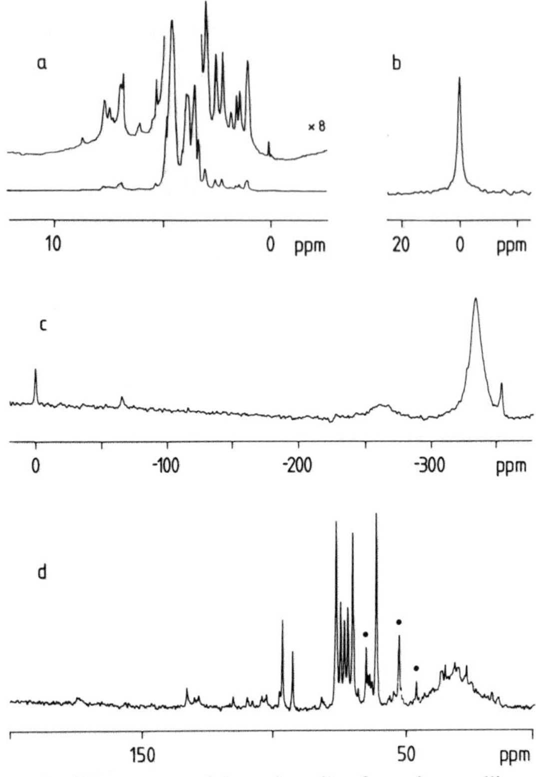

FIG. 2. (a) 270 MHz, ^{1}H spectrum of the coleoptile of a maize seedling, obtained in 1000 scans with a 0.5-sec preirradiation of the water resonance. The large resonance at ~4.5 ppm is the residual water peak and the tissue spectrum is the sum of many overlapping contributions, with only a relatively small number of dominant and potentially assignable resonances [R. G. Ratcliffe, unpublished results (1980)]. (b) 13.99 MHz, ^{39}K spectrum of 5-mm root tips from 9-day-old maize seedlings grown on a potassium-free medium, accumulated in 27 min with a 90° pulse and a 0.2-sec recycle time. The detailed interpretation of this spectrum depends on a careful analysis of the line shape [R. B. Lee and R. G. Ratcliffe, unpublished results (1985)]. (c) 21.68 MHz, ^{14}N spectrum of 5-mm root tips from 10-day-old pea (*Pisum sativum*) seedlings grown on 1 m*M* NO_3^- and supplemented with 5 m*M* NH_4^+ for the 12 hr before excision; accumulated in 1 hr with a 90° pulse and a 0.2-sec recycle time. Most ^{14}N resonances are broad, because of the quadrupolar properties of the nucleus, but the NO_3^- resonance at 0 ppm and the NH_4^+ resonance at −355 ppm are narrow and can be used to investigate inorganic nitrogen metabolism [P. S. Belton, R. B. Lee, and R. G. Ratcliffe, *J. Exp. Bot.* **36,** 190 (1985)]. (d) Natural abundance ^{1}H-decoupled 75.46 MHz ^{13}C NMR spectrum of 2.5-cm root segments from 3-day-old maize seedlings, accumulated in 10 hr with a 90° pulse and a 1.15-sec recycle time. Glucose gives a characteristic pattern of eight resonances which dominate the spectrum between 50 and 100 ppm. The labeled resonances (●) arise from the external buffer solution and the broad hump is caused by the plastic tubing in the system used to circulate the buffer through the NMR tube [R. G. Ratcliffe, unpublished results (1985)].

expected to come from low-molecular-weight metabolites in the tissue. Precipitates, macromolecular assemblies such as cell walls and membranes, and slowly tumbling macromolecules are all undetectable because of the width of the corresponding resonances. Similarly, low-molecular-weight ions and molecules that would usually have short motional correlation times also give undetectably broad lines if the molecular motion is constrained by tight binding to organelles and macromolecules or by the viscosity of the intracellular medium. Other sources of line broadening, including local magnetic field inhomogeneities, paramagnetic ions, and exchange effects, may reduce the number of potentially detectable intracellular components still further. Assigning the molecular and spatial origins of the observed resonances is the key to interpreting the spectra and to using NMR as a noninvasive probe of plant cell metabolism. These two stages in the assignment procedure are considered separately in the following paragraphs.

Depending on the nucleus, the molecular assignment problem ranges from trivial, e.g., in ^{23}Na and ^{39}K NMR where the resonances can come only from the corresponding cations, to very difficult, e.g., in ^{1}H NMR where there are many potential contributors to the spectra that satisfy the mobility and concentration requirements outlined above. In some cases, chemical shifts and the other properties of the resonances may suggest provisional assignments. For example, the characteristic resonance patterns of fructose, glucose, and sucrose are readily identified in natural abundance ^{13}C spectra[35] and only a limited number of well-known ions and molecules, including nitrate and ammonium, are likely to give narrow resonances in ^{14}N spectra.[31] However, in many cases definitive assignments can be made only with a knowledge of the composition of the tissue based on a semiquantitative analysis of tissue extracts. Conventional analytical techniques such as gas chromatography and mass spectroscopy, as well as ^{1}H NMR, can be used to demonstrate the presence of a particular metabolite in the extract. If the ^{1}H spectrum of the extract is recorded before and after the addition of a small quantity of the molecule, it is possible to determine whether the molecule was present at detectable concentrations in the original extract. The resonance positions in the extract can then be used as a guide to the likely resonance positions in the ^{1}H spectrum of the tissue on the assumption that the extract is a reasonable model for the intracellular medium. The ^{13}C, ^{14}N, ^{15}N, and ^{31}P spectra of tissues and their extracts can be correlated in the same way and in some cases it may also be useful to correlate these spectra with the corresponding ^{1}H extract spectra as well. For example, the restricted

[35] T. H. Thomas and R. G. Ratcliffe, *Physiol. Plant.* **63,** 284 (1985).

chemical shift dispersion in the phosphomonoester region of the ^{31}P spectra may make it difficult to make unambiguous assignments. A possible way around this problem is to investigate the pH dependence of the extract spectrum, but an alternative approach is to look for characteristic resonances of the phosphorus-containing metabolites in the ^{1}H spectrum.[36]

Figures 1 and 2 illustrate some of the metabolites that can be detected in plant tissues. Comprehensive descriptions of such spectra are already available elsewhere (see Refs. 1–4), and so it is only necessary to make a few general points here concerning the range of detectable metabolites and the specific problems encountered in the molecular assignment of the spectra of the different nuclei. Starting with the ^{1}H spectra, the difficulty in assigning these spectra arises not only from the relatively high sensitivity of ^{1}H NMR, but also from the restricted chemical shift range and the large line widths of the tissue resonances. This combination of properties leads to severe overlapping in the spectra and, with the notable exception of the work on red blood cells,[37] it is only recently that systematic assignment studies of such systems, including plant tissues,[38] have been undertaken. In contrast, the assignment of the ^{13}C, ^{14}N, ^{15}N, and ^{31}P spectra of plant tissues is made easier by the greater chemical shift range of the resonances and the lower sensitivity of the nuclei which ensures that the pool of detectable metabolites is much smaller. It follows that it is more likely with these nuclei than with ^{1}H that a unique resonance will be resolved corresponding to a particular metabolite. Labeled systems offer some of the best opportunities for both assignment and detection: a knowledge of the metabolic pathways available to a ^{13}C- or ^{15}N-labeled compound simplifies the assignment problem and in addition the existence of heteronuclear scalar coupling to ^{1}H nuclei makes it possible in some cases to detect labeled metabolites with high sensitivity using ^{1}H NMR techniques.[39–41]

Assigning the molecular origin of the plant tissue resonances is only the first stage in the complete assignment of the spectra. In a heterogeneous system the spatial distribution of a molecule may be nonuniform and the NMR properties of the molecule may depend on its location in the sample. Thus the NMR spectra of plant tissues may contain information

[36] S. T. Oxley, R. G. Ratcliffe, and R. J. P. Williams, unpublished results (1982).

[37] F. F. Brown and I. D. Campbell, *Philos. Trans. R. Soc. London, Ser. B* **289,** 395 (1980).

[38] T. W. M. Fan, R. M. Higashi, A. N. Lane, and O. Jardetzky, *Biochim. Biophys. Acta* **882,** 154 (1986).

[39] L. O. Sillerud, J. R. Alger, and R. G. Shulman, *J. Magn. Reson.* **45,** 142 (1981).

[40] K. M. Brindle, R. Porteous, and I. D. Campbell, *J. Magn. Reson.* **56,** 543 (1984).

[41] H. P. Juretschke, *FEBS Lett.* **178,** 123 (1984).

on the metabolite distribution within the tissue and identifying the spatial origin of the resonances is an important aspect of the assignment problem. In particular it is sometimes possible to interpret plant tissue spectra in terms of the subcellular compartmentation of the metabolites making it possible to probe the chemical composition and physical nature of membrane-bound regions, to monitor metabolic changes that occur within such regions, and to investigate processes such as ion transport that are mediated by membranes. These possibilities are all aspects of compartmental analysis and the NMR approach to this problem in plant tissues is examined in the remainder of this chapter.

Compartmental Analysis

The NMR approach to compartmental analysis in biological systems has been discussed[7] and the general principles are summarized in Fig. 3. In principle, NMR may be able to distinguish up to four fractions (A_o, A_a, A_b, and A_c) of a molecule (A) containing a resonant nucleus, by virtue of the spatial dependence of one or more of the usual resonance properties, i.e., chemical shift, intensity, and relaxation behavior. The extracellular fraction (A_o), if present, is usually readily identified but the assignment of the intracellular fractions (A_a, A_b, A_c) may be more difficult. Both cellular heterogeneity and subcellular compartmentation may be responsible for the existence of the different spectroscopic fractions and at the subcellular level the observed fractions do not necessarily correspond to the familiar membrane-bound compartments. Despite these difficulties, a simple two-compartment model of the intracellular space, in which spectroscopic fractions are assigned to cytoplasmic (i.e., extravacuolar) and vacuolar compartments, is often appropriate and it has been used extensively in the interpretation of high-resolution ^{31}P spectra.[1-3] Good evidence in favor of this model can be obtained by comparing tissues that contain different cytoplasmic fractions, for example mature roots and root tips (Fig. 1).

Spectroscopic fractions of A can be distinguished by NMR if differences in the chemical and physical environments of A lead to differences in the NMR properties and the ideal situation occurs when the various fractions are characterized by different chemical shifts, so that A_o, A_a, etc. give rise to resolvable resonances in the spectrum. The magnetic nuclei with useful chemical shift ranges for *in vivo* NMR are ^{1}H, ^{13}C, $^{14,15}N$, and ^{31}P and the spatially dependent property of the intracellular volume which is most likely to influence certain chemical shifts is pH. Thus in the presence of intracellular pH gradients, which are likely to occur in higher plant tissues across the plasmalemma, the tonoplast, the

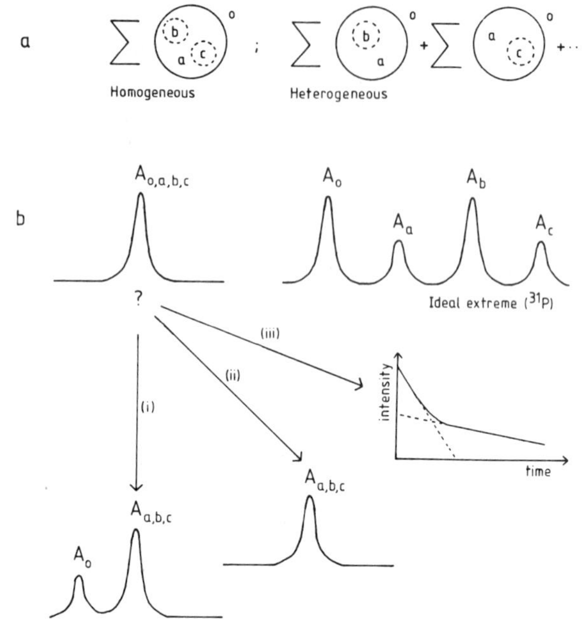

FIG. 3. The principles of compartmental analysis in cellular systems by NMR. (a) The sample may be considered as either a homogeneous collection of similar cells or a heterogeneous collection of different cell types. NMR is unlikely to give evidence for more than four spectroscopic fractions, including the extracellular space (o), and the intracellular fractions (a, b, c) do not necessarily correspond to membrane-bound compartments. (b) Chemical shift differences may make it easy to detect the distribution of A in some cases, but the spectroscopic fractions A_o, A_a, A_b, and A_c may often be superimposed. A_o may be eliminated by changing the composition of the external medium or it may be shifted (i) or broadened (ii) by the addition of paramagnetic agents. The existence of more than one intracellular fraction can often be shown by measuring the time dependence (iii) of a resonance intensity during a relaxation experiment or an exchange experiment.

mitochondrial membrane, and the chloroplast membrane, molecules with pH-dependent shifts arising from the presence of ionizable groups may give spectra which can be interpreted directly in terms of the subcellular distribution of the resonating nucleus. Four conditions need to be satisfied if the subcellular distribution of A is to be revealed by this method: (1) A must have a pH-dependent chemical shift in the pH range encompassed by the different compartments; (2) the size of the pH gradient between the compartments in relation to the pH dependence of the chemical shift must be such as to cause at least a broadening of the intensity from A and preferably a splitting into more than one resonance; (3) A must be present in the compartment(s) of interest at detectable levels; (4) the various fractions of A must be in slow exchange across the membrane boundaries.[7]

The ionizable groups that are most likely to confer a suitable pH dependence on the chemical shift of A are carboxyl groups and phosphate groups. Simple organic acids with pK_a values in the range 4–5, phosphorus compounds with pK_a values of ~6.8, as well as certain amino acids, notably histidine (which has a pK_a of ~6.0 associated with the imidazole ring), are all potential probes for compartmental analysis. For example the ^{13}C resonances of ^{13}C-labeled malate in the leaves of *Kalanchöe tubiflora*[42] and the ^{15}N resonance of the N-1 nitrogen of ^{15}N-labeled histidine in intact mycelia of *Neurospora crassa*[43] have been used to probe the vacuoles of these tissues and to measure the vacuolar pH. The most interesting results are obtained when it is possible to probe more than one intracellular compartment simultaneously and the ubiquity of orthophosphate (P_i) ensures that ^{31}P NMR is potentially the most versatile technique for this purpose. Cytoplasmic and vacuolar contributions to the ^{31}P spectra of plant tissues have been identified in many systems (Fig. 1)[1,2,7] and spectra of this kind can be used to monitor cytoplasmic and vacuolar events and to investigate processes mediated by either the plasmalemma or the tonoplast. Attempts to exploit the small pH gradients across the mitochondrial and chloroplast membranes in the same way have been less successful. Mitochondrial P_i resonances, which have rarely been observed in living systems,[7] have not been resolved in the ^{31}P spectra of the meristematic tissues of plants[26] and the observation of chloroplast resonances in photosynthetic tissues has been hindered by the broadening effects of intercellular air spaces and paramagnetic ions referred to earlier. A stromal P_i resonance, with the expected light-dependent chemical shift, has been observed in wheat leaves[24] and isolated chloroplasts,[23] but the most promising photosynthetic system examined so far by ^{31}P NMR appears to be the green alga *Chlorella*. Stromal P_i resonances have been resolved in two species[44,45] and light-dependent spectral changes have been observed in each case.

In the absence of appreciable chemical shift differences, the contributions to the spectrum of A from the different compartments will be superimposed in a single resonance and ways of determining the number of spectroscopic fractions that contribute to the intensity must be sought

[42] M. A. Stidham, D. E. Moreland, and J. N. Siedow, *Plant Physiol.* **73,** 517 (1983).

[43] T. L. Legerton, K. Kanamori, R. L. Weiss, and J. D. Roberts, *Biochemistry* **22,** 899 (1983).

[44] F. Mitsumori and O. Ito, *FEBS Lett.* **174,** 248 (1984).

[45] J. Sianoudis, A. Mayer, and D. Leibfritz, *Org. Magn. Reson.* **22,** 364 (1984).

(Fig. 3). This problem occurs not only for magnetic nuclei with no useful *in vivo* chemical shift dispersion, principally ^{23}Na and ^{39}K, but also for the other commonly used nuclei when observing intracellular components with effectively pH-independent shifts, such as water and metabolites that lack suitable ionizable groups. There are various ways of tackling this problem, most of which depend on differences in the relaxation behavior of A_a, A_b, and A_c.[7]

In some cases, extreme differences in relaxation behavior, for example between free and bound forms of A, may make the bound spectroscopic fraction undetectably broad. Evidence for this loss of intensity can be obtained by making a quantitative comparison between the resonance intensity and an independent assay of the corresponding metabolite,[7] although the difficulty of such absolute concentration measurements by NMR in living systems should not be underestimated.[13] In addition, exchange effects can complicate the analysis by mixing the intensities of the bound and free fractions and this is particularly important for ^{23}Na and ^{39}K where only a small percentage bound may lead to a large loss in intensity. As an example of this approach, natural abundance ^{13}C NMR studies of *Dunaliella salina* indicated that only 60% of the intracellular glycerol was detectable and it was suggested that the invisible 40% was immobilized in the chloroplasts.[46]

The intrinsic differences in relaxation between spectroscopic fractions are usually much smaller than the extreme case of bound and free and the relaxation behavior of the composite resonance from A may well reflect the contribution of these different fractions. Specifically since it is often, although not invariably, the case that longitudinal and transverse relaxation are simple exponential processes in homogeneous solutions, multiexponential relaxation in heterogeneous systems is frequently attributed to compartmentation.[7] This approach is used in the analysis of the low-resolution ^{1}H spectra of tissues but it is fraught with difficulty as an unambiguous interpretation of the data can be elusive.[7] Examples of the application of this method to plant tissues include a study of Mn^{2+}-treated *Nitella* cells,[47] in which the biphasic longitudinal relaxation of the intracellular water was attributed to the differing relaxation properties of the cytoplasm and the vacuole, and a study of the water relaxation of maize roots,[6] in which the multiphasic uptake of external Mn^{2+} could be followed by its effect on the relaxation of the intracellular water fractions. The *Nitella* system is attractively simple, but even here the data require

[46] R. S. Norton, M. A. Mackay, and L. J. Borowitzka, *Biochem. J.* **202,** 699 (1982).

[47] S. Ratković and G. Bačić, *Bioelectrochem. Bioenerg.* **7,** 405 (1980).

careful interpretation: the distinction between the cytoplasm and the vacuole is an approximate one that arises when Mn^{2+} binds to the cell wall, enhancing the ^{1}H relaxation of the water molecules that diffuse close to the wall. When the diffusion of the water becomes slow in relation to the variation in the relaxation rate, the longitudinal relaxation becomes biphasic and the faster component is assigned to the cytoplasm which is adjacent to the relaxation centers in the cell walls. The sensitivity of the ^{1}H relaxation data to diffusion and exchange effects complicates the compartmental analysis of the water, but it also makes it possible to measure membrane permeabilities to water directly and measurements of this kind were made on both the *Nitella* cells[47] and the maize roots.[6]

Compartmental analysis based on relaxation behavior depends on the time dependence of the resonance intensity following a displacement of the magnetization from equilibrium. Time-dependent resonance intensities may also occur for the resonances of solutes that are normally transported across the plasma membrane and this provides a method for the compartmental analysis of a number of ions. The method has been used in a ^{23}Na NMR study of millet cells in suspension culture under conditions of net influx or efflux,[48] in a ^{39}K NMR study of the potassium composition of maize roots,[49] and in a ^{14}N NMR study of inorganic nitrogen metabolism in plant roots.[31] Experiments of this kind are analogous to the conventional radiotracer techniques for the compartmental analysis of plant tissues,[50] with biphasic kinetics indicating either subcellular compartmentation or cellular heterogeneity within the sample.

Conclusion

Assigning resonances to individual compartments increases the potential usefulness of NMR as a noninvasive technique for investigating plant cell metabolism. It makes it possible, in suitable cases, to investigate the subcellular composition of plant cells, to monitor metabolic processes that occur in particular organelles, and to assess the roles of particular membranes in such processes in functioning tissues. Recurrent themes in the literature include (with references to representative investigations) intracellular pH measurement,[51] with due allowance for the uncertainties in calibrating pH-dependent chemical shifts[52] and for the limitations of ^{31}P

[48] L. O. Sillerud and J. W. Heyser, *Plant Physiol.* **75,** 269 (1984).

[49] R. B. Lee and R. G. Ratcliffe, manuscript in preparation.

[50] N. A. Walker and M. G. Pitman, *Encycl. Plant Physiol., New Ser.* **2A,** 93 (1976).

[51] R. J. Reid, L. D. Field, and M. G. Pitman, *Planta* **166,** 341 (1985).

[52] J. K. M. Roberts, N. Wade-Jardetzky, and O. Jardetzky, *Biochemistry* **20,** 5389 (1981).

NMR as a probe for the vacuolar pH[53]; intracellular pH regulation, particularly during anoxia and metabolic inhibition[54–56]; the partitioning of metabolites between cytoplasmic and vacuolar pools[26,30,57–59]; and the use of ^{13}C and ^{15}N labels to monitor individual pathways.[60,61] The techniques for studying plant tissues by NMR are now well established and a wide range of applications can be anticipated, the more so in view of the limited number of alternative techniques for the compartmental analysis of plant tissues *in vivo*.[62]

Acknowledgment

The author is grateful for the financial support of the Agricultural and Food Research Council.

[53] A. Kurkdjian, H. Quiquampoix, H. Barbier-Brygoo, M. Pean, P. Manigault, and J. Guern, *in* "Biochemistry and Function of Vacuolar Adenosine-Triphosphatase in Fungi and Plants" (B. P. Marin, ed.), p. 98. Springer-Verlag, Berlin and New York, 1985.

[54] J. K. M. Roberts, J. Callis, D. Wemmer, V. Walbot, and O. Jardetzky, *Proc. Natl. Acad. Sci. U.S.A.* **81,** 3379 (1984).

[55] J. K. M. Roberts, J. Callis, O. Jardetzky, V. Walbot, and M. Freeling, *Proc. Natl. Acad. Sci. U.S.A.* **81,** 6029 (1984).

[56] R. J. Reid, B. C. Loughman, and R. G. Ratcliffe, *J. Exp. Bot.* **36,** 889 (1985).

[57] M. J. Kime, B. C. Loughman, and R. G. Ratcliffe, *J. Exp. Bot.* **33,** 670 (1982).

[58] R. J. Robins and R. G. Ratcliffe, *Plant Cell Rep.* **3,** 234 (1984).

[59] F. Rebéillé, R. Bligny, J. B. Martin, and R. Douce, *Biochem. J.* **226,** 679 (1985).

[60] F. Martin, *FEBS Lett.* **182,** 350 (1985).

[61] M. Neeman, D. Aviv, H. Degani, and E. Galun, *Plant Physiol.* **77,** 374 (1985).

[62] M. Thellier, *Physiol. Veg.* **22,** 867 (1984).

Author Index

Numbers in parentheses are footnote reference numbers and indicate that an author's work is referred to although the name is not cited in the text.

A

B

C

E

F

G

H

I

J

K

L

M

N

O

Q

R

S

T

X

Y

Z

Subject Index

A

E

F

H

I

O

P

Q

R

S

T

U

V

W

X

Y

Z